ABC OF ELECTRICAL ENGINEERING

Covers Basic Electrical Engineering and Electrical Machines

For I[st] Year students of
B.E. (all branches), B.Tech. and A.M.I.E.

B.L. Theraja
A.K. Theraja

S. CHAND
PUBLISHING

S Chand And Company Limited
(ISO 9001 Certified Company)

S.CHAND
PUBLISHING

S Chand And Company Limited

(ISO 9001 Certified Company)

Head Office: D-92, Sector–2, Noida – 201301, U.P. (India), Ph. 91-120-4682700

Registered Office: A-27, 2nd Floor, Mohan Co-operative Industrial Estate, New Delhi – 110 044, Phone: 011-49731800

www.**schandpublishing.com**; e-mail: **info@schandpublishing.com**

Marketing Offices:

Chennai : Ph: 23632120; chennai@schandpublishing.com
Guwahati : Ph: 2738811, 2735640; guwahati@schandpublishing.com
Hyderabad : Ph: 40186018; hyderabad@schandpublishing.com
Jalandhar : Ph: 4645630; jalandhar@schandpublishing.com
Kolkata : Ph: 23357458, 23353914; kolkata@schandpublishing.com
Lucknow : Ph: 4003633; lucknow@schandpublishing.com
Mumbai : Ph: 25000297; mumbai@schandpublishing.com
Patna : Ph: 2260011; patna@schandpublishing.com

First Edition 2012
Reprints 2018, 2019, 2020, 2023
Reprint 2024

ISBN: 978-81-219-3909-6 **Product Code:** H3BEE52ELEC10ENAA12R

PRINTED IN INDIA

By Vikas Publishing House Private Limited, Plot 20/4, Site-IV, Industrial Area Sahibabad, Ghaziabad – 201 010 and Published by S Chand And Company Limited, A-27, 2nd Floor, Mohan Co-operative Industrial Estate, New Delhi – 110 044.

Reverentially dedicated to the sacred memory of
Padma Shree Late Sh. Shyam Lal Gupta
Founder of S. Chand & Company Ltd.,
a man of iron will and a golden heart who made
"ordinary men touch the extraordinary"

Preface to the First Edition

The engineer has been, and is, maker of history.
James Kip Finch

The authors feel great pleasure in presenting this unique book to meet the requirements of Ist Year B.E. (all branches) and B. Tech. students of different Technical Universities in India. It attempts to cover the common subject of *Basic Electrical Engineering* introduced recently. The title of the book "*ABC of Electrical Engineering*" itself expresses the philosophy of the book. It gives importance to the basic fundamental concepts and avoids the non-essentials which often confuse the students. The authors believe that everything that can be said can be said clearly. All definitions, derivations and descriptions have been given in a very clear (not confusing) language and in a simple (not complex) manner. How some authors make things difficult and confusing for the students may be judged from the following two examples:

1. Definition of Electrical Force

"*Electrical force is defined as something which causes motion of electrical charge; an electrical charge is something which exerts electrical force.*"

Unfortunately, such roundabout definitions leave students quite confused and clueless.

2. Definition of Norton Theorem

"*Any two-terminal network can be replaced by an equivalent circuit consisting of a perfect current source, I_N — with a current equal to the short-circuit current of the original circuit with the load resistor removed — in parallel with a resistance R_N, equal to the resistance measured back into the original circuit with the load resistance removed and all internal sources in the network removed and replaced with their equivalent internal resistances*".

The above definition is unnecessarily long and is so worded that most of the Ist Year engineering students simply fail to understand it. Now, compare it with the simple definition given by the authors at page 53 of the book.

"*Any given complicated linear circuit can be reduced to a single current source with a parallel resistance.*"

The students will find the above definition quite easy to remember and reproduce in their examination paper.

Plenty of Worked Examples (taken from examination papers of various Technical Universities in India) have been included in the book to help students grasp the basic principle behind them. End-of-chapter problems will enable students to sharpen their newly-acquired problem-solving skills. Similarly, the multiple choice questions (MCQs) given at the end of each chapter are meant to check the students' knowledge of the subject studied by them.

This book would be incomplete without expressing our gratitude to those individuals who motivated us to write it. First and foremost among them is Sh. Navin Joshi, Executive Vice-President (Publishing) who mooted the idea of writing this book and provided us with all the examination material from all over India. Authors offer their sincere thanks to Sh. Amit Gupta, CEO, S. Chand & Company Ltd. whose benevolent guidance and vision inspire not only authors but also contribute to efficient and harmonious working of S. Chand & Company Ltd. Finally, our sincere thanks go to Ms. Sucheta Dogra who has drawn very neat and simple line drawings of rotating electric machinery, single-phase transformer and various electric power generating stations.

In the end, we would request our esteemed readers to kindly send us their valuable suggestions for the improvement of the book and notify us of errors (if any) they may come across while going through the book. Both will be gratefully acknowledged.

B.L. Theraja
A.K. Theraja

Contents

CHAPTER 1 ELECTRIC CURRENT AND RESISTANCE

Q. 1. (*a*) Define the term electric current. What is its unit?

(*b*) The number of electrons flowing per second in a conducting wire is 18.75 × 10^{18}. Calculate the value of the current produced in the wire.

Ans. (*a*) All conductors contain electrons which are free to move among its different atoms. If no voltage is applied across the two ends of the conductor, no electric field is set up in the conductor. That is why the electrons move around freely between different atoms in a random way as seen from Fig. 1(*a*). It is because of this that there is no movement of charge in any *particular* direction within the conductor. However, when a voltage is applied across the conductor, the electrons are attracted towards the positive side of the applied voltage as shown in Fig. 1 (*b*).

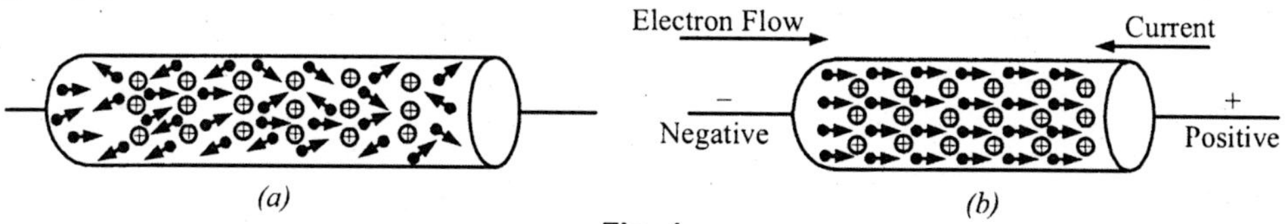

Fig. 1

As a result, there is flow of electronic charge in one *particular* direction. This flow of electronic charge in one particular direction is called an electric current. In other words, an electric current is the rate of flow of electric charge. However, the conventional direction of current is taken opposite to the direction of flow of negatively-charged electrons. The current is measured in terms of the rate of flow of positive charge. If a charge q moves across a given point in a conductor in time t, the current is given by $i = q/t$

If $q = 1$ coulomb (C) and $t = 1$ second, then

$$i = \frac{1 \text{ coulomb}}{1 \text{ second}} = 1 \text{ ampere}$$

Hence, one ampere current flows in a conductor when one coulomb of charge moves across a given point in one second. In simple words, *one ampere equals one coulomb per second.* Now, one coulomb numerically equals the charge of 6.25 × 10^{18} electrons. Hence, for a current of one ampere, 6.25 × 10^{18} electrons per second would have to move across a given point of the conductor.

(*b*) By definition, one ampere is the flow of 6.25 × 10^{18} electrons per second.

$$\therefore \quad i = \frac{18.75 \times 10^{18} \text{ electrons per second}}{6.25 \times 10^{18} \text{ electron per second}} = \mathbf{3\ A}$$

Q. 2. (*a*) State the Laws of Electric Resistance. Define resistivity and give its unit.

(*b*) A coil consists of 2000 turns of copper wire having a cross-sectional area of 0.8 mm^2. The mean length per turn is 80 cm and the resistivity of copper is 0.2 m W-m. Find the resistance of the coil and power absorbed by the coil when connected across 110 V d.c. supply. *(F.Y. Engg. Pune Univ., 2005)*

Ans. (*a*) According to the Laws of Resistance, the resistance R offered by a conductor depends on the following four factors :

1. It varies directly as its length, l.
2. It varies inversely as the cross-section A of the conductor.

3. It depends on the nature of the conductor material.
4. It depends on the temperature of the conductor.

Neglecting the last factor for the time being, we can say that,

$$R \propto \frac{l}{A} \quad \text{or} \quad R = \rho\frac{l}{A} \qquad ...(i)$$

where ρ (rho) is a constant which depends on the nature of the conductor material and is known as its specific resistance (or resistivity).

If in the above equation

$l = 1$ m and $A = 1\text{m}^2$ (*i.e.* 1 m^3 of the material) then, $R = \rho$

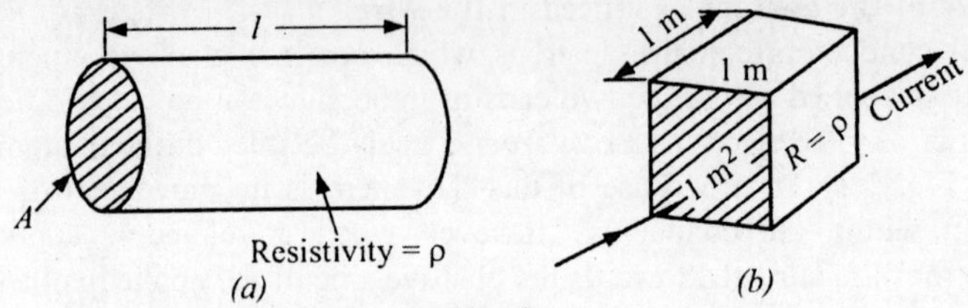

Fig. 2

Hence, the resistivity of the material can be defined as *the resistance between opposite faces of a metre cube of that material.*

From Eq. (*i*), we have

$$\rho = \frac{A\ \text{metre}^2 \times R\ \text{ohm}}{l\ \text{metre}} = \frac{AR}{l}\ \text{ohm} - \text{metre}$$

Hence, the unit of the resistivity is ohm-metre (Ω-m)

(*b*) Here, $l = 0.8 \times 2000 = 1600$ m, $A = 0.8\ \text{mm}^2 = 0.8 \times (10^{-3})^2 = 0.8 \times 10^{-6}\ \text{m}^2$

$$\rho = 0.02 \times 10^{-6}\,\Omega\text{-m}$$

$$\therefore \qquad R = \rho\frac{l}{A} = 0.02 \times 10^{-6} \times 1600/0.8 \times 10^{-6} = \mathbf{40\ \Omega}$$

Power absorbed = $V^2/R = 110^2/40 =$ **302.5 W**

Q. 3 (*a*) A rectangular carbon block has dimensions 1.0 cm × 1.0 cm × 50 cm. (*i*) What is the resistance measured between its two square ends? (*ii*) between two opposing rectangular faces? Resistivity of carbon at 20 °C is 3.5×10^{-5} Ω-m.

(*b*) A current of 5 A exists in a 10 – Ω resistor for 4 minutes.

(*i*) how many coulombs and

(*ii*) how many electrons pass through any section of the resistor in this time? Charge of an electron = 1.6×10^{-19} C. *(MS. Univ. Baroda 2001)*

Ans. (*a*) (*i*) $\qquad R = \rho\dfrac{l}{A}$

Here, $A = (10^{-2}) \times (10^{-2}) = (10^{-2})^2 = 10^{-4}\ \text{m}^2$; $l = 50/100 = 0.5$ m

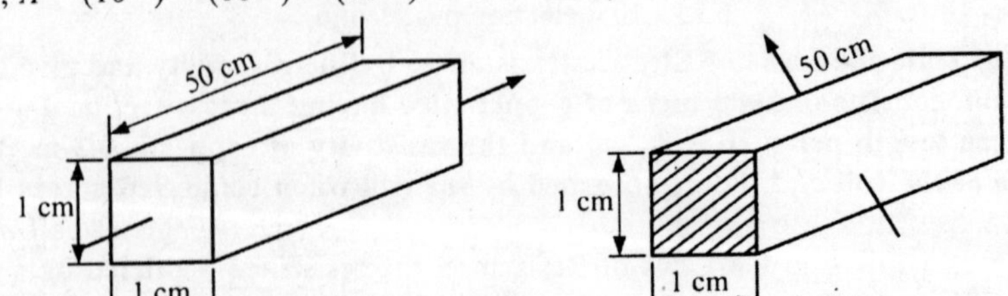

Fig. 3

$$\therefore \qquad R = 3.5 \times 10^{-5} \times 0.5/10^{-4} = \mathbf{0.175\ \Omega}$$

(*ii*) Here, $l = 1 \text{ cm} = 10^{-2} \text{ m}$; $A = 1 \times 50 = 50 \text{ cm}^2 = 5 \times 10^{-3} \text{ m}^2$

$\therefore \quad R = 3.5 \times 10^{-5} \times 10^{-2}/5 \times 10^{-3} = \mathbf{7 \times 10^{-5}\ \Omega}$

(*b*) (*i*) $Q = It = 5 \times (4 \times 60) = \mathbf{1200\ C}$

(*ii*) $Q = ne$ or $n = \dfrac{Q}{e} = \dfrac{1200}{1.6 \times 10^{-19}} = \mathbf{7.5 \times 10^{20}}$

Q. 4 (*a*) A piece of silver wire has a resistance of 1 Ω. What will be the resistance of manganin wire of one-third the length and one-third the diameter, if the specific resistance of manganin is 30 times that of silver. *(Electrical Engg. I, Delhi Univ.)*

(*b*) The resistivity of a ferric-chromium-aluminium alloy is 51 × 10^{-8} Ω-m. A sheet of the material is 15 cm long, 6 cm wide and 0.014 cm thick. Determine resistance between (*i*) opposite ends and (*ii*) opposite sides. *(Electric Circuits, Allahabad Univ.)*

Ans. (*a*) For silver wire, $R_1 = \rho_1 = \dfrac{l_1}{A_1}$; for manganin wire, $R_2 = \rho_2 \dfrac{l_2}{A_2}$

$$\therefore \quad \frac{R_2}{R_1} = \frac{\rho_2}{\rho_1} \times \frac{l_2}{l_1} \times \frac{A_1}{A_2}$$

Now $A_1 = \pi\ d_1^2/4$ and $A_2 = \pi\ d_2^2/4$

$$\therefore \quad A_1/A_2 = d_1^2/d_2^2$$

$$\therefore \quad \frac{R_2}{R_1} = \frac{\rho_2}{\rho_1} \times \frac{l_2}{l_1} \times \left(\frac{d_1}{d_2}\right)^2$$

$R_1 = 1\ \Omega$; $l_2/l_1 = 1/3$, $(d_1/d_2)^2 = (3/1)^2 = 9$; $\rho_2/\rho_1 = 30$

$\therefore \quad R_2 = 1 \times 30 \times (1/3) \times 9 = \mathbf{90\ \Omega}$

(*b*) (*i*) As seen from Fig. 4 (*a*), in this case

$l = 15 \text{ cm} = 0.15 \text{ m}$

$A = 6 \times 0.014 = 0.084 \text{ cm}^2$
$= 0.084 \times 10^{-4} \text{ m}^2$

$$R = \rho\frac{1}{A} = \frac{51 \times 10^{-8} \times 0.15}{0.084 \times 10^{-4}}$$

$= \mathbf{9.1 \times 10^{-3}\ \Omega}$

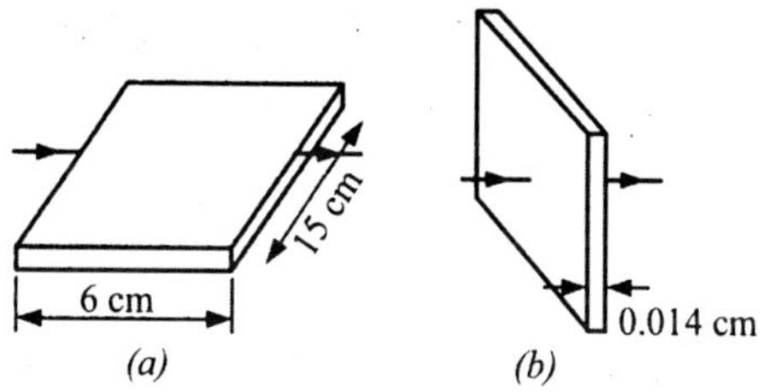

Fig. 4

(*ii*) As seen from Fig. 4(*b*), here

$l = 0.014 \text{ cm} = 14 \times 10^{-5} \text{ m}$

$A = 15 \times 6 = 90 \text{ cm}^2 = 9 \times 10^{-3} \text{ m}^2$

$\therefore \quad R = 51 \times 10^{-8} \times 14 \times 10^{-5}/9 \times 10^{-3}$

$= \mathbf{79.3 \times 10^{-10}\ \Omega}$

Q. 5 (*a*) How do you distinguish between conductance and conductivity of a conductor? What are their units? How are the two related to each other?

(*b*) Find the resistance and conductance of 5000 m of an aluminium wire. The cross-section is circular with a diameter of 2 mm. What will the resistance be if the diameter is doubled? Resistivity of aluminium is 28.3 × 10^{-9} Ω-m.

Ans. (*a*) The conductance of a resistor is the reciprocal of its resistance. The resistance of a conductor measures the *opposition* which it offers to the flow of current. The conductance means how *easily* it lets the current flow through it. The symbol of conductance is G and its unit is siemens. (S).

$$\therefore \qquad G = \frac{1}{R} \text{ siemens}$$

The conductivity (or specific conductance) of a conductor is given by the reciprocal of resistivity. Its symbol is σ (sigma) and its unit is siemens/metre (S/m).

(b) Wire diameter, $d = 2$ mm $= 2 \times 10^{-3}$ m. Hence $A = \pi d^2/4 = \pi\ (2 \times 10^{-3})^2/4$ m²

$$\therefore \qquad R_1 = \rho\,\frac{l}{A} = \frac{28.3 \times 10^{-9} \times 5 \times 10^3}{\pi\,(2 \times 10^{-3})^2/4} = \mathbf{45\ \Omega}$$

$$\therefore \qquad \text{Conductance } G_1 = \frac{1}{R_1} = \frac{1}{45} = 0.0222 \text{ S}$$

$$= 22.2 \times 10^{-3} \text{ S} = \mathbf{22.2\ mS}$$

If the diameter of the wire is doubled, its resistance becomes one-fourth of its original value

$$\therefore \qquad R_2 = \frac{R_1}{4} = \frac{45}{4}\,\Omega$$

$$\therefore \qquad G_2 = \frac{1}{R_2} = \frac{1}{45/4} = \frac{4}{45} = \mathbf{88.8\ mS}$$

Q. 6. What is the effect of the temperature rise on the resistance of (*i*) pure metals (*ii*) alloys and (*iii*) insulators and partial conductors such as carbon?

Ans. (*a*) The effect of temperature rise is as under :

(*i*) It *increases* the resistance of pure metals quite a bit. In fact , temperature/resistance graph is a straight line. Metals are said to have a *positive* temperature-coefficient of resistance.

(*ii*) It *increases* the resistance of alloys although, in their case, the increase in resistance is small and irregular. For some high - resistance alloys like Eureka (60% Cu and 40% Ni) and manganin, the increase in resistance is *negligible* over a large range of temperatures.

(*iii*) It *decreases* the resistance of the insulators (such as paper, rubber, glass, mica etc.) and partial conductors such as carbon. That is why insulators are said to have a *negative* tempeature-coefficient of resistance. In other words, their resistance *decreases* with temperature.

Q. 7 (*a*) What is meant by the term "temperature coefficient" of a resistor? Does it have any unit?

(*b*) A platinum coil has a resistance of 3.146 Ω at 40 °C and 3.767 Ω at 100 °C. Find the resistance at 0 °C and the temperature-coefficient of resistance at 40 °C.

(Electrical Science-II, Allahabad Univ.)

Ans. (*a*) Let a metallic conductor have a resistance of R_0 at 0°C. When heated to t °C, let its resistance be R_t. Obviously, the increase in resistance of the conductor is $\Delta R = R_t - R_0$. This increase depends

(*i*) directly on the initial resistance R_0

(*ii*) directly on rise in temperature, t °C

(*iii*) on the nature of the conductor material

$$\therefore \qquad R_t - R_0 \propto R_0 \times t \quad \text{or} \quad R_t - R_0 = \alpha\, R_0\, t \qquad \ldots(i)$$

where α (alpha) is a constant and is known as the temperature-coefficient of resistance of that conductor.

Re-arranging Eq. (*i*), we get

$$\alpha = \frac{R_t - R_0}{R_0 \times t} = \frac{\Delta R}{R_0 \times t}$$

If $R_0 = 1\ \Omega$, $t = 1$ °C, then $\alpha = \Delta R$

Hence, temperature-coefficient may be defined *as the change in resistance per ohm resistance per degree rise in temperature from 0 °C.*

Equation (*i*) can be written as

$$R_t = R_0 \ (1 \pm \alpha t)$$

The above equation holds good for both rise and fall in temperature.

As seen from above,

$$\alpha = \frac{R_t - R_0}{R_0 \times t} = \frac{\Delta R \text{ ohm}}{R_0 \text{ ohm} \times t\,°\text{C}} = \frac{\Delta R}{R_0 \times t} \text{ per °C}$$

Hence, unit of temperature - coefficient is per °C ($°C^{-1}$).

(*b*) $$R_{100} = R_0 \ (1 + 100 \ \alpha_0) \quad ...(i)$$

$$R_{40} = R_0 \ (1 + 40 \ \alpha_0) \quad ...(ii)$$

$$\therefore \quad \frac{3.767}{3.146} = \frac{1 + 100\,\alpha_0}{1 + 40\,g\alpha_0} \quad \text{or} \quad \alpha_0 = 0.00379 \quad \text{or} \quad 1/264 \text{ °C}$$

From (*i*) we have $3.767 = R_0 \ (1 + 100 \times 0.00379)$ $\therefore$ $R_0 = \mathbf{2.732 \ \Omega}$

Now $$\alpha_{40} = \frac{\alpha_0}{1 + 40\,\alpha_0} = \frac{0.00379}{1 + 40 \times 0.00379} = \frac{\mathbf{1}}{\mathbf{304}} \textbf{ per °C}$$

Q. 8. Derive the relation between temperature-coefficient of a resistor at two different temperatures. Find a general equation between temperature-coefficients at t_1 °C and t_2 °C.

Ans. It has been found that for any given material, the value of α is not constant. It varies and depends on the initial temperature of the resistance. When the increase is based on the resistance measured at 0 °C, then α has the value of α_0. If increase is based on the initial temperature of t °C, then value of α is α_t and so on.

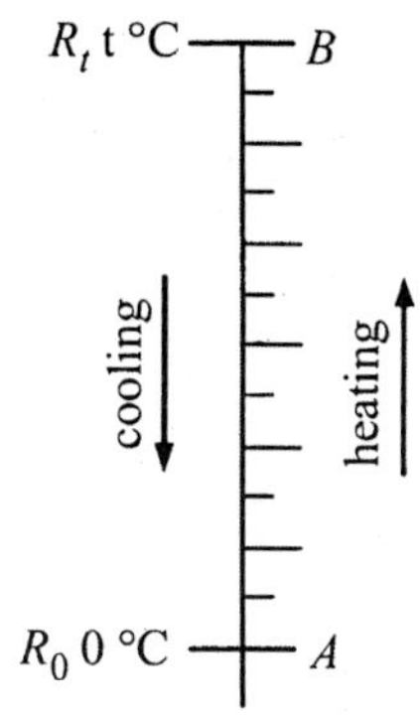

Fig. 5

As seen from Fig. 5, the conductor is heated from point *A* to point *B*. Since the initial temperature is 0 °C, we will take α as α_0. The value of resistance at point *B* is given by

$$R_t = R_0 \ (1 + \alpha_0 \ t) \quad ...(i)$$

Next let the conductor be cooled from initial temperature t °C (point *B*) to final temperature 0 °C (point *A*). Since the initial temperature is t °C, we will take α as α_t. During cooling, there is a decrease in temperature of $(-\ t)$ °C. In this case, the initial temperature is t °C and final temperature is 0 °C. Hence, we will take α as α_t. The value of R_0 in terms of R_t is given by

$$R_0 = R_t \ [1 + \alpha_t \ (-\ t)] = R_t \ (1 - \alpha_t \ t) \quad ...(ii)$$

From equation (*ii*) above, we have

$$\alpha_t = \frac{R_t - R_0}{R_t \times t}$$

Putting the value of R_t from Eq. (*i*) above, we get

$$\alpha_t = \frac{R_0\ (1 + \alpha_0 t) - R_0}{R_0\ (1 + \alpha_0 t) \times t} = \frac{(1 + \alpha_0 t) - 1}{(1 + \alpha_0 t) \times t} = \frac{\alpha_0}{1 + \alpha_0 t}$$

$$\therefore \quad \alpha_t = \frac{\alpha_0}{1 + \alpha_0 t} \quad ...(iii)$$

General relation

In general, let α_1 = tempt – coefficient at t_1 °C

α_2 = tempt – coefficient at t_2 °C

Then, from Eq. (*iii*) above, we get

$$\alpha_1 = \frac{\alpha_0}{1+\alpha_0 t_1} \quad \text{or} \quad \frac{1}{\alpha_1} = \frac{1+\alpha_0 t}{\alpha_0}$$

Similarly, $$\frac{1}{\alpha_2} = \frac{1+\alpha_0 t_2}{\alpha_0}$$

Subtracting one from the other, we get

$$\frac{1}{\alpha_2} - \frac{1}{\alpha_1} = (t_2 - t_1) \quad \text{or} \quad \frac{1}{\alpha_2} = \frac{1}{\alpha_1} + (t_2 - t_1) \quad \therefore \quad \alpha_2 = \frac{1}{1/\alpha_1 + (t_2 - t_1)}$$

Q. 9 (*a*) Derive a relation between the known resistance R_1 at t_1 °C and the unkown resistance R_2 at t_2 °C when α_0 is not known.

(*b*) A potential difference of 250 V is applied to a field winding at 15 °C and the current is 5 A. What will be the mean temperature of the winding when current has fallen to 3.91 A, applied voltage being constant. Assume α_{15} = 1/254.5.

(Elect. Engg., Pune Univ.)

Ans. (*a*) The known resistance R_1 is given by the expression

$$R_1 = R_0\,(1 + \alpha_0\, t_1)$$

Similarly, the unknown resistance R_2 is given by

$$R_2 = R_0\,(1 + \alpha_0\, t_2)$$

Dividing one equation by the other, we get

$$\frac{R_2}{R_1} = \frac{1+\alpha_0\, t_2}{1+\alpha_0\, t_1} = (1+\alpha_0\, t_2)\,(1+\alpha_0\, t_1)^{-1}$$

$$= (1 + \alpha_0\, t_2)\,(1 - \alpha_0\, t_1)$$

$$= 1 + \alpha_0\,(t_2 - t_1) \qquad \text{– approximately}$$

$$\therefore \quad R_2 = R_1\,[1 + \alpha_0\,(t_2 - t_1)]$$

If α_0 is not known, then to a very close approximation, the above equation can be modified as

$$R_2 = R_1\,[1 + \alpha_1\,(t_2 - t_1)]$$

(*b*) Let R_1 = winding resistance at 15 °C

R_2 = winding resistance at unknown mean temperature t_2 °C

$$\therefore \quad R_1 = 250/5 = 50\ \Omega\ ;\ R_2 = 250/3.91 = 63.94\ \Omega$$

Now, $$R_2 = R_1\,[1 + \alpha_{15}\,(t_2 - t_1)]$$

$$\therefore \quad 63.94 = 50\left[1 + \frac{1}{254.5}(t_2 - 15)\right] \quad \therefore \quad t_2 = \mathbf{86\ °C}$$

Q. 10. (*a*) What is the significance of negative temperature-coefficient of resistance?

(*b*) A coil has a resistance of 18 Ω when its mean temperature is 20 °C and of 20 Ω when its mean temperature is 50 °C. Find its mean temperature rise when its resistance is 21 Ω and the surrounding temperature is 15 °C.

(Electrical Technology, Allahabad University 2003)

Ans. (*a*) The temperature coefficient of insulators like paper, rubber, glass, mica and partial conductors like carbon and many semi-conductor materials is *negative*. It means that their resistance *decreases* with temperature. For this reason, these materials are used in temperature control devices (like thermistors) and electrical indicating thermometers.

(*b*) Let R_0 be the resistance of the coil and α_0 its tempt. coefficient at 0 °C.
Then, $18 = R_0\,(1 + \alpha_0 \times 20)$ and $20 = R_0\,(1 + \alpha_0 \times 50)$
Dividing one by the other, we get

$$\frac{20}{18} = \frac{1 + 50\,\alpha_0}{1 + 20\,\alpha_0} \qquad \therefore \quad \alpha_0 = \frac{1}{250} \text{ per °C}$$

If t °C is the temperature of the coil when its resistance is 21 Ω, then,

$$21 = R_0\,(1 + t/250)$$

Dividing this equation by the above equation, we have

$$\frac{21}{18} = \frac{R_0\,(1 + t/250)}{R_0\,(1 + 20\,\alpha_0)}\ ;\ t = 65 \text{ °C} ; \quad \text{tempt. rise} = 65 - 15 = \mathbf{50\ °C}$$

Q. 11. (*a*) Two coils connected in series have resistance of 600 Ω and 300 Ω with tempt. coeff. of 0.1% and 0.4% respectively at 20 °C. Find the resistance of the combination at a tempt. of 50 °C . What is the effective tempt. coefficient of the combination?

(*b*) Two wires *A* and *B* are connected in series at 0 °C and resistance of *B* is 3.5 times that of *A*. The resistance temperature coefficient of *A* is 0.4% and that of the combination is 0.1%. Find the resistance temperature coefficient of *B*.

(Elect. Technology, Hyderabad Univ.)

Ans. (*a*) Resistance of 600 Ω resistor at 50 °C

$$= 600\,[1 + 0.001(50 - 20)] = 618\ \Omega$$

Similarly, resistance of 300 Ω resistor at 50 °C

$$= 300\,[1 + 0.004\,(50 - 20)] = 336\ \Omega$$

Hence, total resistance of combination at 50 °C

$$= 618 + 336 = \mathbf{954\ \Omega}$$

Let β = combination resistance temperature coefficient at 20 °C
Now, combination resistance at 20 °C = 600 + 300 = 900 Ω
Combination resistance at 50 °C = 954 Ω

$\therefore$ $954 = 900\,[1 + \beta\,(50 - 20)]$

$\therefore$ $\beta = \mathbf{0.002\ per\ °C}$

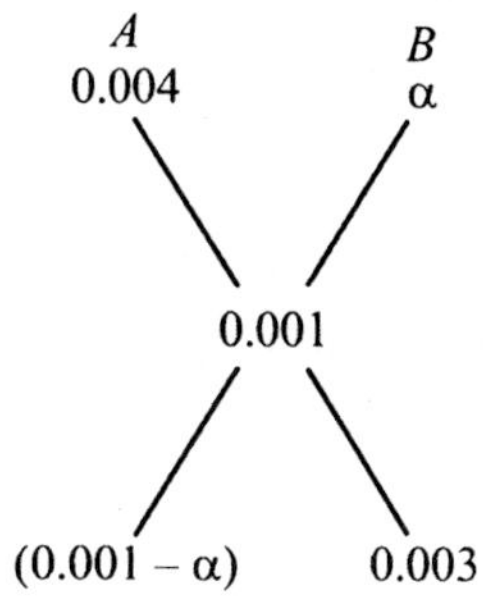

Fig. 6

(*b*) A simple technique which gives quick results in such questions is illustrated by the diagram of Fig. 6.
It is seen that

$$R_B/R_A = 0.003/(0.001 - \alpha)$$

or $3.5 = 0.003/(0.001 - \alpha)$

or $\alpha = 0.000143\ °C^{-1}$ or **0.0143%**

It may be noted that combination temperature coefficient of 0.001 is smaller than the coefficient of wire *A* but larger than the coefficient of wire *B*. That is why we have taken the difference as (0.001 – α) and not (α – 0.001) as shown in Fig. 6.

Q. 12. (*a*) Two conductors, one of copper and the other of iron, are connected in parallel and carry equal currents at 25 °C. What proportion of current will pass through each if the temperature is raised to 100 °C? The temperature coefficient of resistane at 0 °C are 0.0043/°C and 0.0063/°C for copper and iron respectively.

(Principles of Elect. Engg. Delhi Univ. 2000)

(*b*) A semi-circular ring of copper has an inner radius of 6 cm, radial thickness 3 cm and an axial thickness 4 cm. Find the resistance of the ring at 50°C between its two end-faces. Assume specific resistance of Cu at 20 °C = 1.724 × 10^{-6} ohm-cm and resistance tempt. coeff. of Cu at 0 °C = 0.0043/ °C.

Ans. (*a*) Since the copper and iron conductors carry equal currents at 25 °C, their resistances are the same at that temperature. Let each be *R* ohm.

For copper, $R_{100} = R_1 = R\,[1 + 0.0043\,(100 - 25)] = 1.3225\,R$

For iron $R_{100} = R_2 = R\,[1 + 0.0063\,(100 - 25)] = 1.4725\,R$

If I is the current at 100 °C, then as per current divider rule, current in the copper conductor is

$$I_1 = I \times \frac{R_2}{R_1 + R_2} = I \times \frac{1.4725\,R}{1.3225\,R + 1.4725\,R} = 0.5268\,I$$

$$I_2 = I \times \frac{R_1}{R_1 + R_2} = I \times \frac{1.3225\,R}{2.795\,R} = 0.4732\,I$$

Hence, copper conductor will carry **52.68%** of the total current and iron conductor will carry the balance *i.e.* **47.32%.**

(*b*) The semi-circular ring is shown in Fig. 7.

Mean radius of ring = (6 + 9)/2 = 7.5 cm

Mean length between end faces = 7.5 π cm = 23.56 cm

Cross-section of the ring = 3 × 4 = 12 cm^2

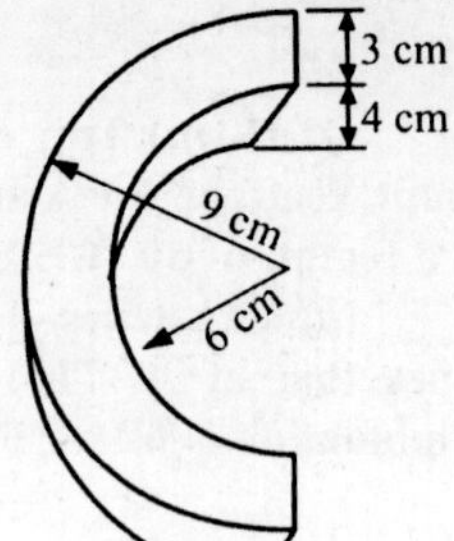

Fig. 7

Now $\alpha_0 = 0043/°C$

$$\alpha_{20} = \frac{0.0043}{1 + 20 \times 0.0043} = 0.00396$$

$$\rho_{50} = \rho_{20}\,[1 + \alpha_0\,(50 - 20)]$$

$$= 1.724 \times 10^{-6}\,(1 + 30 \times 0.00396)$$

$$= 1.93 \times 10^{-6}\ \Omega\text{-cm}$$

$$\therefore \quad R_{50} = \frac{\rho_{50} \times l}{A} = \frac{1.93 \times 10^{-6} \times 23.56}{12} = \mathbf{3.79 \times 10^{-6}\ \Omega}$$

Q. 13 (*a*) Explain Ohm's law and state its limitations.

(*b*) Find the currents in all the resistors of the network shown in Fig. 9(*a*). Also find the voltage across *AB*. *(Basic Elect. Engg. VTU Karnataka, 2008)*

Ans. (*a*) Ohm's law gives the relation between voltage, current and resistance. According to this law, *current in a circuit is directly proportional to voltage provided the resistance remains constant.*

In simple words, it means that voltage and current are linearly related. If one increases or decreases by a certain percentage, the other also increases or decreases by the *same percentage* provided the value of *resistance remains constant.* Such a resistor is called a linear resistor.

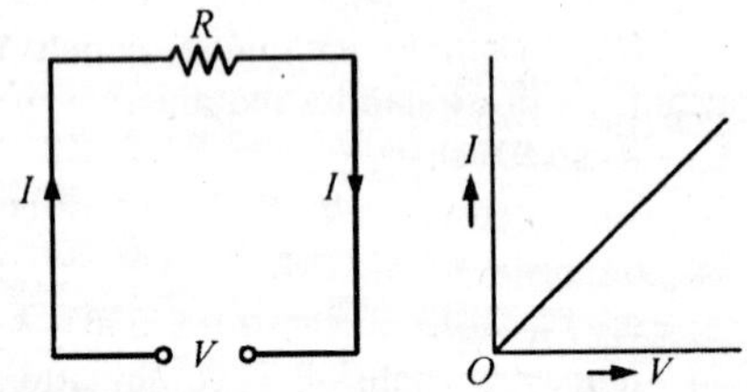

Fig. 8

The graph of voltage V versus current I is a straight line as shown in Fig. 8 (*b*).

Mathematically, Ohm's law can be put as

$$V = IR$$

Other forms of Ohm's law are

$$I = \frac{V}{R} \quad \text{and} \quad R = \frac{V}{I}$$

However, it may be carefully noted that Ohm's law does not mean that resistance varies directly with voltage V and inversely with current I. What it actually means is that the ratio of voltage to current for a given resistance is constant.

Ohm's law is applicable to both dc and ac circuits.

Main limitation of Ohm's law is that it is applicable to only linear resistors. It is so because it is impossible to make an ideal resistor. All actual resistors are non-linear to a lesser or greater extent. When current passes through a resistor, it heats up. This rise in temperature changes its resistance. Hence, the relation $V \propto I$ does not hold good.

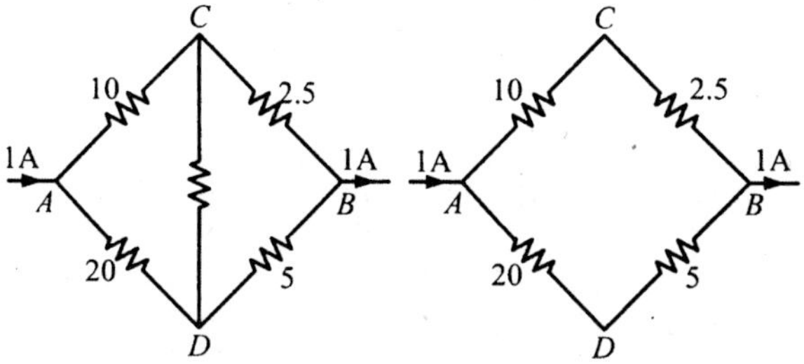

(a) (b)

Fig. 9

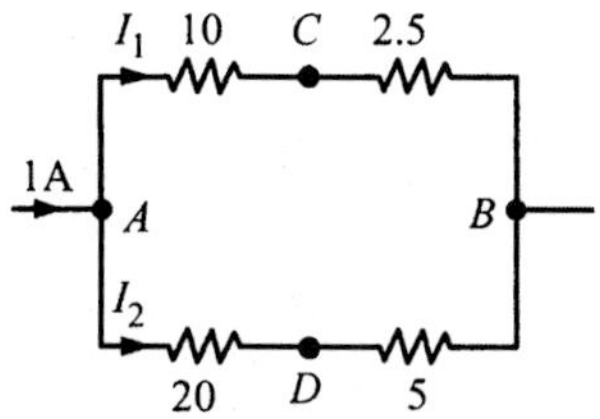

Fig. 10

(b) Since in Fig. 9 (*a*), the products of resistances of the opposite arms of the bridge are equal (each = 50 Ω), hence the bridge is balanced. Therefore, current through the branch *CD* is zero. It means the 5 Ω resistance between *C* and *D* can be removed giving us the circuit of Fig. 9 (*b*).

As seen, the given bridge circuit becomes a parallel circuit. The branch *ACB* has a resistance of $R_1 = (10 + 2.5) = 12.5\ \Omega$. The resistance of parallel branch *ADB* is $R_2 = (20 + 5) = 25\ \Omega$. As per the current divider rule (Fig. 10),

$$I_1 = 1 \times \frac{25}{(25 + 12.5)} = \mathbf{0.67\ A}$$

$$I_2 = 1 \times \frac{12.5}{(25 + 12.5)} = \mathbf{0.33\ A}$$

As seen from Fig.10, $R_{AB} = 12.51125 = 12.5 \times 25 / (12.5 + 25) = 8.33\ \Omega$

$$V_{AB} = I \times R_{AB} = 1 \times 8.33 = \mathbf{8.33\ V}$$

Q. 14 (*a*) Describe in detail the main characteristics of a series circuit.

(*b*) List the main characteristics of a parallel circuit.

Ans. (*a*) Fig. 11 (*a*) shows three resistors R_1, R_2 and R_3 joined in series. Fig. 11(*b*) shows that there is a progressive fall in potential as we go from point *A* to *D*.

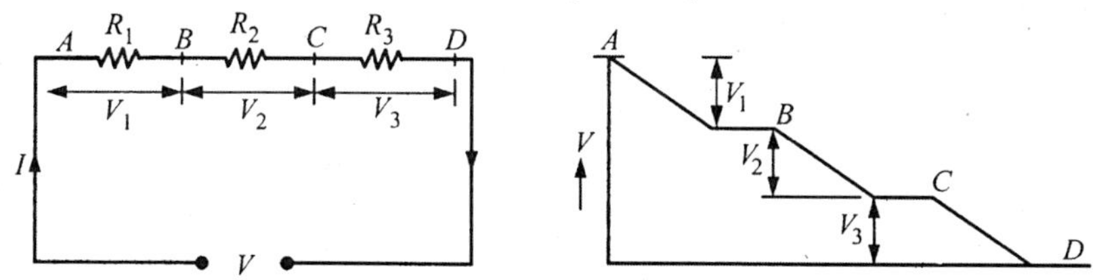

(a) (b)

Fig. 11

The main characteristics of a series circuit are:

(*i*) same current flows through the three resistors
(*ii*) different resistors have different voltage drops
(*iii*) voltage drops are additive
(*iv*) applied voltage equals the sum of the three voltage drops
(*v*) resistances are additive i.e. $R = R_1 + R_2 + R_3$
(*vi*) powers and energies are additive.

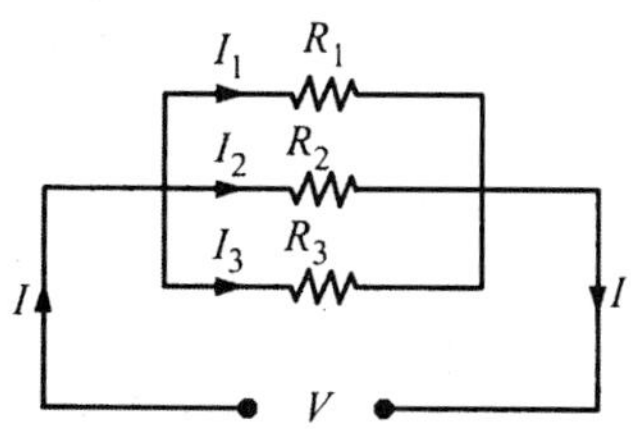

Fig. 12

(*b*) Fig. 12 shows three resistors R_1, R_2 and R_3 connected in parallel across the same voltage source *V*. The three currents are inversely proportional to the resistance of the parallel path as per Ohm's law. The main characteristics of a parallel circuit are:

(*i*) same voltage acts across the three resistors
(*ii*) different resistors have different currents
(*iii*) branch currents are additive i.e. $I = I_1 + I_2 + I_3$
(*iv*) conductances are additive i.e. $G = G_1 + G_2 + G_3$
(*v*) powers and energies are additive.

Q. 15 (*a*) Calculate the current supplied by the battery in the series-parallel circuit of Fig. 13 (*a*). All resistances are in ohm.

(*b*) In the circuit of 14 (*a*), what must the total applied voltage be in order to pass a current of 1 A through the 4 Ω resistor? Also, find the equivalent resistance of the whole combination.

Ans. (*a*) As seen from Fig. 13 (*a*),

$$R_{AB} = 2 \parallel 3 \parallel 6 = 1\ \Omega$$

Hence, the total resistance of the uppermost branch is

$= 1 + 1 = 2\ \Omega.$

This 2 Ω resistance is in parallel with the 2 Ω resistance of branch *CD*. Hence, their equivalent resistance = 2 || 2 = 1 Ω. This 1 Ω resistance is in series with the branch having 10 V battery as shown in Fig. 13 (*b*).

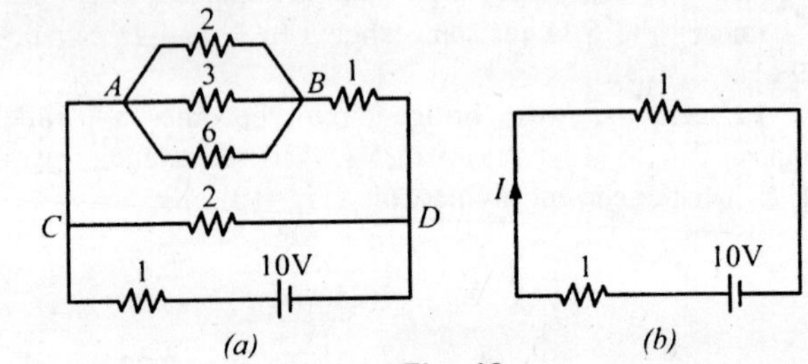

Fig. 13

$$\therefore \quad I = \frac{10}{1+1} = \mathbf{5\ A}$$

(*b*) Since paths *AC* and *AD* are in parallel (points *C* and *D* are connected togther), current divides at point *A* in the inverse ratio of their resistances. Hence, as per current divider rule,

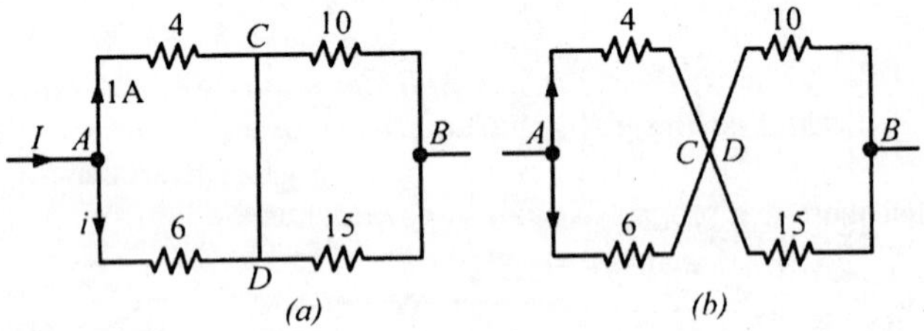

Fig. 14

$$1 = I \times \frac{6}{6+4} = I \times \frac{3}{5}$$

$$\therefore \quad I = \frac{5}{3}\ \text{A}$$

As seen from the Fig. 14 (*b*), combination resistance is given by

$$R_{AB} = 4 \parallel 6 + 10 \parallel 15 = 2.4 + 6 = \mathbf{8.4\ \Omega}$$

Hence, voltage required across the combination to send a current of 1 A through 4 Ω resistor is

$$V_{AB} = I \times R_{AB} = (5/3) \times 8.4 = \mathbf{14\ V}$$

Q. 16 (*a*) Two conductors, one of copper and the other of iron, are connected in parallel and at 20 °C carry equal currents. What proportion of current will pass through each if the temperature is raised to 100 °C? Assume α for copper as 0.0042 and for iron as 0.006 per °C at 20 °C .Find also the values of temperature coefficient at 100 °C.

(Elect. Engg. Madras Univ.)

(*b*) In the circuit of Fig. 16, find the value of supply voltage V so that 20 Ω resistor can dissipate 180 W.

Ans. (*a*) Since they carry equal current at 20 °C, the two conductors have the same resistance at 20 °C *i.e.* R_{20}. As temperature is raised, their resistances increase though unequally.

For Cu, $R_{100} = R_{20}\,(1 + 80 \times 0.0042) = 1.336\,R_{20}$

For iron $R'_{100} = R_{20}\,(1 + 80 \times 0.006) = 1.48\,R_{20}$

As per current divider rule, current through Cu conductor is

$$I_1 = I \times \frac{R'_{100}}{R_{100} + R'_{100}} = I \times \frac{1.48\,R_{20}}{2.816\,R_{20}}$$

$= 0.525$ or **52.5% of *I***

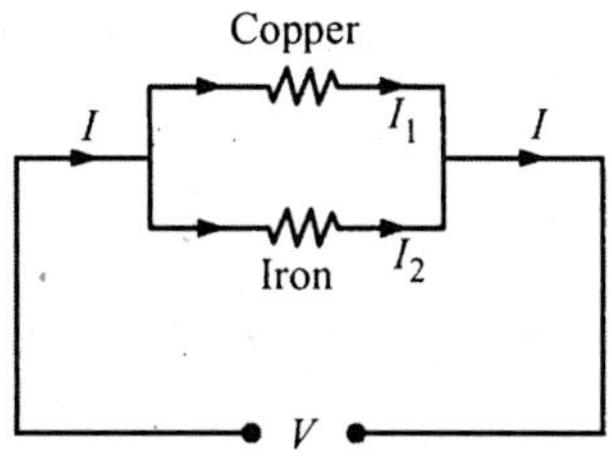

Fig. 15

Hence, current through Cu conductor is 52.5 per cent of the total current. Obviously, the remaining current i.e. 47.5 per cent passes through iron.

Or current through iron conductor is

$$I_2 = I \times \frac{R_{100}}{R_{100} + R'_{100}}$$

$$= I \times \frac{1.336\,R_{20}}{2.816\,R_{20}} \text{ or } \mathbf{47.5\%\ of\ } I$$

For Cu, $\alpha_{100} = \dfrac{1}{(1/0.0042) + 80} = \mathbf{0.00314\ °C^{-1}}$

For iron, $\alpha_{100} = \dfrac{1}{(1/0.006) + 80} = \mathbf{0.0040\ °C^{-1}}$

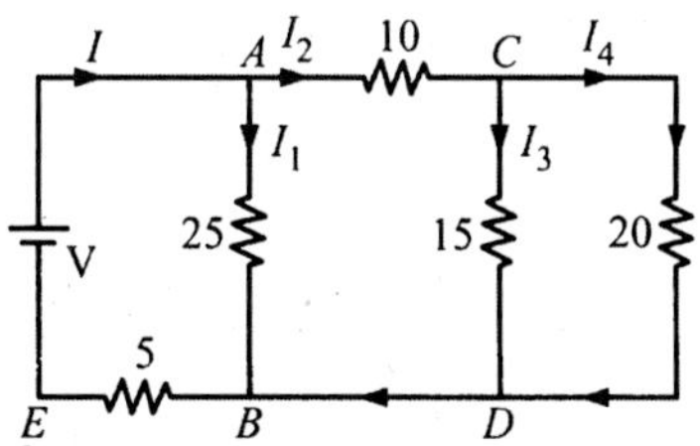

Fig. 16

(*b*) $I_4^2 \times 20 = 180$ W ; $I_4 = 3$ A

Since 15 Ω and 20 Ω are in parallel,

$$I_3 \times 15 = 3 \times 20$$

$\therefore$ $I_3 = 4$ A

$\therefore$ $I_2 = I_3 + I_4 = 4 + 3 = 7$ A

Now, resistance of the circuit to the right of point *A* is

$$= 10 + 15 \times 20/35 = 130/7\ \Omega$$

$\therefore$ $I_1 \times 25 = 7 \times 130/7$

$\therefore$ $I_1 = 26/5$ A $= 5.2$ A

$\therefore$ $I = I_1 + I_2 = 5.2 + 7 = 12.2$ A

Total circuit resistance

$$R_{AE} = 5 + 25 \parallel 130/7 = 955/61\ \Omega$$

$\therefore$ $V = I.R_{AE} = 12.2 \times 955/61 = \mathbf{191\ V}$

Q. 17. A current of 20 A goes through two ammeters *A* ans *B* in the series. The potential difference across *A* is 0.2 V and across *B* is 0.3 V. Find how the same current will divide between *A* and *B* when they are connected in parallel ?

(Basic Elect. Engg. VTU Karnataka Jan. 2008)

Ans. Fig. 17(*a*) shows two ammeters connected in series.Because ammeters also have small amount of internal resistance, hence there is a small voltage drop across them.

$$R_A = \frac{V_A}{20} = \frac{0.2}{20} = 0.01\ \Omega$$

$$R_B = \frac{V_B}{20} = \frac{0.3}{20} = 0.015\ \Omega$$

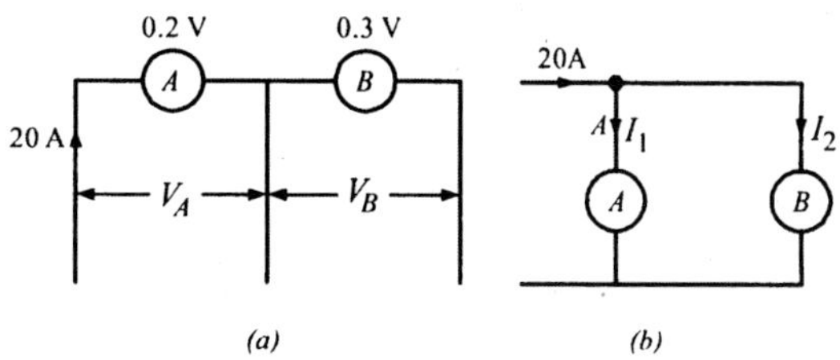

Fig. 17

The ammeters are connected in parallel as shown in Fig. 17(*b*). They act just like small resistances. At point *A*, the 20 A current divides in the inverse ratio of the two ammeter resistances.

$$\therefore \qquad I_1 = 20 \times \frac{0.015}{0.01 + 0.015} = \mathbf{12\ A}$$

$$I_2 = 20 \times \frac{0.01}{0.01 + 0.015} = \mathbf{8\ A}$$

Q. 18. Calculate the battery current *I* in the two circuits of Fig. 18. All resistances are in ohm.

Ans. (*i*) In Fig. 18, all the given three resistors are connected in parallel across points *A* and *D*. For example, 4 Ω resistor is connected to *A* directly and to point *D* via point *B*. Similarly, 16 Ω resistor is connected to *A* and *D* because its two end points *B* and *C* are directly

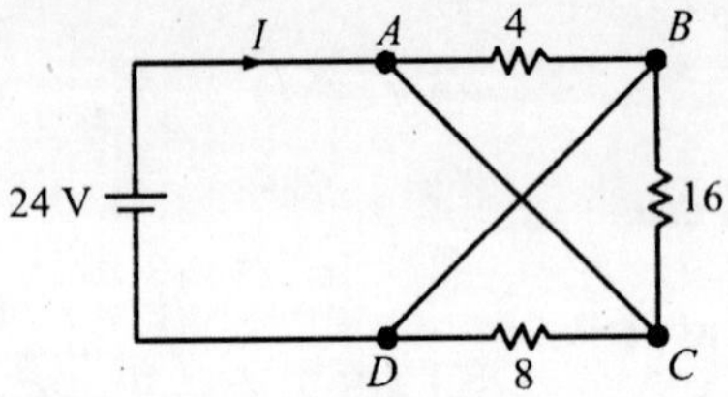

Fig. 18

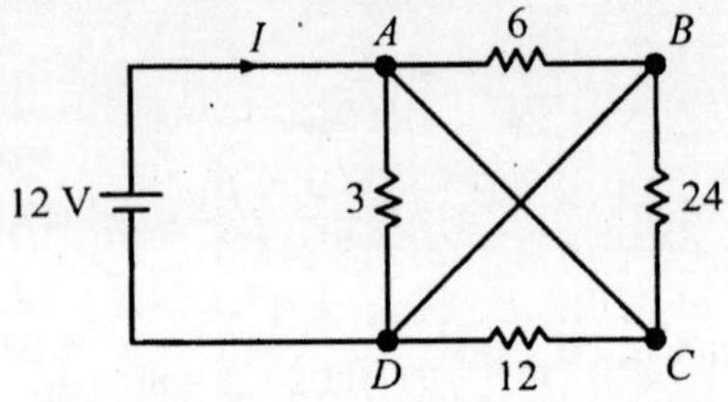

Fig. 19

connected to *D* and *A*. One end of the 8 Ω. resistor is connected directly to *D* and the other to *A* via *C*. Their equivalent resistance is given by

$$\frac{1}{R} = \frac{1}{4} + \frac{1}{8} + \frac{1}{16} = \frac{7}{16}\ \Omega$$

$$\therefore \qquad R = \frac{16}{7}\Omega \qquad \therefore \quad I = \frac{24}{16/7} = \mathbf{10.5\ A}$$

(*ii*) As explained above, the four resistors of Fig. 19 are also connected in parallel.

$$R = 3 \parallel 6 \parallel 12 \parallel 24 = 8/5\ \Omega$$

$$I = \frac{12}{8/5} = \mathbf{7.5\ A}$$

Q. 19 (*a*) What is a voltage divider circuit? What are its applications? Differentiate between a loaded and an unloaded voltage divider.

(*b*) Find the values of different voltages that can be obtained from a 12-V battery with the help of voltage divider circuit of Fig. 20.

Ans. (*a*) A series circuit acts as a voltage divider. As seen from Fig. 20, we can get three different voltages from the three resistors connected in series. Though the three resistors carry the same current, their voltage drops are different as shown below:

Total resistance, $R = R_1 + R_2 + R_3 = 12\ \Omega$

As per the voltage divider rule, the three voltage drops are:

$$V_1 = V \times \frac{R_1}{R} = 24 \times \frac{2}{12} = 4\ \text{V}$$

$$V_2 = V \times \frac{R_2}{R} = 24 \times \frac{4}{12} = 8\ \text{V}$$

$$V_3 = V \times \frac{R_3}{R} = 24 \times \frac{6}{12} = 12\ \text{V}$$

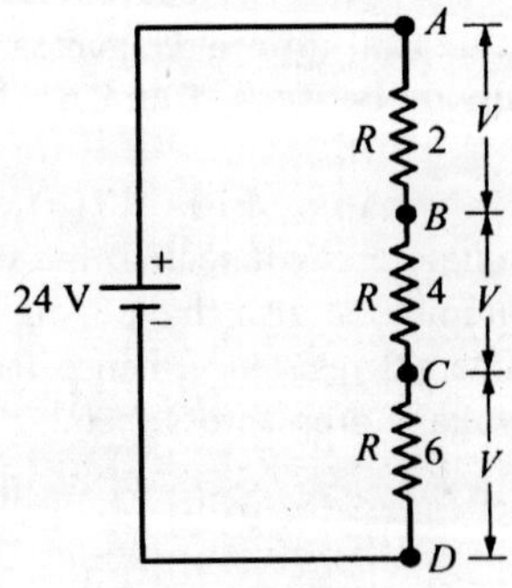

Fig. 20

The volume control of radio or T.V. receivers is a common application of the voltage divider circuit. Another application for the voltage divider is in setting the dc operating voltage (bias) in transistor implifiers.

A simple unloaded voltage divider is shown in Fig. 21(*a*). It gives an output voltage

$$V_0 = 100 \times 30/(20 + 30) = 60 \text{ V}.$$

However, as shown in Fig. 21(*b*), if a load resistor R_L is connected across the output terminals, it is called a loaded voltage divider. In this case, V_0 is reduced by an amount *depending* on R_L.

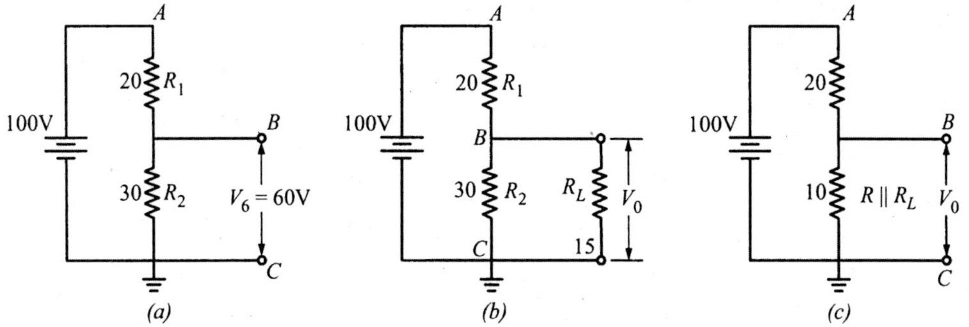

Fig. 21

It is so because resistance across terminals B and C is reduced from R_2 to $R_2 \parallel R_L$. As seen from the Fig. 21(*c*),

$$R_{BC} = 30 \parallel 15 = 10\ \Omega.$$

$$\therefore \quad V_0 = 100 \times 10/(10 + 20) = 33.3 \text{ V}$$

It is found that lesser the value of R_L as compared to R_2, greater the reduction in V_0.

(*b*) $R = R_1 + R_2 + R_3 = 4 + 3 + 1 = \mathbf{8\ \Omega}$.

Drop across $R_1 = 12 \times 4/8 = 6$ V

$\therefore$ $V_B = 12 - 6 = 6$ V above ground

Drop across $R_2 = 12 \times 3/8 = 4.5$ V

$\therefore$ $V_C = V_B - 4.5 = 6 - 4.5 = 1.5$ V

Drop across $R_3 = 12 \times 1/8 = 1.5$ V above ground

Different available load voltages are (Fig. 22)

(*i*) $V_{AB} = V_A - V_B = 12 - 6 = \mathbf{6\ V}$

(*ii*) $V_{AC} = 12 - 1.5 = \mathbf{10.5\ V}$ or

$= 6 + 4.5 = \mathbf{10.5V}$

(*iii*) $V_{AD} = V_A - V_D = 12 - 0 = \mathbf{12\ V}$

(*iv*) $V_{BC} = V_B - V_C = 6 - 1.5 = \mathbf{4.5\ V}$

(*v*) $V_{CD} = V_C - V_D = 1.5 - 0 = \mathbf{1.5\ V}$

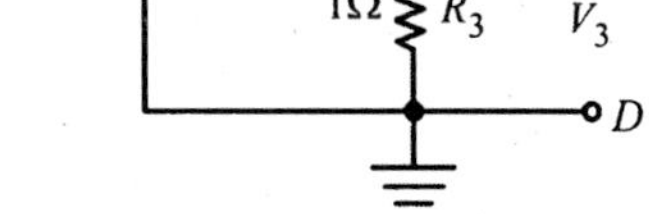

Fig. 22

Q. 20 (*a*) What is a series-parallel circuit ? And parallel-series circuit ?

(*b*) Determine the current supplied by the voltage source in the circuit shown in Fig. 25(*a*). *(Basic Elect. Engg., R.G.P.V. Bhopal 2005)*

Ans. (*a*) A series-parallel circuit is one that contains combination of series-and parallel-connected resistors as shown in Fig. 23 (*a*). It will be seen that the voltage source E and resistor R_1 are in series path and R_2 and R_3 are in parallel. In Fig. 23(*b*), R_2 and R_3 are reduced to a single resistance of $R_2 \parallel R_3$ which is in series with R_1. Hence, the series-parallel circuit is reduced to a single series circuit.

A parallel-series circuit is the one which is a parallel circuit containing series resistors as shown in Fig. 24(*a*).

This circuit can be reduced to an equivalent parallel circuit as shown in Fig. 24(*b*). It is done by reducing the series-connected resistors to a single equivalent resistor in each parallel path.

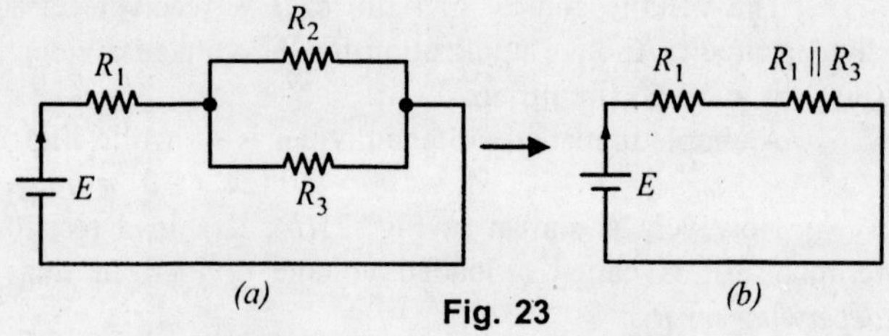

Fig. 23

It should be remembered that a series-parallel circuit is reduced to an equivalent series circuit and a parallel-series circuit to an equivalent parallel circuit. In this way, complex circuits can be simplified by applying this approach.

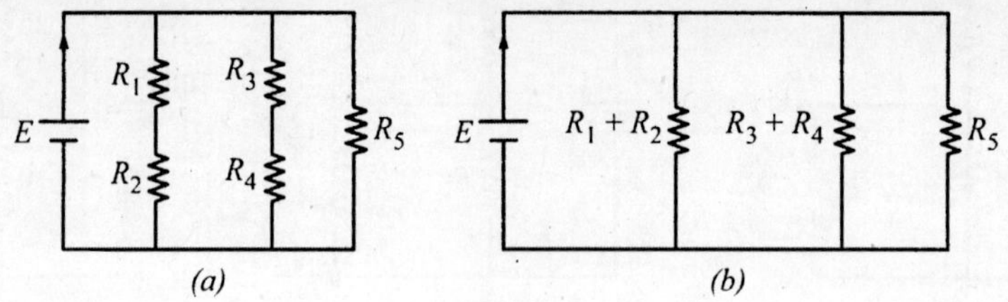

Fig. 24

(***b***) For finding the supply current *I,* we will first simplify the circuit by using the series and parallel resistance laws.

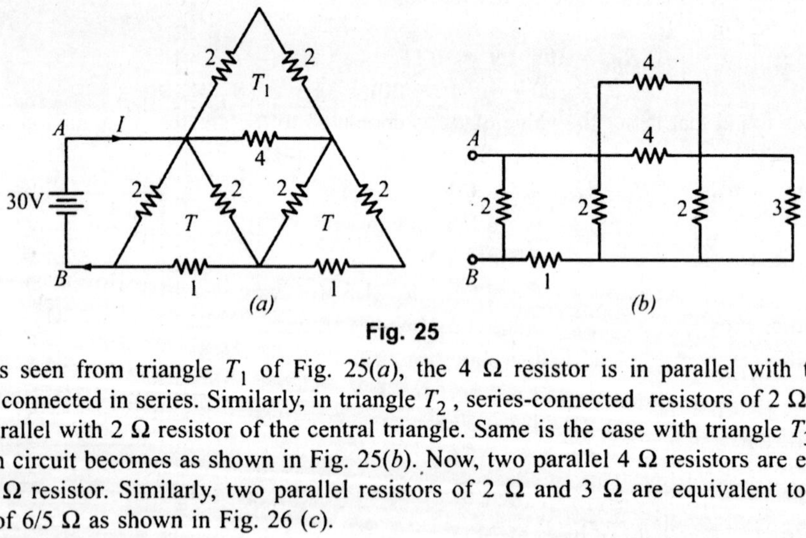

Fig. 25

As seen from triangle T_1 of Fig. 25(*a*), the 4 Ω resistor is in parallel with two 2 Ω resistors connected in series. Similarly, in triangle T_2 , series-connected resistors of 2 Ω and 1 Ω are in parallel with 2 Ω resistor of the central triangle. Same is the case with triangle T_3. Hence, the given circuit becomes as shown in Fig. 25(*b*). Now, two parallel 4 Ω resistors are equal to a single 2 Ω resistor. Similarly, two parallel resistors of 2 Ω and 3 Ω are equivalent to a single resistor of 6/5 Ω as shown in Fig. 26 (*c*).

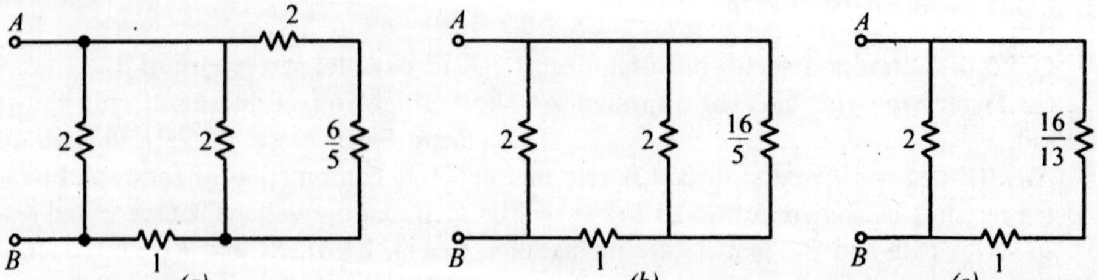

Fig. 26

Combining the two series resistors of 2 Ω and 6/5 Ω gives a single resistor of 16/5 Ω as shown in Fig. 26 (*b*). It is in parallel with 2 Ω resistor.

Their equivalent resistance is $\frac{(2\times16/15)}{(2+16/5)} = \frac{16}{13}\,\Omega$ [Fig. 26(*c*)]

Since, it is in series with 1 Ω resistance, their combined resistance is $\left(\frac{16}{13}+1\right)=\frac{29}{13}\,\Omega$ as shown in Fig. 27 (*a*). The two parallel resistances of 2 Ω and 29/13 Ω are equivalent to a single resistance of 58/55 Ω as shown in Fig. 27(*b*). Hence, in the end, the given circuit of Fig. 25(*a*) is reduced to that shown in Fig. 27(*c*). The circuit current *I* as given by Ohm's law is

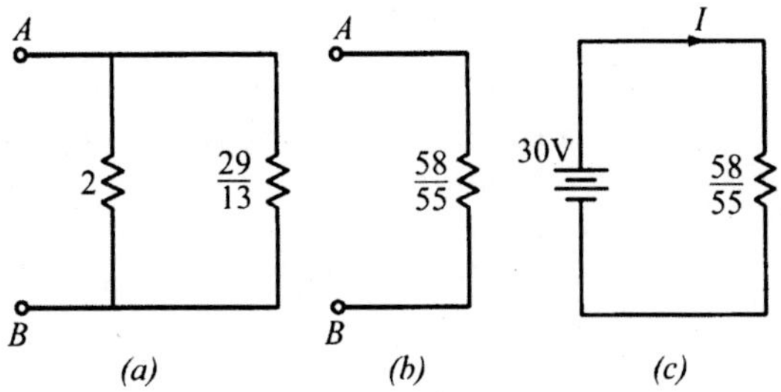

Fig. 27

$$\therefore \qquad I = \frac{V}{R} = \frac{30\times55}{58} = \mathbf{28.45\ A}$$

Q. 21. An 8 Ω resistor is in series with a parallel combination of two resistors 12 Ω and 6 Ω (Fig. 28). If the current in the 6 Ω resistor is 5 A, determine the total power dissipated in the circuit. *(Basic Elect. Engg. VTU, Karnataka 2009)*

Ans. Voltage drop across 6 Ω resistor = 6 × 5 = 30 V

Voltage across 12 Ω resistor is the same i.e. 30 V because it is in parallel with 6 Ω resistor. Current through 12 Ω resistor = 30/12 = 2.5 A.

Current through 8 Ω resistor is equal to the sum of the two currents

i.e. (5 + 2.5) = 7.5 A

Total power dissipated in the circuit is

$= 7.5^2 \times 8 + 5^2 \times 6 + 2.5^2 \times 12$

$= 450 + 150 + 75 =$ **675 Ω**

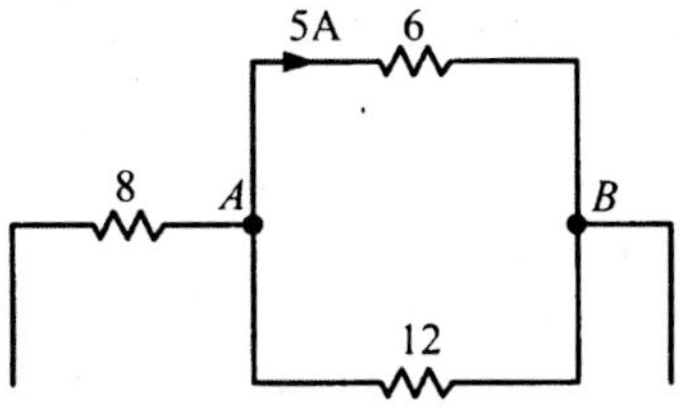

Fig. 28

Q. 22. A circuit consists of two parallel resistors having resistances of 20 Ω and 30 Ω respectively connected in series with a resistance of 15 Ω. If the current through 15 Ω resistor is 3 A , find

(*i*) currents in 20 Ω and 30 Ω resistors
(*ii*) the voltage across the whole circuit
(*iii*) the total power and the power consumed in all resistances.

(Basic Elect. Engg. VTU, Karnataka 2007)

Ans. (*i*) The circuit is shown in Fig. 29. At point *A*, 3 A current will divide in the inverse ratio of the resistances of the two parallel paths.

$$I_1 = 3\times\frac{30}{20+30} = \mathbf{1.8\ A}$$

$$I_2 = 3\times\frac{20}{20+30} = \frac{6}{5} = \mathbf{1.2\ A}$$

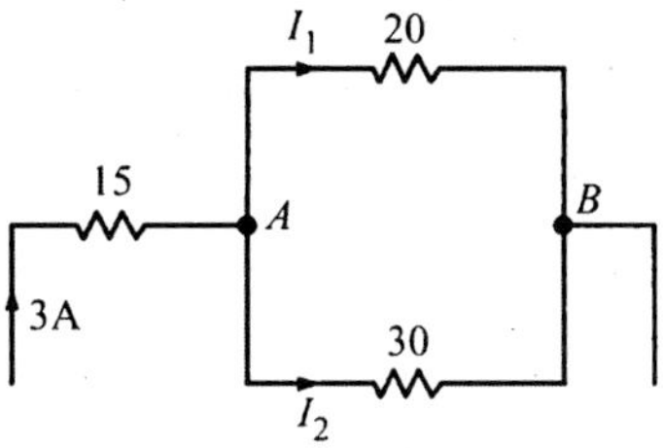

Fig. 29

(*ii*) $V_{AB} = 20 \times 1.8 = 36$ V

or $(30 \times 1.2) = 36$ V

Drop on 15 Ω resistance = 15 × 3 = 45 V

Hence, voltage across the whole circuit = 45 + 36 = **81 V**

(*iii*) Power consumed by 15 Ω resistor = $3^2 \times 15$ = **135 W**

Power consumed by 20 Ω resistor = $1.8^2 \times 20$ = **64.8 W**

Power consumed by 30 Ω resistor = $1.2^2 \times 30$ = **43.2 W**

Total power consumed = 135 + 64.8 + 43.2 = **243 W**

Alternative solution

Total circuit resistance is = 15 + 20 || 30 = 27 Ω

Total power consumed is = $I^2 R = 3^2 \times 27 =$ **243 W**

Q. 23 (*a*) Find the value of resistance *R* if source current is 6 A and source voltage is 66 V as shown in Fig. 30. All resistances are in ohm. *(Elect. Eng.,Nagpur Univ. 2002)*

(*b*) Find the resistance between points *A* and *B* for the circuit shown in Fig. 31 All resistances are in ohm. *(Elect. Eng.,.Nagpur.Univ. 2002)*

Ans. (*a*) The given circuit can be redrawn as shown in Fig. 32. As seen, it is a parallel-series circuit which can be further simplified to the circuit shown in Fig. 33.

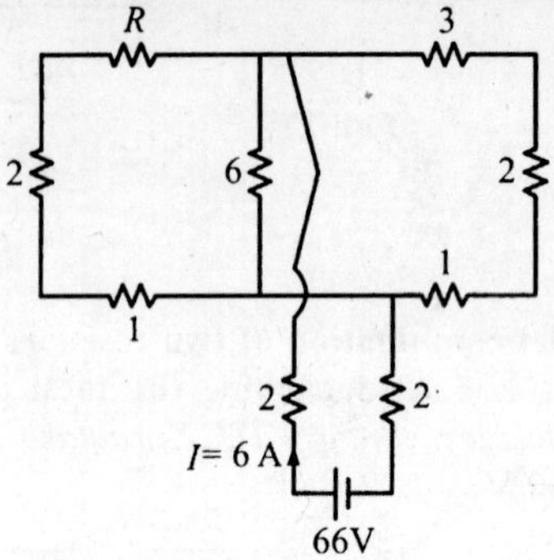

Fig. 30

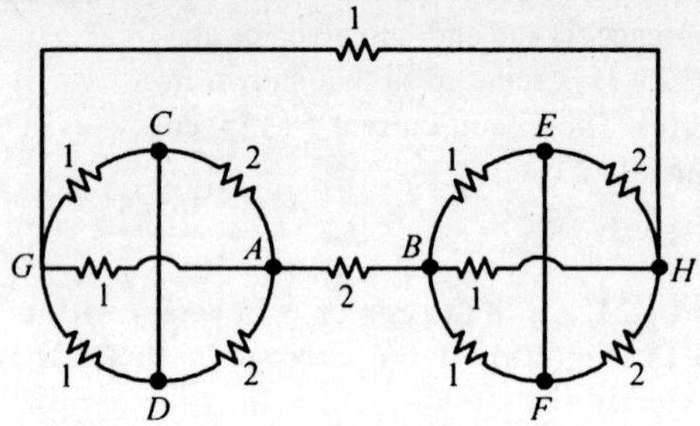

Fig. 31

The combined resistance of the three parallel resistances in Fig. 32 is.

$$= [\, 6 \parallel (3 + 2 + 1)] \parallel (R + 3) = \frac{(3R+9)}{R+6}$$

Hence, we get the simplified circuit of Fig. 33. As seen, this resistance is in series with the two 2 Ω resistors. The total resistance of the series circuit is

$$= \frac{(3R+9)}{R+6} + 2 + 2 = \frac{(7R+33)}{R+6}$$

Since, 66 -V battery supplies a curent of 6 A, it must be seeing a total circuit resistance of 66/6 = 11Ω. Equating the two resistances we get,

$$11 = \frac{(7R+33)}{R+6}$$

$\therefore$ $R =$ **8.25 Ω**

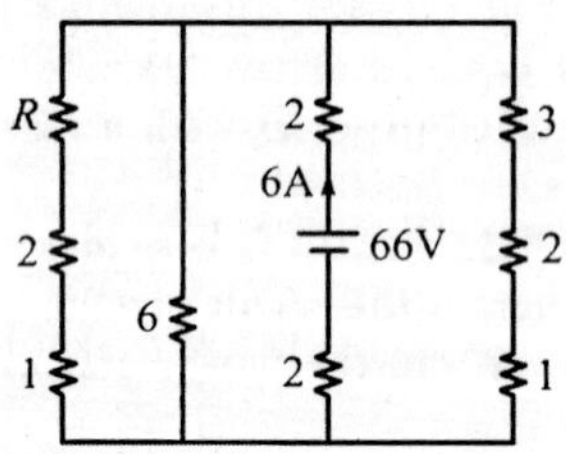

Fig. 32

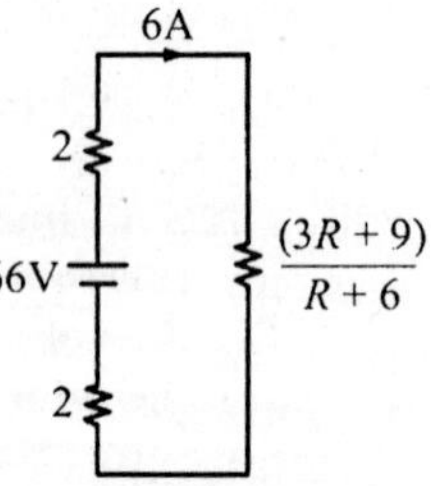

Fig. 33

(*b*) The given circuit of Fig. 31 can be simplified to that shown in Fig. 34. The 2 Ω resistors are in parallel from point *A* to *C*. So are the two 1 Ω resistors connected between points *G* and *C*.

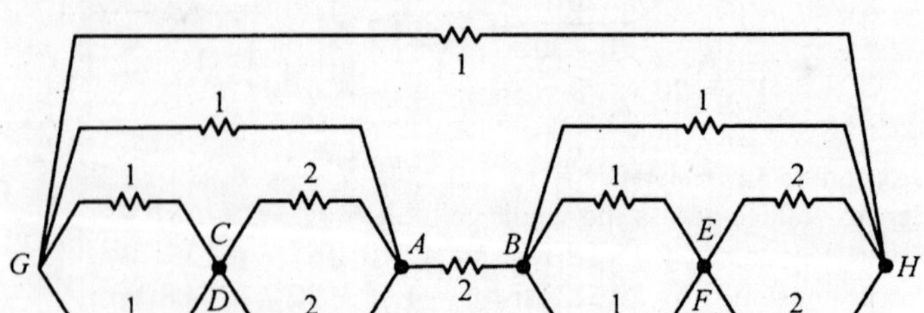

Fig. 34

Similarly, the two 2 Ω resistors connected at point *H* are in parallel between points *H* and *E*. So are the two 1 Ω resistors connected between points *B* and *E*.

Now, $R_{GC} = 1 \parallel 1 = 0.5\ \Omega$

$R_{CA} = 2 \parallel 2 = 1\ \Omega$

∴ $R_{GCA} = 1 + 0.5 = 1.5\ \Omega$

It is in parallel with 1 Ω resistance

∴ $R_{GA} = 1 \parallel 1.5 = 0.6\ \Omega$

In the same, way $R_{BH} = 0.6\ \Omega$

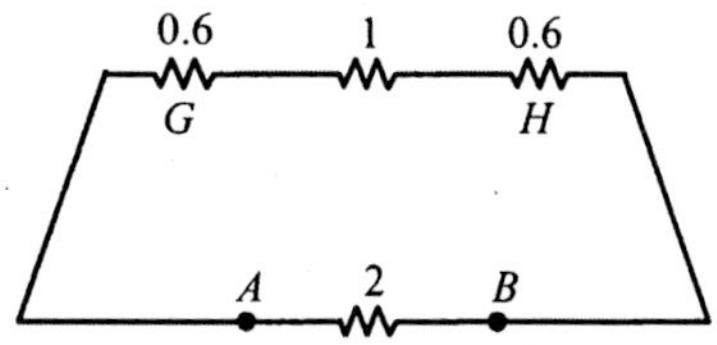

Fig. 35

The circuit of Fig. 34 is reduced to parallel circuit of Fig. 35. As seen, 2 Ω resistor is in parallel with a resistance of (0.6 + 1 + 0.6) = 2.2 Ω across points *A* and *B*.

∴ $R_{AB} = 2 \parallel 2.2 = \mathbf{1.05\ \Omega}$

Q. 24. Determine the value of unknown resistance *R* and the total current drawn from the source in the circuit of Fig. 36. Also compute the total power dissipated in the circuit. All resistances are in ohm. (*Elect. Engg.,Nagpur Univ.2004*)

Ans. As seen from Fig. 36, the given circuit is a series-parallel circuit. The combined resistance of 2 Ω and 8 Ω resistors connected in parallel is = 2 || 8 = 1.6 Ω as shown in Fig. 37.

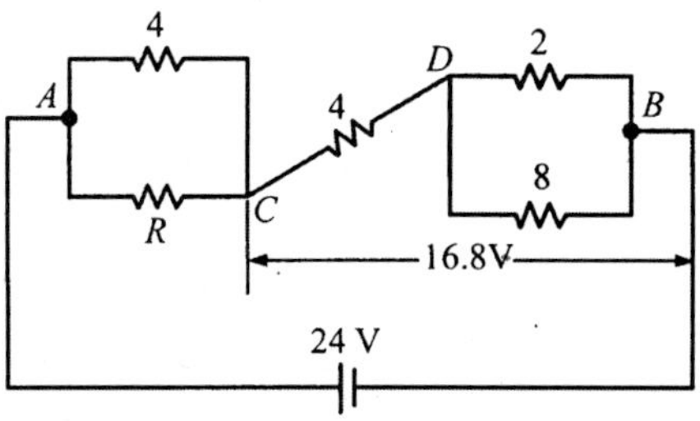

Fig. 36

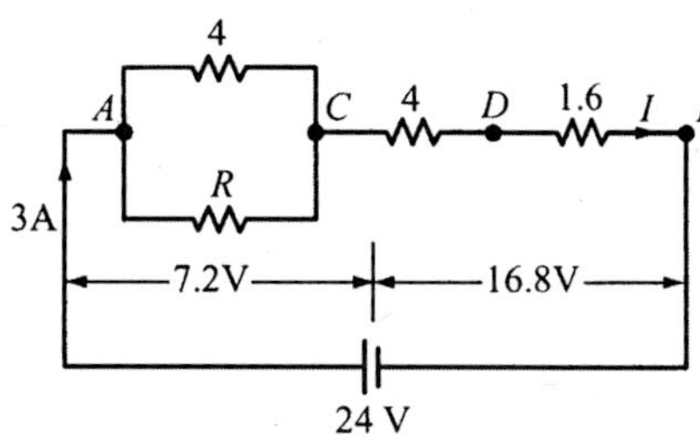

Fig. 37

∴ Obviously, $R_{CB} = 4 + (2 \parallel 8) = 4 + 1.6 = 5.6\ \Omega$

Since, $V_{CB} = 16.8$ V, ∴ $I = 16.8/5.6 = \mathbf{3\ A}$

Now, the supply voltage of 24 V equals the sum of V_{AC} and V_{CB}

∴ $24 = V_{AC} + V_{CB} = V_{AC} + 16.8$

∴ $V_{AC} = 24 - 16.8 = 7.2$ V.

As given by Ohm's law,

$$I \times 4 \parallel R = 7.2 \quad \text{or} \quad 3 \times \frac{4R}{4+R} = 7.2$$

∴ $$\frac{4R}{4+R} = \frac{7.2}{3} = 2.4 \quad \text{or} \quad 4R = 9.6 + 2.4\,R \quad \text{or} \quad 1.6\,R = 9.6$$

∴ $R = 9.6/1.6 = \mathbf{6\ \Omega}$

The total power dissipated in the circuit is

$$P = VI = 24 \times 3 = \mathbf{72\ W}$$

Q. 25. Obtain the potential difference V_{XY} in the circuit of Fig. 38.

(*Basic Elect.Engg. VTU, Karnataka, 2009*)

Ans. Consider the two battery circuits of Fig. 38 seperately. Current in the circuit *ABCX* = $I_1 = 2 / (3 + 2) =$ 0.4 A. Similarly, current in the circuit *EDYF* is $I_2 = 4 / (5 + 3) = 0.5$ A. Voltage drop in the 3 Ω resistor of branch $CX = 3\,I_1 = 3 \times 0.4 = 1.2$ V with polarity as shown in the figure. Similarly, drop on the 3 Ω resistor of branch $DY = 3\,I_2 = 3 \times 0.5 = 1.5$ V with polarity as shown.

For finding V_{XY}, we will start from point X and go to Y via points C and D. We will find the *algebraic* sum of the voltage drops met on the way from point X to Y Sign convention of voltage drops and battery *emfs* would be the same as given in *Q.* 2 of Chapter 5.

$$V_{XY} = 3\,I_1 + 4 - 3\,I_2$$
$$= 1.2 + 4 - 1.5 = \mathbf{3.7\ V}$$

Hence, point X is 3.7 V higher than point Y.

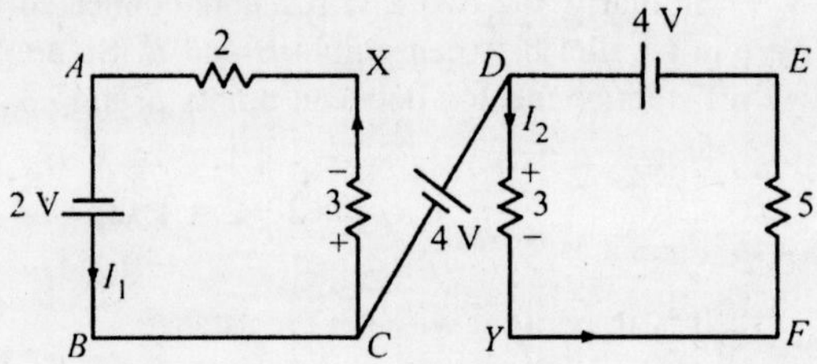

Fig. 38

EXERCISE

1. When a current of 2 A flows for 3 micro-seconds in a copper wire, estimate the number of electrons crossing any cross-section of the wire *(Bombay Univ. 2000)* **(37.5×10^{12})**

   [**Hint :** With 2 A for 3 μs, charge transferred is 6 μC = 6×10^{-6} C]
   No. of electrons crossed = $6 \times 10^{-6} / 1.6 \times 10^{-19} = 37.5 \times 10^{12}$.

2. The resistance of a conductor 1 mm^2 in cross-section and 20 m long is 0.346 Ω. Determine the specific resistance of the conducting material. **(1.73×10^{-8} Ω-m)** *(Elect. Circuits-I, Bangalore Univ. 2002)*

3. Calculate the resistance of 100 m length of a wire having a uniform cross-sectional area of 0.1 mm^2 if the wire is made of manganin having resistivity of 50×10^{-8} Ω-m^2. If the wire is drawn out to three times its original length, by how many times would you expect its resistance to increase? **(500 Ω ; 9 times)**

4. At 20 °C, the field coils of a motor have a resistance of 100 Ω. If, under load, the resistance is increased to 120 Ω, what is the rise in temperature? Temperature coefficient is 0.004/ °C at 0 °C. **(54 °C)**

5. A coil of iron wire takes 2 A at 200 V when the temperature is 14 °C and 1.8 A at 200 V when the temperature is 39 °C. Calculate the temperature-coefficient of iron at 0 °C. **(0.00472/°C)**

6. In the circuit shown in Fig. 39, the total resistance between terminals A and B is 4 Ω. Determine the value of the resistance R. All resistance values are in ohm. **(108/25 Ω)**

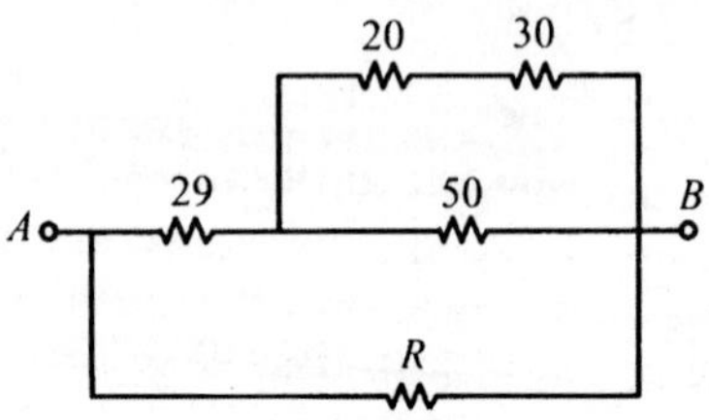

Fig. 39

7. A battery of unknown e.m.f. is connected across resistances as shown in Fig. 40. The voltage drop across the 8 Ω resistor is 20 V. What will be the current reading on the ammeter? What is the e.m.f. of the battery? **(0.7A, 67.3 V)** *(Basic Elect. Engg; Bangladesh Univ.2002)*

8. A length of wire used for overhead transmission line has a resistance of 1.85 Ω at 10 °C and 2.04 Ω at 45 °C. Calculate the temperature coefficient. **(0.00302/ °C)**

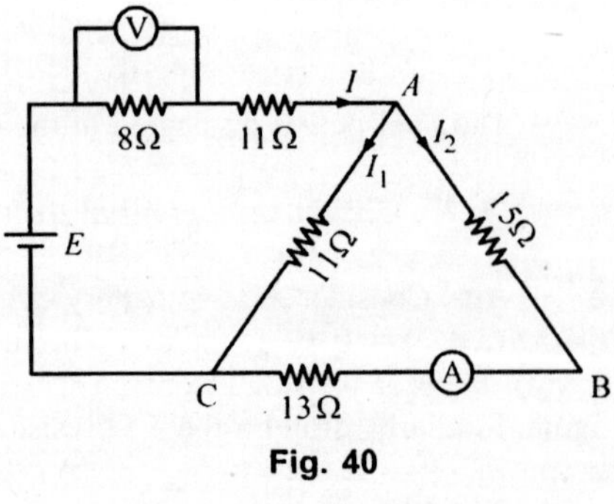

Fig. 40

9. A tungsten lamp filament has a temperature of 2,050 °C and a resistance of 500 Ω when taking normal working current. Calculate the resistance of the filament when it has a temperature of 25 °C. Temperature coefficient at 0 °C is 0.005/ °C. **(50 Ω)**
10. A resistor of 12 Ω is connected in series with a combination of 15 Ω and 20 Ω resistor in parallel. When a voltage of 120 V is applied across the whole circuit, find the current taken from the supply. **(5.83 A)** *(V.T.U. Belgaum, Karnataka Univ. winter 2004)*
11. A 50-ohm resistor is in parallel with 100-ohm resistor. Current in 50-ohm resistor is 7.2 A. How will you add a third resistor and what will be its value if the line-current is to be 12.1 A? **(277 Ω)** *(Elect. Engg., Nagpur Univ. 1997)*
12. An aluminium resistor has a resistance of 43.6 ohm at 20 °C and 47.2 ohm at 40 °C. Calculate the temperature-coefficient of resistance at 0 °C. **(0.045/ °C)**
13. The resistance of a certain length of wire is 4.60 Ω at 20 °C and 5.68 Ω at 80 °C. Find the temperature coefficient of resistance of the material of which the wire is made, referred to 0 °C. What would be the resistance of the wire at 60 °C? **(0.00425/ °C; 5.33 Ω)**
14. Deduce the current I in the circuit of Fig. 41. All resistances are in ohm. **(25 A)**

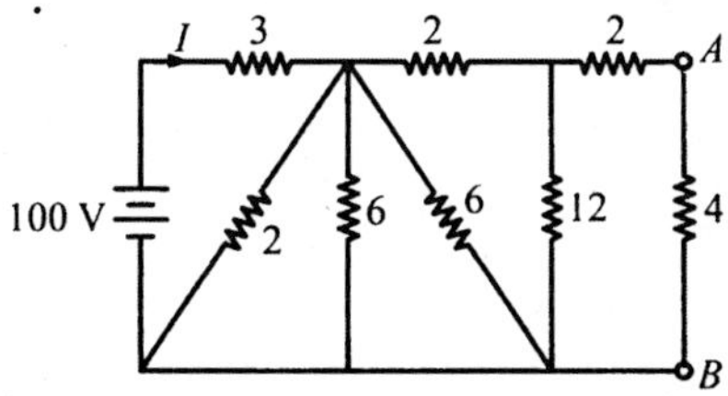

Fig. 41

15. Two resistors R_1 = 2500 Ω and R_2 = 4000 Ω are in series across a 100 V supply. The voltage drops across R_1 and R_2 are successively measured by a voltmeter having a resistance of 50,000 Ω. Find the sum of the two readings. **(97 V)** (*V.T.U.Belgaum, Karnataka Univ. 2004*)
16. A resistance wire 10 m long and cross section area 10 mm^2 at 0 °C passes a current of 10 A when connected to dc supply of 200 volts. Calculate:
(*a*) resistivity of the material
(*b*) current which will flow through the wire when the tempt. rises to 50 °C. Given α_0 = 0.0003 per °C. **(2 × 10^{-15} Ω-m, 9.85 A)**
(*Mumbai Univ. 2003/Gujrat Univ. Summer 2003*)

MULTIPLE CHOICE QUESTIONS

1. When a current of 10 A flows through a wire, the number of electrons that move for second across any given point of the wire is
(*a*) 6.25 × 10^{19} (*b*) 6.25 × 10^{18} (*c*) 6.25 × 10^{17} (*d*) 6.25 × 10^{20}
2. Which of the following material has the least positive temperature coefficient?
(*a*) nickel (*b*) nichrome (*c*) constantan (*d*) aluminium
3. The smallest particle of an element that retains the characteristics of that element is called
(*a*) molecule (*b*) atom (*c*) nucleus (*d*) neutron
4. Electric current is the rate of flow of
(*a*) protons (*b*) neutrons (*c*) electrons (*d*) atoms

5. Resistance is the opposition offered to
   (*a*) energy (*b*) voltage (*c*) charge (*d*) current
6. The resistance of a wire of a given material depends on its
   (*a*) size (*b*) length (*c*) temperature (*d*) all of the above
7. The unit of resistivity of a material is
   (*a*) ohm (*b*) ohm/m (*c*) ohm-metre (*d*) siemens
8. If the length of a given wire is doubled and its cross-section halved, its resistance is
   (*a*) doubled (*b*) tripled (*c*) quadrupled (*d*) none of the above
9. The total voltage across three resistors in series is 24 V. If one of the resistors becomes open, the voltage across it would become
   (*a*) 8 V (*b*) 24 V (*c*) 16 V (*d*) 0 V
10. In a 25 V series circuit having three resistors, the two voltage drops are 5 V and 10 V. The value of the third voltage drop is
    (*a*) 10 V (*b*) 5 V (*c*) 15 V (*d*) 0 V
11. If 100 V is applied across a 200 V, 100 W bulb, the power consumed will be
    (*a*) 100 W (*b*) 50 W (*c*) 25 W (*d*) 12.5 W
    (*Basic Elect. Engg., VTU, Karnataka, Jan.2009*)
12. A series circuit consists of 4.7 kΩ , 5.6 kΩ, 9 kΩ and 10 kΩ. Which resistor has the most voltage across it?
    (*a*) 4.7 kΩ (*b*) 5.6 kΩ (*c*) 9 kΩ (*d*) 10 kΩ
    (*Basic Elect. Engg., VTU Karnataka, July 2009*)
13. The power dissipation in each of the three parallel branches of a circuit is 1 W. The total power dissipation of the circuit is
    (*a*) 1 W (*b*) 4 W (*c*) 3 W (*d*) 9 W
14. Validity of Ohm's law requires that the
    (*a*) voltage should remain constant (*b*) current should remain constant
    (*c*) resistance must remain constant (*d*) power must remain constant

ANSWERS

1. *a*	2. *c*	3. *b*	4. *c*	5. *d*	6. *d*	7. *c*
8. *c*	9. *b*	10. *a*	11. *c*	12. *d*	13. *c*	14. *c*

CHAPTER 2

DC NETWORK THEOREMS

Q. 1. (*a*) Define the following terms used in electrical engineering with the help of suitable diagrams where possible :

(*i*) electric network	**(*ii*) electric circuit**	**(*iii*) closed path**
(*iv*) branch	**(*v*) loop**	**(*vi*) mesh**
(*vii*) node	**(*viii*) active network**	**(*ix*) passive network**

Ans. (*i*) **Electric network :** It is the interconnection of two or more circuit elements either in series or in parallel.

(*ii*) **Electric circuit :** It is an electirc network containing at least one closed path.

(*iii*) **Closed path :** It is a trace which starts from one node, passes through different elements at different nodes and returns to the original node without passing through any node more than once.

(*iv*) **Branch :** It is that part of a network which lies between two junctions or nodes.

For example : $a - f$, $b - e$, $a - b$ etc. are the branches in Fig. 1, their total number is six because *fe* is not a branch as it contains no element.

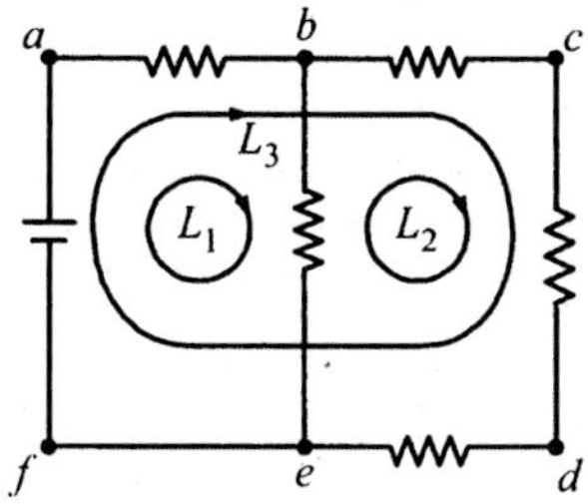

Fig. 1

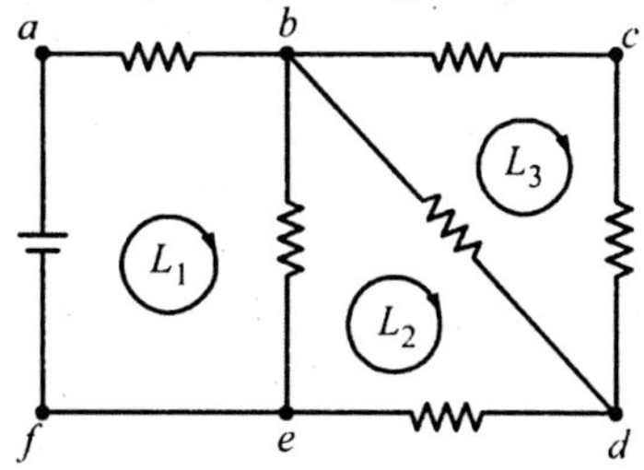

Fig. 2

(*v*) **Loop :** It is any closed path that starts from a given node, passes through different elements connected to different nodes and returns to the starting node.

Fig. 1 shows three loops *abefa*, *bcdeb* and *acdfa* marked as L_1, L_2 and and L_3. Similarly, Fig. 2 shows three loops and seven branches.

(*vi*) **Mesh :** It is a loop that does not contain any other loop within it.

In fact, *a mesh is the shortest possible loop in a network.* It is a special type of loop that does not have a closed path in its interior. For example, in Fig. 1, L_1, L_2 are loops as well as meshes. However, L_3 *is a loop but not mesh.*

(*vii*) **Node :** It is a point or junction in an electric circuit where two or more components are connected together. In Fig. 1, points *a*, *b*, *c*, *d* and *e* are the five nodes because, electrically speaking, point *f* is the same as points *e*.

(*viii*) **Active network :** It is a network which has one or more voltage sources.

(*ix*) **Passive network :** It is a network which has no voltage source.

Q. 2. (*a*) Define ideal voltage and current sources.

(*b*) How do you differentiate between independent and dependent voltage and current sources?

(*Basic Elect. Engg. R.G.P.V, Bhopal 2007*)

Ans. (*a*) An ideal voltage source is one that maintains a constant terminal voltage, *no matter how much current is drawn form it.*

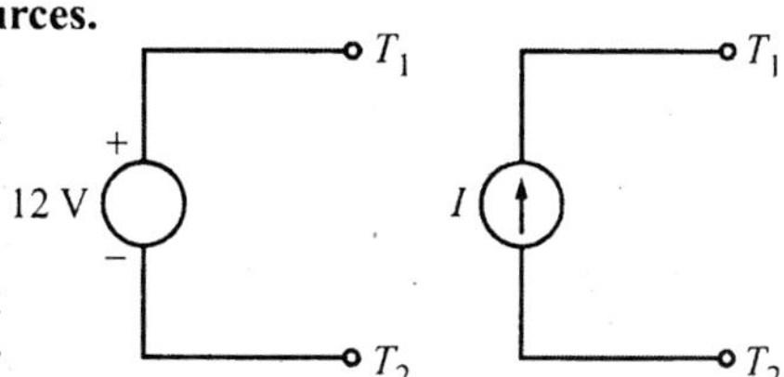

(*a*) (*b*)

Fig. 3

For example, an ideal 12-V battery would, in theory, maintain 12-V across its terminals T_1 and T_2

whether we connect a 1-MΩ or 1-kΩ resistor across its terminals as shown in Fig. 3(*a*). In practice, it is difficult to construct an ideal voltage source. It is due to the fact that all real voltage sources have some *internal resistance* that causes the terminal voltage to *decrease* if the current drawn is large. As shown in Fig. 3(*b*), an ideal current source is the one that supplies the *same current* to any resistance connected across its terminals.

However, it is difficult to construct such a current source because it has some internal resistance that increases the current if the the resistance connected across the terminals is made too large.

(*b*) An independent voltage or current source is one whose voltage or current *does not depend* on other circuit variables. Such examples are shown in Fig. 4.

A dependent voltage or current source is that whose voltage or current depends on other circuit variables.

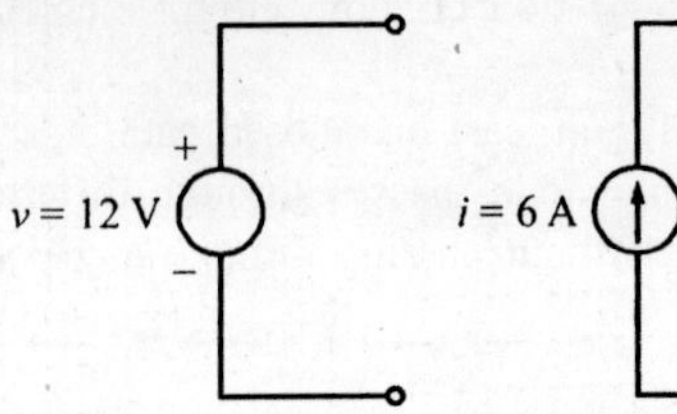

Fig. 4

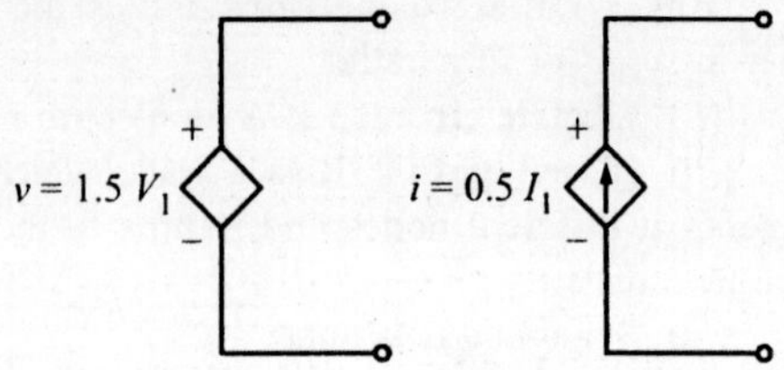

Fig. 5

The symbol for a dependent source is a diamond as shown in Fig. 5. Here, *i* depends on I_1 in some other part of the circuit. Similarly, *v* depends on V_1.

There are four types of dependent sources as shown in Fig. 6.

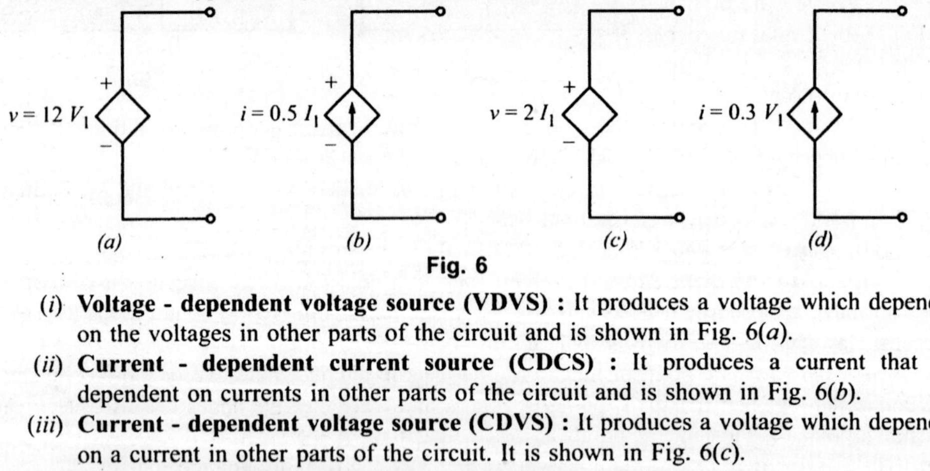

Fig. 6

(*i*) **Voltage - dependent voltage source (VDVS) :** It produces a voltage which depends on the voltages in other parts of the circuit and is shown in Fig. 6(*a*).

(*ii*) **Current - dependent current source (CDCS) :** It produces a current that is dependent on currents in other parts of the circuit and is shown in Fig. 6(*b*).

(*iii*) **Current - dependent voltage source (CDVS) :** It produces a voltage which depends on a current in other parts of the circuit. It is shown in Fig. 6(*c*).

(*iv*) **Voltate - dependent current source (VDCS) :** It produces a current which depends on a voltage in another part of the circuit. It is shown in Fig. 6(*d*).

Q. 3. While applying Kirchhoff's laws, how do you take the algebraic sign of (*i*) battery emfs and (*ii*) voltage or IR drops?

Ans. While applying Kirchhoff's laws for circuit analysis, we must give proper algebraic sign to the emfs and voltages. Otherwise, the results will come out to be wrong. Generally, following sign convention is used:

(*i*) Sign of Battery EMF : A *rise* in voltage is given a +ve sign and a *fall* in voltage a –ve sign. In

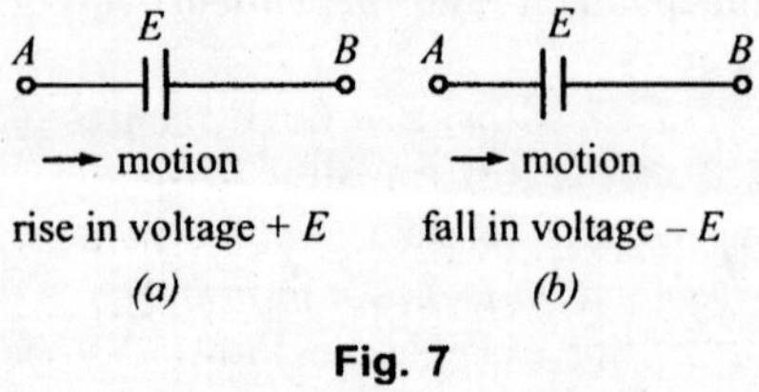

Fig. 7

Fig. 7(*a*), as we go from the negative terminal of the battery to its positive terminal, we experience *a rise* in potential. Hence, battery emf E is given a +ve sign. However, as shown in Fig. 7(*b*) when we go from the positive terminal of the battery to its negative terminal, we experience *a fall* or decrease in potential. Hence, the battery emf E is taken as negative.

It may be noted that sign of battery emf E *does not depend on the direction of the current in that branch.* As seen from Fig. 9, the signs of E_1 and E_2 do not depend on the direction of currents in their branches but only on the *direction of travel i.e.* whether we go from the positive terminal of the battery to its negative terminal or otherwise.

(*ii*) **Sign of IR Drops :** In this case, the sign of IR drops depends on whether we go through the resistor in *the same direction as current* or otherwise. In Fig. 8(*a*), current is flowing from higher voltage point A to lower voltage point B. We are also moving in the *same* direction as the current i.e. we are going down-stream. Obviously, we are experiencing *a fall* in voltage. Hence, this p.d. has a –ve sign *i.e.* $-IR$.

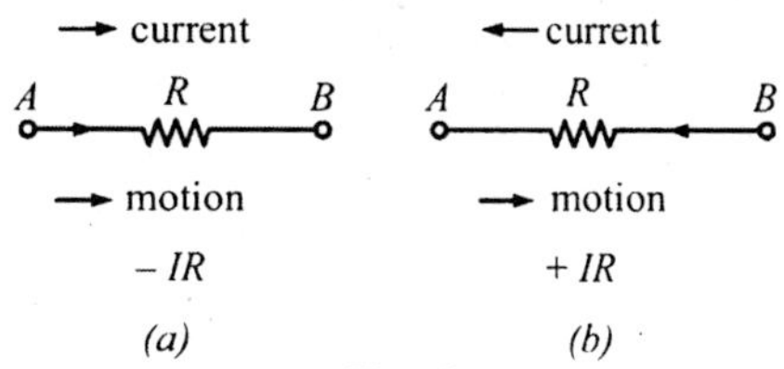

Fig. 8

As seen from Fig. 8(*b*), current is flowing form a high voltage point B to low voltage point A. As we are going from lower potential (point A) to higher potential (point B), we experience a *rise* in voltage. In this case, we are going up-stream. Hence, IR drop is given a +ve sign.

It should be noted that the sign of IR drops *depends on the direction of current through that resistor* but not on the polarity of any other voltage source in the circuit. To further make things clear, consider the closed path $ABCDA$ in Fig. 9. As we travel in the mesh in clockwise direction, different emfs and IR drops will have the following signs :

I_1R_1	is	–ve	fall in potential
I_2R_2	is	–ve	fall in potential
I_3R_3	is	+ve	rise in potential
I_4R_4	is	–ve	fall in potential
E_2	is	–ve	fall in potential
E_1	is	+ve	rise in potential

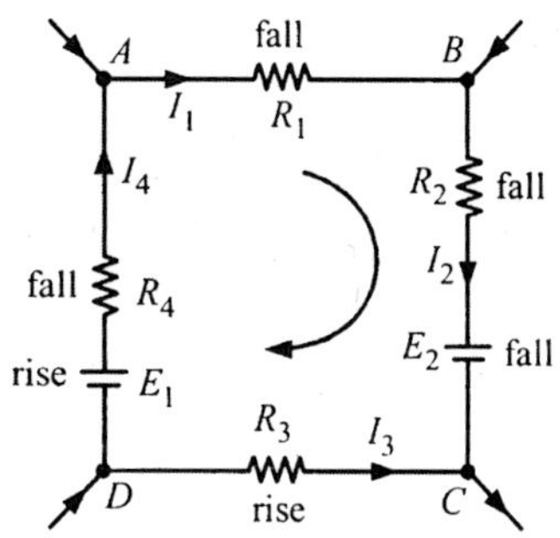

Fig. 9

Q. 4 (*a*) What is meant by circuit analysis? What methods do you use for this purpose?

(*b*) While analysing electrical circuits with the help of Kirchhoff's laws, how do you assume the direction of current flow in the circuit?

Ans. (*a*) As we know, an electric circuit or an electric network consists of electrical elements like voltage and current sources and resistors etc. linked together in a closed path. Circuit analysis is the *process* with which we find the current value in one or more circuit components. Circuits could be simple or complicated. Hence, there are two methods of analysisng electrical circuits or networks.

(*i*) **Direct Method :** In this method, the network is left *in its original form* when finding the voltages and currents in it. This method is meant for very simple circuits only. For analysing such circuits, we take the help of Ohm's law, Loop Current method, Nodal analysis and Superposition theorem etc.

(*ii*) **Network Reduction Method :** This method is used for both simple and complicated circuits. Here, we *reduce* the give network into a much simpler equivalent circuit for rapid calculation of different quantities. For this purpose, we make use of Thevenin's theorem, Norton's theorem, Delta/Star and Star/Delta conversions.

(b) While applying Kirchoff's laws to electrical networks, it is necessary to decide about the *assumed* direction of current flow. Direction of the current flow may be taken either clockwise or anti-clockwise. If the assumed direction of current is not the actual direction, then on solving the question, this current would be found to have *a minus* sign. If the answer turns out to be positive, then the assumed current direction is the same as actual direction.

It should be noted that once we have chosen a certain direction, we must use the same throughout the solution of the question.

Q. 5 (a) What is the role of Cramer's rule in circuit analysis. What is matrix notation?

(b) Find the values of i_1 and i_2 in the following two simultaneous equations

$$i_1 - 3\,i_2 = 6$$
$$2\,i_1 + 4\,i_2 = -5$$

Ans. (a) In circuit analysis, we get two or three simultaneous equations which have to be solved. We can do this by elimination method but often it is long and difficult. Carmer's rule is used to solve simple simultaneous equations with the help of determinants.

However, simultaneous equations having two, three, four or more unknowns are usually solved by using computers.

Suppose, we have to solve the following two equations

$$ax + by = c$$
$$dx + ey = f$$

There are two unknowns x and y, a, b, d and e are coefficients of these unknowns whereas c and f are constants. For solving these two equations by the method of determinants, they are put in the matrix form as

$$\begin{bmatrix} a & b \\ d & e \end{bmatrix} \begin{bmatrix} x \\ y \end{bmatrix} = \begin{bmatrix} c \\ f \end{bmatrix}$$

(b) Here, the coefficient determinant or common determinat Δ is

$$\Delta = \begin{vmatrix} 1 & -3 \\ 2 & 4 \end{vmatrix} = 1 \times 4 - 2 \times (-3) = 10$$

The other determinats Δ_1 and Δ_2 are

$$\Delta_1 = \begin{vmatrix} 6 & -3 \\ -5 & 4 \end{vmatrix} = 6 \times 4 - (-3) \times (-5) = 9$$

$$\Delta_2 = \begin{vmatrix} 1 & 6 \\ 2 & -5 \end{vmatrix} = 1 \times (-5) - 6 \times 2 = -17$$

Hence, according to Cramer's rule,

$$i_1 = \frac{\Delta_1}{\Delta} = \frac{9}{10} = \mathbf{0.9A}$$

$$i_2 = \frac{\Delta_2}{\Delta} = \frac{-17}{10} = \mathbf{-1.7A}$$

Q. 6. Slove the following three simultaneous equations by the use of determinants and Cramer's rule

$$i_1 + 3i_2 + 4i_3 = 14$$
$$i_1 + 2i_2 + i_3 = 7$$
$$2i_1 + i_2 + 2i_3 = 2$$

Ans. As explained earlier, the above equations can be written in the form.

$$\begin{bmatrix} 1 & 3 & 4 \\ 1 & 2 & 1 \\ 2 & 1 & 2 \end{bmatrix} \begin{bmatrix} i_1 \\ i_2 \\ i_3 \end{bmatrix} = \begin{bmatrix} 14 \\ 7 \\ 2 \end{bmatrix}$$

$$\Delta = \begin{bmatrix} 1 & 3 & 4 \\ 1 & 2 & 1 \\ 2 & 1 & 2 \end{bmatrix} = 1\,(4-1) - 1\,(6-4) + (3-8) = -9$$

$$\Delta_1 = \begin{bmatrix} 14 & 3 & 4 \\ 7 & 2 & 1 \\ 2 & 1 & 2 \end{bmatrix} = 14\,(4-1) - 7\,(6-4) + 2\,(3-8) = 18$$

$$\Delta_2 = \begin{bmatrix} 1 & 14 & 4 \\ 1 & 7 & 1 \\ 2 & 2 & 2 \end{bmatrix} = 1\,(14-2) - 1\,(28-8) + 2\,(14-28) = -36$$

$$\Delta_3 = \begin{bmatrix} 1 & 3 & 14 \\ 1 & 2 & 7 \\ 2 & 1 & 2 \end{bmatrix} = 1\,(4-7) - 1\,(6-14) + 2\,(21-28) = -9$$

According to Cramer's rule,

$$i_1 = \frac{\Delta_1}{\Delta} = \frac{18}{-9} = -\mathbf{2A}\ ;\ i_2 = \frac{\Delta_2}{\Delta} = \frac{-36}{-9} = \mathbf{4A}\ ;\ i_3 = \frac{\Delta_3}{\Delta} = \frac{-9}{-9} = \mathbf{1A}.$$

Q. 7. (*a*) State and explain Kirchhoff's Current Law (KCL). Is it based on some physical law? *(Elect. Engg., Nagpur Univ.)*

(*b*) By using Ohm's law and KCL, find the current I_1, I_2 and I_3 and voltage V in the two-node circuit of Fig. 11. The polarities of the unknown voltages and directions of current flow have been *chosen arbitrarily*. All resistances are in ohm.

Ans. (*a*) Kirchnoff's Current Law can be stated in many different ways as given below.

***Statement*-1 :**

The sum of currents entering a node must equal the sum of currents leaving that node.

***Statement*-2 :**

At any node of a circuit, the currents algebraically sum to zero.

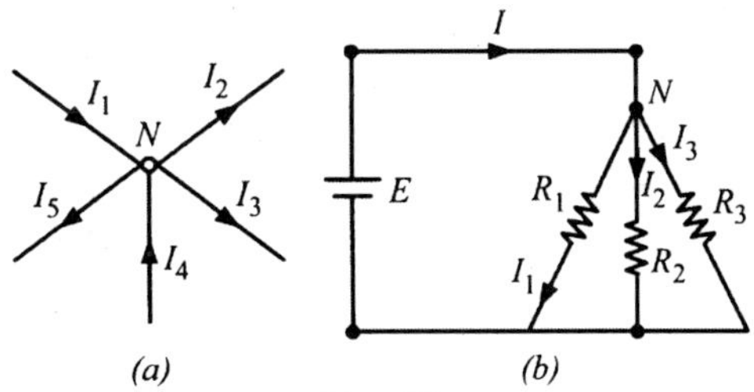

Fig. 10

***Statement*-3 :**

Whatever goes into the node of a circuit must come out of the node.

The algebraic sum of currents means that currents must carry their algebraic sign either +ve or –ve. Consider the case of a few conductors meeting at node N in Fig. 10(*a*). As seen, some conductor currents enter the node whereas some others leave the node. We will assume the *incoming* currents to be positive and the *outgoing* currents negative. It means currents I_1 and I_4 are positive whereas I_2, I_3, I_5 are negative.

$\therefore \quad I_1 + (-I_2) + (-I_3) + I_4 + (-I_5) = 0$

or $\quad I_1 + I_4 - I_2 - I_3 - I_5 = 0 \quad$ or $\quad I_1 + I_4 = I_2 + I_3 + I_5$

Similarly, for node N in Fig. 10(b)

$$I + (-I_1) + (-I_2) + (-I_3) = 0 \quad \text{or} \quad I = I_1 + I_2 + I_3$$

Mathematically, we can say that,

$$\sum_{\text{node}} I = 0 \quad \text{or} \quad \sum_{\text{node}} I_{\text{in}} = \sum_{\text{node}} I_{\text{out}}$$

It is interesting to note that KCL is based on the law of Conservation of Charge which states *that the charge must be conserved.* Since no current can be lost, it is obvious that nothing can cause the amount leaving to be less than the amount entering.

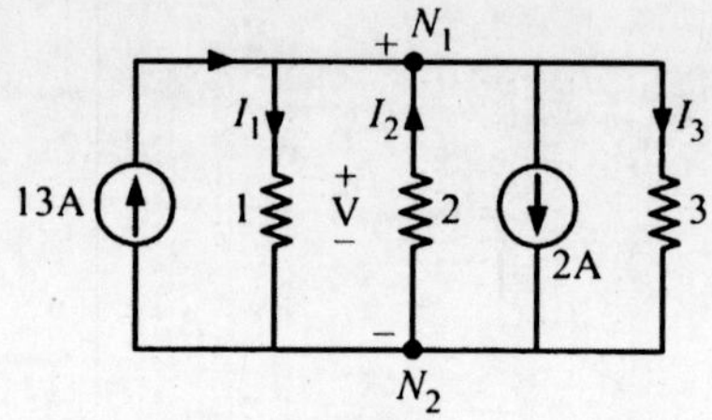

Fig. 11

(***b***) Applying KCL to N_1 of Fig. 11 we have

$13 - I_1 + I_2 - 2 - I_3 = 0$ or $I_1 - I_2 + I_3 = 11$

By Ohm's law $\quad I_1 = V/1,\ I_2 = -V/2$

and $\quad I_3 = V/3$

Please note that I_1 and I_3 have been taken positive because they are being produced from the positive end of the voltage V whereas, I_2 is being driven by the negative end of V. Putting these values in the earlier equation, we get

$$V/1 - (-V/2) + V/3 = 11 \quad \text{or} \quad V = 6\text{ V}$$

Hence, the currents are

$$I_1 = V/1 = 6/1 = \mathbf{6A}\ ;\ I_2 = -V/2 = -6/2 = \mathbf{-3A}$$

$$I_3 = V/3 = 6/3 = \mathbf{2A}$$

Q. 8 (*a*) Using Kirchoff's Current Law, find the unknown currents in Fig. 12. Draw an arrow on each diagram to show the positive direction of each current

(*b*) Find the current I_2 and the voltage V for the 16 Ω resistor of Fig 13. All resistances are in ohm

Ans. (*a*) For node A, we get that current I_1 must equal 80 – (10 + 20) = 50A. It would flow from point A to B. At point B, 30A go to point F and the remaining 20A to point C. A total of (30 + 15) = 45A goes along FD. Hence I_3 = **45A** and flows from F to D. Current I_2 equals **20A** and flows from B to C.

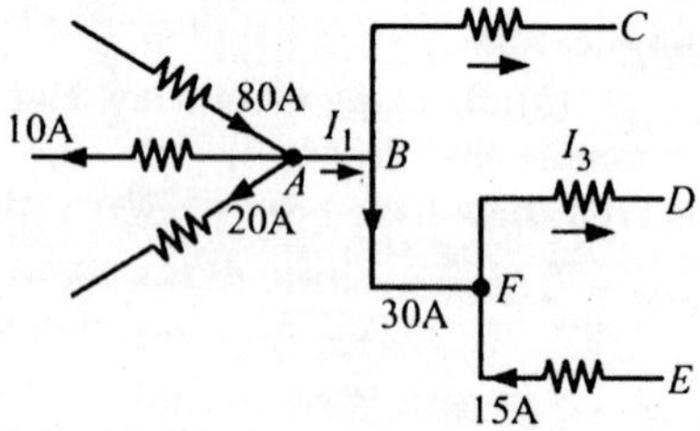

Fig. 12

(***b***) It will be noted that the dependent source is related to the current I_2. Also, the voltage V exsits between node A and node B. Applying KCL to node A, we get

$$4 - I_1 + 3I_2 - I_2 = 0 \qquad \ldots(i)$$

From Ohm's law, we get

$$I_1 = \frac{V}{4+2} = \frac{V}{6} \quad \text{and} \quad I_2 = \frac{V}{16}$$

Putting these values in Eqution (i), we have

$$4 - \frac{V}{6} + 3\left(\frac{V}{6}\right) - \frac{V}{6} = 0$$

or $\quad V = \mathbf{96V}$

$$\therefore \quad I_2 = \frac{V}{16} = \frac{96}{16} = \mathbf{6A}$$

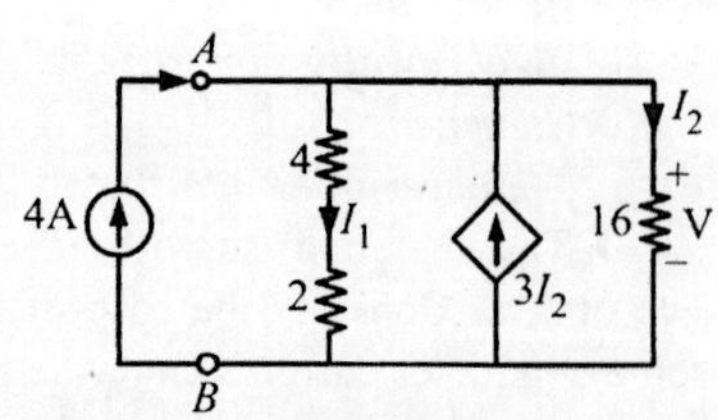

Fig. 13

Q. 9 (*a*) State and explain Kirchhoff's Voltage Law (KVL). Does it apply to any "opens" in the circuit?

(b) Using KVL, find the voltage V_{ab} across the open switch shown in Fig. 15.

Ans. (a) Krichhoff's Voltage Law which is applicable to any *closed path* can be stated in many different ways as given below:

***Statement*-1 :**

Around any loop in a circuit, the voltages algebraically sum to zero.

In other words, $\sum_{\text{Loop}} (V + IR \text{ drops}) = 0$

Here $\sum_{\text{Loop}}$ represents the sum around the closed loop.

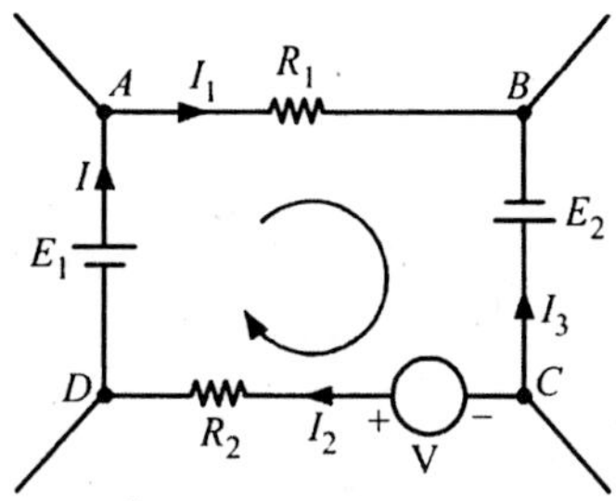

Fig. 14

***Statement*-2 :**

In any closed loop, the algebraic sum of the emfs equals the algebraic sum of the IR drops.

As seen from Fig. 14, if we go around the closed loop *ABCDA*, we meet the following voltages and potential drops. Starting from point *A*, we come across the following voltages and *IR* drops and their sum must be zero

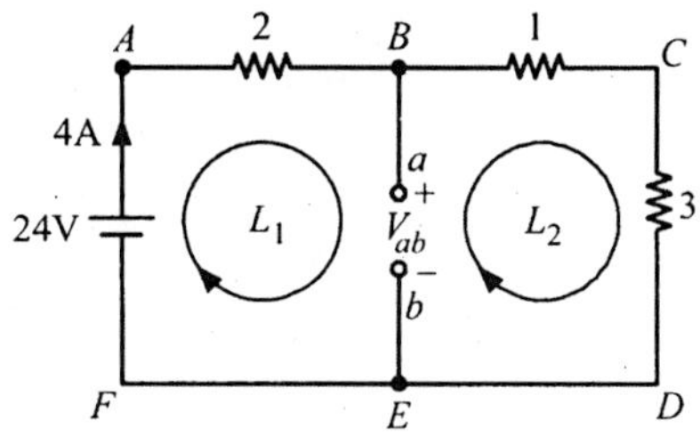

Fig. 15

$$-I_1 R_1 + E_2 + V - I_2 R_2 + E_1 = 0$$

or $$E_1 + E_2 + V = I_1 R_1 + I_2 R_2$$

$\therefore$ $$\sum V_{\text{rises}} = \sum V_{\text{drops}}$$

The basis of Kirchhoff's Voltage Law is as under :

If we start from a particular point in the circuit and go around the loop till we come back to the starting point, then we will be at the *same potential with which we started.* Obviously, whatever voltage rises we met on the way must have been equal to the voltage drops around the loop.

Since KVL does not depend on current, it is applicable even if there is an 'open' in the circuit. It is so because an open switch is treated as just another element of the circuit.

(b) We can find V_{ab} from any one of the two loops. Taking clockwise direction and using KVL, we get

$$-4 \times 2 - V_{ab} + 24 = 0$$

or $$-V_{ab} = -24 + 8 = -16$$

or $$V_{ab} = \mathbf{16V}$$

Q. 10 (a) Using Krichoff's Current Law, find the values of the currents i_1 and i_2 in the circuit shown in Fig. 16(a). (*Elect. Circuit, B.Tech. JNTU Hyderabad, 2006*)

(b) Two batteries *A* and *B* are connected in parallel and load of 10Ω is connected across their terminals. *A* has an e.m.f. of 12V and an internal resistance of 2Ω; *B* has an e.m.f. of 8V and an internal resistance of 1Ω. Use Kirchoff's laws to determine the values and directions of the currents flowing in each of the batteries and in the external resistance. Also determine the potential difference across the external resistance.

(*F.Y. Engg. Pune Univ.*)

Ans. (a) Applying KCL to node *A*, we get

$$2 - i_1 + 4i_1 - i_2 = 0 \quad \text{or} \quad -3i_1 + i_2 = 2$$

As seen from Fig. 16(*a*), the voltage across the three parallel branches is *v*.

Using Ohm's law, we get

$$i_1 = v/3 \quad \text{and} \quad i_2 = v/2$$

Putting these values in the above equation, we get $-3\,(v/3) + v/2 = 2$ or $v = -4\text{V}$

$\therefore \qquad i_1 = -\,\mathbf{4/3\ A}$ and $i_2 = -\,4/2 = -\,\mathbf{2\ A}$

The value of the dependent current source is

$$= 4i_1 = 4 \times (-\,4/3) = -\,16/3\text{A}.$$

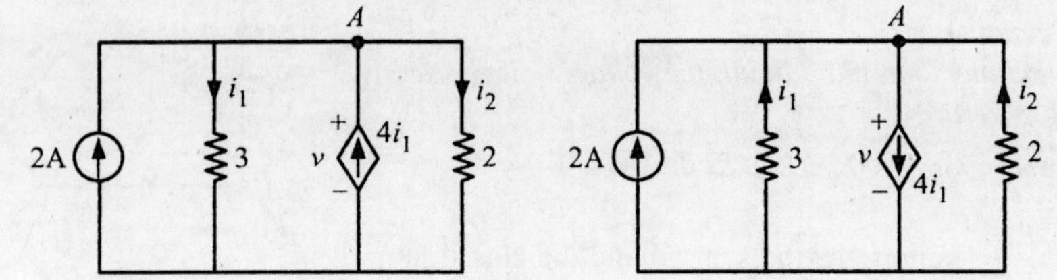

(*a*) (*b*)

Fig. 16

Since i_1 and i_2 come out to be negative, it means that they flow *upwards* as shown in Fig. 16(*b*) and not downwards as presumed. Similarly, the current of the dependent source flows *downwards* as shown in Fig. 16(*b*). It may also be noted that the sum of the upward currents equals that of the downwards currents.

(*b*) Applying KVL to the closed cicuit *ABCDA* of Fig. 17, we get.

$$-\,12 + 2x - 1y + 8 = 0 \quad \text{or} \quad 2x - y = 4 \qquad ...(i)$$

Similarly, from the closed circuit *ADCEA*, we get

$$-\,8 + 1y + 10\,(x + y) = 0$$

or $\qquad 10x + 11y = 8 \qquad ...(ii)$

From Eqn. (*i*) and (*ii*), we get

$$x = \mathbf{1.\,625\ A} \quad \text{and} \quad y = -\,\mathbf{0.75\ A}$$

The negative sign of y shows that the current is flowing into the 8-V battery and not out of it. In other words, it is a charging current and not a discharging current.

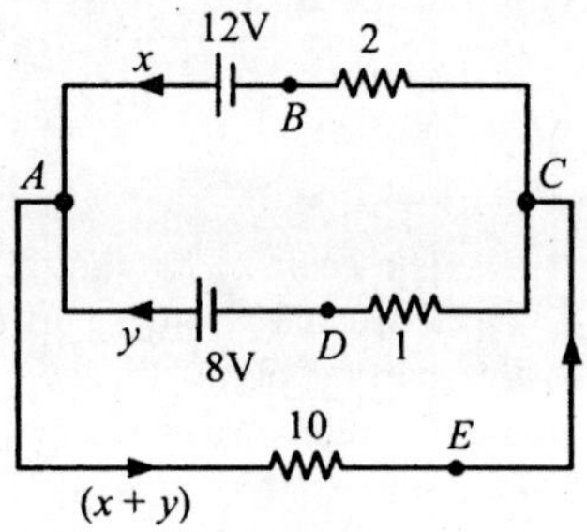

Fig. 17

Current flowing in the external resistance

$$= x + y = 1.625 - 0.75 = \mathbf{0.875\ A}$$

P.D. across the external resistance

$$= 10 \times 0.875 = \mathbf{8.745\ V}$$

Q. 11. (*a*) Applying Krichhoff's laws to different loops in Fig. 18, find the values of V_1 and V_2.

(*b*) In the network shown in Fig. 19(*a*), calculate the values of the unknown resistance *R* and the current flowing through it when the current in branch *OC* is zero.

Ans. (*a*) Starting from point *A* and applying Kirchhoff's voltage law to loop No. 3, we get

$$-\,V_3 + 5 = 0$$

or $\qquad V_3 = 5\text{V}$

Starting from point *A* and applying Kirchhoff's voltage law to loop No. 1 we get

$$10 - 30 - V_1 + 5 = 0$$

or $\qquad V_1 = -\,\mathbf{15\ V}$

The negative sign of V_1 denotes that its polarity is oppsite to that shown in the figure.

Starting from point *B* in loop No. 3, we get

$$-\,(-\,15) - V_2 + (-\,15) = 0$$

or $\qquad V_2 = \mathbf{0}$

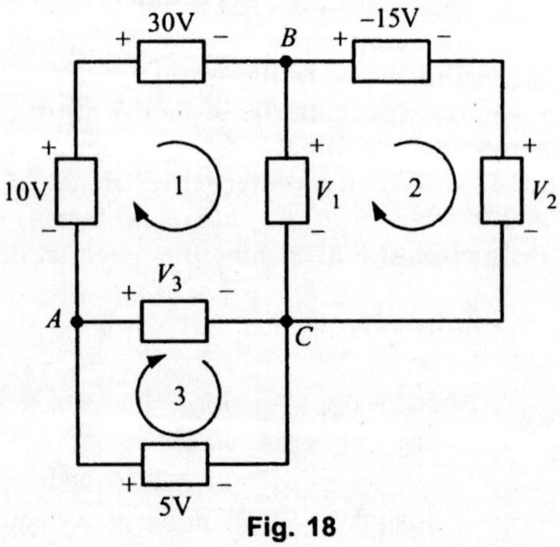

Fig. 18

(*b*) Since there is no current in branch *OC*, it acts as an open circuit. Hence, the given Fig. 19(*a*) can be redrawn as the balanced bridge circuit shown in Fig. 19(*b*). Under such conditions, the products of the resistances of the opposite sides of the bridge are equal.

$\therefore \quad 1 \times R = 4 \times 1.5 \quad \text{or} \quad R = 6\ \Omega$

Applying KVL to outer loop *ACBDA*, we get

$$(-I_1 \times 1) - (1.5 \times I_1) - 2\,(I_1 + I_2) + 10 = 0$$

or $\quad 4.5\, I_1 + 2\, I_2 = 10 \quad$ ...(*i*)

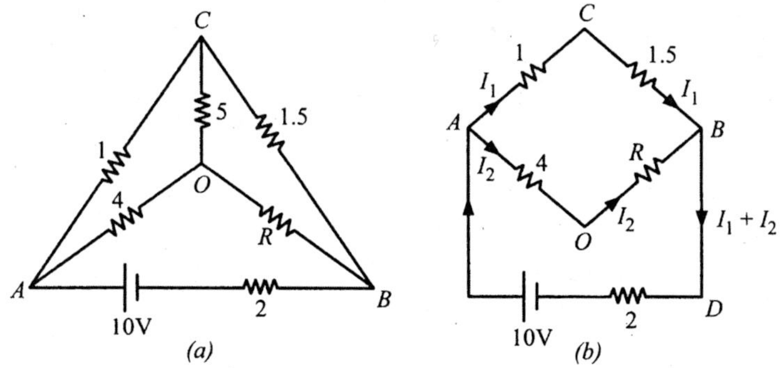

Fig. 19

Applying KVL to loop *AOBDA*, we get

$$-4I_2 - 6I_2 - 2\,(I_1 + I_2) + 10 = 0$$

or $\quad 2I_1 + 12\, I_2 = 10 \quad$ ...(*ii*)

Multiplying Eq. (*i*) by 12 and Eq. (*ii*) by 2, we get

$$54\, I_1 + 24\, I_2 = 120 \quad ...(iii)$$

$$4\, I_1 + 24\, I_2 = 20 \quad ...(iv)$$

Subtracting Eq. (*iv*) form Eq. (*iii*), we get

$$50\, I_1 = 100 \quad \text{or} \quad I_1 = 2\text{A}$$

Putting this value of I_1 in Eq. (*ii*), we get

$$2 \times 2 + 12\, I_2 = 10 \quad \text{or} \quad 12\, I_2 = 6$$

or $\quad I_2 = 0.5$ A

$\therefore$ Current through $\quad R = I_2 =$ **0.5 A**

Q. 12. A Wheatstone bridge circuit has $R_{AB} = 60\Omega = R_{CD}$, $R_{BC} = R_{AD} = 40\Omega$, $R_{BD} = 100\Omega$ (Fig. 20). Supply is connected to points *A* and *C*. If the current drawn from the supply is 100 mA, find the currents through R_{BC}, R_{CD} and R_{BD}.

Ans. The current of 100 mA = 0.1A gets distributed as shown in Fig. 20(*b*). Applying Kirchhoff's voltage law to the closed circuit *ABDA*, we get

$$-60i_1 - 100i_2 + 40\,(0.1 - i_1) = 0$$

or $\quad 25i_1 + 25i_2 = 1 \quad$ ...(*i*)

Similarly, applying Kirchhoff's voltage law to loop *BCDB*, we get

$$-40\,(i_1 - i_2) + 60\,(0.1 - i_1 + i_2) + 100i_2 = 0$$

or $\quad 50i_1 - 100i_2 = 3 \quad$ ...(*ii*)

Solving from Eq. (*i*) and Eq. (*ii*) , we get

$$i_1 = \frac{7}{150}\ \text{A} \quad \text{and} \quad i_2 = -\frac{1}{150}\ \text{A}$$

The negative sign of i_2 shows that it flows from *D* to *B* and not form *B* to *D* as assumed. The actual flow of currents in various branches is shown in Fig. 20(*c*).

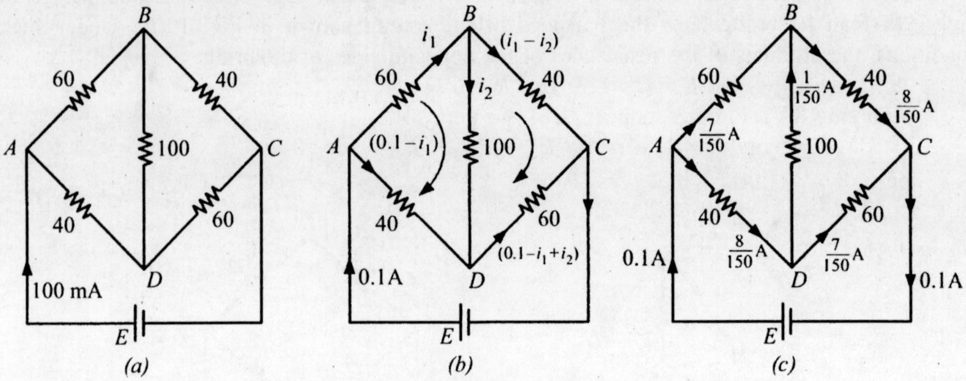

Fig. 20

$$\text{Current in branch } BC = \frac{8}{150}\text{ A}$$

$$\text{Current in branch } DC = \frac{7}{150}\text{ A}$$

$$\text{Current in branch } DB = \frac{1}{150}\text{ A}$$

Q. 13 (*a*) Explain the process of source conversion. How is it helpful in solving electrical networks? (*Basic Elect. Engg., R.G.P.V. Bhopal, 2007*)

(*b*) Find the current in the 6Ω resistor of Fig. 22(*a*) by converting the current source to a voltage source. All resistances are in ohm.

Ans. (*a*) A given independent voltage source with a *series* resistance can be converted into a current source with a *parallel* resistance. Also, a current source with a parallel resistance can be converted into a voltage source with a series resistance.

Suppose, we want to convert the voltage source of Fig. 21(*a*) into an equivalent current source. First, we put a "short" across terminals *A* and *B* of Fig. 21(*b*) to find the value of short circuit current I_{SC}. Obviously, $I_{SC} = V/R$.

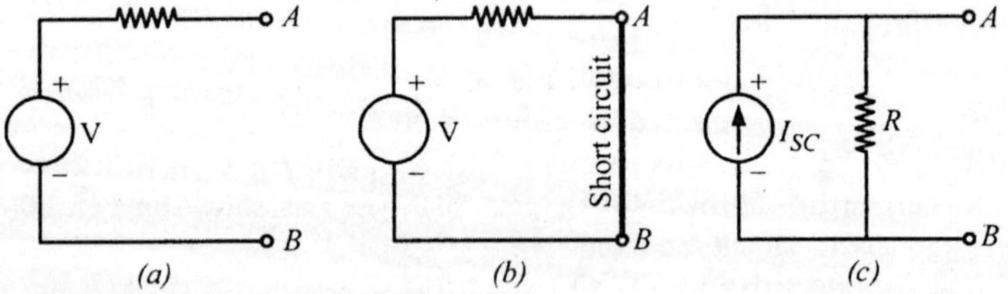

Fig. 21

The equivalent parallel current source would be the one supplying current of I_{SC} and having the same resistance *R* connected in *parallel* as shown in Fig. 21(*c*).

Now, consider the current source shown in Fig. 22(*a*). It is equivalent to a voltage source with a series resistance as shown in Fig. 22(*b*).

The reason why one type of source is converted to the other type is that it often simplifies the analysis of a circuit containing both types. For example, it is generally easier to use mesh current analysis when all the sources are voltage sources. Similarly, it is usually easier to use node analysis when all the sources are current sources. When we have both voltage and current sources in a circuit, it is helpful if all the sources are of *one type*. It can be achieved with the help of source conversion technique explained above.

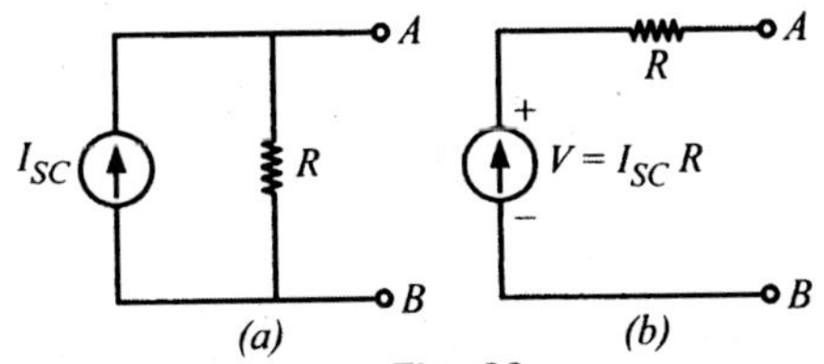

Fig. 22

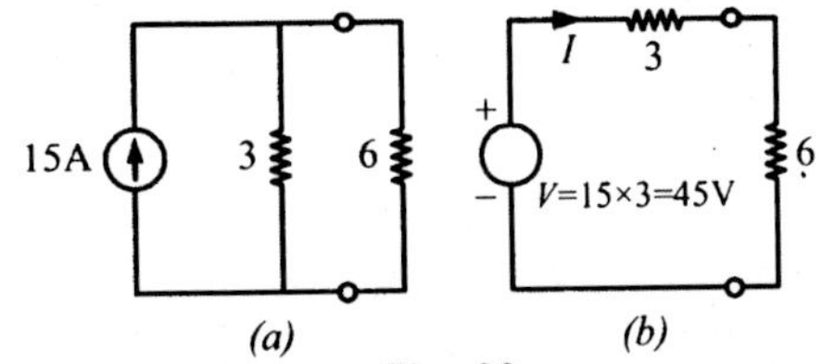

Fig. 23

(*b*) The current source of 15A with its parallel resistance of 3Ω shown in Fig. 23(*a*) will be converted into its equivalent voltage source with *same* resistance in series.

As shown in Fig. 23(*b*), $V = 15 \times 3 = 45$V and it is connected in series with 3Ω resistance.

The circuit current is $I = 45/(6 + 3) = 5$A

Hence, current through 6Ω resistor is **5A**.

Q. 14. Use source conversion method to find the load current I_L in the circuit of Fig. 24(*a*). All resistances are in ohm.

Ans. As shown in Fig. 24(*a*), 6-V voltage source with a series resistance of 3Ω has been converted into an equivalent 2A current source with 3Ω resistance in parallel. [Fig. 24(*b*)]

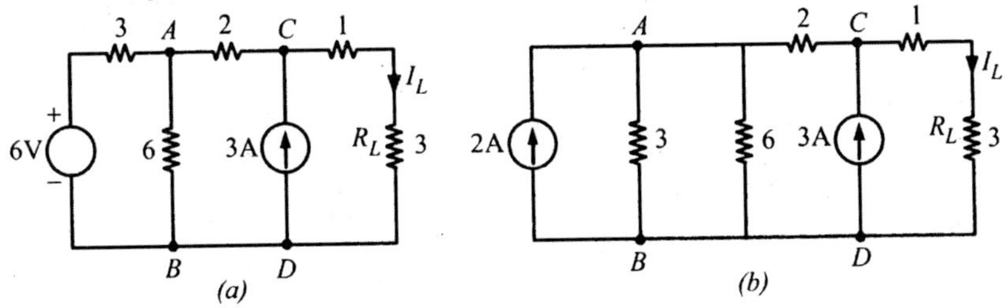

Fig. 24

The two parallel resistances of 3Ω and 6Ω can be combined into a single resistance of 2Ω as shown in Fig. 25(*a*).

The two current sources cannot be combined together because of the 2Ω resistance present between points *A* and *C*. To get over this problem, we convert the 2A current source into the equivalent 4V source as shown in Fig. 25(*b*).

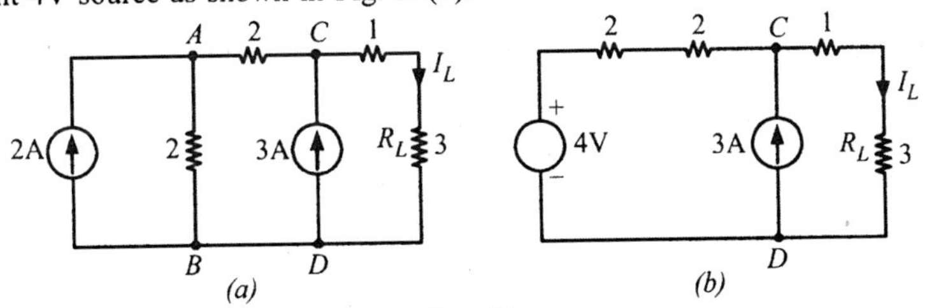

Fig. 25

Now, this 4V source with a total series resistance of (2 + 2) = 4Ω can again be converted into the equivalent current source as shown in Fig. 26(*a*). Finally, the two current source of 1A and 3A can be combined into a single 4A source as shown in Fig. 26(*b*).

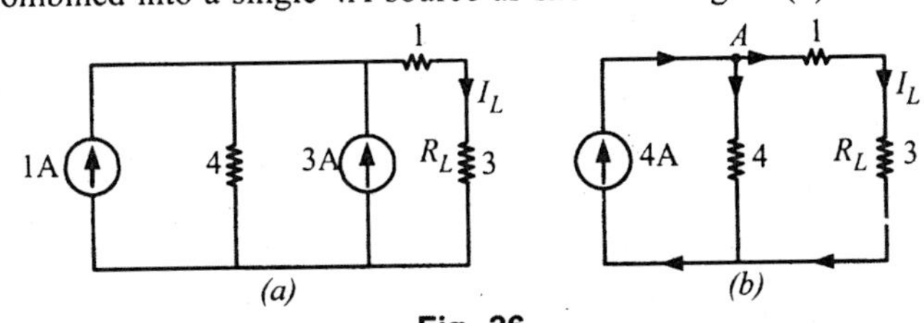

Fig. 26

This 4A current is divided into two equal parts at point A because each parallel path has a resistance of 4Ω.

Hence $\quad I_L = 4/2 = \mathbf{2A}$.

Q. 15. Fully explain with diagram the Loop (or Mesh) current method. Which Kirchhoff's Law does it depend on?

For which type of electrical networks does this method suit the most?

Ans. Consider the network of Fig. 27(a) which contains two meshes or loops marked L_1 and L_2. It has four nodes marked 1, 2, 3 and 4 and five branches. This method uses loop currents I_1 and I_2 instead of *branch* currents. For convenience, the two loop currents are considered to flow in the clockwise direction. The number of independent equations required to find the values of the two loop currents I_1 and I_2 is given by the equation

$$n = B - (N - 1)$$

where B is the number of branches and N is the number of nodes. In the present case

$$n = 5 - (4 - 1) = 2$$

These equations are found by applying Kirchhoff's voltage law to loops L_1 and L_2.

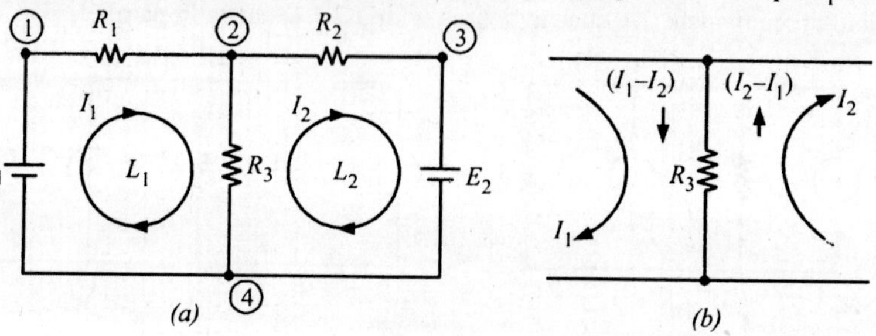

Fig. 27

It should be noted that R_3 is present in both loops. Current I_1 flows through R_3 *downwards* but I_2 flows upwards. When using KVL around L_1, net current in R_3 is $(I_1 - I_2)$ flowing *downwards* as shown in Fig. 27(b). But when we write KVL around L2, the net current through R_3 is $(I_2 - I_1)$ flowing *upwards*. The values of I_1 and I_2 can be obtained from the following two KVL equations

Loop L_1 : $\quad E_1 - I_1 R_1 - (I_1 - I_2) R_3 = 0$

Loop L_2 : $\quad - I_2 R_2 - E_2 - (I_2 - I_1) R_3 = 0$

Obviously, loop current method depends on KVL.

Loop method is very suitable for solving electrical circuits which have the following conditions:

(i) two or more energy sources

(ii) most of the energy sources are voltage sources (any current source if present, may be converted into voltage source)

(iii) current values are to be found

(iv) number of loops is less than the number of nodes.

Q. 16 (*a*) Using Mesh analysis, find the value of current in each resistor in Fig. 28(*a*).

(*b*) Find the loop currents I_1, I_2 and I_3 in the network of Fig. 29 by using Mesh analysis. *(Basic Elec. Engg. R.G.P.V., Bhopal 2008)*

Ans. Fig. 28(a) shows two loops and the two loop currents I_1 and I_2 flowing clockwise. The 4Ω resistor is in both loops. When considered in L_1, current through it is $(I_1 - I_2)$ flowing *downwards* as shown in Fig. 28(b). However when considering L_2, current through the 4Ω resistor is $(I_2 - I_1)$ flowing upwards.

Applying KVL to loop L_1, we get

$$16 - 2I_1 - 4\,(I_1 - I_2) = 0 \quad \text{or} \quad 3\,I_1 - 2\,I_2 = 8 \qquad ...(i)$$

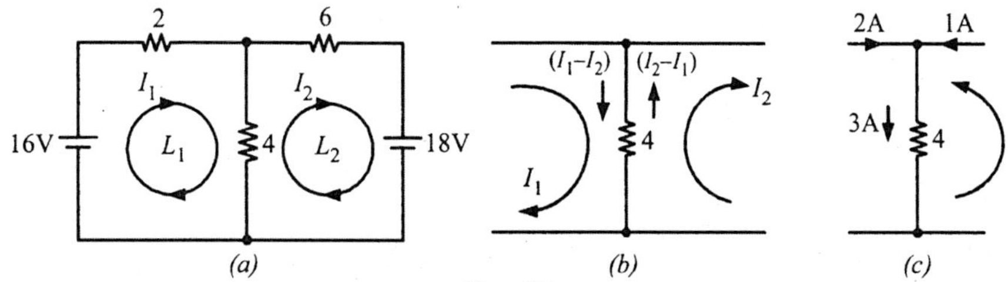

Fig. 28

Similarly, from loop L_2, we get

$$-6I_2 - 18 - 4(I_2 - I_1) = 6 \quad \text{or} \quad 2I_1 - 5I_2 = 9 \qquad ...(ii)$$

Using determinant method, we have

$$\Delta = \begin{vmatrix} 3 & -2 \\ 2 & -5 \end{vmatrix} = 3 \times (-5) - (-2) \times (2) = -11$$

$$\Delta_1 = \begin{vmatrix} 8 & -2 \\ 9 & -5 \end{vmatrix} = 8 \times (-5) - (-2) \times 9 = -22$$

$$\Delta_2 = \begin{vmatrix} 3 & 8 \\ 2 & 9 \end{vmatrix} = 3 \times 9 - 2 \times 8 = 11$$

$$\therefore \qquad I_1 = \frac{\Delta_1}{\Delta} = \frac{-22}{-11} = \mathbf{2A}$$

$$I_2 = \frac{\Delta_2}{\Delta} = \frac{11}{-11} = \mathbf{-1A}$$

Since I_2 turns out to be negative, it means that it flows counter-clockwise. Hence, current through 4Ω resistor adds up to (2 + 1) = **3A** as shown in Fig. 28(*c*).

(***b***) Let V be the voltage of node N in Fig. 29.

By applying KVL to the first loop, we get

$$5 - 1 \times I_1 - 2 \times (I_1 - I_2) = 0 \quad \text{or} \quad 3I_1 - 2I_2 = 5 \qquad ...(i)$$

Similarly, from second loop, we get

$$-2(I_2 - I_1) - 1 \times I_2 - V = 0 \quad \text{or} \quad 2I_1 - 3I_2 = V \qquad ...(ii)$$

Finally, from loop 3 we have,

$$-2I_3 - 4I_1 + V = 0 \quad \text{or} \quad 4I_1 + 2I_3 = V \qquad ...(iii)$$

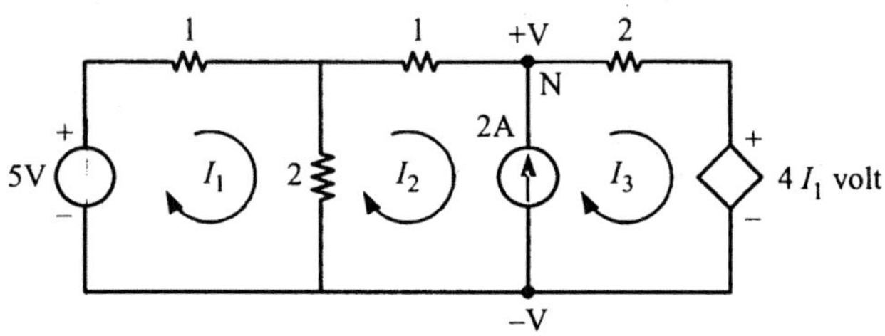

Fig. 29

Putting the value of V from Eq. (*iii*) into Eq. (*ii*), we get

$$2I_1 - 3I_2 = 4I_1 + 2I_3 \quad \text{or} \quad 2I_1 + 3I_2 + 2I_3 = 0 \qquad ...(iv)$$

It will be noted from loop 3 that

$$I_3 - I_2 = 2 \quad \text{or} \quad I_3 = 2 + I_2 \qquad ...(v)$$

Putting the value of I_3 from Eq. (*v*) into Eq. (*iv*), we get

$$2I_1 + 3I_2 + 2(2 + I_2) = 0 \quad \text{or} \quad 2I_1 + 5I_2 = -4 \qquad ...(vi)$$

By solving Eq. (*i*) and (*vi*), we have

$$I_1 = \frac{17}{19}\text{A}, \; I_2 = -\frac{22}{19}\text{ A and } I_3 = \frac{16}{19}\text{A}.$$

Q. 17 (*a*) Find the value of current in 10Ω resistance for the circuit shown in Fig. 30(*a*) by using loop current method. All resistances are in ohm.

(*Basic Elect. Engg., R.G.P.V., Bhopal 2008*)

(*b*) Using mesh analysis, find the current supplied by the 31V source and the current in 4Ω resistance of circuit shown in Fig. 31.

Ans. (*a*) Applying KVL to the first loop of Fig. 30(*b*), we get

$$20 - 5\, I_1 - 5\,(I_1 - I_2) = 0 \quad \text{or} \quad 2\, I_1 - I_2 = 4 \qquad ...(i)$$

Similarly, from second loop, we get

$$-\,10\, I_2 - 5\,(I_2 - I_3) - 5\,(I_2 - I_1) = 0 \quad \text{or} \quad I_1 - 4\, I_2 + I_3 = 0 \qquad ....(ii)$$

Also, loop 3 gives the equation,

$$-\,5\, I_3 - 10 - 5\,(I_3 - I_2) = 0 \quad \text{or} \quad I_2 - 2\, I_3 = 2 \qquad ...(iii)$$

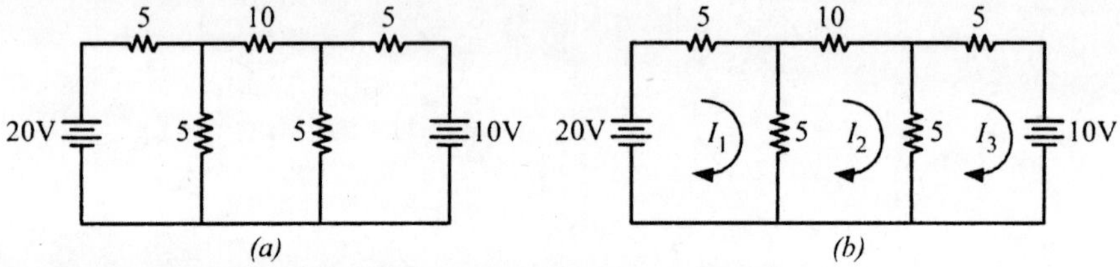

Fig. 30

Solving the above three equations, we get

$$I_1 = 2.17\text{A}, \; I_2 = 0.33\text{A} \quad \text{and} \quad I_3 = -\,0.83\text{A}$$

Obviously, current flowing through 10Ω resistance is **0.33A.**

(*b*) Let I_1 and I_2 be the mesh currents as shown in Fig. 31.

Form the first loop, we get

$$31 - 15I_1 - 1 \times (I_1 - I_2) + 10V_1 = 0$$

or $$16I_1 - I_2 - 10V_1 = 31 \qquad ...(i)$$

Similarly from second loop, we get

$$-\,4I_2 - 10V_1 - 1 \times (I_2 - I_1) = 0$$

or $$I_1 - 5\, I_2 - 10V_1 = 0 \qquad ...(ii)$$

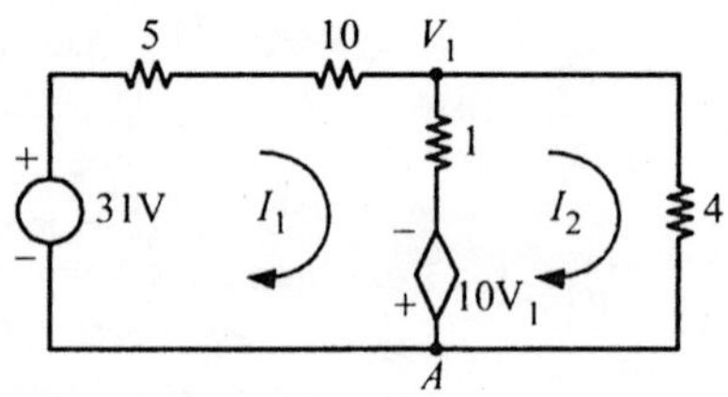

Fig. 31

It should be noted that V_1 is +ve with respect to bottom point *A* of Fig. 31.

As seen from Fig. 31 and moving upward from the voltage dependent source, we find that

$$-\,10V_1 - 1\,(I_2 - I_1) = V_1 \quad \text{or} \quad I_1 - I_2 = 11V_1 \qquad ...(iii)$$

Form Eq. (*iii*) we have,

$$11V_1 = I_1 - I_2 \quad \text{or} \quad V_1 = \frac{I_1 - I_2}{11} \qquad ...(iv)$$

Putting this values of V_1 in Eq. (*i*) and (*ii*), we get

$$166\, I_1 - I_2 = 341 \qquad ...(v)$$

and $$I_1 - 45\, I_2 = 0 \qquad ....(vi)$$

Solving Eq. (*v*) and (*vi*), we get

$$I_1 = \mathbf{2.05\ A} \quad \text{and} \quad I_2 = \mathbf{0.046A.}$$

Q. 18 (*a*) Explain in detail Nodal Analysis as applied to a network having a number of voltage sources in parallel. Also, derive the nodal equation for finding the node voltage.

(*b*) Find the node voltage V_A in Fig. 33. All resistances are in ohm.

Ans. (*a*) Nodal analysis is based on finding voltages at each node using Kirchhoff's Current Law (KCL). The current equations are written in terms of voltages and resistance. These equations are used for finding the value of voltages.

Let us consider the circuit shown in Fig. 32(*a*). The general steps for analysing the circuit are as under:

1. Select one of the nodes as a *reference* or *zero-potential* node. It should be the node having most of the components connected to it. All other node voltages are measured with respect to this node.
2. Mark all other node voltages as V_1, V_2, V_3 etc. It will be assumed that $V_1 > V_2 > V_3 > \ldots\ldots > 0$.
3. Take direction of currents through resistors from a higher to a lower voltage.
4. Write KCL for each node and solve the equations thus obtained for finding V_1, V_2, V_3 etc. Once voltages are known, branch currents can be found.

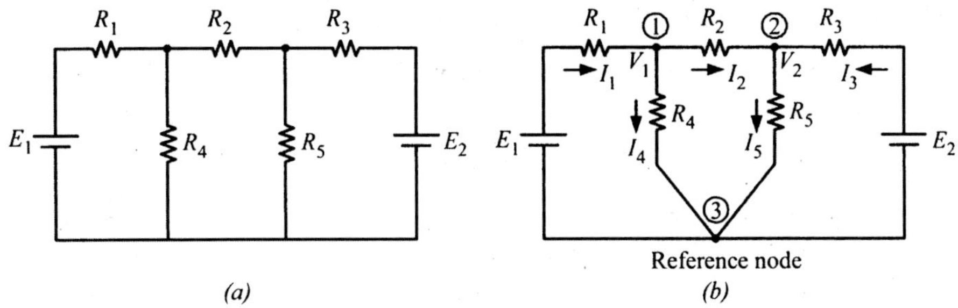

Fig. 32

Let us now draw circuit of Fig. 32(*a*) in a different way as in Fig. 32(*b*). The directions of flow of current have been drawn as per Step 3 above.

For node 1, we have as per KCL,

$$I_1 = I_2 + I_4 \qquad \ldots(i)$$

Now,
$$I_1 = (E_1 - V_1) / R_1$$

$$I_4 = \frac{V_1}{R_4} \quad \text{and} \quad I_2 = \frac{V_1 - V_2}{R_2} \qquad (\because V_1 > V_2)$$

Putting these values in Eq. (*i*), we get

$$\frac{E_1 - V_1}{R_1} = \frac{V_1}{R_4} + \frac{V_1 - V_2}{R_2}$$

or
$$V_1\left(\frac{1}{R_4} + \frac{1}{R_1} + \frac{1}{R_2}\right) - \frac{V_2}{R_2} - \frac{E_1}{R_1} = 0 \qquad \ldots(ii)$$

For node 2, we have as per KCL

$$I_5 = I_2 + I_3$$

or
$$\frac{V_2}{R_5} = \frac{V_1 - V_2}{R_2} + \frac{E_2 - V_2}{R_3}$$

or
$$V_2\left(\frac{1}{R_2} + \frac{1}{R_3} + \frac{1}{R_5}\right) - \frac{V_1}{R_2} - \frac{E_2}{R_3} = 0 \qquad \ldots(iii)$$

These nodal equations look complicated but are easy to write simply by inspection as given below :

(*i*) the product of node potential V_1 and $(1/R_1 + 1/R_2 + 1/R_4)$ *i.e.* the sum of the reciprocals of the branch resistances connected to this node.

(*ii*) minus the ratio of adjacent potential V_2 and the interconnecting resistance R_2.
(*iii*) minus the ratio of adjacent battery voltage E_1 and interconnecting resistance R_1.
(*iv*) all the above set to zero.

Using conductances instead of resistances, Eq. (*ii*) and (*iii*) can be written as

$$V_1 (G_1 + G_2 + G_4) - V_2G_2 - E_1G_1 = 0 \quad ...(iv)$$

and

$$V_2 (G_2 + G_3 + G_5) - V_1G_2 - E_2G_3 = 0 \quad ...(v)$$

It would be seen from above that number of equations required is $(n - 1)$ *i.e.* one less than the number of nodes. Nodal analysis is particularly suited for networks having many parallel circuits with a common ground. Also, nodal analysis is *complementary* to mesh analysis.

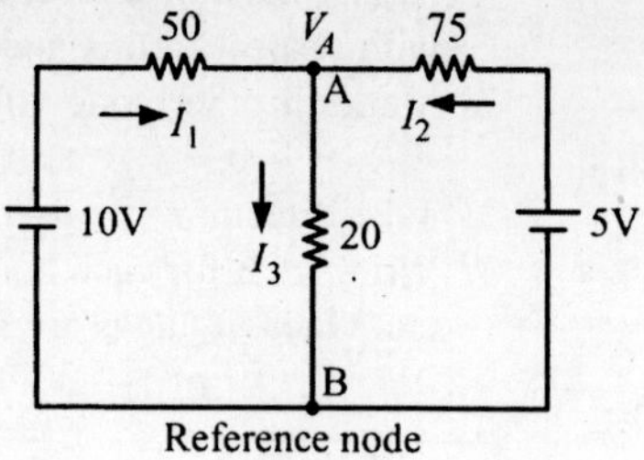

Fig. 33

(*b*) In Fig. 33, node *B* has been chosen as the reference node. There is only one unknown voltage at node *A i.e.* V_A. The current directions have been chosen as per Step 3 given above. The current equation at node *A* is

$$I_3 = I_1 + I_2 \quad ...(i)$$

or

$$\frac{V_A}{20} = \frac{10 - V_A}{50} + \frac{5 - V_A}{75} \qquad \therefore \quad V_A = \mathbf{3.2\ V}$$

Q. 19 (*a*) Using nodal analysis, write node voltage equation for the circuit shown in Fig. 34. Find the current through 25Ω branch. All the resistances are in ohm.

(Basic Elect. Engg. R.G.P.V., Bhopal, 2006)

(*b*) Use node analysis to find current in the 4Ω resistor of Fig. 35(*a*).

Ans. (*a*) In Fig. 34, the current directions have been assumed as per Step No. 2 of Q. 18(*a*). Applying KCL to node *A*, we get

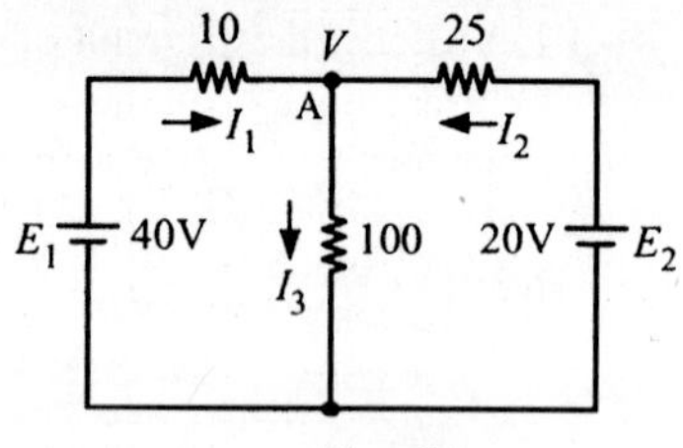

Fig. 34

$$I_1 + I_2 = I_3$$

or

$$\frac{40 - V}{10} + \frac{20 - V}{25} = \frac{V}{100}$$

or

$$15V = 480 \quad \text{or} \quad V = 32V$$

$$\therefore \qquad I_2 = \frac{20 - 32}{25} = -\mathbf{0.48\ A}$$

Since the answer comes out to be negative, the actual direction of flow of I_2 is opposite to that shown in Fig. 34. It means the current is actually flowing from node *A* into battery E_2.

(*b*) The given circuit of Fig. 35(*a*) has been redrawn with its node voltages and currents as shown in Fig. 35(*b*). We will now apply KCL to both nodes.

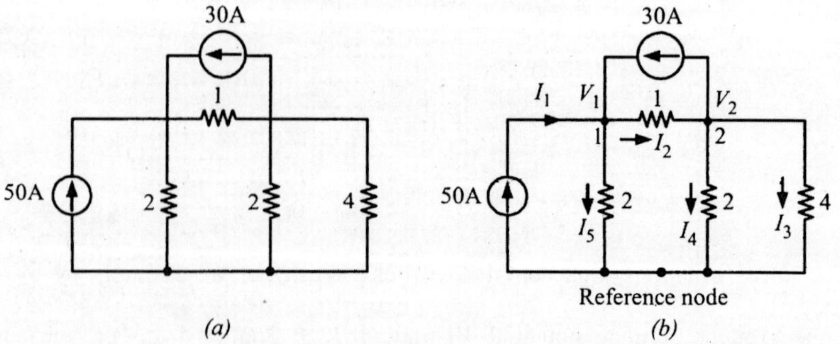

Fig. 35

For node 1, $I_1 + 30 = I_2 + I_5$ or $80 = \dfrac{V_1 - V_2}{1} + \dfrac{V_1}{2}$

$\therefore \quad 3V_1 - V_2 = 160$ ...(*i*)

For node 2, $I_2 = 30 + I_3 + I_4$

or $$\frac{V_1 - V_2}{1} = 30 + \frac{V_2}{4} + \frac{V_2}{2}$$

or $4V_1 - 7V_2 = 120$ ...(*ii*)

Solving the above two equations, we get

$$V_2 = 21.54\text{V} \quad \therefore \quad I_3 = \frac{21.54}{4} = \mathbf{5.39A}$$

Q. 20 (*a*) Using node voltage method, find the current in the 3Ω resistance for the network shown in Fig. 36. All resistances are in ohm.

(*b*) Use nodal analysis to determine the voltage across *BC* and the current in the 12-V source in Fig. 37. All resistances are in ohm.

(*Basic Elect. Engg., R.G.P.V., Bhopal, 2007*)

Ans. (*a*) It will be seen from Fig. 36 that the two battries on the left side of the figure are connected in *additive* series. Applying KCL. to node 1, we have

$$I_1 + I_2 = I_3$$

or $$\frac{6 - V_1}{(3+2)} + \frac{4 - V_1}{2} = \frac{V_1}{2}$$

$\therefore \quad 12V_1 = 32$ or $V_1 = \dfrac{8}{3}$ volt

$$\therefore \quad I_1 = \frac{6 - (8/3)}{(3+2)} = \mathbf{\frac{2}{3}\ A}$$

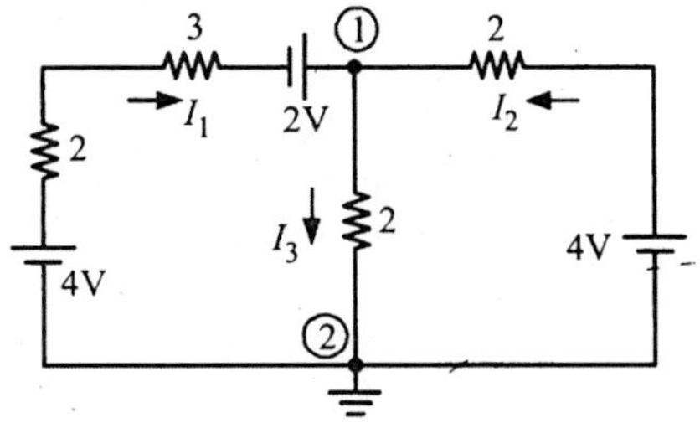

Fig. 36

(*b*) As shwon in Fig. 37, we have taken the three nodes as *A*, *B* and *C*. The resistor currents have been assumed as per Step 2 of Q. 18(*a*). We will now apply KCL to each node.

For node *A*,

$$\frac{12 - V_A}{4} = \frac{V_A - V_B}{2} + \frac{V_A - V_C}{4} + 9$$

or $4V_A - 2V_B - V_C = -24$ ...(*i*)

For node *B*

$$\frac{V_A - V_B}{2} = \frac{V_B - V_C}{4} + \frac{V_B}{4}$$

or $2V_A - 4V_B + V_C = 0$ ...(*ii*)

For node *C*

$$\frac{V_B - V_C}{4} + \frac{V_A - V_C}{4} + 9 = \frac{V_C}{2}$$

or $V_A + V_B - 4V_C = -36$ ...(*iii*)

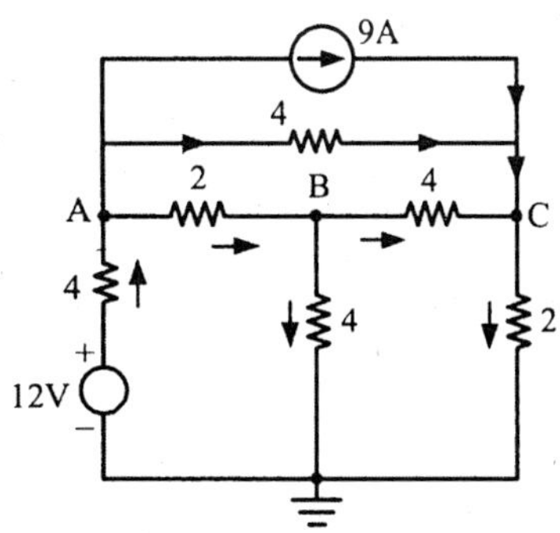

Fig. 37

Solving the above three equations, we get

$$V_A = -4\text{V},\ V_B = 0\text{V},\ V_C = 8\text{V}$$

Voltage across *BC* is

$$V_{BC} = V_B - V_C = 0 - 8 = \mathbf{-8V}$$

Since, V_A is lesser than the 12V source voltage, the potential difference between the two is 12 – (– 4) = 16V. Current supplied by the voltage source is 16/4 = **4A**.

It may be noted that current flowing from *B* to *A* is [0 – (–4)]/2 = 2A and current from *C* to *A* is [8 – (– 4)]/4 = 3A.

Hence, total current coming towards node *A* = 4 + 2 + 3 = 9A. The same current leaves the node *A*. The assumed directions of currents in branches *AB*, *BC* and *CA* are opposite to the actual directions.

Q. 21. (*a*) Define a linear circuit. Fully explain the superposition theorem for solving linear networks. *(Basic Elect. Engg. R.G.P.V., Bhoopal 2006)*

(*b*) Using the superposition theorem, find the current I in 15Ω resistor in Fig 38(*a*). All resistors are in ohm.

Ans. (*a*) A linear circuit is one which contains *linear* components like resistors, independent sources and linear dependent sources. A resistor is called a *linear element* if it obeys Ohm's law $V = IR$. The graph of voltage across it versus the current through it is a straight line. Superposition principle applies only to linear circuits. While applying superposition principle to multi-source electrical networks, we calculate the contribution made by each source when *acting alone*. For this purpose, we remove (or disable or kill) all other sources by

(*i*) short circuiting other voltage sources and

(*ii*) open-circuiting other current sources.

After calculating the response of each source when acting alone, we algebraically add them to get the total response *when all sources act simultaneously*. If a *dependent* source is present, it is *not disabled* during superposition process.

The procedure for analysing a circuit by superposition principle is as under :

1. Select any one source in the circuit and remove all other sources by shorts or opens as explained above.
2. Calculate the required voltage or current in a component when only one source acts at a time.
3. Repeat the above steps 1 and 2 for each source in turn.
4. Algebraically add the above values to get the *actual* voltage or current values when all source act simultaneously.

(*b*) We will first short 54V battery and then find I_1 produced by 30V battery as shown in Fig. 38(*b*). Next, we will short 30V battery as shown in Fig. 38(*c*) and find current I_2 produced by 54V battery. When both batteries act simultaneously, current through the 15Ω resistor would be $(I_1 + I_2)$ as shown in Fig. 38(*d*).

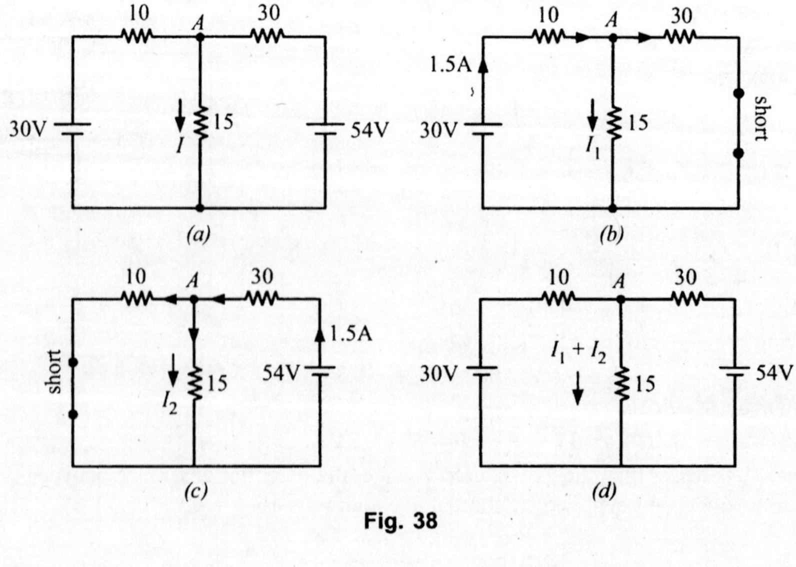

Fig. 38

As seen from Fig 38(*b*), the total resistance seen by the 30V battery is

$$10 + 15 \parallel 30 = 10 + 10 = 20\Omega$$

$\therefore$ battery current = 30/20 = 1.5A

This current divides at point *A* in the inverse ratio of the resistances of the two parallel paths.

$$\therefore \quad I_1 = 1.5 \times \frac{30}{30+15} = 1 \text{ A}$$

Now, the 30V battery is shorted and current is supplied by 54V battery only as shown in Fig. 38(*c*).

Total resistance offered to 54V battery is

$$= 30 + 15 \parallel 10 = 36\Omega$$

Hnece, current supplied by the battery is

$$= 54/36 = 1.5\text{A}$$

$$\therefore \quad I_2 = 1.5 \times 10/25 = 0.6\text{A}$$

Since I_1 and I_2 flow in the same direction, current through 15Ω resistance produced by both batteries is

$$I = I_1 + I_2 = 1 + 0.6 = \mathbf{1.6\ A.}$$

Q. 22 (*a*) Find the current flowing in the branch *XY* of the network shown in Fig. 39(*a*) by superposition theorem. All resistances are in ohm. (*Elect. Engg. Nagpur Univ.*)

(*b*) Use superposition theorem to find current in various branches of the circuit shown in Fig. 40(*a*). All resistances are in ohm. (*Elect. Engg. Nagpur Univ.*)

Ans. (*a*) First consider the current due to 20V battery of internal resistance 1Ω when the other 20V battery is shorted. There are two parallel paths across terminal *X* and *Y*. One has a resistance of 3Ω and the other has resistance of (1 + 3 + 2) = 6Ω. Their combined resistance is 3 || 6 = 2Ω. Hence, total resistance seen by 20V battery is (2 + 2 +1) = 5Ω.

Hence, current supplied by the battery is

$$I = 20/5 = 4\text{A}$$

This current divides at point *X* in the inverse ratio of the resistance of the two parallel circuits.

$$\therefore \quad I_1 = 4 \times 6/9 = 2.67\text{A} \qquad \text{– from } X \text{ to } Y$$

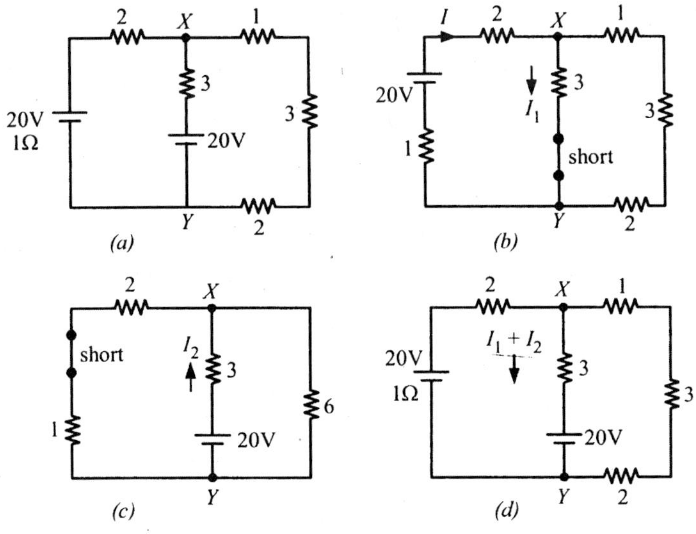

Fig. 39

Next, consider the case when 20V, 1Ω battery is shorted as shown in Fig. 39(*c*). Now, the current will be produced by the second 20V battery. The total resistance seen by the battery is

$$= 3 + (2 + 1) \parallel 6 = 3 + 2 = 5\Omega$$

Therefore, current produced by the battery is

$$I_2 = 20/5 = 4\text{A} \qquad \text{– from } Y \text{ to } X$$

∴ total current flowing in the branch *XY* is

$$I_{XY} = 2.67 - 4 = -\ \mathbf{1.33A} \qquad \text{– from } X \text{ to } Y$$

The –ve sign shows that current actually flows form *Y* to *X*.

(*b*) As shown in Fig. 40(*b*), 24V battery has been replaced by a short because it has zero internal resistance.

Here,

$$I_1' = 6\text{A} \qquad \text{– from } C \text{ to } B$$

$$I_2' = 6 \times \frac{4}{4+4} = 3\text{A} \qquad \text{– from } B \text{ to } A$$

$$I_3' = I_1' - I_2' = 6 - 3 = 3\text{A} \qquad \text{– from } B \text{ to } D$$

In Fig. 40(*c*), 6A current source has been replaced by an open.

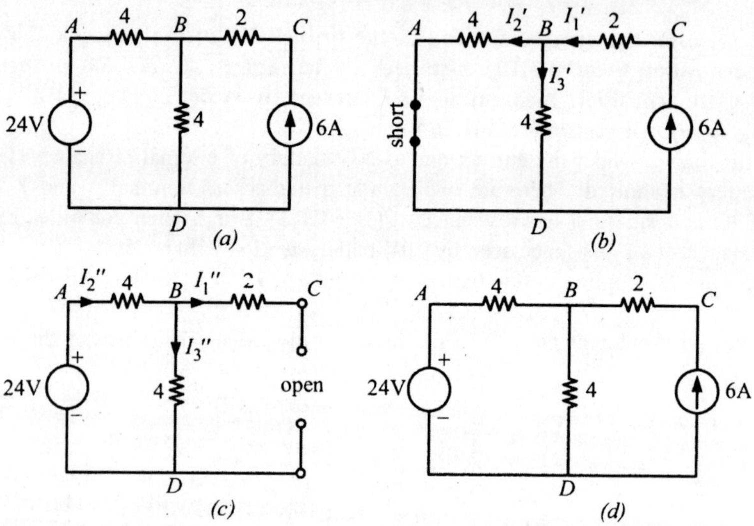

Fig. 40

The different currents are :

$$I_1'' = 0\text{ A} \qquad \text{– open circuit}$$

$$I_2'' = 24/(4 + 4) = 3\text{A} \qquad \text{– form } A \text{ to } B$$

$$I_3'' = I_2' = 3\text{A} \qquad \text{– form } B \text{ to } D$$

The net resultant current through different branches when both sources are acting is

$$I_{CB} = I_1' - I_1'' = 6 - 0 = \mathbf{6\ A}$$

$$I_{AB} = I_2'' - I_2' = 3 - 3 = \mathbf{0\ A}$$

$$I_{BD} = I_3' + I_3'' = 3 + 3 = \mathbf{6\ A}$$

Q. 23. Calculate the current through 3Ω resistor (Fig. 41) by superposition theorem. All resistances are in ohm. *(Elect. Engg. West Bengal Tech. Univ., 2008)*

Ans. (*i*) **When 6V battery acts alone :** As shown in Fig. 41(*b*), the 12V battery has been removed leaving behind its internal resistance of 1Ω. The total resistance seen by 6V battery is

$$= (0.5 + 2.5) + (2 + 1) \parallel 3 = 3 + 1.5 = 4.5\ \Omega$$

Hence, circuit current $I = 6/4.5 = 4/3$ A.

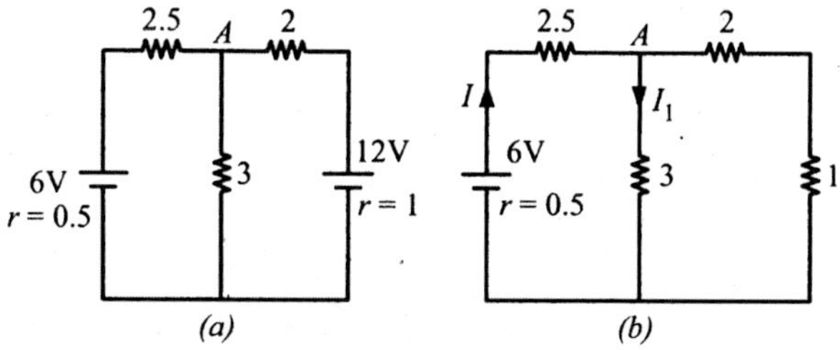

Fig. 41

This current divides equally between the two parallel paths at point A because each has a resistance of 3Ω.

$$\therefore \qquad I_1 = \frac{1}{2} \times \frac{4}{3} = \frac{2}{3} \text{ A.}$$

(*ii*) **When 12V battery acts alone :** Now, 6-V battery is removed leaving behind its internal resistance of 0.5Ω as shown in Fig. 42(*a*).

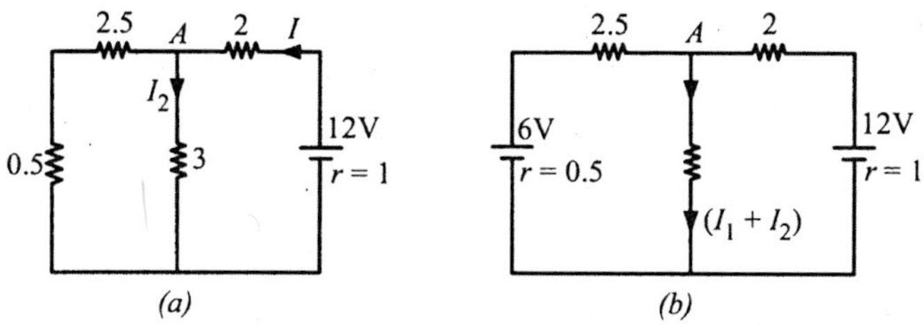

Fig. 42

The total resistance seen by the 12V battery is

$$= (1 + 2) + 3 \parallel (2.5 + 0.5) = 3 + 1.5 = 4.5\Omega$$

∴ Current I in Fig. 42(*a*) is

$$= 12/4.5 = 8/3\text{A}$$

This current is divided into two parallel paths at point A. Since each path has the same resistance, hence $I_2 = 8/2 \times 3 = 4/3$A.

(*iii*) **When both voltage sources act simultaneously :** As shown in Fig. 42(*b*), current through 3Ω resistance is

$$= I_1 + I_2 = \frac{2}{3} + \frac{4}{3} = \mathbf{2A.}$$

Q. 24 (*a*) Compute the power dissipated in the 9Ω resistor in Fig. 43(*a*) by applying superposition theorem. The voltage and current sources should be treated as ideal. All resistances are in ohm. (*Elect. Engg. Nagpur Univ. 2005*)

(*b*) Find the value and direction of current in the 15Ω resistance of Fig. 44(*a*) by using superposition theorem. All resistances are in ohm.

(*Basic Elect. Engg. R.G.P.V., Bhopal, 2008*)

Ans. (*a*) It should be kept in mind that an ideal voltage source has zero internal resistance whereas an ideal current source has an infinite internal resistance.

(*i*) **When voltage source acts alone :** As shown in Fig. 43(*b*), the constant current source has been replaced by an open circuit *i.e.* infinite resistance. Total resistance seen by the 32V battery is

$$= 4 + 12 \parallel 15 = \frac{32}{3} \ \Omega$$

∴ current supplied by the battery is

$$= \frac{32}{32/3} = 3\text{A}$$

At point A, this current divides into two parts. The current going to 9Ω resistance is

$$I_1 = 3 \times \frac{12}{12+15} = \frac{4}{3} \text{ A.}$$

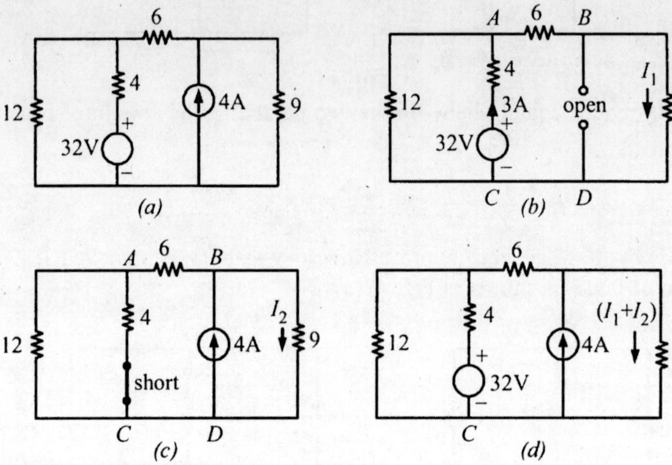

Fig. 43

(*ii*) **When current source acts alone :** In this case, 32V battery has been short-circuited as shown in Fig. 43(*c*). Here, 4A current divides at point B into two parts. Current I_2 goes to 9Ω resistor and the rest $(4 - I_2)$ goes from B towards A.

The resistance of the series-parallel circuit to the left of the line BD is

$$= 6 + (4 \parallel 12) = 9\ \Omega$$

Hence, at point B the current divides into two equal parts.

$$\therefore \qquad I_2 = 4/2 = 2\text{A}$$

∴ total current through 9Ω resistor is

$$= I_1 + I_2 = \frac{4}{3} + 2 = \frac{10}{3} \text{ A}$$

Power dissipated through 9Ω resistor

$$= I^2 R = \left(\frac{10}{3}\right)^2 \times 9 = \mathbf{100\ W}$$

(*b*) (*i*) When 10V source acts alone : As shown in Fig. 44(*b*), the circuit resistance as seen by 10V battery is

$$= 5 + (3 + 3) \parallel 15 = 9.28\ \Omega$$

$$\therefore \qquad I = 10/9.28 = 1.07\text{A}$$

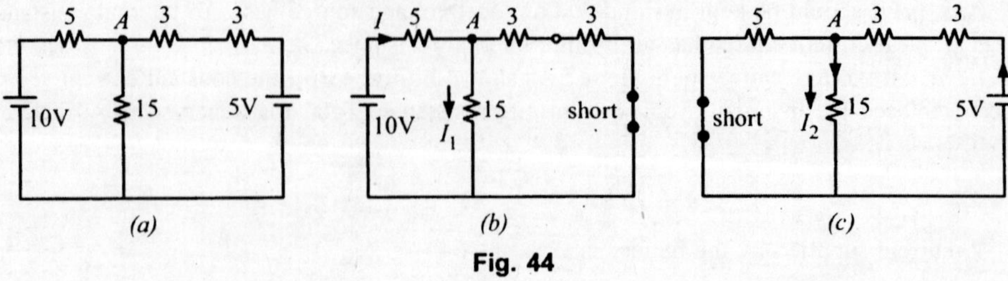

Fig. 44

This current is divided into two parallel paths at point A. The current through 15Ω resistance is

$$I_1 = 1.07 \times \frac{3+3}{3+3+15} = 1.07 \times \frac{6}{21} = 0.31\text{A}$$

(*ii*) **When 5V source acts alone :** As seen from Fig. 44(*c*), the 5 V battery sees a total resistance of (3 + 3) + 5 || 15= 9.75 Ω

$$\therefore \qquad I = 5/9.75 = 0.51 \text{ A}$$

This current divides between the two parallel paths at point A in the inverse ratio of the resistance of the two parallel paths.

$$\therefore \qquad I_2 = 0.51 \times \frac{5}{5 \times 15} = 0.13 \text{ A}$$

Hence, according to superposition theorem, current through 15Ω resistance when both batteries act simultaneously, is

$$= I_1 + I_2 = 0.31 + 0.13 = 0.44 \text{ A}$$

Q. 25. (*a*) Find the current *I* in the circuit shown in Fig. 45.

(*Elect. Circuits B. Tech. JNTU, Hyderabad, 2006*)

(*b*) Using superposition theorem, determine the current through 3Ω resistor (all resistances are in ohm) as shown in Fig. 46.

(*Basic Elect. Engg. B. Tech. JNTU, Hyderabad, 2007*)

Ans. (*a*) We will use superposition theorem to solve this question. Let us first remove the 1A source and replace it by an open circuit as shown in Fig. 45(*b*). Obviously, no current flows through branch *BA*. The two branches *CD* and *CBE* are in parallel across the 1V battery. Hence, battery sees a resistance equal to the combined resistance of branches *CD* and *CBE*.

The combined resistance = 3 || (4 + 1) = 15/8 Ω

$$\therefore \qquad i = \frac{1}{15/8} = \frac{8}{15}\text{A}$$

At point *C*, current i divides into two parts, one going along *CD* and the other I_1 going along *CBE*. As per the current divider rule.

$$I_1 = \frac{8}{15} \times \frac{3}{3+5} = \frac{8}{15} \times \frac{3}{8} = 0.2\text{A}$$

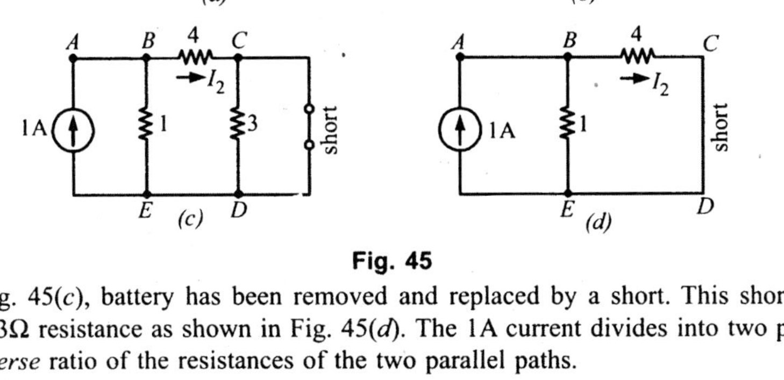

Fig. 45

In Fig. 45(*c*), battery has been removed and replaced by a short. This short also short-circuits the 3Ω resistance as shown in Fig. 45(*d*). The 1A current divides into two parts at point *B* in the *inverse* ratio of the resistances of the two parallel paths.

$$\therefore \qquad I_2 = 1 \times \frac{1}{(1+4)} = 1 \times \frac{1}{5} = 0.2\text{A}$$

It is seen from the above that the two currents I_1 and I_2 are each equal to 0.2A but flow in the opposite directions. Hence, currents I = 0.2 – 0.2 = **0 A.**

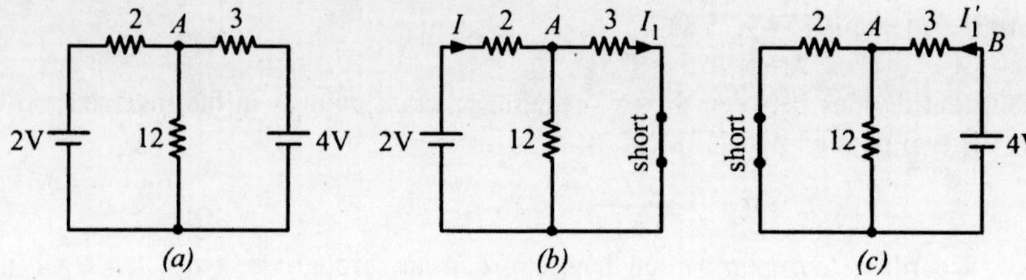

Fig. 46

(*b*) In Fig. 46(*b*), 4V battery has been removed and replaced by a short. The current I is divided into two parts at point A as per the current divider rule

The total resistance seen by the 2V battery is

$$= 2 + 3 \parallel 12 = 2 + 36/15 = 4.4\Omega$$

$$\therefore \qquad I = 2/4.4 = 0.45 \text{ A}$$

$$I_1 = 0.45 \times 12/(3 + 12) = 0.36\text{A}$$

In Fig. 46(*c*), 2V battery has been replaced by a short. The total resistance seen by the 4V battery is

$$= 3 + 2 \parallel 12 = 3 + \frac{12}{7} = \frac{33}{7}\,\Omega$$

$$\therefore \qquad I_1' = 4/(33/7) = 0.85\text{A}$$

Hence, current through the 3Ω resistance

$$= I_1' - I_1 = 0.85 - 0.36 = \mathbf{0.49A}. \qquad \text{– from point } B \text{ to } A$$

Q. 26 (*a*) State and explain Thevenin's theorem with suitable diagrams. How does it differ from Norton's theorem? (*Elect. and Mech. Tech., Osmania Univ.*)

(*b*) With reference to the network of Fig. 50(*a*) by applying Thevenin's theorem find the following.

(*i*) the equivalent e.m.f. of the network when viewed from terminals *A* and *B*.

(*ii*) the equivalent resistance of the network when looked into from terminals *A*&*B*.

(*iii*) current in the load resistance R_L of 15Ω. (*Basic Circuit Analysis, Nagpur Univ.*)

Ans. (*a*) It is a very useful method of reducing complex linear circuits into simple circuits. According to this theorem, *any complex linear circuit can be reduced to a single voltage source in series with a single resistance.*

This *two-element* circuit is called Thevenin's equivalent circuit for the given network (Fig. 47). Different steps for finding Thevenin's equivalent are as follows:

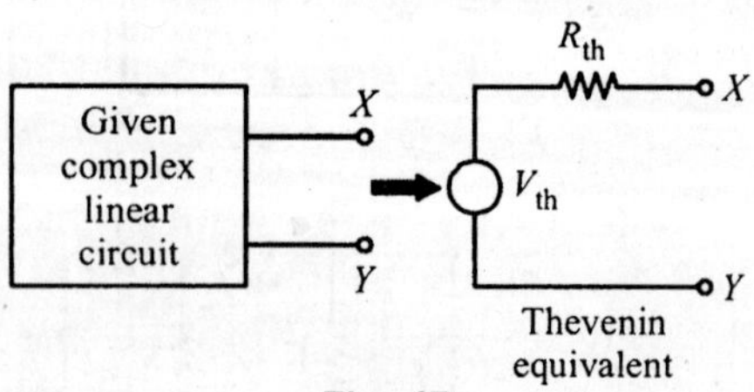

Fig. 47

1. Open-circuit the terminals of the given circuit with respect to which Thevenin's equivalent is required.
2. The voltage which appears across the open terminals is called Thevenin's voltage, V_{th}.
3. Next, voltage sources in the given circuit are shorted and current sources opened. The resistance of the given circuit as viewed from the two open terminals is calculated. It is called Thevenin's resistance R_{th}.

4. Voltage source V_{th} connected in series with R_{th} gives the Thevenin equivalent for the given network as shown in Fig. 47.

Explanation : Suppose we want to 'thevenise' the circuit of Fig. 48(*a*) to the left side of the terminals $X - Y$.

1. First we remove the load resistance of 50Ω as shown in Fig. 48(*b*). This way terminals $X - Y$ become open.
2. Next, we find R_{th}. For this purpose, 24V battery is shorted and the resistance of the given network as viewed from the terminals X-Y is calculated. As seen from Fig. 48(*b*).
$$R_{th} = 100 + (60 \parallel 120) = 140\Omega$$
3. Now, we find V_{th} from Fig. 48(*c*). It is equal to the voltage drop across 120Ω resistance. It is so because there is no current through the 100Ω resistor and, hence, no voltage drop across it

$$\therefore \qquad V_{th} = 24 \times \frac{120}{120+60} = 16\text{V}$$

Obviously, the given circuit is equivalent to Thevenin's circuit shown in Fig. 48(*d*). The complete circuit is shown in Fig. 48(*e*).

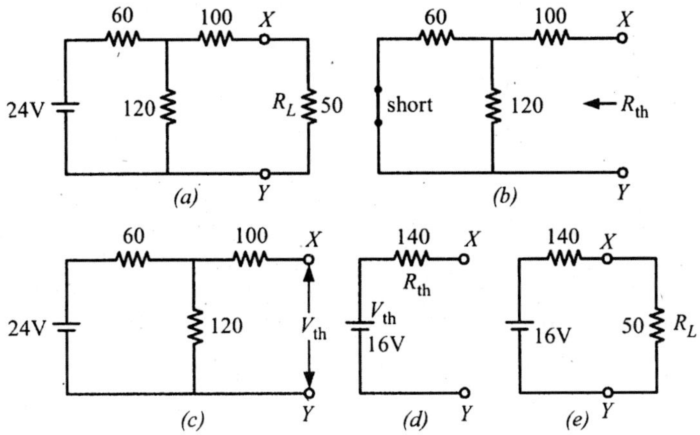

Fig. 48

Thevenin theorem reduces a given network between any two terminals to an equivalent circuit having *voltage* source with a *series* resistance. However, Norton theorem reduces the given circuit to an ideal *current* source with a *parallel* resistance as shown in Fig. 49.

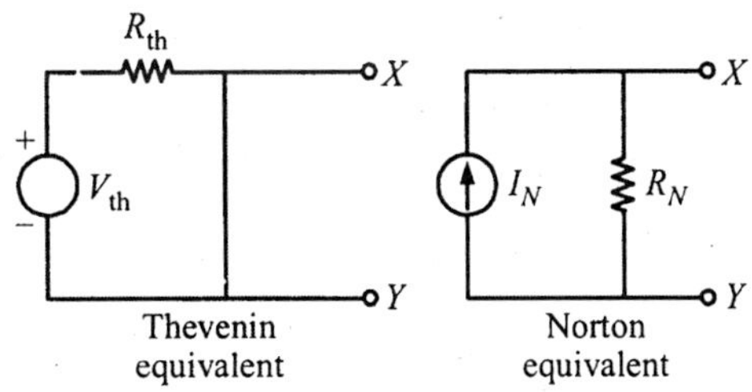

Fig. 49

(*b*) (*i*) Current in the network before load resistance is connected [Fig. 50(*a*)].

$$= 24/(12 + 3 + 1) = 1.5\text{A}$$

∴ voltage across terminals *AB*

$$= V_{oc} = V_{th} = 12 \times 1.5 = \mathbf{18V}$$

Hence, so for as terminals *A* and *B* are concerned, the network has an e.m.f. of 18 volt (and not 24 V).

(*ii*) There are two parallel paths between points *A* and *B*. Imagine that battery of 24V is removed but not its internal resistance. Then, resistance of the circuit as looked into from points *A* and *B* is [Fig. 50(*c*)].

$$R_i = R_{th} = 12 \times 4/(12 + 4) = 3\Omega$$

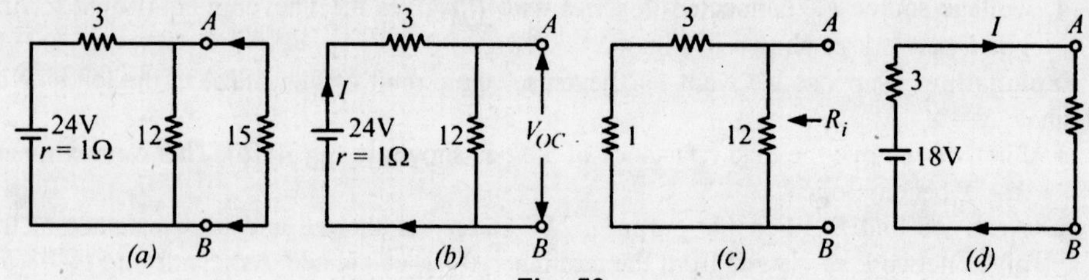

Fig. 50

(*iii*) When load resistance of 15Ω is connected across the terminals, the network is reduced to the structure shown is Fig. 50(*d*).

$$I = V_{th}/(R_{th} + R_L) = 18/(15 + 3) = \mathbf{1\ A.}$$

Q. 27. State and explain superposition theorem.

(*b*) Find the current through 5Ω resistor of Fig. 51(*a*) using Thevenin's Theorem.

(Elec. Engg. West Bengal Tech. Univ. 2009)

Ans. (*a*) Please see Q. 21.

(*b*) First step : Open the terminals $T_1 - T_2$ as shown in Fig. 51(*b*).

Second Step : Now, find the voltage available across terminals $T_1 - T_2$. It is called V_{th}. As there is no current through 3Ω resistace connected to T_1, hence there is no voltage drop across it.

$\therefore \quad V_{th} = V_{BC}$ or V_{AD}.

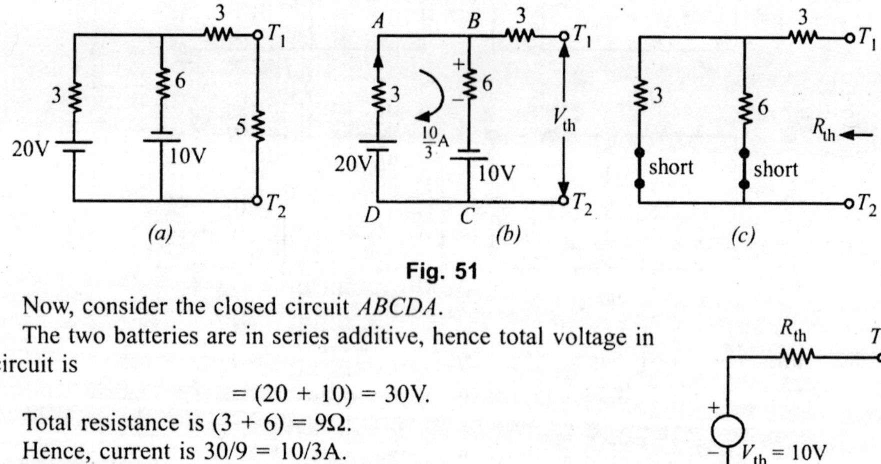

Fig. 51

Now, consider the closed circuit *ABCDA*.

The two batteries are in series additive, hence total voltage in the circuit is

$$= (20 + 10) = 30\text{V}.$$

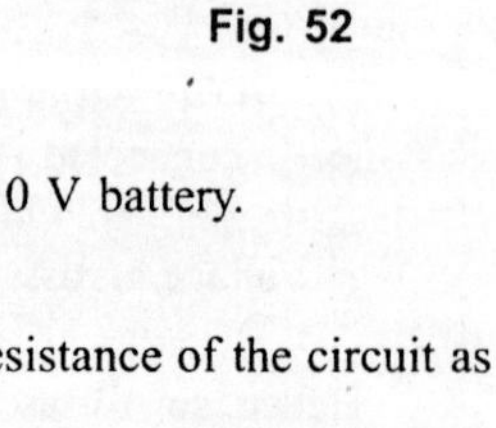

Fig. 52

Total resistance is (3 + 6) = 9Ω.

Hence, current is 30/9 = 10/3A.

Now, $\quad V_{AD} = 20 - 3 \times \dfrac{10}{3}$

$= 10$ V $\quad$ with point *A* positive.

Alternatively, we may consider the branch *BC*.

Drop on 6 Ω resistance $= 6 \times \dfrac{10}{3} = 20$ V. It is *opposed* by the 10 V battery.

Hence, $\quad V_{BC} = 20 - 10 = 10$ V as before.

Third Step : For finding R_{th}, the two batteries are shorted. The resistance of the circuit as viewed form terminals $T_1 - T_2$ is

$$= 3 + 6 \parallel 3 = 5\ \Omega$$

The Thevenin equivalent consists of a voltage source of 10V and a series resistance of 5Ω as shown in Fig. 52.

Fourth step : Finally connect the load resistance of 5Ω back to terminals $T_1 - T_2$.

$\therefore \quad I = 10/(5 + 5) = \mathbf{1\ A}$

Q. 28. (a) State and explain Thevenin's theorem.

(b) The galvanometer in Fig. 53(a) has a resistance of 5Ω. Find the current through the galvanometer using Thevenin's theorem. All resistances are in ohm.

(*Basic Elect. West Bengal Tech, Univ., 2009*)

Ans. (a) Please see Q. 26.

(b) (*i*) When galvanometer is removed form Fig. 53(*a*), we get the circuit of Fig. 53(*b*).

(*ii*) Let us now find the open-circuit voltage $V_{oc} = V_{th}$ between terminals *B* and *D*. Keeping in mind that *ABC* as well as *ADC* is a voltage divider on which a voltage drop of 10V takes place, we get

Potential of *B* w.r.t *C* *i.e.* $V_{BC} = 10 \times \frac{15}{10+15} = 6 \text{ V}$

Potential of *D* w.r.t *C* *i.e.* $V_{DC} = 10 \times \frac{16}{16+12} = 5.7 \text{ V}$

∴ p.d. between *B* and *D* *i.e.* $V_{th} = 6 - 5.7 = 0.3\text{V}$

(*iii*) Now, remove the 10-V battery. Since, battery has no internal resistance, it amounts to short-circuiting points *A* and *C* as shown in Fig. 53(*d*).

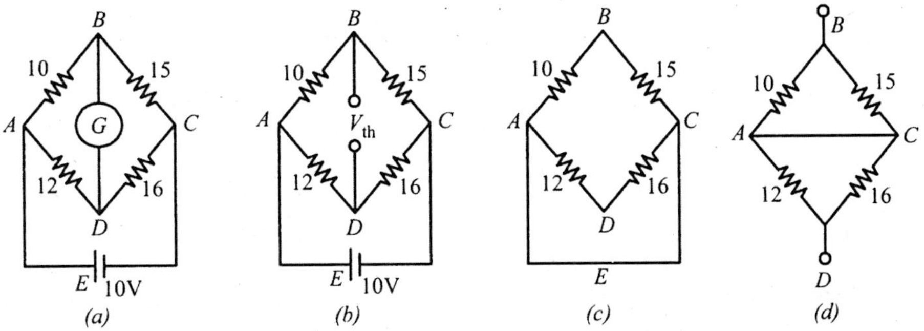

Fig. 53

(*iv*) Next, we will find resistance of the whole network as viewed from points *B* and *D*. It will equal R_{th}. It will be noted that electrically speaking , points *A* and *C* have become one as shown in Fig. 54(*a*).

Obviously, $R_{BD} = 10 \parallel 15 + 12 \parallel 16 = 6 + 6.85 = 12.85\Omega$ as shown in Fig. 54(*c*).

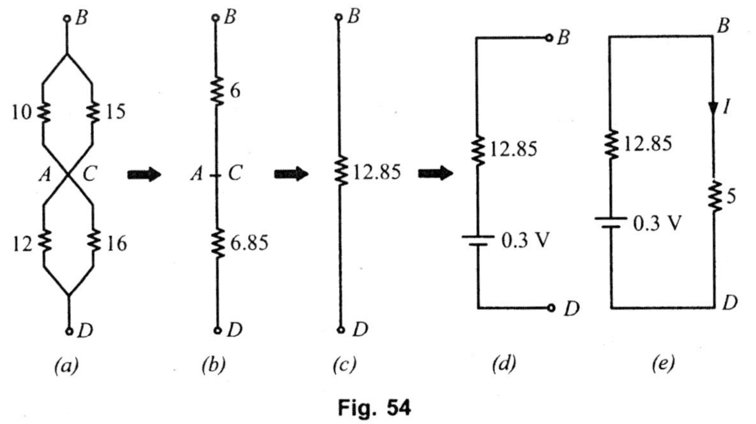

Fig. 54

(*v*) Now, as for as points *B* and *D* are concerned the network has a voltage source of 0.3V and a series resistance of 12.85Ω as shown in Fig. 54(*d*). It is called Thevenin's equivalent.

(*vi*) Finally let us connect the galvanometer back to the Thevenin equivalent and calculate the current I going through it. As seen from Fig,. 54(*e*).

$$I = \frac{0.3}{5 + 12.85} = 0.0168\text{A} = \mathbf{6.8\ mA}$$

Q. 29 (*a*) Give different forms of resistive delta (Δ) network and wye (*Y*) network. What is their role in network analysis?

(*b*) Explain Δ/*Y* and *Y*/Δ conversion formulae with suitable diagrams.

(Basic Elect. Engg. R.G.P.V Bhopal, 2006)

Ans. (*a*) Fig. 55(*a*) shows three different forms of three resistors connected in Δ. Usually, such resistors carry letter subscripts *A*, *B* and *C* etc.

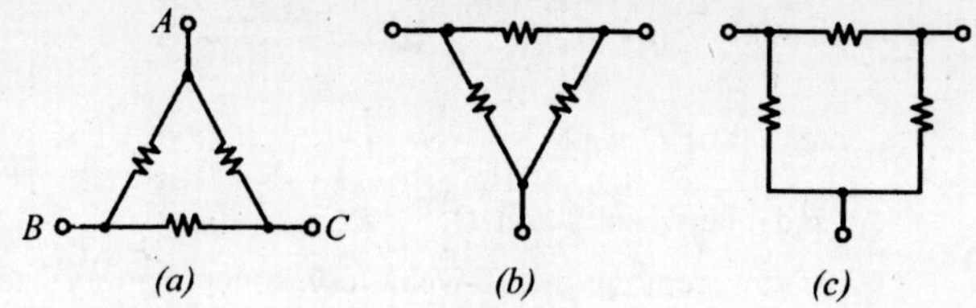

Fig. 55

Fig. 56 shows three resistors connected in *Y* or star. These resistors carry numerical subscripts like 1, 2 and 3.

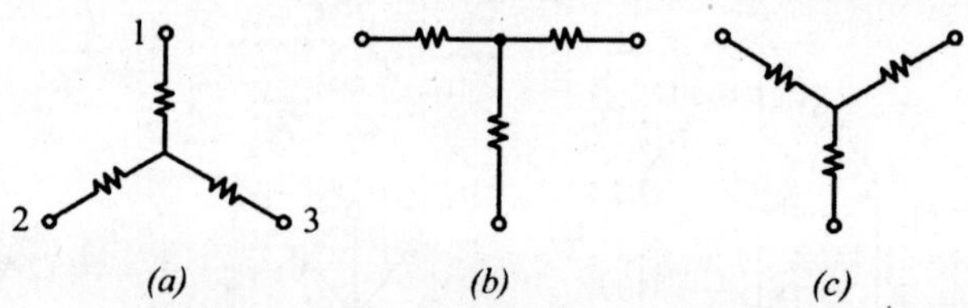

Fig. 56

In most of the complicated networks, we come across many resistors connected in either delta or wye. It is very difficult to solve such circuits. However, it becomes easy to solve most circuits if we use Δ/*Y* or *Y*/Δ transformations (or conversions).

(*b*) Δ/*Y* Conversion : It is convenient to think of *Y* placed within the delta as shown in Fig. 57. To convert from Δ to *Y*, we need R_1, R_2 and R_3 in terms of R_A, R_B and R_C. The conversion rule is as follows:

Each resistor in *Y* is equal to the product of resistors in two adjacent delta branches, divided by the sum of all three delta resistors.

In Fig. 57, R_A and R_C are adjacent to R_1

$$\therefore \qquad R_1 = \frac{R_A R_C}{R_A + R_B + R_C}$$

Also, R_A and R_B are adjacent to R_2

$$\therefore \qquad R_2 = \frac{R_A R_B}{R_A + R_B + R_C}$$

Similarly, R_B and R_C are adjacent to R_3

$$\therefore \qquad R_3 = \frac{R_B R_C}{R_A + R_B + R_C}$$

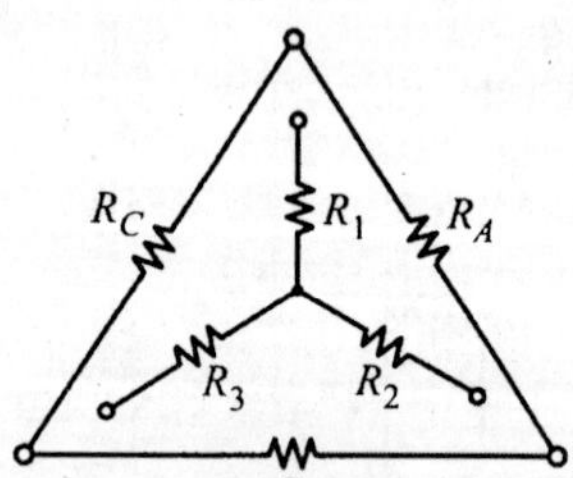

Fig. 57

***Y*/Δ Conversion :** To convert form *Y* to Δ, we need R_A, R_B and R_C in terms of R_1, R_2 and R_3. The conversion rule is as follows :

Each resistor in delta is equal to the sum of all possible products of *Y* resistors taken two at a time, divided by the opposite *Y* resistor.

In Fig. 57, R_A is opposite to R_3.

$$\therefore \qquad R_A = \frac{R_1 R_2 + R_2 R_3 + R_3 R_1}{R_3} = R_1 + R_2 + \frac{R_1 R_2}{R_3}$$

Also, $R_B = \dfrac{R_1R_2 + R_2R_3 + R_3R_1}{R_1} = R_2 + R_3 + \dfrac{R_2R_3}{R_1}$

Similarly, $R_C = \dfrac{R_1R_2 + R_2R_3 + R_3R_1}{R_2} = R_1 + R_3 + \dfrac{R_1R_3}{R_2}$

Q. 30 (*a*) Find the input resistance of the circuit between the points *A* and *B* of Fig. 58. (*AMIE, Sec. B Network Anlysis Summer 1992*)

(*b*) Find the voltage to be applied across *AB* in order to drive a current of 5A into the circuit of Fig. 59(*a*) by using star-delta transformation.

(*Elect. Circuits, B. Tech JNTU, Hyderabad 2006*)

Sol. (*a*) For finding R_{AB}, we will be convert the delta *CDE* of Fig. 58(*a*) into equivelent star as shown in Fig. 58(*b*).

$R_{CS} = 8 \times 4/18 = 16/9\ \Omega$; $R_{ES} = 8 \times 6/18 = 24/9\ \Omega$; $R_{DS} = 6 \times 4/18 = 12/9\Omega$.

The two parallel resistances between *S* and *B* can be reduced to a single resistance of 35/9Ω.

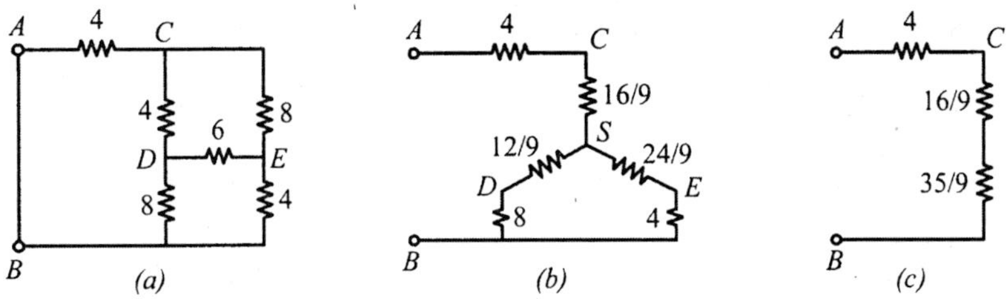

Fig. 58

As seen from Fig. 58(*c*), $R_{AB} = 4 + (16/9) + (35/9) = \mathbf{87/9\Omega}$.

(*b*) The two deltas *C*, *D*, *E* and *L*, *M*, *N* have been converted to the equivalent stars as shown in Fig. 59(*c*). Using the delta-star converstion formula, we get from Fig. 59(*b*).

$$R_1 = \frac{3 \times 2}{2 + 5 + 3} = \frac{6}{10} = 0.6\ \Omega$$

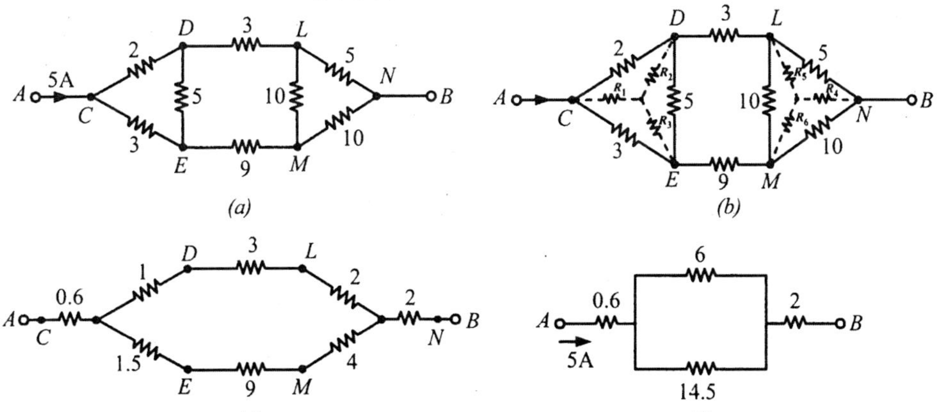

Fig. 59

Similarly $R_2 = 5 \times 2/10 = 1\Omega$

$R_3 = 5 \times 3/10 = 1.5\Omega$

Fig. 59(*c*) has been simplified to Fig. 59(*d*). In Fig. 59(*d*).

$$R_{AB} = 0.6 + 6 \parallel 14.5 + 2 = 6.84\Omega$$

Hence, voltage to be applied across *A* and *B* is = 5 × 6.84 = **34.2 V**

Q. 31 (*a*) Calculate the current *I* in the circuit and the resistance between the terminals *A* and *B* (all resistances are in ohm) as shown in Fig. 60.

(Basic Elect. Engg. B. Tech. JNTU, Hyderabad 2006)

(*b*) Find the current in 11-Ω resistor in Fig. 61(*a*) using star/delta conversion. All resitances are in ohm. *(Elect. Engg., Nagpur Univ. 2003)*

Ans. (*a*) The given circuit of Fig. 60(*a*) has been redrawn in Fig. 60(*b*). When delta *CDE* is replaced by its equivalent star, the circuit becomes as shown in Fig. 60(*c*). Now the circuit is reduced to a simple series-parallel circuit shown in Fig. 60(*d*).

Resistance $R_{AB} = 30 + 100 \parallel 100 = 30 + 50 = 80\Omega$

$\therefore \quad I = 160/80 = \mathbf{2A}$

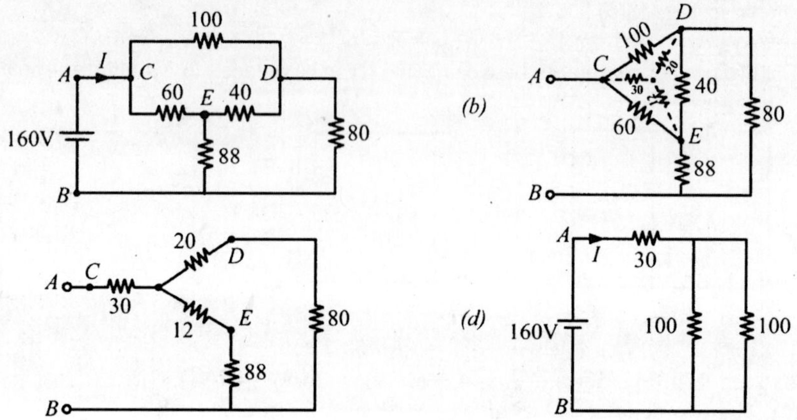

Fig. 60

(*b*) The given figure can be redrawn as shown in Fig. 61(*b*). Let us convert Δ*ABC* into its equivalent star. The value of each star resistance is $= \dfrac{6 \times 6}{6+6+6} = 2\Omega$.

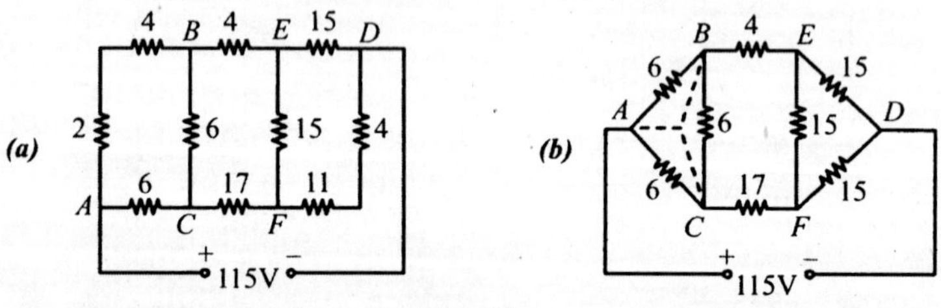

Fig. 61

These are shown in Fig. 62(*a*) which is redrawn as Fig. 62(*b*).

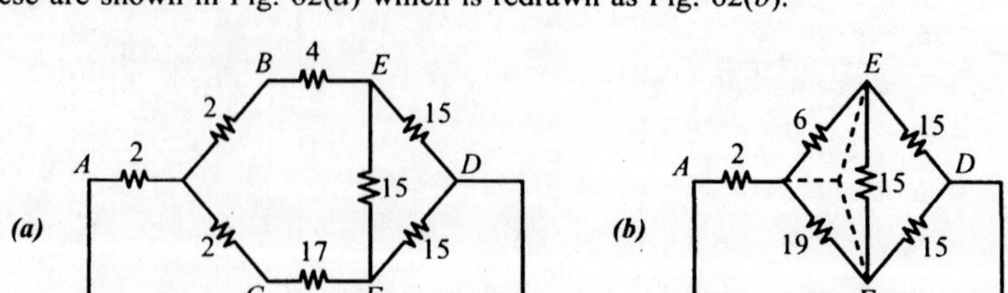

Fig. 62

When (6-19-15) delta of Fig. 62(*b*) is converted to its equivalent star, we get Fig. 63(*a*).

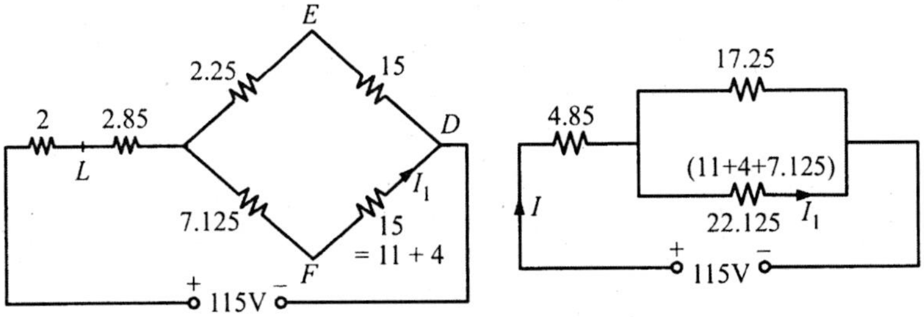

(*a*) (*b*)

Fig. 63

The circuit of Fig. 63(*a*) is a simple series parallel circuit as shown is Fig. 63(*b*). The total resistance of this series-parallel circuit is

$$= 4.85 + (17.25) \parallel (22.125) = 14.53\Omega$$

$\therefore$ $$I = 115/\ 14.53 = 7.91\text{A}$$

By current division rule, we get

$$I_1 = I \times \frac{17.25}{17.25 + 22.125} = 7.\ 91 \times \frac{17.25}{39.375} = \mathbf{3.46A}$$

This is the current which flows through 11Ω resistance.

Q. 32 (*a*) Find the total current flowing through the circuit of Fig. 64(*a*) using *Y*/Δ transformation. Value of each resistor is 4Ω. (*Elect. Engg. Nagpur Univ., 2006*)

(*b*) Find the voltage drop on 10Ω resistor in Fig. 67(*a*). All resistances are in ohm. (*Elect. Engg. Nagpur Univ., 2003*)

Ans. (*a*) The given circuit of Fig. 64(*a*) is redrawn as shown in Fig. 64(*b*). When we convert Δ*DBC* of Fig. 64(*b*), we get the circuit of Fig. 65(*a*).

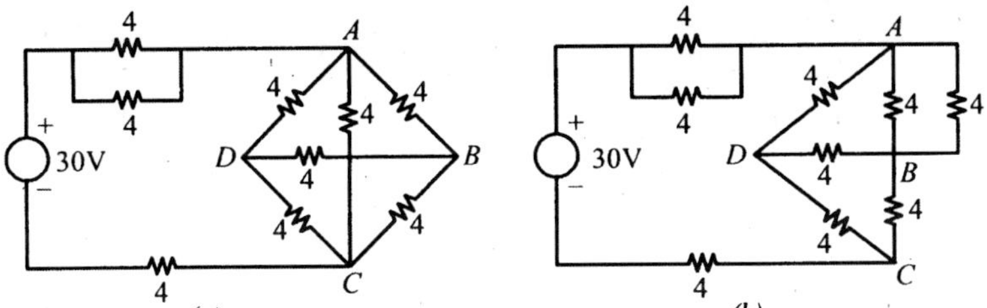

(*a*) (*b*)

Fig. 64

$$\text{Each star resistance} = \frac{4 \times 4}{(4+4+4)} = 1.34\ \Omega \text{ as shown in Fig. 65}(a).$$

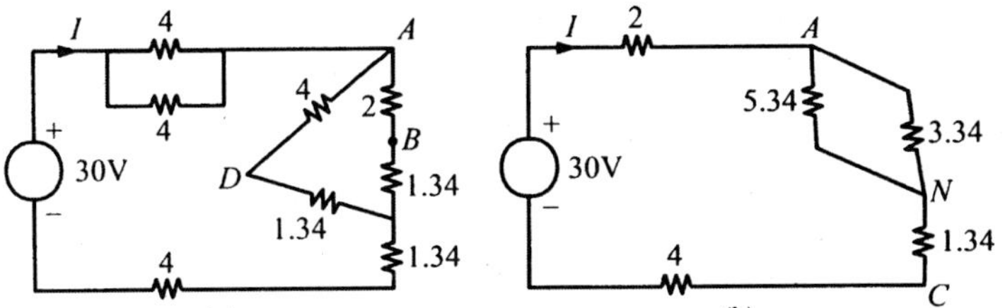

(*a*) (*b*)

Fig. 65

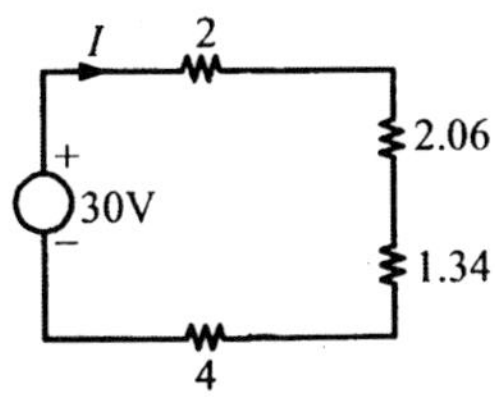

Fig. 66

Circuit of Fig. 65(*a*) can be simplified to that shown in Fig. 65(*b*).

$$R_{AN} = 5.34 \parallel 3.34 = 2.06\Omega$$

Hence, the given circuit of Fig. 64(*a*) is finally reduced to that shown in Fig. 66.

$$\therefore \qquad I = \frac{30}{(2 + 2.06 + 1.34 + 4)} = \mathbf{3.19A}$$

(*b*) In Fig. 67(*a*), the two 16Ω parallel resistors can be combined to give one 8Ω resistor. When combined with the 8Ω series resistor, we get a total resistance of 16Ω as shown in Fig. 67(*b*).

When we convert star (8–8–16) of Fig. 67(*b*), we get delta resistances as calculated below :

$$R_A = 8 + 16 + \frac{8 \times 16}{8} = 40\Omega$$

$$R_B = 8 + 16 + \frac{8 \times 16}{8} = 40\Omega$$

$$R_C = 8 + 8 + \frac{8 \times 8}{16} = 20\Omega$$

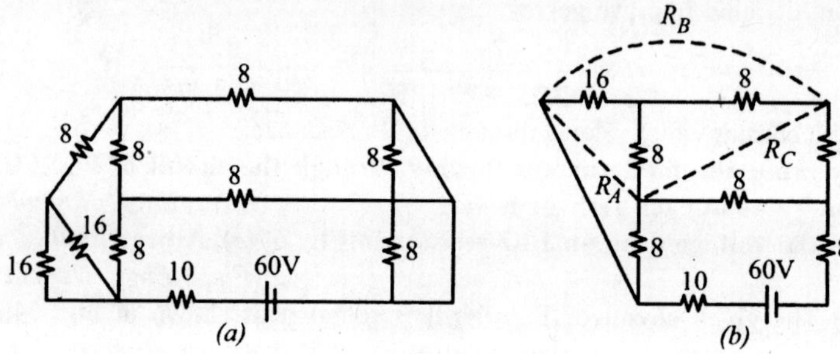

Fig. 67

We will convert the delta (20–8–8) of Fig. 68(*a*) into an equivalent star.

$$R_1 = \frac{20 \times 8}{20 + 8 + 8} = 4.45\Omega$$

$$R_2 = \frac{20 \times 8}{20 + 8 + 8} = 4.45\Omega$$

$$R_3 = \frac{8 \times 8}{20 + 8 + 8} = 1.78\Omega$$

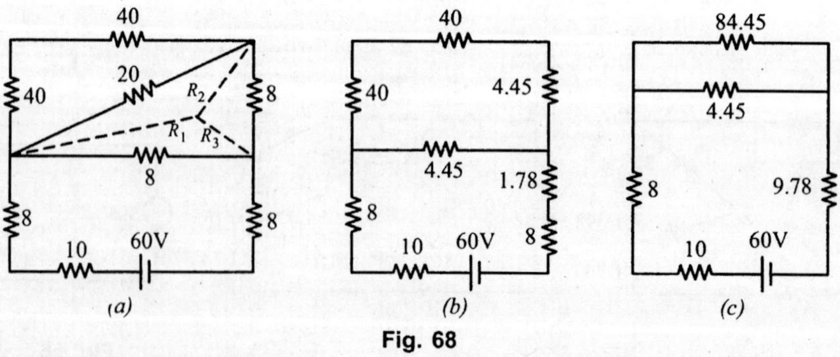

Fig. 68

Hence, circuit of Fig. 68(*a*) becomes that of Fig. 68(*b*) which further reduces to the circuit of Fig. 68(*c*) which is a series-parallel circuit. As seen from Fig. 69, as per the voltage divider rule,

Voltage drop across 10Ω resistor

$$= \frac{60 \times 10}{(10 + 8 + 4.22 + 9.78)}$$

$$= \mathbf{18.75\ V}$$

Note : We have avoided calculating I by using voltage divider rule.

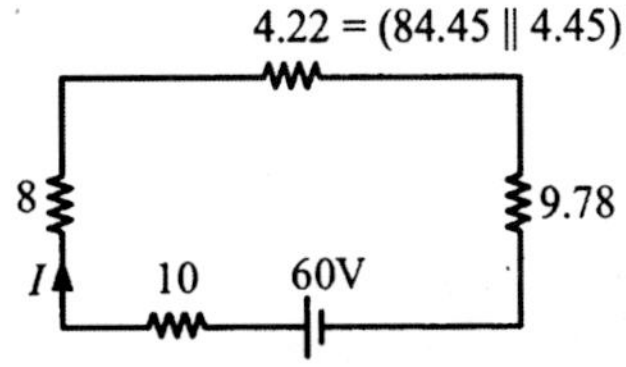

Fig. 69

Q. 33. State and explain Norton's theorem. What are the two elements of a Norton equivalent circuit?

Ans. In simple words, Norton's theorem states that *any given complicated circuit can be reduced to a single current source with a parallel resistance.*

This fact has been shown in Fig. 70.

Different steps required for finding Norton's equivalent circuit are as under :

1. Open the circuit terminals of the given circuit with respect to which Norton's equivalent is required.
2. Short all voltage sources and open all current sources of the given circuit. And calculate the value of the circuit resistance as viewed from the open terminals. It is called Norton's resistance R_N.

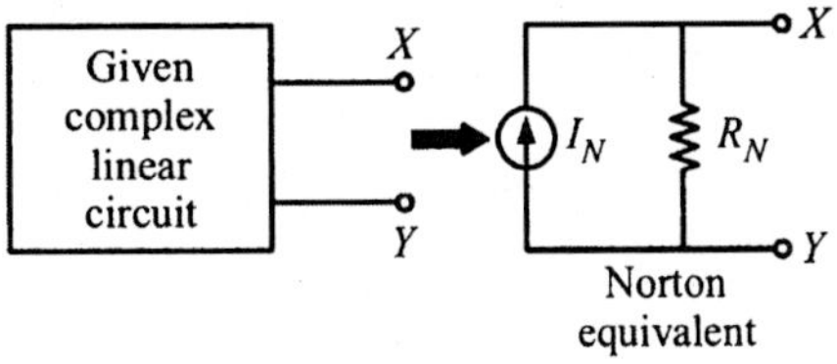

Fig. 70

3. Next, short circuit the two terminals of the given circuit. Calculate the short-circuit current I_{SC} through the shorted terminals. It is called Norton's current I_N.
4. The resistance R_N is connected in parallel with the current source I_N to get the Norton equivalent for the original circuit.

As seen, the Norton equivalent circuit consists of two components. One is an ideal current source I_N and the other is a resistor R_N both connected in parallel.

Explanation : Suppose, we want to find the Norton equivalent across terminals X–Y of the circuit given in Fig. 71(a).

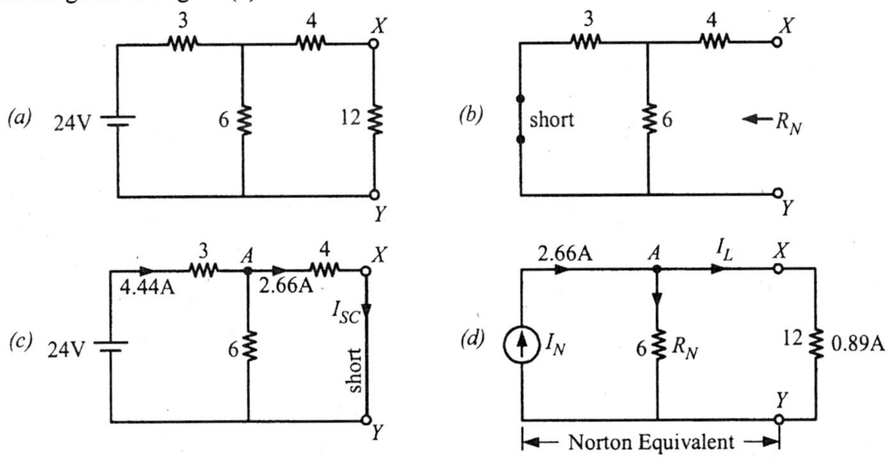

Fig. 71

First, we remove the 12Ω load resistance and short circuit the 12V battery as shown in Fig. 71(b). Then, we find circuit resistance as viewed from the terminals X–Y as shown in Fig. 71(b)

$$R_N = 4 + 6 \parallel 3 = 6\Omega$$

Now, we short circuit the terminals X–Y and find current flowing through this short as shown in Fig. 71(c).

The total resistance as seen by the battery

$$= 3 + 6 \parallel 4 = 5.4\Omega$$

Hence, current supplied by the battery is 24/5.4 = 4.44A. This current divides at point A in the inverse ratio of the two parallel resistances.

$$\therefore \quad I_{SC} = 4.44 \times \frac{6}{6+4} = 2.66\text{A}$$

Hence, Norton's current is 2.66A.

As seen from the Norton equivalent of Fig. 71(d), I_N divides into two parts. The load current is given by (Fig. 72).

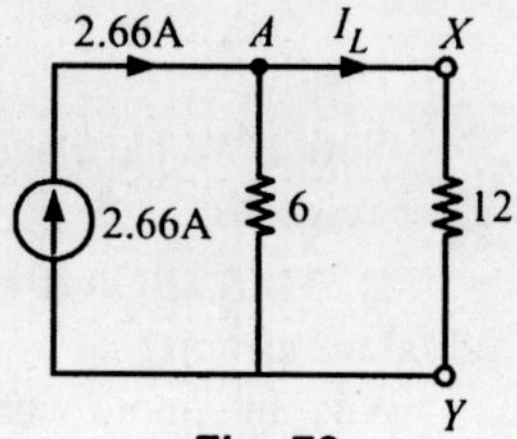

Fig. 72

$$I_L = 2.66 \times \frac{6}{6+12} = \mathbf{0.89A}$$

Q. 34. Find the Norton's equivalent for the network to the left of terminals X–Y in Fig. 73(a).

Ans. For finding R_N, we short circuit the 30V battery and open circuit 6A current source as shown in Fig. 73(b).

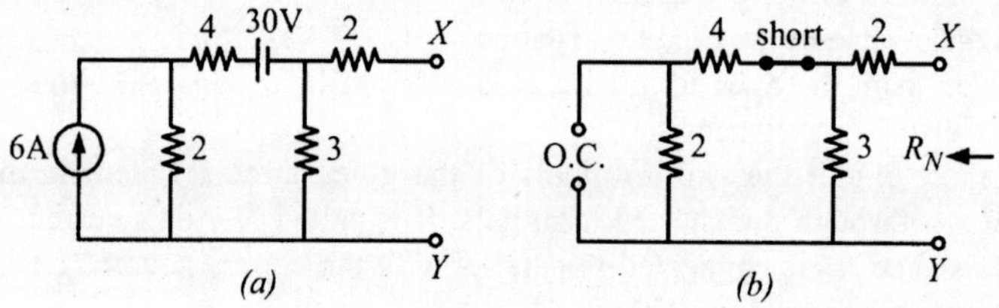

Fig. 73

$$R_N = 2 + 3 \parallel (4 + 2) = \mathbf{4\ \Omega}$$

Now, we convert the 6A current source into voltage source as shown in Fig. 74(a). The equivalent voltage source has $V = 2 \times 6 = 12$ V and 2Ω series resistance as shown in Fig. 74(a).

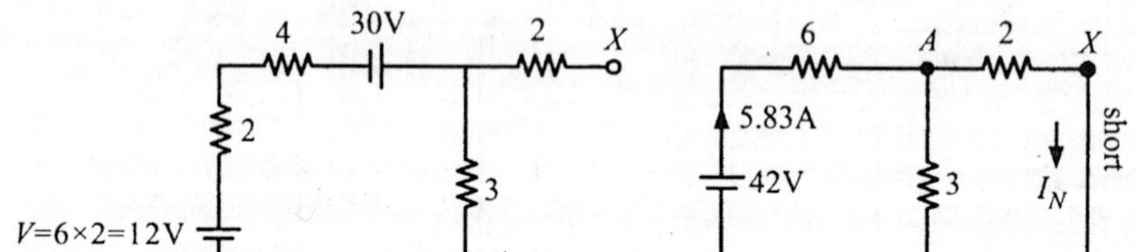

Fig. 74

The circuit of Fig. 74(a) is simplified to that shown in Fig. 74(b). For finding I_N, we will short circuit terminals X–Y. The resistance seen by the (34 + 12) = 42V battery is

$$= 6 + 3 \parallel 2 = 7.2\ \Omega$$

Battery current = 42 / 7.2 = 5.83A

This current divides at point A into two parallel paths. According to current divider rule,

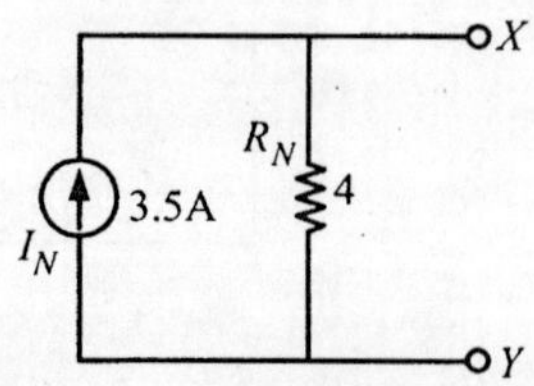

Fig. 75

$$I_N = 5.83 \times \frac{3}{3+2} = \mathbf{3.5\ A}$$

Hence, Norton equivalent consists of a 3.5 A current source in parallel with 4Ω resistance as shown in Fig. 75.

Q. 35 (*a*) State and prove maximum power transfer theorem for DC networks.
(*Elect. Engg., West Bengal Tech. Univ., 2008*)

(*b*) Determine current through the 15Ω resistor of the network given in Fig. 76(*a*) by using Norton's Theorem. All resistances are in ohm.

Ans. (*a*) Please see Q. 38.

(*b*) First Step : Open the terminals $T_1 - T_2$ as shown in Fig. 76(*b*).

Second step : Short circuit the 10V battery. The resistance of the network as viewed from $T_1 - T_2$ is R_N

$$R_N = (20 \parallel 20) \parallel 10 = 10 \parallel 10 = 5\Omega$$

Third Step : Now, short circuit the terminals as shown in Fig. 76(*c*). This short-circuits both 20Ω resistances. Hence, 10V battery has to drive current through the series resistance of 10Ω and the short between terminals $T_1 - T_2$.

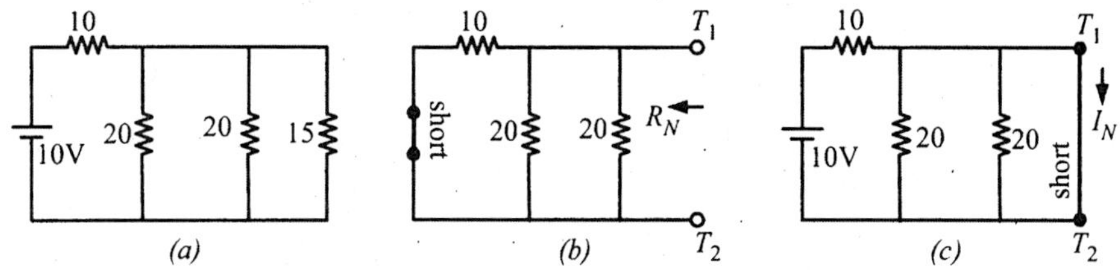

Fig. 76

$\therefore \quad I_{SC} = 10/10 = 1\text{A}$

It is also the Norton current I_N.

Hence, Norton equivalent consists of a current source of 1A in parallel with a resistance of 5Ω.

Forth Step : The 15Ω resistor earlier removed from the terminals is connected back to them, (Fig. 77).

Now, the current of 1A is divided between the two parallel paths in the inverse ratio of their resistances. By current divider rule, we have

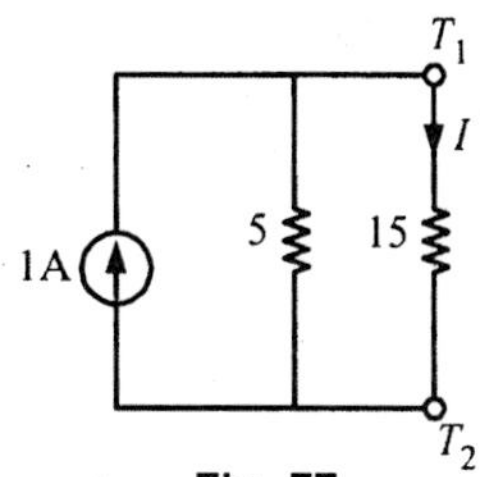

Fig. 77

$$I = 1 \times \frac{5}{5+15} = \mathbf{0.25A.}$$

Q. 36 (*a*) State and explain Maximum Power Transfer (MPT) theorem as applied to resistive loads. What is its special field of application? (*Basic. Elect. Engg., Bhopal, 2007*)

(*b*) Find the value of R_L in Fig. 80(*a*) for

(*i*) getting maximum power

(*ii*) value of maximum power in R_L.

Ans. (*a*) This theorem concerns itself with the transfer of electric power from a *source to a load* (Fig. 78). According to this theorem

maximum power is transferred to a load resistance if it equals the Thevein's resistance of the source to which it is connected.

It may be noted that when the source and load have the same resistance, they are said to *matched.*

Explanation : In Fig. 79(*a*) is shown the source circuit which has Thevenin's voltage V_{th} = 36 and R_{th} = 18Ω. The load resistance is 2Ω.

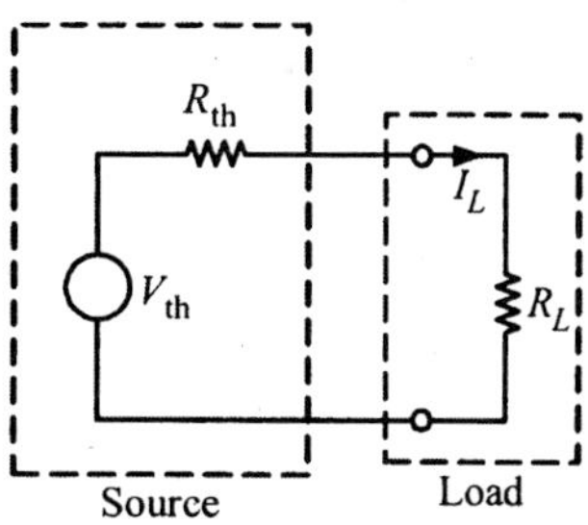

Fig. 78

As seen, $\quad I_L = 36/(18 + 2) = 1.8$ A

Power delivered to the load is

$$P_L = I_L^2 R_L = 1.8^2 \times 2 = 6.48 \text{ W}$$

Now, let the load resistance be increased to 18Ω.

Obviously, $R_L = R_{th} = 18\Omega$ as shown in Fig. 79(*b*).

$I_L = 36/(18 + 18) = 1A$

$P_L = 1^2 \times 18 = 18$ W

Fig. 79(*c*) shows the graph between load power P_L and load resistance R_L. It is seen that power transfer is maximum when $R_L = R_{th}$.

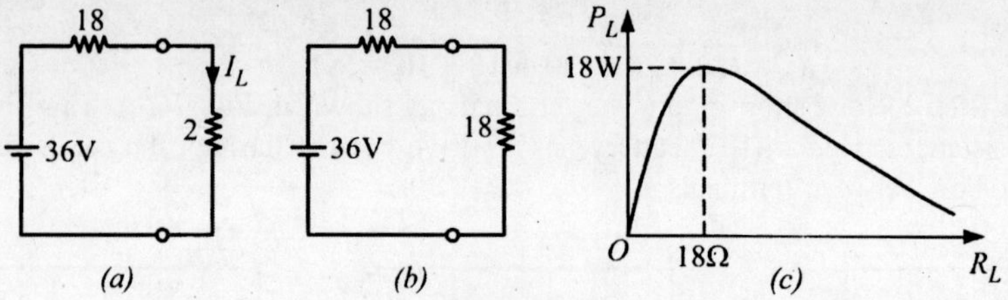

Fig. 79

However, it is found that although power transferred to R_L is maximum, the power transfer efficiency (η) is only 50%. MPT theorem is very useful for analysing electronic and communication networks where the main purpose is to either transmit or receive maximum power *even at reduced efficiency.* It is so because the power involved is extremely small. Hence, the main goal is to get maximum power transfer and not maximum efficiency.

(*b*) (*i*) For finding R_{th}, we will remove R_L temporarily as shown in Fig. 80(*b*).

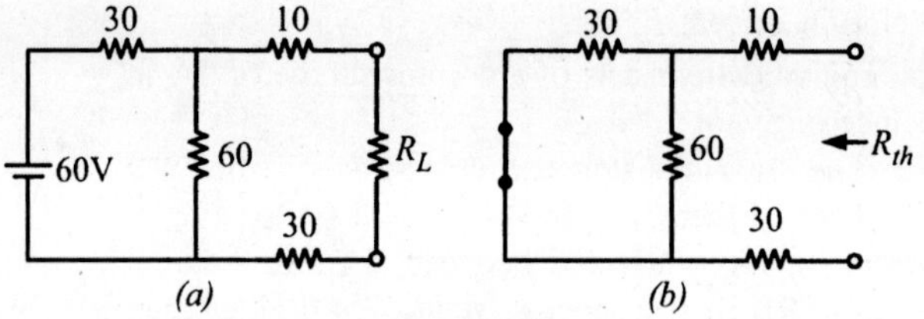

Fig. 80

$$R_{th} = 10 + (30 \parallel 60) + 30 = 60\Omega$$

Hence, for getting maximum power R_L = **60Ω.**

(*ii*) For finding V_{th}, we will use the circuit of Fig. 81(*a*).

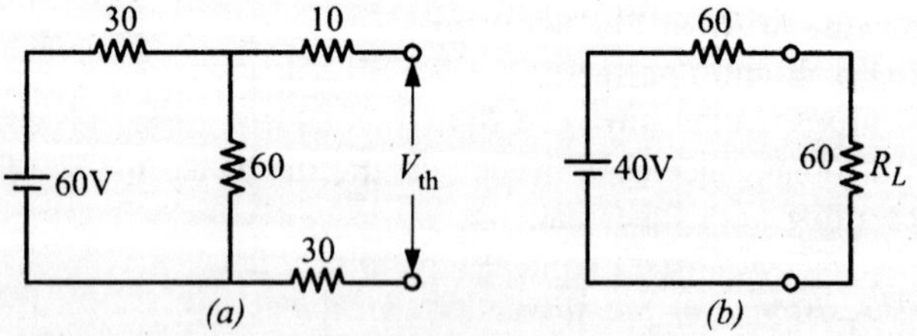

Fig. 81

There are no voltage drops on the 10Ω and 30Ω resistors because they carry no current. The 60 volt will be shared by 30Ω and 60Ω resistors. As per voltage divider rule,

$$\text{drop across } 60\Omega = 60 \times \frac{60}{60 + 30} = 40\text{V}$$

This Thevenin's voltage source is connected in series with R_L.

Current through $R_L = 40/(60 + 60) = 1/3$A

Maximum power transferred is $= (1/3)^2 \times 60 =$ **6.7 W.**

Q. 37. A source delivers 5A when the load connected to it is 5Ω and 2A when load is increased to 20Ω. Find the maximum power available from this source.

Ans. As shown in Fig. 82, let V_{th} and R_{th} be the Thevenin's equivalent of the given source. The values of V_{th} and R_{th} can be found from the given data.

(*i*) When $R_L = 5\Omega$, $I_L = 5A$

∴ voltage across load is

$$V_L = 5 \times 5 = 25V$$

As seen, $V_L = V_{th} - I_L R_{th}$ or $25 = V_{th} - 5 R_{th}$ ...(*i*)

(*ii*) When $R_L = 20\Omega$, $I_L = 2A$

∴ $V_L = I_L R_L = 2 \times 20 = 40V$

∴ $40 = V_{th} - 2 R_{th}$ ...(*ii*)

Solving from (*i*) and (*ii*), we get

$$15 = 3 R_{th} \quad \text{or} \quad R_{th} = 5\Omega$$

Hence, power delivered to the load will be maximum in the first case because

$$R_L = R_{th} = 5\Omega$$

Since, under the above condition,

$$I_L = 5A \quad \text{and} \quad R_L = 5\Omega$$

∴ $P_{L\,max} = I_L^2 R_L = 5^2 \times 5 = \mathbf{125W}$

Alternatively, $P_{L\,max} = V_L I_L = 25 \times 5 = \mathbf{125W}$

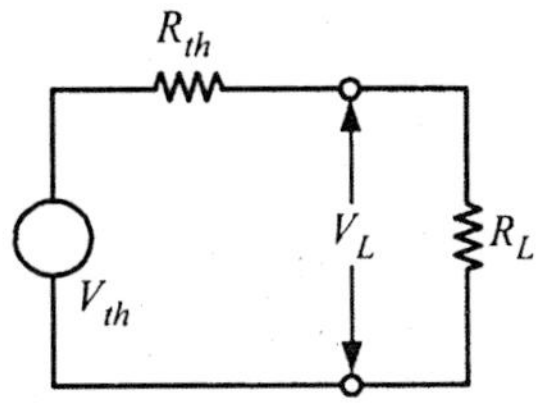

Fig. 82

Q. 38. State and prove maximum power transfer (MPT) theorem.

(*Elect. Engg., West Bengal Tech. Univ., 2009*)

Ans. Statement of MPT : According to MPT theorem, maximum power is transferred to a load resistance if it equals the Thevenin's resistance of the source to which it is connected (Fig. 78).

Proof of MPT theorem : As shown in Fig. 83, the variable load resistor R_L is connected to the Thevenin equivalent of a power source.

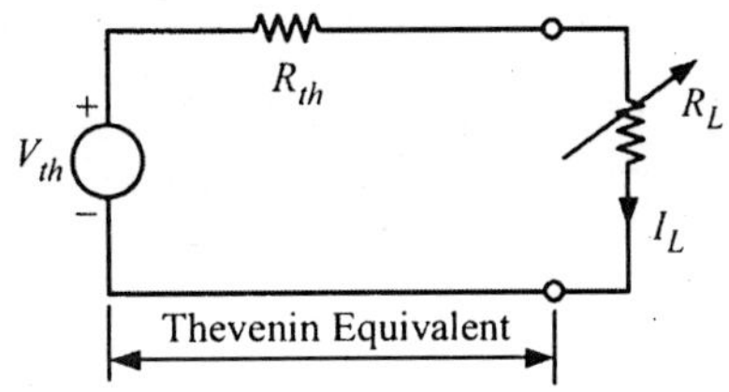

Fig. 83

Load current $\quad I_L = \dfrac{V_{th}}{R_{th} + R_L}$

Power in the load, $\quad P_L = I_L^2 R_L = \dfrac{V_{th}^2 R_L}{(R_{th} + R_L)^2}$ ...(*i*)

Differentiating with respect to the variable R_L, we get

$$\frac{dP_L}{dR_L} = V_{th}^2 \cdot \frac{a}{dR_L}\left[\frac{R_L}{(R_{th} + R_2)^2}\right]$$

$$= V_{th}^2\left[\frac{1}{(R_{th} + R_L)^2} - \frac{2R_L}{(R_{th} + R_L)^3}\right] = V_{th}^2 \frac{R_{th} - R_L}{(R_{th} + R_L)^3}$$

The load power P_L will be maximum when this derivative equals zero.

or, when $\quad \dfrac{dP_L}{dR_L} = 0 \quad$ or $\quad V_{th}^2 \dfrac{R_{th} - R_L}{(R_{th} + R_L)^3} = 0$

or $\quad R_{th} - R_L = \dfrac{(R_{th} + R_L)^3}{V_{th}^2} \times 0$

or $\quad R_{th} - R_L = 0 \quad$ or $\quad R_{th} = R_L$

Incidentally, the maximum power $P_{L \cdot max}$ as given by Eq. (*i*) above is

$$P_{L.\max} = \frac{V_{th}^2 R_L}{(R_L + R_L)^2} = \frac{V_{th}^2}{4R_L} \quad \text{or} \quad \frac{V_{th}^2}{4R_{th}}$$

Q. 39. What percentage of maximum possible power is delivered to load R_L when $R_L = 2 R_{th}$?

Ans. In Fig. 84, as per voltage divider rule, voltage V_L across load is

$$V_L = V_{th} \frac{2R_{th}}{R_{th} + 2R_{th}} = \frac{2}{3} V_{th}$$

$$P_L = \frac{V_L^2}{R_L} = \frac{\left(\frac{2}{3}V_{th}\right)^2}{2R_{th}} = \frac{2\,V_{th}^2}{9\,R_{th}}$$

Now under maximum power condition, $R_L = R_{th}$. The voltage V_{th} divides equally between R_{th} and R_L. Hence, $V_L = V_{th}/2$.

$$\therefore \quad P_{L.\max} = \frac{V_L^2}{R_L} = \frac{(V_{th}/2)^2}{R_{th}} = \frac{V_{th}^2}{4\,R_{th}}$$

$$\therefore \quad \frac{P_L}{P_{L.\max}} = \frac{2\,V_{th}^2/9\,R_{th}}{V_{th}^2/4\,R_{th}} = \frac{8}{9}$$

$$\therefore \quad P_L = \frac{8}{9} \times P_{L.\max} \quad \text{or} \quad \left(\frac{8}{9} \times 100\right) \times P_{L.\max}$$

$$= 88.89\%\ P_{L.\max}$$

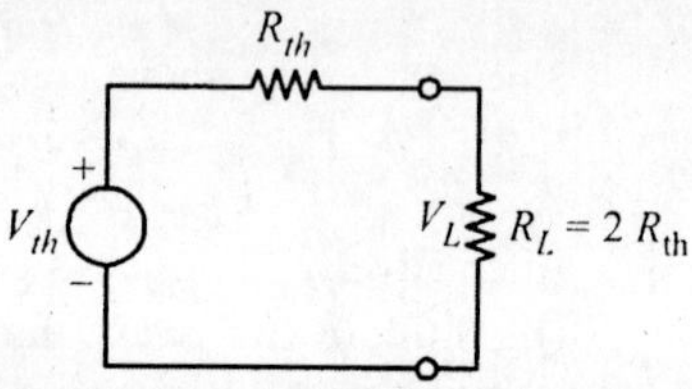

Fig. 84

Note : If, in general, $R_L = n\,.\,R_{th}$, then as thumb rule

$$\frac{P_L}{P_{L.\max}} = \frac{4n}{(n+1)^2}$$

If, $n = 10$, then $$\frac{P_L}{P_{L.\max}} = \frac{4 \times 10}{(10+1)^2} = \frac{40}{121} = 0.3306 \quad \text{or} \quad 33.06\%$$

EXERCISE

Note : All resistances in the undergiven problems are in ohms unless stated otherwise.

Kirchhoff's Current Law :

1. The circuit of Fig. 85 shows a typical interconnection in a telephone system. Find I_1 and I_6. **[I_1 = 12 mA, I_6 = –11 mA]**

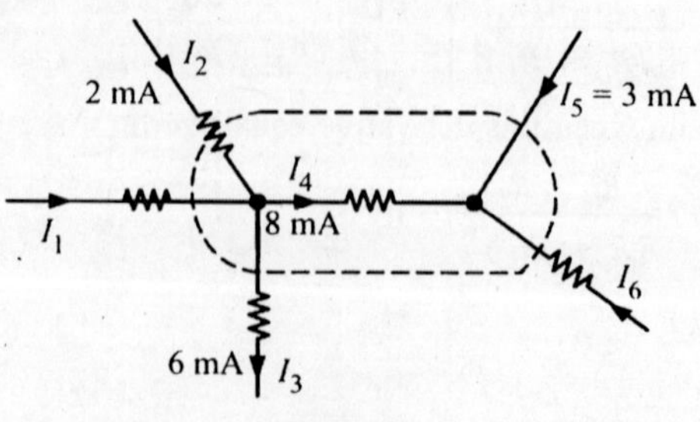

Fig. 85

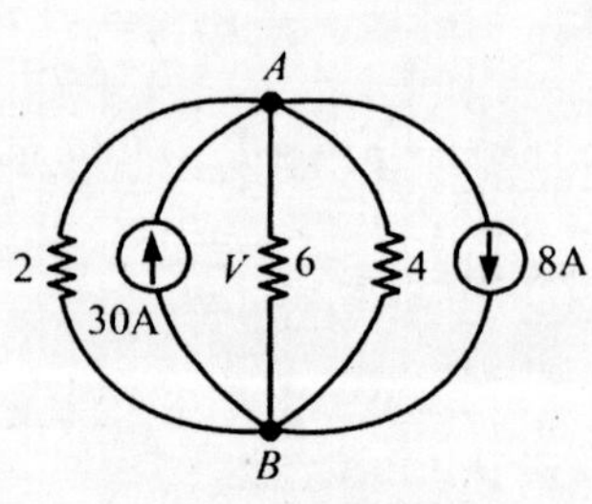

Fig. 86

2. Using Kirchhoff's current law and Ohm's law, find the magnitude and polarity of voltage V in Fig. 86. **[24 V, Point A is positive]**

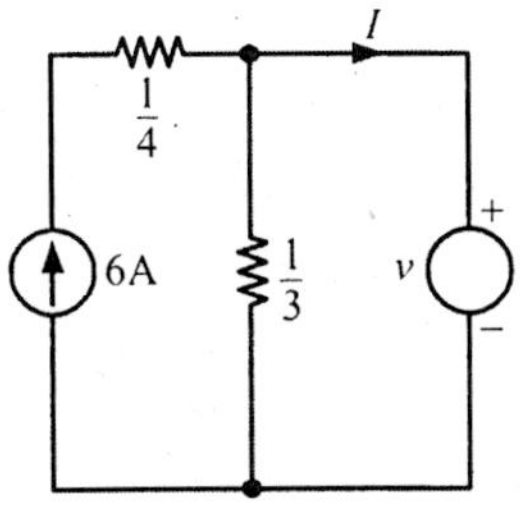

Fig. 87

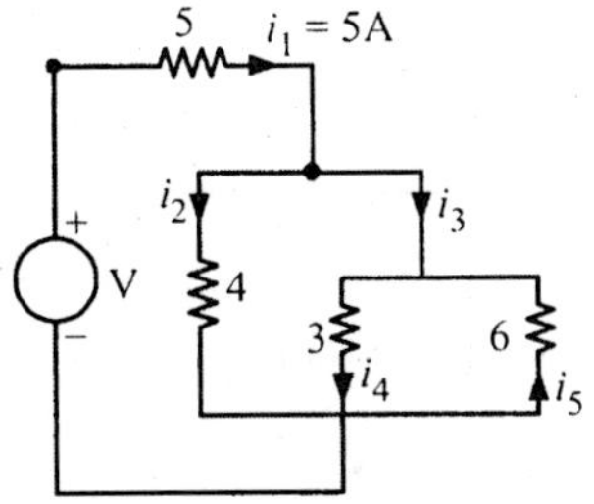

Fig. 88

3. For the circuit shown in Fig. 87, find the value of current I when v = 1V. **[3A]**
4. For the circuit given in Fig. 88, find i_2, i_3, i_4 and i_5. Assume i_1 = 5A.

$$\left[i_2 = \frac{5}{3}\text{A},\ i_3 = \frac{10}{3}\text{A},\ i_4 = \frac{20}{9}\text{A},\ i_5 = -\frac{10}{9}\text{A} \right]$$

Kirchhoff's Voltage Law

1. For the circuit shown in Fig. 89, find the value of V and its polarity. **[8V, with A +ve]**

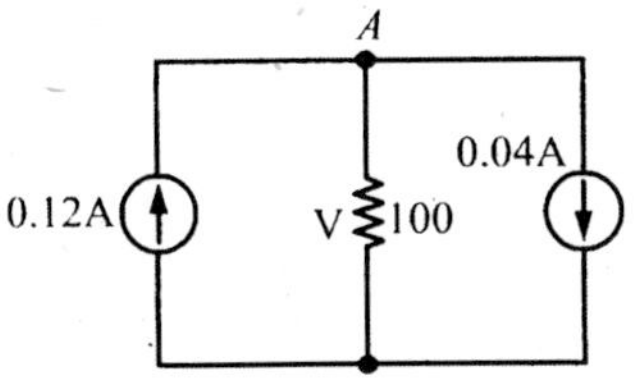

Fig. 89

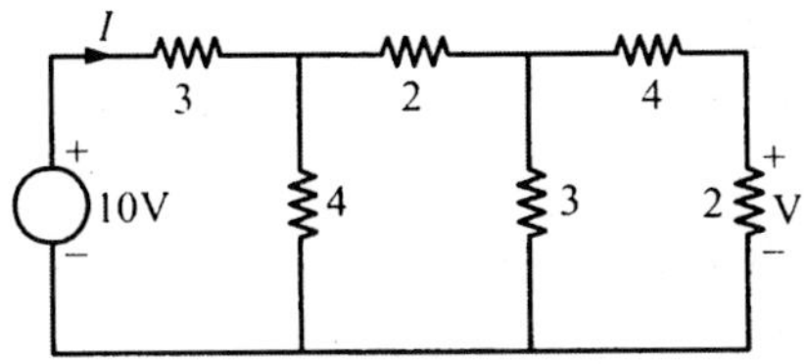

Fig. 90

2. For the series-parallel circuit given in Fig. 90, find V and I. $\left[\frac{2}{3}\text{V},\ 2\text{A} \right]$

3. For the circuit of Fig. 91, find I and V. **[I = − 1A, V = 7V]**

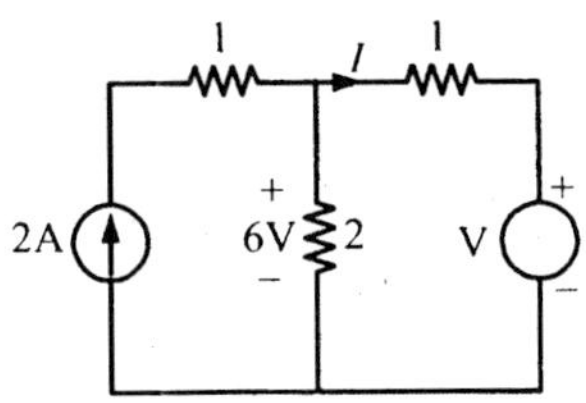

Fig. 91

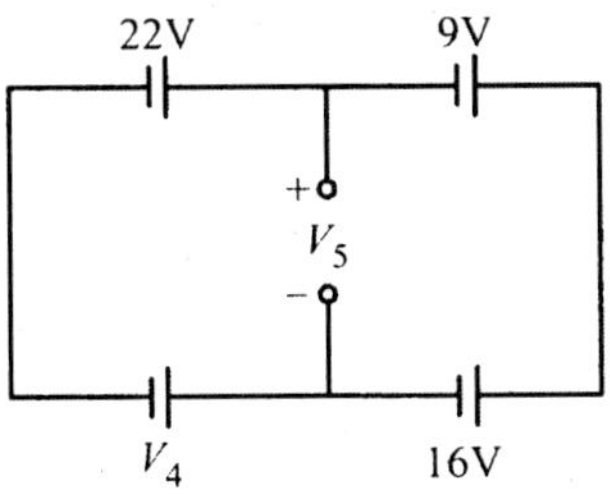

Fig. 92

4. In the circuit of Fig. 92, find the unknown voltages V_4 and V_5. **[V_4 = 15V, V_5 = 7V]**

Source Conversion

1. Find the source conversions for the circuits of Fig. 93(a) and (b).

[15V, 18Ω series resistor ; 8A, 3Ω parallel resistor]

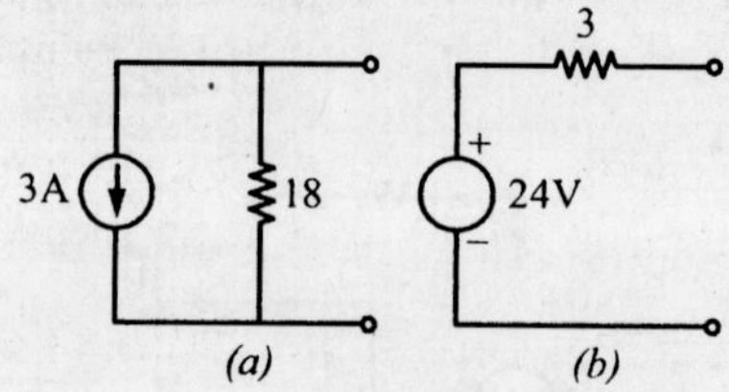

Fig. 93

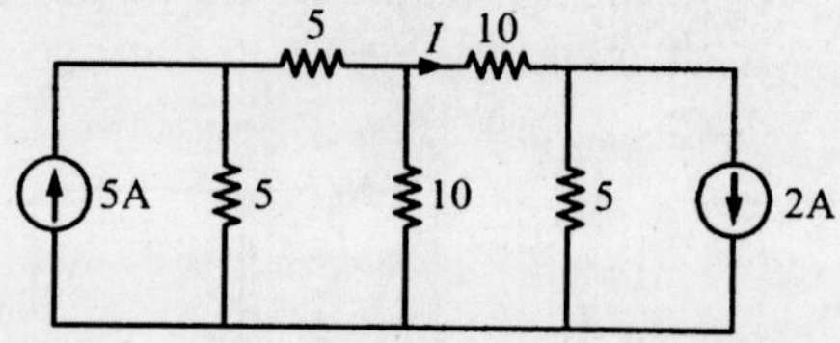

Fig. 94

2. Use source conversion method to find the current I for the circut of Fig. 94. **[1.125A]**

Mesh Analysis

1. Solve the network shown in Fig. 95 by Mesh current method.

(Basic Elect. Engg., R.G.P.V. Bhopal, 2006)

[I_1 = 3.82A, I_2 = 0.545A]

2. Find I_1, I_2 and I_3 in the network shown in Fig. 96 using loop-current method.

[I_1 = 1A, I_2 = 2A, I_3 = 3A]

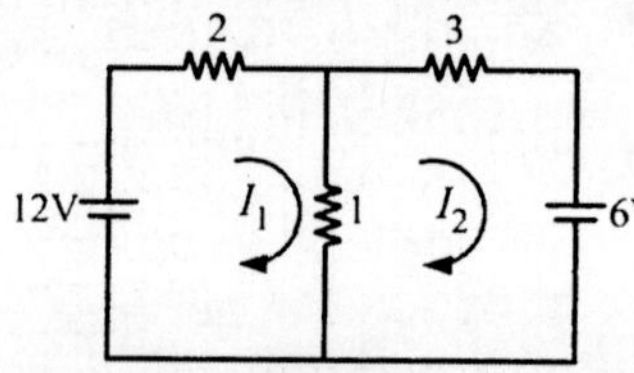

Fig. 95

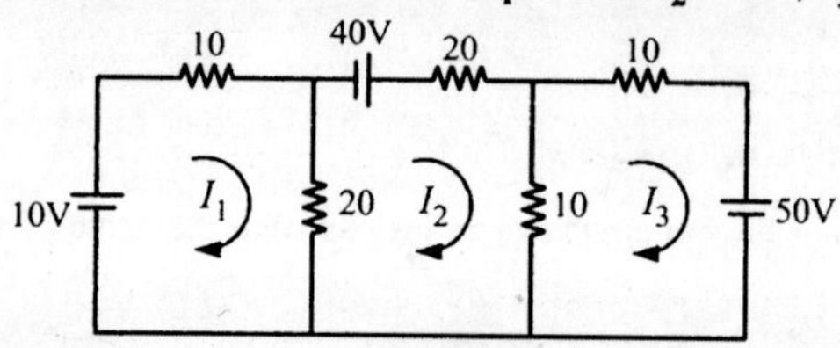

Fig. 96

Nodal Analysis

1. Use node voltage method to find the voltage of node A in Fig. 97. **[V_A = – 13.33V]**

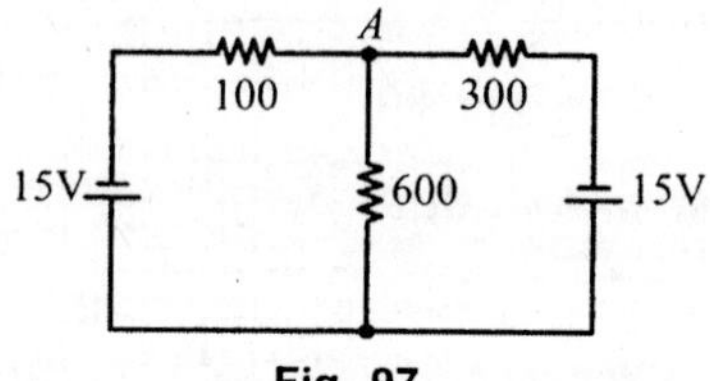

Fig. 97

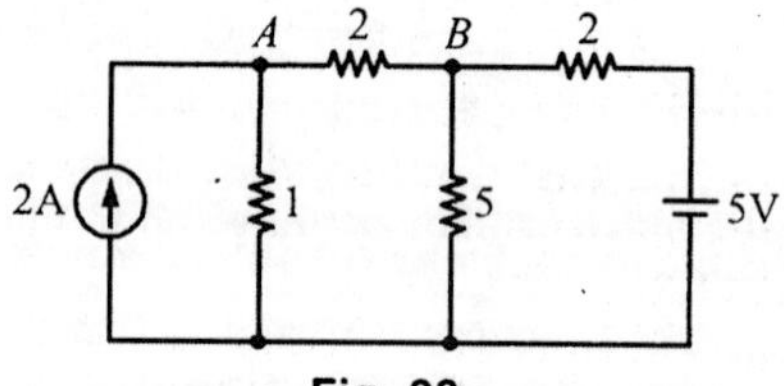

Fig. 98

2. Determine the current through the branch AB for the circuit shown in Fig. 98 using nodal analysis.

(Basic Elect. Engg., R.G. P. V., Bhopal, 2006)

[V_A = 2.35V, V_B = 3.06V]

Superposition Theorem

1. For the circuit of Fig. 99, determine current I through the 5Ω resistor using superposition theorem.

(Elect. Engg. Nagpur Univ. 2004)

[2.104A]

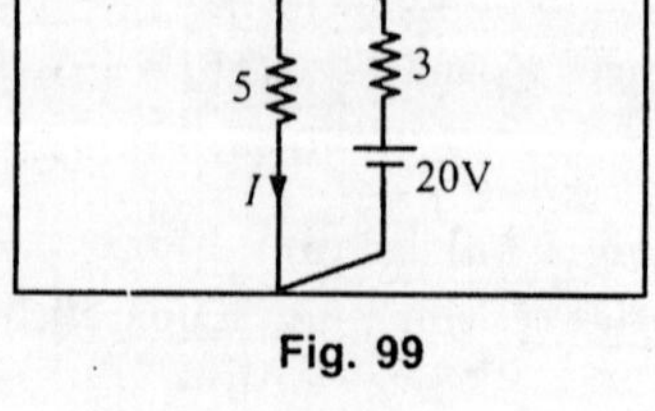

Fig. 99

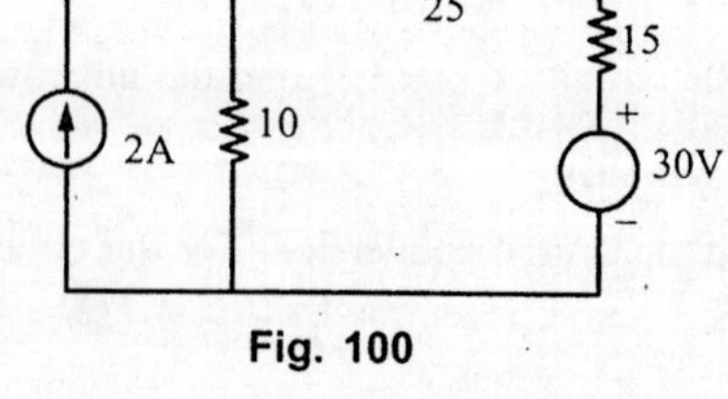

Fig. 100

2. Apply the superposition theorem and find the current through 25Ω resistance of the circuit shown in Fig. 100. *(Elect. Engg., Nagpur Univ., 2003)*
[0.2A]
3. Calculate the current flowing through the 2-Ω resistance in Fig. 101 by using superposition theorem. *(Elect. Engg., Nagpur Univ., 2004)*
[0.092A]

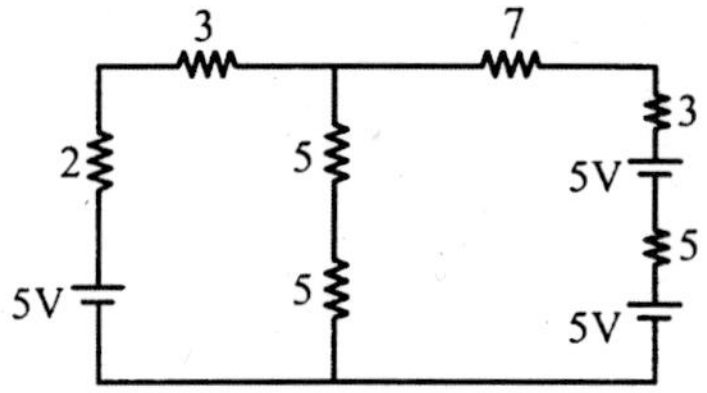

Fig. 101

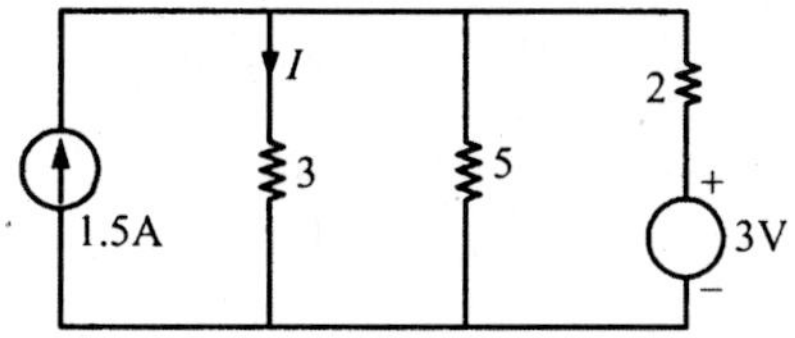

Fig. 102

4. Use the superposition theorem to find current I in the circuit shown in Fig. 102.
(Elect. Engg., Nagpur Univ., 2004)
[0.968A]
5. Using superposition theorem, find the value of the output voltage V_0 in the circuit of Fig. 103. *(Elect. Engg., Nagpur Univ., 2004)*
[10V]

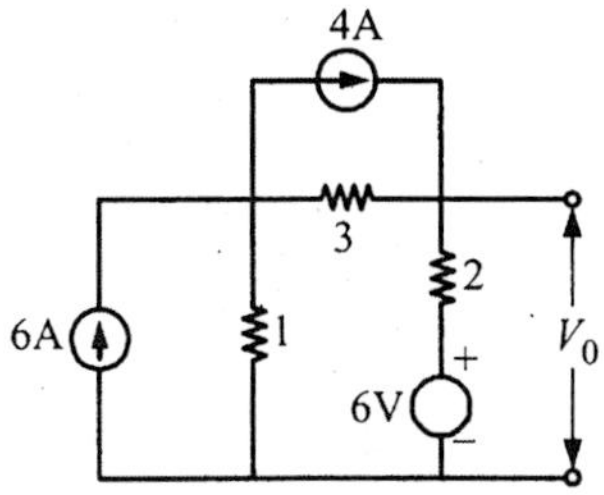

Fig. 103

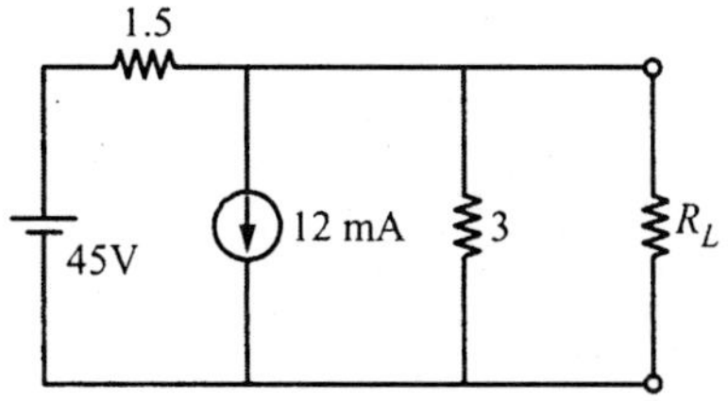

Fig. 104

Thevenin's Theorem

1. Use Thevenin's theorem to find voltage across R_L in Fig. 104. All resistances are in kilohm (kΩ). **[2V]**
2. Find the Thevenin equivalent circuit between terminals A and B for the network shown in Fig. 105 and then calculate the value of the current I. **[–1.25A, *i.e.* from *B* to *A*]**

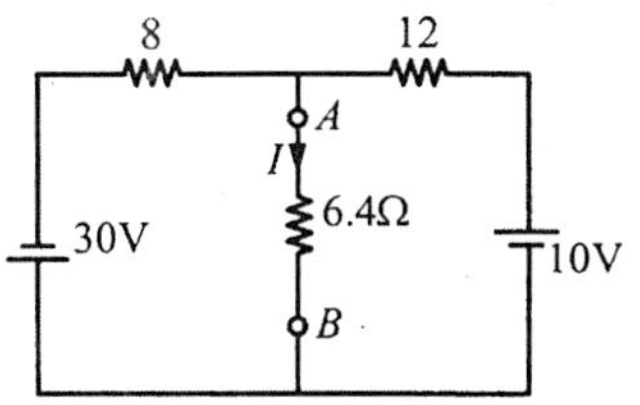

Fig. 105

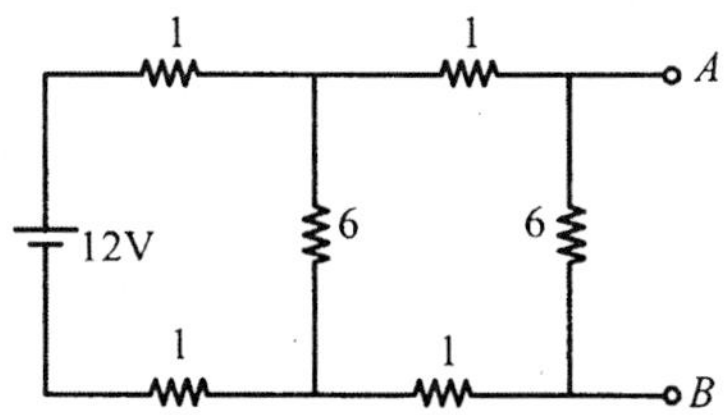

Fig. 106

3. Determine Thevenin equivalent circuit wihich may be used to represent the given network of Fig. 106 at the terminals A and B. **[V_{th} = 5.68V, R_{th} = 2.21Ω]**

Δ/*Y* Conversions

1. Convert the delta network of Fig. 107 into a wye network.
[R_1 = 125Ω, R_2 = 25Ω, R_3 = 62.5Ω]

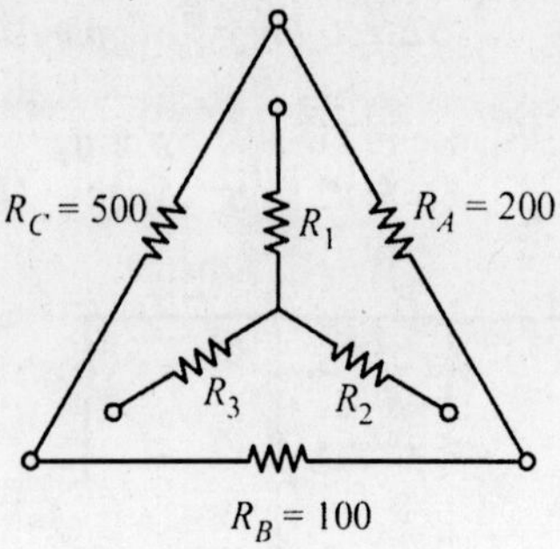

Fig. 107

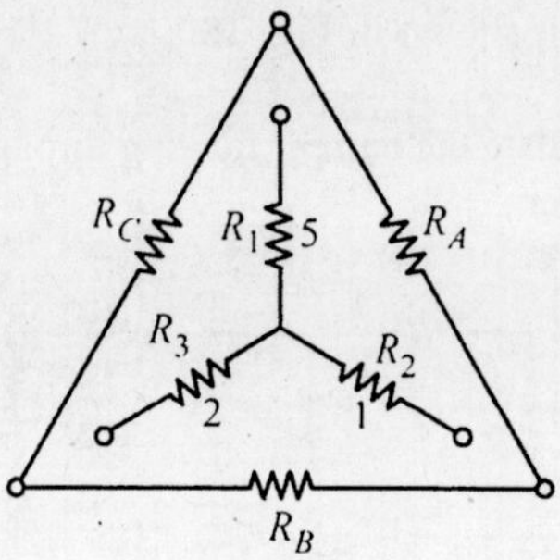

Fig. 108

2. Convert the wye network of Fig. 108 into a delta network.

[R_A = 8.5Ω, R_B = 3.4Ω, R_C = 17Ω]

3. Using star-delta transformation, determine resistance between terminals *A* and *C* of Fig. 109. (*Elect. Engg., Nagpur Univ., 2005*)

[R_{AC} = 0.58Ω]

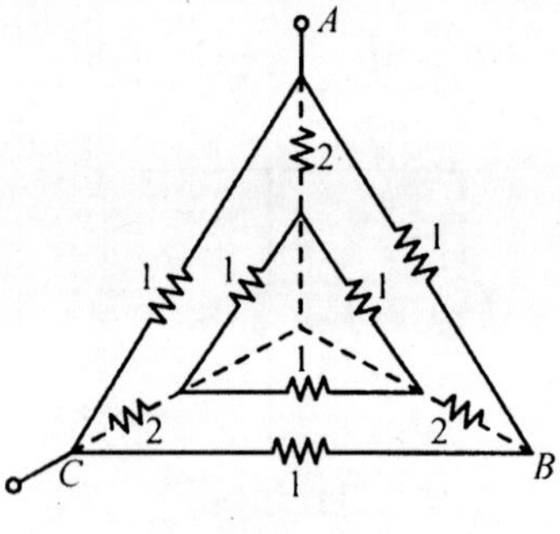

Fig. 109

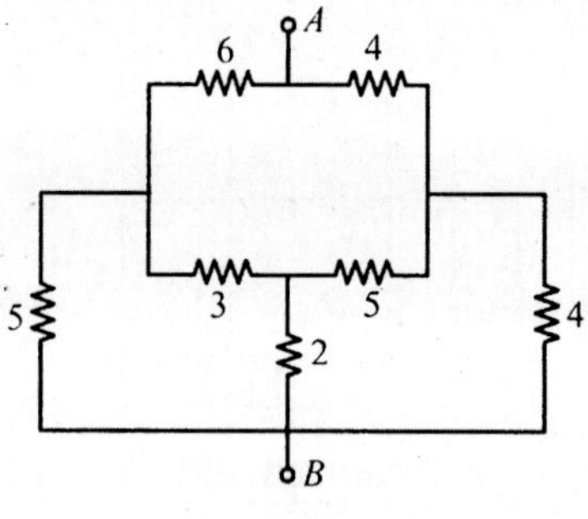

Fig. 110

4. Determine the resistance between points *A* and *B* in the network of Fig. 110. **[3.85Ω]**

Norton Theorem

1. Using Norton's theorem, calculate the current flowing through the 12Ω resistor of Fig. 111. **[1.33A]**

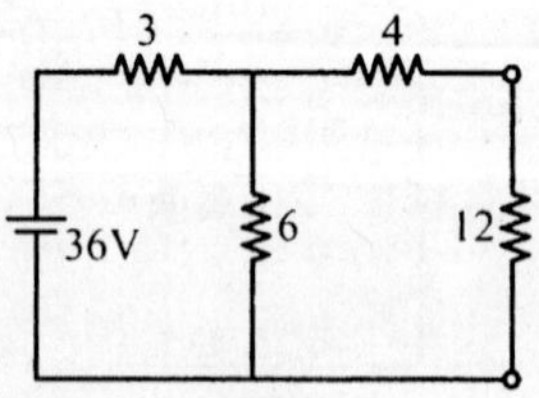

Fig. 111

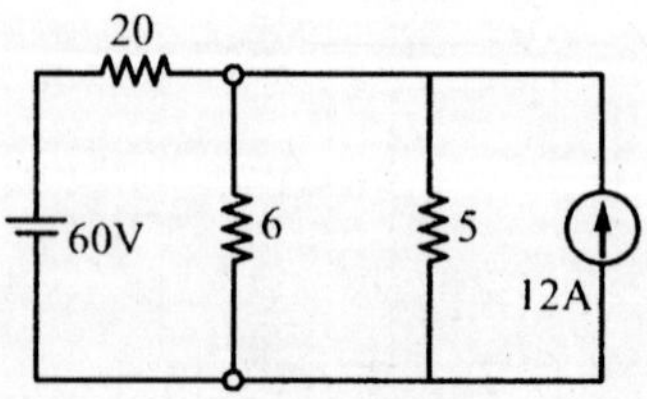

Fig. 112

2. Use Norton's theorem to calculate the current passing through the 6 Ω resistor of Fig. 112. **[6A]**

3. Using Norton's current-source equivalent circuit of the network shown in Fig. 113, find the current that will flow through the resistor *R* when it takes the values of 12, 24 and 36 ohm respectively. (*Elect. Circuits, South Gujarat Univ.*)

[4A, 3A, 2.4A]

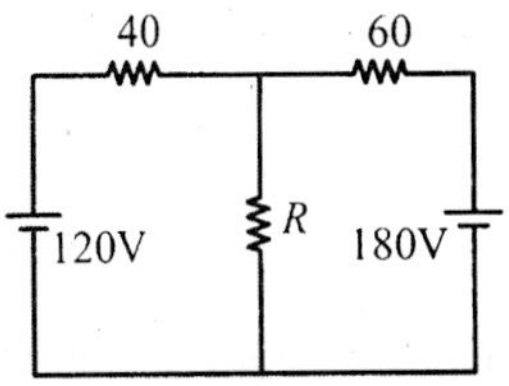

Fig. 113

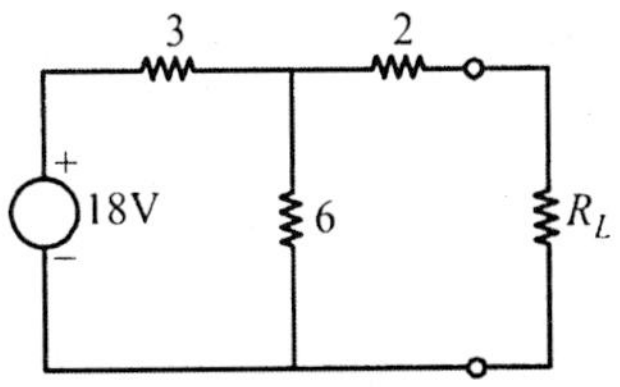

Fig. 114

MPT Theorem

1. Calculate the maximum power delivered to load resistor R_L of the circuit of Fig. 114 using maximum power transfer theorem. **[9W]**
2. In the network shown in Fig. 115, find the value of load resistor R_L so that maximum power is developed across it. What is the maximum power? **[6.477Ω, 0.3W]**

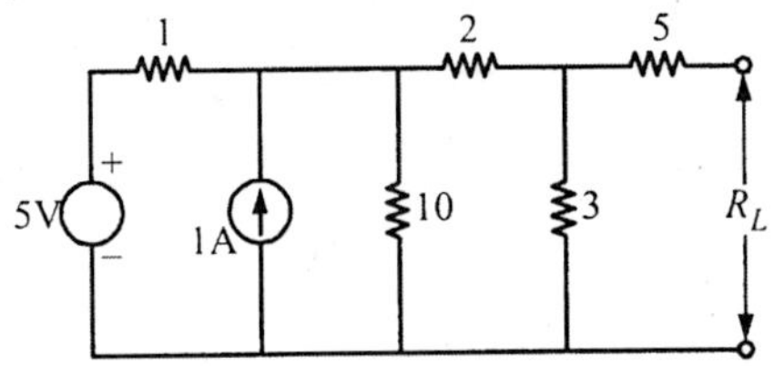

Fig. 115

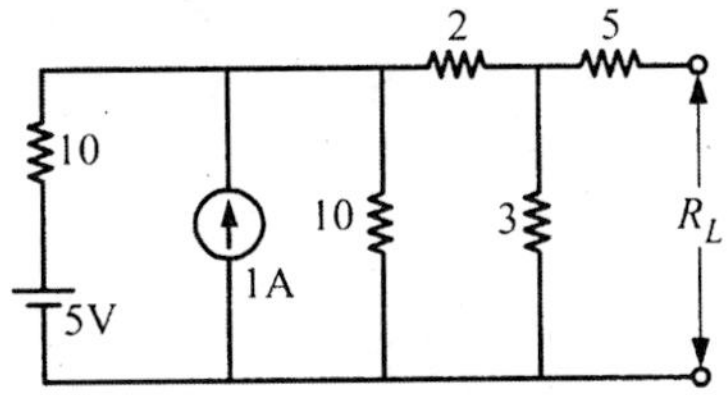

Fig. 116

3. In the circuit shown in Fig. 116, obtain the condition for maximum power transfer to the load R_L. Hence determine the maximum power transferred. **[R_L = 7.1 Ω, 178 mW]**

MULTIPLE CHOICE QUESTIONS

1. In an electrical network, mesh is defiend as
   (*a*) any closed path (*b*) a loop containing other loops
   (*b*) the shortest possible loop (*d*) the longest loop
2. An ideal voltage source in one that
   (*a*) provides constant voltage irrespectve of current.
   (*b*) has small internal resistance
   (*c*) supplies constant current
   (*d*) delivers constant power
3. An ideal current source has
   (*a*) zero internal resistance (*b*) infinite internal resistance
   (*c*) low value of voltage (*d*) large value of current
4. According to Kirchhoff's current law
   (*a*) at any node, currents algebraically sum to zero
   (*b*) all that goes into a node, need not come out of it
   (*c*) arithmatic sum of currents at a node is zero
   (*d*) incoming currents need not equal outgoing currents
5. Kirchhoff's voltage law is concerned with
   (*a*) IR drops (*b*) battery e.m.fs
   (*c*) junction voltages (*d*) both (*a*) and (*b*)
   (*Basic Elect. Engg. R.G.P.V., Bhopal, 2006*)
6. Kirchhoff's voltage law is applicable to
   (*a*) any closed path (*b*) any node (*c*) every branch (*d*) passive networks

7. A current source of 15A and 3Ω parallel resistance is equivalent to a voltage source of .... volt and 3Ω series resistor
(*a*) 5 (*b*) 45 (*c*) 10 (*d*) 30
8. Mesh analysis is based on
(*a*) KVL (*b*) KCL
(*c*) Ohm's law (*d*) law of conservation of energy
9. Maxwell's loop current method of solving electrical networks
(*a*) uses branch currents (*b*) utilizes Kirchhoff's voltage law
(*c*) is confined to single-loop circuits (*d*) is a network reduction method
(*Basic Elect. Engg., R.G.P.V. Bhopal, June 2007*)
10. Nodal analysis depends on
(*a*) applying KVL (*b*) selecting reference node
(*c*) comparing node voltages (*d*) using minimum number of equations.
11. In n node analysis, if there are n nodes in the circuit, then number of equations needed to slove the network is
(*a*) $n - 1$ (*b*) $n + 1$ (*c*) n (*d*) $n - 2$
(*Basic Elect. Engg. R.G.P.V., Bhopal 2006*)
12. Point out the ***wrong*** statement.
In the node voltage techinque of solving complex networks, choice of a refernce node does not
(*a*) affect the operation of the circuit
(*b*) change the voltage across any elements
(*c*) alter the potential difference between any pair of nodes.
(*d*) affect the voltage of various nodes.
13. Superposition theorem is meant for solving .... circuits
(*a*) linear (*b*) non-linear (*c*) non-resistive (*d*) simple
14. Thevenin's theorem reduces any complex linear circuit to a
(*a*) voltage source with a series resistance (*b*) simple series circuit
(*c*) parallel circuit (*d*) series-parallel circuit.
15. Practical applications of MPT theorem include
(*a*) electronic networks (*b*) communication networks
(*c*) audio systems (*d*) all of the above
16. According to MPT theorem, power transferred to a load resistance R_L is maximum when
(*a*) $R_L < R_{th}$ (*b*) $R_L > R_{th}$ (*c*) $R_L = R_{th}$ (*d*) $R_L = 2R_{th}$
17. What percentage of the maximum power is delivered to a load if load resistance is 10 times greater than the Thevenin resistance of the source to which it is connected?
(*a*) 25% (*b*) 40% (*c*) 35.39% (*d*) 33.06%
(*Basic Elect. Engg. West Bengal Tech. Univ. 2007*)
18. The efficiency in the case of maximum power transfer theorem is
(*a*) 100% (*b*) 50% (*c*) less than 50% (*d*) 50% - 100%
19. An electric circut is an electric network which contains at least one
(*a*) node (*b*) closed path (*c*) branch (*d*) mesh
20. Circuit analysis is the process of finding .... in one or more circuit components
(*a*) voltage (*b*) current (*c*) power (*d*) voltage drop
21. Kirchhoff's current law is based on the law of conservation of
(*a*) energy (*b*) power (*c*) charge (*d*) none of the above

22. The Thevenin's equivalent of any complex linear circuit consists of
    (*a*) current source with series resistance (*b*) current source with parallel resistance
    (*c*) voltage source with parallel resistance (*d*) voltage source with series resistance
23. Source conversion process helps in converting a
    (*a*) voltage source into a current source (*b*) current source into a voltage source
    (*c*) series resistance to a parallel resistance (*d*) both (*a*) and (*b*)
24. According to MPT, maximum power is transferred to a load resistance if it is equal to .... resistance of the supply source.
    (*a*) Thevenin (*b*) Norton (*c*) ohmic (*d*) none of the above

ANSWERS

1. *c*	2. *a*	3. *b*	4. *a*	5. *d*	6. *a*	7. *b*
8. *a*	9. *b*	10. *b*	11. *a*	12. *b*	13. *a*	14. *a*
15. *d*	16. *c*	17. *d*	18. *b*	19. *b*	20. *b*	21. *c*
22. *d*	23. *d*	24. *a*				

3 CHAPTER ELECTROSTATICS AND CAPACITORS

Q. 1 (*a*) What is the difference between 'current' electricity and 'static' electricity?

(*b*) What is meant by the term 'charge of a body?

Ans. (*a*) The subject of 'current' electricity deals with electricity in *motion* i.e. when electrons move from one point of a circuit to another. In other words, movement of charge is *current* which produces its own effects.

Electrostatics deals with the behaviour of charge when it is *at rest* i.e. static. A charge at rest has its own laws which are different from the laws of moving charge.

(*b*) When some negatively-charged electrons from a body are removed, then it is left with a *net* positive charge. Hence, it is said to be positively charged. If, on the other hand, some electrons are added to it, then it has a net negative charge. Hence, the body is said to be negatively charged. *The total deficiency or excess of electrons in a body is called its charge.*

Q. 2 (*a*) Explain clearly the term 'permittivity' as used in Electrostatics. Distinguish between the relative and absolute permittivity of a medium.

(*b*) Relative permittivity of mica is 5. What is its absolute permittivity?

Ans. (*a*) Every charge placed in a medium gives out electrostatic flux through the medium. Some media focus this flux more strongly than others. Similar thing happens in the case of ferromagnetic materials which focus magetic flux. This property of the medium to focus electrostatic flux is called *permittivity*.

Each dielectric medium has two permittivities

(*i*) absolute permittivity ($\in$) and

(*ii*) relative permittivity ($\in_r$)

The relative permittivity of a medium is measured with respect to the permittivity of vacuum or free air.

The absolute permittivity of air is denoted by $\in_0$ and has been *allotted* a value of 8.854×10^{-12} F/m.

$$\therefore \qquad \in_r = \frac{\text{absolute permittivity of a meadium}}{\text{absolute permittivity of vacuum}}$$

$$\text{or} \qquad \in_r = \frac{\in}{\in_0}$$

It is seen from above that absolute permittivity of a medium is given by $\in = \in_0 \in_r$.

It should be obvious that relative permittivity of vacuum with respect to *itself* is one

i.e. $\in_r = 1$.

Generally, it is the relative permittivity of a medium which is given in the literature.

(*b*) Absoloute permittivity,

$$\in = \in_0 \in_r = 8.854 \times 10^{-12} \times 5$$
$$= \mathbf{44.27 \times 10^{-12}\ F/m}$$

Q. 3 (*a*) State and explain the laws of Electrostatics. How do you define one coulomb?

(*b*) The small identical conducting spheres have charges of 2.0×10^{-9} C and -0.5×10^{-9} respectively. When they are placed 4 cm apart, what is the force between them? If they are brought into contact and then separated by 4 cm, what is the force between them?

(*Electromagnetic Theory, A.M.I.E. Sec. B*)

Ans. (*a*) There are the two laws of Electrostatics

(*i*) **First Law :** According to this law,

like charges repel each other whereas unlike charges attract each other.

This fact is shown in Fig. 1.

(*ii*) **Second Law :** It is also known as Coulomb's Law. According to this law, the force between two point charges is

(*a*) directly proportional to the strength of the two charges Q_1 and Q_2 (Fig. 2).

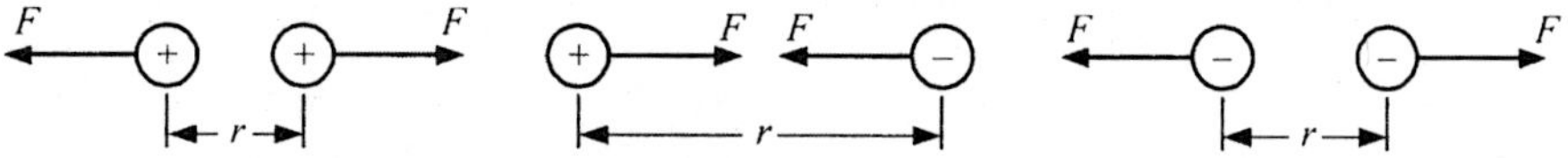

(*a*) (*b*) (*c*)

Fig. 1

(*b*) inversely proportional to the square of the distance between them i.e. $1/r^2$ and

(*c*) inversely proportional to the **absolute** permittivity ($\in$) of the surrounding medium.

It can be expressed mathematically **as**

$$F \propto \frac{Q_1\, Q_2}{\in d^2} \quad \text{or} \quad F = k\, \frac{Q_1\, Q_2}{\in d^2}$$

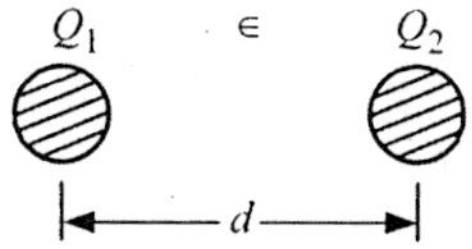

Fig. 2

where k is a constant of proportionality whose value is $1/4\pi$.

$$\therefore \qquad F = \frac{Q_1\, Q_2}{4\,\pi \in d^2}$$

Since, $\in\, = \,\in_0\, \in_r$, hence the above equation becomes

$$F = \frac{Q_1\, Q_2}{4\,\pi \in_0\, \in_r\, d^2} \qquad \text{.... in a medium}$$

$$= \frac{Q_1\, Q_2}{4\,\pi \in_0\, d^2} \qquad \text{.... in air}$$

Now,
$$\frac{1}{4\pi \in_0} = \frac{1}{4\pi \times 8.854 \times 10^{-12}} = 9 \times 10^9 \text{ (approx)}$$

The Coulomb's Law can be put in the simple form as

$$F = 9 \times 10^9\, \frac{Q_1\, Q_2}{\in_r\, d^2} \qquad \text{.... in a medium}$$

$$= 9 \times 10^9\, \frac{Q_1\, Q_2}{d^2} \qquad \text{.... in air} \qquad \text{... } (i)$$

If in Eq. (*i*) above,

$$Q_1 = Q_2 = Q,\ d = 1 \text{ metre}, \quad F = 9 \times 10^9 \text{ newton (N)}$$

Then,
$$9 \times 10^9 = 9 \times 10^9\, \frac{Q^2}{1^2}$$

$$\therefore \qquad Q^2 = 1 \quad \text{or} \quad Q = \pm 1 \text{ coulomb}$$

Hence, one coulomb of charge may be defined as :

the charge which when placed in air from an equal and similar charge at a distance of one metre repels it with a force of 9×10^9 *N.*

(*b*) $F = 9 \times 10^9 Q_1\, Q_2/d^2 = 9 \times 10^{-9} \times (9 \times 10^{-8} \times -0.5 \times 10^{-9})/0.04^2 = -56.25 \times 10^{-7}$ N.

The negative sign shows that it is a force of attraction.

When two identical spheres are brought into contact with each other and then separated, each gets half of the total charge. Hence,

$$Q_1 = Q_2 = [2 \times 10^{-9} + (-0.5 \times 10^{-9})/2] = 0.75 \times 10^{-9} \text{ C}$$

When they are seprated by 4 cm,

$$F = 9 \times 10^{-9} \times (0.75 \times 10^{-9})^2/0.04^2 = \mathbf{0.316 \times 10^{-5}\ N.}$$

Q. 4 (*a*) How do you define an electric (or electrostatic) field? Give different formulae for field strength.

(*b*) A positive point charge of 400 μC is placed in air. Find the electric field intensity at a distance of 2 m from the charge.

(*c*) Three equal charges each of magnitude 3×10^{-6} C are placed at three corners of a right-angled triangle of sides 3 cm and 4 cm. Find the force on the charge at the apex corner if 4 cm side is the base of the triangle.

(Elect. Engg. West Bengal Tech. Univ., 2009)

Ans. (*a*) It is found that a positive charge gives out lines of force in space which extend *upto infinity* unless stopped by some negative charge (Fig. 3). These lines of force produce stress in the surrounding medium. Any other positive or negative charge when placed in the space around the fixed positive charge experiences a certain amount of force. This spacial spread of lines of force is called electric field or electrostatic field.

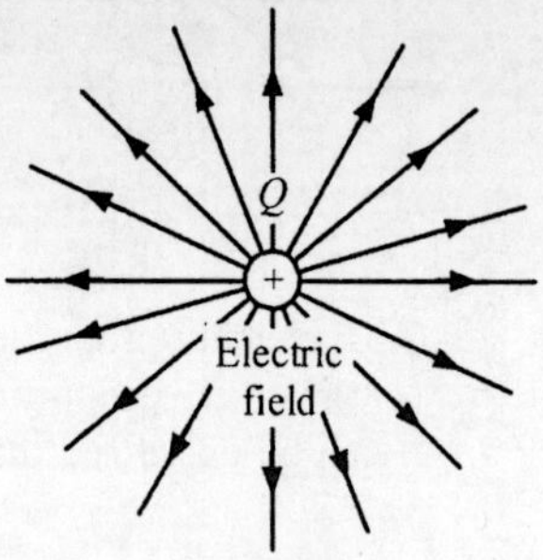

Fig. 3

The field strength or electric intensity of the electric field at any point may be defined in the following two ways.

(*i*) *it is equal to the lines of force passing normally through a unit cross-section at that point.*

Suppose, we have a point charge of Q coulomb (Fig. 4). The number of lines of force produced by it is $Q/\in$. If these lines fall perpendicularly on an area of A m^2, then

$$E = \frac{Q/\in}{A} = \frac{Q}{\in A} \text{ volt/metre}$$

$$E = \frac{Q}{\in_0 \in_r A} \text{ V/m} \qquad \text{.... in a medium}$$

$$= \frac{Q}{\in_0 A} \text{ V/m} \qquad \text{.... in vacuum or air}$$

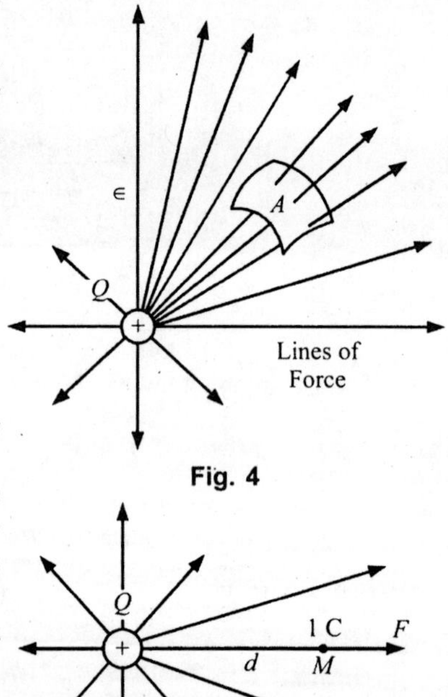

Fig. 4

(*ii*) *it is numerically equal to the force felt by a unit positive charge placed at that point.*

As shown in Fig. 5, a charge of 1 coulomb has been placed at point M in the electric field produced by charge Q. The force experienced by this unit charge gives the electric field intensity E at point M.

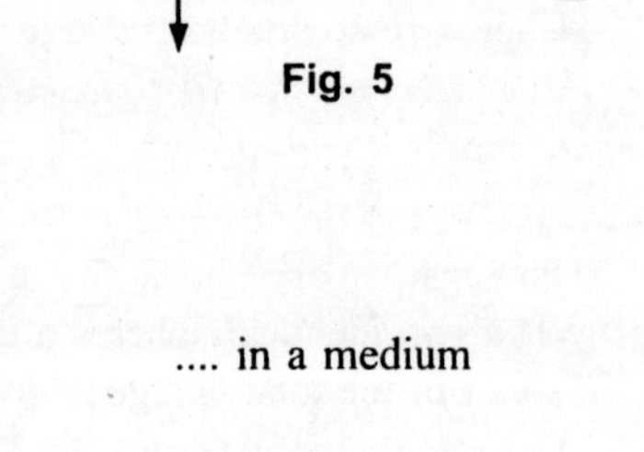

Fig. 5

In general, if a charge of q coulomb experiences a force of F newtons, then

$$\text{electric field intensity, } E = \frac{F}{q} \text{ newton/coulomb}$$

The value of E at point M of Fig. 5 can be calculated from Coulomb's law

$$F = \frac{Q \times 1}{4\pi \in_0 \in_r d^2} \text{ N/C} = 9 \times 10^9 \frac{Q \times 1}{\in_r d^2} \text{ N/C}$$

$$\therefore \quad E = \frac{Q}{4\pi \in_0 \in_r d^2} \text{ N/C} = 9 \times 10^9 \frac{Q}{\in_r d^2} \text{ N/C} \qquad \text{.... in a medium}$$

$$= \frac{Q}{4\pi \in_0 d^2} \text{ N/C} = 9 \times 10^9 \frac{Q}{d^2} \text{ N/C} \qquad \text{.... in air}$$

(*b*) As seen from above,

$$E = 9 \times 10^9 = \frac{Q}{d^2} \text{ N/C}$$

$$\therefore \qquad E = 9 \times 10^9 \, \frac{400 \times 10^{-6}}{2^2} = \mathbf{9 \times 10^5 \ N/C}$$

(*c*) The charge of 3×10^{-6} C experiences force of repulsion from similar charges placed at points. *A* and *B* of Fig. 6(*a*).

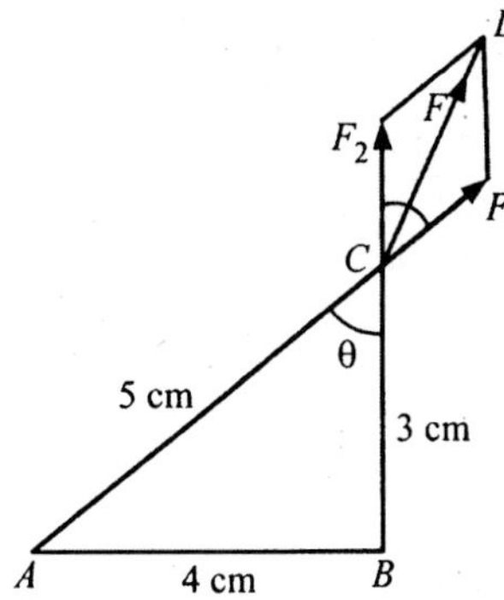

Fig. 6(a)

As per Coulomb's law,

$$F_1 = 9 \times 10^9 \ Q_1Q_2/d^2 \text{ newton}$$
$$= 9 \times 10^9 \times (3 \times 10^{-6})^2/(5 \times 10^{-2})^2$$
$$= 32.4\text{N}$$

Similarly, $F_2 = 32.4$ N

The resultant of the two equal forces F_1 and F_2 is F given by

$$F = \sqrt{F_1^2 + F_2^2 + 2F_1F_2 \cos\theta}$$

Since $F_1 = F_2 = 32.4$ N, hence

$$F = \sqrt{2 \times 32.4^2 + 2 \times 32.4^2 \times 0.6}$$

$$= \sqrt{2099.5 + 2099.5 \times 0.6} = \mathbf{58 \ N}$$

As seen from Fig. 6(*a*), $\cos\theta = BC/AC = 3/5 = 0.6$.

Q. 5. Define electric flux and electric flux density. Give their units. What is the relation between permittivity, field strength and flux density?

Ans. Charges are *induced* in *conducting* materials when placed within an electric field. The total *induced charge* is called the electric flux. Its unit is the same as that of the electric charge *i.e.* coulomb. If a charge of Q coulomb is induced, then the flux produced is Ψ coulomb. In other words, *flux and charge are numerically equal.*

i.e. $\Psi = Q$

The charge-inducing capability of an electric field is called its *electric flux density*. If a flux of Ψ coulomb is produced in a surface area of A m^2, then the electric flux density is given by

$$D = \frac{Q}{A} = \frac{\Psi}{A} \text{ C/m}^2$$

The factor which controls the relation between the field strength and the flux density is the permittivity $\in$ of the surrounding material or medium.

The exact relationship between permittivity, field strength and flux density is

$$\in = \frac{D}{E} \quad \text{or} \quad D = \in E = \in_0 \in_r E$$

Q. 6 (*a*) Find an expression for the potential of a point within an electric field.

(*b*) An energy of 30 joules is used to move a charge of 2 C from infinity to a point *A*. An energy of 40 joules is required to bring the same charge to a point *B* from infinity. Calculate the :

(*i*) potential of point *A*

(*ii*) potential of point *B*

(*iii*) potential difference between points *B* and *A*

(*iv*) which of the two points is at the higher potential and why?

Ans. (*a*) Since force on an electric charge at infinity is zero, hence 'infinity' is chosen as the place of *zero electric potential.* We define the potential at a point in an electric field as

numerically equal to the work done in bringing a unit positive charge from infinity to that point against the electric field E.

If we bring one coulomb of charge from infinity to a certain point in an electric field and the work done is one joule, then potential of that point is one volt.

$$\therefore \qquad 1 \text{ volt} = \frac{1 \text{ joule}}{1 \text{ coulomb}} = 1 \text{ J/C}$$

In simple words, *potential is work per unit charge.* If W is the work done in shifting a charge of Q coulombs from infinity to a certain point in an electric field of strength E, then potential of that point is

$$V = \frac{W}{Q} \text{ volts}$$

Similarly, a potential difference of 1 volt exists between two points in an electric field if 1 joule of work is done in shifting one coulmb of charge from one point to the other.

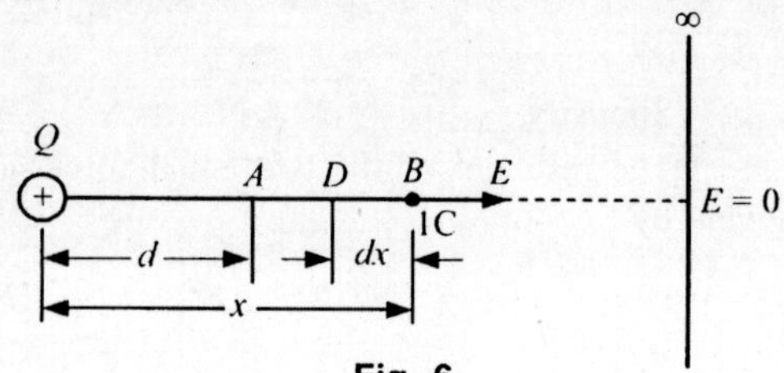

Fig. 6

Potential at a point : In Fig. 6, let the positive charge of 1 C be moved from point B to point D against the repulsive force of charge Q. The work done would be

$$dW = \text{force} \times \text{distance moved}$$

$$\text{or} \qquad dW = \frac{Q}{4\pi \epsilon_0 x^2} \times (-dx)$$

The small distance dx has been taken as negative because it is measured along the negative direction of x.

The total work done in bringing this one coulomb of positive charge from infinity to point A within the electric field E is given by

$$\int dW = \int_{x=\infty}^{x=d} Q \cdot \frac{(-dx)}{4\pi \epsilon_0 x^2}$$

$$\text{or} \qquad W = \frac{-Q}{4\pi \epsilon_0} \int_{\infty}^{d} \frac{dx}{x^2} = \frac{-Q}{4\pi \epsilon_0} \left| \frac{-1}{x} \right|_{\infty}^{d}$$

$$= \frac{-Q}{4\pi \epsilon_0} \left[-\frac{1}{d} - \left(-\frac{1}{\infty} \right) \right] = \frac{Q}{4\pi \epsilon_0 d} \text{ joules}$$

By definition, this work is equal to the potential of that point expressed in volts.

$$\therefore \qquad V = \frac{Q}{4\pi \epsilon_0 d} = 9 \times 10^9 \frac{Q}{d} \text{ volts} \qquad \text{.... in air}$$

$$= \frac{Q}{4\pi \epsilon_0 \epsilon_r d} = 9 \times 10^9 \frac{Q}{\epsilon_r d} \text{ volts} \qquad \text{.... in a medium}$$

(*b*) (*i*) $V_A = W_A / Q = 30/2 =$ **15 volt**

(*ii*) $V_B = W_B / Q = 40/2 =$ **20 volt**

(*iii*) $V_{BA} = V_B - V_A = 20 - 15 =$ **5 volt**

(*iv*) Point B is at the higher potential because more work is done against the field to reach point B.

Q. 7 (*a*) What is meant by the term 'dielectric' and the dielectric strength of a medium?

(*b*) A roll of rubber tape having thickness of 0.5 mm is used by an electrician to insulate soldered splices. What maximum voltage can the splices withstand if the electrician wraps them with (*i*) one layer of tape (*ii*) three layers of tape.

Dielectric strength of rubber is 18 kV/mm.

Ans. (*a*) A dielectric is an insulator material which is used to separate positive and negative charges. For example, a dielectric material is placed between the oppositely-charged plates of a capacitor. This dielectric material is able to concentrate the lines of force of the electric field and thus increase the capacity of a capacitor to store energy. The relative permittivity ($\in_r$) of the dielectric material is also called dielectric strength.

The dielectric strength of a medium is given by the potential difference which when applied across one metre thickness of the medium breaks down its insulation. It is expressed in volts /metre. For example, when we say that dielectric strength of air is 3×10^6 V/m, then we mean that the maximum potential difference which one metre thickness of air can withstand across it is 3×10^6 volts. If the voltage is greater than this value, the air insulation breaks down and air becomes a conductor.

It is obvious that the dielectric strength of a medium is the maximum value of the electric intensity E that can be safely set up in that medium. In fact, the dielectric strength and breakdown voltage of a medium *mean the same thing.*

(*b*) (*i*) V = dielectric strength × tape thickness

$$= 18 \frac{\text{kV}}{\text{mm}} \times 0.5 \text{ mm} = \mathbf{9 \text{ kV}}$$

(*ii*)
$$V = 18 \frac{\text{kV}}{\text{mm}} \times (3 \times 0.5 \text{ mm}) = \mathbf{27 \text{ kV}}$$

Q. 8. (*a*) What is a parallel-plate capacitor? How do you define its capacitance?

(*b*) A parallel-plate capacitor has plates 0.15 mm apart, a plate area of 0.1 m² and a dielectric of relative permittivity 3. Find electric flux density, electric field intensity and the p.d. between the plates if capacitor charge is 0.5 μC.

Ans. (*a*) The basic structure of a parallel-plate capacitor is shown in Fig. 7. It is consists of two parallel plates P_1 and P_2 made of conducting material and seprated by an insulator. The insulator could be air whose permittivity is $\in_0$ or any other dielectric medium of absolute permittivity $\in = \in_0 \in_r$.

When this capacitor is connected across a battery (Fig. 8), then

(*i*) a charging current i is set up by the battery.

(*ii*) hence, positive charge is transferred to the plate connected to the positive terminal of the battery.

(*iii*) a potential difference is set up between the two parallel plates.

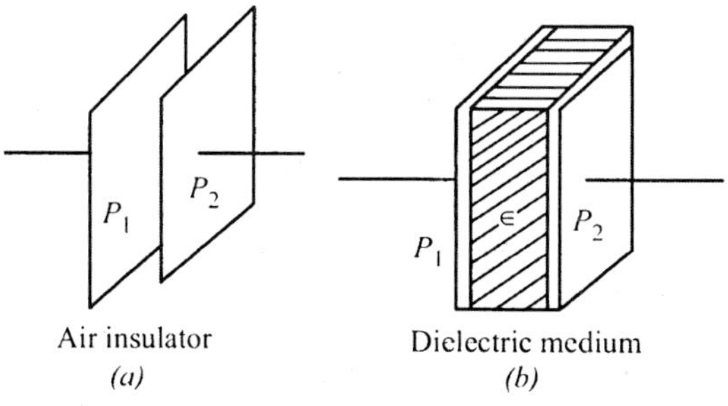

Fig. 7

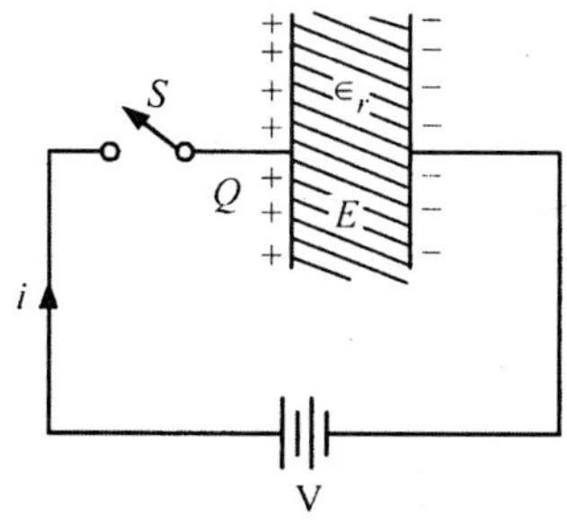

Fig. 8

(*iv*) electric intensity E is developed in the dielectric medium.

(*v*) electric energy is stored in electric field between the two plates.

Capacitance means the *ability of a capacitor to store charge.* In fact, capacitance may be defined as

the amount of charge required to create a potential difference of 1 volt between the two plates of a capacitor.

Suppose, the positive plate of the capacitor is given a charge of Q coulomb which produces a potential difference of V volt between the two plates. Then the capacitance of this capacitor is

$$C = \frac{Q}{V} \text{ coulomb/volt}$$

$$= \frac{Q}{V} \text{ farad}$$

As seen, the unit of capacitance is farad (F) which may be defined as

the capacitance of the capacitor which requires a charge of one coulomb to create a p.d. of one volt between its plates.

Capacitance Calculation : Consider the parallel capacitor shown in Fig. 9. We will use the following three relations to calculate the value of capacitance :

(*i*) electric flux density, $D = \frac{Q}{A}$

(*ii*) absolute permittivity, $\epsilon = \frac{D}{E}$

(*iii*) electric intensity $E = \frac{V}{d}$

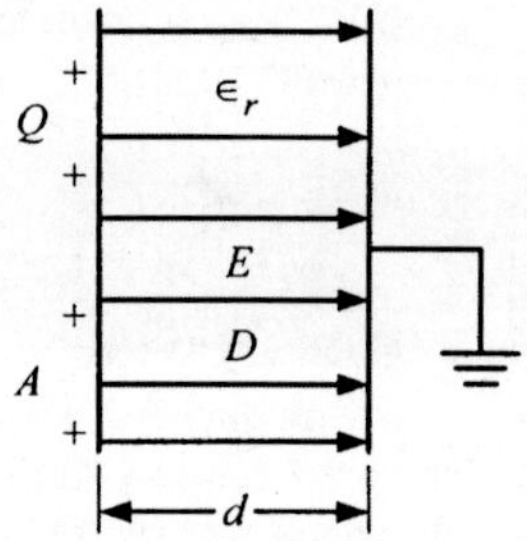

Fig. 9

Now, $$\epsilon = \frac{D}{E} = \frac{Q/A}{V/d} = \frac{Q}{V} \times \frac{d}{A} = C \times \frac{d}{A}$$

or $$C = \frac{\epsilon A}{D} \text{ farad}$$

$$= \epsilon_0 \epsilon_r \frac{A}{d} \text{ farad} \quad \text{.... with a medium}$$

$$= \epsilon_0 \frac{A}{d} \quad \text{.... with air}$$

(*b*) $$C = \epsilon_0 \epsilon_r A/d = 8.854 \times 10^{-12} \times 3 \times 0.1/0.15 \times 10^{-3}$$

$$= \mathbf{1.77 \times 10^{-8} \text{ F}}$$

$$D = Q/A = 0.5 \times 10^{-6}/0.1 = \mathbf{0.5 \times 10^{-5} \text{ C/m}^2}$$

$$Q = CV$$

∴ $$V = Q/C = 0.5 \times 10^{-6}/1.77 \times 10^{-8} = \mathbf{28 \text{ V}}$$

$$E = V/d = 28/0.15 \times 10^{-3} = \mathbf{186.667 \text{ V/m}}$$

Q. 9. (*a*) Derive an expression for the capacitance of a parallel-plate capacitor whose dielectric medium consists partially of air.

(*b*) A capacitor consists of two similar square aluminium plates, each 10 cm × 10 cm mounted parallel and opposite each other. What is their capacitance in μμF when the distance between them is 1 cm and the dielectric air? If the capacitor is given a charge of 500 μμC, what will be the difference of potential difference between plates? How will this be affected if the space between the plates is filled with wax which has a relative permittivity of 4?

Ans. **(*a*)** As shown in Fig. 10, the medium consists partly of air and partly of a medium slab of thickness t and relative permittivity $\in_r$. The electric flux density $D = Q/A$ is the same *in both media.* But electric intensities are different.

In air, $$E_1 = \frac{D}{\in_0} = \frac{Q}{\in_0 A}$$

In the medium, $$E_2 = \frac{D}{\in_0 \in_r} = \frac{Q}{\in_0 \in_r A}$$

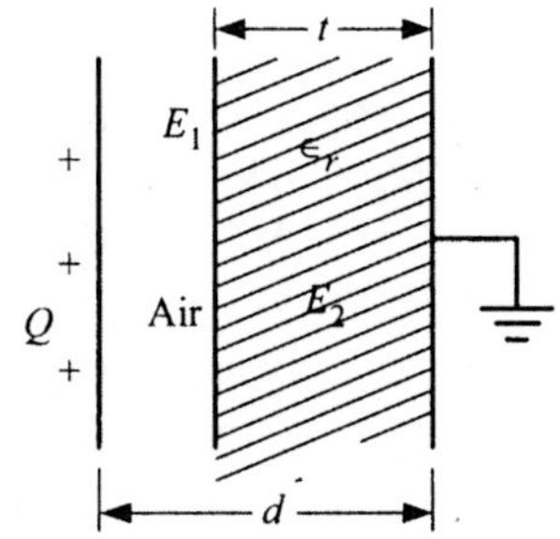

Fig. 10

Now, we will use the relation, $E = V/d$ or $V = E.d.$

P.d. between plates,

$$V = E_1 (d - t) + E_2 \cdot t$$

$$= \frac{D}{\in_0}(d-t) + \frac{D}{\in_0 \in_r} t$$

$$= \frac{Q}{\in_0 A}\left(\frac{t}{\in_r} + d - t\right) \quad \left(\because D = \frac{Q}{A}\right)$$

$$= \frac{Q}{\in_0 A}\left[d - \left(t - \frac{t}{\in_r}\right)\right]$$

or $$C = \frac{Q}{V} = \frac{\in_0 A}{\left[d - \left(t - \frac{t}{\in_r}\right)\right]} \text{ farad}$$

(*b*) $$C = \frac{\in_0 A}{d} \text{ farad}$$

$$A = 10 \times 10 = 100 \text{ cm}^2 = 10^{-2} \text{ m}^2 \; ; \; d = 1 \text{ cm} = 10^{-2} \text{ m}$$

$$\therefore \quad C = \frac{8.854 \times 10^{-12} \times 10^{-2}}{10^{-2}} = 8.854 \times 10^{-12} \text{ farad} = \mathbf{8.854\ \mu\mu F}$$

Now, $$C = \frac{Q}{V} \quad \therefore \quad V = \frac{Q}{C} = \frac{500 \times 10^{-12}}{8.854 \times 10^{-12}} = \mathbf{56.5\ Volts}$$

Since, $\in_r = 4$ for wax, the capacitance will be increased 4 times.

New capacitance = 4 × 8.854 = **35.4 μμF**

The p.d. will decrease to one-fourth value

∴ New $V = 56.5/4 =$ **14.1 volts**

Q. 10 (*a*) Derive an expression for the energy stored by a charged capacitor.

(*b*) A parallel-plate capacitor has 20 plates each of 0.5 m² area, the separation between the plates being 0.2 cm with a dielectric of relative permittivity 2.5. Calculate (*i*) the capacitance (*ii*) the value of the stored energy if it is charged to 1000 volts.

Ans. (*a*) Whenever a capacitor is charged, some energy is spent by the charging battery. This energy is stored up in the electrostatic field E set up in the dielectric medium. When the capacitor is discharged, the field collapses and the stored energy is released.

To start with when the capacitor is uncharged, no work is done in transferring *first charge* from the battery to the positive plate of the capacitor. But further charges have to be carried against *the repulsion of the positive charge already present on the plates.* Hence, more energy is spent in further charging the capacitor.

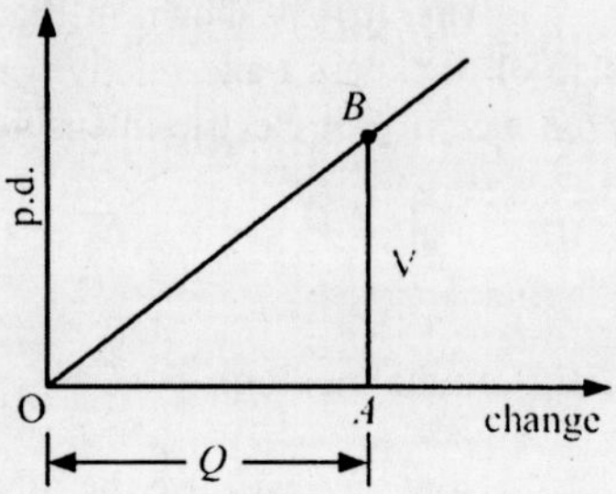

Fig. 11

Let C be the capacitance and Q the charge on the plate which produces a potential difference of V between the two plates.

Now, $C = Q/V$ or $V = Q/C$

For a capacitor of given capacity C, $V \propto Q$

Hence, graph between V and Q is a straight line passing through the origin as shown in Fig. 11. The energy required to charge the capacitor is equal to the area of the triangle OAB.

$$\therefore \quad \text{Energy stored} = \frac{1}{2}\,(\textit{base} \times \textit{height})$$

$$= \frac{1}{2}\,QV \text{ joules}$$

$$= \frac{1}{2}\,CV^2 \text{ joules} \qquad (\text{putting } Q = CV)$$

$$= \frac{Q^2}{2C} \text{ joules} \qquad \left(\text{putting } V = \frac{Q}{C}\right)$$

(*b*)

$$C = \frac{(n-1)\,\epsilon_0\,\epsilon_r\,A}{d}$$

$$= \frac{(20-1)\times 8.854\times 10^{-12}\times 2.5\times 0.5}{2\times 10^{-3}} = \mathbf{105 \times 10^{-9}\ F}$$

$$\text{Energy stored} = \frac{1}{2}\,CV^2 = \frac{1}{2} \times 105 \times 10^{-9} \times 1000^2 = \mathbf{0.0524\ J}$$

Q. 11 (*a*) What is the combined capacitance of a few capacitors connected in series?

(Gujrat Univ. Summer 2003)

(*b*) A 10 μF, a 20 μF and a 40 μF capacitor are connected in series to a 399 volt source of e.m.f.

(*i*) What is the equivalent capacitance?

(*ii*) What is the magnitude of charge across each capacitor?

(*iii*) What is the potential difference across each capacitor?

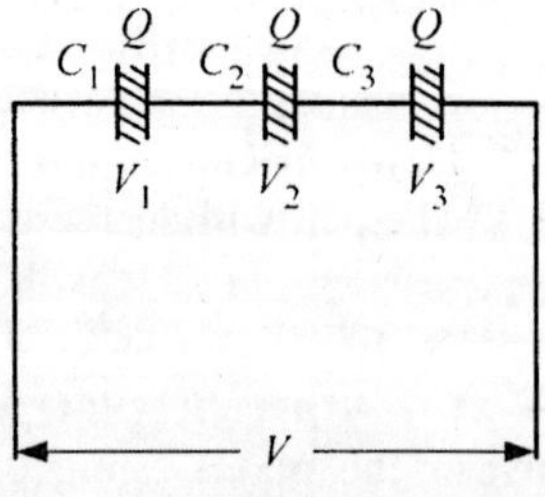

Fig. 12

Ans. (*a*) Fig. 12 shows three capacitors connected in series across a voltage supply of V (volt). In series combination, every capacitor *has the same charge* regardless of its capacitance. Moreover, though charge is the same, yet voltage across each is different because capacitances are different.

$$\therefore \quad V = V_1 + V_2 + V_3$$

$$\text{or} \quad \frac{Q}{C} = \frac{Q}{C_1} + \frac{Q}{C_2} + \frac{Q}{C_3}$$

$$\therefore \quad \frac{1}{C} = \frac{1}{C_1} + \frac{1}{C_2} + \frac{1}{C_3}$$

It will be found that the combined capacitance C will be less than the *smallest* capacitor.

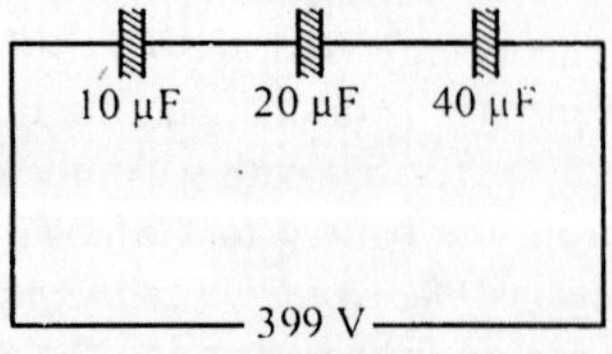

Fig. 13

(*b*) The three capacitors are connected in series as shown in Fig. 13. The equivalent capacitance is given by :

$$(i) \quad \frac{1}{C} = \frac{1}{10} + \frac{1}{20} + \frac{1}{30} \quad \therefore \quad C = \frac{40}{7}\ \mu\text{F}$$

(*ii*) The total charge across the combined capacitance is given by

$$Q = CV = (40/7) \times 399$$
$$= 15{,}960/7 = 2280\ \mu C$$

Charge across each capacitor would be the same i.e **2280 μF.**

(*iii*)
$$V_1 = Q/C_1 = 2280/10 = \mathbf{228\ V}$$
$$V_2 = Q/C_2 = 2280/20 = \mathbf{114\ V}$$
$$V_3 = Q/C_3 = 2280/40 = \mathbf{57\ V}$$

Q. 12 (*a*) Find an expression for the combined capacitance of a few capacitors connected in parallel. *(Gujarat Univ., Summer 2003)*

(*b*) Three capacitors *A*, *B*, *C* have capacitances of 10, 50, and 25 μF respectively. Calculate (*i*) charge on each when connected in parallel to a 250 V supply (*ii*) total capacitance and (*iii*) p.d. across each when connected in series. *(Elect. Tech. Gwalior Univ.)*

Ans. (*a*) Fig. 14 shows three capacitors connected in parallel across a voltage supply of V volts. It will be seen that p.d. across each capacitor is the same but *charge on each is different* because their capacitances are different.

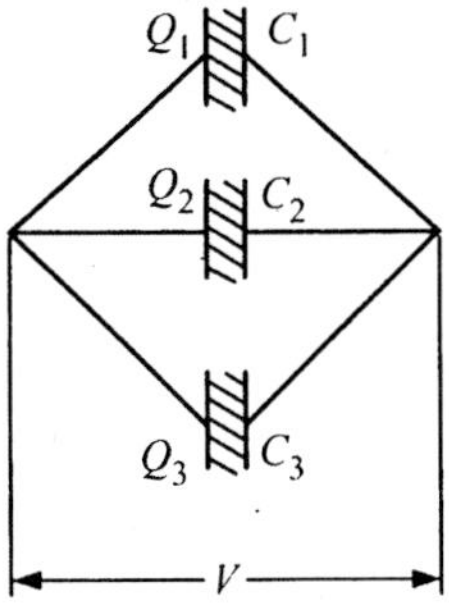

Fig. 14

The sum of three charges is equal to the total charge Q supplied by the voltage source. The value of $Q = CV$ where C is the *combined* capacitance.

$\therefore$
$$Q = Q_1 + Q_2 + Q_3$$
or
$$CV = C_1V + C_2V + C_3V$$
or
$$C = C_1 + C_2 + C_3$$

It is obvious that the combined capacitance is *larger* than the largest capacitor.

(*b*) (*i*) Parallel connection is shown in Fig. 15(*a*). Each capacitor has a p.d.of 250 V across it.

$$Q_1 = C_1V = 10 \times 250 = \mathbf{2500\ \mu C}$$
$$Q_2 = 50 \times 250 = \mathbf{12{,}500\ \mu C}$$
$$Q_3 = 25 \times 250 = \mathbf{6{,}750\ \mu C}$$

(*ii*) $C = C_1 + C_2 + C_3$

$$= 10 + 50 + 25 = \mathbf{85\ \mu F}$$

(*iii*) Series connection is shown in Fig. 15(*b*). Here charge on each capacitor is the same and is equal to that on the equivalent single capacitor.

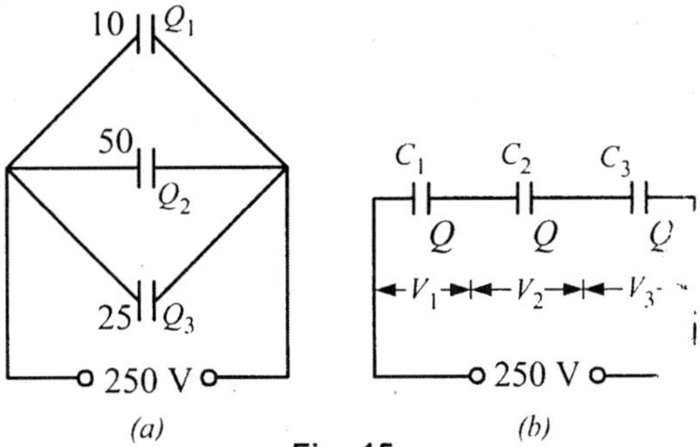

Fig. 15

$$1/C = 1/10 + 1/50 + 1/25,\ C = 25/4\ \mu F$$
$$Q = CV = (25/4) \times 250 = 1562.5\ \mu C$$
$$Q = C_1V_1,\ V_1 = Q/C_1 = 1562.5\ \mu C/10\ \mu F = \mathbf{156.25\ V}$$
$$V_2 = 1562.5/50 = \mathbf{31.25\ V},\ V_3 = 1562.5/25 = \mathbf{62.5\ V}$$

Q. 13. Find the voltage and charge across each capacitor in Fig. 16(*a*). All capacitances are in μF.

Ans. The combined capacitance of the series-parallel capacitor is

$$C = \frac{12 \times (10+8)}{12 + (10+8)} = 7.2\ \mu F \qquad \text{– Fig. 16}(b)$$

Total charge delivered by the 10 V battery is

$$Q = 10 \times 7.2 = 72\ \mu C \qquad \text{– Fig. 16}(c)$$

As shown in Fig. 17(*a*), 12 μF and 18 μF capacitors are in series. Hence, they have the same charge i.e. 72 μC. Hence, voltage across 12 μF capacitor is

$$V_1 = \frac{\text{charge}}{\text{capacitance}} = \frac{72\ \mu C}{12\ \mu F} = \mathbf{6\ V}$$

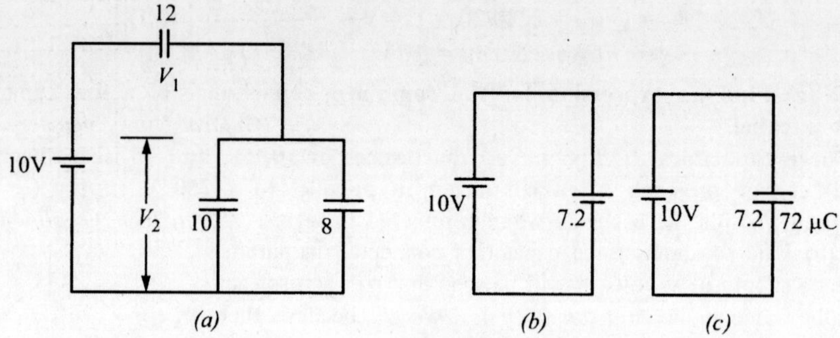

Fig. 16

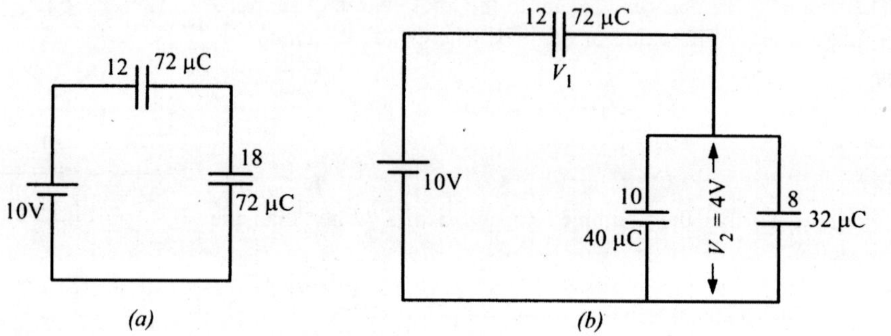

Fig. 17

Since, $10\ V = V_1 + V_2$ $\therefore$ $V_2 = 10 - 6 = \mathbf{4\ V}$

The 18 μF capacitor represents two capacitors connected in parallel as shown in Fig. 17(*b*). They have the same potential V_2 across them.

The charge on 8 μF capacitor = 8 μF × 4 V = **32 μC**

Similarly, charge on 10 μF capacitor = 10 μF × 4 V = **40 μC.**

Q. 14. Fully explain the variations in the voltage, charging current and charge of a capacitor when it is charged from a dc voltage source. Draw the charging curves of the above three quantities.

Ans. In Fig. 18(*a*) is shown an arrangement by which a capacitor *C* may be charged or discharged through a high resistance *R* from a battery of V volts. The voltage across *C* can be measured by a voltmeter and charging current by an ammeter. When the switch *S* is closed, the

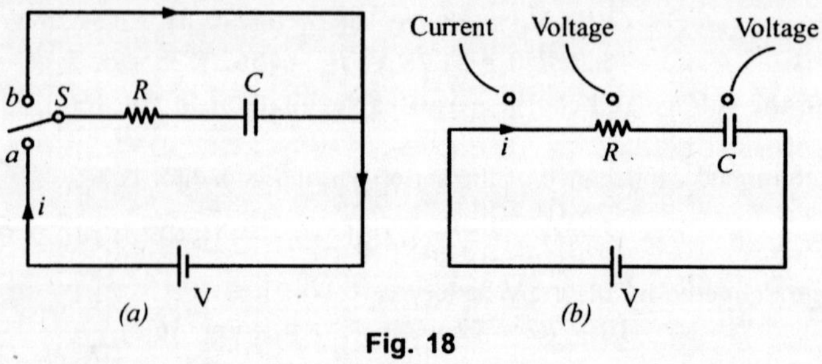

Fig. 18

voltage v across C does not *rise to V immediately*. It builds up gradually and non-linearly. The charging current i is maximum at the start when C is uncharged. At that time, whole of V drops on R so that maximum charging current is $I_m = V/R$. It then decreases exponentially and finally stops when p.d. across capacitor becomes equal to the battery voltage V. Same is the case with the capacitor charge which builds up gradually. These changes are shown in Fig. 18(b).

At any time during charging, let

v = p.d. across C, i = charging current, q = charge on capacitor plates

As seen from Fig. 19, the applied voltage V is always equal to the sum of (a) resistive drop, iR and (b) voltage v developed across capacitor.

$$\therefore \quad V = iR + v$$

Now, $$q = Cv$$

$$\therefore \quad i = \frac{dq}{dt} = \frac{d}{dt}(Cv) = C\frac{dv}{dt}$$

Hence, $$V = v + CR\frac{dv}{dt} \quad \text{or} \quad \frac{dv}{V - v} = \frac{dt}{CR}$$

or $$-\frac{dv}{V - v} = -\frac{dt}{CR}$$ (putting –ve sign on both sides)

Integrating both sides, we get $$-\int\frac{dv}{V - v} = -\frac{1}{CR}\int dt$$

$$\therefore \quad \log_e(V - v) = -\frac{t}{CR} + K \qquad ...(i)$$

where K is the constant of integration.

Since, $v = 0$ when $t = 0$, putting these values in Eq. (i) above, we get $\log_e V = K$

Substituting this value of K in Eq. (i), we get

$$\log_e(V - v) = -\frac{t}{CR} + \log_e V$$

or $$\log_e(V - v) - \log_e V = \frac{-t}{CR}$$

or $$\log_e\frac{V - v}{V} = \frac{-t}{CR} = \frac{-t}{\lambda}$$

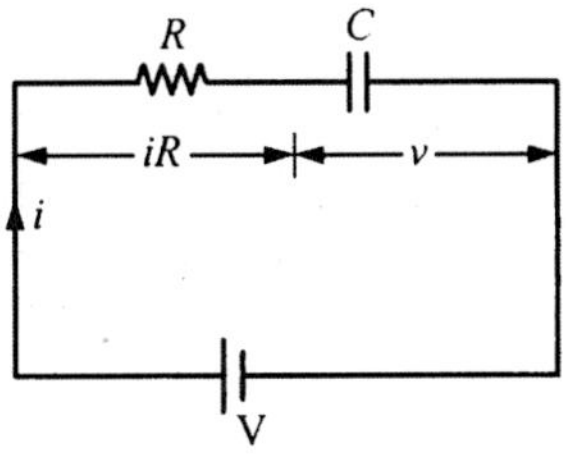

Fig. 19

Here, $\lambda = CR$ is called the *time constant*.

$$\therefore \quad \frac{V - v}{V} = e^{-t/\lambda} \quad \text{or} \quad v = V(1 - e^{-t/\lambda}) \qquad ...(ii)$$

The above equation gives the variations with time of voltage across the capacitor plates and is shown in Fig. 20(a) and 20(b).

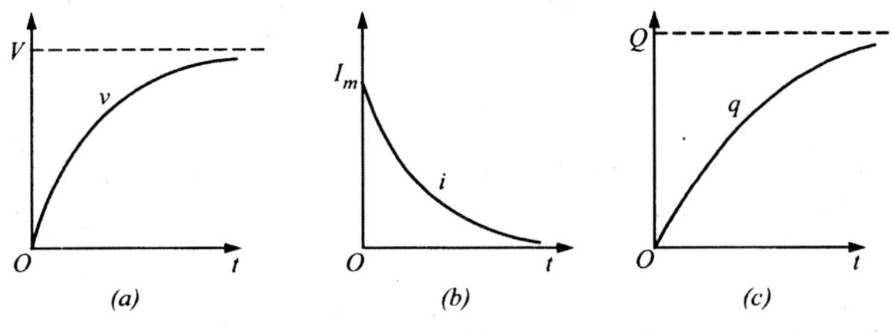

Fig. 20

Now $\qquad v = \dfrac{q}{C}$ and $V = \dfrac{Q}{C}$

Hence, Eq. (*ii*) becomes

$$\frac{q}{C} = \frac{Q}{C}(1 - e^{-t/\lambda}) \quad \text{or} \quad q = Q(1 - e^{-t/\lambda}) \qquad ...(iii)$$

As seen, increase of charge also follows an exponential law and reaches maximum value after *infinite time* [Fig. 20(*c*)].

The charging current is given by

$$i = \frac{dq}{dt}$$

Putting the value of q from Eq. (*iii*) above, we get

$$i = \frac{d}{dt}(q) = \frac{d}{dt}\left[Q\left(1 - e^{-t/\lambda}\right)\right]$$

$$= Q\frac{d}{dt}\left(1 - e^{-t/\lambda}\right)$$

$$= Q\left(\frac{1}{\lambda}e^{-t/\lambda}\right) = \frac{Q}{\lambda}e^{-t/\lambda}$$

$$\therefore \qquad i = \frac{Q}{\lambda}e^{-t/\lambda}$$

Now, $\quad Q = CV$ and $\lambda = CR \qquad \therefore \quad i = \dfrac{CV}{CR}e^{-t/\lambda}$

or $\qquad i = \dfrac{V}{R}e^{-t/\lambda}$ or $i = I_m e^{-t/\lambda} \qquad ...(iv)$

where, I_m = maximum charging current *at the start*. As charging continues, charging current decreases according to Eq. (*iv*) as shown in Fig. 20(*b*). However, for all practical purposes, charging current reaches its maximum value in about *five time constants* of the circuit.

Q. 15. (*a*) What is meant by time constant of a series *RC* network? What is its significance?

(*b*) A capacitor is charged from a d.c. source through a resistor of 0.5 megohm. If the p.d. across it reaches 75% of its initial value in half a second, find its capacitance.

Ans. (*a*) The quantity '*RC*' is called the time constant (λ) of a series *RC* network. If *R* is in ohm and *C* is in farad, then time constant is given in seconds. Hence, $\lambda = RC$ seconds. The time constant is a measure of how long it takes the capacitor to charge. The larger the value of *R* or *C* or both, the longer it takes for the capacitor to charge. It is due to the fact that *R* limits the charging current and *C* determines how much charge can be stored.

The variation with time of voltage across the capacitor plate is given by the equation.

$$v = V(1 - e^{-t/\lambda})$$

If $t = \lambda$, then $\qquad v = V \times (1 - e^{-\lambda/\lambda}) = V(1 - e^{-1})$

$$= V\left(1 - \frac{1}{2.718}\right) = 0.632 \text{ V}$$

Hence, time constant may be defined as

the time during which capacitor voltage rises to 63.2% of its final steady value.

Similarly, the charging current of a capacitor is given by the equation.

$$i = I_m e^{-t/\lambda}$$

Putting $t = \lambda$, we get

$$i = I_m e^{-\lambda/\lambda} = I_m e^{-1} = \frac{I_m}{2.718} = 0.37\, I_m$$

Hence, the time constant of a series *RC* circuit may be defined as

the time during which the charging current of a capacitor falls to 37% of its initial maximum value.

(**b**) Using the relation $v = V(1 - e^{-t/\lambda})$, we have

$v = 0.75$ V — V being the applied voltage

$t = 0.5$ second

$\therefore \quad 0.75\text{ V} = V(1 - e^{-1/2\lambda}) \quad \text{or} \quad e^{-1/2\lambda} = 0.25$

$\therefore \quad \frac{-1}{2\lambda} = \log_e 0.25 \quad \text{or} \quad \lambda = 0.3612$ second (approx.)

Now $\lambda = CR$

When C is in farads and R in ohms, then λ is in seconds.

$\therefore \quad C \times 0.5 \times 10^6 = 0.3612$

$C = 0.3612/500{,}000$ farad

$= 0.7224 \times 10^{-6}$ farad $=$ **0.7224 μF**

Q. 16. A capacitor of 0.1 μF is charged from a 100-V battery through a series resistance of 1000 Ω. Find :

(*a*) the time for the capacitor to receive 63.2% of its final charge.

(*b*) the charge received in this time

(*c*) the rate of charging when the charge is 63.2% of the final charge.

(Elect. Engineering, Bombay Univ. 2001)

Ans. (*a*) A capacitor receives 63.2% of its final charge in a time equal to one time constant of the circuit.

$\therefore$ time required $= \lambda = CR = 0.1 \times 10^{-6} \times 1000 =$ **10^{-4} second**

(**b**) Final charge, $Q = CV = 0.1 \times 100 = 10$ μC

Charge received during this time $= 0.632 \times 10 =$ **6.32 μC**

(**c**) Here, $v = 0.632\,V = 0.632 \times 100 = 63.2$ volts

$$\frac{dv}{dt} = \frac{100 - 63.2}{10^{-4}} = \textbf{368 kV/s}$$

Q. 17. An 8 μF capacitor is connected in series with a 0.5 MΩ resistor across a 200 V d.c. supply. Calculate (*i*) the time constant (*ii*) the initial charging current (*iii*) the time taken for the p.d. across the capacitor to grow to 160 V and (iv) the current and the p.d. across the capacitor in 4 seconds after it is connected to the supply.

(Electrical Science, AMIE Summer)

Ans. (*i*) Time constant $\lambda = CR = 8 \times 10^{-6} \times 0.5 \times 10^6 =$ **4 second**

(*ii*) Initial charging current is maximum because at the start, there is no voltage across the capacitor to oppose the applied voltage. Initial current is restricted by R alone

$\therefore \quad I_m = V/R = 200/0.5 \times 10^6 = 0.4 \times 10^{-3}$ A $=$ **0.4 mA**

(*iii*) $\quad v = V(1 - e^{-t/\lambda}) \quad \text{or} \quad 160 = 200(1 - e^{-t/4})$

$\therefore \quad 1 - e^{-t/4} = 0.8 \quad \text{or} \quad e^{t/\lambda} = 5 \quad \text{or} \quad (t/4)\log_e e = \log_e 5$

or $\quad t/4 = 1.61 \quad \text{or} \quad t =$ **6.44 seconds**

(*iv*) The given time of 4 seconds represents the time constant of the circuit. Hence, capacitor voltage will raise to 63.2% of its maximum voltage and its charging current will fall to 37% of its maximum initial value. Hence, after 4 seconds.

$v = 0.632 \times 200 =$ **126.4 V**

$i = 0.37\,I_m = 0.37 \times 0.4 =$ **0.148 mA**

Q. 18. A resistance *R* and a 4 μF capacitor are connected in series across 200 V d.c. supply. Across the capacitor is a neon lamp that strikes (glows) at 120 V. Calculate the value of *R* to make the lamp strike (glow) 5 seconds after the switch has been closed.

(*Electrotechnics-I.M.S. Univ. Baroda*)

Ans. Obviously, the capacitor voltage has to rise to 120 V in 5 seconds.

$\therefore \quad 120 = 200\,(1 - e^{-5/\lambda}) \quad \text{or} \quad e^{5/\lambda} = 2.5 \quad \text{or} \quad \lambda = 5.464 \text{ second.}$

Now, $\quad \lambda = CR \quad \therefore \quad R = 5.464/4 \times 10^{-6} = \mathbf{1.366\ M\Omega}$

Q. 19 (*a*) Discuss the discharging of a capacitor. Derive equations for the discharge current and the potential difference across the capacitor.

(*b*) A 20 μF capacitor is found to have an insulation resistance of 50 MΩ measured between the terminals. If this capacitor is charged off a d.c. supply of 230-V, find the time required after disconnection from the supply for the p.d. across the capacitor to fall to 60 V.

Ans. (*a*) When switch *S* of Fig. 18(a) is shifted to terminal '*b*', the capacitor is discharged through *R*. To being with, discharge current is maximum but then it starts decreasing and becomes zero when capacitor is fully discharged. The voltage equation across the capacitor at any time during the discharge is given as

$$V = v + iR \qquad \text{...}(i)$$

Here, i is the discharge current and v is the voltage across the capacitor at any time during discharge.

Now, $\quad q = Cv$

$$\therefore \quad i = \frac{dq}{dt} = \frac{d}{dt}(Cv) = C\frac{dv}{dt}$$

Hence, Eq. (*i*) becomes

$$V = v + CR\frac{dv}{dt}$$

Since battery is cut out of the circuit, $V = 0$

∴ the above equation becomes

$$0 = CR\frac{dv}{dt} + v \quad \text{or} \quad v = -\,CR\frac{dv}{dt}$$

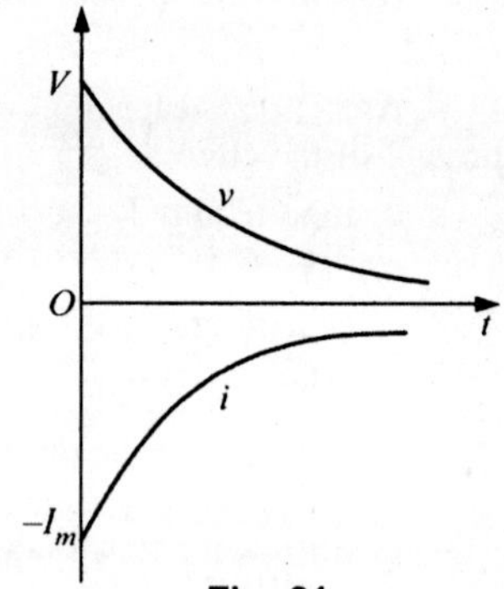

Fig. 21

$$\text{or} \quad \frac{dv}{v} = \frac{-dt}{CR}$$

$$\therefore \quad \int \frac{dv}{v} = -\frac{1}{CR}\int dt \quad \text{or} \quad \log_e v = -\frac{t}{CR} + k$$

At the start of discharge when $t = 0$, $v = V$

$$\therefore \quad \log_e V = 0 + K \quad \text{or} \quad \log_e V = K$$

Putting this value above, we get

$$\log_e v = -\frac{t}{\lambda} + \log_e V \quad \text{or} \quad \log_e v/V = -\frac{t}{\lambda}$$

$$\therefore \quad \frac{v}{V} = e^{-t/\lambda} \quad \text{or} \quad v = V e^{-t/\lambda}$$

Similarly, $\quad q = Q\,e^{-t/\lambda}$ and $i = I_m\,e^{-t/\lambda}$

The fall of potential and the discharge current are shown in Fig. 21.

(*b*) Obviously, the fully-charged capacitor discharges through its insulation resistance.

Here $\quad V = 230 \text{ V}\ ;\ v = 60 \text{ V},\ \lambda = CR = 20 \times 10^{-6} \times 50 \times 10^{6} = 1000 \text{ seconds}$

Now, $v = V e^{-t/\lambda}$ or $60 = 230\, e^{-t/1000}$

or $e^{t/1000} = \frac{23}{6}$ or $\frac{t}{1000} \cdot \log_e e = \log_e 23/4 = 1.343$

$t =$ **1343 second**

EXERCISE

1. A positive charge of 40 μC is placed at a distance of 5 cm from a negative charge of 100 μC. Find the force of attraction between the two charges. **(14.4×10^3 N)**
2. A flux of 100 μC penetrates vertically through an area 10 mm x 10 mm. Find the value of flux density. **(1 C/m^2)**
3. When a charge of 1000 μC is placed at a certain point in an electric field, it experiences a force of 28N. If the medium is air, find
   (*i*) field strength at the point where the charge is placed.
   (*ii*) flux density at the same point. **(28,000 N/C, 24.76 C/m^2)**
4. Two parallel plates are separated by a distance of 10 mm. A charge of 600 μC experiences a force of 9 N when placed between them. Calculate the voltage difference between the two plates. **(150 V)**
5. Two oppositely-charged plates are separated by a 0.12 mm thickness of mica. Voltage across the two plates is gradually increased. At what voltage will mica break down if dielectric strength of mica is 20×10^7 V/m. **(24,000 V)**
6. An energy of 18×10^{-3} joules is required to move a positive charge of 300 μC from a negatively-charged plate to a positively charged plate (Fig. 22). If the two plates are separated by a distance of 15 mm, calculate
   (*a*) potential difference between the two charged plates
   (*b*) electric field intensity *E* in the region between the two plates. **(60 V, 4 kV/m)**

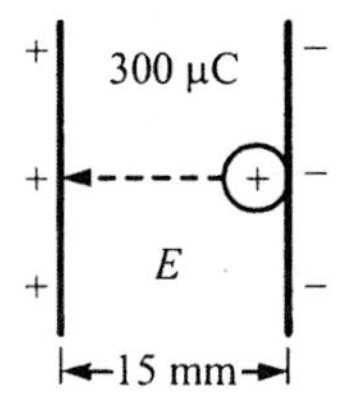

Fig. 22

7. Calculate the total flux passing through 10 cm × 5 cm surface in a region where electric flux density is 300 μC/m^2. **(1.5×10^{-5} C)**
8. At a certain point in a dielectric medium, the flux density is 0.09 C/m^2 and electric field intensity is 1.2×10^9 V/m. What is the relative permittivity of the medium? **(8.5)**
9. How much charge must be placed on a 60 μF capacitor in order to make voltage across it equal to 20 V? **(1200 μC)**
10. For the circuit shown in Fig. 23, find
    (*i*) equivalent capacitance and
    (*ii*) voltage developed across each capacitor
    All capacitances are in μF. **[(*i*) 6 μF (*ii*) V_{AB} = 50 V, V_{BC} = 40 V]**

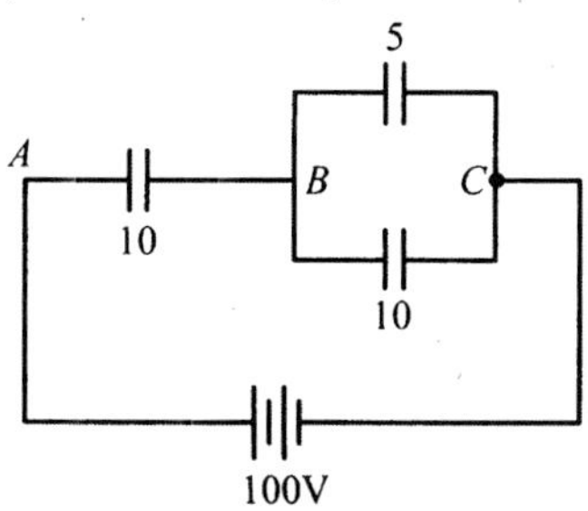

Fig. 23

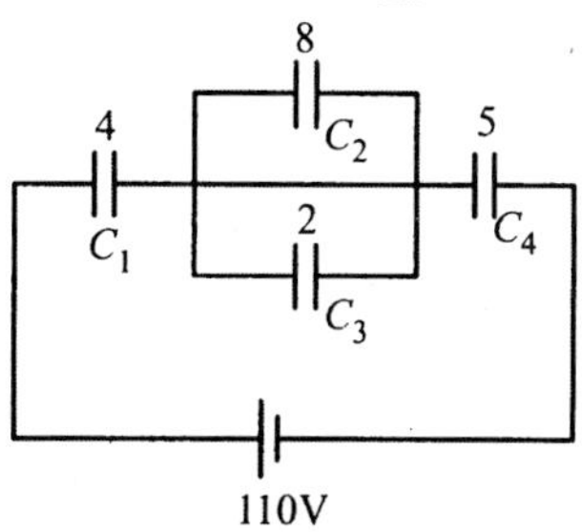

Fig. 24

11. In the circuit of Fig. 24, find
   (*i*) equivalent capacitance
   (*ii*) voltage across each capacitor and
   (*iii*) charge on each capacitor
   All capacitances are in μF. **[(*i*)1.82 μF (*ii*) V_1 = 50 V, V_2= V_3 = 20 V, V_4 = 40 V (*iii*) Q_1 = 200 μC, Q_2 = 160 μC, Q_3 = 40 μC, Q_4 = 200 μC]**
12. The total capacitance of two capacitors is 0.03 μF when joined in series and 0.16 μF when connected in parallel. Calculate the capacitance of each capacitor. *(Gujarat Univ., Summer 2003)* **(0.12 μF, 0.04 μF)**
13. Three capacitors of 1, 2, 3 μF are connected in series across a supply voltage of 100 V. Find the equivalent capacitance of the combination and energy stored in each capacitor. *(V.T.U. Belgaum Karnataka Univ. 2003)* **(6/11 μF, 14.88 μJ, 17.44 μJ, 4.96 μJ)**
14. Find the equivalent capacitance between points *A* and *B* of the network shown in Fig. 25. All capacitances are in μF. *(V.T.U. Belgaum Karnataka Univ. 2008)* **(2.14 μF)**

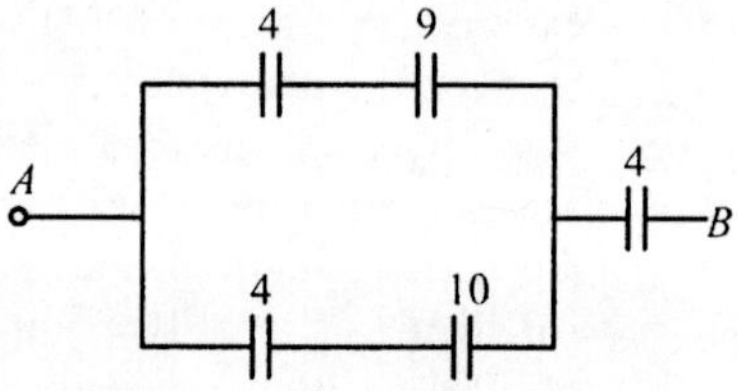

Fig. 25

15. When a capacitor, charged to a p.d. of 400 V, is connected to a voltmeter having a resistance of 25 MΩ, the voltmeter reading is found to have fallen to 50 V at the end of an interval of 2 minutes. Find the capacitance of the capacitor. *(App. Elect. London Univ.)* **(2.31 μF)**
16. A 500 μF capacitor in series with 2Ω resistor and a switch is charged by a 10 V source. Calculate
   (*i*) initial charging current (I_m) when switch is closed
   (*ii*) time constant of the circuit
   (*iii*) time for the voltage across the capacitor to reach 9.95 V
   [(*i*) 5 A (*ii*) 1 ms (*iii*) 5.3 ms]

MULTIPLE CHOICE QUESTIONS

1. Capacitance means the ability of a capacitor to store
   (*a*) energy (*b*) charge (*c*) flux (*d*) power
2. A flux of 96 μC falls perpendicularly on a surface area of 48 mm^2. Neglecting fringing, the flux density *D* is .... C/m^2
   (*a*) 2 (*b*) 0.2 (*c*) 0.02 (*d*) 20
3. In a capacitor, energy is stored in its
   (*a*) plates (*b*) dielectric (*c*) electric field (*d*) connecting wires
4. The amount of energy stored in a parallel plate capacitor varies as the square of its
   (*a*) capacitance (*b*) plate charge (*c*) plate separation (*d*) plate potential
5. Lighting strikes Earth from a charged cloud 100 m above its surface. If breakdown strength of air is 3 × 10^6 V/m, the voltage between the cloud and Earth surface is .... million volts.
   (*a*) 3 (*b*) 30 (*c*) 300 (*d*) 3000

6. An ideal capacitor does not
   (*a*) store energy (*b*) dissipate energy (*c*) contain flux (*d*) have electric field
7. Farad is the unit of
   (*a*) charge (*b*) capacitance (*c*) flux (*d*) field intensity
8. The unit of electric flux density *D* is
   (*a*) V/m (*b*) F/m (*c*) N/C (*d*) C/m^2
9. The capacitance of a parallel plate capacitor can be increased by
   (*a*) increasing plate area (*b*) decreasing plate separation
   (*c*) choosing dielectric of high permittivity (*d*) all of the above
10. The capacitance of a capacitor is NOT influenced by
    (*a*) plate thickness (*b*) plate separation
    (*c*) plate area (*d*) nature of dielectrics (*West Bengal Tech. Univ. 2007*)
11. With the introduction of a dielectric, the capacity of a capacitor
    (*a*) increases (*b*) decreases (*c*) does not change (*d*) changes arbitrarily
12. A capacitor blocks
    (*a*) ac current (*b*) dc current
    (*c*) sinusoidal current (*d*) triangular current
13. The relative permittivity of a medium has the unit of
    (*a*) volt/metre (*b*) coulomb/metre^2 (*c*) farad/metre (*d*) none of the above
14. The property of a medium to focus electrostatic flux is called its
    (*a*) dielectric strength (*b*) flux density (*c*) permittivity (*d*) charge density
15. .... time constants are required for a capacitor to charge or discharge fully
    (*a*) 5 (*b*) 3 (*c*) 2 (*d*) 7
16. During charging or discharging of a capacitor, no current passes through the
    (*a*) connecting wires (*b*) battery (*c*) dielectric (*d*) circuit switch
17. A capacitor in a series *RC* circuit is charged to 100 V. When allowed to discharge for one time constant, the capacitor voltage would be .... volts.
    (*a*) 37 (*b*) 75 (*c*) 63 (*d*) 0
18. The time constant of an *R-C* series circuit is equal to
    (*a*) R/C (*b*) C/R (*c*) RC (*d*) $1/RC$
19. The smallest capacitor in a series combination has the largest ....
    (*a*) charge (*b*) voltage (*c*) capacitance (*d*) insulation
20. Charging and discharging of a capacitor follow .... curves
    (*a*) linear (*b*) no-linear (*c*) exponential (*d*) sinusoidal
21. The time constant of a series *RC* circuit during charging and discharging
    (*a*) remains the same (*b*) is different
    (*c*) depends on the ratio *R*/*C* (*d*) depends on the circuit current
22. When a capacitor is charged in an *RC* circuit, after five time periods
    (*a*) capacitor voltage is maximum (*b*) voltage drop across *R* is zero
    (*c*) charging current is zero (*d*) all of the above.

ANSWERS

1. (*b*)	2. (*a*)	3. (*c*)	4. (*d*)	5. (*c*)	6. (*b*)	7. (*b*)
8. (*d*)	9. (*d*)	10. (*a*)	11. (*a*)	12. (*b*)	13. (*d*)	14. (*c*)
15. (*a*)	16. (*c*)	17. (*a*)	18. (*c*)	19. (*b*)	20. (*c*)	21. (*a*)
22. (*d*)						

CHAPTER 4 MAGNETISM AND MAGNETIC CIRCUITS

4.1 MAGNETISM

Q. 1 (*a*) Explain the statement that 'magnetism is the property of the atoms of a material'.

(*b*) How do you explain that some materials like paper, wood and glass etc. cannot be magnetised even when placed in a magnetic field?

Ans. (*a*) It has been found that the magnetic properties of a material are due to the normal electronic activity of its atoms. In fact, each atom is a tiny electromagnet because it contains charge (electrons) *in motion.* These moving electrons *within* the atom produce magnetic fields which add up or cancel each other out. As is well known, the electrons have two types of motion:

(*i*) **Orbital motion** - electrons move in orbits around the nucleus of the atom.

(*ii*) **Spinning motion** - electrons spin around their *own axis* either clockwise or counter-clockwise.

The resultant magnetic effect of the atoms is due to the above two motions. However, it has been found that the electron spin is the more important of the two factors.

In most of the materials, all electron spins are *paired* or *compensated.* For each electron spinning in a clockwise direction, there is a second electron which produces a compensating (or balancing) spin in a counterclockwise direction. The same paired situation applies to the orbital motion of the electrons as well. Hence, the net magnetic field produced is zero and such materials are called *nonmagnetic* materials. However, if there is an unbalanced electron spin in a material (like iron), it is called a *magnetic* material.

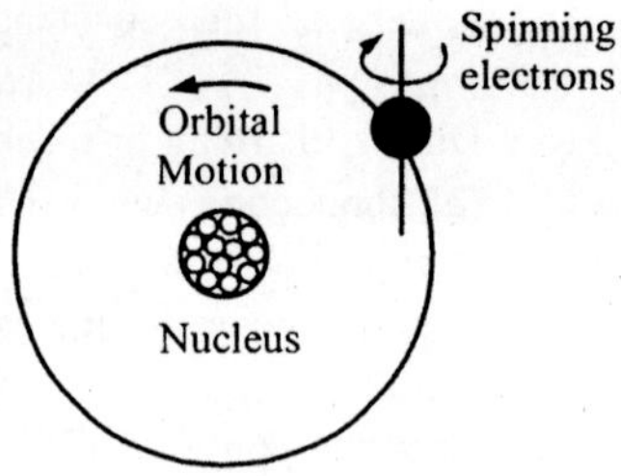

Fig. 1

Consider the case of iron which is a magnetic material. Let us find out why it is so. There are 26 electrons in an iron atom. The two electrons in the first shell neutralise each other because of opposite spins. Same is the case with second shell–4 electrons spin in one direction and 4 electrons in the opposite direction. The third shell has 14 electrons–9 spin in one direction and 5 in the other. The outermost shell has only 2 electrons which spin in the opposite directions and so cancel out each other's magnetic effects. Obviously, the magnetic properties of iron are determined by the $(9 - 5) = 4$ unbalanced electrons in the *third shell*. Hence, when a piece of iron is placed in an external magnetic field, each of its 4 uncompensated electrons aligns its axis of spin parallel to the external field. As a result, The magnetic field of iron increases and adds to the applied external field. The group of atoms that have been so aligned is called a **magnetic domain.** Each domain contains from 10^{12} to 10^{15} atoms. Each domain produces a magnetic field in a *particular* direction. In unmagnetised materials, adjacent domains produce magnetic fields in different directions as shown in Fig. 2(*a*).

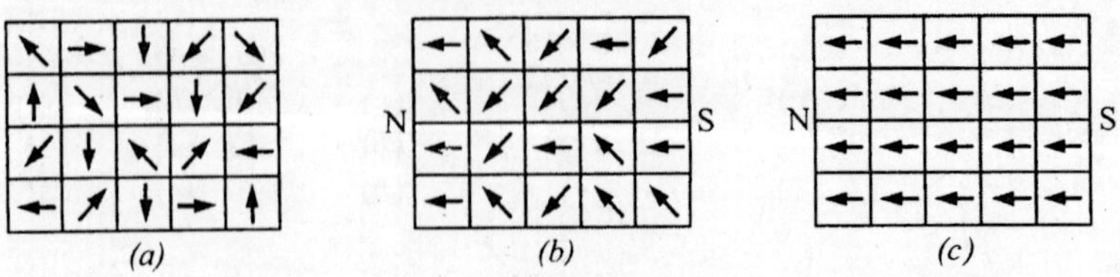

Fig. 2

When an external magnetic field is applied, the domains in the unmagnetised magnetic material start aligning themselves parallel to the external field as shown in Fig. 2(*b*). However, not every domain is perfectly aligned as yet. When the external field is strong enough, all domains get aligned and the material becomes highly magnetised. When all domains are perfectly aligned, the magnetic material is said to be *saturated* because it cannot be magnetised further as shown in Fig. 2(*c*).

(*b*) Materials like paper, wood and glass are called nonmagnetic materials because they cannot be magnetised. The reason is that they have as many electrons spinning in clockwise direction as those spinning in counterclockwise direction. Hence, their magnetic fields cancel out.

Q. 2. How do you define a magnetic field? Are the lines of force or magnetic flux real or imaginary? Give the characteristics of these magnetic lines of force. Is the field affected by the presence of magnetic and non-magnetic materials in its neighbourhood?

Ans. Every magnet is surrounded by an energy field. The strength and direction of the energy field is represented by lines of force. These lines of force leave the *N*-pole and enter the *S*-pole as shown in Fig. 3. When a nonmagnetic material such as glass, wood or plastic is placed in the magnetic field, the lines of force are unaltered as shown in Fig. 3. However, when a magnetic material such as soft iron is placed in the field, the lines of force tend to change course and pass through the iron rather than through the surrounding air. They do so because the iron provides a magnetic path that is more easily established than that of air.

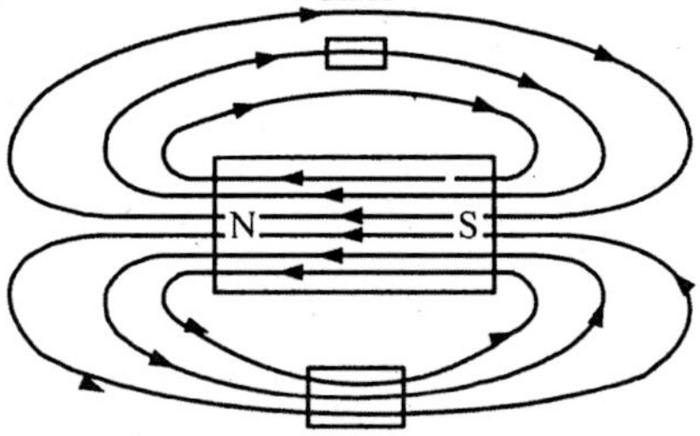

Fig. 3

The space around a magnet in which a magnetic material experiences force of attraction or repulsion is called the *magnetic field* of the magnet.

The lines of force, in fact do not exist at all (if they did, we would see them!). These lines are only *imaginary* lines along which magnetic force acts. Some of the main characteristics of magnetic lines of force are as under:

1. They always go from *N*-pole to *S*-pole and always form closed loops.
2. They never intersect each other.
3. They tend to repulse each other and, hence, diverge as one moves away from the pole.
4. They exhibit tension along their length and try to become as short as possible.
5. They tend to take up the path of least resistance. It means they are easier to set up in a magnetic material (like iron) than in a non-magnetic material (like plastic or glass).

Q. 3. How do you classify solid materials with respect to their magnetic properties? Give two examples of each material.

Ans. There are *four* groups of solid materials depending on their magnetic properties:

1. **Ferromagnetic materials** — these materials have unbalanced electron spins in their atomic shells as explained in Q. 1. Hence, in the presence of an external magnetic field, they *increase* the total magnetic flux greatly. In other words, they can be strongly magnetised by the magnetic field. Such materials are used to manufacture permanent magnets. These materials are: iron, steel, nickel, cobalt and alloys such as ALNICO. The relative permeability of such materials is *very high.*

2. **Paramagnetic materials** — such materials are only *slightly* attracted by a magnetic field. In other words, when placed in a magnetic field, they increase the total flux *only slightly*. Examples are: aluminium, platinum and maganese. Their relative permeability is slightly greater than *unity.*

3. **Diamagnetic materials** — such materials produce a slight *decrease* in the total flux when placed in a magnetic field. Example are: bismuth, copper, gold and silver. Their relative permeability is *less than unity.*

4. **Non-magnetic materials** — these materials have no effect on a magnetic field. When placed in air, they neither increase nor decrease the magnetic field. Examples are: wood, rubber, paper and plastic. Such materials have the same number of electrons spinning in each direction which cancel the atomic fields.

Q. 4. State the laws of magnetic force. Use these laws to define a magnetic pole.

Ans. Let us consider two magnetic poles having pole strengths of m_1 and m_2 webers placed in a medium of permeability μ and d metre apart (Fig. 4). The force of attraction or repulsion between them is

(*i*) directly proportional to their pole strengths.

(*ii*) inversely proportional to the square of the distance between them and

(*iii*) inversely proportional to the absolute permeability of the surrounding medium.

$$\therefore \quad F \propto \frac{m_1 m_2}{\mu d^2} = k\frac{m_1 m_2}{\mu d^2}$$

where k is a constant whose value in the SI system of units is $= 1/4\pi$

$$\therefore \quad F = \frac{m_1 m_2}{4\pi\mu d^2} = \frac{m_1 m_2}{4\pi\mu_0\mu_r d^2} \text{ newton}$$

If the two poles are placed in vacuum or air, then

$$F = \frac{m_1 m_2}{4\pi\mu_0 d^2} \text{ newton}$$

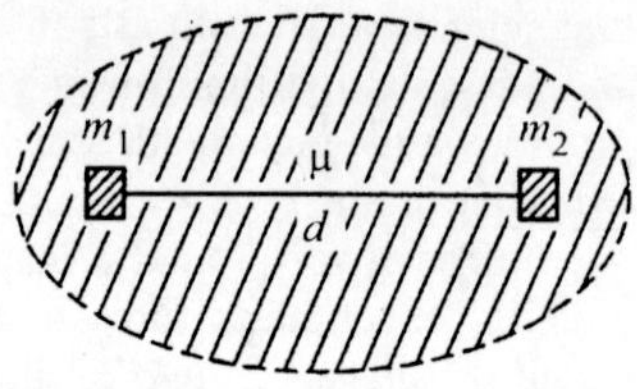

Fig. 4

Definition of Unit Pole

If in the above equation,

$$m_1 = m_2 = m \text{ (say) ; } d = 1 \text{ metre}$$

and $\quad F = \dfrac{1}{4\pi\mu_0}$ newton

Then, $\quad m^2 = 1 \quad$ or $\quad m = \pm 1$ weber

Hence, a unit magnetic pole may be defined as

the pole which when placed in air at a distance of one metre from a similar and equal pole repels it with a force of $1/4\pi\mu_0$ *newtons.*

Q. 5 (*a*) What are soft and hard magnetic materials? Give examples and applications of these materials.

(*b*) What are ferrites? Give their main magnetic properties and field of application.

Ans. (*a*) Broadly speaking *magnetic* materials can be divided into two classes:

(*i*) **Soft magnetic materials** and (*ii*) **hard magnetic materials :** Soft magnetic materials are those which are magnetised easily but lose their magnetism **readily** after magnetising force is removed. Such materials have high permeability, low coercivity and small hysteresis loss which makes them suitable for transformer cores, motor and generator armatures and electromagnets etc. Examples are: silicon steel, nickel-iron alloys and alloys of iron, nickel and cobalt.

Hard magnetic materials are those which are difficult to magnetise but once magnetised, they *retain* their magnetism when the magnetising force is removed.

Such magnetic materials have low permeability, high coercivity and large hysteresis loss. Hence, they are suitable for making permanent magnets. Examples are: cobalt steel, alloys like ALNI (aluminium, nickel and iron) and ALNICO (aluminium, nickel and cobalt).

(*b*) Ferrites are ferromagnetic ceramics which are usually produced by powdered metalllurgical techniques using iron oxide. They have almost constant permeability and low hysteresis loss. Due to their extremely high electrical resistivity, they have negligible eddy current loss. For this reason, they are well-suited for high frequency applications. Example are: cores for radio frequency transformers and for switching and memory devices in high-speed computers.

4.2 ELECTROMAGNETISM

Q. 6. (*a*) What is meant by electromagnetism and electromagnets? Mention some of the uses of electromagnets.

(*b*) What is a solenoid? What is it used for?

Ans. (*a*) Production of a magnetic field around a current-carrying conductor is known as *electromagnetism.*The field is called an electromagnetic field. The invisible lines of force of the magnetic field form concentric circles around the conductors as shown in Fig. 5. The strength of the magnetic field is directly proportional to the current.

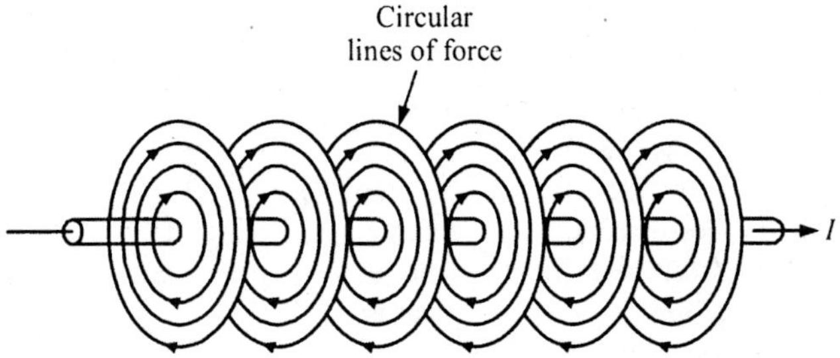

Fig. 5

The direction of the lines of force surrounding a current-carrying conductor can be found with the help of either right-hand rule or Maxwell's corkscrew rule. Electromagnets are *temporary artificial magnets.* They are produced with the help of electric current. An ordinary soft iron rod can be made a magnet if insulated wire is would around the rod and a current is passed through the wire as shown in Fig. 6.

The magnetic properties of this electromagnet will be similar to those of a permanent bar magnet but they would be *much stronger.*

Uses of Electromagets :

Electromagnets are commonly used in

1. motors and generators
2. transformers
3. relay coils
4. for lifting heavy loads
5. electromagnetic deflection in oscilloscopes which includes (*i*) television (*ii*) radar (*iii*) electric bells and electrodynamic speakers.

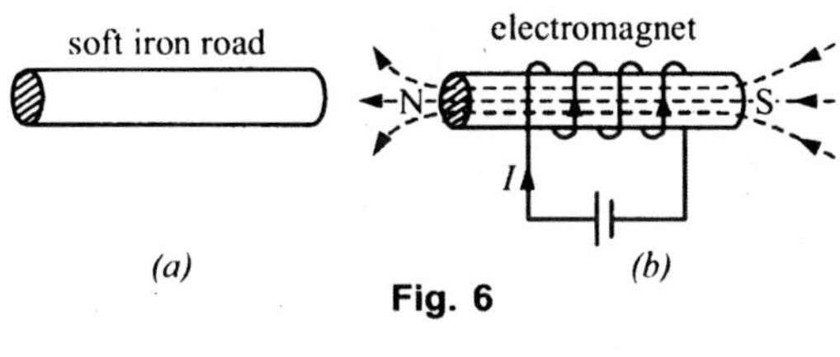

Fig. 6

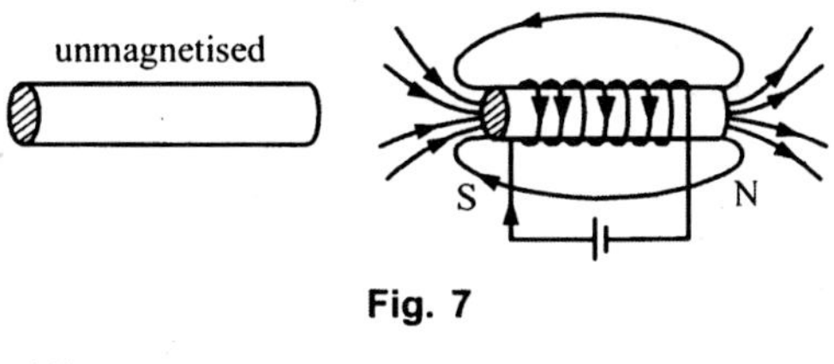

Fig. 7

(*b*) An electromagnet formed by winding coils around a cylindrical iron alloy core is known as a solenoid. It is shown in Fig. 7. One end of the solenoid has N polarity while the other has S polarity. By using a large number of turns of wire, we can increase the magnetic flux density in the core of the solenoid. The flux density also depends on the current carried by the coil. Hence, the final flux density is proportional to the number of turns and the current in each tern *i.e.* NI. The product 'NI' gives the *magnetomotive force* (mmf) of the solenoid which is responsible for producing magnetic flux. Electrical relays, bell ringers and electrical hammers make use of the principle of the solenoid.

Q. 7 (*a*) Give an expression for the force experienced by a current-carrying conductor placed in a uniform magnetic field.

(*b*) A single conductor 10 cm long carries a current of 5 A and lies perpendicular to a uniform magnetic field of 1.4 tesla.

Calculate:

(*i*) force on the conductor produced by the field and

(*ii*) force on the conductor when it makes an angle of 60° with the field.

Ans. (*a*) Whenever a current-carrying conductor is placed in a magnetic field, it experiences a force which acts in a direction perpendicular both to the direction of the current and the field. Fig. 8 shows a conductor *XY* which lies in a uniform magnetic field of flux density *B* produced by the permanent magnet. If *l* is the length of the conductor lying *within* this field and *I* the current carried by it, then the value of the force experienced by the conductor is

$F = BIl = \mu_0 \, \mu_r \, HIl$ newton $\quad (\because \; B = \mu_0 \, \mu_r \, H)$

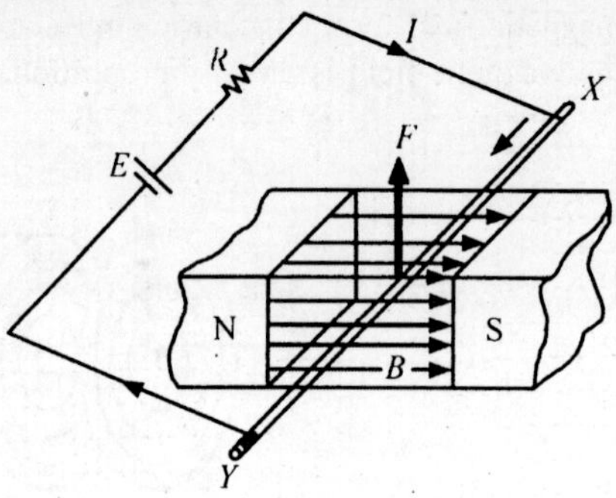

Fig. 8

This force is due to the interaction between the magnetic field set up by the current and the permanent magnet field. The direction of the force can be found by Fleming's left hand rule (or motor rule) as shown in Fig. 9. In Fig. 8. force is upwards as shown. If current is reversed, the force *F* would be downwards.

If the conductor lies at an angle of θ with the field, then force is given by

$F = BIl \sin \theta$ newton

It is obvious that no force is exerted on the conductor if it lies *parallel* to the magnetic field.

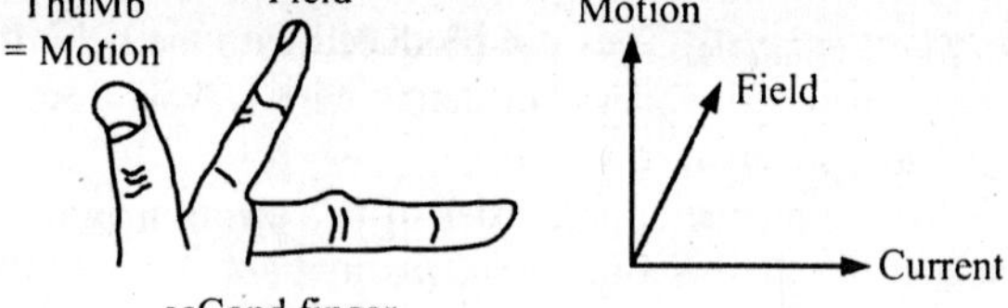

Fig. 9

(*b*) (*i*) $F = BIl = 1.4 \times 5 \times 0.1$

$= \mathbf{0.7 \; N}$

(*ii*) $F = BIl \sin\theta$

$= 1.4 \times 5 \times 0.1 \times \sin 60°$

$= 0.7 \times 0.866 = \mathbf{0.60 \; N}$

Q. 8. Derive the formula for the field strength *H* at a distance of *r* for a long straight current-carrying conductor. (*Basic Elect. Engg., West Bengal Tech. Univ., 2001*)

Ans. In Fig. 10 is shown a straight conductor which extends to infinity in either direction. Let it carry a current of *I upwards*. The magnetic field consists of circular lines of force having their plane perpendicular to the conductor and their centres at the centre of the conductor.

Suppose, the field strength at point *C* is *H*. It means if a unit *N*-pole is placed at *C*, it will experience a force of *H*. The direction of this force would be tangential to the circular lines of force passing through point *C*. If this unit *N*-pole is taken once round the conductor *against* this force, then work done is $H \times 2 \pi r$. It must equal the magnetomotive force (mmf) produced by the conductor *i.e.* *NI*. Since there is only one conductor, hence

$$\text{mmf} = NI = 1 \times I = I.$$

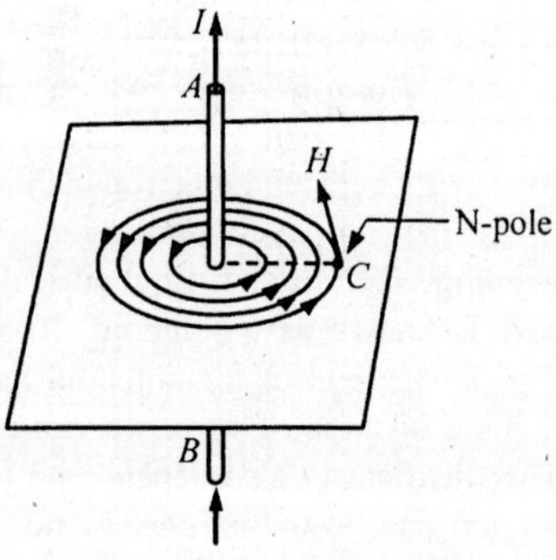

Fig. 10

Hence, $H \times 2\pi r = I$ or $H = \dfrac{NI}{2\pi r}$

Also, $B = \mu_0 H = \mu_0 \dfrac{NI}{2\pi r}$ ... in air

Q. 9. Derive an expression for the mutual force of attraction or repulsion between two current-carrying conductors lying in vacuum. Hence, define one ampere.

Ans. In Fig. 11(*a*) are shown two parallel conductors L and M carrying currents of I_1 and I_2 in the *same* direction *i.e.* upwards.

The field strength in the space between the two conductors is *decreased* because the two fields there oppose each other. Hence, the two conductors are attracted towards each other.

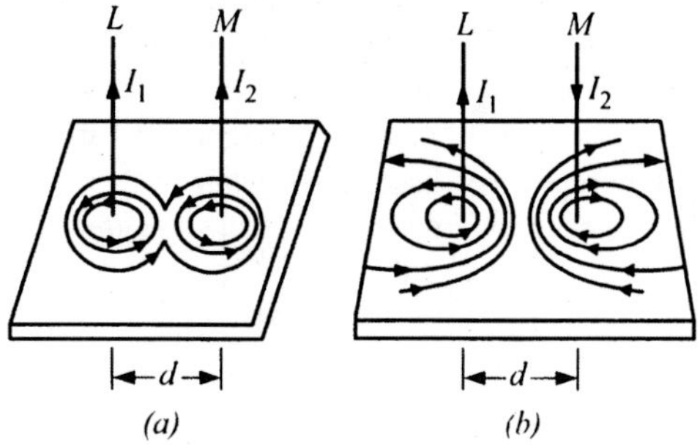

Fig. 11

Now, let the two parallel conductors carry currents in *opposite* directions as shown in Fig. 11(*b*). In this case, the field strength is *increased* in the space between the two conductors because the two fields are in the *same* direction there. Due to the lateral repulsion of the lines of force, the two conductors experience a mutual force of repulsion as shown.

Magnitute of Mutual Force

It is obvious that each of the two parallel conductors lies in the magnetic field of the other conductor. Conductor L lies in the magnetic field of M and M lies in the magnetic field of L. If 'd' is the distance between them, then flux density at Q due to L is

$$B = \frac{\mu_0 I_1}{2\pi d} \text{ Wb/m}^2$$

If l is the length of the conductor M lying in this flux density, then the force (either of attraction or repulsion) is given by

$$F = BI_2 l = \frac{\mu_0 I_1 I_2 l}{2\pi d} \text{ newton}$$

Obviously, conductor L will experience an equal force in the opposite direction.

Definition of one ampere :

Now, $F = \dfrac{\mu_0 I_1 I_2 l}{2\pi d}$ or $F = \dfrac{4\pi \times 10^{-7} I_1 I_2 l}{2\pi d} = 2 \times 10^{-7} \dfrac{I_1 I_2 l}{d}$ newton

The force *per metre* run of the conductor is

$$F = 2 \times 10^{-7} \frac{I_1 I_2}{d} \text{ newton/metre}$$

If $I_1 = I_2 = 1\text{A}$ and $d = 1$ m, then

$$F = 2 \times 10^{-7} \text{ N}$$

Hence, one ampere current is that current which *when flowing in each of the two long parallel conductors situated in vacuum and separated 1 metre between centres, produces on each conductor a force of 2×10^{-7} newton per metre length.*

Q. 10. (*a*) Explain the meaning of induced magnetism.

(*b*) Explain the terms residual magnetism and retentivity.

Ans. (*a*) The magnetism produced in a material by bringing it very close to a magnet or inside a magnetic field is called *magnetic induction* or magnetising by induction. For example, soft iron becomes temporarily magnetised when brought in contact with or close to a permanent magnet or any magnetic field. As an example, small iron nails become temporarily magnetised when attracted to the ends of a magnet. Each nail becomes a small magnet by induction. When removed away from the magnet, they lose their magnetism very rapidly.

(b) Residual magnetism - it is the magnetism which remains behind after the external magnetising force is removed.

Retentivity - it is the ability of a magnetic material to retain its magnetism after the external magnetic field is removed.

When a strong magnetic field is applied to a piece of hard steel or alloy such as ALNICO (combination of aluminium, nickel and cobalt), its atoms align themselves with the external field. Hence, hard steel becomes a magnet. When the piece of hard steel is removed from the external field, nearly all of its atoms *remain* in their magnetised position. This magnetism which remains even after removing the external magnetising field is called *residual magnetism*. And the property of a magnetic material to keep back its induced magnetism is called its *retentivity*.

Permanent magnets have high retentivity and, therefore, a large amount of residual magnetism. On the other hand, soft iron has very little retentivity and can be only temporarily magnetised.

Q. 11. What is meant by the permeability of a medium? What is its electrical counterpart? Why do magnetic materials have two permeabilities? Mention two highly magnetisable materials.

Ans. The *ease* with which a magnetic field can be established in a given material is called the permeability (μ) of that material. The higher the value of μ, the more easily a magnetic field can be established in the material. Alternatively, the ability of a material to *concentrate magnetic flux is called permeability*. Any material that is easily magnetised tends to concentrate magnetic flux. Because soft iron is easily magnetised, we say that it has a high permeability. Numerical values of μ for differnt materials are allotted by comparing their permeability with the permeability of air or vacuum. It is similar to conductivity (σ) in electrical circuits.

Since the permeability of every magnetic medium is compared with the permeability of air, each material has two permeabilities:

(*i*) **Relative permeability (μ_r)** : It is given by the ratio of the permeability of the material and permeability of air.

$$\therefore \qquad \mu_r = \frac{\mu}{\mu_0}$$

where μ_0 is the permeability of air or vacuum. It may be noted that μ_r has no units.

Materials like iron, steel, cobalt and nickel have values of μ_r ranging from 100 to 1,00,000.

(*ii*) **Absolute permeability (μ) :**

It is given by $\mu = \mu_0\mu_r$

It is also given by $\mu = B/H$

where H is the magnetising force and B is the flux density in the material.

It is called the absolute perbeability of a material. Since relative permeability of air with respect to itself is unity, hence its absolute permeability is μ_0. Its value is $4\pi \times 10^{-7}$ henry/metre.

4.3 MAGNETIC CIRCUITS

Q. 12 (*a*) What is a magnetic circuit? Define magnetomotive force (mmf) and magnetising force or magnetic field intensity (H).

(*b*) Find the magnetic field intensity in the magnetic circuit of Fig. 13. The mean path of the magnetic circuit is as indicated in the figure.

Ans. (*a*) The closed path of the magnetic lines of force or flux is called a *magnetic circuit*. Fig. 12 shows a magnetic circuit made up of soft iron in the form of a toroid. The magnetic flux is produced by a coil of N turns carrying a current of I wound on the toroid. The length of the magnetic path is given by the mean circumference l of the toroid. Magnetomotive force is given by the product NI. It is the *driving force* which produces the magnetic flux Φ. The magnetic field intensity (or magnetic field strength) is given by mmf per unit length.

$$\therefore \quad H = \frac{\text{mmf}}{l} = \frac{NI}{l} \text{ ampere/metre (since } N \text{ has no units)}$$

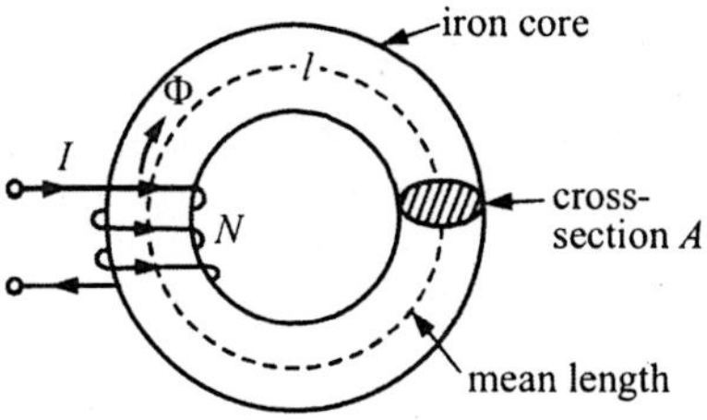

Fig. 12

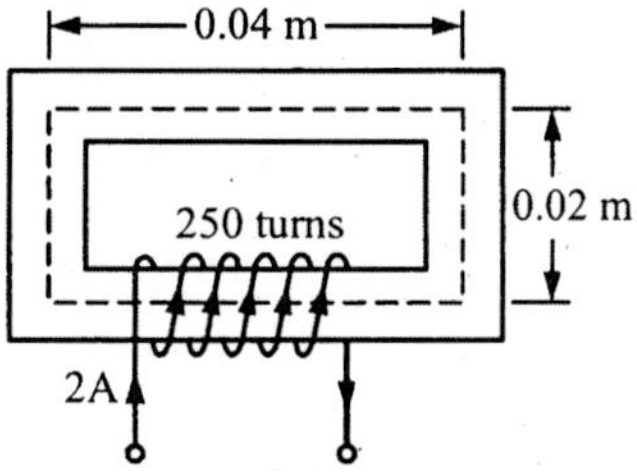

Fig. 13

(*b*) $\text{mmf} = NI = 250 \times 2 = 500$ ampere-turns (AT)

Length of the mean magnetic path (Fig. 13).

$$= 2\,(0.04 + 0.02) = 0.12 \text{ m}$$

$$\therefore \quad H = \frac{\text{mmf}}{l} = \frac{500}{0.12} = \mathbf{41.7 \times 10^2 \ AT/m}$$

Q. 13 (*a*) State Ohm's law for a magnetic circuit.

(*b*) The simple magnetic circuit of Fig. 14 has a cross-sectional area of 50 cm^2 and mean length of 2 m. The relative permeability of the core is 100. The coil has 250 turns and the flux produced is 100 μWb. Find:

(*i*) reluctance of the magnetic circuit

(*ii*) current flowing in the coil.

Ans. (*a*) Fig. 12 of Q. 12 shows a simple magnetic circuit. A magnetic circuit is exactly like an electric circuit. An electric circuit has three things:

(*i*) a source of emf

(*ii*) circuit current and

(*iii*) circuit resistance

Similarly, a magnetic circuit has three things

(*i*) Magnetomotive force (mmf) provided by a current-carrying coil.

(*ii*) Magnetic flux Φ produced by the magnetomotive force.

(*iii*) reluctance of the magnetic circuit which opposes the production of flux in the core.

The relation between the above three magnetic quantities is known as Ohm's law of magnetic circuit. It is given by

$$\Phi = \frac{\text{mmf}}{\text{reluctance}} \quad \therefore \quad \Phi = \frac{NI}{S} = \frac{NI}{l/\mu_0\mu_r A} \text{ weber}$$

As seen, $\text{mmf} = NI$

$$\text{reluctance, } S = \frac{l}{\mu A} = \frac{l}{\mu_0\,\mu_r\,A}$$

(*b*) $\mu = \mu_r\,\mu_0 = 100 \times 4\pi \times 10^{-7} = 12.56 \times 10^{-5}$

$$\therefore \text{ Reluctance, } S = \frac{l}{\mu A} = \frac{2}{12.56 \times 10^{-5} \times (50 \times 10^{-4})}$$

$$= \mathbf{3.18 \times 10^6 \ AT/Wb}$$

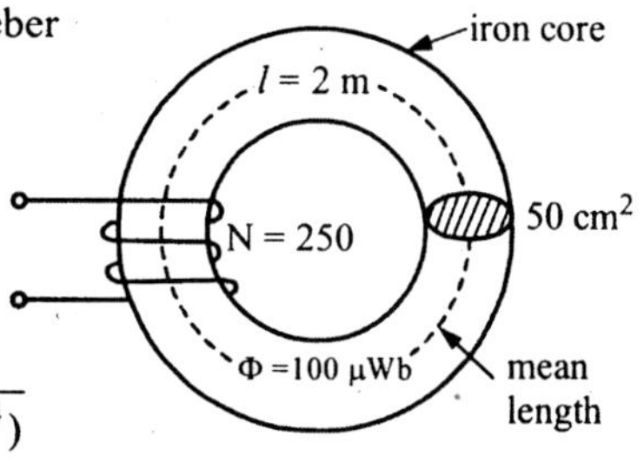

Fig. 14

Now, $\Phi = \frac{\text{mmf}}{S}$

$\therefore$ $\text{mmf} = \Phi \times S = 100 \times 10^{-6} \times 3.18 \times 10^6 = 318$ AT

Also, $\text{mmf} = NI$ or $318 = 250 \times I$ $\therefore$ $I =$ **1.27 A.**

Q. 14. The air gap in a magnetic circuit is 1.5 mm long and 2500 mm² in cross-sectional area. Calculate

(*i*) the reluctance of the air gap.

(*ii*) the mmf required to set up a flux of 800 × 10⁻⁶ Wb. in the air gap.

(*Basic Elect. Engg., B. Tech, JNTU, Hyderabad 2007*)

Sol. (*i*) $S = \frac{l}{\mu_0 \mu_r A} = \frac{l}{\mu_0 A}$ — $\because$ for air $\mu_r = 1$

$$= \frac{1.5 \times 10^{-3}}{4\pi \times 10^{-7} \times 2500 \times 10^{-6}} = \frac{1.5 \times 10^6}{\pi} = \mathbf{4.77 \times 10^5 \text{ AT/Wb}}$$

(*ii*) $\text{mmf reqd.} = \Phi S = 800 \times 10^{-6} \times 4.77 \times 10^5 =$ **381.6 AT**

Q. 14. Define the following terms along with their units:

(*i*) magnetic field strength (*ii*) magnetic flux (*iii*) flux density (*iv*) magnetomotive force (*v*) reluctance (*vi*) permeance.

Ans. (*i*) Magnetic field strength (*H*) - it is the magnetising force that produces flux in a given magnetic circuit. It is defined as magnetomotive force per unit length of the magnetic circuit. If in a given magnetic circuit, mmf = NI and l, the average length of the magnetic path then.

$$H = \frac{\text{mmf}}{\text{length}} = \frac{NI}{l} \text{ ampere/metre (or ampere – turns/metre)}$$

(*ii*) **Magnetic flux (Φ)** - it equals the entire number of lines of force produced by a magnet. The more the lines of force, the greater the flux and stronger the magnetic field. The unit of magnetic flux is weber (Wb). One weber equals 10^8 lines.

(*iii*) **Flux density (*B*)** - the flux density B is given by the flux Φ passing normally through a surface area A of the magnetic field

$\therefore$ $B = \frac{\Phi}{A}$ — Fig. 15

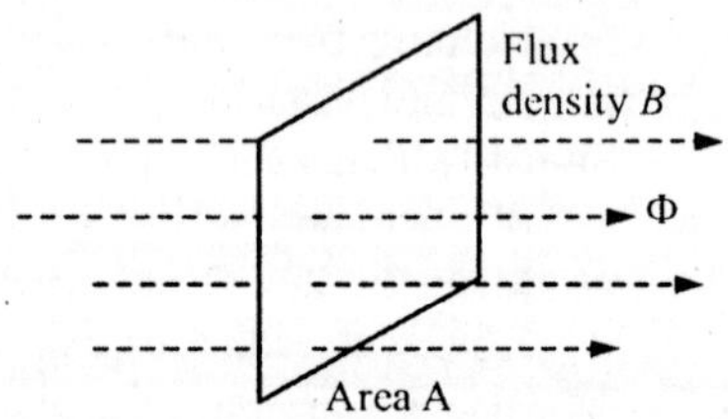

Fig. 15

Its unit is Wb/m² or tesla (T).

(*iv*) **Magnetomotive force** - it is the force produced by current flowing through a coil of wire. Greater the mmf of a circuit, greater the flux produced by it.

$$\text{mmf} = NI \text{ AT}$$

For example, if a coil has 5000 turns and carries a current of 10 mA, then mmf produced

$$= 5000 \times 10 \times 10^{-3} = 50 \text{ AT.}$$

(*v*) **Reluctance (*S*)** - it is the *opposition* that a magnetic circuit offers to the passage of magnetic flux through it. It is given by the ratio of the mmf and the amount of flux produced

$\therefore$ $S = \frac{\text{mmf}}{\Phi} = \frac{NI}{\Phi}$ AT/Wb

In terms of the magnetic quantities, it is given by,

$$S = \frac{l}{\mu A} = \frac{l}{\mu_0 \mu_r A} \text{ AT/Wb}$$

Since, 1 AT/Wb = 1/ henry, hence unit of reluctance is also 'reciprocal' henry.

(*vi*) **Permeance** - it is reciprocal of reluctance and means the *ease* with which magnetic flux is produced. Its unit is henry or Wb/AT.

Q. 15. Compare magnetic and electric circuits giving their similararities and differences. (*Basic Elec. Engg., West Bengal Tech. Univ., 2009*)

Ans. All magnetic quantities used in magnetic circuits are related to each other. The two circuits have both similarities and differences as given below:

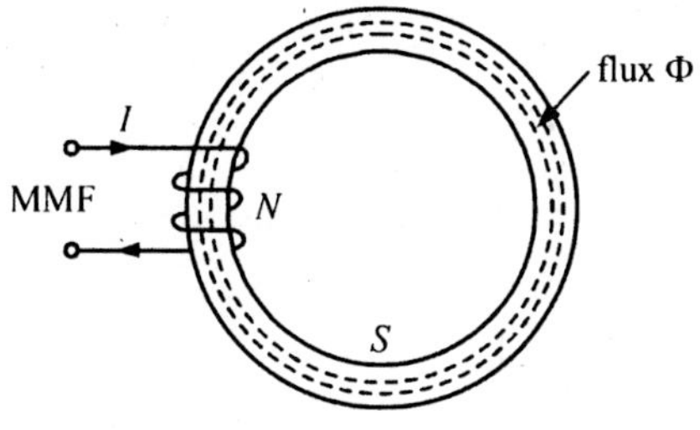

Fig. 16

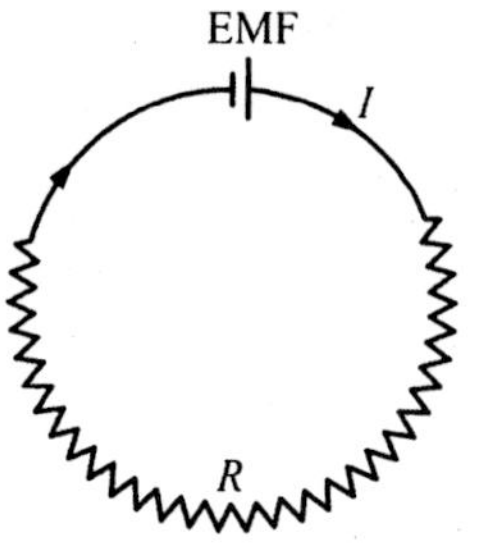

Fig. 17

SIMILARITIES

Magnetic Circuit	Electric Circuit
1. flux = $\dfrac{\text{mmf}}{\text{reluctance}}$	1. current = $\dfrac{\text{e.m.f.}}{\text{resistance}}$
2. mmf (ampere-turn)	2. e.m.f. (volt)
3. flux Φ (Weber)	3. current (ampere)
4. flux density B (Wb/m^2)	4. current density (A/m^2)
5. reluctance $S = \dfrac{l}{\mu A} = \dfrac{l}{\mu_0 \mu_r A}$ (AT/Wb)	5. resistance $R = \rho \dfrac{l}{A} = \dfrac{l}{\sigma A}$ (ohm)
6. magnetic field intensity (AT/m)	6. electrical field intensity (V/m)
7. permeance = 1/reluctance (henry)	7. conductance = 1/resistance (siemens)
8. reluctivity (AT-m/Wb)	8. resistivity (ohm-metre)
9. permeability = 1/reluctivity (Wb/AT-m)	9. conductivity = 1/resistivity (siemens/metre)

DIFFERENCES

1. In a magnetic circuit, flux is just created. It does not 'flow' in the sense in which an electric current flows.
2. If temperature is kept constant, the conductivity of an electric circuit remains constant and is independent of the current flowing through the circuit. On the other hand, the magnetic conductivity *i.e.* permeability of the magnetic circuit depends on the total magnetic flux established through the core.
3. When current flows in an electric circuit, energy is continuously lost in the form of heat. But in a magnetic circuit, energy is needed only for *creating* the flux intially but not for maintaining it.
4. It is not possible to confine magnetic flux to a definite path as accurately as current is confined to a conductor.

Q. 16. Define the terms 'leakage flux' and 'fringing flux' in a magnetic circuit. In which part of the magnetic circuit do they occur?

Ans. When a magnetic circuit has a concentrated winding as shown in Fig. 18, leakage of flux occurs. If air-gaps are also present in the circuit, the leakage of flux is increased. We may define leakage flux as the flux which follows a path *not meant for it.* Even in the best designed magnetic circuits, it is found that 15 to 20% of the total flux produced leaks away without being utilized.

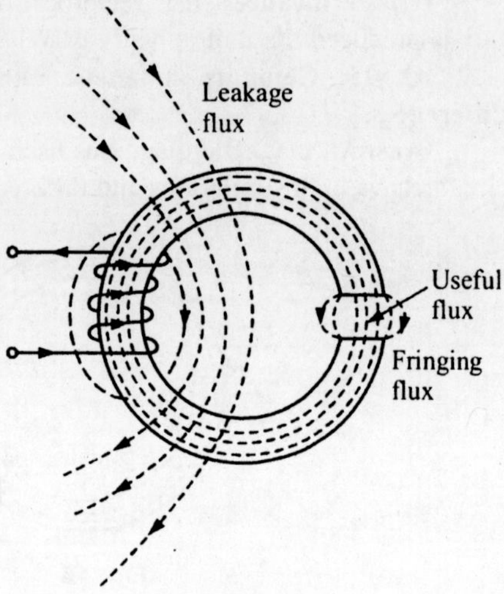

Fig. 18

If Φ_t = total flux produced

Φ = useful flux available in the air-gap

then $\lambda = \dfrac{\text{total flux}}{\text{useful flux}} = \dfrac{\Phi_t}{\Phi}$

In electric machines like motors and generators, leakage flux is undesirable. It lowers their power efficiency and also increases their weight and cost of manufacturing. Magnetic leakage can be minimised by placing the windings very close to the air-gap or where flux is to be utilised.

It is also seen from Fig. 18 that there is *fringing* or *spreading* of flux lines at the edges of the air-gap. It increases the effective area of the air-gap thereby reducing the flux density in it.

Naturally, every effort is made to keep air-gaps in magnetic circuits as short as possible in order to reduce leakage, fringing and the total reluctance of the circuit.

Leakage flux takes place throughout the length of the core. However, fringing of flux takes place at the air-gaps only.

Q. 17 (*a*) Give one example each of

(*i*) series magnetic circuit

(*ii*) parallel magnetic circuit and

(*iii*) series-parallel magnetic circuit

(*b*) A rectangular shaped core is made of mild steel plates 15 mm × 20 mm cross-section. The mean length of the magnetic path is 18 cm. The exciting coil has 300 turns and current 0.7 A. Calculate :

(*i*) Magnetizing force **(*ii*) Flux density**

(*iii*) Reluctance **(*iv*) Flux of the magnetic circuit.**

Assuming relative permeability of mild steel is 940.

(*Basic Elect. Engg., R.G.P.V., Bhopal, 2006.*)

Ans. (*a*) (*i*) **Series magnetic circuit** - a simple series magnetic circuit and its equivalent electrical circuit are shown in Fig. 19(*a*). The average length of a magnetic circuit is shown by the dashed line. The mmf of the magnetic circuit is '*NI*', flux produced is Φ and the magnetic path has a reluctance of $l/\mu A = l/\mu_0\mu_r A$. The mmf of the magnetic circuit is similar to the emf of the electric circuit, flux Φ is like electric current *I* and reluctance *S* is like resistance *R* of the electric circuit.

(*ii*) **Parallel magnetic circuit** - Fig. 19(*b*) shows a parallel magnetic circuit having two paths of equal reluctance (reluctance of the central core has been neglected). Hence, the total flux Φ divides into two equal parts. In its equivalent circuit, *I* divides into two equal parts because the two parallel paths have equal resistance of *R*.

(*iii*) **Series-parallel magnetic circuit** - it is shown in Fig. 19(*c*). Each parallel path is identical. The central core consists of an air gap which has high reluctance. Again, magnetic flux is divided equally between the two parallel paths.

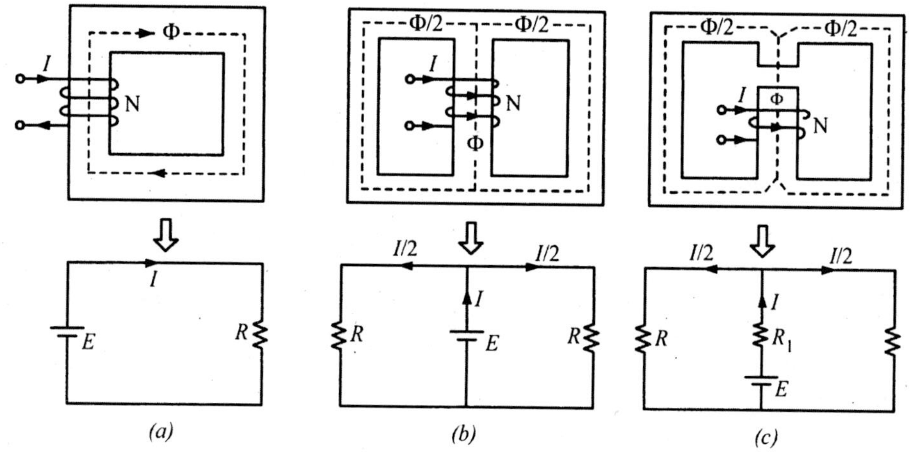

Fig. 19

(b) (*i*) Magnetizing force, $H = NI/l$

Now, $N = 300,\ I = 0.7$ A, $l = 18$ cm $= 0.18$ m

$\therefore$ $H = 300 \times 0.7/0.18 =$ **116.7 AT/m**

(*ii*) Flux density, $B = \mu_0 \mu_r H = 4\pi \times 10^{-7} \times 940 \times 1166.7$

$=$ **1.38 Wb/m²**

(*iii*) Reluctance,

$$S = \frac{l}{\mu_0 \mu_r A}$$

$$= \frac{18 \times 10^{-2}}{4\pi \times 10^{-7} \times 940 \times 15 \times 20 \times 10^{-6}}$$

$= \mathbf{5.08 \times 10^5}$ **AT/Wb**

(*iv*) $\Phi = B \times A$

$A = 15 \times 20 = 300 \text{ mm}^2 = 300 \times 10^{-6} \text{ m}^2 = 3 \times 10^{-4} \text{ m}^2$

$\therefore$ $\Phi = 1.38 \times 3 \times 10^{-4} = \mathbf{0.414 \times 10^{-3}}$ **Wb**

Q. 18. A mild steel ring having a cross-sectional area of 500 mm² and a mean circumference of 400 mm has a coil of 200 turns wound uniformly around it.

Calculate :

(*i*) the reluctance of the ring and

(*ii*) the current required to produce a flux of 800 μWb in the ring.

Take relative permeability of mild steel as 400 at the given flux density.

(*Elect. Engg. 1, Punjab Univ.*)

(*b*) A ring of 30 cm mean diameter is made up of round iron rod 2.5 cm in diameter. At one end, a saw-cut 1 mm wide is made through it. It is uniformly wound with 500 turns of wire. Calculate the current required by the exciting coil to produce a total flux of 4 mWb. Assume a relative permeability of iron at this flux density as 800. Neglect leakage and fringing. (*Basic Elect. Engg. R.G.P.V., Bhopal, 2008*)

Sol. (*i*)

$$S = \frac{1}{\mu_0 \mu_r A} = \frac{400 \times 10^{-3}}{4\pi \times 10^{-7} \times 400 \times 500 \times 10^{-6}} = \mathbf{1.6 \times 10^6\ AT/Wb}$$

(*ii*)

$$\Phi = \frac{MMF}{S} = \frac{NI}{S}$$

$\therefore$ $NI = \Phi \times S = 800 \times 10^{-6} \times 1.6 \times 10^6 = 1280$

$I = 1280/200 =$ **6.4 A**

(*b*) The magnetic circuit is shown in Fig. 20.

$$\Phi = 4 \text{ mWb} = 4 \times 10^{-3} \text{ Wb}$$

$$A = \frac{\pi}{4} \times (2.5)^2 = 4.91 \text{ cm}^2$$

$$= 4.91 \times 10^{-4} \text{ m}^2$$

$$B = \Phi/A = 4 \times 10^{-3}/4.91 \times 10^{-4}$$

$$= 8.15 \text{ Wb/m}^2$$

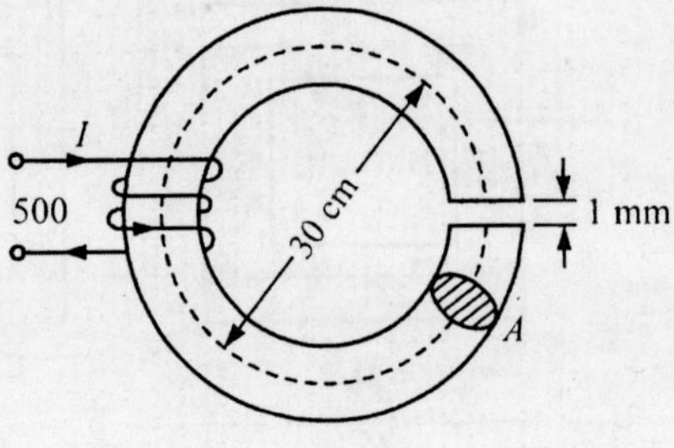

Fig. 20

Air gap

$$H = B/\mu_0 = 8.15/4\pi \times 10^{-7}$$

$$= 6487 \times 10^3 \text{ AT/m}$$

$$l = 1 \text{ mm} = 1 \times 10^{-3} \text{ m}$$

$$\text{AT reqd. for air gap} = \text{H} \times l = 6487.4 \times 10^3 \times 1 \times 10^{-3} = 6487$$

Iron ring

$$H = B/\mu_0\mu_r = 8.15/4\pi \times 10^{-7} \times 800 = 8107$$

Length of the iron ring = $\pi \times 30$ = 94.25 cm = 0.9425 m

Net length of the magnetic flux path in the ring

$$= 0.9425 - 0.001 = 0.9415 \text{ m}$$

$$\text{AT reqd.} = H \times l = 8107 \times 0.9415 = 7633$$

$$\text{Total AT reqd.} = 7633 + 6487 = 14{,}120$$

No. of turns in the coil = 500

Hence, current in the coil, $I = 14120/500 =$ **28.24 A**

Q. 19 (*a*) An iron ring of mean circumference 50 cm has an air-gap of 0.1 cm and a winding of 300 turns. If the relative permeability of iron is 400 when a current of 1 ampere flows through the coil, find the flux density in the air-gap.

(*Basic Elect. Engg., West Bengal Tech. Univ., 2001*)

(*b*) A cast steel structure is made of a rod of square section 2.5 cm × 2.5 cm as shown in Fig. 22. What is the current that should be passed in a 500 turn coil on the left limb, so that a flux of 2.5 mWb is made to pass in the right limb. Assume permeability as 750 and neglect leakage. (*Elect. Circuits, B.Tech JNTU, Hyderabad 2006*)

Ans. (*i*) (*i*) Iron Ring :

Let B be the flux density in the iron ring (Fig. 21). Its value would also be the same in the air gap. The value of the magnetizing force is

$$H = B/\mu_0 \mu_r$$

$$= B/4\pi \times 10^{-7} \times 400$$

$$= 1990 B$$

$$\text{AT reqd.} = H \times l = 1990 B \times 50 \times 10^{-2}$$

$$= 995 B$$

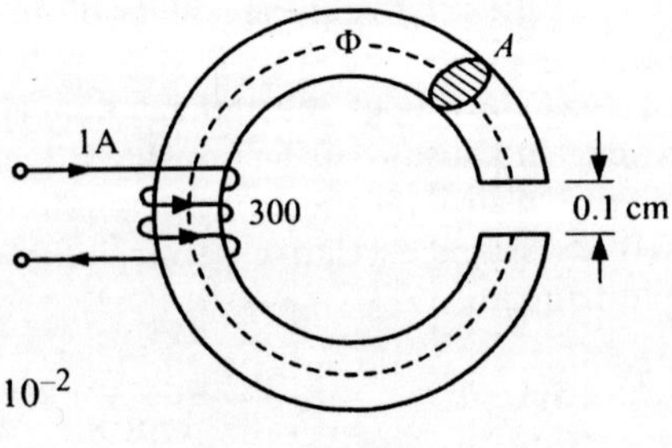

Fig. 21

(*ii*) **Air Gap :**

$$H = B/\mu_0 = B/4\pi \times 10^{-7}$$

$$= 796 \times 10^3 B$$

$$\text{AT reqd.} = H \times l = 796 \times 10^3 B \times 0.1 \times 10^{-2}$$

$$= 796 B$$

Total No. of AT reqd.

$$= 995 B + 796 B$$

$$= 1791 B$$

No. of given AT = 300 × 1 = 300

∴ $1791 B = 300$, $B =$ **0.165 Wb/m²**

(*b*) As seen from Fig. 22, the two magnetic paths *C* and *D* are in parallel with the path *E*. Obviously, the same mmf ats across the paths *C* and *D* and produces flux Φ_1 in path *C* and Φ_2 in path *D*.

Now, $\text{mmf} = \Phi \times S$ – where *S* is the reluctance

Considering the two parallel paths *C* and *D*, we have

$$\Phi_1 S_1 = \Phi_2 S_2$$

Now $$S_1 = \frac{l}{\mu A} = \frac{25}{\mu A} \text{ and } S_2 = \frac{40}{\mu A}$$

$$\therefore \quad \Phi_1 \times \frac{25}{\mu A} = \Phi_2 \times \frac{40}{\mu A} \text{ or } \Phi_1 \times \frac{25}{\mu A} = 2.5 \times \frac{40}{\mu A}$$

$$\therefore \quad \Phi_1 = 4 \text{ mWb}$$

Hence, the flux Φ through the magnetic path *E* is given by the sum of Φ_1 and Φ_2

$$\therefore \quad \Phi = \Phi_1 + \Phi_2 = 4 + 2.5 = 6.5 \text{ m Wb}$$

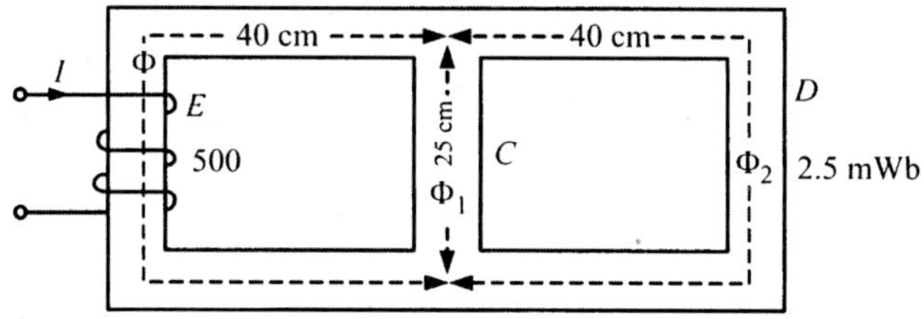

Fig. 22

The mmf required for the whole magnetic circuit is equal to the sum of

(*i*) the mmf-1 required for path *E*

(*ii*) the mmf-2 for the parallel path *C* or *D*

$$\therefore \quad \text{mmf reqd.} = \text{mmf-1} + \text{mmf-2} = \Phi S + \Phi_2 S_2$$

Since the lengths of the path *E* and *D* is the same, hence

$$S = S_2$$

Now $$S = \frac{l}{\mu_0 \mu_r A} = \frac{0.4}{4\pi \times 10^{-7} \times 750 \times 6.25 \times 10^{-4}} = 679 \times 10^3$$

Since $S = S_2$, hence $S_2 = 679 \times 10^3$

$$\therefore \quad \text{mmf} = 6.5 \times 10^{-3} \times 679 \times 10^3 + 2.5 \times 10^{-3} \times 679 \times 10^3 = 6111 \text{ AT}$$

$$NI = 6111$$

$$\therefore \quad NI = 6111, \quad \therefore \quad I = 6111/500 = \mathbf{12.2\ A}$$

Q. 20 (*a*) An iron ring of mean length 50 cm has an air gap of 3 mm and a winding of 200 turns. If the permeability of iron core is 400 and the winding carries a current of 1.5 A, calculate the value of the flux density.

(*Basic. Elect. Engg., B. Tech, JNTU, Hyderabad 2007*)

(*b*) A cast steel ring of mean circumference 50 cm has a cross-section of 0.52 cm². (Fig. 24). It has a saw-cut of 1 mm at one place. Given the following data –

***B* (Wb/m²)**	**1.0**	**1.25**	**1.46**	**1.60**
μ_r	**714**	**520**	**360**	**247**

Calculate how many ampere-turns are required to produce a flux of 0.052 mWb in air gap if leakage factor is 1.2. (*Basic. Elect. Engg, R.G.P.V, Bhopal, June 2007*)

Ans. (*a*) Let *A* be the area of iron path as well as that of the air-gap. (Fig. 23)

If S_1 is the reluctance of the iron path and S_2 that of the air gap, then total reluctance is $S = (S_1 + S_2)$.

Now, $\quad \text{mmf} = \Phi S = \Phi\,(S_1 + S_2)$,

$$\therefore \quad \Phi = \frac{\text{mmf}}{(S_1 + S_2)}$$

Now, $\quad S_1 = \dfrac{l_1}{\mu_0 \mu_r A}$

and $\quad S_2 = \dfrac{l_2}{\mu_0 A} \quad - \because$ for air, $\mu_r = 1$

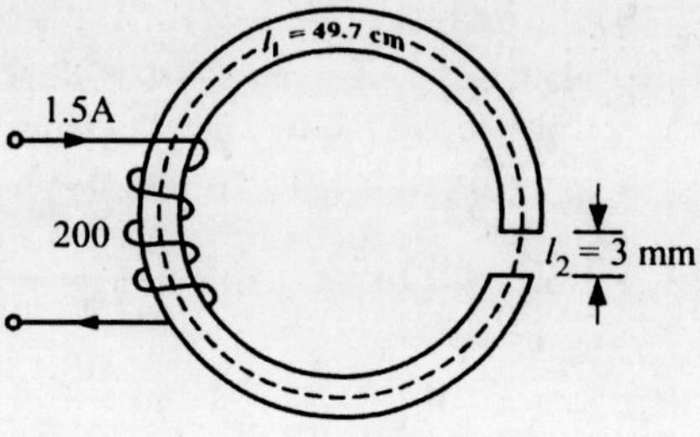

Fig. 23

$$\therefore \quad \Phi = \frac{\text{mmf}}{\dfrac{l_1}{\mu_0\mu_r A} + \dfrac{l_2}{\mu_0 A}} = \mu_0\mu_r A \times \frac{\text{mmf}}{l_1 + \mu_r l_2}$$

Now $\quad \Phi = BA \quad$ – where B is the flux density

$$\therefore \quad BA = \mu_0\mu_r A \times \frac{\text{mmf}}{l_1 + \mu_r l_2}$$

or

$$B = \mu_0\mu_r \times \frac{\text{mmf}}{l_1 + \mu_r l_2}$$

$$= 4\pi \times 10^{-7} \times 400 \times \frac{200 \times 1.5}{(50 - 0.003) + 400 \times 3 \times 10^{-3}}$$

$$= \frac{48\pi \times 10^{-3}}{12.497} = 3.84 \times 10^{-3} \text{ Wb/m}^2 = \mathbf{3.84 \ mWb/m^2}$$

(*b*) Air gap

$\Phi = 0.052$ mWb

$= 0.052 \times 10^{-3}$ Wb

Area of cross-section

$= 0.52 \times 10^{-4}$ m^2

$B = \Phi/A = 0.052 \times 10^{-3}/0.52 \times 10^{-4}$

$= 1$ Wb/m^2

$\therefore \quad H = B/\mu_0 = 1/4\pi \times 10^{-7}$

$= 796 \times 10^3$ AT/m

$l = 1$ mm $= 1 \times 10^{-3}$ m

AT reqd. $= H \times l = 796 \times 10^3 \times 1 \times 10^{-3} = 796$

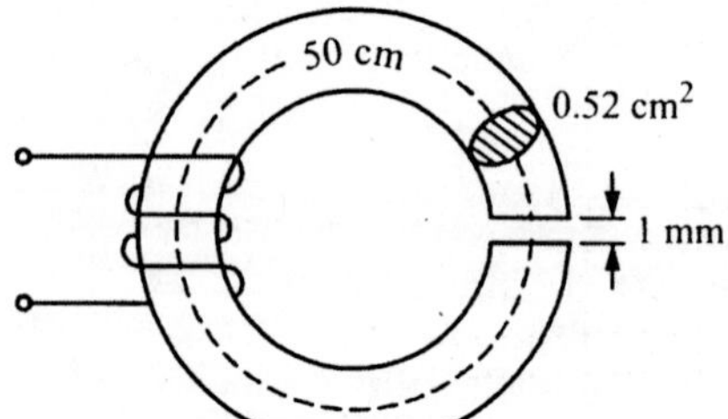

Fig. 24

Cast steel ring :

$\Phi = 1.2 \times$ air-gap flux $= 1.2 \times 0.052 = 0.0624$ mWb

$= 6.24 \times 10^{-5}$ Wb

$A = 0.52$ cm^2 $= 0.52 \times 10^{-4}$ m^2

$\therefore \quad B = \Phi/A = 6.24 \times 10^{-5}/0.52 \times 10^{-4} = 1.2$ Wb/m^2

If we draw the B/μ_r curve from the given data, we find that corresponding to $B = 1.2$ Wb/m^2, value of $\mu_r = 565$.

$$\therefore \quad H \text{ in cast steel} = \frac{B}{\mu_0\mu_r} = \frac{1.2}{4\pi \times 10^{-7} \times 565} = 1690$$

$l = 50$ cm $= 0.5$ m $\quad$ – neglecting length of the air gap

AT reqd. $= H \times l = 1690 \times 0.5 = 845$

Total AT reqd. $= 845 + 796 =$ **1641**

Q. 21 (*a*) An iron ring of mean circumference 50 cm has an air-gap of 1 mm (Fig. 25). It is uniformly wound with a coil having 200 turns. If the relative permeability of iron is 300 when a current of 1A flows through the coil, calculate the flux density in the air-gap. Neglect leakage and fringing. (*Basic Elect. Engg., R.G.P.V., Bhopal 2008*)

(*b*) The magnetic circuit of Fig. 26 has cast steel core with dimensions as shown.

Mean length from *A* to *B* through either outer limb = 0.5 m

Mean length from *A* to *B* through the central limb = 0.2 m.

In the magnetic circuit shown, it is required to establish a flux of 0.75 mWb in the air-gap of the central limb. Determine the mmf of the exciting coil if for the core material relative permeability, μ_r = 5000. Neglect fringing.

(*Basic. Elect. Engg., R.G.P.V., Bhopal, Dec. 2003*)

Ans. (*a*) Air gap :

$$H = \frac{B}{\mu_0\mu_r} = \frac{B}{4\pi \times 10^{-7}}$$

$$= B \times 0.796 \times 10^6 \text{ AT/m}$$

$$\text{AT reqd.} = H \times l$$

$$= B \times 0.796 \times 10^6 \times 1 \times 10^{-3}$$

$$= 796 \text{ B}$$

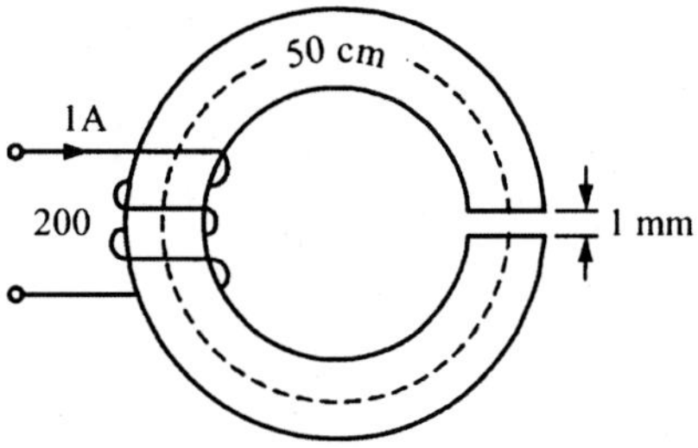

Fig. 25

Iron ring :

$$H = \frac{B}{\mu_0\mu_r A} = \frac{0.5}{4\pi \times 10^{-7} \times 300}$$

$$= B \times 2.65 \times 10^3 \text{ AT/m}$$

$$l = 50 - 0.1 = 49.9 \text{ cm} = 49.9 \times 10^{-2} \text{ m}$$

$$\text{AT reqd.} = H \times l = B \times 2.65 \times 10^3 \times 49.9 \times 10^{-2} = 1322\, B$$

$$\text{Total AT reqd.} = 796\, B + 1322\, B = 2118\, B \text{ AT}$$

AT provided by the coil = $NI = 200 \times 1 = 200$ AT

$\therefore \quad 200 = 2118\, B \quad \therefore \quad B = 200/2118 =$ **0.094 Wb/m²**

(*b*) As seen from Fig. 26, the flux Φ in the central core divides into two equal parts at point *A* because the two parallel paths have equal reluctances. Total AT required for this magnetic circuit will be equal to

(*i*) AT required for central core

(*ii*) AT required for one parallel path and

(*iii*) AT required for three air gaps.

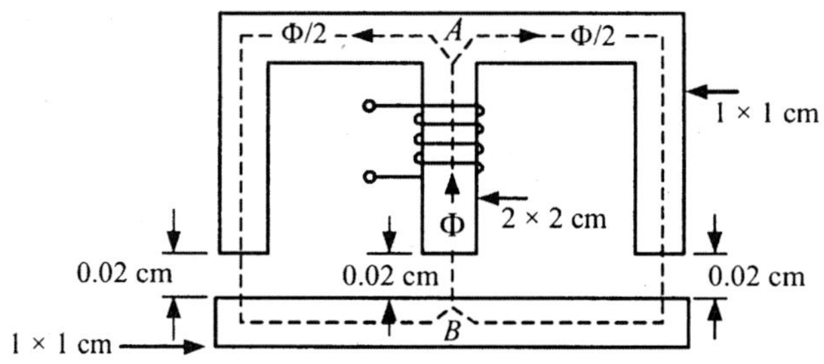

Fig. 26

Flux in the central core = 0.75×10^{-3} Wb

Flux in the outer cores = $0.75 \times 10^{-3}/2 = 3.75 \times 10^{-4}$ Wb

(*i*) **Central Core :**

$$A = 2 \times 10^{-2} \times 2 \times 10^{-2} = 4 \times 10^{-4} \text{ m}^2$$

$$S = \frac{l}{\mu_0 \mu_r A} = \frac{0.2}{4\pi \times 10^{-7} \times 5000 \times 4 \times 10^{-4}} = 79.58 \times 10^3 \text{ AT/Wb}$$

(*ii*) **Outer Limb :**

We need consider only one outer limb.

Here, $l = 0.5$ m, $A = 1 \times 10^{-2} \times 1 \times 10^{-2} = 10^{-4}$ m^2, $\mu_r = 5000$

$$S = \frac{0.5}{4\pi \times 10^{-7} \times 10^{-4} \times 5000} = 795.77 \times 10^3 \text{ AT/Wb}$$

(*iii*) **Air gaps :**

Length of the mean path of flux in the three air gaps

$= 0.02 + 0.02 + 0.02 = 0.06$ cm $= 6 \times 10^{-4}$ m

$$S = \frac{l}{\mu_0 A} = \frac{6 \times 10^{-4}}{4\pi \times 10^{-7} \times 4 \times 10^{-4}} = 1193.66 \times 10^3 \text{ AT/Wb}$$

Total reluctance of the magnetic circuit is

$= 79.58 \times 10^3 + 795.77 \times 10^3 + 1193.66 \times 10^3$

$= 2069 \times 10^3$ AT/Wb

Now, $\Phi = \frac{\text{mmf}}{S}$ or mmf = flux × reluctance

$\therefore$ mmf $= \Phi \times S = 0.75 \times 10^{-3} \times 2069 \times 10^3 =$ **1551.75 AT**

Q. 22. (*a*) The magnetic circuit shown in Fig. 27 is built-up of iron of square cross-section 2 cm wide. Each air-gap is 2 mm wide, μ for part *A* may be taken as 900 and for *B* as 1250. Each of the exciting coil is wound with 1200 turns and the exciting coil is carrying current of 0.8 A. Find :

(*i*) Reluctance of *A* **(*ii*) Reluctance of *B***

(*iii*) Reluctance of air-gaps **(*iv*) Total reluctance**

(*v*) Total flux **(*vi*) Flux density**

(*Basic. Elect. Engg., R.G. P.V., Bhopal, June 2007*)

(*b*) An iron ring of mean length 30 cm is made of three pieces of iron, each has the same length but their respective diameters are 3.9, 3.2 and 2.5 cm (Fig. 28). An air gap of length 0.5 mm is cut in the 2.5 cm diameter piece. If a coil of 1000 turns is wound on the ring, find the value of current it has to carry to produce a flux density of 0.5 Wb/m^2 in the air gap.

The *B-H* characteristics of iron is as follows :

B (Wb/m^2)	**0.1**	**0.2**	**0.3**	**0.4**	**0.5**	**0.6**
μ_r	**280**	**620**	**990**	**1400**	**2000**	**2800**

(*Elect. Engg., Nagpur Univ. 2006*)

Ans. (*a*) (*i*) Reluctance of part *A*, $S_A = \frac{l_A}{\mu_0 \mu_r A}$

Now, $l_A = 22 - (1 + 1) = 20$ cm $= 0.2$ m

$A = 2 \times 2 = 4$ cm^2 $= 4 \times 10^{-4}$ m^2

$$\therefore \quad S_A = \frac{0.2}{4\pi \times 10^{-7} \times 900 \times 4 \times 10^{-4}} = \mathbf{44.21 \times 10^4 \text{ AT/Wb}}$$

(*ii*) Reluctance of part *B*, $S_B = \frac{l_B}{\mu_0 \mu_r A}$

$l_B = 22 - (2 \times 1) + (11 - 1) + (11 - 1) = 40$ cm $= 0.4$ m

$$S_B = \frac{0.4}{4\pi \times 10^{-7} \times 1250 \times 4 \times 10^{-4}}$$

$= \mathbf{63.66 \times 10^4}$ **AT/Wb**

(*iii*) Length of the two air gaps
$= 2 + 2 = 4$ mm
$= 4 \times 10^{-3}$ m

$\mu_r = 1$

$$\therefore \quad S_g = \frac{4 \times 10^{-3}}{4\pi \times 10^{-7} \times 4 \times 10^{-4} \times 1}$$

$= \mathbf{795.77 \times 10^4}$ **AT/Wb**

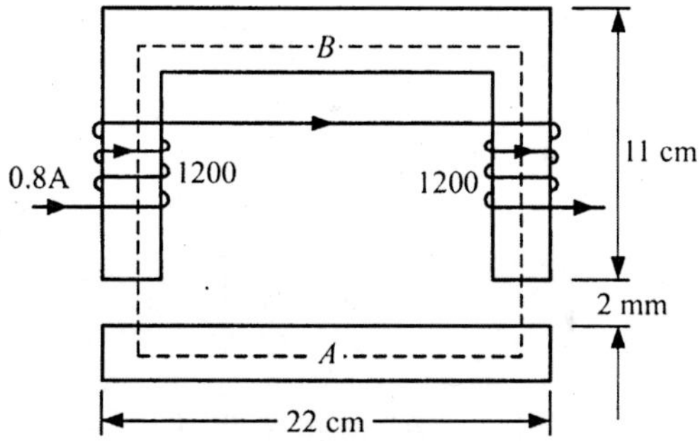

Fig. 27

(*iv*) Total reluctance of the complete circuit is given by

$$S = S_A + S_B + S_g$$
$$= (44.21 + 63.66 + 795.77) \times 10^4$$
$$= \mathbf{903.64 \times 10^4 \text{ AT/Wb}}$$

(*v*) Total flux, $\Phi = \dfrac{\text{total mmf}}{\text{total reluctance}}$

Now, total mmf $= 1200 \times 0.8 + 1200 \times 0.8 = 1920$ AT

$\therefore$ Total flux, $\Phi = \dfrac{1920}{903.64 \times 10^4} = \mathbf{2.124 \times 10^{-4}}$ **Wb**

(*vi*) Flux density, $B = \dfrac{\Phi}{A} = \dfrac{2.124 \times 10^{-4}}{4 \times 10^{-4}} = \mathbf{0.531}$ **Wb/m²**

(***b***) It is seen from Fig. 28 that the three iron paths have the same length of 10 cm *i.e.* 0.1 m but have different diameters. From the given flux density in the air gap, we can find the value of magnetic flux which will be the same in all parts of the iron ring.

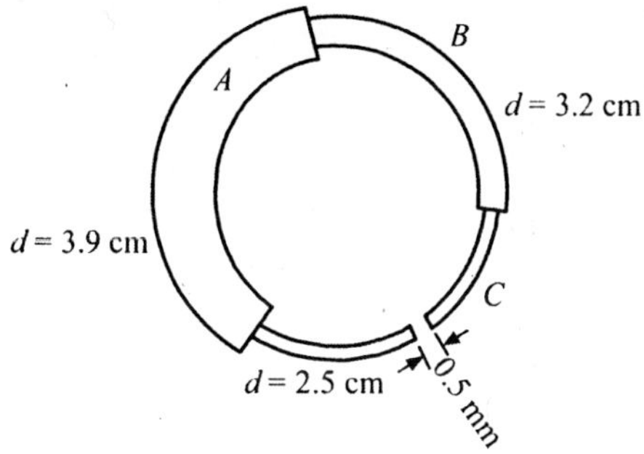

Fig. 28

For air gap :

$$B = \mu_0 H$$

or $$H = \frac{B}{\mu_0} = \frac{0.5}{4\pi \times 10^{-7}}$$
$$= 397.89 \times 10^3 \text{ AT/m}$$

AT reqd. $= H \times l = 397.89 \times 10^3 \times 0.5 \times 10^{-3} = 199$ AT

Now, $$A = \frac{\pi (d)^2}{4} = \frac{\pi (2.5 \times 10^{-2})^2}{4} = 4.9 \times 10^{-4} \text{ m}^2$$

$$\Phi = B \times A = 0.5 \times 4.9 \times 10^{-4} = 2.45 \times 10^{-4} \text{ Wb}$$

This flux would be the same throughout the ring

For material *A* :

$$A = \frac{\pi (d)^2}{4} = \frac{\pi (3.9 \times 10^{-2})^2}{4} = 1.19 \times 10^{-3} \text{ m}^2$$

$$B = \frac{\Phi}{A} = \frac{2.45 \times 10^{-4}}{1.19 \times 10^{-3}} = 0.2 \text{ Wb/m}^2$$

As seen from the table, for $B = 0.2$ Wb/m^2, the value of $H = 620$

$\therefore$ AT reqd. $= H \times l = 620 \times 0.1 = 62$ AT

For Material *B* :

$$A = \frac{\pi (d)^2}{4} = \frac{\pi (3.2 \times 10^{-2})^2}{4} = 8.04 \times 10^{-4} \text{ m}^2$$

$$B = \frac{\Phi}{A} = \frac{2.45 \times 10^{-4}}{8.04 \times 10^{-4}} = 0.3 \text{ Wb/m}^2$$

From the table, the corresponding value of $H = 990$

$\therefore$ AT reqd. $= 990 \times 0.1 = 99$ AT

For material *C* :

$$A = 4.9 \times 10^{-4} \text{ m}^2 \quad \text{– as calculated earlier for air-gap}$$

$$B = \frac{\Phi}{A} = \frac{2.45 \times 10^{-4}}{4.9 \times 10^{-4}} = 0.5 \text{ Wb/m}^2$$

From the table, for $B = 0.5$ Wb/m^2, $H = 2000$ AT/m

$\therefore$ AT reqd. $= H \times l = 2000 \times 0.1 = 200$ AT

Total AT reqd. $= 199 + 62 + 99 + 200 = 560$ AT

$\therefore$ $NI = 560$ $\quad\therefore\quad$ $I = 560/1000 =$ **0.56 A**

Q. 23. State and explain Ampere's Circuital law for a magnetic circuit.

(*Basic. Elect. Engg., West Bengal Tech. Univ., 2008*)

Ans. This law is similar to Kirchhoff's voltage law for the electrical circuit which states that *the algebraic sum of all voltage rises and drops in any electric circuit is zero.* Consider the magnetic circuit shown in Fig. 29 which consists of three different materials. Like a battery in an electric circuit, the coil provides the mmf for the entire circuit. This mmf is distributed over the three different parts of the circuit. If the three mmf drops over these three parts are mmf_1, mmf_2, mmf_3, then according to this law.

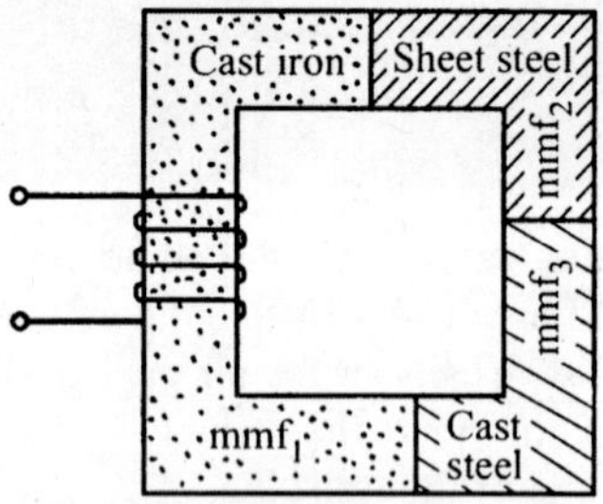

Fig. 29

Sum of the three mmf *drops around magnetic circuit equals the applied* MMF.

$$\therefore \quad MMF = mmf_1 + mmf_2 + mmf_3$$

$$\text{or} \quad MMF - mmf_1 - mmf_2 - mmf_3 = 0 \quad \text{or} \quad \Sigma\, mmf = 0$$

It may be noted that while going around a closed magnetic circuit, one may come across more than one winding. If the windings produce flux in the same direction, then the mmfs due to each are simply added. Otherwise, their algebraic sum is taken when applying Ampere's Circuital law.

4.4 ELECTROMAGNETIC INDUCTION

Q. 24. Explain the phenomenon of Electromagnetic Induction. Name the law which governs it.

Ans. When a moving conductor cuts magnetic lines of flux, an emf is induced in it. This emf is called *induced* emf. The process is shown in Fig. 30.

It is found that whenever conductor *AB* is moved up or down *perpendicular* to the lines of flux, an emf is induced in it *momentarily*. The direction of the induced emf is opposite to each other in the two cases as shown. It may be noted that this emf is produced only if *AB cuts the lines of flux* perpendicularly or at some other angle. For example, if *AB* moves parallel to the lines of flux, it does not cut the magnetic lines, hence no emf is induced in it. It will also be seen that the moving conductor is simply an *open-circuited voltage source* which produces no induced current because there is no closed path.

As shown in Fig. 30(*a*), when *AB* moves *upwards*, induced emf is such that end *A* becomes positive with respect to end *B*.

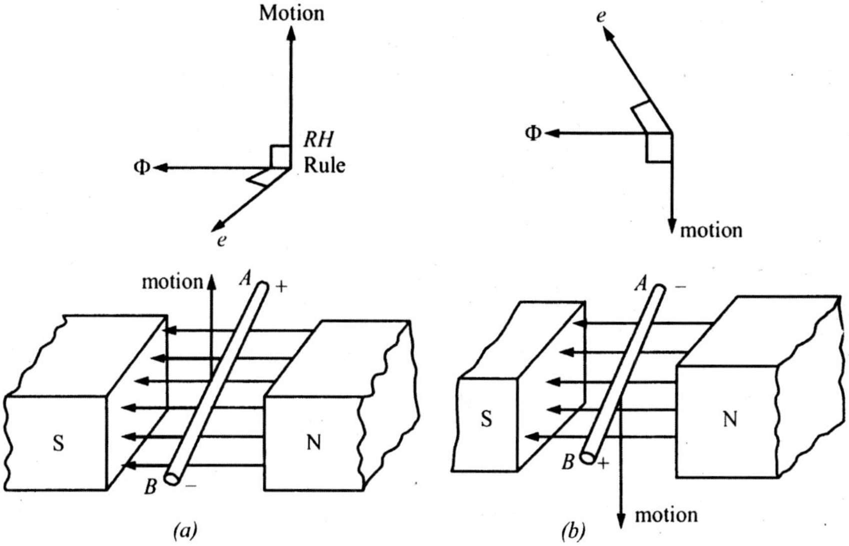

Fig. 30

Fig. 30(*b*) shows the conductor *AB* moving *downwards* at right-angles to the lines of flux. In this case, the induced voltage is such that it makes point *B* positive with respect to point *A*.

The magnitude and polarity of the induced emf is governed by Faraday's laws of Electromagnetic Induction and Lenz's law respectively.

Q. 25. State Faraday's laws of Electromagnetic Induction. Derive an expression for the value of the induced emf.

Ans. The two Faraday's laws of electromagnetic induction are as follows :

1. **First Law :** It may be stated in the following two ways—
   (*i*) Whenever the magnetic flux linked with a circuit changes, an emf is induced in it.

   or

   (*ii*) Whenever a conductor cuts magnetic lines of flux, an emf is induced in that conductor.
2. **Second Law :** It states that

   the magnitude of the induced emf is equal to the rate of change of flux linkages.

Explanation :

Suppose a coil has N turns and flux through it changes from an initial value of Φ_1 weber to the final value of Φ_2 weber in time t seconds. Now, by flux linkages is meant the product of number of turns by the flux linked with the coil.

$\therefore$ initial flux linkages $= N\Phi_1$; final flux linkages $= N\Phi_2$

$\therefore$ induced emf, $\quad e = \dfrac{N\Phi_2 - N\Phi_1}{t}$ volt

The above equation may be put in the differential form as

$$e = \frac{d}{dt}(N\Phi) \quad \text{or} \quad e = N\frac{d\Phi}{dt} \text{ volt}$$

Q. 26 (*a*) A coil of 500 turns is linked by a flux of 0.4 mWb. If the flux is reversed in 0.01 second, find the e.m.f. induced in the coil.

(*b*) A coil of resistance 100 Ω is placed in a magnetic field of 1 mWb. The coil has 100 turns and a galvanometer of 400 Ω resistance is connected in series with it. Find the average e.m.f. and the current if the coil is moved in 1/10th second from the given field to a field strength of 0.2 mWb.

Ans. (*a*) Induced e.m.f. is

$$e = N\frac{d\Phi}{dt} \text{ volt ; } N = 500, \text{ flux} = 0.4 \text{ mWb}$$

$$d\Phi = 0.4 - (-0.4) = 0.8 \text{ mWb} = 8 \times 10^{-4} \text{ Wb}$$

$$dt = 0.01 \text{ second ; } e = ?$$

$$e = 500 \times \frac{8 \times 10^{-4}}{0.01} = \mathbf{40 \text{ volt}}$$

(*b*) Induced e.m.f. $e = N \cdot \frac{d\Phi}{dt}$ volt

Here, $d\Phi = 1 - 0.2 = 0.8 \text{ mWb} = 0.8 \times 10^{-3} \text{ Wb}$

$dt = 1/10 \text{ second} = 0.1 \text{ s ; } N = 100$

∴ $e = 100 \times 0.8 \times 10^{-3}/0.1 = 0.8 \text{ V}$

Total circuit resistance = 100 + 400 = 500 Ω

∴ induced current = $0.8/500 = 1.6 \times 10^{-3}$ A = 1.6 mA

Q. 27. State and explain the rules by which the direction of induced emf and current in a conductor can be found when it cuts magnetic flux.

Ans. There exists a definite relationship between the direction of the induced emf, the direction of the magnetic lines of flux and the direction of motion of the conductor. The direction of the induced emf may be found by applying

(*i*) Right-hand rule (RHR)

(*ii*) Right Flat-hand rule and

(*iii*) Lenz's law

Fig. 31(*a*) is used when induced emf is due to flux cutting as in dc generators. It is called *dynamically-induced* emf or *motionally-induced* emf. The right-hand rule (or generator rule) is as under :

When the first three fingures of the right hand are extended at right angles to each other with the thumb pointing in the direction of motion and the forefinger pointing in the direction of the magnetic flux, the second fingure points in the direction of current flow.

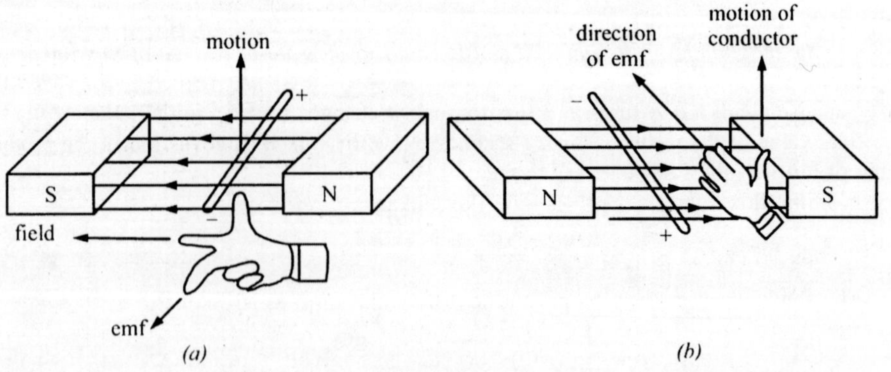

Fig. 31

Fig. 31(*b*) shows the use of right flat-hand rule. Here, the palm of the right-hand is held perpendicular to the incident flux with the thumb pointing in the direction of motion of the conductor. The direction of the fingures gives the direction of the induced emf and current.

Lenz's law is used when induced emf is due to changes in flux-linkages i.e. for *statically* induced emf.

Q. 28. State and explain Lenz's law. Show that it obeys the Law of Conservation of Energy.

Ans. The Lenz's law helps us to find the direction of the induced emf when it is due to *change in flux linkages* as in the case of transformers. It may be stated in slightly three different ways as given below :

(*i*) **First statement :** The direction of the current induced in a coil is always such that it opposes the *change* in the magnetic flux that produces it.

(*ii*) **Second statement :** The direction of flow of induced current is such that the flux produced by it opposes the change in original flux which causes this current.

This statement is based on Newton's third law of motion which states that for every action, there is an equal and opposite reaction.

(*iii*) **Third statement :** Perhaps the easiest way to state Lenz's law is to say that *the polarity of the induced emf is opposite to that of the driving emf.*

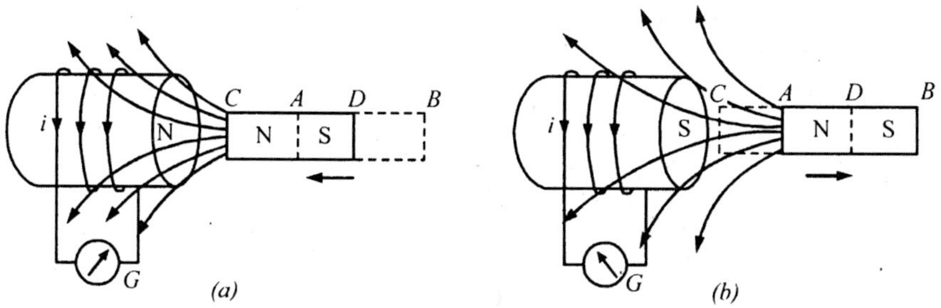

Fig. 32

Explanation :

The statements of Lenz's law will be further clarified with the help of Fig. 32. In Fig. 32(*a*) when *N*-pole of the bar magnet approaches the coil from position *AB* to position *CD*, the flux linked with the coil *increases*.

Hence, an emf is induced in the coil which sets up an induced current. This induced current flows in the counter-clockwise direction in the coil as seen from the *magnet side*. Hence, face of the coil opposite to the magnet becomes a *N*-pole and so tends to stop the approach of the magnet towards itself (like poles repel each other). The mechanical energy spent in overcoming this repulsive force is converted into electrical energy which appears in the coil.

When the magnet is withdrawn back to its original position *AB*, the flux linked with the coil *decreases* (Fig. 30*b*). As a result, induced current now flows in the clockwise direction. This makes the face of the coil opposite to the magnet a *S*-pole. This *S*-pole of the coil *attracts* the *N*-pole of the magnet. Obviously, mechanical energy is required to move away the *N*-pole of the magnet against the attractive force of the *S*-pole of the coil. Again, the mechanical energy spent is converted into electrical energy. It will now be shown that Lenz's law obeys the Law of Conservation of Energy. Suppose for a moment that when *N*-pole of the magnet in Fig. 32(*a*) approaches the coil, induced current flows *clockwise* and makes the coil face a *S*-pole. In that case, the magnet would be *automatically* pulled towards the coil without spending any mechanical energy. It will mean that we will be able to create electrical energy *without* spending mechanical or any other energy. It is something which is denied by the Law of Conservation of Energy.

Q. 29. What are the different types of induced emfs? What practical use is made of these emfs? (*Basic Elec. Engg., VTU, Karnataka, 2008*)

Ans. An induced emf is either (*i*) dynamically induced (*i.e.* motionally induced) or (*ii*) statically induced. In the first case, usually the field is stationary and conductors cut the flux. All dc generators make use of this induced emf. However, in the second case, usually the conductors or the coil remain stationary and the flux linked with them is changed by simply increasing or decreasing the current producing this flux. This fact is made use of in the working of a transformer.

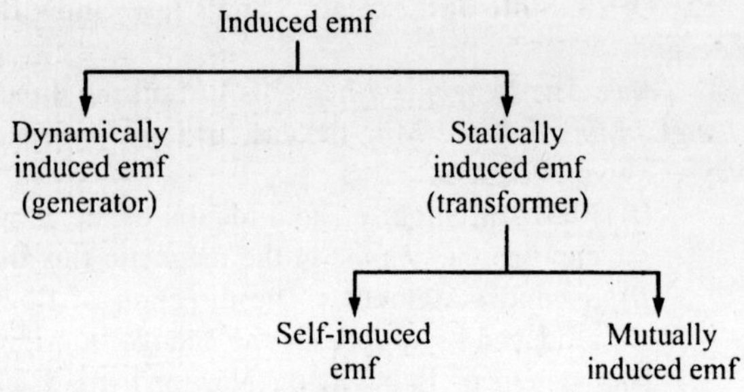

Fig. 33

The statically induced emf can be further sub-divided into (*i*) mutually-induced emf and (*ii*) self-induced emf as shown in Fig. 33.

Consider the coil shown in Fig. 34(*a*). As the current I starts to increase from zero, passing first in turn A, a magnetic flux Φ_{orig}, is produced in this turn and it links with the turns B and C because they are close to A. Hence, an induced flux Φ_{ind} is produced in both turns B and C which opposes the increase in Φ_{orig}. The direction of Φ_{ind} is opposite to Φ_{orig} as shown. The induced flux is due to induced current i produced by the induced emf. The induced emf, e which is opposite to the applied emf E is called self-induced emf. It is so called because it is produced by the change of magnetising current in the circuit itself. This phenomena is called self-induction.

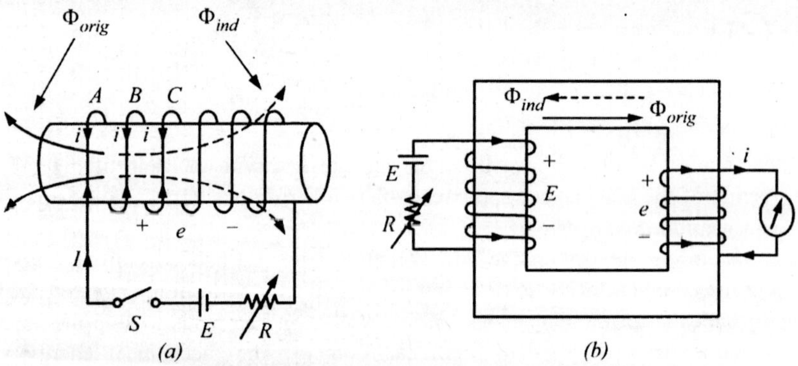

Fig. 34

Fig. 34(*b*) shows two coils wound on the same magnetic core. The first winding is energised by a battery and a variable resistance, R. When current is set up through the first coil, magnetic flux Φ_{orig} is produced in it which links with the second coil. As current is increased. Φ_{orig} increases. This increase in current and hence flux produces an induced emf, e in the second coil. The induced current i produced by the induced emf, e sets up the flux Φ_{ind} in the second coil which opposes the change in Φ_{orig}. This generation of voltage in the secondary coil due to a changing current in the first coil is called *mutual induction*. The emf e produced in the second coil is called *mutually induced emf*. As mentioned earlier, the working of a transformer is based on the phenomenon of mutual induction.

Q. 30(*a*) Derive an expression for the value of dynamically induced emf in a conductor lying in a uniform magnetic field.

(b) A straight conductor has an active length of 20 cm and moves in uniform field of 0.5 T at the rate of 5 m/s. Find the generated emf if the motion of the conductor is (i) parallel (ii) perpendicular (iii) at an angle of 30° to the direction of the field.

(*Elect. Engg. A.M.Ae. S.I. 2000*)

Ans. Fig. 35(*a*) shows the cross-section of a conductor A lying perpendicularly within a uniform field of flux density B. Suppose l is the length of the conductor lying *within* the field and let it move a distance dx in time dt to the right side. The area of the magnetic field swept by the conductors is $= l \,.\, dx$. Hence, magnetic flux cut by it $= l \,.\, dx \times B$.

$\therefore$ change in flux,

$$d\Phi = Bldx \text{ ; time taken } = dt$$

Hence, according to Faraday's laws of Electromagnetic Induction, the emf induced in the conductor (also called dynamically induced emf) is

= rate of change of flux linkages

$$= \frac{d\Phi}{dt} = \frac{Bldx}{dt} = Bl\frac{dx}{dt} = Blv \text{ volt}$$

where $\frac{dx}{dt}$ = conductor velocity v

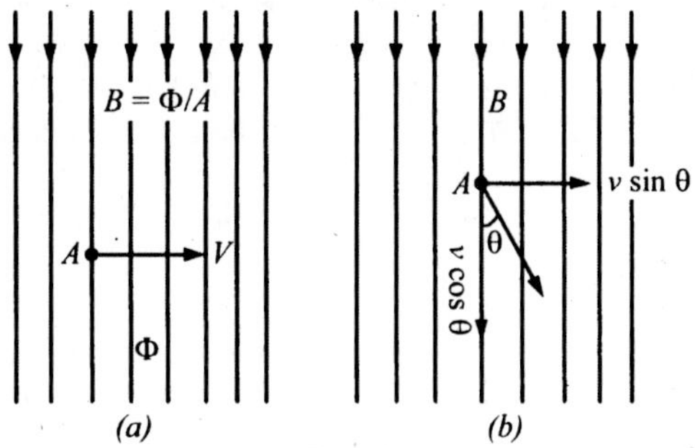

Fig. 35

If the conductor A moves at an angle θ with the direction of the lines of flux as shown in Fig. 33(*b*), then the induced emf is

$$e = Blv \sin\theta$$

As stated earlier, generators work on the production of dynamically induced emf in the conductors housed in a revolving armature which lies within a strong magnetic field.

(b) (*i*) Since the conductor moves parallel to the field, it does not cut it. Hence, no dynamically induced emf is produced. Alternatively,

$$e = Blv \sin\theta = 0.5 \times 0.2 \times 5 \times \sin 0° = \mathbf{0}$$

(*ii*) Here, $\theta = 90°$

$$e = Blv \sin\theta = 0.5 \times 0.2 \times 5 \times \sin 90° = \mathbf{0.5\ V}$$

(*iii*) Here, $\theta = 30°$, $e = 0.5 \times 0.2 \times 5 \times \sin 30° = \mathbf{0.25\ V}$

Q. 31 (a) Calculate the e.m.f. generated in the axle of a car travelling at 72 km/h assuming the length of the axle to be 2 m and vertical component of the Earth's magnetic field to be 40 μWb/m^2.

(b) A conductor 10 cm long and carrying a current of 50 A lies perpendicular to a field of strength 1000 AT/m, Calculate

(i) the force acting on the conductor

(ii) the mechanical power to move this conductor against this force with a speed of 1 m/s and

(iii) e.m.f. induced in the conductor.

(*Magnetic Circuits & Materials, Punjab Univ. 2001*)

Ans. (a) Car velocity = 72 km/h = 72 × 5/18 = 20 m/s

Area of the magnetic field swept by the axle

= axle length × car velocity

$\therefore$ $A = 2 \times 20 = 40 \text{ m}^2$

flux cut, $d\Phi = BA = 40 \times 10^{-6} \times 40 = 1600 \times 10^{-6}$ Wb

time taken = 1 second

$$e = N\frac{d\Phi}{dt} = 1 \times \frac{1600 \times 10^{-6}}{1} = 1600 \times 10^{-6} \text{ V} = \mathbf{1.6\ mV}$$

(b) (i) $F = BIl$ newton

Now, $H = 1000$ AT/m

$\therefore$ $B = \mu_0 H = 4\pi \times 10^{-7} \times 10^3 = 4\pi \times 10^{-4}$ Wb/m^2

$F = 4\pi \times 10^{-4} \times 50 \times 0.1 = \mathbf{6.28 \times 10^{-3}}$ **N**

(ii) $P = F \times v = 6.28 \times 10^{-3} \times 1 = \mathbf{6.28 \times 10^{-3}}$ **W**

(iii) $e = Blv = 4\pi \times 10^{-4} \times 0.1 \times 1 = \mathbf{4\pi \times 10^{-5}}$ **V**

Q. 32. How do you define the property of self-inductance? Derive three different expressions for the coefficient of self induction of a coil.

Ans. The property of an inductor or coil that opposes any change of current (and, hence, flux) through it is called *inductance*. The ability of a coil to induce voltage *into itself* due to change in its *own current* (and, hence, flux) is called self-inductance or self-induction. Its unit is henry (H).

The coefficient of self-induction (L) may be defined in any one of the three ways given below :

(i) **First definition for *L* :** The coefficient of self-induction of a coil is defined as its *weber-turns produced per ampere current in the coil.* By 'weber-turns' is meant the product of flux and the number of coil turns with which the flux is linked.

In other words, it means the flux linkages of the coil.

Consider a coil having N turns and carrying a current of I ampere (Fig. 36). As seen, the weber-turns of the coil are $N\Phi$.

Hence, weber-turns per ampere are $= N\Phi/I$.

By definition, self-nnductance of the coil,

$$L = N\Phi/I.$$

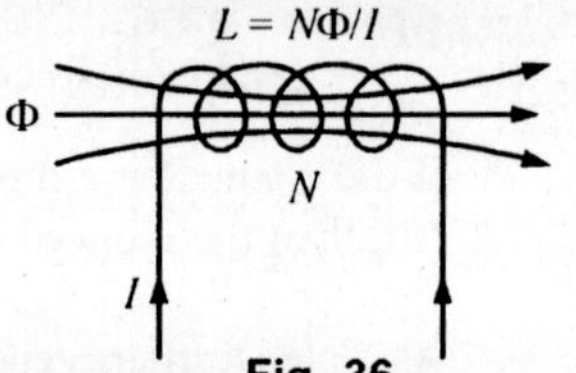

Fig. 36

The unit of self-inductance is henry (H).

If $N\Phi = 1$ Wb-turn, $I = 1$ ampere, then $L = 1$ henry (H).

Hence, *a coil has a self-inductance of one henry if a current of one ampere flowing through it produces flux linkages of one weber-turn in it.*

(ii) **Second definition for *L* :** As proved in Q. 13, the flux produced in an iron-cored coil or solenoid is

$$\Phi = \frac{\text{mmf}}{S} = \frac{NI}{l/\mu_0\mu_r A} \qquad \therefore \quad \frac{\Phi}{I} = \frac{N}{l/\mu_0\mu_r A}$$

Now, $$L = \frac{N\Phi}{I} = N \cdot \frac{\Phi}{I} = N \cdot \frac{N}{l/\mu_0\mu_r A} \text{ henry}$$

or $$L = \frac{N^2}{l/\mu_0\mu_r A} = \frac{N^2}{S} \text{ henry}$$

(iii) **Third definition for *L* :**

$$L = \frac{N\Phi}{I} \quad \text{or} \quad N\Phi = LI$$

Differentiating both sides, we get

$$\frac{d}{dt}(N\Phi) = \frac{d}{dt}(LI) \quad \text{or} \quad N \cdot \frac{d\Phi}{dt} = L\frac{dI}{dt}$$

Now, $$N \cdot \frac{d\Phi}{dt} = \text{self-induced e.m.f, } e_L$$

$\therefore$ $$e_L = L\frac{dI}{dt} \quad \text{or} \quad L = \frac{e_L}{dI/dt}$$

The above relation may be used to define unit of self-inductance.

If dI/dt = 1 ampere/second and e_L = 1 volt, then L = 1 henry.

Hence, *a coil has a self-inductance of one henry if one volt is induced in it when current through it changes at the rate of one ampere per second.*

Q. 33 (*a*) A magnetic field is produced by a coil of 300 turns which is wound on a closed iron ring. The ring has a cross-section of 20 cm^2 and mean length of 120 cm. The permeability of the iron is 800. If the current in the coil is 10A, find the energy stored in the magnetic field.

(*b*) An air-cored toroidal coil has 450 turns and a mean diameter of 30 cm and a cross-sectional area of 5 cm^2. Calculate the inductance of the coil and the average induced emf if a current of 4A is reversed in 60 milliseconds.

(*Basic. Elect. Engg., B.Tech., JNTU, Hyderabad 2007*)

Ans. (*a*) $L = \mu_0\mu_r AN^2/l$ henry

$= 4\pi \times 10^{-7} \times 800 \times 20 \times 10^{-4} \times 300^2/1.2 = 0.151$ H

$$E = \frac{1}{2} LI^2 = \frac{1}{2} \times 0.151 \times 10^2 = \mathbf{7.55\ J}$$

(*b*) $$L = \frac{N^2}{S} = \frac{N^2}{l/\mu_0\mu_r A} = \frac{\mu_0\mu_r AN^2}{l}$$

Now, $l = \pi d = 30\pi = 94.3$ cm $= 0.943$ m, $A = 5$ cm^2 $= 5 \times 10^{-4}$ m^2

∴ $L = 4\pi \times 10^{-7} \times 1 \times 5 \times 10^{-4} \times 450^2/0.943$

$= 20\pi \times 450^2 \times 10^{-11}/0.943 = \mathbf{134.9\ mH}$

$e_L = LdI/dt$

Now, $dI = 4 - (-4) = 8$A, $dt = 60$ ms $= 60 \times 10^{-3}$ second $= 0.06$ s

$e_L = 134.9 \times 10^{-3} \times 8/0.06 = \mathbf{18\ V}$

Q. 34. An iron ring 30 cm mean diameter is made of square iron of 2 cm × 2 cm cross-section and is uniformly wound with 400 turns of wire of 2 mm^2 cross-section. Calculate the value of the self-inductance of the coil. Assume μ_r = 800.

(*b*) A circuit has 1000 turns enclosing a magnetic circuit 20 cm^2 in cross-section. With 4A, the flux density is 1.0 Wb/m^2 and with 9A, it is 1.4 Wb/m^2. Find the mean value of the inductances between these current limits and the induced e.m.f. if the current falls from 9A to 4A in 0.05 seconds. (*Elect. Engg.-1, Delhi Univ.*)

Ans. (*a*) $L = \mu_0\mu_r AN^2/l$.

Here $N = 400$; $A = 2 \times 2 = 4$ cm^2 $= 4 \times 10^{-4}$ m^2 ; $l = 0.3\ \pi$ m ; $\mu_r = 800$

∴ $L = 4\pi \times 10^{-7} \times 800 \times 4 \times 10^{-4}\ (400)^2/0.3\ \pi = \mathbf{68.3\ mH}$

(*b*) $$L = N\frac{d\Phi}{dI} = N\frac{d}{dI}(BA) = NA\frac{dB}{dI} \text{ henry}$$

$= 1000 \times 20 \times 10^{-4}\ (1.4 - 1)/(9 - 4) = \mathbf{0.16\ H}$

Now, $e_L = L \cdot dI/dt$; $dI = (9 - 4) = 5$ A, $dt = 0.05$ s

∴ $e_L = 0.16 \times 5/0.05 = \mathbf{16\ V}$

Q. 35. What is meant by the term mutual inductance? What is its unit?

Derive an expression for the coefficient of mutual inductance between two mutually-coupled coils. (*Basic Elect. Engg., West Bengal Tech. Univ., 2009*)

Ans. When two coils are placed close to each other such that flux from one coil links with the turns of the other, they are said to possess mutual inductance. Due to this mutual inductance, a *varying* current in one coil produces an induced emf in the other coil. The coils are said to be mutually-coupled. In fact, the property of mutual inductance is basically the same as the property of self-inductance. The main difference between the two is that self-inductance is the property of a *single coil* while mutual inductance is the property of *two or more coils* acting together. Like self-inductance; mutual inductance is also measured in henrys.

Coefficient of mutual inductance, M betweem two coils may be defined in any of the three ways given below :

(*i*) **First definition of M :** Consider the two mutually-coupled coils of Fig. 37. Their coefficient of mutual inductance may be defined as *the weber-turns in one coil due to one ampere current in the other coil.*

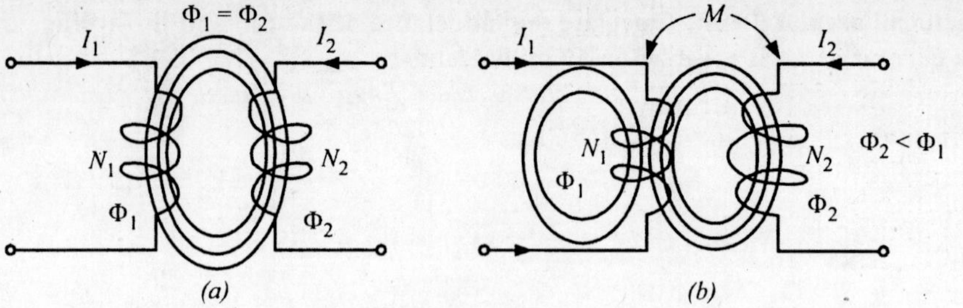

Fig. 37

Let a current of I_1 ampere when flowing in the first coil produce a flux of Φ_1 weber in it. Suppose that *whole* of this flux links with the turns of the second coil. Then, flux linkages in the second coil = $N_2\,\Phi_1$. Hence, flux linkages in second coil for unit current in the first coil

$$= N_2\Phi_1/I_1$$

By definition $$M = \frac{N_2\Phi_1}{I_1}$$

If $N_2\Phi_1/I_1 = 1$, then as seen from above $M = 1$ H.

Hence, two coils possess a mutual inductance of 1 henry if one ampere current flowing in one coil producess flux linkages of one Wb-turn in the other.

Note : In general, $M = N_2\Phi_2/I_1$ where Φ_2 is the flux *actually* linked with the second coil. It is some percentage of Φ_1 as shown in Fig. 37(*b*).

(*ii*) **Second definition of M :** This definition gives the value of M in terms of the dimensions of the two coils.

Flux in the first coil, $$\Phi_1 = \frac{N_1 I_1}{l/\mu_0\mu_r A} \text{ weber}$$

$$\therefore \quad \text{Flux/ampere} = \frac{\Phi_1}{I_1} = \frac{N_1}{l/\mu_0\mu_r A}$$

Assuming that *whole* of this flux (in reality, it is some percentage of it) is linked with the other coil of N_2 turns, then weber-turns in it due to one ampere in the first coil is

$$M = \frac{N_2\Phi_1}{I_1} = \frac{N_2 N_1}{l/\mu_0\mu_r A} \text{ henry} \qquad \text{– putting value of } \frac{\Phi_1}{I_1}$$

$$\therefore \quad M = \frac{\mu_0\mu_r A\, N_1 N_2}{l}$$

or $$M = \frac{N_1 N_2}{l/\mu_0\mu_r A} = \frac{N_1 N_2}{S} \text{ henry}$$

(*iii*) **Third definition of M :** Assuming $\Phi_2 = \Phi_1$, we have

$$M = \frac{N_2\Phi_2}{I_1} = \frac{N_2\Phi_1}{I_1} \quad \text{or} \quad N_2\Phi_1 = M I_1$$

Differentiating both sides, we get

$$\frac{d}{dt}(N_2\Phi_1) = M \cdot \frac{dI_1}{dt}$$

Now, $\frac{d}{dt}(N_2\Phi_1)$ = mutually-induced emf. e_M in the second coil.

$$\therefore \qquad e_M = M\frac{dI_1}{dt}$$

If dI_1/dt = 1ampere/second and e_M = 1 volt, then M = 1 henry. Hence, *two coils have a mutual-inductance of one henry if current changing at the rate of 1 ampere/second in one coil induces an* emf *of one volt in the other coil.*

Q. 36 (*a*) Two identical coils *X* and *Y* of 1,000 turns each lie in parallel planes such that 80% of flux produced by one coil links with the other. If a current of 5A flowing in *X* produces a flux of 0.05 mWb in it find the mutual inductance between *X* and *Y*.

(*b*) Two coils having 30 and 60 turns are wound side-by-side on a closed iron circuit of cross-section 0.01 m^2 and of mean length 1.5 m (*i*) calculate the mutual inductance between the coils if the relative permeability of iron is 2000 (*ii*) a current in the first coil grows steadily from 0 to 10 A in 0.01 second, find the emf induced in the coil.

(*Magnetic Circuits and Materials, Punjab Univ. 2002*)

Ans. (*a*) Formula used :

$$M = \frac{N_2\,\Phi_2}{I_1} \text{ henry}$$

Flux produced in X = 0.05 mWb = 0.05×10^{-3} Wb

Flux linked with Y = $0.05 \times 10^{-3} \times 0.8 = 0.04 \times 10^{-3}$ Wb

$$\therefore \qquad M = \frac{1000 \times 0.04 \times 10^{-3}}{5} = \mathbf{8 \times 10^{-3}\ H}$$

(*b*) (*i*)

$$M = \frac{N_1\,N_2}{l/\mu_0\,\mu_r\,A} = \frac{\mu_0\,\mu_r\,AN_1\,N_2}{l}$$

$$\therefore \qquad M = \frac{4\pi \times 10^{-7} \times 2000 \times 0.01 \times 30 \times 60}{1.5} = \mathbf{0.03\ H}$$

(*ii*) $e_M = MdI_1/dt = 0.03 \times 10/0.01 = \mathbf{30\ V}$

Q. 37 (*a*) Two coils, *A* of 5,000 turns and *B* of 3,000 turns lie in parallel planes. A current of 6 A in coil A produces a flux of 0.1 mWb. If 60% of the flux produced by coil *A* links with the turns of coil *B*, calculate the e.m.f. induced in coil *B* when the current in coil *A* changes from 5 A to – 5 A in 0.01 second.

(*b*) Two identical 2000-turn coils *A* and *B* lie in parallel plane such that 70% of flux produced by one links with the other. A current of 4 A in coil A produces in it a flux of 0.06 mWb. If the current in the coil A changes from +5A to –5A in 0.01 second, what emf is induced in the coil *B*?

Calculate the self-inductance of each coil and their mutual inductance.

Ans. (*a*) Flux/ampere in coil $A = \Phi_1/I_1 = 0.1 \times 10^{-3}/6 = 1.667 \times 10^{-5}$ Wb

Flux linked with coil $B = 0.6 \times 1.667 \times 10^{-5}$ Wb

Now, $$M = \frac{N_2\,\Phi_2}{I_1} = 3000 \times 10^{-5} = 0.03 \text{ H}$$

Mutually-induced e.m.f. in coil B

$$e_M = M\frac{dI_1}{dt} = 0.03 \times 10/0.01 = \mathbf{30\ V}$$

(*b*) The self inductance of coil *A* is given by

$$L_1 = \frac{N_1 \Phi_1}{I_1} = \frac{2000 \times 0.06 \times 10^{-3}}{4} = \mathbf{0.03\ H}$$

Flux linked with coil *B*,

$$\Phi_2 = 70\% \text{ of } 0.06 = 0.042 \text{ mWb}$$

$$\therefore \quad M = \frac{N_2 \Phi_2}{I_1} = \frac{2000 \times 0.042 \times 10^{-3}}{4} = \mathbf{0.021\ H}$$

Change in current, $dI_1 = 5 - (-5) = 10\text{A}$

The emf induced in coil *B* is given by

$$e_2 = \frac{M dI_1}{dt} = \frac{0.021 \times 10}{0.01} = \mathbf{21\ V}$$

Q. 38 (*a*) Define coefficient of coupling and establish relation between self-inductance and mutual inductance with coefficient of coupling.

(*Basic. Elect. Engg. VTU karnataka, July 2009*)

(*b*) Two identical coils *X* and *Y* of 1000 turns each lie in parallel planes. A current of 5A in coil *X* produces 0.5×10^{-3} weber of flux in it. The coupling coefficient between the coils is 0.8. Determine the self-inductance and mutual inductance of the arrangement.

Ans. (*a*) Two coils are said to be magnetically coupled if full or part of the flux produced by one coil links with the other. Let L_1 and L_2 be the self-inductances of the two coils and *M* their mutual inductance. The coefficient of coupling is given by

$$k = \frac{M}{\sqrt{L_1 \cdot L_2}} \quad \text{or} \quad M = k \cdot \sqrt{L_1 \cdot L_2}$$

When all the flux produced by one coil links with the other, then $k = 1$. Hence,

$$M = \sqrt{L_1 \cdot L_2}$$

When there is no common flux between the two coils, they are said to be magnetically *isolated*. In that case,

$$k = 0 \text{ and hence } M = 0.$$

Since $\sqrt{L_1 \cdot L_2}$ represents the value of *maximum* possible mutual inductance, hence coefficient of coupling is given by

$$k = \frac{\text{actual mutual inductance}}{\text{maximum possible mutual inductance}}$$

(*b*) Self-inductance of coil $X = N_1 \Phi_1 / I_1 = 1000 \times 0.5 \times 10^{-3}/5 = \mathbf{0.1\ H}$

Since the two coils are identical, self-inductance of *Y* is the same *i.e.* 0.1 H.

Now, $$M = \frac{k}{\sqrt{L_1 L_2}} = \frac{0.8}{\sqrt{0.1 \times 0.1}} = 0.8 \times 0.1 = \mathbf{0.08\ H}$$

Q. 39 (*a*) The windings of a transformer have an inductance $L_1 = 6\text{H}$, $L_2 = 0.06$ H and a coefficient of coupling $k = 0.9$. Find the emf induced in both windings when the primary current increases at the rate of 1000 A/s.

(*b*) The number of turns in a coil is 250. When a current of 2 A flows in this coil, the flux in the coil is 0.3 mWb. When this current is reduced to zero in 2 milliseconds, the voltage induced in a coil lying in the vicinity of coil is 63.75 volts. If the coefficient of coupling between the coils is 0.85, find self inductances of the two coils, mutual inductance and the number of turns in the second coil. (*Electrical Engg. A:M.Ae.S.I. 2003*)

Ans. (*a*) It is obvious that self-induced emf e_L is produced in primary but mutually-induced e.m.f. e_M is produced in secondeary.

$$e_L = L \cdot dI_1/dt = 6 \times 1000 = \mathbf{6000\ V}$$

Now, $$M = k\sqrt{L_1 L_2}$$

$$= 0.9\sqrt{6 \times 0.06} = 0.54$$

$\therefore$ $$e_M = M \cdot dI_1/dt = 0.54 \times 1000 = \mathbf{540\ V}$$

(*b*) $$L_1 = \frac{N_1 \Phi_1}{I_1} = \frac{250 \times 0.3 \times 10^{-3}}{2} = 37.5 \times 10^{-3}\ \text{H}$$

$$e_M = M \cdot dI_1/dt \quad \text{or} \quad 63.75 = M \times (2 - 0)/2 \times 10^{-3},$$

$$M = 63.75\ \text{mH}$$

$$k = M/\sqrt{L_1 L_2} \quad \text{or} \quad L_1 L_2 = M^2/k^2$$

or $$L_2 = (63.75 \times 10^{-3})^2/0.85^2 \times 37.5 \times 10^{-3}$$

$$= \mathbf{150\ mH}$$

Now, $$\Phi_2 = k\Phi_1 = 0.85 \times 0.3 = 0.255\ \text{mWb}$$

$$M = N\,\Phi_2/I_1 \quad \text{or} \quad 63.75 \times 10^{-3} = N_2 \times 0.255 \times 10^{-3}/2$$

or $$N_2 = \mathbf{500}$$

Q. 40 (*a*) Calculate the total inductance of two coils when joined in (*i*) series aiding and (*ii*) series opposing.

(*b*) The combined inductance of two coils when connected in series are 0.42 H and 0.096 H for series-aiding and series-opposing connections respectively. If one of the coils when isolated has an inductance 0.1 H, calculate (*i*) the self-inductance of the other coil (*ii*) mutual inductance between the coils (*iii*) the coupling coefficient.

(*Elect. Engg., A.M.Ae.S.I., 2000*)

Ans. (*a*) (*i*) Series aiding : In Fig. 38(*a*), the two coils *A* and *B* have been joined in series in such a way that their fluxes (or mmfs) are additive *i.e.* their currents flow in the *same direction.*

Let L_1 = self-inductance of coil *A*

L_2 = self-inductance of coil *B*

M = mutual inductance between *A* and *B*

Self-induced emf in *A* $= e_1 = -L_1 \dfrac{dI}{dt}$

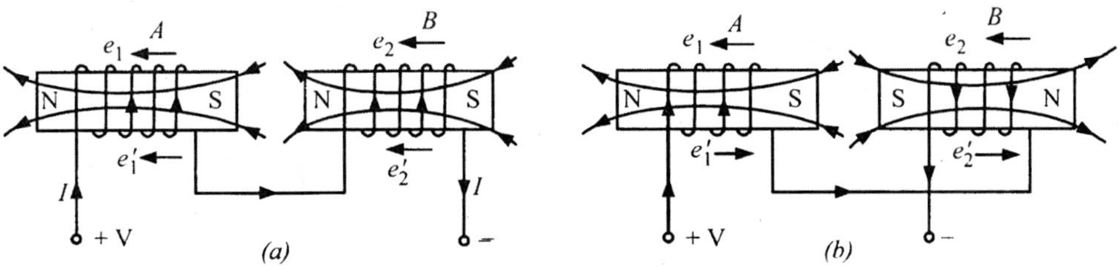

Fig. 38

Mutually-induced emf in *A* due to change of current in *B*

$$= e_1' = -M\frac{dI}{dt}$$

Self-induced emf in *B* $= e_2 = -l_2 \dfrac{dI}{dt}$

Mutually-induced emf in *B* due to change of current in *A* = e_2'

$$e_2' = -M\frac{dI}{dt}$$

All emfs have –ve sign because they all are in opposition to the applied voltage V.
Total induced emf in the combination $= e_1 + e_2 + e_1' + e_2'$

$$= -\left(L_1\frac{dI}{dt} + L_2\frac{dI}{dt} + M\frac{dI}{dt} + M\frac{dI}{dt}\right)$$

$$= -\frac{dI}{dt}(L_1 + L_2 + 2M) \qquad ...(i)$$

If L is the *equivalent* inductance of the single coil, then total induced emf in that single coil would be

$$= -L \cdot \frac{dI}{dt} \qquad ...(ii)$$

Equating (*i*) and (*ii*), we have

$$L = L_1 + L_2 + 2M$$

(*ii*) **Series opposing :** In Fig. 38(*b*) are shown two series-connected coils A and B such that their fluxes are in opposite directions.

As before $e_1 = -L_1 \cdot \frac{dI}{dt}$ – opposite to applied voltage

$e_1' = +M\frac{dI}{dt}$ – in the same direction as applied voltage

$e_2 = -L_2\frac{dI}{dt}$ – opposite to applied voltage

$e_2' = +M\frac{dI}{dt}$ – in the same direction as applied voltage

$\therefore$ Total induced emf $= -e_1 + e_1' - e_2 + e_2'$

$$= -L_1\frac{dI}{dt} + M\frac{dI}{dt} - L_2\frac{dI}{dt} + M\frac{dI}{dt}$$

$$= -\frac{dI}{dt}(L_1 + L_2 - 2M)$$

$\therefore$ equivalent inductance, $L = L_1 + L_2 - 2M$

To sum up, we have

$L = L_1 + L_2 + 2M$ – if mmfs are aiding

$L = L_1 + L_2 - 2M$ – if mmfs are opposing

If we add the two equations, we get

$$M = \frac{L_a - L_0}{4}$$

where L_a = equivalent inductance in series-aiding
L_0 = equivalent inductance in series-opposing

(*b*) (*i*) For series-aiding $L = L_1 + L_2 + 2M$ or $0.42 = L_1 + L_2 + 2M$

For series-opposing, $L = L_1 + L_2 - 2M$ or $0.096 = L_1 + L_2 - 2M$

From the above we get, $M =$ **0.081 H**

(*ii*) Now $L_1 = 0.1$ H. Substituting this value above, we get,

$L_2 =$ **0.482 H**

(*iii*) Coupling coefficient, $k = \frac{M}{\sqrt{L_1 L_2}} = \frac{0.081}{\sqrt{0.1 \times 0.482}} =$ **0.4 H**

Q. 41. Two coils are packed on the same core. Their combined inductance is 0.9H when current through them flows in the same direction. The combined inductance falls to 0.2H when currents through them flow in opposite directions. The self-inductance of one coil is 0.3H. Calculate the values of (*i*) self-inductance of other coil (*ii*) mutual inductance (*iii*) the coupling coefficient. (*Basic Elect. Engg. West Bengal Tech. Univ. 2007*)

Ans. Series-aiding – currents in the same direction

$$0.9 = L_1 + L_2 + 2M \quad ...(i)$$

Series-opposing – currents in opposite directions

$$0.2 = L_1 + L_2 - 2M \quad ...(ii)$$

Subtracting (*ii*) from (*i*), we get

$$0.7 = 4\text{M} \quad \text{or} \quad M = 0.175 \text{ H}$$

or $M = (L_a - L_0)/4 = (0.9 - 0.2)/4 = 0.175$ H

Now, $L_1 = 0.3$H. Substituting this value of L_1 and M in (*i*) above,

we have $0.9 = 0.3 + L_2 + 2 \times 0.175 \quad \therefore \; L_2 = \mathbf{0.25H}$

(*i*) self-inductance of the other coil = **0.25 H**

(*ii*) As found above, $M = 0.175$ H

(*iii*) $$k = \frac{M}{\sqrt{L_1 L_2}} = \frac{0.175}{\sqrt{0.3 \times 0.25}} = \mathbf{0.63 \ H}$$

Q. 42. Two similar coils connected in series gave a total inductance of 600 mH and when one of the coils is reversed, the total inductance is 300 mH. Determine the mutual inductance between the coils and coefficient of coupling?

(*Elect. Circuits. B.Tech., JNTU, Hyderabad 2006*)

Ans. Since the two coils are similar, hence $L_1 = L_2$.

Now, $600 = L_1 + L_2 + 2M$ – series-aiding

$300 = L_1 + L_2 - 2M$ – series-opposing

or $600 = 2L + 2M \quad ...(i)$

and $300 = 2L - 2M \quad ...(ii)$

Adding (*i*) and (*ii*), we get

$$900 = 4L \quad \text{or} \quad L = \mathbf{225 \ mH}$$

Hence, each coil has a self-inductance of 225 mH.

Putting this value in (*i*) above, we get

$$600 = 2 \times 225 + 2M \quad \therefore \; M = \mathbf{75 \ mH}$$

or $M = (L_a - L_0)/4 = (600 - 300)\ 4 = \mathbf{75 \ mH}$

Coefficient of coupling is given by

$$k = \frac{M}{\sqrt{L_1 L_2}} = \frac{M}{\sqrt{L \times L}} = \frac{M}{L} = \frac{75}{225} = \mathbf{0.33}$$

Q. 43. The combined inductance of two coils connected in series is 1.2 H or 0.2 H depending on the relative directions of the current in the coils. If one of the coils has a self inductance of 0.4 H ; Calculate:

(*i*) self-inductance of the second coil

(*ii*) mutual inductance between the coils

(*iii*) the coupling coefficient (*Basic Elect. Engg. B.Tech, JNTU, Hyderabad 2007*)

Ans. We are given that $L_1 = 0.4$ mH

Now, $1.2 = L_1 + L_2 + 2M \quad ...(i)$

and $0.2 = L_1 + L_2 - 2M \quad ...(ii)$

(*i*) Subtracting (*ii*) from (*i*), we get

$$1 = 4\ M \quad \text{or} \quad M = \mathbf{0.25 \ mH}$$

or $M = (L_a - L_0)/4 = 1/4 = 0.25$ H

(*ii*) From (*i*) above, we have

$$1.2 = 0.4 + L_2 + 2 \times 0.25 \quad \therefore \; L_2 = \mathbf{0.3\ mH}$$

(*iii*)
$$k = \frac{M}{\sqrt{L_1 L_2}} = \frac{0.25}{\sqrt{0.4 \times 0.3}} = \frac{0.25}{0.346} = \mathbf{0.72}$$

Q. 44 (*a*) Explain the term magnetic hysteresis. What is it due to? What is meant by hysteresis loop and what is its significance?

(*b*) The hysteresis loop of a sample of sheet subjected to a maximum flux density of 1.3 Wb/m² has an area of 93 cm², the scales being 1 cm = 0.1 Wb/m² and 1 cm = 50 AT/m. Calculate the hysteresis loss in watts when 1500 cm² of the same material is subjected to an alternating flux density of 1.3 Wb/m² peak value at a frequency of 65 Hz.

(*Magnetic Circuits and Materials, Punjab Univ. 2002*)

Ans. Let us take an unmagnetised bar of iron and magnetise it by placing it within the magnetic field of a solenoid.

We will magnetise the bar first in one direction and then in the opposite direction by reversing the direction of current through the coil. The field H produced by the solenoid is called the magnetising field or force H. It produces magnetic flux density B in the iron bar. The value of H can be increased or decreased by increasing or decreasing the current. We will now plot values of B for different values of H. Let us start from point O where $H = 0$.

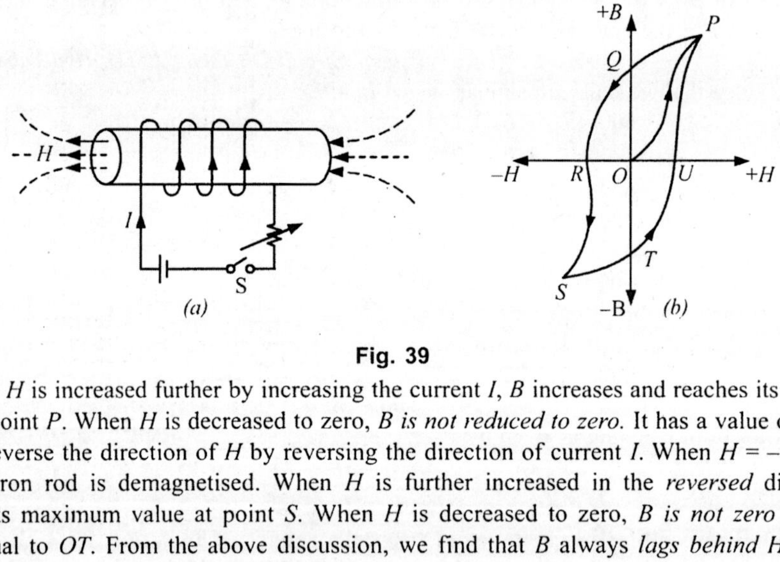

Fig. 39

As H is increased further by increasing the current I, B increases and reaches its maximum value at point P. When H is decreased to zero, B *is not reduced to zero.* It has a value of OQ. Let us now reverse the direction of H by reversing the direction of current I. When $H = -OR$, $B = 0$ and the iron rod is demagnetised. When H is further increased in the *reversed* direction, B reaches its maximum value at point S. When H is decreased to zero, B *is not zero* and has a value equal to OT. From the above discussion, we find that B always *lags behind H.* The two never become zero or maximum at the same time.

The value of B becomes zero *much after H has become zero.* This lagging of flux density B behind the magnetising force H is called hysteresis.

This hysteresis (or lagging) is produced because the magnetic domains inside the iron bar take time to align with the external magnetising force H. Also, the domains do not return exactly to their original positions when H is removed.

The B/H curve shown in Fig. 39 is called the *hysteresis loop.* The shape of the hysteresis loop shows the qualities of the magnetic material from which the bar is made. A material having a high retentivity produces a nearly square hysteresis loop. However, the area of the hysteresis loop represents the loss of energy when a unit volume of the magnetic material is carried through one reversal of magnetisation. Due to this loss of energy, the magnetic bar is heated up if carried through many cycles of reversals. This heat is produced due to friction between the elementary magnets in the bar as they turn from one direction to another.

(*b*) Loss = xy (area of B/H loop) J/m^3/cycle

$= 0.1 \times 50 \times 93 = 465$ J/m^3/cycle

Volume = 1500 cm^3 $= 15 \times 10^{-4}$ m^3

Number of reversals per second = 65

$\therefore \quad W_h = 465 \times 15 \times 10^{-4} \times 65$ J/s = **45.3 W**

Q. 45 (*a*) Derive an expression for the energy stored in the magnetic field of an inductor. (*Basic. Elect. Engg. VTU Karnataka, July 2009*)

(*b*) A solenoid 1 m long and 10 cm in diameter has 500 turns. Calculate (*i*) its approximate inductance in henrys (*ii*) The energy in joules stored in it when a current of 2A passes through it. (*Basic. Elect. Engg. R.G.P.V, Bhopal, 2007*)

Ans. (*a*) Suppose in an inductor, the current grows uniformally from zero value to its maximum value of I. Obviously, average value of the current is $I_{av} = (0 + 1)/2 = I/2$. If L is the self-inductance of the inductor, then self-induced emf, $e_L = L\, I_{av}/t$ where t is the time for the current to change from zero to I.

$\therefore$ average power absorbed during time t is

$$= \text{induced emf} \times \text{average current}$$

$$= e \times I_{av}$$

$$= L \cdot \frac{I}{t} \times \frac{1}{2} I = \frac{1}{2} \cdot \frac{LI^2}{t}$$

$$\text{Total energy absorbed} = \text{Power} \times \text{time}$$

$$= \frac{1}{2} \frac{LI^2}{t} \times t = \frac{1}{2} LI^2$$

$$\therefore \quad \text{energy stored} = \frac{1}{2} LI^2 \text{ joules}$$

(*b*) (*i*) As shown in Fig. 40.

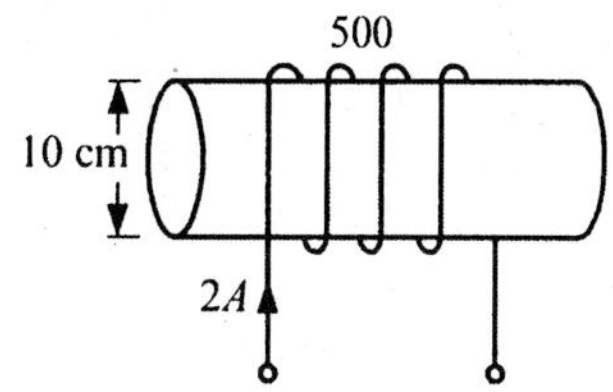

Fig. 40

$$L = \frac{N^2}{S} \text{ henry} \qquad \text{– Q. 32}$$

Reluctance, $\quad S = l/\mu_0\mu_r A$

$$\therefore \quad L = \frac{N^2}{l/\mu_0\mu_r A} = \frac{\mu_0\mu_r A\, N^2}{l} \text{ henry}$$

Now, $\quad l = 1$ m, $A = \pi\,(10)^2 = 100\,\pi$ cm^2 $= \pi \times 10^{-2}$ m^2

$N = 500$, $\mu_0 = 4\pi \times 10^{-7}$, $\mu_r = 1$

$$\therefore \quad L = \frac{4\pi \times 10^{-7} \times \pi \times 10^{-2} \times (500)^2}{1} = \mathbf{9.87\ mH}$$

(*ii*) $\quad$ Energy stored $= \frac{1}{2} LI^2 = \frac{1}{2}(9.87 \times 10^{-3}) \times 2^2 =$ **19.74 × 10^{-3} joules.**

Q. 46. Derive an expression for the energy stored in a magnetic field.

(*Basic. Elect. Engg. VTU Karnataka, July 2009*)

Ans. It is well-known that to establish a magnetic field, energy must be supplied. We know that when current through an inductor rises from zero to I, the flux produced gives rise to an induced emf in the inductor. This induced emf opposes the battery voltage. Hence, battery has to spend energy in overcoming this opposing emf in the inductor. This energy is stored in the magnetic field of the inductor. Suppose an inductor has a magnetic field of length l and cross-section A. As proved in Q. 45, the energy stored is

$$E = \frac{1}{2} LI^2 = \frac{1}{2} \times \frac{\mu_0 \mu_r AN^2}{l} I^2 \qquad \text{– putting the value of } L \text{ from Q. 32}$$

Now, $H = \frac{NI}{l}$

$$\therefore \quad E = \left(\frac{NI}{l}\right)^2 \times \frac{1}{2} \mu_0 \mu_r Al$$

$$= \frac{1}{2} \mu_0 \mu_r H^2 \times Al$$

Now, Al = volume of the magnetic field

$$\therefore \text{ energy stored per unit volume} = \frac{1}{2} \times \mu_0 \mu_r H^2$$

$$= \frac{1}{2} BH$$

Now, $H = B/\mu$. Putting the value of H, we get

$$\text{energy stored per unit volume} = \frac{B^2}{2\mu_0 \mu_r} \qquad \text{– in a medium}$$

$$= \frac{B^2}{2\mu_0} \qquad \text{– in air}$$

If B is in Wb/m^2, H in AT/m and volume in m^3, then the above expressions give the energy stored in joules per m^3 of the magnetic field.

Q. 47. A coil consists of 600 turns and a current of 10 A in the coil gives rise to a magnetic flux of 1 mWb.

Calculate :

(*i*) self-inductance

(*ii*) the induced emf and

(*iii*) the energy stored when the current is reversed in 0.01 second.

(*Basic. Elect. Engg. VTU Karnataka, 2009*)

Ans. (*i*) $L = \frac{N\Phi}{I} = \frac{600 \times 1 \times 10^{-3}}{10} = \mathbf{0.06\ H}$

(*ii*) $e_L = L \cdot \frac{dI}{dt}$

$dI = 10 - (-10) = 20\text{A}, \ dt = 0.01$ second

$$\therefore \quad e_L = 0.06 \times \frac{20}{0.01} = \mathbf{120\ V}$$

(*iii*) Energy stored $= \frac{1}{2} LI^2 = \frac{1}{2} \times 0.06 \times 10^2 = \mathbf{3\ joules}$

Q. 48. (*a*) Derive an expression for the tractive or lifting power of an electromagnet.

(*Basic Elect. Engg., West Bengal Tech., Univ., 2003*)

(*b*) Determine the force required to separate two iron surfaces each of 0.3 m^2 section when the flux normal to the surface is 8 × 10^{-2} Wb.

Ans. (*a*) Consider two magnetic poles of cross-sectional area A and separated by a distance x as shown in Fig. 41. Further, suppose that the flux density in the air gap is B and the attractive force between the two poles is F. If one of the poles (say N-pole) is pulled apart *against* the attractive force F through a distance of dx, then work done is = $F \times dx$.

This work goes to provide energy for the *additional* volume of the magnetic field created. Additional volume of the magnetic field created

$$= A \times dx$$

Energy required per unit volume $= \dfrac{B^2}{2\mu_0}$

$\therefore$ energy required for the additional volume

$$= \frac{B^2}{2\mu_0} \times Adx$$

Equating the work done and the energy stored, we get

$$F \,.\, dx = \frac{B^2 \times A \,.\, dx}{2\mu_0} \quad \text{or} \quad F = \frac{B^2 A}{2\mu_0} \text{ newton}$$

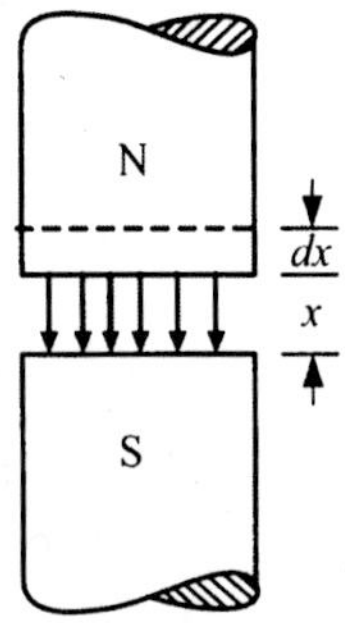

Fig. 41

or $$F = \frac{B^2 A}{9.81 \times 2\mu_0} = \frac{B^2 A}{19.62\,\mu_0} \text{ kg-wt} \qquad (\because 1 \text{ kg-wt} = 9.81 \text{ N})$$

(*b*) The pulling force required is given by

$$F = \frac{B^2 A}{2\mu_0} \text{ newton} = \frac{B^2 A}{19.62\,\mu_0} \text{ kg-wt}$$

$$B = 8 \times 10^{-2}/0.3 = 0.267 \text{ Wb/m}^2$$

$$\therefore \quad F = \frac{(0.267)^2 \times 0.3}{19.62 \times 4\pi \times 10^{-7}} = \mathbf{867.4 \text{ kg-wt}}$$

EXERCISE

1. A 10 cm long straight wire carries a current of 100 A and lies at right angles to a uniform magnetic field of strength 1.5 Wb/m^2. Find the mechanical force on the conductor. **(15N)**
2. Two bus bars 100 mm apart are supported by insulators every metre along their length. If each busbar carries a current of 20 kA, find the force acting on each insulator. **(800N)**
3. A conductor 1 m in length is moved at a velocity of 50 cm/s perpendicular to a magnetic field of 20 tesla (Fig. 42). What is the emf generated? **(12.5 V)**
4. A straight conductor 20 cm long carries a current of 50A at right angles to a magnetic field of flux density 0.3 Wb/m^2. Calculate the force on the conductor. **(3 N)**
5. A current of 3A flows through a coil of 1000 turns uniformly wound on an iron ring having a mean circumference of 40 cm and a cross-sectional area of 4 cm^2. The relative permeability of the iron is 80. Calculate : (*a*) the magnetomotive force (*b*) the magnetising force (*c*) the flux density (*d*) the flux.

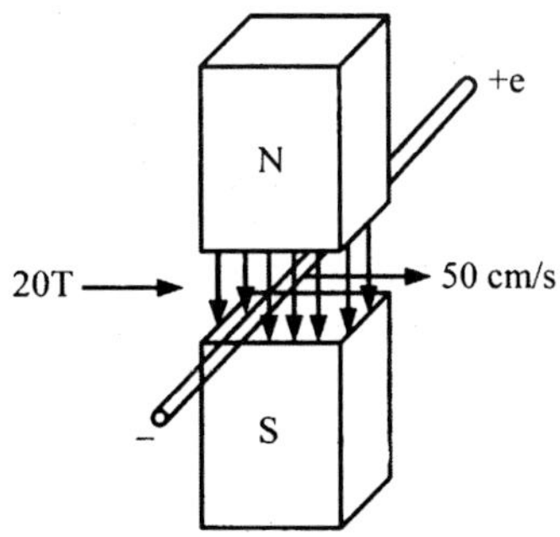

Fig. 42

(3000AT, 7500AT/m, 0.754Wb/m^2, 0.3mWb/m^2)

6. A coil of wire consists of 50 turns and carries a current of 5 A. If the length of the coil is 20 cm, calculate : (*i*) magnetic field intensity and (*ii*) flux density within the coil. **(1250 AT/m, 1.57T)**

7. A conductor of effective length 150 mm moves with a velocity of 5 m/s perpendicular to a magnetic field of uniform flux density 0.4T. Calculate :
   (*a*) the emf induced in the conductor.
   (*b*) the force acting on the conductor when it carries a current of 20A.
   (*c*) the power required to drive the conductor. **(0.3V. 1.2N, 6.0W)**
8. A coil having 200 turns and carrying a current of 3A is wound uniformally over a toroidal ring with a mean circumference of 20 cm and cross-section of 5 cm^2. The material of the ring has $\mu_r = 1000$. Calculate : (*i*) magnetic field strength (*ii*) flux density (*iii*) total flux. **(3000 AT/m, 3.77T, 1.89 mWb)**
9. A coil of 600 turns has a flux of 0.03 Wb linked with it. The flux is reversed in 0.2 second. Calculate the average value of the emf induced in the coil. **(180 V)**
10. An inductor having 50 turns of wire is placed in a magnetic field that is changing at the rate of 25 Wb/s. What is the induced voltage? **(1250 V)**
11. Calculate the reluctance and permeance of a ferromagnetic circuit of mean length 20 cm, having area 10 cm^2 and a relative permeability 100. If the circuit is uniformly wound with 1000 turns of wire carrying 0.1 A, determine the magnetic flux in the core. **(62.9×10^{-6} Wb)**
12. The mean length of a magnetic circuit is 40 cm, cross-sectional area 8 cm^2 and relative permeability of the core material 200. When the magnetising coil carries a current of 6A, the magnetic flux produced in the core is 3 mWb. Find : (*i*) reluctance of the magnetic circuit (*ii*) flux density in the core (*iii*) magnetic field intensity (*iv*) number of turns of the coil. **(1.989×10^6AT/Wb, 3.75 Wb/m^2, 1.492×10^4 AT/m, 995)**
13. The magnetic circuit of Fig. 43 has a cross-sectional area of 50 cm^2 and a mean length of 2 m. The relative permeability of core material is 80. Calculate the reluctance of the magnetic circuit if the magnetising coil has 150 turns and the core flux is 80 μWb. What is the value of the current flowing in the coil? **(3.98×10^6AT/Wb, 2.12A)**

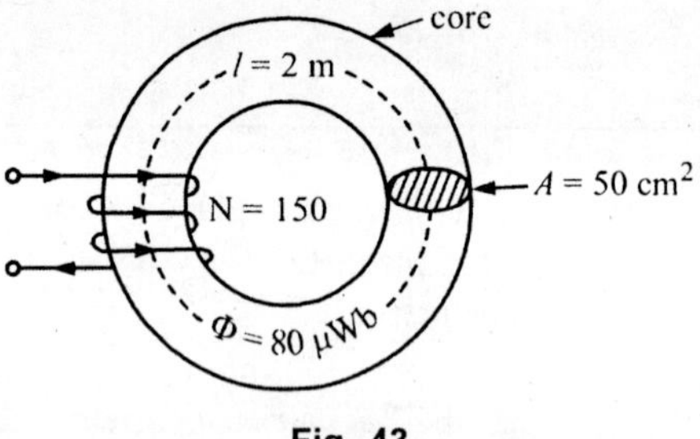

Fig. 43

14. A mild steel ring having a cross-sectional area of 5 cm^2 and a mean circumference of 40 cm has a coil of 200 turns wound uniformly around it. Calculate (*i*) the reluctance of the ring (*ii*) the current required to produce a flux of 800 μWb in the ring. Assume relative permeability of mild steel to be 380 at the flux density developed in the core. (*Elect. Engg.-I, Punjab Univ.*) **(1.675×10^6 AT/Wb, 6.7A)**
15. A magnetic ring has a mean circumference of 1.5 m and is of 0.01 m^2 in cross-section and is wound with 250 turns. A saw-cut of 4 mm wide is made in the ring. Calculate the magnetizing current required to produce a flux of 0.8×10^{-3} Weber in the air-gap. Assume relative permeability of iron as 400 and leakage facor as 1.25. **(2.21A)**
16. A wooden ring has a circular cross-section of 300 mm^2 and a mean diameter of 200 mm (Fig. 44). It is uniformly wound with 800 turns. Calculate :
   (*i*) the field strength produced in the coil by a current of 2A : (assume $\mu_r = 1$)
   (*ii*) the magnetic flux density produced by this current and

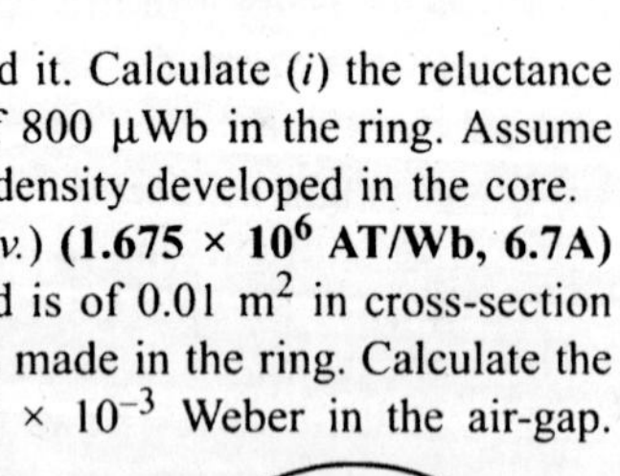

Fig. 44

(*iii*) the current required to produce a flux density of 0.02Wb/m^2.

(*Elect. Engg. Nagpur Univ., 2000*) **(2546.5AT/m, 3.2 × 10^{-3} Wb/m^2, 12.5A)**

17. An electromagnet as shown in Fig. 45 has area of cross-section of 12 cm^2. Mean length of iron path is 50 cm. It is excited by two coils each having 400 turns. When the current in the coil is 1.0 A, the resulting flux density gives a relative permeability of 1300. Calculate :
    (*i*) reluctance of iron part of magnetic circuit
    (*ii*) reluctance of air-gap
    (*iii*) total flux
    (*iv*) flux density in the air-gap
    Neglect leakage and fringing.

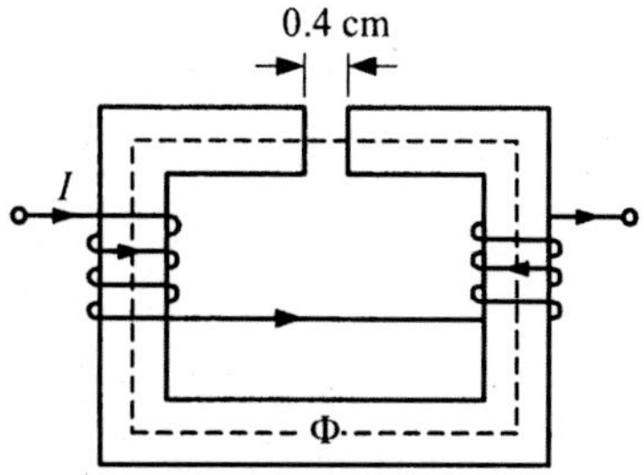

Fig. 45

(*Basic. Elect. Engg. R.G.P.V, Bhopal Univ., 2006*)
(253 × 10^{3}AT/Wb, 2653 × 10^3 AT/Wb, 0.275mWb, 0.23 Wb/m^2)

18. A cast iron ring of 40 cm mean length and circular cross-section of 5 cm diameter is wound with a coil. The coil carries a current of 3A and produces a flux of 3 mWb in the air gap. The length of air gap is 2 mm. The relative permeability of the cast iron is 800. The leakage coefficient is 1.2. Calculate number of turns of the coil.
(*Elect. Engg. Nagpur Univ.,2004*)
(1054 turns)

19. An iron magnetic circuit has uniform cross-sectional area of 5 cm^2 and a length of 25 cm. A coil of 120 turns is wound uniformly over the magnetic circuit. When the current in the coil is 1.5 A, the total flux is 0.3 mWb; when the current is 5A, the total flux is 0.6 mWb.
    For each value of current, calculate :
    (*i*) the magnetising force
    (*ii*) the relative permeability of the iron. (*Elect. Engg. Nagpur Univ. 2004*)
(For *I* = 1.5A : 720AT/m, 663 ; *I* = 5A : 2400 AT/m, 398)

20. Two coils have mutual-inductance of 0.75 H and their coefficient of coupling is 0.9. If one coil has an inductance of 4.5 H, what is the inductance of the other? **(0.32H)**

21. Two coils of 270 and 180 mH are connected in series-aiding. If their mutual inductance is 100 mH, calculate total inductance between the two coils. **(650 mH)**

22. When connected in series-aiding, the total inductance of two inductors is 560 mH. When connected in series-opposing, their combined inductance is 320 mH. What is the value of mutual inductance between them? **(60 mH)**

23. The combined inductance of two coils connected in series is 0.6H or 0.1H, depending upon the relative directions of the currents in the coils. If one of the coils when isolated has a self inductance of 0.2 H, calculate : (*i*) mutual inductance and (*ii*) the coefficient of coupling. (*Elect. Circuits. B.Tech., JNTU Hyderabad 2006*)
(0.125 mH, 0.72)

MULTIPLE CHOICE QUESTIONS

1. Iron is called a magnetic material because it has
   (a) lot of free electrons
   (*b*) unbalanced electron spins
   (*c*) no valence electrons
   (*d*) no spinning electrons in its atoms

2. Mark the **WRONG** statement :
Magnetic lines of force
(*a*) are elastic (*b*) repel each other
(*c*) never intersect each other (*d*) actually do exist
3. Materials which when placed in a magnetic field increase its flux are called .... materials.
(*a*) diamagnetic (*b*) paramagnetic
(*c*) ferromagnetic (*d*) non-magnetic
4. Silicon steel is called soft magnetic material because it has
(*a*) high permeability (*b*) low coercivity
(*c*) high retentivity (*d*) both (*a*) and (*b*)
5. Production of a magnetic field around a current-carrying conductor is called
(*a*) electromagnetism (*b*) ferromagnetism
(*c*) inductance (*d*) electrodynamism
6. The force experienced by a current-carrying conductor when placed in a magnetic field depends on
(*a*) flux density of the field (*b*) conductor current
(*c*) conductor length (*d*) all of the above
7. The unit of electric intensity is
(*a*) joule/coulomb (*b*) newton/coulomb
(*c*) volt/metre (*d*) both (*b*) and (*c*)
(*Basic. Elect. Engg., R.G.P.V., Bhopal Univ, 2007*)
8. The unit of flux density is
(*a*) Tesla (*b*) A/mm^2
(*c*) N/m^2 (*d*) Wb/m
(*Basic. Elect. Engg., R.G.P.V., Bhopal Univ, 2007*)
9. The principle of dynamically induced emf is utilized in
(*a*) transformers (*b*) chokes
(*c*) generators (*d*) thermocouples
10. The ability of a magnetic material to retain its magnetism even after the magnetising force is removed is called its
(*a*) residual magnetism (*b*) retentivity
(*c*) coercivity (*d*) permeance
11. Which of the following media has a relative permeability of unity?
(*a*) iron (*b*) air
(*c*) vacuum (*d*) water
12. Energy stored by a coil is doubled when its current is increased ....: percent
(*a*) 100 (*b*) 41.4
(*c*) 50 (*d*) 35
(*Basic Elect. Engg. RGPV, Bhopal Univ. 2007*)
13. Reluctance offered by the magnetic circuit depends upon
(*a*) nature of magnetic material (*b*) length of magnetic flux path
(*c*) cross-sectional area of the material (*d*) all of the above
(*Basic. Elect. Engg. RGPV, Bhopal Univ. 2007*)
14. A cast-steel electromagnet has an air-gap length of 0.3 cm. Find the ampere-turns for the air-gap to produce a flux density of 0.7 Wb/m^2 in the gap.
(*a*) 2100 AT (*b*) 1671 AT
(*c*) 1447 AT (*d*) 167 AT
(*Basic. Elect. Engg. RGPV, Bhopal Univ, 2006*)

15. A magnetic circuit requires 800 AT to produce a certain amount of flux in a magnetic circuit. If its exicitation coil has 100 turns and 5 Ω resistance, the voltage to be applied to exciting coil is
    (*a*) 60 volt (*b*) 40 volt
    (*c*) 80 volt (*d*) 8 volt
    (*Basic. Elect. Engg. RGPV, Bhopal Univ. 2006*)
16. Reluctivity of a magnetic circuit is similar to .... of an electric circuit.
    (*a*) resistivity (*b*) conductivity
    (*c*) resistance (*d*) conductance
17. The conductance of an electrical circuit is similar to .... in a magnetic circuit.
    (*a*) flux (*b*) reluctance
    (*c*) permeance (*d*) relative permeability
    (*Basic Elect. Engg. RGPV, Bhopal Univ. 2007*)
18. Leakage flux in generators and motors reduces their
    (*a*) weight (*b*) cost of manufacture
    (*c*) power efficiency (*d*) hysteresis loss
19. The direction of magnetic field produced by a linear conductor is given by
    (a) Ampere's law (b) Fleming's left hand rule
    (*c*) Right hand thumb rule (*d*) none of the above
    (*Basic Elect. Engg. VTU, Karnatka. 2009*)
20. When a magnet is suddenly plunged into an open-circuited coil, ... is induced in the coil.
    (*a*) current (*b*) emf
    (*c*) power (*d*) energy
21. The direction of induced emf in a conductor can be deduced by
    (*a*) Fleming's left hand rule (*b*) Fleming's right hand rule
    (*c*) Maxwell's corkscrew rule (*d*) Lenz's law
    (*Basic Elect. Engg. VTU, Karnatka. 2009*)
22. Making a wire into a coil increases its
    (*a*) length (*b*) resistance
    (*c*) mmf (*d*) current
23. A magnetic circuit, like an electric circuit, must be
    (*a*) long (*b*) short
    (*c*) complete (*d*) circular
24. Hysteresis loss least depends on
    (*a*) volume of the material (*b*) frequency
    (*c*) Steinmetz coefficient of material (*d*) ambient temperature
    (*Basic Elect. Engg. RGPV, Bhopal Univ. 2007*)
25. In a magnetic material, hysteresis loss takes place primarily due to
    (*a*) flux density lagging behind magnetizing force
    (*b*) molecular friction
    (*c*) its high retentivity
    (*d*) rapid reversals of its magnetization
    (*Basic Elect. Engg. RGPV, Bhopal Univ, 2007*)
26. The area of hysteresis loop is a measure of
    (*a*) magnetic flux (*b*) permeance
    (*c*) mmf per cycle (*d*) energy loss per cycle/m^3
    (*Basic Elect. Engg. RGPV, Bhopal Univ, 2007*)

27. Which of the following inductors will have the least eddy current loss?
    (*a*) air-cored (*b*) iron-cored
    (*c*) wooden-cored (*d*) laminated iron-cored
    (*Basic Elect. Engg. RGPV, Bhopal Univ, 2007*)
28. Hysteresis loss occurrs only in .... materials.
    (*a*) paramagnetic (*b*) non-magnetic
    (*c*) diamagnetic (*d*) ferromagnetic
29. The combined inductance of two coils is 0.8H in series-aiding and 0.2H in series-opposing. Their mutual inductance is .... henry.
    (*a*) 0.6 (*b*) 1.0
    (*c*) 0.15 (*d*) 0.5
30. In case all the flux from the current in one coil links with another coil, the coefficient of coupling will be
    (*a*) 2.0 (*b*) 1.0
    (*c*) 0.5 (*d*) zero
    (*Basic Elect. Engg. RGPV, Bhopal Univ, 2007*)

ANSWERS

1. *b*	2. *d*	3. *c*	4. *d*	5. *a*	6. *d*	7. *d*
8. *a*	9. *c*	10. *b*	11. *c*	12. *b*	13. *d*	14. *b*
15. *b*	16. a	17. c	18. c	19. c	20. b	21. *b*
22. *c*	23. *c*	24. *d*	25. *a*	26. *d*	27. *a*	28. *d*
29. *c*	30. *b*.					

CHAPTER 5

AC FUNDAMENTALS

Q. 1. What is the difference between a direct current (dc) and an alternating current (ac). Give examples of different alternating currents?

Ans. A direct current always flows in the same *direction* and a dc voltage has always the same *polarity*. Fig. 1(*a*) shows a steady direct current which has a constant value of 2A. Some of the alternating current waves are also shown in the figure. Alternating current is one which reverses its direction of flow again and again at regular intervals. Its magnitude also changes regularly. Fig. 1(*b*) shows an alternating current which first flows in one direction and then in the opposite direction. It is called a sine wave because it varies according to the sine function. Fig. 1(*c*) shows an alternating square wave. Its magnitude remains the same but it reverses its direction of flow at regular intervals.

Fig. 1(*d*) shows a sawtooth wave which rises and falls in straight lines. Both its magnitude and direction of flow change regularly.

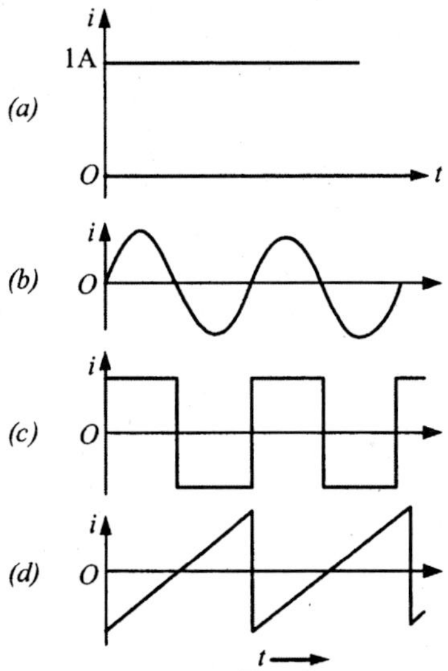

Fig. 1

The main differences between dc and ac are tabulated below:

S. No.	Alternating Current	Direct Current
1.	It is easy to generate from turbines.	1. It is obtained by converting ac to dc.
2.	Its frequency is 50 or 60 Hz.	2. Its frequency is zero.
3.	It can be easily changed to lower or higher values as per the needs of electrical apparatus.	3. It is not easy to change dc currents or voltages.
4.	It changes both its magnitude and direction regularly.	4. It keeps flowing in one direction only.
5.	AC transmission at high voltages using wire conductors over great distances is less efficient. It is so because the wire offers more impedance to it than to dc. Moreover, there is loss of energy in the magnetic fields around the wires.	5. DC transmission at extremely high voltages is more efficient. It is so because the wire offers less *impedance* with dc than with ac. Moreover, much less energy is lost in the magnetic fields around the wires.
6.	It is more dangerous than dc at the same voltage rating.	6. It is less dangerous than ac at the same voltage rating.

Q. 2. What is a sine wave. What is its importance?

Ans. The sine wave is a very common type of alternating current and alternating voltage. It is also known as *sinusoidal* wave. The electric supply provided by the power companies is in the form of sinusoidal voltages and currents. Fig. 2. shows the general shape of a sine wave which can be either current or voltage. Let us see how the voltage or current varies with time. Starting form zero, it increases to a positive maximum value, returns to zero and then reverses its direction and increases to a negative maximum value before returning to zero again.

Obviously, a sine wave changes its polarity *i.e.* its direction of flow at its *zero values.* In other words, it alternates between positive and negative values at regular intervals.

The combined positive and negative alternations make up one cycle of a sine wave.

More than 90 percent of electric energy is generated as alternating current though it may also be converted into direct current. The sine wave is important in ac circuits because

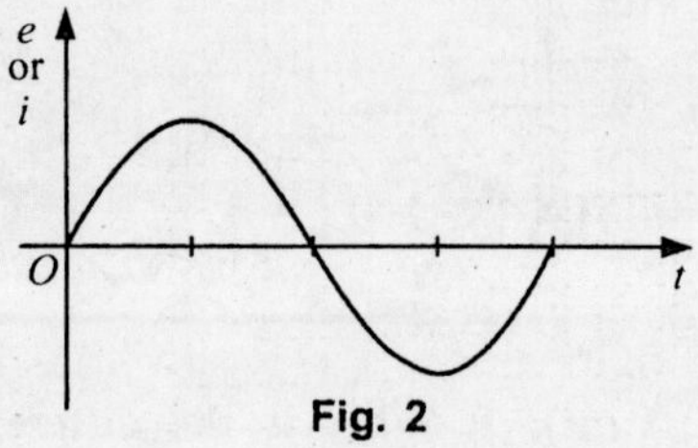

Fig. 2

(*i*) it is the easiest waveform to generate
(*ii*) it is the easiest waveform to distribute and use
(*iii*) all other periodic complex waveforms (called harmonics) can be produced by combining many sine waves.

Q. 3. Mention three methods for producing ac voltages and currents.

Ans. A voltage can be developed in a coil of wire in one of the three ways given below:

1. by changing the flux linked with a coil
2. by moving a coil through a magnetic field so that flux is cut
3. by changing the *direction* of the flux linked with the coil. In the first case, the voltage is said to be an *induced* e.m.f. and as per Faraday's law, its value is given by

$$e = -N \cdot \frac{d\Phi}{dt} \text{ volt}$$

In this method, there is no motion of the coil or magnet. The value of the current through the magnetizing coil is increased or decreased to change the flux linked with the coil. The voltage so produced is called emf.

In the other two methods, there is actual physical motion of the coil or conductor and the magnet. A voltage produced in either of the above ways is called generated e.m.f. Its value is given by

$$e = B\ l\ v \text{ volt}$$

where B = flux density ; l = conductor length ; v = conductor velocity.

Q. 4. With the help of a suitable diagram, describe the working of a simple ac generator?

Ans. All ac generators are based on Farday's Laws of Electromagnetic Induction. Two very simple forms of such an ac generator are shown in Fig. 3. The ac generator of Fig. 3(*a*) consists of a conducting rectangular loop of wire which is rotated within a magnetic field in counter- clockwise direction . As the loop rotates, its two *longer sides* cut the magnetic flux. Hence, an e.m.f. is induced in them. As the two sides cut the lines of flux in opposite directions, the induced voltage is *alternating*. Hence, it appears across the two ends of the loop as an alternating voltage *e*. It has a sinusoidal shape as shown in Fig. 3.

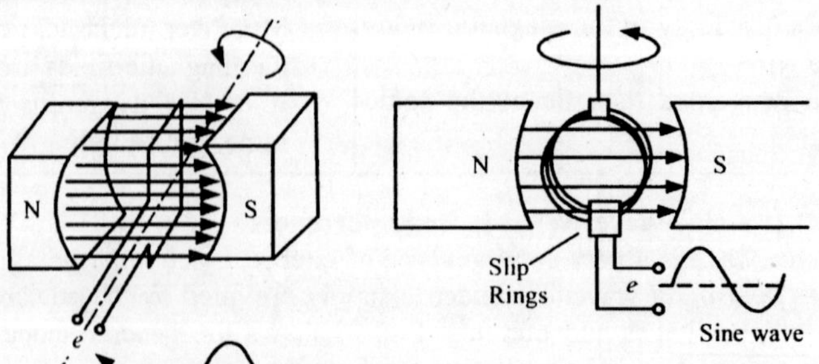

Fig. 3

As seen from Fig. 4, the two loop sides move *parallel* to the lines of magnetic flux at angles of 0° or 360°. They *do not cut* any magnetic flux. Hence, no e.m.f. is induced in them. However, when the loop sides move at a right-angle *i.e.* at 90° or 270° to the lines of magnetic flux, their *rate* of cutting the flux is *maximum.*

Hence, maximum e.m.f. E_m is induced in them.

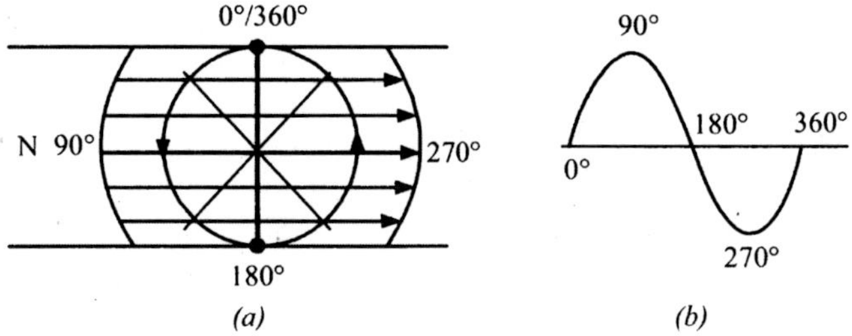

Fig. 4

As the loop rotates from points of maximum induced voltage to the points of zero induced voltage, the voltage changes according to the formula

$$e = E_m \sin \theta \text{ volt}$$

Here, E_m is the maximum induced e.m.f. and e is the *instantaneous* e.m.f. at any angle θ of the rotation. As the loop conductors rotate through a full revolution (360°), one full cycle of the sinusoidal e.m.f. is induced as shown in Fig. 4(*b*).

It may be noted that Fig. 3 only shows the basic idea of an ac generator. In practice, there must be means of mechanically rotating the loop and of electrically connecting the ends of the loop across a load in order to utilize the generated voltage. For this purpose, the ends of the loop are connected to sliprings that rotate along with the loop. The sliprings, in turn, make continuous contact with stationary brushes across which electric load is connected.

Q. 5. Derive an expression for the ac voltage induced in a coil when it rotates in a magnetic field in terms of its angular velocity and flux linked with it.

Ans. Fig. 5 shows a rectangular coil of N turns rotating in a magnetic field with an angular velocity of ω radian/second. Let time be measured from the positive direction of X-axis. Maximum flux Φ is linked with the coil when its plane is *horizontal i.e.* when it coincides with the X-axis. In time, t seconds, the coil rotates through an angle $\theta = \omega t$. In this deflected position, the part of the flux which is perpendicular to the plane of the coil is $\Phi \cos \theta = \Phi \cos \omega t$. It is this flux which is linked with the coil. Hence, flux linkages of the coil in this postion are

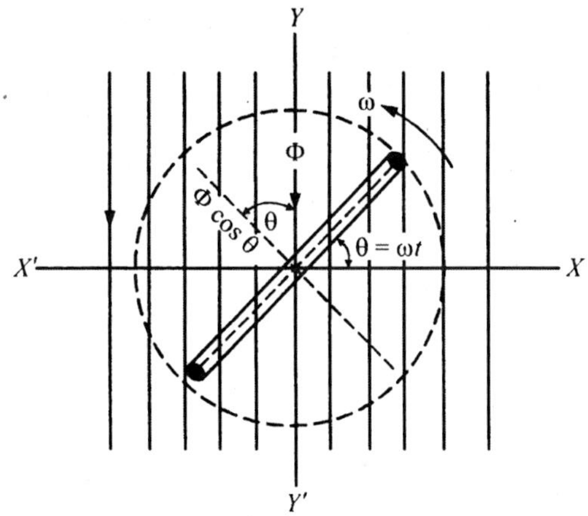

Fig. 5

$$= N\Phi \cos \omega t$$

The value of induced e.m.f. is

$$e = -\frac{d}{dt}(N\Phi) = -N\frac{d}{dt}(\Phi \cos\omega t) = -\omega N\Phi(-\sin \omega t)$$

$$= \omega N \Phi \sin \theta \text{ volt} \qquad \ldots(i)$$

When the coil has turned through 90° *i.e.* when $\theta = 90°$, then $\sin\theta = 1$. Hence, e has the maximum value, say, E_m. In that case from Eqn. (*i*), we get

$$E_m = \omega N\Phi = \omega NBA \text{ volt}$$

where B is the flus density and A is the area of the coil. Substituting this value of E_m in Eqn. (*i*), we have

$$e = E_m \sin\theta = E_m \sin \omega t$$

Similarly, the equation of the induced current is

$$i = I_m \sin \omega t$$

Since $\omega = 2\pi f$, we get

$$e = E_m \sin 2\pi ft = E_m \sin\left(\frac{2\pi}{T}\right)t$$

and $$i = I_m \sin 2\pi ft = I_m \sin\left(\frac{2\pi}{T}\right)t \qquad \text{where } f = 1/T.$$

Note : $E_m = \omega N\Phi = \omega NB\,A = 2\pi f\,NB\,A$ volt $\quad \therefore\ e = 2\pi f\,N\,B\,A \sin\theta$.

Q. 6. Derive the equation of the ac voltage induced in a rectangular coil when it rotates in a magnetic field in terms of its dimensions and flux density,

Ans. Fig. 6. shows a rectangular coil AC having N turns and rotating in a magnetic field of flux density. B Wb/m^2 with an angular velocity of ω rad/s. Let l be the length and b be the breadth of the coil in metres and their linear velocity be v m/s. When in the horizontal position, the two sides A and C move *parallel* to the lines of magnetic flux. Hence, no flux is cut and therefore, no emf is induced in the coil.

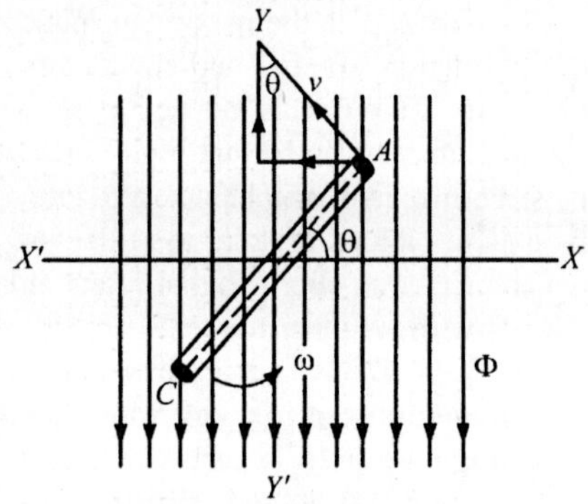

Fig. 6

When the coil has turned through angle θ, its velocity v can be resolved into two mutually perpendicular components.

(*i*) $v \cos\theta$ parallel to the direction of magnetic flux and

(*ii*) $v \sin\theta$ perpendicular to the direction of flux.

The e.m.f. generated is due entirely to the *perpendicular* component *i.e.* $v \sin\theta$. Hence, e.m.f. generated in one side of the coil which contains N conductors is given by e

$$e = N \times B\,l\,v \sin\theta$$

Total e.m.f. generated in both sides of the coil is

$$e = 2\,B\,N\,l\,v \sin\theta \qquad ...(i)$$

Now, e has maximum value *of* E_m when $\theta = 90°$.

Hence, form Eqn. (*i*) above, we get

$$E_m = 2\,B\,N\,l\,v \qquad ...(ii)$$

Therefore, Eqn. (*i*) can be written as

$$e = E_m \sin\theta$$

If b = breadth of the coil AC in metres

f = frequency of rotation of the coil in rev/s

Then, $v = (\pi b) \times f$ metre/second

Hence, from Eqn. (*ii*), we get

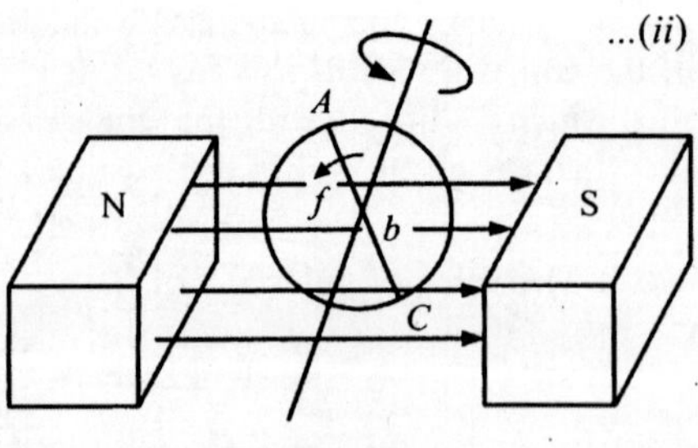

Fig. 7

$$\therefore\ E_m = 2\,BNl \times \pi\,b\,f = 2\,\pi\,f\,BNlb = 2\pi f\,BNA$$

$$\therefore\ e = 2\pi f\,BNA \sin\theta \text{ volt}$$

or $$e = E_m \sin\theta \text{ volt}$$

where E_m = maximum value of generated e.m.f.

$A = l \times b$ – area of the coil and

N = No. of the conductors in the coil.

Q. 7. A coil of 1000 turns is rotated at 3000 rpm in a magnetic field having a uniform flux of 0.2 tesla. The average area of each turn is 50 cm^2 and axis of rotation is at right angles to the direction of the flux. Calculate:

(*a*) frequency (*b*) time period (*c*) angular velocity (*d*) maximum value of the generated voltage.

Write down the expression for the instantaneous voltage and calculate its value when the coil has rotated 7 π/6 radian from the position of zero voltage.

Ans. (*a*) frequency, $f = 3000/\ 60 =$ **50 Hz**

(*b*) time period, $T = 1/f = 1/50$ second = **20 millisecond**

(*c*) angular velocity, $\omega = 2\pi f = 2\pi \times 50 =$ **314 radians/second**

(*d*) maximum voltage,

$$E_m = \omega NBA = 314 \times 1000 \times 0.2 \times (50 \times 10^{-4}) = \mathbf{314\ volt}$$

∴ instantaneous voltage is

$$e = E_m \sin \omega t = 314 \sin\theta$$

Now $$\theta = \frac{7\pi}{6} \text{ radian} = \left(\pi + \frac{\pi}{6}\right) \text{ radian}$$

∴ $$e = -\ 314 \sin\left(\pi + \frac{\pi}{6}\right) = -\ 314 \sin\left(\frac{\pi}{6}\right) = -\ 314 \sin 30°$$

$$= -\ 314 \times 0.5 = \mathbf{-\ 157\ V}$$

Q. 8. A square coil of 10 cm side and with 100 turns is rotated at the uniform speed of 500 rpm about an axis at right angles to a uniform field of 0.5 Wb/m^2. Calculate the instantaneous value of the induced emf when the plane of the coil is (*a*) at right angles to the plane of the field (*b*) at 30° to the plane of the field and (*c*) in the plane of the field.

(*Elect. Engg. A.M.Ae.S.I. 2000*)

Ans. The instantaneous value of the e.m.f. is

$$e = 2\pi f NBA \sin\theta \text{ volt}$$ – refer Q. 6.

where θ is measured from the position of the coil when its plane is at right angles to the flux (Fig. 7A).

Here, $f = 500/60 = 25/3$ rps

$N = 100;\ B = 0.5$ Wb/m^2

$A = 10 \times 10 = 100$ cm^2 $= 10^{-2}$ m^2

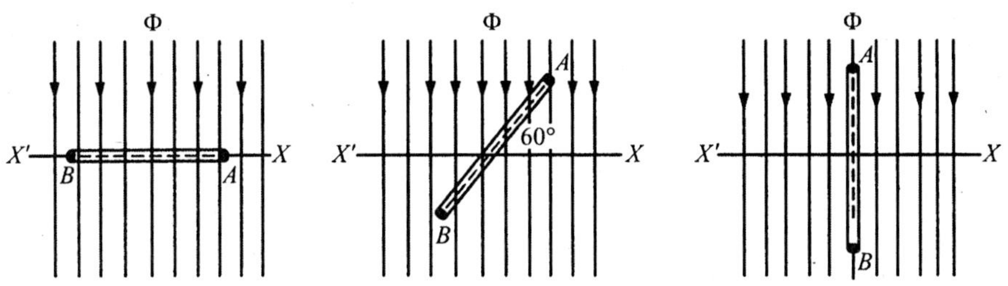

Fig. 7A

(*a*) $\theta = 0°, \sin\theta = 0$ ∴ $e = \mathbf{0}$

(*b*) $\theta = 60°, \sin 60° = 0.866$

$$e = 2\pi \times (25/3) \times 0.5 \times 100 \times 10^{-2} \times 0.866 = \mathbf{22.6\ V}$$

(*c*) $\theta = 90°$, $\sin 90° = 1$

$\therefore$ $e = 2\pi \times (25/3) \times 0.5 \times 100 \times 10^{-2} = \mathbf{26.18\ V}$

Q. 9. Define the following terms with reference to a sine wave : (*i*) cycle (*ii*) period (*iii*) frequency (*iv*) wavelength and (*v*) velocity.

Ans. (*i*) Cycle : One complete set of positive and negative values of an alternating quantity is known as its cycle. Simply put, **a cycle is one complete alternation.** Fig. 8 shows one cycle of a sinusoidal wave. It occupies 2π radian or 360°.

(*ii*) **Period :** The length of time needed for one cycle is called the period of the wave as shown in Fig. 8. Its symbol is *T*. Also, $T = 1/f$

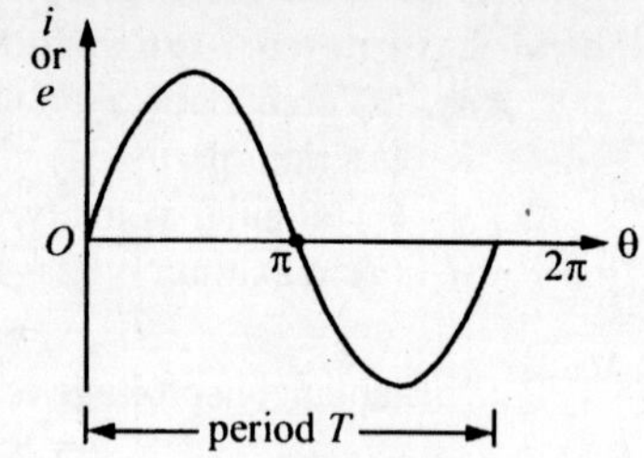

Fig. 8

(*iii*) **Frequency :** It is the number of cycles that a sine wave completes in one second. Its unit is cycle per second or hertz (Hz).

There is a reciprocal relationship between frequency and period of a sine wave *i.e.*

$$f = 1/T \quad \text{or} \quad T = 1/f$$

(*iv*) **Wavelength :** It is the distance occupied by one complete cycle of a sine wave.

(*v*) **Velocity :** The velocity of a wave is given by the product of its frequency and wavelength *i.e.* $v = f\lambda$.

Q. 10. The FM radio band covers the frequency range 88 to 108 MHz. What is the equivalent wavelength range? The velocity of propagation of radio waves is 3×10^8 m/s.

Ans. For f = 88 MHz

$$v = f\lambda \quad \text{or} \quad \lambda = \frac{v}{f} = \frac{3 \times 10^8}{88 \times 10^6} = \mathbf{3.409\ m}$$

For f = 108 MHz

$$v = f\lambda \quad \text{or} \quad \lambda = \frac{v}{f} = \frac{3 \times 10^8}{108 \times 10^6} = \mathbf{2.778\ m}$$

Q. 11. Explain the following terms for a sine wave with the help of suitable diagrams

(*a*) phase **(*b*) phase difference** **(*c*) in-phase**
(*d*) anti-phase **(*e*) lag and lead** **(*f*) peak-to-peak value**
(*g*) peak amplitude

Ans. (*a*) Phase : It is given by the amount of time (or angle) since the wave *last passed through its positive zero value.*

As shown in Fig. 9(*a*), the phase of the sine wave at point *A* is *T*/4 second or $\pi/2$ radian or 90°.

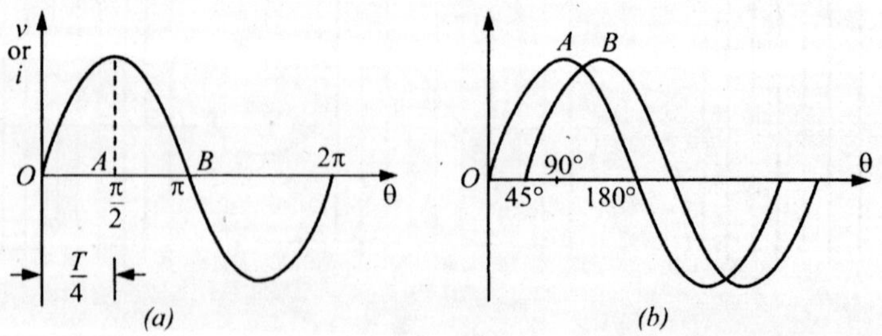

Fig. 9

(*b*) Phase difference : It is the *time* difference or *angle* difference between two sine waves since they *last* passed through their zero value.

For example, the phase difference between the two sine waves shown in Fig. 9(*b*) is 45° or π/4 radian or *T*/8 seconds.

(*c*) In-phase : When two sine waves achieve their zero value or maximum value at the *same time*, they are said to be in phase (or in-step) with each other. The phase difference of the two sine waves shown in Fig. 10(*a*) is zero

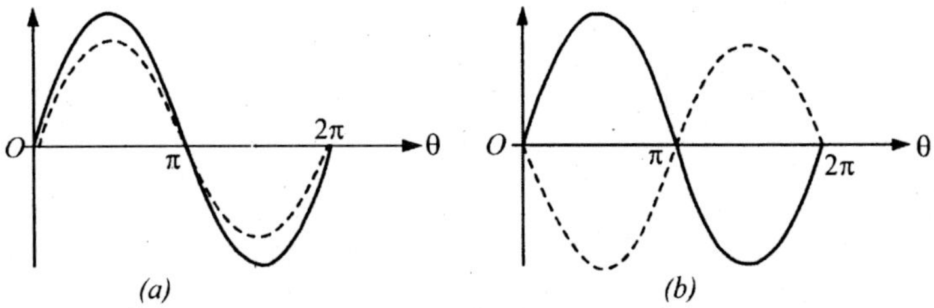

Fig. 10

(*d*) Anti-phase : If two sine waves achieve their zero values at the same time but one swings positive when the other swings negative, they are said to be in antiphase. Their phase difference is 180° or π radians. They are said to be 180° out of phase with each other as shown in Fig. 10(*b*). Their net resultant is zero.

(*e*) Lag and lead : A lagging sine wave is one which reaches its zero or maximum value *later* than the other wave.

As shown in Fig. 11(*a*), the dotted sine wave reaches its zero value 1ms later than the other wave. Also, it reaches its maximum value 2 ms later. Hence, it is said to lag behind the other wave by 1ms or *T*/4 seconds. In terms of angle, it lags behind by π/2 radian or 90°.

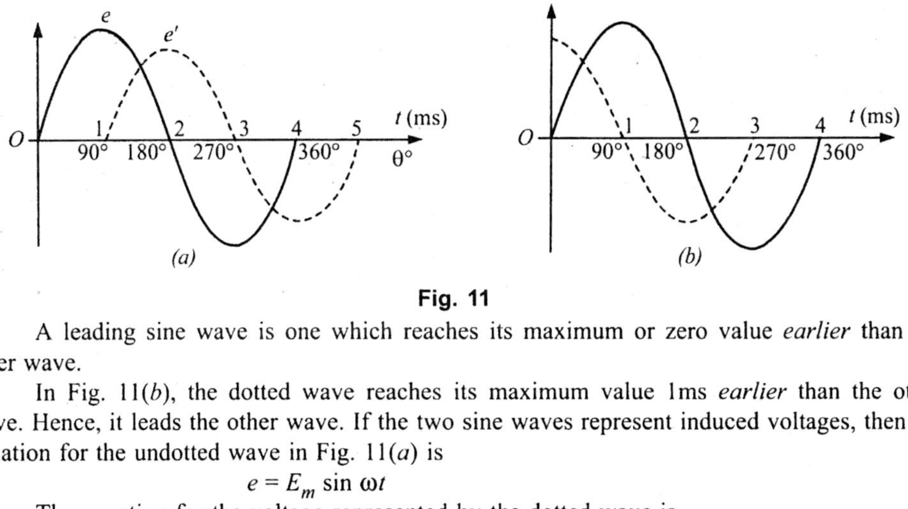

Fig. 11

A leading sine wave is one which reaches its maximum or zero value *earlier* than the other wave.

In Fig. 11(*b*), the dotted wave reaches its maximum value 1ms *earlier* than the other wave. Hence, it leads the other wave. If the two sine waves represent induced voltages, then the equation for the undotted wave in Fig. 11(*a*) is

$$e = E_m \sin \omega t$$

The equation for the voltage represented by the dotted wave is

$$e' = E'_m \sin (\omega t - \pi/2)$$

In Fig. 11(*b*), the voltage equation for the undotted sine wave is

$$e = E_m \sin \omega t$$

The value of voltage of the dotted sine wave is

$$e' = E'_m \sin (\omega t + \pi/2)$$

It may be noted that the phase angle of +π/2 means a *leading* phase and a phase angle of – π/2 means a *lagging* phase.

(*f*) Peak-to-Peak value : In the symmetrical sine wave shown in Fig. 12, the *p-p* value is twice the peak value of the wave *i.e.* (5 + 5) = 10 V. It is seen that *p-p* value of a sine wave is the total waveform magnitude taken from the positive peak to the negative peak.

The *p-p* value indicates how much the wave level or amplitude swings during the cycle.

(*g*) Peak Amplitude : The peak amplitude of a sine wave is its maximum value *either side of zero.*

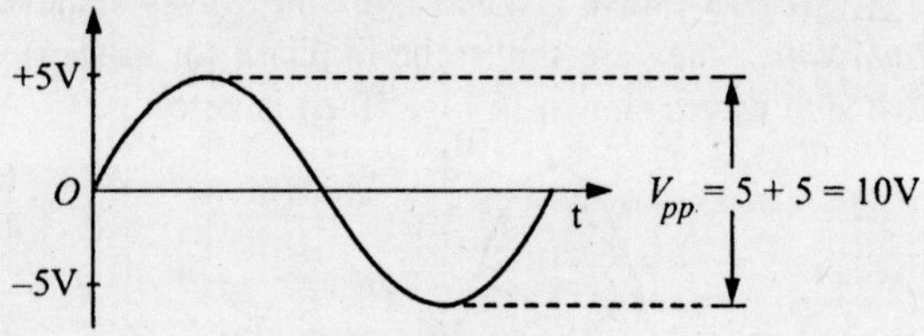

Fig. 12

As shown in Fig. 12, the peak vaule of the wave is 5V and *p-p* value is 10V.

Q. 12. Explain with suitable diagrams, how the instantaneous equation of an ac sine wave changes with change in the reference point?

Ans. Suppose there is a sinusoidal ac current of maximum value 15A, frequency of 50Hz and hence an angular velocity of ω = 314 radian/second (Fig. 13). If the angle is measured from point *O* where *i* = 0 and is becoming positive, then the equation of the current is

$$i = 15 \sin\theta = 15 \sin 2\pi ft$$
$$= 15 \sin 100\pi t = 15 \sin 314t$$

Suppose we now want to find the equation of the wave with reference to a point other than the origin point *O*. Then, we simply *add the angle at the new reference point to the angle* 314*t radian.* If at the new refrence point θ = 90° as shown is Fig. 13. then the equation of the current wave becomes.

$$i = 15 \sin (314\, t + 90°) = \mathbf{15 \cos 314} t$$

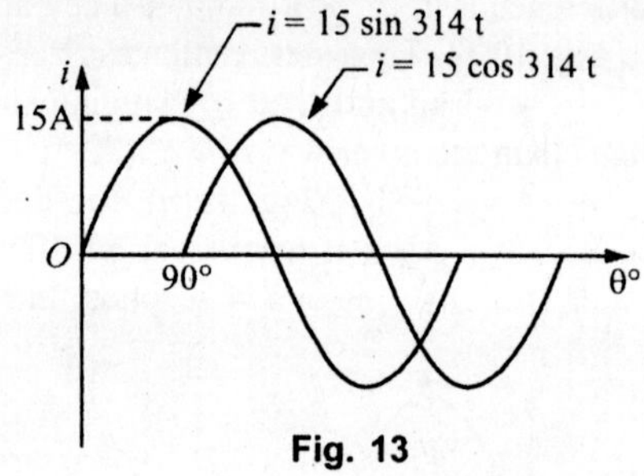

Fig. 13

This wave is shown in Fig. 13.

Q. 13 (*a*) Calculate the instantaneous value at the 90° reference point *D* on the horizontal axis for each voltage sine wave shown in Fig. 14.

(*b*) An alternating current is given by the expression

$$i\,(t) = 300 \sin (157\, t + \pi/3)$$

Find : (*i*) maximum value of current (*ii*) frequency and (*iii*) periodic time.

(*Elect. Engg. Punjab Univ.*)

Ans. As seen in Fig. 14, sine wave *A* is the normal reference wave since it starts from the origin *O*. Since wave *B leads A* by 20° and wave *C lags A* by 45°. The instantaneous values of the three sine waves with reference to point *D* are as under:

$$e_A = E_m \sin\theta$$
$$= 20 \sin (90° + 0°) = 20 \times 1 = \mathbf{20\ V}$$
$$e_B = 10 \sin (\theta + \phi_B)$$
$$= 10 \sin 90° + 20°)$$
$$= 10 \cos 20°$$
$$= 10 \times 0.9397 = \mathbf{9.4\ V}$$
$$e_C = 16 \sin (\theta - \phi_C)$$
$$= 16 \sin (90° - 45°)$$
$$= 16 \times 0.707$$
$$= \mathbf{11.32\ V}$$

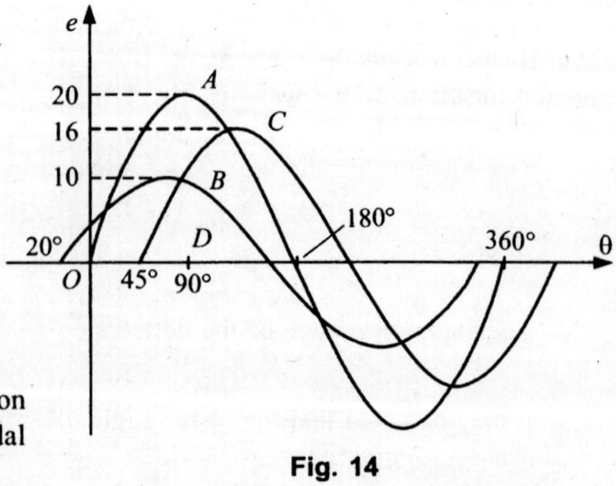

Fig. 14

(*b*) Comparing the given current equation with the standard equation for a sinusoidal alternating current, we get the following:

(*i*) I_m = **300 A**

(*ii*) ω = 157 or 2πf = 157 or f = **25 Hz**

(*iii*) Time period, $T = 1/f = 1/25$ = **0.04 second.**

Q. 14 (*a*) Sketch and explain a phasor diagram to represent a sine wave. Use the peak value of the wave.

(*b*) an alternating current varying sinusoidally with a frequency of 50 Hz has r.m.s value of 10A. Write down the equation for instantaneous value and find this value at

(*i*) 0.0025 second and

(*ii*) 0.0125 second after passing through a positive maximum value

(*iii*) at what time measured from the positive maximum value, will the instantaneous current be 7.07A? Draw this current wave. (*Elect. Engg, Nagpur Univ. 2000*)

Ans. A full cycle of a sine wave can be represented by the rotation of a phasor (or vector) through 360°. The instantaneous value of the sine wave at any point is equal to the *vertical* distance from the *tip* of the phasor to the horizontal axis.

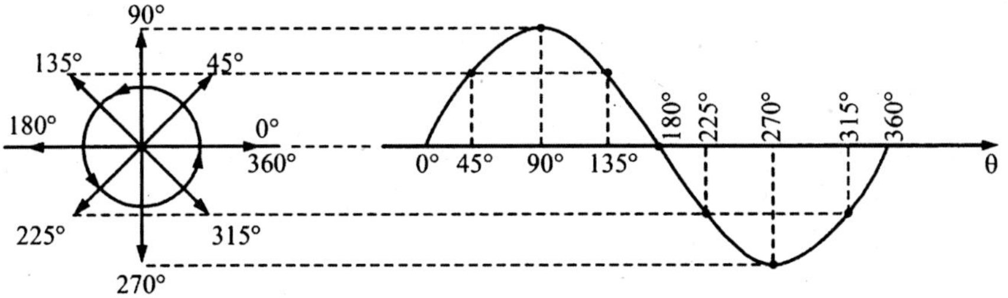

Fig. 15

Fig. 15 shows how the phasor traces out the sine wave as it goes from 0° to 360° in the counter-clockwise direction with a velocity of ω radian/second. It may be noted that the length of the phasor is equal to the peak or the *maximum* value of the sine wave. The angle of the phasor measured form 0° is the corresponding angular point on the sine wave. If the sine wave represents ac voltage of maximum value E_m, then its instantaneous value is given by the perpendicular projection of E_m on *X*-axis

$$e = E_m \sin \theta = E_m \sin \omega t$$

(*b*) $I_{rms} = 10\text{A}, f = 50\text{Hz}, I_{rms} = I_m/\sqrt{2}$

∴ $I_m = \sqrt{2} \times 10 = 14.14\text{A}$

$\omega = 2\pi f = 100\pi$ rad/sec

The instantaneous equation of the given ac current wave with reference to point *O* in Fig. 16 is

$$\mathbf{i = 14.14 \sin \omega t}$$
$$\mathbf{= 14.14 \sin 100\,\pi\, t}$$

Since time value is given from point *A* where current has positive and maximum, we will refer the given current equation itself to point *A*. In that case, the equation of the ac current becomes

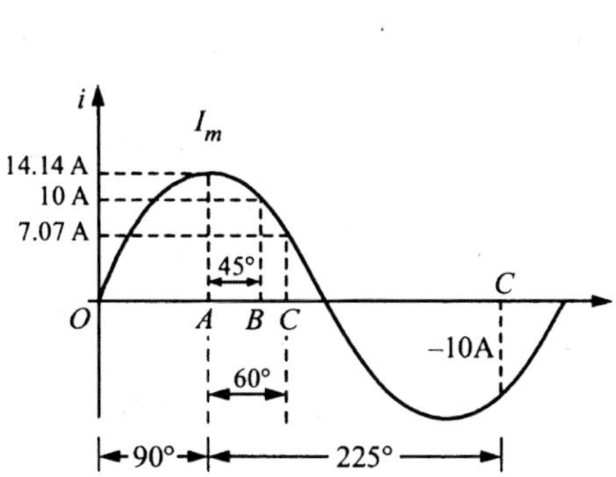

Fig. 16

$$i = 14.14 \sin (\omega t + 90°)$$
$$= 14.14 \cos \omega t$$
$$= 14.14 \cos 100\,\pi\, t \quad \text{– refer Q. 12}$$

(*i*) **When *t* = 0.0025 second**

$$i = 14.14 \cos 100\,\pi \times 0.0025 \quad \text{– angle in radian}$$

$= 14.14 \cos 100 \times 180 \times 0.0025$ – angle in degrees.

$= 14.14 \cos 45° = \mathbf{10\ A}$ – point *B*

(*ii*) **When $t = 0.0125$ second**

$i = 14.14 \cos 100 \times 180° \times 0.0125 = 14.14 \cos 225°$

$i = 14.14 \times (-1/\sqrt{2}) = -10$ A – point *D*

(*iii*) Here, $i = 7.07$ A. Substituting this value in the equation, we get

$7.07 = 14.14 \cos 100 \times 180 \times t$

$\therefore \quad 100 \times 180\ t = \cos^{-1}(0.5) = 60°$ – point *C*

$\therefore \quad t = \mathbf{1/300\ second}$

Q. 15. Differentiate between the average and root-mean square (rms) values of a current sine wave. What do they signify?

Ans. Fig. 17(*a*) shows positive half-cycle of a current sine wave. It can be divided into a large number of n instantaneous currents $i_1, i_2, i_3 \ldots i_n$. The average value of these currents over one half-cycle is given by the mean of these currents.

$$\therefore \quad I_{av} = \frac{i_1 + i_2 + i_3 + \ldots\ldots\ldots i_n}{n}$$

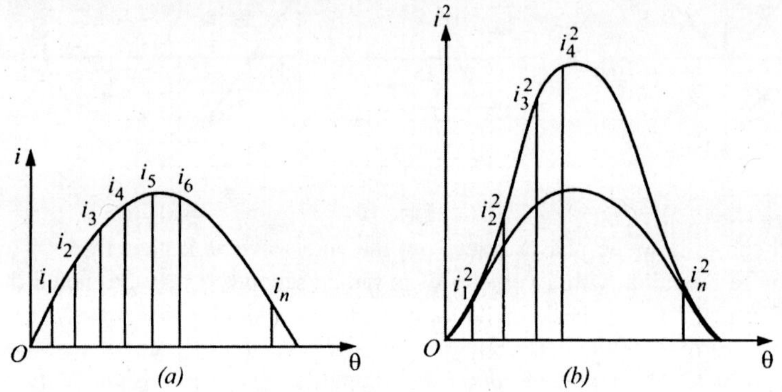

Fig. 17

Fig. 17(*b*) shows the half- cycle for the square values of the instantaneous values *i.e.*

$$i_1^2,\ i_2^2,\ i_3^2\ \ldots\ i_n^2.$$

The mean value of these squares is

$$= \frac{i_1^2 + i_2^2 + i_3^2\ \ldots\ldots\ldots i_n^2}{n}$$

The square root of this mean value is known as the root-mean-square (rms) value of the current sine wave. It is denoted by I_{rms} or simply by *I*.

$$\therefore \quad I_{rms} = \sqrt{\left(\frac{i_1^2 + i_2^2 + i_3^2\ \ldots\ldots\ldots i_n^2}{n}\right)}$$

The significance of these currents is as under.

(*i*) the I_{av} of an ac current wave represents that value of a direct current which transfers the same amount of *charge* through a given circuit for a given time as done by the ac current wave.

(*ii*) The rms value of an ac current wave represents that value of direct current which produces the same current *heat* through a given resistance for a given time as produced by the varying ac current wave.

Q. 16. Can the average and RMS values of a sine current wave be found from the area of its half-cycle? Does it apply only to sine waves?

Ans. In Fig. 18, let the area of the half-cycle of the current sine wave be A. I_{av} is the equivalent *dc* current which transfers the same *charge* as the given varying *ac* current. Area of the dotted rectangle is $= I_{av} \times \pi$. It must equal the area A of the half-cycle.

$$\therefore \quad I_{av} \times \pi = A \quad \text{or} \quad I_{av} = A/\pi$$

Since the ac current wave is given by $i = I_m \sin\theta$

$$\therefore \quad A = \int_0^{\pi} i\, d\theta = \int_0^{\pi} I_m \sin\theta\, d\theta$$

$$\therefore \quad I_{av} = \frac{1}{\pi}\int_0^{\pi} i\, d\theta = \frac{1}{\pi}\int_0^{\pi} I_m \sin\theta\, d\theta$$

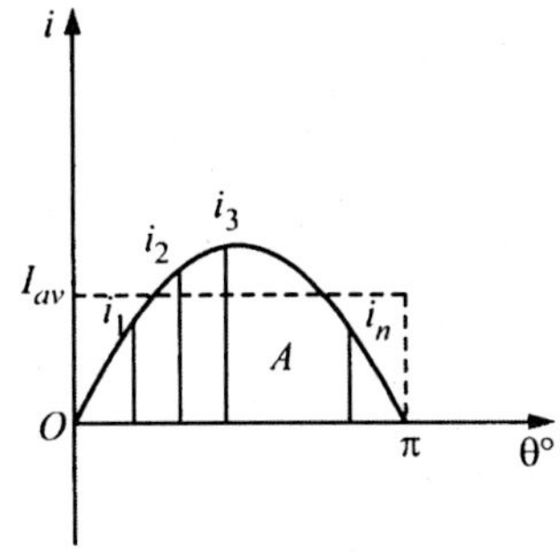

Fig. 18

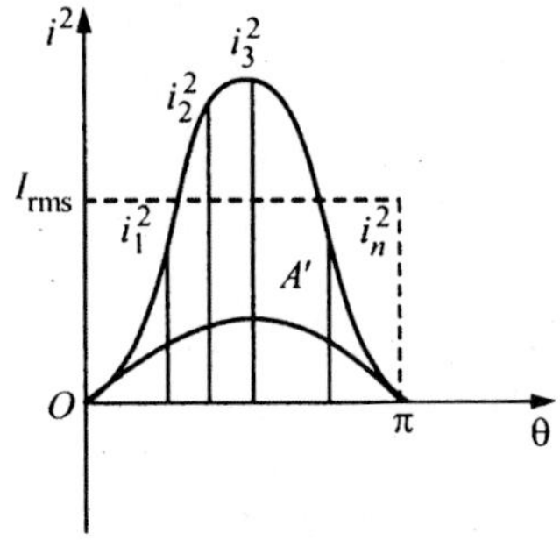

Fig. 19

If the equation of the sine wave is not given, then A can be found from the geometry of the wave. Fig. 19 shows the power half-cycle of the given current sine wave. It represents the *heat* produced during the half-cycle. Let the area of the power half-cycle be A'. Also, let I_{rms} be the equivalent *dc* current which produces the *same* heat. Then, area of the dotted rectangle equals A'.

$$\therefore \quad I_{rms}^2 \times \pi = A' \quad \text{or} \quad I_{rms} = \sqrt{A'/\pi}$$

$$\text{Now,} \quad A' = \int_0^{\pi} i^2\, d\theta \qquad \therefore \quad I_{rms} = \sqrt{\frac{1}{\pi}\int_0^{\pi} i^2\, d\theta}$$

It applies to all waves whether sinusoidal or non-sinusoidal, symmetrical or asymmetrical.

Q. 17. What is meant by the average value of an alternating current? Calculate its value for a sinusoidal current wave.

Ans. The average value of a current wave is equal to the *mean* of its instantaneous values in a full cycle or half cycle. In the case of a symmetrical sine wave, the average value over a full cycle is zero. It is so because the average of its positive half-cycle is cancelled by the average of its negative half-cycle. Hence, in the case of *symmetrical* waveforms, finding the average value over one half-cycle is sufficient. Let the equation of a sine wave be $i = I_m \sin\theta$.

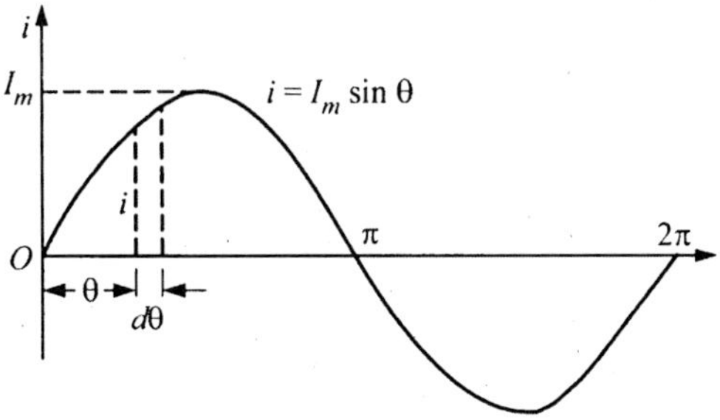

Fig. 20

$$I_{av} = \frac{1}{\pi}\int_0^{\pi} i\, d\theta = \frac{1}{\pi}\int_0^{\pi} I_m \sin\theta . d\theta \qquad \text{– Fig. 20}$$

$$= \frac{I_m}{\pi}\int_0^{\pi} \sin\theta . d\theta = \frac{I_m}{\pi}\left| -\cos\theta \right|_0^{\pi} = \frac{I_m}{\pi}\left| -\cos\pi - (-\cos 0) \right|$$

$$= \frac{I_m}{\pi} \mid -(-1)-(-1) \mid = \frac{2I_m}{\pi}$$

$$\therefore \quad I_{av} = \frac{2I_m}{\pi} = 0.637\ I_m$$

Note **:** If sine wave is drawn on a time base, then

$$I_{av} = \frac{1}{T/2} \int_0^{T/2} i dt = \frac{I_m}{T/2} \int_0^{T/2} \sin \omega t . dt$$

$$= \frac{2I_m}{T} \int_0^{T/2} \sin \frac{2\pi}{T} t\, dt = \frac{2I_m}{\omega T} \left| -\cos \frac{2\pi}{T} . t \right|_0^{T/2}$$

$$= \frac{2I_m}{\frac{2\pi}{T} . T} \mid -\cos \pi - \cos 0 \mid = \frac{2I_m}{2\pi} \mid -(-1)-(-1) \mid$$

$$= \frac{I_m}{\pi} (2) = \frac{2I_m}{\pi} = 0.637\ I_m \qquad \text{– as above}$$

Q. 18. Define the root mean square value of a symmetrical current waveform. Derive its value in the case of a standard sine wave whose equation is $i = I_m \sin\theta$.

Ans. The rms value of a current waveform is equal to the that dc current which produces the same amount of *heat* as the given waveform.

The term *rms* stands for (square) root of the mean of the squares of the instantaneous current values.

$$\therefore \quad I_{rms} = \sqrt{\frac{1}{\pi} \int_0^{\pi} i^2 . d\theta}$$

$$I_{rms}^2 = \frac{1}{\pi} \int_0^{\pi} I_m^2 \sin^2 \theta\, d\theta = \frac{I_m^2}{\pi} \int_0^{\pi} \sin^2 \theta . d\theta$$

$$= \frac{I_m^2}{\pi} \int_0^{\pi} \left(\frac{1-\cos 2\theta}{2} \right) .d\theta = \frac{I_m^2}{2\pi} \int_0^{\pi} (1-\cos 2\theta) . d\theta$$

$$= \frac{I_m^2}{2\pi} \left| \theta - \frac{1}{2} \sin 2\theta \right|_0^{\pi}$$

$$= \frac{I_m^2}{2\pi} \left| \left(\pi - \frac{1}{2} \sin 2\pi \right) - \left(0 - \frac{1}{2} \sin 2 \times 0 \right) \right|$$

$$= \frac{I_m^2}{2\pi} \mid \pi - 0 - 0 + 0 \mid = \frac{\pi}{2\pi} I_m^2 = \frac{I_m^2}{2}$$

$$\therefore \quad I_{rms} = \sqrt{\frac{I_m^2}{2}} = \frac{I_m}{\sqrt{2}} = 0.707\ I_m$$

The rms value is *always* greater than the average value except for a reactangular wave when these are equal. Its value can *never* be negative.

Points to Note :

1. The average value I_{av} of an alternating quantity over a half-cycle may be found by dividing the half-cycle area by the length of its base. The area itself may be found by using the equation of the alternating quantity (if given) or geometrically
2. The procedure for finding the rms value is as under

   (*i*) find the area under the curve of the square of the alternating quantity.

(*ii*) divide the area by the length of the base

(*iii*) take the square root of the result.

3. If the given sine waveform contains both dc and ac currents, then

   (*i*) average value of the given waveform equals the sum of dc current and the average of the alternating current.

   (*ii*) the *rms* value of a given waveform containing ac and dc is given by

$$I_{rms} = \sqrt{I_{dc}^2 + I_{rms}^2}$$

where, I_{dc} is the dc component and I_{rms} is the rms value of the ac component.

Q. 19. Define (*i*) Form Factor and (*ii*) Peak Factor of a sine wave. What is their importance?

Ans. (*i*) Form factor is defined as the ratio,

$$K_f = \frac{\text{rms value}}{\text{average value}} = \frac{0.707\ I_m}{0.637\ I_m} = 1.1$$

As seen, with the help of form factor, we can find the rms value from the average value and vice-versa. This factor shows the "peakiness" of the waveform. A square wave, which is flatter than a sine wave, has a form factor of 1.0. A triangular wave, on the other hand, is more "peaky" than a sine wave and has a form factor of 1.15

(*ii*) Peak factor is given by the ratio,

$$K_a = \frac{\text{maximum value}}{\textit{rms}\text{ value}} = \frac{I_m}{I_{m/\sqrt{2}}} = \sqrt{2} = 1.414$$

This factor is also known as Crest Factor or Amplitude factor.

This factor has importance in dielectric insulation testing. It is so because the dielectric stress on the material being tested depends on the peak value of the applied voltage. This factor is also important when measuring iron loss because iron loss depends on the value of maximum flux.

Q. 20. For the current waveform shown in Fig. 21, find

(*i*) peak current **(*ii*) average value**

(*iii*) frequency **(*iv*) periodic time**

(*v*) instantaneous value at *t* = 3 ms

(*Basic Elect. Engg. VTU Karnataka 2007*)

Ans. (*i*) Peak or maximum current = **20 A**

(*ii*) average value, $I_{av} = 0.637\ I_m = 0.637 \times 20 =$ **12.74 A**

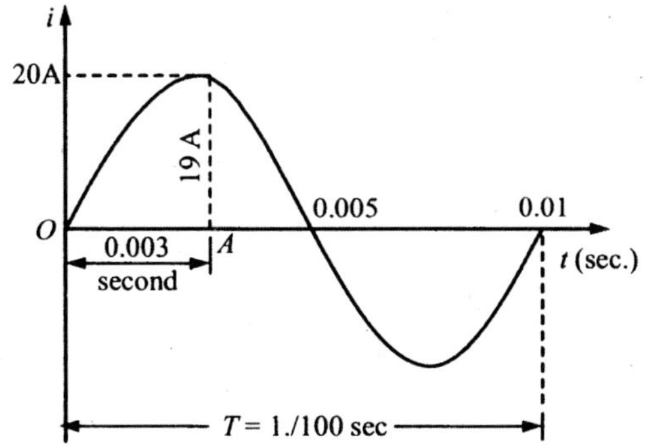

Fig. 21

(*iii*) frequency, $f = \frac{1}{T} = \frac{1}{1/100} =$ **100 Hz** (*iv*) periodic time, $T = 1/100$ second = **0.01 second**

(*v*) $i = I_m \sin\theta = I_m \sin \omega t = I_m \sin 2\pi \times 100 \times t =$ **20 sin 628 t**

Now, putting $t = 3$ ms $= 3 \times 10^{-3}$ second = 0.003 second (point *A*)

$\therefore$ $i = 20 \sin 628 \times 0.003 = 20 \sin 1.884$ – angle in radian

$= 20 \sin 1.884 \times 57.3° = 20 \sin 108° = 20 \cos 18°$

$= 20 \times 0.95 =$ **19 A**

Q. 21. A resultant current wave is made up of two components 5A dc component and a 50 Hz sinusoidal component of maximum value 5A

(*i*) draw a sketch of the resultant wave

(*ii*) write the equation for the current wave taking $t = 0$ at a point where ac component is zero and becoming positive.

(*iii*) calculate the average value and rms value of the resultant current over one cycle.

Ans. (*i*) The sketch of the resultant wave is shown in Fig. 22.

(*ii*) $i = 5 + 5 \sin \theta$

(*iii*) Since the wave is symmetrical, the average value of the ac component over one cycle is zero. Hence, the average value of the resultant current is equal to the value of the dc component *i.e.* **5A**.

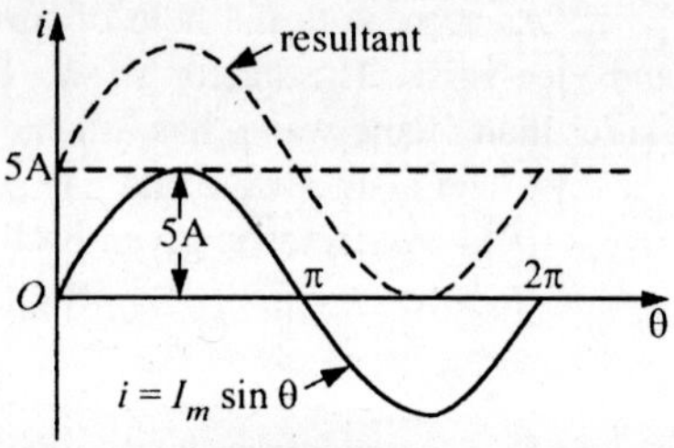

Fig. 22

Mean value of i^2 over one complete cycle is

$$= \frac{1}{2\pi}\int_0^{2\pi} i^2\, d\theta = \frac{1}{2\pi}\int_0^{2\pi} (5 + 5\sin\theta)^2\, d\theta$$

$$= \frac{1}{2\pi}\int_0^{2\pi} (25 + 50\sin\theta + 25\sin^2\theta)\, d\theta$$

$$= \frac{1}{2\pi}\int_0^{2\pi}\left[25 + 50\sin\theta + 25\left(\frac{1-\cos 2\theta}{2}\right)\right] d\theta$$

$$= \frac{1}{2\pi}\int_0^{2\pi} (37.5 + 50\sin\theta - 12.5\cos 2\theta)\, d\theta$$

$$= \frac{1}{2\pi}\left| 37.5\,\theta - 50\cos\theta - \frac{12.5}{2}\sin 2\theta \right|_0^{2\pi} = 37.5$$

$\therefore$ RMS value $= \sqrt{37.5} =$ **6.12 A**

Note : Alternatively, combined $I_{rms} = \sqrt{(I_{dc}^2 + I_{rms}^2 \text{ of } ac)}$

I_{rms} of ac component $= 5/\sqrt{2}$.

$\therefore$ combined $I_{rms} = \sqrt{5^2 + (5/\sqrt{2})^2} = \sqrt{37.5} =$ **6.12A.**

Q. 22. An ac current is given by $i = 10 \sin \omega t + 3 \sin 3\omega t + 2 \sin 5\omega t$. Find the rms value of the current. (*Basic Elect. Engg. VTU Karnataka 2008*)

Ans. The given ac current is a *complex* current consisting of three different sinusoidal currents. The rms value of this complex ac current wave is

$$= \sqrt{\left(I_{1m}/\sqrt{2}\right)^2 + \left(I_{2m}/\sqrt{2}\right)^2 + \left(I_{3m}/\sqrt{2}\right)^2}$$

$$\therefore \quad I_{rms} = \sqrt{\left(\frac{10}{\sqrt{2}}\right)^2 + \left(\frac{3}{\sqrt{2}}\right)^2 + \left(\frac{2}{\sqrt{2}}\right)^2} = \sqrt{56.5} = \mathbf{7.52\ A.}$$

Q. 23. Define RMS value, average value and form factor of a sinusoidal waveform. Find the average and rms values of the sawtooth waveform shown in Fig. 23.

(*Basic Elect. Engg. Kerala Univ. 2004*)

Ans. For definitions, please see Q. 18 and 19.

In the sawtooth waveform of Fig. 23, the voltage increases linearly *i.e.* in a straight line.

$$\therefore \quad V_{av} = \frac{V_{max} + V_{min}}{2}$$

$$= \frac{50 + 0}{2} = \mathbf{25\ V}$$

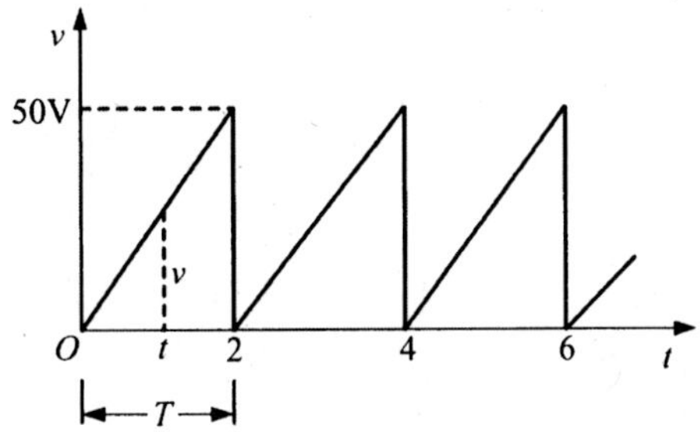

Fig. 23

For finding the equation of the voltage, we will consider voltage v at time t. From the two similar triangles of Fig. 23, we get

$$\frac{v}{t} = \frac{50}{2} \quad \text{or} \quad v = 25t$$

The rms value of the given voltage is

$$V_{rms} = \sqrt{\frac{1}{T}\int_0^T v^2 dt} = \sqrt{\frac{1}{2}\int_0^2 (25t)^2 dt} \qquad - (\because T = 2)$$

$$= \sqrt{\frac{1}{2}\int_0^2 625t^2 . dt} = \sqrt{\frac{625}{2}\left|\frac{t^3}{3}\right|_0^2}$$

$$= \sqrt{\frac{625}{2}\left|\frac{2^3}{3} - 0\right|} = \sqrt{833.33} = \mathbf{28.9\ V}$$

Q. 24. Find the average and rms values of the sawtooth current waveform given in Fig. 24(*a*).

(*Basic Elect. Engg. Kerala Univ. 2004*)

Ans. We will use the *area* method rather than the *integration* method for calculating the required current values.

Average value : The area of the right-angled triangle OBC is in Fig. 24(a)

$$A = \frac{1}{2} \times BC \times OB$$

$$= \frac{1}{2} \times 2 \times 1 = 1$$

Base length = 1

$$\therefore \quad I_{av} = \frac{A}{\text{base length}} = \frac{1}{1} = \mathbf{1\ A}$$

Alternatively, we can find the average value by taking the mean of the maximum and minimum values of the current. It is due to the fact that the relation between time and current is linear.

$$\therefore I_{av} = \frac{I_{max} + I_{min}}{2} = \frac{2 + 0}{2} = \mathbf{1\ A}$$

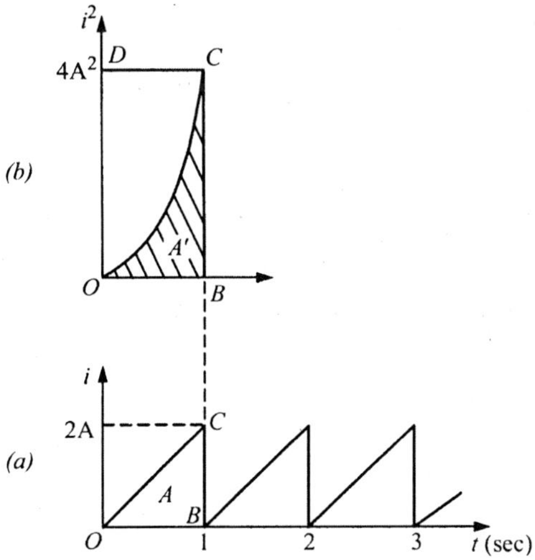

Fig. 24

RMS Value : The square curve is drawn in Fig. 24(*b*).

Here, the shaded area A' is equal to one-third of the area of rectangle *OBCD*.

Area of rectangle $OBCD = OD \times OB = 4 \times 1 = 4$

$$\therefore \quad A' = \frac{1}{3} \times 4 = \frac{4}{3}$$

$$\therefore \quad I_{rms} = \sqrt{\frac{\text{area of shaded curve}}{\text{base length}}}$$

$$= \sqrt{\frac{4/3}{1}} = \frac{2}{\sqrt{3}} = \mathbf{1.15\ A}$$

Q. 25. Determine the rms value, average value and form factor of the waveform shown in Fig. 25. (*Basic Elect. Engg. Nagpur Univ. 2006*)

Ans. The equation of the waveform can be found from Fig. 25 as shown below.

$$\frac{DE}{AE} = \frac{BC}{AC} \quad \text{or} \quad \frac{(y-10)}{t} = \frac{10}{T} \quad \therefore \quad y = 10 + (10/T)\, t$$

Mean square value of the function is

$$= \frac{1}{T}\int_0^T y^2 . dt = \frac{1}{T}\int_0^T \left(10 + \frac{10}{T}t\right)^2 dt$$

$$= \frac{1}{T}\int_0^T \left(100 + \frac{100}{T^2}t^2 + \frac{200}{T}t\right) dt$$

$$= \frac{1}{T}\left| 100\,t + \frac{100\,t^3}{3\,T^2} + \frac{200\,t^2}{2\,T}\right|_0^T$$

$$= \frac{1}{T}\left(100\,T + \frac{100\,T}{3} + 100\,T\right) = \frac{700}{3}$$

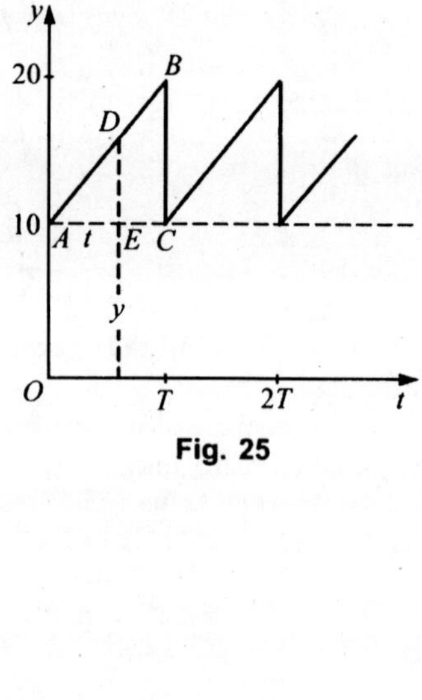

Fig. 25

Square root of the mean square values of function y

$$= \sqrt{700/3} = 10\sqrt{7/3} = 15.2$$

$\therefore$ rms value of y $= \mathbf{15.2}$

An easy way to find the average value of y is as under:

Maximum value of $y = 20$, Minimum value of $y = 10$

$$\therefore \quad \text{average value of } y = \frac{(10+20)}{2} = \mathbf{15}$$

$$\therefore \quad \text{form factor} = \frac{\text{rms value}}{\text{average value}} = \frac{15.2}{15} = \mathbf{1.01}$$

Alternatively, $$y_{av} = \frac{1}{T}\int_0^T y\, dt = \frac{1}{T}\int_0^T \left(10 + \frac{10}{T}t\right) dt$$

$$= \frac{1}{T}\int_0^T \left[10 . dt + \frac{10}{T}. t.dt\right]$$

$$= \frac{1}{T}\left| 10t + \frac{5t^2}{T}\right|_0^T = \mathbf{15}$$

Q. 26. Find the rms value, average value, form factor and peak factor for the trapezoidal waveform shown in Fig. 26. (*Elect. Engg. Nagpur Univ, 2003*)

Ans. For time interval $0 < t < 3T/20$, the equation of the current can be found by equating the two slopes in the right-angled triangle *OAB*.

$$\therefore \qquad \frac{i}{t} = \frac{I_m}{3T/20} \quad \text{or} \quad i = \left(\frac{20 I_m}{3T}\right) t$$

Also, for interval $3T/20 < t < 7T/20$, the equation of the current is $i = I_m$ *i.e.* current remains constant. Moreover, ΔOAB is identical with ΔDCE. Hence, expression for ΔOAB will be multiplied by 2 to include area of ΔDCE.

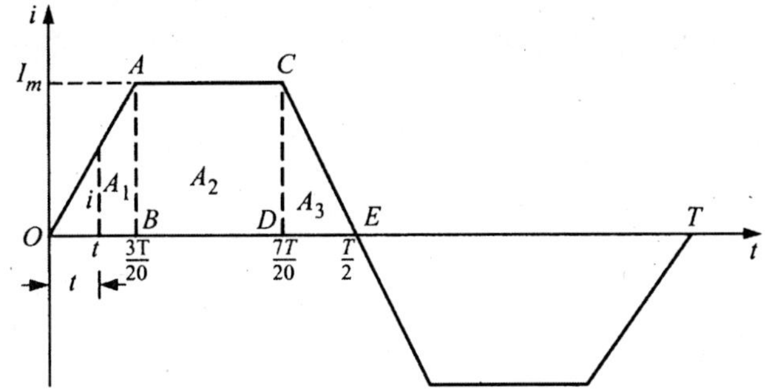

Fig. 26

Now, rms value is given by the square root of the mean of i^2 values in the trapezium *OACE*.

$$\therefore \quad I_{rms} = \sqrt{\frac{1}{T/2}\left[2\int_0^{3T/20} i^2 dt + \int_{3T/20}^{7T/20} I_m^2 \, dt\right]}$$

$$= \sqrt{\frac{2}{T}\left[2\left(\frac{20\, I_m}{3T}\right)^2 \int_0^{3T/20} t^2 dt + I_m^2 \int_{3T/20}^{7T/20} dt\right]} = \sqrt{(3/5)\, I_m^2} = \mathbf{0.775}\, \boldsymbol{I_m}$$

$$I_{av} = \frac{1}{T/2}\left[2\int_0^{3T/20} i dt + \int_{3T/20}^{7T/20} I_m \, dt\right]$$

$$= \frac{2}{T}\left[2\int_0^{3T/20}\left(\frac{20\, I_m}{3T}\right) t dt + I_m \int_{3T/20}^{7T/20} dt\right]$$

$$= \frac{2}{T}\left[2\left(\frac{20\, I_m}{3T}\right)\left|\frac{t^2}{2}\right|_0^{3T/20} + I_m \left|t\right|_{3T/20}^{7T/20}\right] = \frac{7}{10} I_m = \mathbf{0.7}\, \boldsymbol{I_m}$$

$$\text{From factor} = \frac{\text{rms value}}{\text{average value}} = \frac{0.775\, I_m}{0.7\, I_m} = \mathbf{1.107}$$

$$\text{Peak factor} = \frac{\text{max. value}}{\text{rms value}} = \frac{I_m}{0.775\, I_m} = \mathbf{1.29}$$

Q. 27. Find the average and rms values of the trapezoidal waveform shown in Fig. 27. (*Elect. Engg., Nagpur Univ. 2001*)

Ans. We will use the area method rather than the integration method for finding the current values.

$$I_{av} = \frac{\text{area under one half-cycle}}{\text{base length}}$$

Area under one half-cycle [Fig. 27(*a*)]

$$A = A_1 + A_2 + A_3$$

$$= \frac{1}{2} V_m \times 1 + V_m \times 1 + \frac{1}{2} V_m \times 1$$

$$= 2 V_m$$

Base length = 3

$$\therefore \quad I_{av} = \frac{A}{3} = \frac{2V_m}{3} = \mathbf{\frac{2}{3} V_m}$$

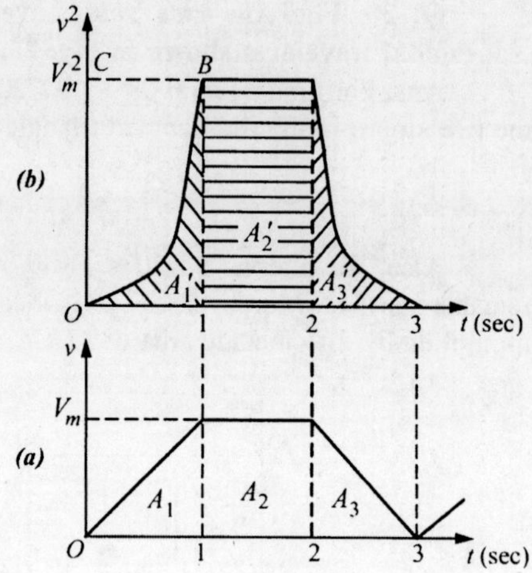

Fig. 27

For finding the rms value, we will have to draw squared curve as shown in Fig. 27(*b*). It may be noted that the shaded area in rectangle *OABC* of Fig. 27(*b*) is one-third the area of the whole rectangle *OABC*.

$$\therefore \text{ shaded area,} \qquad A' = \frac{V_m^2 \times 1}{3} = \frac{V_m^2}{3}$$

The entire area of the shaded curve is

$$A' = A_1' + A_2' + A_3'$$

$$= \frac{V_m^2}{3} + V_m^2 \times 1 + \frac{V_m^2}{3} = \frac{5\,V_m^2}{3}$$

Now, $$I_{rms} = \sqrt{\frac{\text{area of the squared curve}}{\text{base length}}}$$

$$= \sqrt{\frac{A'}{\text{base}}} = \sqrt{\frac{5\,V_m^2/3}{3}} = \sqrt{\frac{5}{9}}\,.V_m = \mathbf{0.74\ V_m}$$

Q. 28. A varying current with a periodic waveform as shown in Fig. 28 flows through an 8Ω resistor. Determine the

(*i*) average value
(*ii*) rms value
(*iii*) heat dissipated in 5 minutes.

(*Elect. Engg., Nagpur Univ. 2001*)

Ans. Instead of integration, we will use the area method for this question.

$$I_{av} = \frac{A}{T}$$

where T = time period = 0.1 second

$$I_{rms} = \sqrt{\frac{A'}{T}}$$

For average value, A stands for the area of Fig. 28(*a*) but for rms value, A' stands for the area of Fig. 28(*b*) because it is the area of i^2 values.

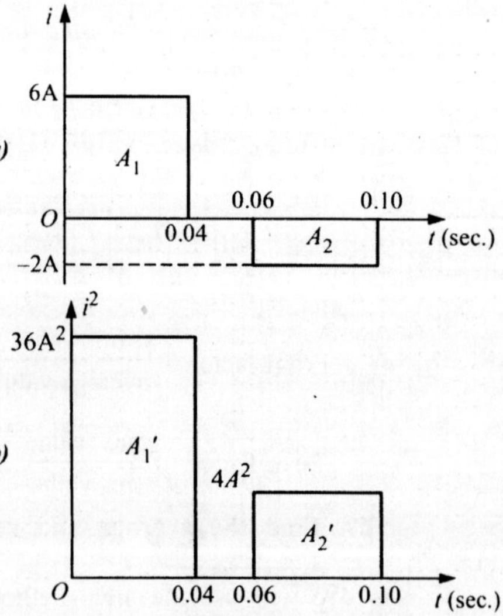

Fig. 28

(i) Average value

$$\text{Area } A_1 = 6 \times 0.04 = 0.24 \text{ A-s}$$
$$\text{Area } A_2 = -2 \times 0.04 = -0.08 \text{ A-s}$$

Total area of one-cycle $A = A_1 + A_2 = 0.24 + (-0.08) = 0.16$ A-s

$$\therefore \quad I_{av} = \frac{\text{total area}}{\text{base length}} = \frac{0.16 \text{ A-s}}{0.1 \text{s}} = \mathbf{1.6\ A}$$

(ii) Squaring the values of i, we get Fig. 28(b).

$$A_1' = 36 \times 0.04 = 1.44 \text{ A}^2\text{-s}$$
$$A_2' = 4 \times 0.04 = 0.16 \text{ A}^2\text{-s}$$

Total area of the whole cycle

$$A' = A_1' + A_2' = 1.44 + 0.16 = 1.6 \text{ A}^2\text{-s}$$

$$\therefore \quad I_{\text{rms}} = \sqrt{\frac{A'}{T}} = \sqrt{\frac{1.6 \text{ A}^2\text{-s}}{0.1 \text{ s}}} = \sqrt{16 \text{ A}^2} = \mathbf{4\ A}$$

(iii) Heat dissipated in 5 minutes *i.e.* 300 seconds

$$= I_{rms}^2 Rt = 4^2 \times 8 \times 300 = \mathbf{38{,}400\ joules}$$

Q. 29. What is meant by half-wave rectified ac? Find its average and rms value and form factor.

Ans. It is an alternating current whose one half-cycle has been removed *i.e. one which flows for half the time during one cycle.* It is shown in Fig. 29.

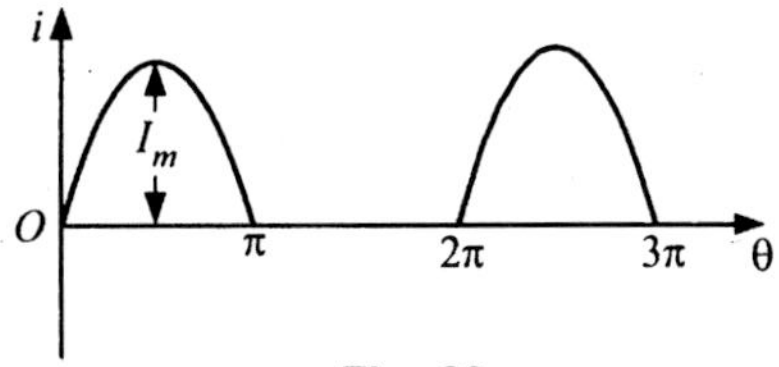

Fig. 29

For finding the average and rms values, summation of i values and i^2 values would be done over the period for which current *actually* flows *i.e.* from 0 to π, though it would be *averaged over the whole cycle i.e.* from 0 to 2π.

Let the equation of the sinusoidal current be, $i = I_m \sin\theta$

$$I_{av} = \frac{1}{2\pi}\int_0^{\pi} i\, d\theta = \frac{I_m}{2\pi}\int_0^{\pi} \sin\theta\, d\theta \qquad (\because\ i = I_m \sin\theta)$$

$$= \frac{I_m}{2\pi}\left| -\cos\theta \right|_0^{\pi} = \frac{I_m}{2\pi}\left| (-\cos\pi) - (-\cos 0) \right|$$

$$= \frac{I_m}{2\pi}\left| 1 + 1 \right| = \frac{\mathbf{I_m}}{\mathbf{2\pi}}$$

$$I_{rms}^2 = \frac{1}{2\pi}\int_0^{\pi} i^2 d\theta = \frac{I_m^2}{2\pi}\int_0^{\pi} \sin^2\theta\, d\theta$$

$$= \frac{I_m^2}{4\pi}\int_0^{\pi}(1 - \cos 2\theta)\, d\theta = \frac{I_m^2}{4\pi}\int_0^{\pi} d\theta - \cos 2\theta . d\theta$$

$$= \frac{I_m^2}{4\pi}\left| \theta - \frac{\sin 2\theta}{2} \right|_0^{\pi} = \left(\frac{I_m^2}{4\pi} \times \pi \right) = \frac{I_m^2}{4}$$

$$\therefore \quad I_{\text{rms}} = \sqrt{\frac{I_m^2}{4}} = \frac{I_m}{2} = \mathbf{0.5\ I_m}$$

$$\text{From factor} = \frac{\text{rms value}}{\text{average value}} = \frac{I_m/2}{I_m/\pi} = \frac{\pi}{2} = \mathbf{1.57}$$

Q. 30. A voltage of $e = 150 \sin 314\, t$ is applied to a device which offers a resistance of 20Ω to the flow of current in one direction while preventing the flow of current in opposite direction. Calculate the rms, average value and form factor for the current over one cycle.

(*Basic Elect. Engg. West Bengal Univ. 2001*)

Ans. Since 150 represents the maximum voltage of the source, hence maximum current through the device = 150/20 = 7.5A. Obviously, it is a half-wave rectified ac.

$$\therefore \qquad I_{av} = \frac{I_m}{\pi} = \frac{7.5}{\pi} = \mathbf{2.39A}$$

$$I_{rms} = \frac{I_m}{2} = \frac{7.5}{2} = \mathbf{3.75A} \qquad \therefore \quad \text{form factor} = 3.75/2.39 = \mathbf{1.57}$$

Q. 31. Briefly discuss the vector (or phasor) representation of sinusoidal voltages and currents. Will it make any difference if the rotating vector represents rms value rather than the maximum value?

Ans. For easy handling of alternating quantities, they are represented by rotating vectors rather than waves. These vectors represent the *maximum* value of the quantity and rotate counter-clockwise with the same angular velocity as that of the alternating quantity. (Fig. 30)

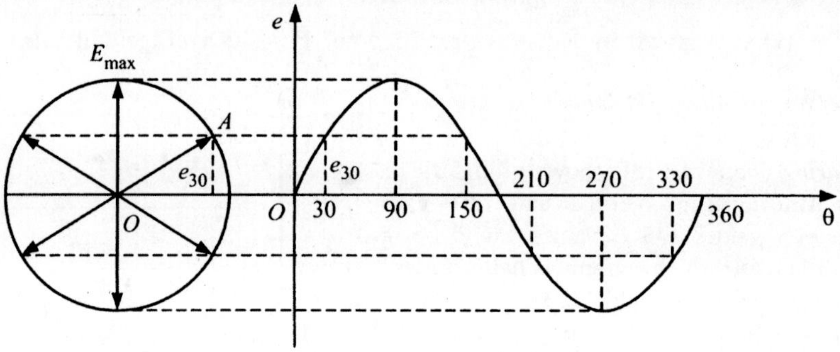

Fig. 30

It is seen that the projection of *OA* on the *vertical* axis represents the instantaneous value of the sinewave. When *OA* rotates through one revolution or 2π radians, one cycle of the sine wave is completed.

The vector *OA* can represent the sinusoidal voltage or current only when it satisfies the following conditions :

1. Its length should be equal to the *peak or maximum* value.
2. It should be in the horizontal position at the instant when the alternating quantity is zero and increases *positively*.
3. Its angular velocity should equal that of the alternating quantity.

A rotating vector will give the instantaneous value of an alternating quantity only *when it represents the maximum* value of that quantity. If it is made to represent the rms values, its projections on the *Y*-axis will not give the instantaneous values on the *Y*-axis.

Q. 32. Can vectors be used to represent two sine waves of the same frequency in the same vector diagram? What will happen if the two sine waves have different frequencies?

Ans. Yes, they can be as shown in Fig. 31. Since their frequency is the same, the two vectors will rotate counter-clockwise at the same frequency and hence *keep a fixed position relative to each other.* Suppose we have two sine waves of equations

$$i = I_m \sin \omega t \quad \text{and} \quad e = E_m \sin (\omega t + \phi)$$

As seen, voltage leads the current by an angle ϕ. The two vectors representing E_m and I_m rotate with same frequency and maintain a phase difference of ϕ degree throughout. If the two

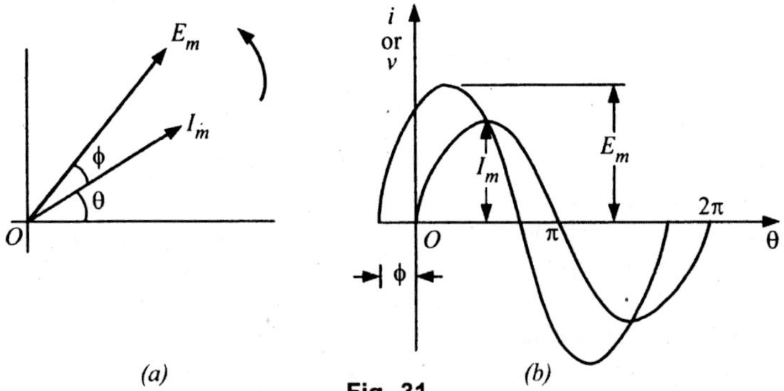

Fig. 31

sine waves have different frequencies f_1 and f_2, *they cannot be represented* on the same vector diagram. It is so because due to different speeds, their phase angle will not remain constant but keep changing continuously.

Q. 33 (*a*) Add the following currents as waves and as vectors.

$$i_1 = 7 \sin \omega t \text{ and } i_2 = 10 \sin (\omega t + \pi/3)$$

(*b*) In a parallel circuit with branches, the instantaneous branch currents are represented by

$$i_1 = 10 \sin \omega t$$
$$i_2 = 20 \sin (\omega t + \pi/3)$$
$$i_3 = 12 \sin (\omega t - \pi/6)$$

Write down the expression for the total instantaneous current in the form of

$$i = I_m \sin (\omega t + \phi).$$

Ans. (*a*) As waves :

Let i be the instantaneous value of the resultant current.

$$\therefore \quad i = i_1 + i_2 = 7 \sin \omega t + 10 \sin (\omega t + 60°)$$
$$= 7 \sin \omega t + 10 \sin \omega t \cos 60° + 10 \cos \omega t \sin 60°$$
$$= 12 \sin \omega t + 8.66 \cos \omega t$$

Dividing and multiplying both sides by $\sqrt{(12^2 + 8.66^2)} = 14.8$, we get

$$i = 14.8 \left(\frac{12}{14.8} \sin \omega t + \frac{8.66}{14.8} \cos \omega t \right)$$
$$= 14.8 (\cos\alpha \sin\omega t + \sin\alpha \cos\omega t)$$

where $\cos\alpha = \frac{12}{14.8}$ and $\sin\alpha = \frac{8.66}{14.8}$ as seen from impedance ΔODC of Fig. 32(*a*).

$$\therefore \quad i = 14.8 \sin (\omega t + \alpha)$$

where $\tan\alpha = 8.66/12$ or $\alpha = \tan^{-1} (8.66/12) = 35.8°$.

$$\therefore \quad \mathbf{i = 14.8 \sin (\omega t + 35.8°)}$$

As vectors :

Vector diagram is shown in Fig. 32(*b*). Resolving the current vectors into their horizontal and vertical components, we have

X-component = 7 + 10 cos 60° = 12

Y-component = 0 + 10 sin 60° = 8.66

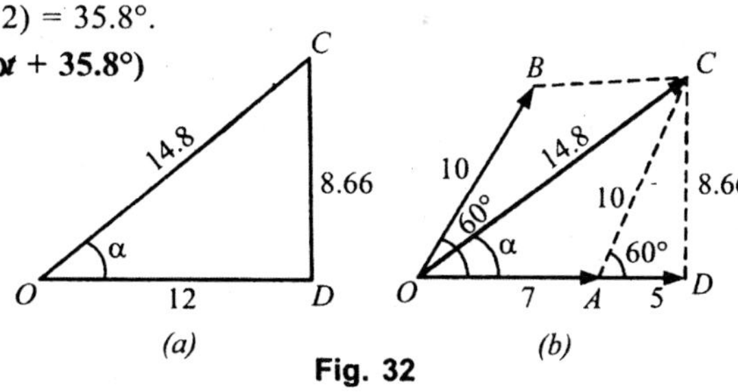

Fig. 32

$$\text{Resultant} = \sqrt{(12^2 + 8.66^2)} = 14.8 \text{ A}$$

and $\alpha = \tan^{-1}(8.66/12) = 35.8°$

Hence, the resultant equation can be written as

$$i = \mathbf{14.8 \sin (\omega t + 35.8°)}$$

(*b*) The vectors representing the maximum values of the three currents in their proper phases are shown in Fig. 33(*a*). Resolving them into their *X*-and *Y*-components, we have

$$X\text{-component} = 10 + 20 \cos 60° + 12 \cos 30° = 30.39 \text{ A}$$

$$Y\text{-component} = 20 \sin 60° - 12 \sin 30° = 11.32 \text{ A}$$

As shown in Fig. 33(b), the maximum value of the resultant current is

$$I_m = \sqrt{(X\text{-comp})^2 + (Y\text{-comp})^2}$$

$$= \sqrt{30.39^2 + 11.32^2} = 32.4 \text{ A}$$

Its angle with *X*-axis is given by

$\tan\phi = Y\text{-comp}/X\text{-comp}$

$= 11.32/30.39 = 0.3726$

$\phi = \tan^{-1}(0.3726)$

$= 20° \, 26' = 0.357$ radian

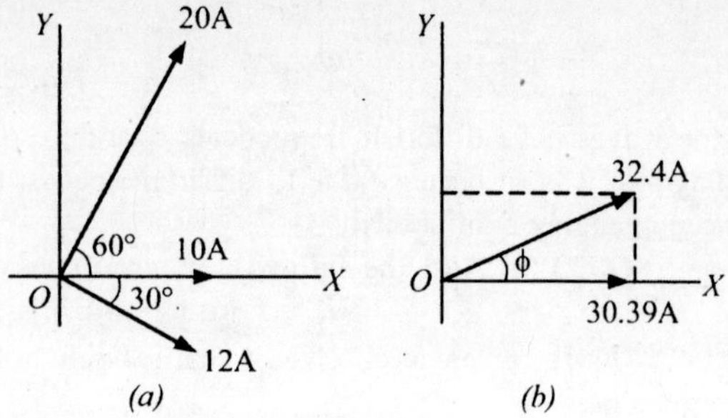

Fig. 33

Hence, the expression for the total instantaneous current is

$$i = \mathbf{32.4 \sin (\omega t + 0.357)}$$

Q. 34. Find the vector sum of the following instantaneous emfs

$$\mathbf{e_1 = 20 \sin \omega t}$$
$$\mathbf{e_2 = 10 \sin (\omega t + \pi/6)}$$
$$\mathbf{e_3 = 15 \cos \omega t}$$
$$\mathbf{e_4 = 10 \sin (\omega t - \pi/3)}$$
$$\mathbf{e_5 = 25 \cos (\omega t + 2\pi/3)}$$

(*Basic Elect. Engg. RGPV, Bhopal 2001*)

Ans. We will draw the different vectors by taking *X*-axis as reference vector [Fig. 34(*a*)]

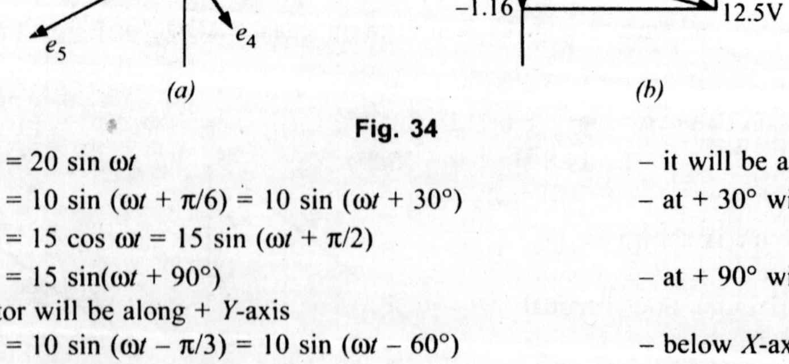

Fig. 34

$e_1 = 20 \sin \omega t$ – it will be along *X*-axis

$e_2 = 10 \sin (\omega t + \pi/6) = 10 \sin (\omega t + 30°)$ – at + 30° with *X*-axis

$e_3 = 15 \cos \omega t = 15 \sin (\omega t + \pi/2)$

$= 15 \sin(\omega t + 90°)$ – at + 90° with *X*-axis

Its vector will be along + *Y*-axis

$e_4 = 10 \sin (\omega t - \pi/3) = 10 \sin (\omega t - 60°)$ – below *X*-axis by 60°

$e_5 = 25 \cos (\omega t + 2\pi/3)$

$= 25 \sin (\omega t + 2\pi/3 + \pi/2) = 25 \sin (\omega t + 210°)$ – in the 3rd quadrant

Fig. 34(*b*) shows the vector diagram for the resultant voltage and its components. Resolving the vectors into their *X*-and *Y*-components, we get

X-component, $e_x = 20 + 10 \cos 30° + 10 \cos 60° - 25 \cos 30°$
$= 20 + 8.66 + 5 - 21.65 = 12$

Y-component, $e_y = 10 \sin 30° + 15 - 10 \sin 60° - 25 \sin 30°$
$= 10 \times 1/2 + 15 - 10 \times \sqrt{3}/2 - 25/2$
$= 5 + 15 - 8.66 - 12.5 = -1.16$

Resultant e.m.f., $e = \sqrt{e_x^2 + e_y^2} = \sqrt{(12)^2 + (-1.16)^2} = \mathbf{12.05\ V}$

Phase angle of resultant e.m.f. *e* is given by

$$\tan \phi = \frac{Y\text{-component}}{X\text{-component}} = \frac{12}{1.16} = 10.34$$

$\therefore$ $\phi = \tan^{-1} (10.34) = \mathbf{5.5°}$ – Fig. 34(*b*)

Q. 35. Derive an expression for the current produced in a pure ohmic resistance when a sinusoidal voltage is applied across it. What is the power consumed by the resistor? Is the frequency of the power waveform the same as that of the applied voltage?

Ans. As shown in Fig. 35, a resistor *R* is connected across an ac voltage source given by the equation

$$v = V_m \sin \theta = V_m \sin \omega t \quad \text{... } (i)$$

If *i* is the instantaneous value of the current produced, then by Ohm's law, we have

$$v = i R$$

Substituting the value of *v* from above, we get

$$V_m \sin \omega t = i R$$

or

$$i = \frac{V_m}{R} \sin \omega t \quad \text{... } (ii)$$

Current *i* is maximum when sin ωt is unity

$\therefore$ $I_m = V_m/R$

Hence, from Eq. (*ii*), we get

$$i = I_m \sin \omega t \quad \text{...} (iii)$$

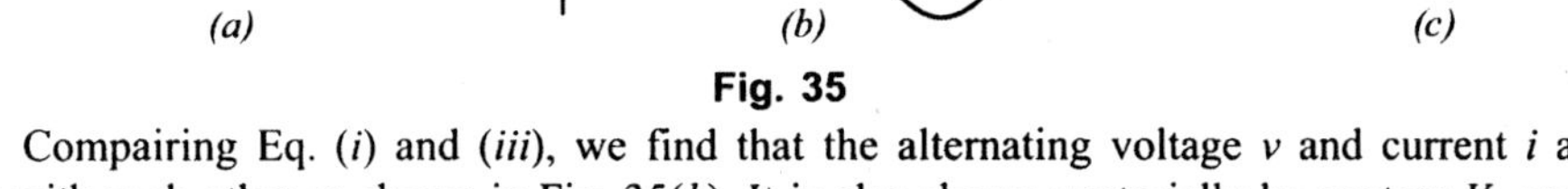

Fig. 35

Compairing Eq. (*i*) and (*iii*), we find that the alternating voltage *v* and current *i* are in phase with each other as shown in Fig. 35(*b*). It is also shown vectorially by vectors V_R and *I* in Fig. 35(*c*).

Power :

Instantaneous power, $p = vi$
$= V_m \sin \omega t \times I_m \sin \omega t$
$= V_m I_m \sin^2 \omega t$
$= \frac{V_m I_m}{2} (1 - \cos 2 \omega t)$

$$= \frac{V_m I_m}{2} - \frac{V_m I_m}{2} \cos 2\omega t \quad \text{– Fig. 36}$$

As seen, power consists of two parts

(*i*) constant power $= \frac{V_m I_m}{2}$ and

(*ii*) fluctuating power $= \frac{V_m I_m}{2} \cos 2\omega t$

For a complete cycle, the average value of *fluctuating power* is zero. Hence, the value of real power over one cycle is

$$P = \frac{V_m I_m}{2} = \frac{V_m}{\sqrt{2}} \times \frac{I_m}{\sqrt{2}} = V \times I \text{ watts}$$

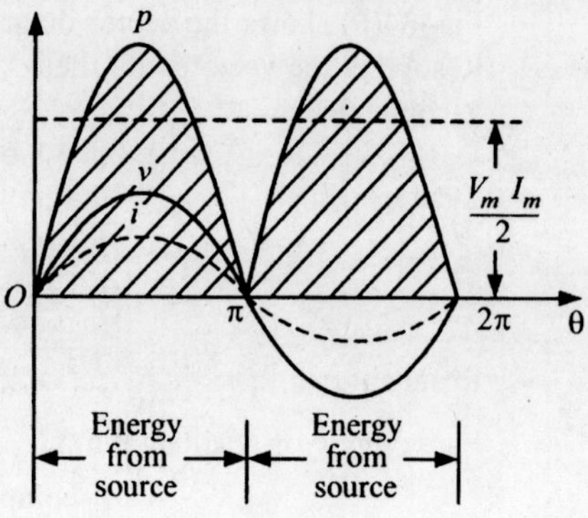

Fig. 36

where V and I are the rms values

Following points should be noted :

1. In a purely resistive circuit, power is *never* zero. It means no part of the power cycle becomes negative at any time.
2. Power is *absorbed* during the positive half cycle as well as during the negative half-cycle. It is so because the voltage and current are in phase with each other in both half-cycles.
3. In the positive half-cycle, both v and i are positive. Hence, their product is *positive.* It means power is *drawn* from the supply source. In the negative half-cycle, both v and i are negative. Hence, their product is *positive* which means again power is *drawn from the source.*
4. Power drawn by the resistor is wasted as heat. It is *not given back* to the source at any time.
5. The frequency of power cycle is *double* that of the voltage and current waves.

Q. 36. Prove that in a pure inductor, the applied sinusoidal voltage and the current are 90° out of phase with each other. Does the inductor absorb any power over one cycle of the applied voltage?

Ans. When sinusoidal voltage is applied to a pure inductive coil, a back or counter-emf is produced due to the sef-inductance (L) of the inductor. This counter emf opposes the rise and fall of current through the inductor at *every step.* Since it is a pure inductor *i.e.* it has no ohmic resistance, the applied voltage v has to overcome this self-induced emf e_L only. Value of

$$e_L = L di/dt.$$

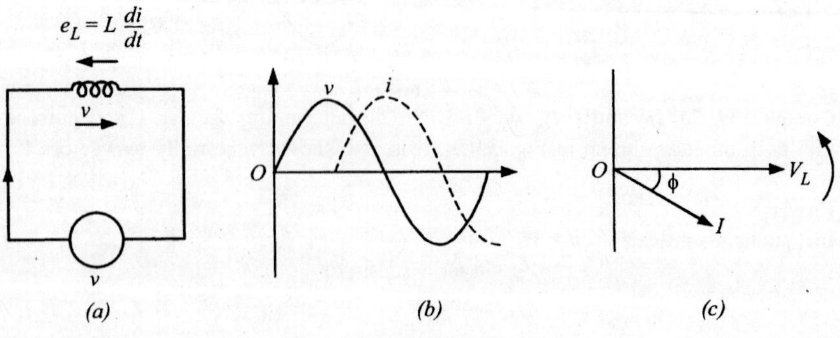

Fig. 37

Hence, at every step of the applied voltage cycle

$$v = L\frac{di}{dt}$$

Now, $v = V_m \sin \omega t$

$\therefore \quad V_m \sin \omega t = L\dfrac{di}{dt} \quad \text{or} \quad di = \dfrac{V_m}{L}\sin \omega t\, dt$

Integrating both sides, we get

$$i = \frac{V_m}{L}\int \sin \omega t\, dt = \frac{V_m}{\omega L}(-\cos \omega t) = -\frac{V_m}{\omega L}\cos \omega t$$

$\therefore \quad i = \dfrac{V_m}{\omega L}\sin\left(\omega t - \dfrac{\pi}{2}\right) \qquad — \because \sin\left(\theta - \dfrac{\pi}{2}\right) = -\cos\theta$

Maximum value of i is $I_m = \dfrac{V_m}{\omega L}$ when $\sin\left(\omega t - \dfrac{\pi}{2}\right)$ is unity.

Hence, equation of the current through the inductor is

$$i = I_m \sin\left(\omega t - \frac{\pi}{2}\right) = I_m \sin(\omega t - 90°).$$

Hence, we find that if applied voltage is given by

$$v = V_m \sin \omega t$$

then, current in a pure inductor is given by

$$i = I_m \sin(\omega t - \pi/2)$$

Obviously, current *lags* behind the applied voltage by $\pi/2$ radian or 90° *i.e.* one quarter cycle as shown in Fig. 37(*b*). Vectors are shown in Fig. 37(*c*).

As seen, $I_m = \dfrac{V_m}{\omega L}$.

Here, 'ωL' plays the part of "resistance" or opposition. It is called the (inductive) *reactance* of the inductor. It is denoted by X_L and is given in ohms if L is in henrys and ω in radians/second.

Power :

The power cycle is shown in Fig. 38.

The instantaneous power,

$$p = vi = V_m I_m \sin \omega t \sin\left(\omega t - \frac{\pi}{2}\right)$$

$$= -V_m I_m \sin \omega t \cos \omega t$$

$$= -\frac{V_m I_m}{2}\, 2 \sin \omega t \cos \omega t$$

$$= -\frac{V_m I_m}{2}\sin 2\omega t$$

Power for the whole cycle is

$$P = -\frac{V_m I_m}{2}\int_0^{2\pi} \sin 2\omega t\, dt = 0$$

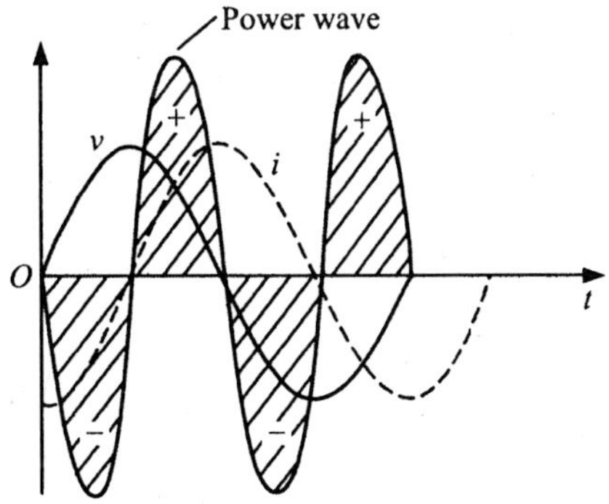

Fig. 38

It means that the *net* power drawn by an ideal or pure inductor from the supply for a complete cycle *is zero*. It may also be noted that power wave is also a sine wave of frequency *double* that of the voltage and current waves.

Q. 37. Even with voltage and current being present in a pure inductor, the average power drawn by it over one cycle is zero. How do you explain this fact? Does it mean that pure inductor draws no power at all from the supply source?

Ans. Normally, when voltage and current are present in a circuit, it is expected that power will also be present. But in the case of a pure inductor, power drawn during one complete cycle is zero. It is due to the fact that current drawn by the inductor has a phase difference of 90° with its voltage. However, it does not mean that a pure inductor draws no power at all from the source in *any part of the cycle*. In fact, it draws power when its v and i are either both positive or negative. As seen from Fig. 38 or 39, both v and i have similar sign in the second and fourth half-cycles. Hence, inductor draws power from the supply source. But in first and third quarter-cycles when v is positive, i is negative and vice versa. Hence, the same power is *returned back* to the source. So, net power drawn by an inductor in one cycle is zero.

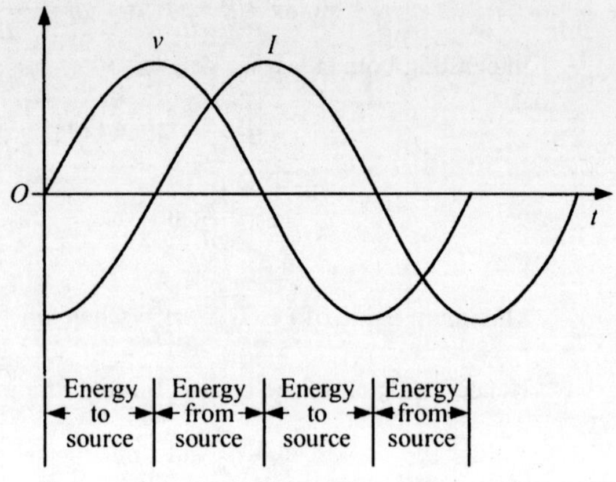

Fig. 39

Incidentally, the power drawn by the inductor is stored in its magnetic field. In the first and third quater-cycles, when magnetic field collapses, power is returned back to the supply source.

This give and take of power between the supply source and the inductor goes on so long as they are connected together. There is no *loss of energy* in a pure inductor. Whatever power the inductor draws from the source in one quater cycle is given back to the source in the next quater-cycle (Fig. 39).

Q. 38. Derive an expression for the current drawn by a pure capacitor when connected across a sinusoidal voltage. Explain with the help of a power diagram that the value of average power drawn by the capacitor during one cycle is zero.

Ans. Let the equation of the applied sinusoidal voltage be $v = V_m \sin \omega t$. At any time, the charge on the capacitor is given by

$$q = Cv = C\, V_m \sin \omega t$$

Now, the value of the capacitor charging current is given by rate of flow of charge.

$$\therefore \quad i = \frac{dq}{dt} = \frac{d}{dt}(C\, V_m \sin \omega t) = \omega\, C\, V_m \cos \omega t$$

$$\text{or} \quad i = \frac{V_m}{1/\omega C} \cos \omega t = \frac{V_m}{1/\omega C} \sin\left(\omega t + \frac{\pi}{2}\right) \qquad ...(i)$$

$$\text{Now,} \quad i = I_m \text{ when } \sin\left(\omega t + \frac{\pi}{2}\right) \text{ is unity.}$$

$$\therefore \quad I_m = \frac{V_m}{1/\omega C} = \omega\, C\, V_m$$

Hence, from Eq. (i), we get

$$i = I_m \sin\left(\omega t + \frac{\pi}{2}\right)$$

The denominator '$1/\omega C$' is known as 'capacitive' reactance. Its value is in ohm if C is in farads and ω in radian per second. It is denoted by X_C.

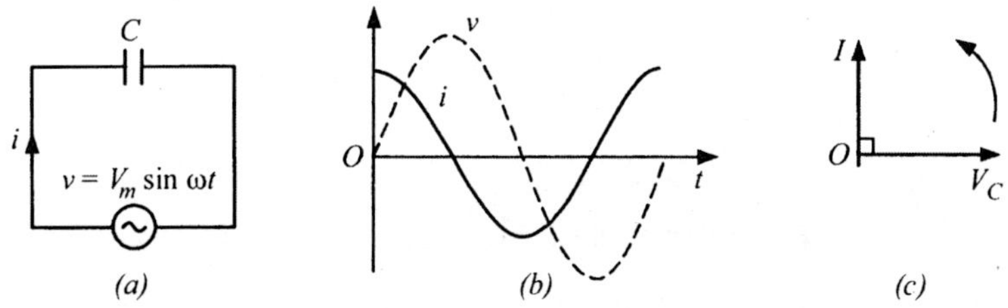

Fig. 40

It is seen from above as well as Fig. 40(*b*) that when the applied voltage is

$$v = V_m \sin \omega t$$

the current is given by

$$i = I_m \sin\left(\omega t + \frac{\pi}{2}\right)$$

Obviously, current drawn by a pure capacitor *leads* its voltage by 90° as shown graphically in Fig. 40(*b*) and vectorially in Fig. 40(*c*).

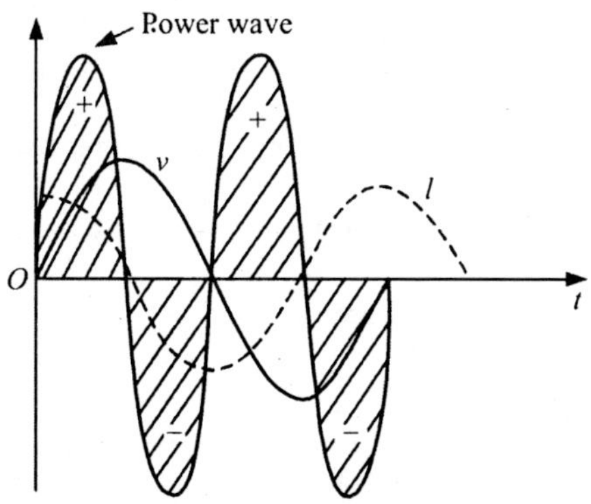

Fig. 41

Power :

Instantaneous power

$$p = vi = V_m \sin \omega t \,.\, I_m \sin(\omega t + 90°)$$

$$= V_m I_m \sin \omega t \cos \omega t$$

$$= \frac{1}{2} V_m I_m \sin 2\,\omega t$$

Power for the whole cycle

$$P = \frac{V_m I_m}{2} \int_0^{2\pi} \sin 2\,\omega t\, dt = 0$$

The capacitor draws power from the source in the first and third quarter-cycles when v and i both have the same algebraic sign either positive or negative. This power goes to set up electric field between the plates of the capacitor. In the second and fourth quarter-cycles, v and i have opposite signs. Hence, the electric field collapses and power is returned to the supply source. To put it in simple words, power is drawn by the capacitor during charging and given back during discharging. Hence, net power drawn by the pure capacitor during one cycle of the applied voltage is zero.

Q. 39. Explain the meaning of the terms inductive reactance and capacitive reactance?

Ans. Inductive reactance : It is the *opposition* to the flow of an alternating current offered by the inductance of a circuit.

In fact, it is a property of *all inductors*. It is denoted by X_L.

$$X_L = \omega L = 2\pi f L \text{ ohm}$$

where f = frequency in Hz ; L = inductance in henrys.

As seen, inductive reactance of a given inductor is directly propotional to the frequency of the applied voltage (Fig. 42). It means X_L increases linearly with frequency.

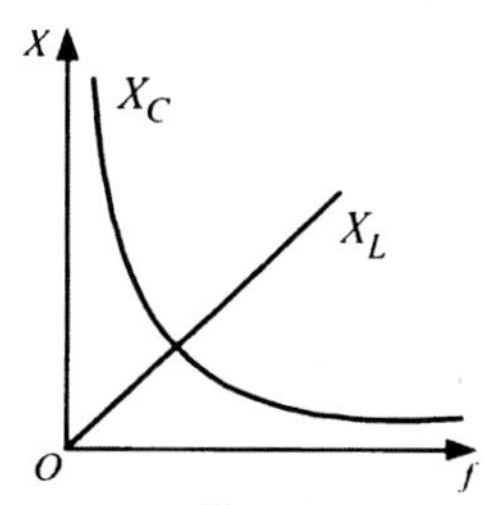

Fig. 42

Capacitive reactance : It is the opposition to the flow of an alternating current offered by the capacitance of a capacitor.

In fact, it is a property of all capacitive circuits. It is denoted by X_C.

$$X_C = \frac{1}{\omega C} = \frac{1}{2\pi f C}$$

where, f = frequency in Hz ; C = capacitance in farad.

As seen from above, capacitive reactance varies inversely as the frequency of the applied voltage. Hence, X_C does not vary linearly with frequency. (Fig. 42)

Q. 40 (*a*) A small inductor used in a TV reciever has a reactance of 5 kΩ at 50 MHz. What is its inductance?

(*b*) What is the reactance of an 8 μF filter capacitor at a frequency of 120 Hz?

Ans. (*a*) $X_L = 2\pi f L$ or $L = \frac{X_L}{2\pi f}$

$$\therefore \quad L = \frac{5\times10^3}{2\pi\times50\times10^6} = 1.59 \times 10^{-5}\text{ H}$$

$$= 15.9 \times 10^{-6}\text{ H} = \mathbf{15.9\ \mu H}$$

(*b*) $$X_C = \frac{1}{2\pi f C} = \frac{1}{6.28\times120\times8\times10^{-6}} = \mathbf{165.9\ \Omega}$$

Q. 41. The voltage and current sinusoidals of an element are respectively

$i(t) = 10 \sin 314\, t$ and $v(t) = 10 \cos 314\, t$

What is the frequency of sinusoidals and the element value?

(*Basic Elect. Engg. Kerala Univ. 2004*)

Ans. We are given that

$$i(t) = 10 \sin 314\, t$$

$$v(t) = 10 \cos 314\, t = 10 \sin\left(314t + \frac{\pi}{2}\right)$$

It is seen from the above that voltage *leads* the current by π/2 radian. Put in an other way, it means that *current* lags behind the voltage by 90°. Hence, the element is a *pure inductor*. Its inductance L can be found from the relation

$$X_L = \omega L$$

Now, $X_L = V_m/I_m = 10/10 = 1\Omega$, $\omega = 314$ rad/second

$\therefore$ $1 = 314 \times L$ or $L = 1/314 = \mathbf{3.18\ mH}$

Q. 42. A sinusoidal 50 Hz current of maximum value of 100 A flows through a capacitor of 318 μF capacitance. Calculate

(*i*) the expression for instantaneous current

(*ii*) reactance of the capacitor

(*iii*) equation of the applied emf

(*iv*) rms values of applied emf and current. (*Elect. Engg., Nagpur Univ. 2001*)

Ans. $I_m = 100$ A, $C = 318$ μF, $f = 50$ Hz

$$\omega = 2\pi f = 2\pi \times 50 = 314 \text{ rad/second}$$

(*i*) The equation of the instantaneous current is given by

$$i = I_m \sin\left(\omega t + \frac{\pi}{2}\right) = \mathbf{100\ sin\left(314t + \frac{\pi}{2}\right)}$$

(*ii*) Reactance of the capacitor is given by

$$X_C = \frac{1}{\omega C} = \frac{1}{2\pi f C} = \frac{1}{2\pi\times50\times318\times10^{-6}} = \mathbf{10\ \Omega}$$

(*iii*) Equation of the applied emf is

$v = V_m \sin \omega t$ where $V_m = I_m X_C = 100 \times 10 = 1000$ V

$\therefore$ $v = \mathbf{1000\ sin\ \omega t}$

(iv) $$V_{rms} = \frac{V_{max}}{\sqrt{2}} = \frac{1000}{\sqrt{2}} = \mathbf{707.1\ volt}$$

$$I_{rms} = \frac{I_m}{\sqrt{2}} = \frac{100}{\sqrt{2}} = \mathbf{70.71\ A}$$

Q. 43. How do you explain the puzzling fact that in a pure inductor, sinusoidal current is maximum when applied voltage is zero?

Ans. When a sinusoidal current flows through a pure inductor, a very surprising thing happens. It is found that current is maximum when voltage across the inductor is zero! And the current is zero when voltage is maximum! It is something which never happens with dc voltages and currents.

We know that a sinusoidal current always induces a sinusoidal emf e_L in an inductor. This induced emf is always *equal and opposite* to the applied voltage v_L at every instant.

Since the value of

$$e_L = L \cdot di/dt,$$

hence $$v_L = e_L = L \cdot di/dt$$

Obviously, value of v_L depends on the rate of change of current *i.e.* di/dt at different points of the wave. It is seen from Fig. 43(*a*) that when the current is zero at point *O* and is going positive, its di/dt is *maximum*. Similarly di/dt is maximum at point *C* although in the negative direction. Similarly, di/dt will be positive maximum at point *D*. Hence, at all these points, v_L will be maximum though current is zero.

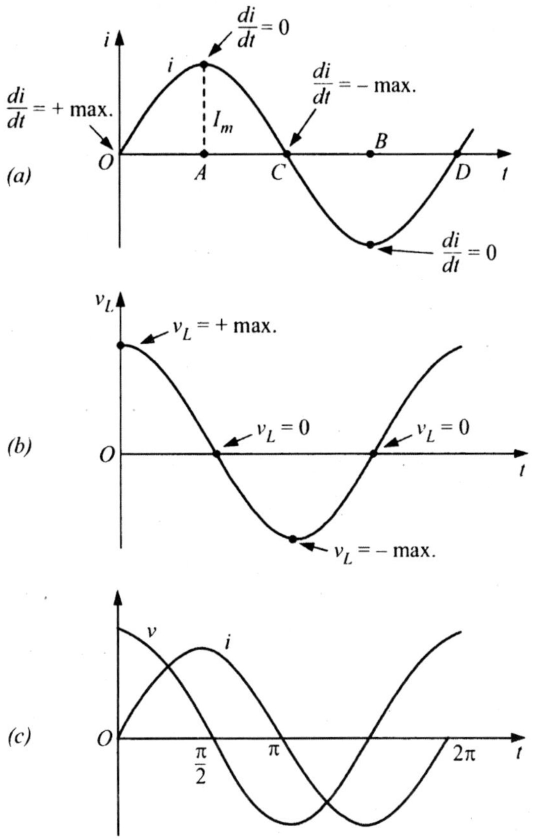

Fig. 43

The value of di/dt is zero at points *A* and *B* because after having achieved maximum value of I_m, current remains constant for a moment. In other words, di/dt is zero. Hence, the value of v_L is zero at these points as shown in Fig. 43(*b*). Hence, current is maximum when v_L is zero. This explains why the voltage and current do not achieve maximum or zero values at the some time. As shown in Fig. (*c*), the two waves have a phase difference of 90° – with voltage wave leading by a quater cycle with respect to the current wave. When one has maximum value, the other has zero value.

Q. 44. Briefly explain why current drawn by a pure capacitor leads the applied voltage by 90°?

Ans. A sinusoidal voltage always produces a sinusoidal current in a capacitor. The charging current *i* of the capacitor is given by

$$i = C \times \text{rate of change of p.d.}$$

or $$i = C \times dv/dt$$

Obviously, i will achieve maximum value I_m when dv/dt becomes maximum. As seen from Fig. 44(a), dv/dt has maximum positive rate of change at point O. Hence, current achieves maximum value of + I_m at that instant as shown in Fig. 44(b). Again, at point A, voltage changes at the maximum negative rate. Consequently, current achieves – I_m value at that instant.

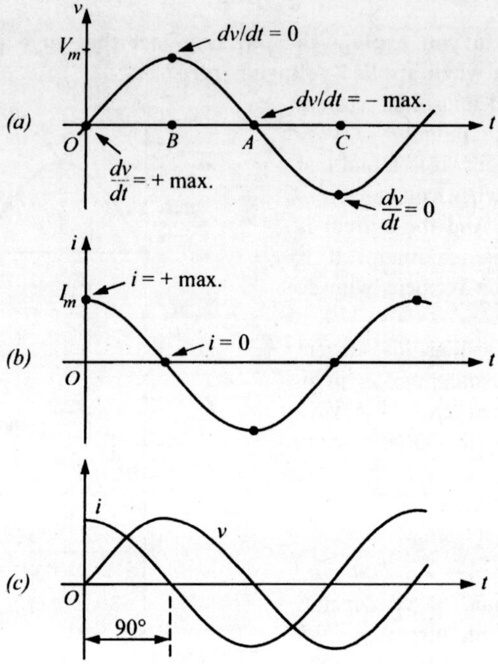

Fig. 44

As seen from Fig. 44(a), $dv/dt = 0$ at points B and C. Hence, the value of charging current is also zero at these points. The voltage and current waves have been shown together in Fig. 44(c). As seen, current leads the voltage by 90° or a quater-cycle.

EXERCISE

1. A coil of 4000 turns is rotated at 24,000 rpm at right angles to a magnetic field having a uniform flux density of 0.25 tesla (Wb/m^2). The average area of each turn is 25 cm^2. Calculate : (i) frequency (ii) period (iii) angular velocity (iv) maximum value of the generated voltage. **(400 Hz, 2.5 ms, 1600 π rad/s, 6283 volt)**
2. A rectangular coil of mean area of 600 cm^2. and having 40 turns is rotated about an axis coinciding with one of its longer sides at a speed of 1500 rpm in a uniform magnetic field of flux density 0.075 T. Find an expression for the instantaneous emf induced in the coil if the flux is at right angle to the axis of rotation.
   Evaluate this emf at an instant 0.002 second after the plane of the coil has been perpendicular to the field. **(28.27 sin 157t volt; 8.74V)**
3. A square coil of 10 cm side and with 100 turns is rotated at the uniform speed of 500 r.p.m. about an axis at right angles to a uniform field of 0.5 Wb/m^2. Calculate the instantaneous value of the induced emf when the plane of the coil is (a) at right angles to the plane of the field (b) at 30° to the plane of the field and (c) in the plane of the field. ($Elect. Engg. A.M.Ae.S.I.$) **(0 volt, 22.6 volt, 26.18 volt)**

4. An alternating current of frequency 60 Hz has a maximum value of 120A. Write down the equation for its instantaneous value. Reckoning time from the instant the current is zero and is becoming positive, find (*a*) the instantaneous value after 1/360 second and (*b*) the time taken to reach 96 A for the first time. **(103.9A, 0.00245 second)**
5. For a certain 0-degree reference sinusoidal current, the peak value is 100 mA. Determine the instantaneous values at each of the following points
(*a*) 35° (*b*) 95° (*c*) 215° (*d*) 275° (*e*) 360°.
(57.36 mA ; 99.62 mA ; – 57.36 mA ; – 99.62 mA ; 0 mA)
6. With reference to Fig. 45, determine the instantaneous values of sine wave *B* at the following angles taking sine wave *A* as the reference.
(*a*) 30°
(*b*) 90°
(*c*) 180°
(*d*) 300°
(12.99 V, 12.99 V, –7.5 V, -7.5 V)

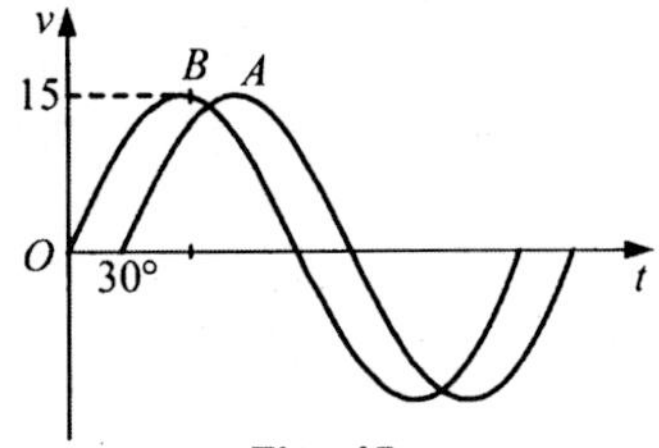

Fig. 45

7. Write the time-domain expressions of the two sine voltage waveforms shown in Fig. 46.

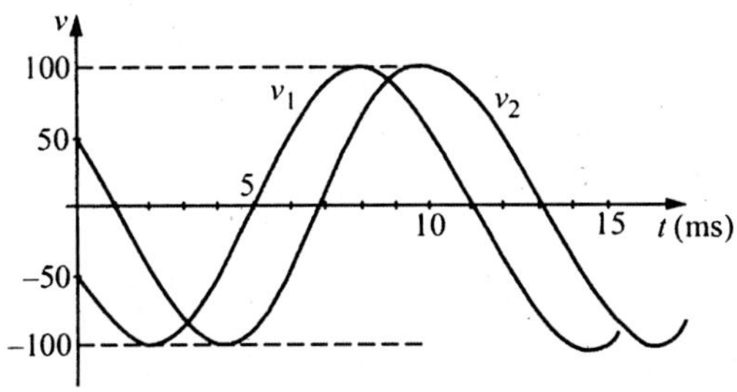

Fig. 46

[(v_1 = 100 sin (523.6t - 150°), v_2 = 100 sin (523.6t + 150°)]

8. Calculate the average value of the non-sinusoidal waveform called a stairstep shown in Fig. 47. **(3 volt)**

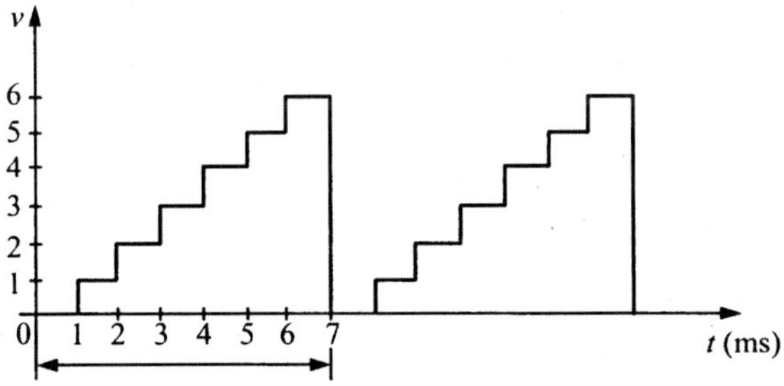

Fig. 47

9. A complex voltage waveform has the equation of $v = 20 + 10 \sin 200\,\pi t$ volt. Calculate its :
(*i*) average value (*ii*) rms value of the ac component
(*iii*) rms value of the waveform (*iv*) form factor
(*v*) crest factor **(20 V ; 7.07 V ; 21.21 V ; 1.06 ; 1.414)**
10. Write the mathematical expression for the current waveform shown in Fig. 48.
(30 sin ($2\pi \times 25 \times 10^3\, t$ – 90°)

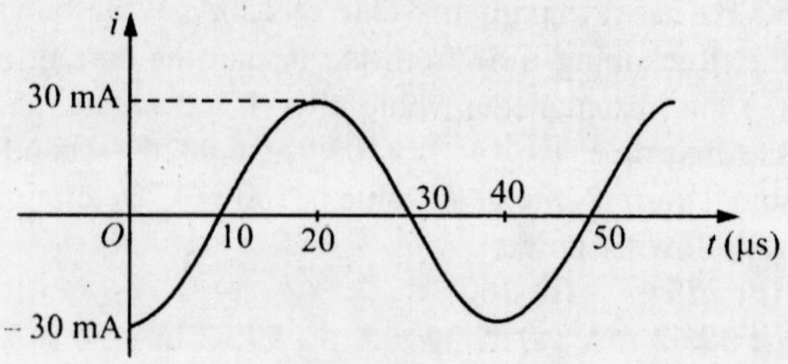

Fig. 48

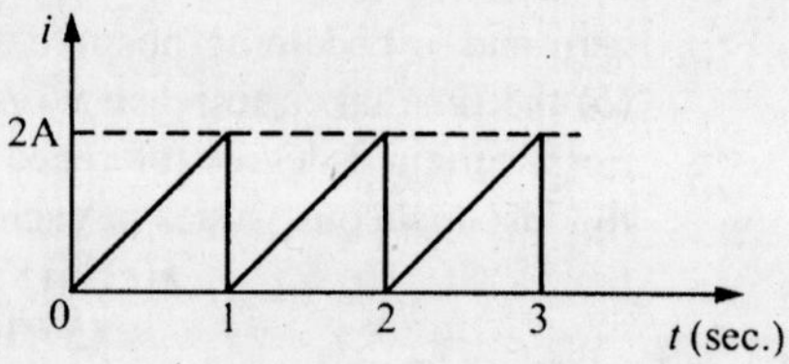

Fig. 49

11. In the sawtooth waveform shown in Fig. 49, the current increases linearally from 0 to 2 A and then suddenly drops to zero. Find the average and rms values.

(*Basic. Elect. Engg. Kerala Univ. 2002*)

(1A, $2/\sqrt{3}$ A)

12. A triangular voltage wave has the following values over one half-cycle, both half-cycles being symmetrical about the zero axis:

Time (ms)	0	10	20	30	40	50	60	70	80	90	100
Voltage	0	2	4	6	8	10	8	6	4	2	0

Plot half-cycle of the waveform and hence determine (*a*) the average value (*b*) the rms value (*c*) the form factor. **(5 volt, 5.77 volt, 1.154)**

13. The current in a circuit is given by i = 100 sin 728 t. Find the maximum value and frequency of the current. (*Basic Elect. Engg. Kerala Univ. 2002*)

(100 A, 116 Hz)

14. Calculate from first principles, the rms value and form factor of an alternating voltage having the following values over half a cycle, both half-cycles being symmetrical about the zero axis :

Time, in milliseconds	0	1	2	3	4
Voltage, in volts	0	100	100	100	0

These voltage values are joined by straight lines. **(81.5 V, 1.087)**

MULTIPLE CHOICE QUESTIONS

1. A standard sinusoidal voltage wave changes its polarity at
(*a*) maximum value (*b*) minimum value (*c*) zero value (*d*) none of the above
2. The period of a certain sine wave is 10 milliseconds. Its frequency is
(*a*) 10 MHz (*b*) 10 kHz (*c*) 10 Hz (*d*) 100 Hz
3. Two sine waves are said to be in phase with each other if they achieve their
(*a*) zero value at the same time (*b*) maximum value at the same time
(*c*) minimum value at the same time (*d*) all of the above
4. The distance occupied by one complete cycle of the wave is called its
(*a*) time period (*b*) wave length (*c*) velocity (*d*) frequency
5. The rms value of a sine wave of peak value I_m is given by
(*a*) $I_m/\sqrt{2}$ (*b*) I_m (*c*) $I_m/2$ (*d*) I_m/π
6. The average value of a sine wave of maximum value I_m over one cycle is
(*a*) I_m/π (*b*) $2\,I_m/\pi$ (*c*) zero (*d*) $I_m/2$

7. The rms value of a sine wave of maximum value 10 A equals a dc current of .... ampere.
(*a*) 7.07 (*b*) 6.37 (*c*) 5 (*d*) 5.77
8. The rms value of a sinusoidal voltage with peak-to-peak value of 240 volt is .... volt.
(*a*) 84.84 (*b*) 77.82 (*c*) 94.68 (*d*) 89.15
9. The rms value of a sawtooth waveform of maximum value E_m is
(*a*) $E_m/\sqrt{3}$ (*b*) (3/5) E_m (*c*) $E_m/\sqrt{2}$ (*d*) $E_m/2$
10. The phase difference between sinusoidal voltage and current in a pure capacitor
(*a*) is 90° (*b*) depends on the value of *C*
(*c*) increases with frequency (*d*) is zero
11. The average power in a purely inductive or capacitive circuit
(*a*) depends on X_L (*b*) depends on X_C (*c*) is zero (*d*) is positive
12. Power is dissipated regardless of the direction in which current flows through a/an
(*a*) resistor (*b*) inductor (*c*) capacitor (*d*) none of these
13. Pure inductors are
(*a*) costly (*b*) used in T.V. (*c*) hypothetical (*d*) bulky
14. A certain inductor has a reactance of 4 kΩ at 5 kHz. Its reactance at 15kHz is .... kilo-ohm.
(*a*) 8 (*b*) 10 (*c*) 12 (*d*) 20
15. In a purely inductive circuit, if the frequency is doubled and the applied voltage is halved, the resulting current becomes.
(*a*) one-fourth (*b*) one-fifth (*c*) one-half (*d*) one-third
16. Which of the following combination of *f* and *C* presents the greatest capacitive reactance.
(*a*) 1μF at 1 kHz (*b*) 0.005 μF at 50 kHz
(*c*) 250 ρF at 10 MHz (*d*) 0.01μF at 5 kHz
17. The time period of a sinusoidal waveform witth 200 Hz frequency is .... second.
(*a*) 0.05 (*b*) 0.005 (*c*) 0.0005 (*d*) 0.5
(*Basic Elect. Engg. VTU Karnataka, 2009*)
18. The peak value of a sine wave is 400 V. Its average value is .... volt.
(*a*) 254.6 (*b*) 282.6 (*c*) 400 (*d*) 565.5
(*Basic Elect. Engg. VTU Karnataka, 2009*)
19. The form factor of a sine wave is
(*a*) 1.01 (*b*) 1.11 (*c*) 1.21 (*d*) none of the above
(*Basic Elect. Engg. West Bengal Tech. Univ. 2003*)

ANSWERS

1. *c*	2. *d*	3. *d*	4. *b*	5. *a*	6. *c*	7. *a*
8. *a*	9. *a*	10. *a*	11. *c*	12. *a*	13. *c*	14. *c*
15. *a*	16. *d*	17. *b*	18. *a*	19. *b*		

6 CHAPTER COMPLEX ALGEBRA

Q. 1. What is the significance of the *j*-operator. Why is it called an 'imaginary' quantity?

Ans. In Electronics and electrical calculations, it is often necessary to show quantities that are rotated exactly 90° in either clockwise (CW) or counterclockwise (CCW) direction from the *X*-axis. Engineers use the *j*-operator for this purpose. When the letter *j* is placed in front of an electrical quantity like voltage *V*, it means that the vector for *V* has been rotated 90° from *X*-axis in CCW direction. Now, it lies on the *Y*-axis as seen from Fig. 1(*a*). Similarly, when *V* is operated on by – *j*, it is rotated 90° in the clockwise direction as shown.

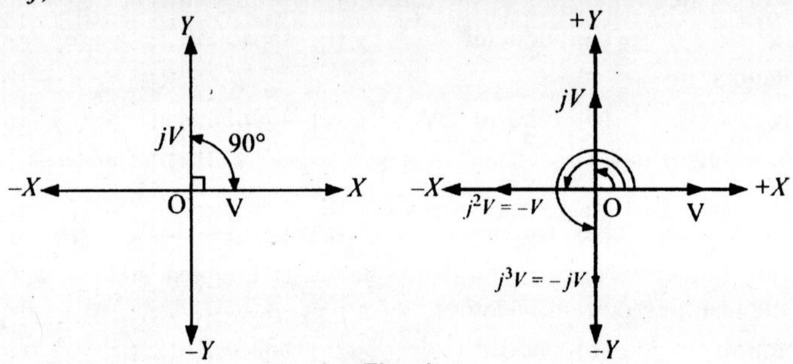

Fig. 1

Obviously, the operator *j* gives the instruction, "rotate by 90° or π/2 radian". Often it means, "change the direction from horizontal to vertical".

Unlike most other operators, *j* has a numerical value of $\sqrt{-1}$.

Hence, $$j^2 = \left(\sqrt{-1}\right)^2 = -1, j^3 = \left(\sqrt{-1}\right)^3 = -j, j^4 = \left(\sqrt{-1}\right)^4 = 1$$

As seen from Fig. 1(*b*).

(*i*) When *j* operates once on *V*, we get jV along *Y*-axis.

(*ii*) When *j* operates twice on *V*, we get $j^2V = -V$ along negative *X*-axis.

(*iii*) When *j* operates thrice on *V*, we get $j^3V = -jV$ along the negative *Y*-axis.

(*iv*) When *j* operates on *V* for the fourth time, we get $j^4V = V$. In fact, we come back to the starting point after rotating through 4 × 90° = 360°. In short, when *V* is operated on by the *j*-operator continuously, it rotates *counterclockwise* in steps of 90°.

It may also be noted that

$$\frac{1}{j} = \frac{j}{j^2} = \frac{j}{-1} = -j$$

It should be carefully noted that

(*i*) jV does not mean that *j* is *multiplied* by *V*. Placing *j* in front of *V* simply means that it has been rotated 90° in CCW direction.

(*ii*) In the term (3 + *j*4), it does not mean that we have simply to add *j*4 to 3. It needs *vector* addition.

(*iii*) Mathematically, $j = \sqrt{-1}$ and is regarded as an *imaginary* number because *its square gives a negative number*. The squares of all *real* numbers, whether plus or minus, *never* give a minus number. But electrically speaking, *j* as an operator does not change the *reality* of the quantities involved. It only changes their *direction* but not their magnitude.

Q. 2. What is meant by a complex number? Is it really complicated? What are the two common forms of representing complex numbers?

Ans. A number which consists of both real and imaginary numbers is called a complex number. The real number is usually along the *X*-axis and the imaginary number along the *Y*-axis. In electrical calculations, they are often called *in-phase* and quadrature components respectively. The term *imaginary* is generally used for numbers containing the *j*-operator such as *j*5. However, since long, the term *imaginary* has lost its meaning of *unreality* Now-a-days, the terms 'real' and 'imaginary' are only used as technical terms like positive and negative.

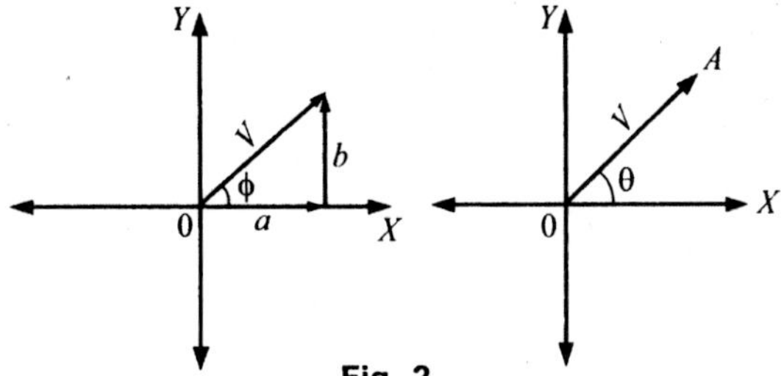

Fig. 2

Suppose we have an ac current of 5 A having a real component of 3 A and an *imaginary* component of *j*4 A. When this current is passed through a 10 Ω resistor, power due to imaginary component = $4^2 \times 10 = 160$ W and that due to real component = $3^2 \times 10 = 90$ W. The total power = 160 + 90 = 250 W.

It is the same as produced by current of 5A because $5^2 \times 10 = 250$ W.

Complex numbers are not *complicated* ones. They are called 'complex' simply because they contain so-called real and imaginary parts. Such numbers are very useful in the analysis of ac circuits. With their help, we can add, subtract, multiply and divide different electrical quantities that have both magnitude and phase angle.

There are two common forms of complex numbers each having certain advantages depending on the particular application. We will use letters such as *V* and *I* to represent magnitude only. The boldface letters such as **V** and **I** will be used to represent phasor quantities.

(*i*) **Rectangular form :** In this case, the complex number is represented by its horizontal component '*a*' and vertical component '*b*'. As shown in Fig. 2(*a*), $\mathbf{V} = a + jb$. The magnitude of the voltage is given by $\sqrt{a^2 + b^2}$. Its angle with *X*-axis is given by $\phi = \tan^{-1}$ (b/a).

(*ii*) **Polar form :** In this form, we express the phasor magnitude and its angle with the positive *X*-axis. As shown in Fig. 2(*b*), *OA* represents the magnitude of the voltage and its phase angle with respect to *X*-axis which is taken as the reference axis.

Q. 3. Briefly discuss the rectangular form of representing complex numbers. Is this form suitable for multiplication and division of such numbers?

Ans. In this form, a complex number can be expressed in terms of its *X*-component and *Y*-component. For example, the phasor voltage $\mathbf{V_1}$ of Fig. 3 may be described by its real component 'a_1' and imaginary component 'b_1' both of which are positive.

$$\mathbf{V_1} = a_1 + jb_1$$

Similarly, other voltages can be expressed as follows:

$$\mathbf{V_2} = -a_2 + jb_2,$$

$$\mathbf{V_3} = -a_3 - jb_3, \quad \mathbf{V_4} = a_4 - jb_4$$

This form of representation is not well-suited for multiplication and division of complex numbers.

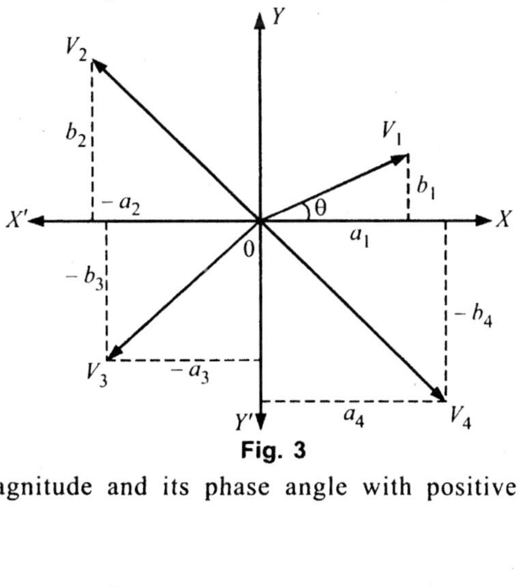

Fig. 3

A phasor expressed in rectangular form can be easily changed into polar form. For this purpose, we need the absolute value of its magnitude and its phase angle with positive *X*-axis.

Absolute value of $V_1 = \sqrt{a_1^2 + b_1^2}$

Phase angle with X-axis,

$\tan\theta = (b_1/a_1) \quad \therefore \quad \theta = \tan^{-1}(b_1/a_1) \quad \therefore \quad \mathbf{V}_1 = \sqrt{a_1^2 + b_1^2}\ \angle\tan^{-1}(b_1/a_1)$

Q. 4. What is the polar form of representing complex numbers? What are its advantages over the rectangular form?

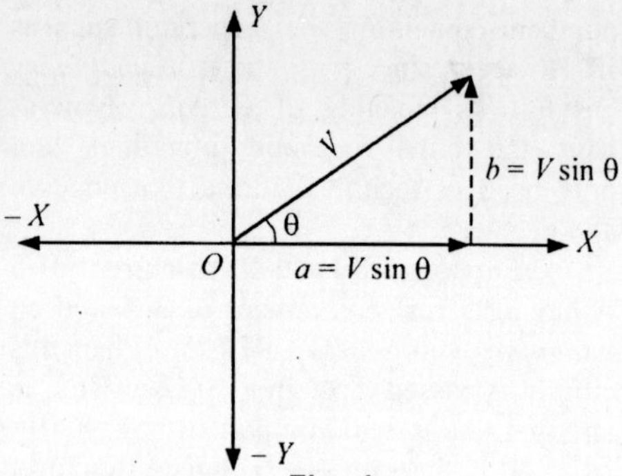

Fig. 4

Ans. This form gives the magnitude and angle of a phasor quantity as shown in Fig. 4. To get the rectangular form, the X-component and Y-component must be found from the right-angled triangle using the rule of trigonometry. For example,

real component, $a = V\cos\theta$

imaginary component, $b = V\sin\theta$

$\therefore$ polar form is,

$$\mathbf{V}\angle\theta = V\cos\theta + j\,V\sin\theta = a + jb$$

Polar form is ideal of multiplication and division of complex numbers.

Q. 5. Define conjugate complex numbers. How can they be used for calculating the power consumed by a circuit?

Ans. Two complex numbers are said to be *conjugate* if they differ in the algebraic sign of their *imaginery* components. For example, the numers $(a + jb)$ and $(a - jb)$ are conjugate numbers. Their product gives the volt-ampere (VA) of the circuit. The real part of the product gives the actual or real power consumed by the circuit whereas the j-part gives the reactive volt-amperes (VAR). It should be noted that the active volt-ampere or real power is the *same* whether we use conjugate of the voltage or of the current. If voltage conjugate is used, then capacitive VARs are positive and inductive VARs negative. It will be the other way around if current conjugate is used.

Q. 6. Describe the method of solving a series *R-L* circuit with the help of complex algebra.

Ans. The series *R-L* circuit is shown in Fig. 5(*a*). As seen from the impedance triangle of Fig. 5(*c*),

$$\text{impedance, } \mathbf{Z} = R + j\,X_L = R + j\omega L \qquad - X_L \text{ is taken +ve}$$

The numerical value of impedance is $|Z| = \sqrt{R^2 + X_L^2}$

Its phase angle with respect to the reference axis is

$$\tan\phi = X_L/R \quad \text{or} \quad \phi = \tan^{-1}(X_L/R)$$

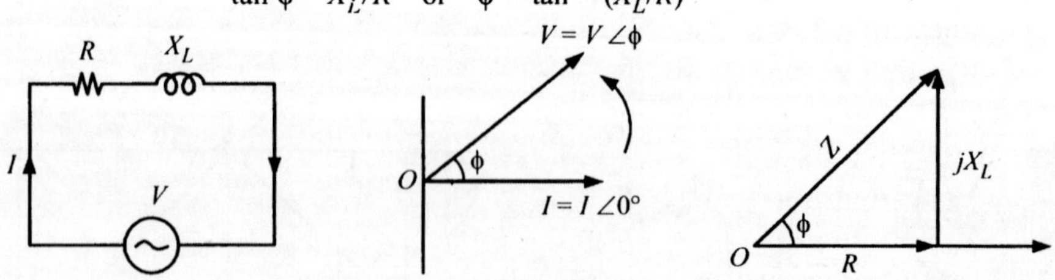

Fig. 5

We have taken **I** as the reference vector because current is *common to both R and X_L*.

$\therefore \qquad \mathbf{I} = I\angle 0°$

and $\qquad \mathbf{V} = \mathbf{IZ} = I\angle 0° \times Z\angle\phi = IZ\angle\phi$

It shows that current *lags* behind the voltage by $\phi°$

Q. 7. Analyse a series *R-C* circuit with the help of complex numbers?

Ans. The series *R-C* circuit is shown in Fig. 6(*a*) and impedance triangle is shown in Fig. 6(*c*). X_C is taken *negative* because *I* has been taken as the reference quantity as shown in Fig. 6(*c*).

$$\mathbf{Z} = R - jX_C = R - j\,\frac{1}{\omega C}$$

The numerical value of impedance is

$$|Z| = \sqrt{R^2 + (-X_C)^2} = \sqrt{R^2 + X_C^2}$$

$$\tan\phi = -X_C/R \quad \text{or} \quad \phi = \tan^{-1}(-X_C/R)$$

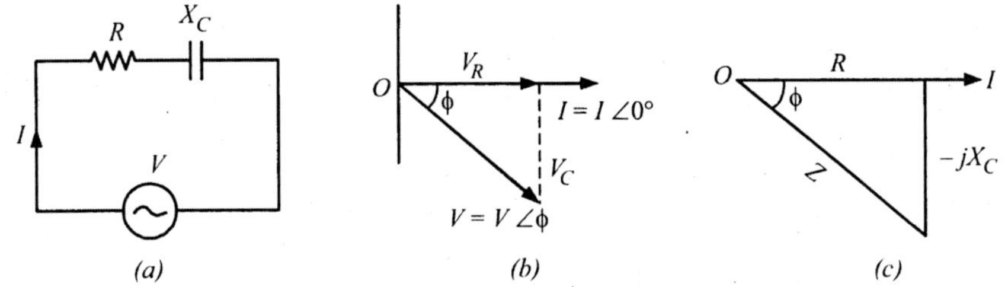

Fig. 6

Since current vector has been taken along the reference *X*-axis, hence, it may be written as

$$\mathbf{I} = I\angle 0°$$

$$\mathbf{V} = \mathbf{IZ} = I\angle 0° \times Z\angle -\phi = IZ\angle -\phi$$

It shows that voltage *lags* behind the current or current leads the voltage by $\phi°$

Q. 8. In the figures given below, each vector has been shown with its positive angle *i.e.* angle measured from positive *X*-axis in the counter-clockwise direction. Find their equivalent negative angles measured in the clockwise direction. Also, find the values of their in-phase and quadrature components.

Ans. (*a*) $\theta = 30° = -(360° - 30°) = -\mathbf{330°}$ — Fig. 7(*a*)

in-phase component = 10 cos 30° = 10 × 0.866 = **8.66 V**

quadrature component = 10 sin 30° = 10 × 0.5 = **5 V**

(*b*) $\theta = 135° = -(360° - 135°) = -\mathbf{225°}$ — Fig. 7(*b*)

in-phase component = 10 cos 135° = 10 cos (90° + 45°) = 10 sin 45°

10 × (– 0.707) = – **7.07 V**

quadrature component = 10 sin 135° = 10 sin (90° + 45°) = 10 cos 45°

= 10 × (0.707) = + **7.07 V**

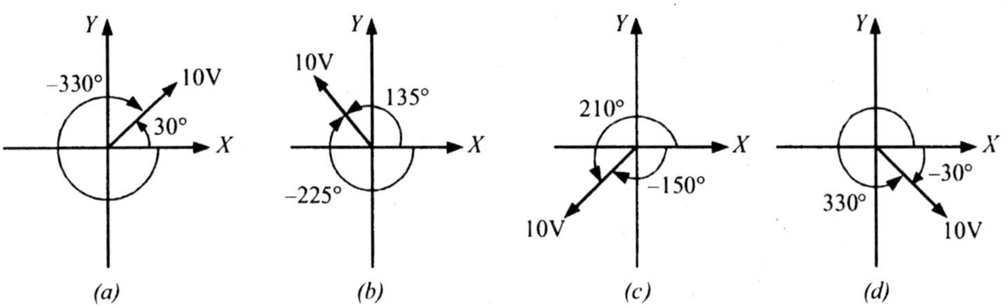

Fig. 7

(c) $\theta = 210° = -(360° - 210°) = -\mathbf{150°}$ — Fig. 7(*c*)

in-phase component = 10 cos 210° = 10 cos (180° + 30°)

$= 10 \times - \cos 30° = - 10 \times 0.866 = \mathbf{-8.66\ V}$

quadrature component $= 10 \sin 210° = 10 \sin (180° + 30°)$

$= 10 \times - \sin 30° = \mathbf{- 5\ V}$

(*d*) $\theta = 330° = - (360° - 330°) = \mathbf{- 30°}$

in-phase component $= 10 \cos 330° = 10 \cos (2\pi - 30°)$

$= 10 \cos 30° = 10 \times 0.866 = \mathbf{8.66\ V}$

quadrature component $= 10 \sin 330° = 10 \sin (2\pi - 30°)$

$= 10 \times - \sin 30° = 10 \times - 0.5 = \mathbf{- 5\ V}$

Note : The best and easy procedure for solving such questions is as under.

(*i*) Find the sine and cosine components of the given vector along +ve *X*-axis and –ve *X*-axis.

(*ii*) Give proper sign whether positive or negative depending on the direction of the axis along which the components lie.

Q. 9. Find the absolute values of the following complex numbers

(*i*) – 25 – *j*0 (*ii*) 10 – *j* 30 (*iii*) – 3 – *j*4

Ans. **(*i*)** Here, $a = -25$ and $b = 0$

$\therefore$ absolute value $= \sqrt{(-25)^2 + 0^2} = \mathbf{25}$

(*ii*) Here, $a = 10$ and $b = -30$

$\therefore$ absolute value $= \sqrt{(10)^2 + (-30)^2} = \mathbf{31.6}$

(*iii*) Here, $a = -3$ and $b = -4$

$\therefore$ absolute value $= \sqrt{(-3)^2 + (-4)^2} = \mathbf{5}$

Q. 10. Add the following vectors given in rectangular form and illustrate the process graphically.

$\boldsymbol{A = 16 + j12, B = - 6 + j10.4}$

Ans. A + B = C $= (16 + j12) + (- 6 + j10.4)$

$= 10 + j22.4$

$\therefore$ Magnitude of $C = \sqrt{(102 + 22.4)^2}$

$= 24.5$ units

Slope of $C = \theta = \tan^{-1} (22.4/10)$

$= 65.95°$

The vector addition is shown in Fig. 8.

$\alpha = \tan^{-1} (12/16) = 36.9°$

$\beta = \tan^{-1} (-10.4/6)$

$= - 240°$ or $120°$

The resultant vector is found by using parallelogram law of vectors (Fig. 8).

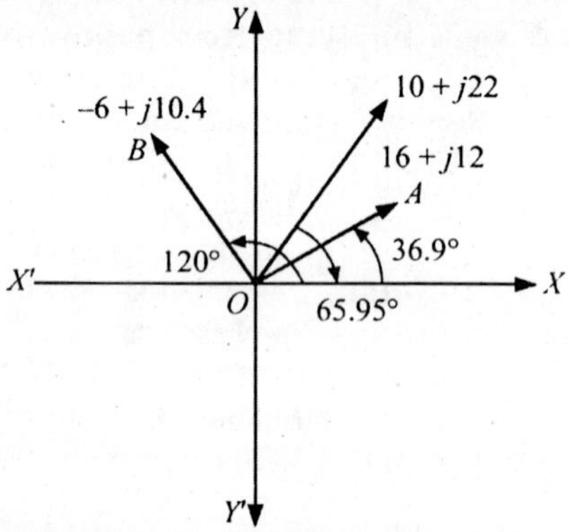

Fig. 8

Q. 11. Find polar form for the complex numbers given in rectangular form.

(*i*) – 4 + *j*12 and (*ii*) –3 – *j*6.

Ans. It must have been found by many students that in this type of conversion, finding the angle can sometimes be a problem. This can be avoided by drawing a rough freehand sketch of the two components of the complex number before attempting the solution. Here, we find that the real component is negative but imaginary component is positive. Obviously, the vector lies in the second quadrant as shown in Fig. 9(*a*). One can guess that the angle θ would be more than 90° and the amplitude would be slightly greater than 12.

$$\text{vector amplitude} = \sqrt{(-4)^2 + 12^2} = 12.6$$

$\tan \alpha = 12/4 = 3,$

$\therefore \quad \alpha = \tan^{-1} 3 = 72.5°$

$\therefore \quad \theta = 180° - 72.5° = 107.5°$

$\therefore$ polar form = $\mathbf{12.6\angle 107.5°}$

(*ii*) Here, both components are negative which is possible only if vector lies in the third quadrant as shown in Fig. 9(*b*). Again rough sketch would be helpful.

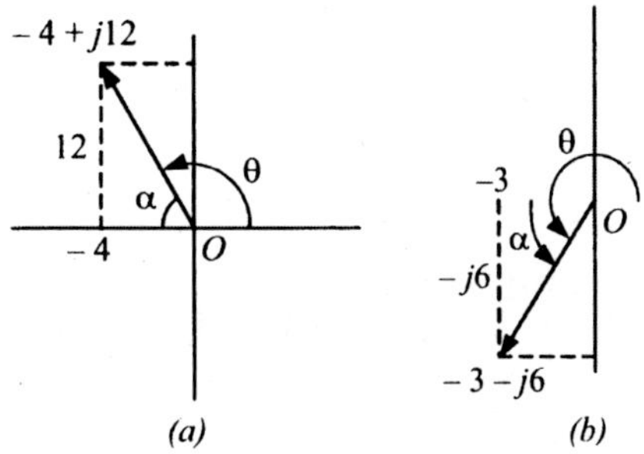

Fig. 9

vector amplitude = $\sqrt{(-3)^2 + (-6)^2} = 6.7$

$\tan \alpha = (-6/-3) = 2,$

$\therefore \quad \alpha = \tan^{-1} 2 = 63.5°$

$\therefore \quad \theta = 180° + 63.5° = 243.5°$

$\therefore$ polar form is $\mathbf{6.7\angle 243.5°}$ or $\mathbf{6.7\angle -116.5°}$

Q. 12. Draw the vector and impedance diagrams and calculate the circuit constants for the following applied voltage and resultant current.

$$v = 100 \sin(2500t + 45°) \quad \text{and} \quad i = 2 \sin(2500t - 15°).$$

Ans. From the phase angles of voltage and current, it is obvious that vector for *V* lies in the first quadrant and that for *I* lies in the fourth quadrant. Their phase difference is 45° – (–15°) = 60° with current *lagging*. Hence, it is an *RL* series circuit.

Since 100 V and 2A represent the maximum values, their rms values are $100/\sqrt{2}$ and $2/\sqrt{2}$ respectively.

$$\mathbf{Z} = \frac{\mathbf{V}}{\mathbf{I}} = \frac{(100/\sqrt{2})\angle 45°}{(2/\sqrt{2})\angle -15°} = 50\angle 60°$$

$$= 25 + j43.3$$

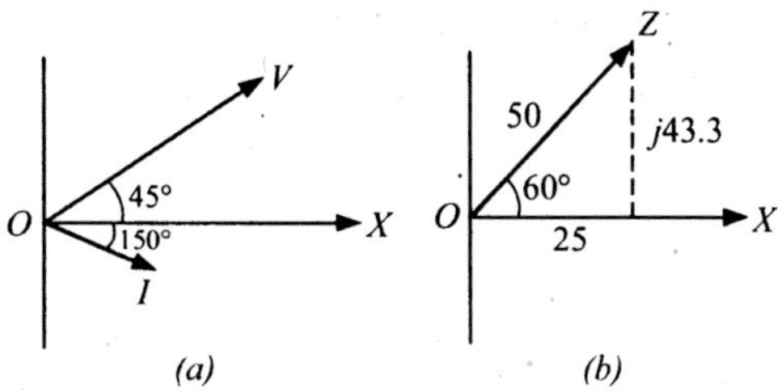

Fig. 10

Also, $\quad Z = R + jX_L = 25 + j\,43.3$

$R = \mathbf{25\ \Omega}$ and $X_L = 43.3\ \Omega$

Now, $\quad X_L = \omega L$

$\therefore \quad \omega L = 43.3$

or $\quad 2500\,L = 43.3,\ L = 0.0173$ H

$= \mathbf{17.3\ mH}$

Q. 13. Draw the vector and impedance diagrams and calculate the circuit constants for the following applied voltage and current.

$$v = 300 \sin(2000t + 170°) \text{ and}$$
$$i = 15 \sin(2000t - 145°)$$

Ans. It should be kept in mind that 300 V and 15 A represent the maximum values. The rms values are $300/\sqrt{2} = 212$ V and $15/\sqrt{2} = 10.6$ A.

Hence,

$V = 212\angle 170°$

and $\quad I = 10.6\angle -145°$

$$\mathbf{Z} = \frac{\mathbf{V}}{\mathbf{I}} = \frac{212\angle 170°}{10.6\angle -145°}$$

$= 20\angle 315° = 20\angle -45°$

$= 14.14 - j14.14$

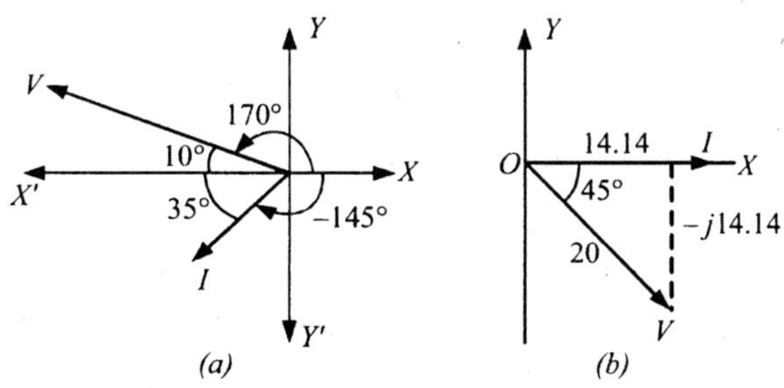

Fig. 11

As seen from the phasor diagram of Fig. 11(*a*), the current leads the voltage by (10° + 35°) = 45° indicating that it is an *RC* circuit. From the expression for *Z*, we find that R = **14.14 Ω** and X_C = 14.14 Ω.

Now, $$X_C = \frac{1}{\omega C} \quad \text{or} \quad 14.14 \frac{1}{2000 \times C} \qquad \therefore \quad C = \mathbf{35.36\ \mu F}$$

Q. 14. In a given *R-L* circuit, *R* = 35 Ω and *L* = *IH*. Find the current flowing through the series circuit if a 50-Hz voltage V = 110∠30° volts is applied across the circuit.

Ans. $R = 35\ \Omega,\ X_L = \omega L$

$$X_L = 1 \times 314 = 314\ \Omega$$

$$\mathbf{Z} = 35 + j314$$

$$= \sqrt{(35^2 + 314^2)}\angle\tan^{-1}(314/35)$$

$$= 315.9\angle 83.67°$$

$$\therefore \quad \mathbf{I} = \frac{\mathbf{V}}{\mathbf{Z}} = \frac{110\angle 30°}{315.9/83.67°}$$

$$= \mathbf{0.35\angle{-53.67°}}$$

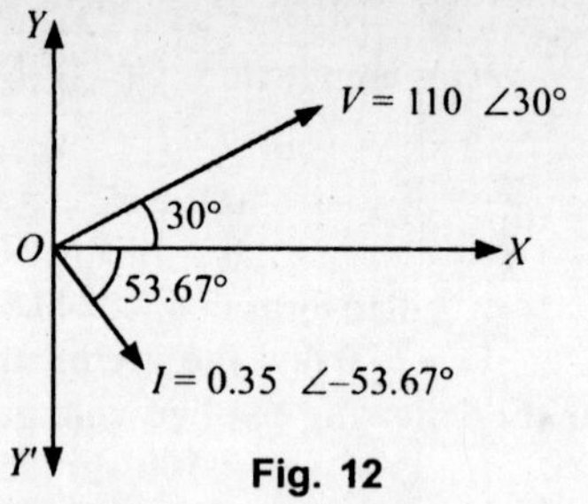

Fig. 12

Q. 15. A coil of inductance 0.08 H and negligible resistance is connected in series with a 15Ω non-inductive resistance. The combined circuit is energised from a 240 V, 50 Hz supply.

Calculate : (*i*) reactance of the coil (*ii*) impedance of the circuit (*iii*) current in the circuit (*iv*) voltage across the resistance (*v*) voltage across the coil (*vi*) power absorbed by the circuit (*vii*) power factor of the circuit. (*Basic Elect. Engg. R.G.P.V, Bhopal, 2008*)

Ans. (*i*) $X_L = 2\pi f L = 2\pi \times 50 \times 0.08 = \mathbf{25.13\ \Omega}$

(*ii*) Impedance of the circuit is

$$\mathbf{Z} = R + j\,X_L = 15 + j\,25.13 = \mathbf{29.27\angle 59.17°}$$

(*iii*) Taking voltage as the reference quantity, circuit, current is

$$\mathbf{I} = \frac{\mathbf{V}}{\mathbf{Z}} = \frac{240\angle 0°}{29.27\angle 59.17°} = \mathbf{8.2\angle{-59.17°}\ A}$$

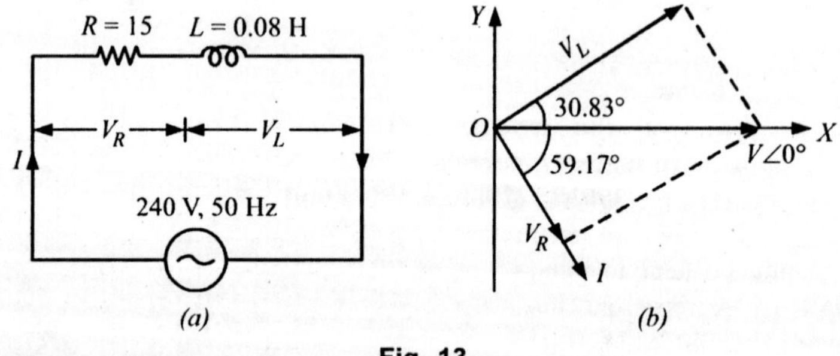

Fig. 13

Obviously, current *lags* behind the voltage by 59.17°

(*iv*) $V_R = \mathbf{IR} = (8.2\angle{-59.17°}) \times 15\angle 0° = \mathbf{123\angle{-59.17°}}$ volt

(*v*) voltage across the coil is

$$\mathbf{V_L} = I \times jX_L = (8.2\angle 59.17°) \times 25.13\angle 90° = \mathbf{206\angle 30.83°\ V}$$

(*vi*) power absorbed by the circuit is,

$$P = VI\cos\phi = 240 \times 8.2 \times \cos(-59.17°) = \mathbf{1008\ W}$$

(*vii*) Power factor of the circuit is given by,

$$\text{power factor} = \cos 59.17° = \mathbf{0.51\ (lagging)}$$

Voltage vector diagram is shown in Fig. 13(*b*).

Q. 16. In a particular ac circuit, three impedances are connected in parallel. Currents as shown in Fig. 14(*a*) are flowing through its parallel branches.

(*i*) Write the equations for the currents in terms of sinusoidal variations and draw the waveforms (*ii*) find the total current supplied by the source (*iii*) power consumed.

(*Elect. Engg. Nagpur Univ.*)

Ans. **(*ii*)** Since I_1 lags behind V by 30°, it must be a *R-L* series circuit. However, I_2 leads V by 45° indicating that it is an *R-C* circuit. Since I_3 leads V by 90°, it must be a pure capacitor.

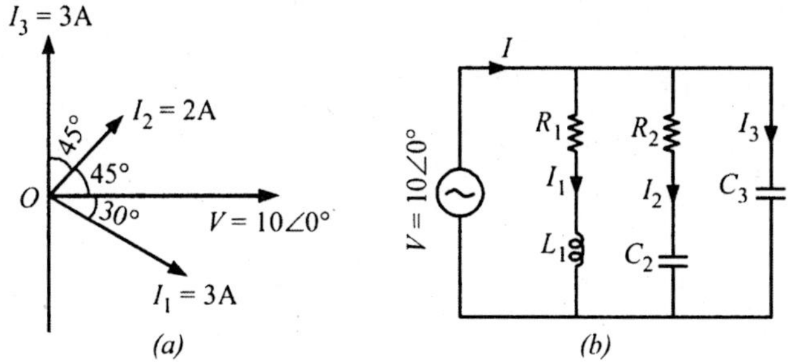

Fig. 14

$$\mathbf{I} = \mathbf{I_1} + \mathbf{I_2} + \mathbf{I_3}$$
$$= 3\,(\cos 30° - j \sin 30°) + 2\,(\cos 45° + j \sin 45°) + j\,3$$
$$= 4.0123 + j\,2.9142 = \mathbf{4.96\angle 36°}$$

Hence, the total circuit current is 4.96 A and leads V by 36°. Current waveforms have been shown in Fig. 15.

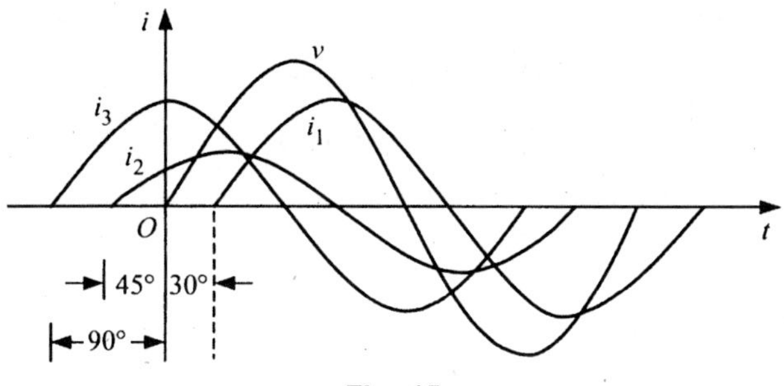

Fig. 15

(*i*) Assuming f = 50 Hz and as seen from Fig. 15(*b*), expressions for the voltage and currents using maximum values are as under:

$$v = \mathbf{10\sqrt{2}\ \sin(314t + 0°)};\ i_1 = \mathbf{3\sqrt{2}\ \sin(314t - 30°)}$$
$$i_2 = \mathbf{2\sqrt{2}\ \sin(314t + 45°)};\ i_3 = \mathbf{3\sqrt{2}\ \sin(314t + 90°)}$$
$$i = \mathbf{4.96\sqrt{2}\ \sin(314t + 36°)}$$

(*iii*) Total power consumed

$$P = \text{voltage} \times \text{in-phase component of } I$$
$$= 10 \times 4.0123 = \mathbf{40.12\ W}$$

Q. 17. A series R-C circuit has R = 20Ω and C = 125μF. A voltage of 240V, 50 Hz is applied to the circuit. Calculate current I, V_R, V_C and power lost.

(*Elect. Engg. Nagpur Univ. 2003*)

Ans.
$$X_C = \frac{1}{\omega C} = \frac{1}{2\pi f C} = \frac{1}{2\pi \times 50 \times 125 \times 10^{-6}} = \frac{10^6}{314 \times 125} = 25.5\ \Omega$$

$$\therefore \quad \mathbf{Z} = R - j\,X_C = 20 - j\,25.5 = 32.4\angle{-51.85°}$$

Let $\mathbf{V} = 240\angle 0°$

$$\therefore \quad \mathbf{I} = \frac{\mathbf{V}}{\mathbf{Z}} = \frac{240\angle 0°}{32.4\angle -51.85°} = \mathbf{6.2\angle 51.85°}$$

$$\mathbf{V_R} = \mathbf{IR} = (6.2\angle 51.85°) \times 20\angle 0° = \mathbf{124\angle 51.85°\ volt}$$

$$\mathbf{V_C} = \mathbf{IX_C} = (6.2\angle 51.85°) \times 25.5\angle{-90°} = \mathbf{158\angle{-38.15°}\ volt}$$

It should be noted that power lost in the circuit is due to the power consumed by the resistor only which produces heat.

$$\text{power lost} = I^2 R = (6.2)^2 \times 20 = \mathbf{769\ W}$$

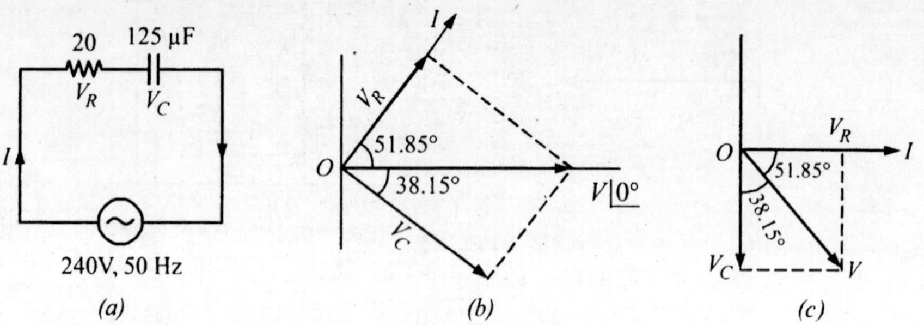

Fig. 16

The vector diagram is shown in Fig. 16(*b*). As seen, I leads V_C by (51.85° + 38.15°) = 90°.

Incidentally, it may be observed that we have taken V as the reference vector in Fig. 16(*b*). If I is taken as the reference vector, then the phasor diagram would be as shown in Fig. 16(*c*).

Q. 18. The potential difference and the current in a circuit are represented by (100 + *j* 200) V and (10 + *j*5) A, respectively. Calculate the power and the reactive volt-ampers.

(*Elect. Engg. A.M.Ae. S.I.*)

Ans. We will use the method of conjugates to calculate the active and reactive powers. Using current conjugate, we have

$$\begin{aligned} \mathbf{P_{VA}} &= (100 + j\,200)\,(10 - j\,5) \\ &= 100 \times 10 - j\,5 \times 100 + j\,200 \times 10 + 200 \\ &= 2000 + j\,1500 \end{aligned}$$

Active power = 2000 W = **2 kW**

Reactive power = **1.5 kVAR**

Q. 19. What is meant by the complex power of a circuit? What are its two rectangular components?

Draw power triangle for (*i*) an inductive load and (*ii*) a capacitive load.

Ans. Suppose a given circuit is connected across an ac voltage source of rms value V and draws an rms current of I. Then, the apparent power drawn by the circuit is VI volt-ampere. It is denoted by S.

The two rectangular components of S are :

(*i*) true or active power, $P = VI \cos \phi$ *i.e.* voltage × in-phase component of current as shown in Fig. 17(*d*).

(*ii*) imaginary or reactive power $Q = VI \sin \phi$ = voltage × quadrature component of current.

As seen from the power triangle of Fig. 17(*d*), $S = \sqrt{P^2 + Q^2}$.

(*i*) **Inductive load :** An inductive load shown in Fig. 17(*a*) draws a lagging current at a phase angle of ϕ as shown. The in-phase and quadrature components are shown in Fig. 17(*c*).

The power triangle is shown in Fig. 17(*d*). It will be seen that the quadrature component of the circuit current *I* (*i.e.* *I* sin ϕ) as well as the quadrature power (*S*) of the circuit are drawn in the fourth quadrant. It is so because we have taken *V* along the *X*-axis.

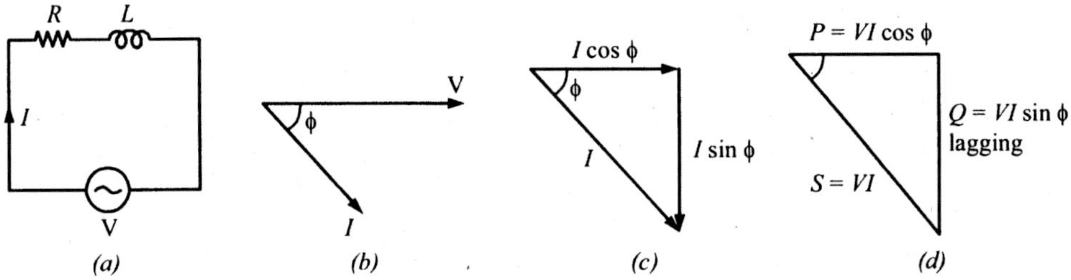

Fig. 17

(*ii*) **Capacitive load :** The capacitive circuit shown in Fig. 18(*a*) draws a current *I* leading the voltage *V* by an angle ϕ. The in-phase component of *I* is *I* cos ϕ and quadrature component is *I* sin ϕ as shown in Fig. 18(*c*).

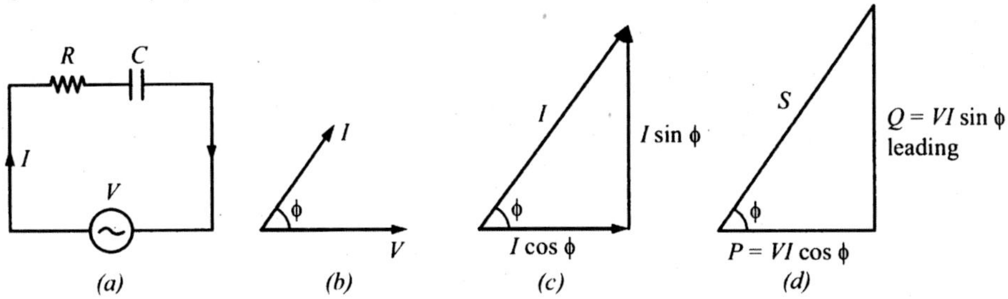

Fig. 18

The corresponding power triangle for the circuit is shown in Fig. 18(*d*). The apparent power *S* has two rectangular components *P* and *Q* as shown.

It will be noted that *V* has been taken along positive *X*-axis. Hence, the leading current *I* lies in the first quadrant. For this reason, the vector for reactive power *Q* in Fig. 18(*d*) lies in the first quadrent.

Q. 20. An *RL* circuit has an impedance of (6 + *j*8) Ω and its applied voltage is 200∠30°. Draw its power triangle.

Ans. $\mathbf{Z} = 6 + j8 = 10\angle 53.1°$

$\mathbf{I} = \mathbf{V}/\mathbf{Z} = (200\angle 30°)/(10\angle 53.1°) = 20\angle -33.1°$

The circuit power can be found by the following three methods:

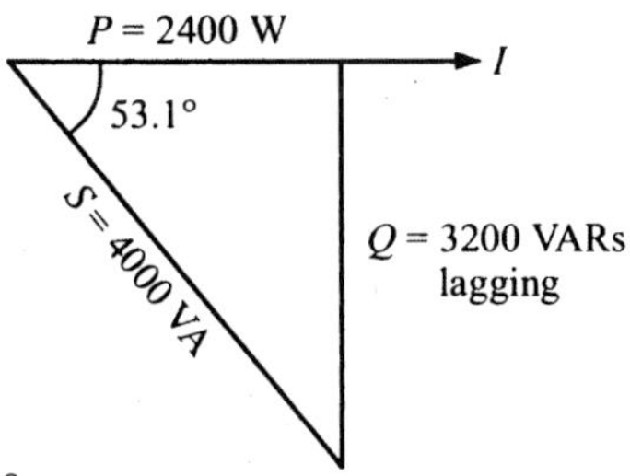

Fig. 19

First method : $P = I^2 R = 20^2 \times 6 = \mathbf{2400\ W}$

$Q = I^2 X_L = 20^2 \times 8 = \mathbf{3200\ VARs}$

$S = I^2 Z = 20^2 \times 10 = \mathbf{4000\ VA}$

Second method : $S = VI = 200 \times 20 = \mathbf{4000\ VA}$

$P = VI\cos\phi = 200 \times 20 \times \cos 53.1°$

$= 200 \times 20 \times 0.6 = \mathbf{2400\ W}$

$Q = VI\sin\phi = 200 \times 20 \times \sin 53.1°$

$= 4000 \times 0.8 =$ **3200 VARs (lagging)**

Third method : For this purpose, we will use the current conjugate **I***.

Here $\mathbf{I} = 20\angle -23.1°$ and its conjugate $\mathbf{I}^* = 20\angle 23.1°$ *i.e.* the sign of the imaginary component has been reversed.

$$\mathbf{S} = \mathbf{VI}^* = (200\angle 30°)(20\angle 23.1°) = 4000\angle 53.1° = 2400 + j3200$$

∴ $P = $ **2400 W**, $Q = $ **3200 VARs lagging**

$$S = \sqrt{2400^2 + 3200^2} = \mathbf{4000\ VA}$$

Q. 21. Determine the power triangle of the circuit shown in Fig. 20(*a*).

Ans. As seen from Fig. 20(*a*),

$\mathbf{Z} = 6 + j10 - j2 = 6 + j8 = 10\angle 53.1°$

$\mathbf{I} = \mathbf{V}/\mathbf{Z} = (100Ð{-90°})/(10\angle 53.1°)$

$= 10\angle{-143.1°}$

Hence,

$\mathbf{S} = \mathbf{VI^*} = (100\angle{-90°})\ (10\angle 143.1°)$

$= 1000\angle 53.1° = 600 + j800$

The angle of the current I has been changed because we have used the current conjugate. The power triangle is shown in Fig. 20(*b*). Its components are P = **600 W**, Q = **800 VAR lagging**, S = **1000 VA** and P.f. = cos 53.1° = **0.6 lagging.**

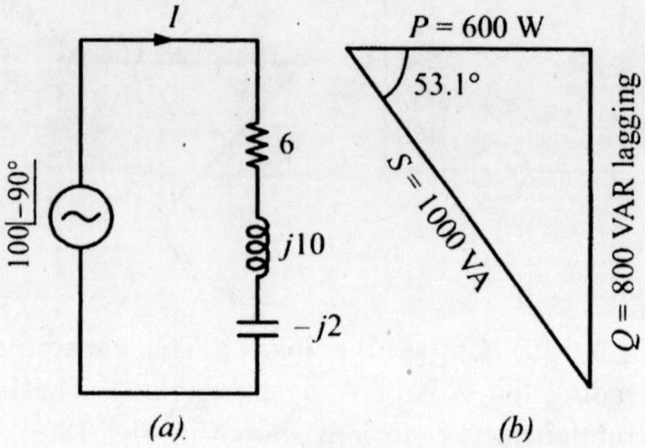

Fig. 20

Q. 22. An alternating voltage (80 + *j*60) volt is applied to a circuit and current flowing is (–4 + *j*10) A. Find :

(*i*) the impedance of the circuit (*ii*) power consumed and (*iii*) phase angle

(*Basic Elect. Engg., CSV Tech Univ. Bhilai, 2009*)

Ans. Students should note that when voltages and currents are given in rectangular form, positive *X*-axis is taken as the reference vector.

Magnitude of $V = \sqrt{80^2 + 60^2} = 100$ volt

As seen, $\tan\alpha = 60/80 = 0.75$

∴ $\alpha = \tan^{-1}(0.75) = 37°$ ∴ $\mathbf{V} = 100\underline{|37°}$

Its vector lies in the first quadrant because both its components are positive as shown in Fig. 21.

Magnitude of

$$I = \sqrt{(-4)^2 + (10)^2} = 10.77\ \text{A}$$

Since its real component is negative and *j* component is positive, it lies in the second quadrant as shown in Fig. 21. As seen, $\beta = \tan^{-1}(10/4) = \tan^{-1}(2.5) = 68.3°$

∴ $\theta = 180° - 68.3° = 111.7°$

∴ $\mathbf{I} = 10.77\angle 111.7°$

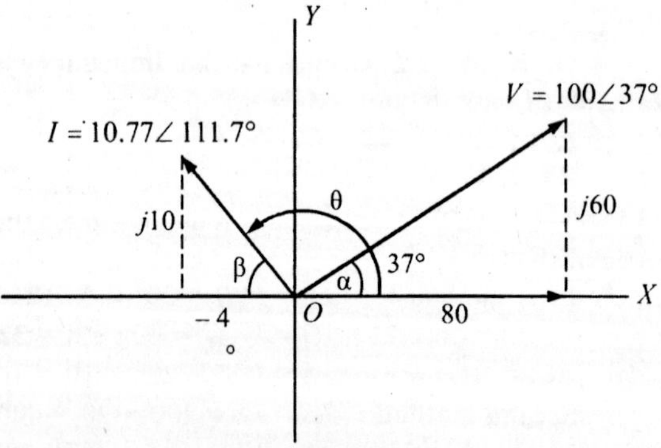

Fig. 21

(*i*) $\mathbf{Z} = \dfrac{\mathbf{V}}{\mathbf{I}} = \dfrac{100\angle 37°}{10.77\angle 111.7°}$

$= \mathbf{9.28\angle{-74.7°}}$

(*ii*) We will use the current conjugate to find volt-ampere of the circuit

∴ $\mathbf{P_{VA}} = (80 + j60) \times (-4 - j10) = 280 - j1040$

∴ power consumed = **280 W**

(*iii*) The phase difference between the voltage and current is

$= \theta - \alpha = 111.7° - 37° = \mathbf{74.7°}$

Incidentally, power factor = cos 74.7° = 0.264 (leading).

Q. 23. Find the value of R and C so that $V_b = 3\ V_a$ and V_b and V_a are in quadrature (Fig. 22). Find also the phase relation between V and V_b as well as V_a and I.

(*Basic Elect. Engg. Tech Univ. Bhilai, 2007*)

Ans. Since I is common to both impedances, it is taken as the reference vector as shown in Fig. 22(*b*).

$$X_L = 2\pi f L = 2\pi \times 50 \times 25.5 \times 10^{-3} = 8\ \Omega$$

$$\therefore \quad \mathbf{Z}_b = 6 + j8 = 10\angle 53.13^\circ$$

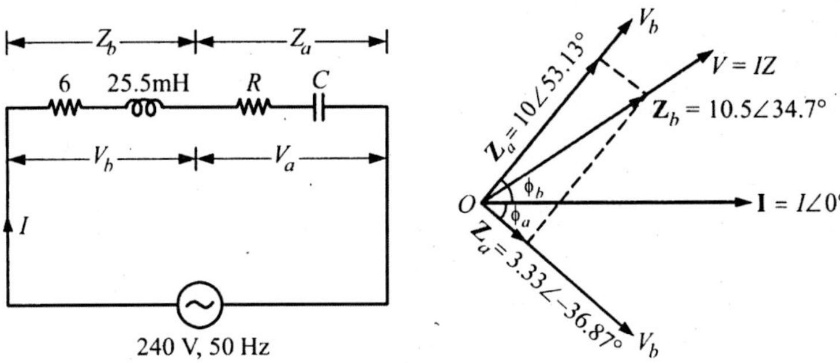

(*a*) (*b*)

Fig. 22

Now, $|\mathbf{Z}_b| = 10$ and we are given that $|\mathbf{Z}_b| = 3\ |\mathbf{Z}_a|$

$$\therefore \quad 3\ |\mathbf{Z}_a| = 10 \quad \text{or} \quad |\mathbf{Z}_a| = 3.33$$

We are also given that V_b and V_a are in quadrature. Since $\mathbf{Z}_a$ is capacitive, hence it is behind $\mathbf{Z}_b$ by 90° and similarly, $\mathbf{V}_a$ lags $\mathbf{V}_b$ by 90° as shown in Fig. 22(*b*). It is so because capacitive voltage lags current whereas inductive voltage is ahead of current.

$$\therefore \quad \phi_b + \phi_a = 90^\circ \quad \text{or} \quad 53.13^\circ + \phi_a = 90^\circ \quad \therefore \quad \phi_a = 36.87^\circ$$

$$\therefore \quad \mathbf{Z}_a = 3.33\angle -36.87^\circ$$

$$\mathbf{Z}_a = 3.33\ [\cos(-36.87^\circ) + j\ 3.33 \sin(-36.87^\circ)] = 2.66 - j2$$

Hence, $R = 2.66\ \Omega$ and $X_C = 2\ \Omega$

Now, $$X_C = \frac{1}{2\pi f C} \quad \text{or} \quad C = \frac{1}{2\pi \times 50 \times 2}\ \text{F} = \frac{10^6}{314 \times 2} = 1590\ \mu\text{F}$$

Circuit impedance, $\mathbf{Z} = \mathbf{Z}_a + \mathbf{Z}_b = 6 + j8 + 2.66 - j2$

$$= 8.66 + j6 = 10.5\angle 34.7^\circ$$

As seen from the vector diagram, angle between V and V_b = 53.13° – 34.7° = 18.43°. Similarly, angle between V_a and $I = \phi_a = 36.87^\circ$

Note : Students should note that in the vector diagram of Fig. 22(*b*), voltage vectors are the same as the corresponding impedances, multiplied by I (since the phase of I is zero).

Q. 24. Two impedance $Z_1 = (6 + j8)\ \Omega$ and $Z_2 = (8 - j6)\ \Omega$ are connected in parallel across 100 V supply.

Determine :

(*i*) current and power factor of each branch

(*ii*) overall current and power factor

(*iii*) power consumed by each branch and total power.

(*Elect. Engg. Nagpur Univ. 2003*)

Ans. We are given that,

$$\mathbf{Z}_1 = (6 + j8) = 10\angle 53.13^\circ$$

$$\mathbf{Z}_2 = (8 - j6) = 10\angle -36.86^\circ$$

Since it is a parallel circuit [Fig. 23(*a*)] so that voltage is common to both, we will take $\mathbf{V} = 100\angle 0°$.

(*i*) $I_1 = \dfrac{\mathbf{V}}{\mathbf{Z}_1} = \dfrac{100\angle 0°}{10\angle 53.13°}$

$= \mathbf{10\angle -53.13°}$

$I_2 = \dfrac{\mathbf{V}}{\mathbf{Z}_2} = \dfrac{100\angle 0°}{10\angle -36.86°}$

$= \mathbf{10\angle 36.86°}$

$\cos\phi_1 = R_1/Z_1 = 6/10 = $ **0.6 lagging**

$\cos\phi_2 = R_2/Z_2 = 8/10 = $ **0.8 leading**

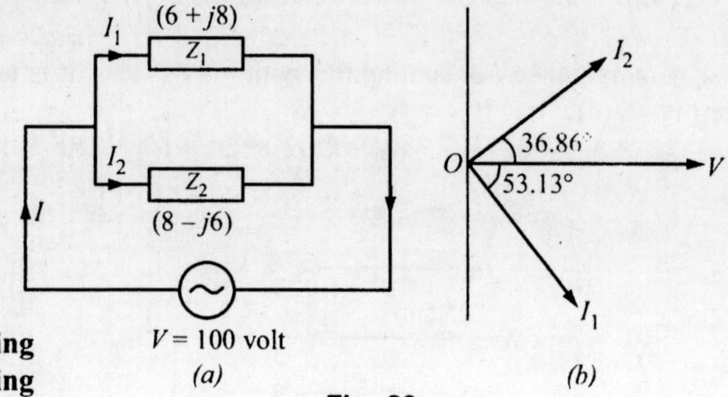

Fig. 23

(*ii*) Combined $\mathbf{Z} = \mathbf{Z}_1 \parallel \mathbf{Z}_2 = \dfrac{\mathbf{Z}_1\mathbf{Z}_2}{\mathbf{Z}_1 + \mathbf{Z}_2}$

$= \dfrac{100\angle 16.27°}{14 + j2} = \dfrac{100\angle 16.27°}{14.14\angle 8.13°} = 7.07\angle 8.14° = (7 + j1)$

$\mathbf{I} = \dfrac{100\angle 0°}{7.07\angle 8.14°} = \mathbf{14.14\angle -8.14°}$

It is seen from the combined **Z** value, that combined resistance is 7 Ω

∴ $\cos\phi = 7/7.07 = $ **0.99 lag**

(*iii*) $P_1 = I_1^2 \times R_1 = 10^2 \times 6 = 600$ W

$P_2 = I_2^2 \times R_2 = 10^2 \times 8 = 800$ W

∴ total power, $P = P_1 + P_2 = 600 + 800 = $ **1400 W**

Alternatively, we can find P from the circuit current and total resistance.

As found above, $I = 14.14$ A and total resistance = 7 Ω

∴ $P = I^2R = 14.14^2 \times 7 = $ **1400 W**

EXERCISES

1. Evaluate the following :
(*i*) $(+j15)(-j20)$, (*ii*) $(-j15)(+j20)(+j30)$, (*iii*) $(-j50)(-j30)$
(300; $j9 \times 10^3$; $-j1.5 \times 10^3$)

2. Convert the following into polar form :
(*i*) $(5 + j25).(20 - j10)$ (*ii*) $(15 - j30).(-10 - j20)$ (*iii*) $(30 + j30).(20 - j10)$.
(570∠52.1°, –750, 949∠18.4°)

3. Convert the following quatities to rectangular form :
(*i*) 240∠48° (*ii*) 100∠–32° **(160.6 + j178.4 ; 84.8 – j53)**

4. Convert the following voltages into polar form:
(*i*) $(707 + j707)$ V (*ii*) $(40 - j360)$ V (*iii*) $(-40 - j30)$ V.
(1000∠45°; 362∠–83.4; 25∠21.7°)

5. A series *R-C* circuit consists of a resistance of 1 kΩ and a capacitor of 0.022 μF connected to 10-V, 5 kHz ac source. Calculate V_R, V_C and phase angle of the circuit.
(5.685∠55.35°, 8.226Ð–34.65°, 55.35°)

6. The current drawn by an unknown impedance is 5∠20°A and the voltage across the impedance is 100∠60° V. Calculate

(*i*) impedance (*ii*) Circuit phase angle and power factor (*iii*) values of R and X (*iv*) drop across R (*v*) drop across X.

($20\angle 40°$ Ω; 40°, 0.766; 15.32 Ω; j1286 Ω; $76.6\angle 20°$ V, $V_L = 64.28\angle 110°$V)

7. The current in a circuit is $(10 - j12)$ A when the applied voltage is $(50 + j30)$ V. Find: (*a*) impedance of the circuit (*b*) power consumed and (*c*) power factor.

(*Basic Elect. Engg. Kerala Univ, 2004*)

($3.7\angle 81.5°$, 140 W, 0.148 lag)

8. A voltage of 125 V at 60 Hz is applied across a non-inductive resistor connected in series with a capacitor. The current is 2.2 A. The power loss in the resistor is 96.8 W and that in the capacitor is negligible. Calculate the resistance and the capacitance.

(20 Ω, 50 μF)

9. A circuit consists of a 100 Ω resistor connected in series with a pure inductor of 0.1 H. The voltage drops across resistor and inductor are 10 V each. Assuming current to be the reference vector, find:
(*a*) frequency (*b*) value of current (*c*) circuit impedance (*d*) applied voltage.

(159 Hz ; $141.4\angle 45°$ Ω ; $14.14\angle 45°$)

10. In a series RC circuit, voltage across the resistor is 15 V and that across the capacitor is 10 V. The circuit current is 5 A and the angular frequency of the applied source is 10,000 rad/s. Find the values of R and C taking current as the reference vector and also find V. **(3 Ω, 50 μF, $18\angle -337°$ V)**

11. Two impedances $(14 + j5)$ Ω and $(18 + j10)$ Ω are connected in parallel across a 220 V, 50 Hz supply. Determine:
   (*i*) impedance of the entire circuit.
   (*ii*) current, power and power factor.
   (*iii*) Draw the phasor diagram for above circuit indicating all the electrical quantities.

(*Elect. Engg, Nagpur Univ. 2004*)

($8.66\angle 23.59°$, $25.4\angle -23.59°$, 5122 W, 0.917)

MULTIPLE CHOICE QUESTIONS

1. The square of an imaginary number
(*a*) can never be negative (*b*) can never be positive
(*c*) could be either positive or negative (*d*) is equal to j
2. A complex number
(*a*) is the same thing as imaginary number (*b*) has a real part and an imaginary part
(*c*) is negative number (*d*) is merely a technical term
3. The sum of $(3 + j\,6)$ and $(-3 - j\,6)$ is
(*a*) $0 + j0$ (*b*) $6 + j12$ (*c*) $-6 - j12$ (*d*) $0 - j12$
4. The product of $(-4 - j7)(6 - j2)$ is
(*a*) $-24 + j14$ (*b*) $24 - j14$ (*c*) $-38 - j34$ (*d*) $-24 - j14$
5. The magnitude of the vector $(6 - j8)$ is
(*a*) 10 (*b*) 100 (*c*) -48 (*d*) $-j^2$
6. The impedance vector $(0 - j20)$ represents
(*a*) a pure resistor (*b*) a pure inductance
(*c*) a pure capacitance (*d*) an inductance combined with resistance.
7. An inductive reactance is written as
(*a*) jX_L (*b*) $X_L j$ (*c*) $-j\,X_L$ (*d*) $-X_L j$

8. The total impedance of a series circuit with $R = 1000\ \Omega$ and $j\,X_L = 1000\ \Omega$ is ...... ohm.
(*a*) 2000 (*b*) 1414 (*c*) 1000 (*d*) 707
9. If a series *RLC* circuit has $R = 20\ \Omega$, $X_L = 30\ \Omega$ and $X_C = 40\ \Omega$ its impedance in rectangular form is
(*a*) $20 + j70$ (*b*) $-j20 + 10$ (*c*) $20 + j10$ (*d*) $20 - j10$
10. The capacitive susceptance of a 100 μF capacitor at 50 Hz is
(*a*) j 0.0314 (*b*) $-j$ 0.0314 (*c*) j0.314 (*d*) $-j$ 0.314
11. The impedance of a series ac circuit is $50\angle{-60°}\ \Omega$. The circuit is
(*a*) resistive (*b*) capacitive
(*c*) inductive (*d*) none of the above

(*Basic Elect. Engg. VTU Karnataka, 2009*)

ANSWERS

1. *b*	2. *b*	3. *a*	4. *c*	5. *a*	6. *c*	7. *a*
8. *b*	9. *d*	10. *a*	11. *b*.			

7 CHAPTER AC SERIES CIRCUITS

7.1 SERIES R-L CIRCUITS

Q. 1 (*a*) Discuss the ac voltage-current relationship in a series *R-L* circuit. Is volt-ampere (VA) the same thing as watt (W)?

Ans. A *pure* resistor *R* and a *pure* inductance *L* are shown connected in series in Fig. 1. We will take *I* as the reference vector because it is common to both elements.

Let V = rms value of the applied voltage

I = rms value of the resultant current

$V_R = IR$ = voltage drop across *R* (in phase with *I*)

$V_L = IX_L$ = voltage drop across inductor (at right angle to *I*)

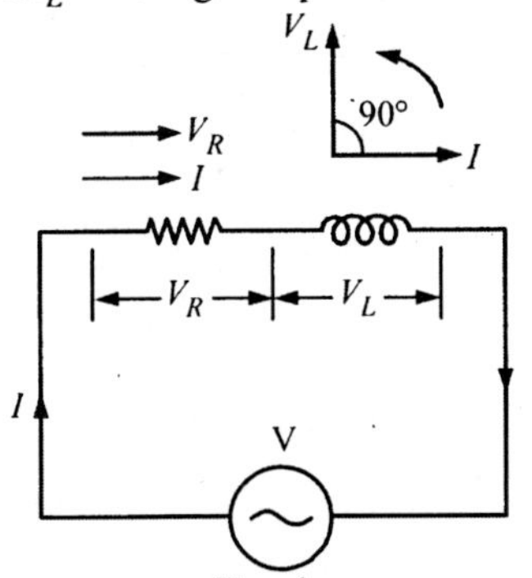

Fig. 1

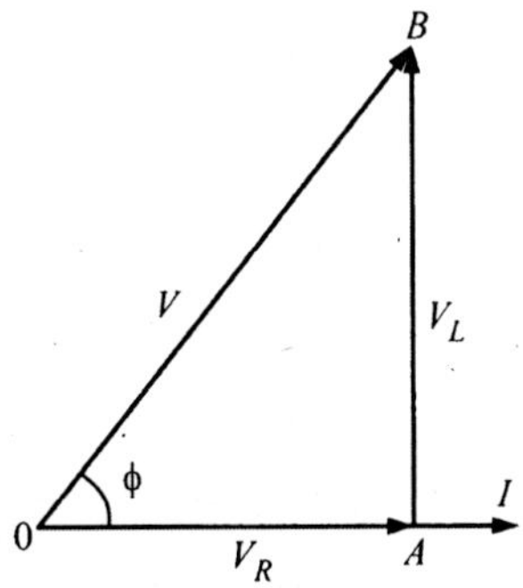

Fig. 2

These voltage drops are shown in the voltage triangle *OAB* in Fig. 2. The applied voltage *V* is the *vector* sum of the two voltage drops. It equals vector *OB*.

$$\therefore \quad V = \sqrt{V_R^2 + V_L^2}$$

$$= \sqrt{(IR)^2 + (IX_L)^2} = I\sqrt{(R^2 + X_L^2)}$$

$$\therefore \quad I = \frac{V}{\sqrt{(R^2 + X_L^2)}} = \frac{V}{Z}$$

The quantity $\sqrt{(R^2 + X_L^2)}$ is known as the *impedance* (*Z*) of the circuit.

As seen from the impedance triangle of Fig. 3.

$$Z^2 = R^2 + X_L^2 \quad \text{or} \quad Z = \sqrt{R^2 + X_L^2}$$

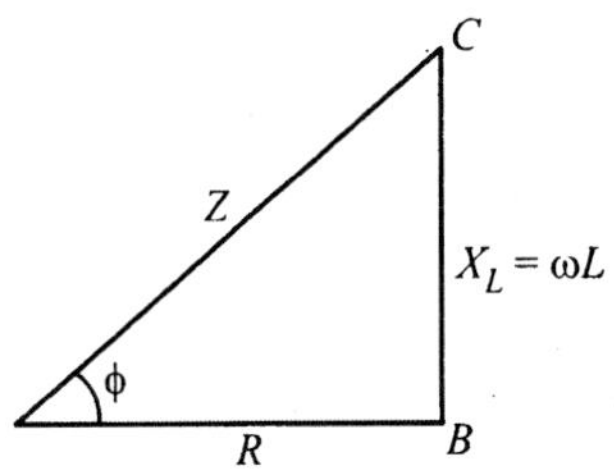

Fig. 3

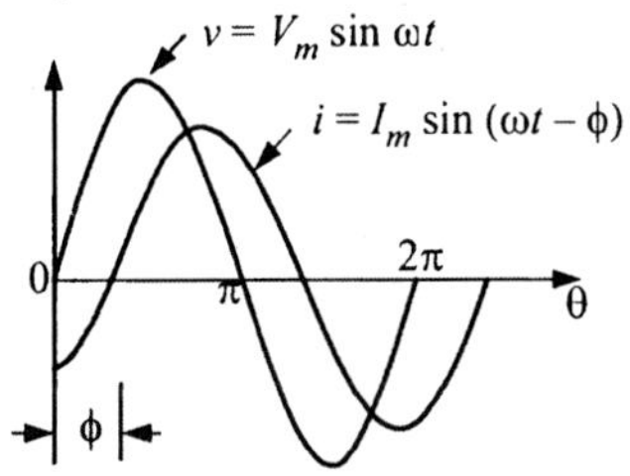

Fig. 4

It is clear from Fig. 2 that current I lags behind the applied voltage V by an angle ϕ such that

$$\tan \phi = \frac{X_L}{R} = \frac{\omega L}{R} = \frac{\text{reactance}}{\text{resistance}}$$

$$\therefore \quad \phi = \tan^{-1} (X_L/R)$$

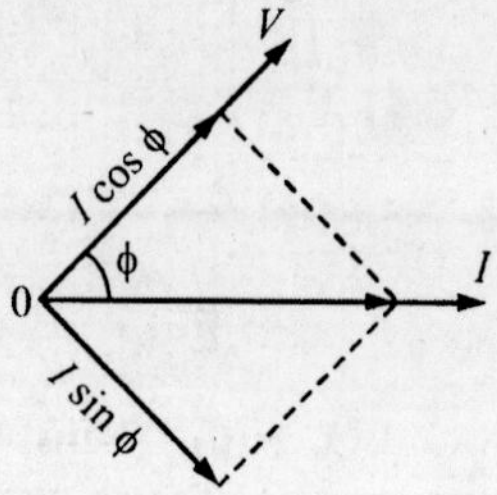

Fig. 5

The same fact is shown graphically in Fig. 4.

Hence, we find that if the applied voltage is given by

$$v = V_m \sin \omega t$$

Then $\quad i = I_m \sin (\omega t - \phi)$

where, $\quad I_m = V_m/Z$

Power :

As seen from Fig. 5, the lagging current I can be resolved into two rectangular components,

(*i*) $I \cos \phi$ along the applied voltage V

(*ii*) $I \sin \phi$ perpendicular to V

The mean power consumed by *R-L* circuit is given by the product of V and that part of the current which is in phase with V.

$$\therefore \quad P = V \times I \cos \phi$$

$$= \text{rms volts} \times \text{rms current} \times \cos \phi$$

The term '$\cos \phi$' is known as the *power factor* of the circuit.

Also, $\quad P = VI \cos \phi = VI \times (R/Z) \qquad$ — ($\because \cos \phi = R/Z$)

$$= (V/Z) \times I \times R = I \times I \times R = I^2 R$$

In fact, power is consumed by the *resistor and not by the inductor.*

In an ac circuit, the product VI does not give the true power in a circuit (because V and I are not in-phase with each other as in d.c. circuits). It gives apparent power.

Hence, it is called volt-ampere. the true power in watts is given by

$$\text{Watts} = VA \times \cos \phi$$

Q. 2. Give different expressions for the power factor of a series *R-L* circuit. Also, define the apparent power, active power and reactive power drawn by such a circuit.

Ans. The power factor of an ac circuit may be defined in the following different ways:

1. It is given by the cosine of the angle of lead or lag

$$\therefore \quad \text{p.f.} = \cos \phi$$

2. It is given by the ratio, $\dfrac{\text{resistance}}{\text{impedance}}$

$$\therefore \quad \text{p.f.} = \frac{R}{Z}$$

3. It is given by the ratio, $\dfrac{\text{watts}}{\text{volt} - \text{amperes}}$

$$\therefore \quad \text{p.f.} = \frac{\text{W}}{\text{VA}} \text{ or } \frac{\text{kW}}{\text{kVA}}$$

In dc circuits, the product VI gives the power in watts drawn by the circuit. But in ac circuits, things are different. In ac circuits, the product 'VI' gives the *apparent* power and not the *real* power because the voltage and current are *not in phase with each other.* The VA has two components: (*i*) active component which gives the true power in watts and (*ii*) reactive component which gives volt-ampere-reactive (VAR).

These facts are shown in the power triangles or kVA triangles of Fig. 6 and Fig. 7 for lagging and leading power factors respectively.

Talking in terms of kVA, we have the following three powers:

(*i*) **apparent power (S)**

It is given by S = kilovolt-ampere (kVA)

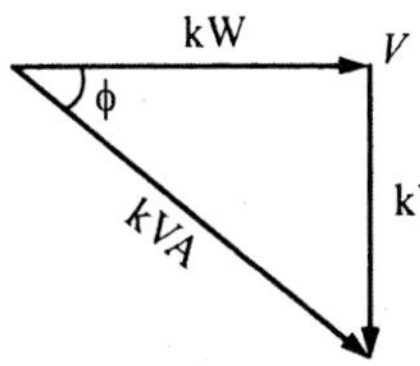

lagging p.f.

Fig. 6

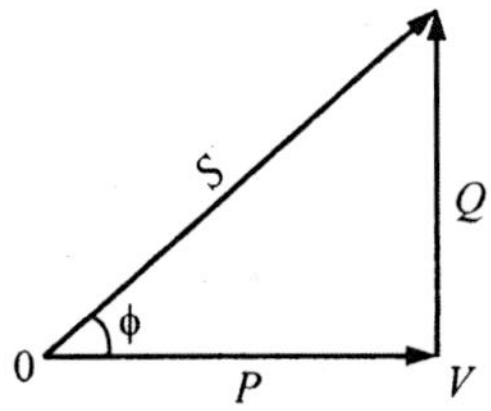

leading p.f.

Fig. 7

(*ii*) **active power (P or W)**

It is given by P = kVA × cos ϕ or killowatts (kW).

It is the power which is actually dissipated in the circuit resistance R and is equal to $I^2 R$. It is also called true power.

(*iii*) **reactive power (Q)**

It is given by Q = kVA sin ϕ. It is written as kVAR. In fact, it is the power developed in the inductive reactance of the circuit. It is also called *imaginary* power. As seen from the kVA triangle of Fig. 7 for a lagging power factor.

$$S^2 = P^2 + Q^2 \quad \text{or} \quad S = \sqrt{P^2 + Q^2}\,.$$

Q. 3. A single-phase series circuit consists of 10 Ω resistance and 0.1 H inductance. A voltage of 230 volts, 50 Hz is applied across the circuit. Find the voltage drop across the resistance (*Basic Elect. Engg., Kerala Univ. 2003*)

Ans. Let us first find the rms current I in the circuit.

For this purpose, we find

$$Z = \sqrt{R_L^2 \times X_L^2}$$

Now, $R = 10\ \Omega,\ X_L = 2\pi f L = 2\pi \times 50 \times 0.1 = 31.4\ \Omega$

∴ $Z = \sqrt{10^2 + 31.4^2} = \sqrt{100 + 986} = \sqrt{1086} = 33\ \Omega$

∴ $I = \dfrac{V}{Z} = \dfrac{230}{33} = 6.97\ \text{A}$

Now, $V_R = I \times R = 6.97 \times 10 = \mathbf{69.7\ V}$

Q. 4. An inductive coil takes 10 A and dissipates 1000 W when connected to a supply of 250 V, 50 Hz. Calculate:

(*i*) the impedance **(*ii*) the effective resistance**

(*iii*) reactance **(*iv*) the inductance**

(*v*) power factor

Also, draw the vector diagram. (*Elect. Circuits B. Tech JNTU, Hyderabad 2006*)

Sol. (*i*) $Z = V/I = 250/10 = \mathbf{25\ \Omega}$

(*ii*) $I^2R = 1000$ or $10^2 \times R = 1000$, ∴ $R = \mathbf{10\ \Omega}$

(*iii*) $Z^2 = R^2 + X_L^2$ or $X_L = \sqrt{Z^2 - R^2} = \sqrt{25^2 - 10^2} = \mathbf{23\ \Omega}$

(*iv*) Now, $X_L = \omega L = 314\ L$ — (Fig. 8)

∴ $23 = 314 \times L$ ∴ $L = \mathbf{0.073\ mH}$

(*v*) $\cos\phi = R/Z = 10/25 = \mathbf{0.4}$

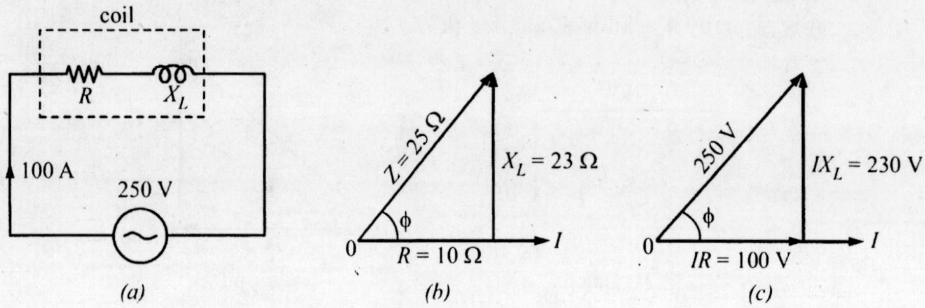

Fig. 8

Fig. 8 (c) shows the vector diagram. Here, the resistive drop and inductive drop add vectorially to give the applied voltage of 250 V as shown.

Q. 5. A coil having a resistance of 7 Ω and an inductance of 31.8 mH is connected to 230 volts, 50 Hz supply. Calculate:

(*i*) the circuit current (*ii*) phase angle (*iii*) power factor (*iv*) power consumed.

(*Basic. Elect. Engg., RGPV, Bhopal 2008*)

Ans. First, let us find the value of impedance, *Z*.

$$X_L = 2\pi fL = 2\pi \times 50 \times (31.8 \times 10^{-3}) = 10\ \Omega$$

$$Z = \sqrt{R^2 + X_L^2} = \sqrt{7^2 + 10^2} = \sqrt{149} = 12.2\ \Omega$$

(*i*) $I = V/Z = 230/12.2 = \mathbf{18.85\ A}$

(*ii*) $\tan\phi = 10/7,\ \phi = \tan^{-1}(10/7) = \mathbf{55° \text{ (lag)}}$

(*iii*) power factor, $\cos\phi = R/Z = 7/12.2 = \mathbf{0.574 \text{ (lag)}}$

(*iv*) power consumed $= VI\cos\phi = 230 \times 18.85 \times 0.574 = \mathbf{2488\ W}$

or $P = I^2R = (18.85)^2 \times 7 = 2488$ W

Q. 6. A coil dissipates 10 W at 0.1 power factor lagging when connected to a 240 V, 50-Hz sinusoidal supply. Calculate the resistance, impedance and inductance of the coil.

Ans. Power, $P = V \times I \times \cos\phi$

$$\therefore \quad I = \frac{P}{V \times \cos\phi} = \frac{10}{240 \times 0.1} = 0.417\ \text{A}$$

Also, $P = I^2R \quad \therefore \quad R = \dfrac{P}{I^2} = \dfrac{10}{0.417^2} = \mathbf{57.5\ \Omega}$

Now, $\cos\phi = \dfrac{R}{Z} \quad \therefore \quad Z = \dfrac{R}{\cos\phi} = \dfrac{57.5}{0.1} = \mathbf{575\ \Omega}$

$$X_L = \sqrt{Z^2 - R^2} = \sqrt{575^2 - 57.5^2} = 572\ \Omega$$

Now, $X_L = 2\pi fL \ \therefore\ L = X_L/2\pi f = 572/2\pi \times 50 = \mathbf{1.82\ H}$

Q. 7. A series *R-L* circuit draws a current of 1 A when connected across a 10 V, 50 Hz supply. Assuming the resistance to be 5 Ω, find the inductance of the circuit. What is its power factor? Draw the phasor diagram of the circuit.

(*Basic Elect. Engg., Kerala Univ. 2005*)

Ans. The circuit is shown in Fig. 9.

$$Z = \frac{V}{I} = \frac{10}{1} = 10\ \Omega$$

$$X_L = \sqrt{Z^2 - R^2}$$

$$= \sqrt{10^2 - 5^2} = 8.66\ \Omega$$

$$X_L = 2\pi f L$$

or $8.66 = 2\pi \times 50 \times L,$

$$L = 0.0276\ \text{H} = \mathbf{27.6\ mH}$$

$$\text{p.f.} = R/Z = 5/10 = \mathbf{0.5\ lag.}$$

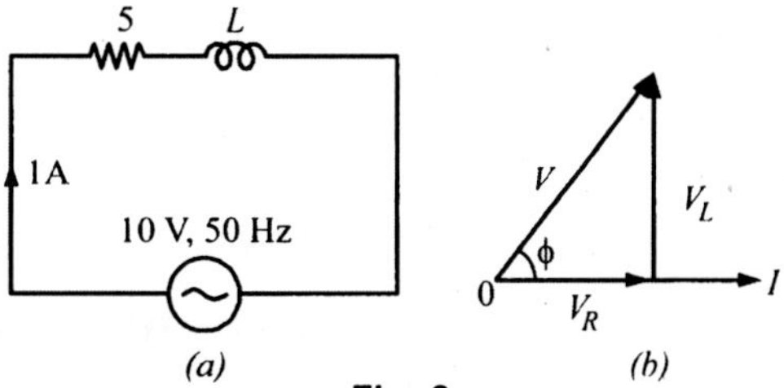

Fig. 9

Q. 8. A coil has an inductance of 50 mH and an impedance of 20 Ω. If the power factor of the circuit is 0.5 lag, find the values of angular frequency and resistance of the coil. *(Basic Elect. Engg. RGPV, Bhopal 2005)*

Ans. Given, $L = 50$ mH, $Z = 20\ \Omega$, p.f., $\cos\phi = 0.5$ lag

Now, $\cos\phi = \frac{R}{Z}$ $\quad\therefore\ 0.5 = \frac{R}{20}$ $\quad\therefore\ R = 0.5 \times 20 = \mathbf{10\ \Omega}$

Now, $Z^2 = R^2 + X_L^2 = R^2 + (\omega L)^2$

$\therefore$ $20^2 = 10^2 + \omega^2 \times (50 \times 10^{-3})^2$

or $300 = \omega^2 \times 25 \times 10^{-4},$ $\qquad \omega^2 = 300/25 \times 10^{-4} = 12 \times 10^4$

$$\omega = \sqrt{12 \times 10^4} = \mathbf{346.4\ rad/s}$$

Q. 9. A coil of resistance 24 Ω has a reactance of 32 Ω when connected across a supply voltage, $v = 566 \sin 314\, t$. Calculate :

(*i*) frequency (*ii*) rms value of current (*iii*) power factor (*iv*) equation of the resultant current.

Ans. **(*i*)** As seen from voltage equation, $\omega = 314$ rad/second

$\therefore$ $$f = \frac{\omega}{2\pi} = \frac{314}{6.28} = \mathbf{50\ Hz}$$

(*ii*) It should be noted that, $V_m = 566$ volt.

Hence, $V_{rms} = 0.707\ V_m = 0.707 \times 566 = 400$ volt

$$Z = \sqrt{R^2 + X_L^2} = \sqrt{24^2 + 32^2} = 40\ \Omega$$

$\therefore$ $$I_{rms} = \frac{V_{rms}}{Z} = \frac{400}{40} = \mathbf{10\ A}$$

(*iii*) $$\cos\phi = \frac{R}{Z} = \frac{24}{40} = \mathbf{0.6\ lag}$$

(*iv*) Now, $I_{max} = \sqrt{2} \times I_{rms} = \mathbf{14.1\ A}$

$\cos\phi = 0.6,\ \phi = 53°8' = 53.13°$

$\therefore$ $\mathbf{i = 14.1 \sin (314t - 53.13°)}$

Q. 10. A current of 5 A flows through a non-inductive resistance connected in series with a choke coil when supplied at 250 V, 50 Hz. If voltage across resistance is 125 V and across coil is 200 V, calculate : (*i*) impedance, resistance and reactance of the coil (*ii*) power in coil (*iii*) total power consumed in the circuit (*iv*) Draw phasor diagram.

(Elect. Engg. Nagpur Univ. 2003)

Ans. The vector diagram of Fig. 10(*b*) shows the relationship between different voltages. The voltage drop of 200 V on the choke coil consists of two rectangular components:

(*i*) drop V_r across the choke resistance and

(*ii*) drop V_L across choke inductance.

The angle θ gives the phase angle of the *choke* and *not of the entire circuit.* The angle ϕ gives the power factor angle of the *entire circuit.* From the vector diagram, we get

$$AC^2 + CD^2 = 250^2$$

or $$(125 + BC)^2 + CD^2 = 250^2 \quad ...(i)$$

Also $$BC^2 + CD^2 = 200^2 \quad ...(ii)$$

Subtracting (*ii*) from (*i*), we get

$$(125 + BC)^2 - BC^2 = 250^2 - 200^2$$

$$\therefore \quad BC = V_r = 27.5 \text{ V}$$

$$CD = \sqrt{(200^2 - 27.5^2)} = 198.1 \text{ V}$$

$$\therefore \quad CD = V_L = 198.1 \text{ V}$$

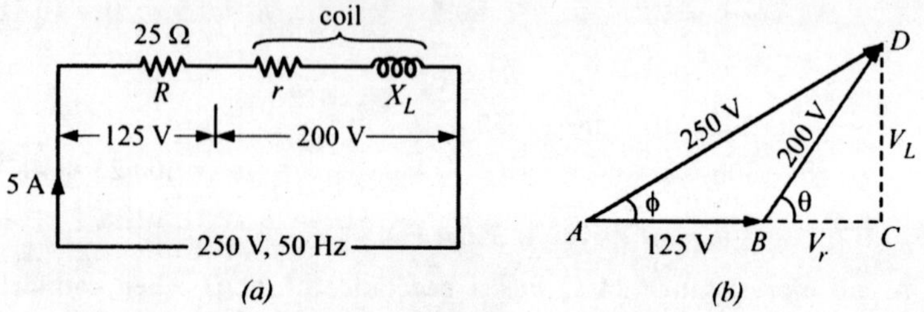

Fig. 10

(*i*) Coil impedance = 200/5 = **40 Ω**

$$V_r = I \times r \quad \text{or} \quad 27.5 = 5 \times r \qquad \therefore \; r = \mathbf{5.5\ \Omega}$$

$$V_L = I \cdot X_L = CD = 198.1$$

$$\therefore \quad X_L = 198.1/5 = \mathbf{39.62\ \Omega}$$

(*ii*) Power absorbed by the coil = $I^2 . r = 5^2 \times 5.5$ = **137.5 W**

(*iii*) $R = 125/5 = 25\ \Omega$

Total circuit resistance = 25 + 5.5 = 30.5 Ω

Total power consumed = $5^2 \times 30.5$ = **762.5 W**

(*iv*) Phasor diagram is shown in Fig. 10(*b*).

Q. 11. Given *v* = 200 sin 377*t* volts and *i* = 8 sin (377*t* – 30°) ampere for an ac circuit, determine

(*i*) power factor **(*ii*) true power**

(*iii*) apparent power **(*iv*) reactive power**

Indicate the unit of power calculated. (*Basic Elect. Engg. VTU Karnataka 2007*)

Ans. As seen form the two given equations, current lags behind the voltage by 30°. Obviously, it is an *R-L* circuit. Also, V_m = 200 volt and I_m = 8 A.

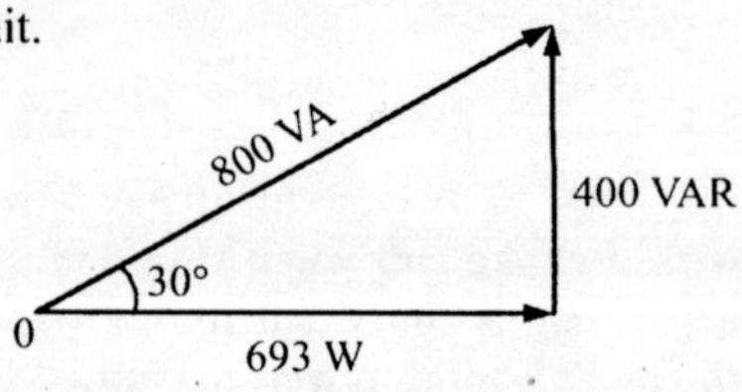

Fig. 11

(*i*) power factor = cos ϕ = cos 30°

= **0.866 lag**

(*ii*) True power = *VI* cos ϕ

$$= \frac{200}{\sqrt{2}} \times \frac{8}{\sqrt{2}} \times 0.866 = \mathbf{693\ W}$$

(*iii*) Apparent power = $VI = \frac{200}{\sqrt{2}} \times \frac{8}{\sqrt{2}}$ = **800 volt-ampere**

(*iv*) reactive power = $VI \sin\phi = \frac{200}{\sqrt{2}} \times \frac{8}{\sqrt{2}} \times \frac{1}{2}$ = **400 volt-ampre-reactive (VAR).**

Q. 12. The apparent power drawn by an ac circuit is 10 kVA and active power is 8 kW. What is the reactive power in the circuit? What is the power factor of the circuit?
(*Basic Elect. Engg. Kerala Univ. 2003*)

Ans. We are given, S = 10 kVA, P = 8 kW. As seen from Fig. 7 of *Q*. 2, the reactive power Q is given by

$$Q = \sqrt{S^2 - P^2} = \sqrt{10^2 - 8^2} = \sqrt{36} = \mathbf{6\ kVAR}$$

$$\text{Power factor} = \cos\phi = \frac{\text{active power}}{\text{apparent power}} = \frac{P}{S} = \frac{8}{10} = \mathbf{0.8}$$

Q. 13. A coil takes 1.3 kVA and 1.2 kVAR when connected to a 240-V, 50 Hz sinusoidal supply. Calculate:

(*a*) power dissipated (*b*) current (*c*) inductance of the coil.

Ans. We are given, S = 1.3 kVA, Q = 1.2 kVAR

(*a*) Now, $P = \sqrt{S^2 - Q^2}$ — Fig. 7

$= \sqrt{1.3^2 - 1.2^2}$ = **0.5 kW**

(*b*) $kVA = 1.3$ or $VA = 1.3 \times 1000 = 1300$

or $VI = 1300 \quad \therefore \quad I = \frac{1300}{240}$ = **5.42 A**

(*c*) As seen from Fig. 3, $X_L = Z \times \sin\phi$

Now, $Z = V/I = 240/5.42 = 44.3\ \Omega$

$$\sin\phi = \frac{\text{kVAR}}{\text{kVA}} = \frac{1.2}{1.3} = 0.923 \qquad \text{— Fig. 6}$$

$\therefore \quad X_L = Z\sin\phi = 44.3 \times 0.923 = 40.9\ \Omega$

$L = X_L/2\pi f = 40.9/2\pi\times 50$ = **0.13 H**

Q. 14. In a series circuit containing pure resistance and a pure inductance, the current and the voltage are expressed as :

$$\boldsymbol{i(t) = 5\sin(314t + 2\pi/3) \text{ and } v(t) = 15\sin(314t + 5\pi/6)}$$

(*a*) What is the impedance of the circuit? (*b*) What is the value of the resistance? (*c*) What is the inductance in henrys? (*d*) What is the average power drawn by the circuit? (*e*) What is the power factor? (*Elect. Technology, Indore Univ.*)

Ans. Phase angle of the current = $2\pi/3 = 2 \times 180°/3 = 120°$

Phase angle of the voltage = $5\pi/6 = 5 \times 180°/6 = 150°$

Obviously, there is a phase difference of (150° – 120°) = 30° between voltage and current with current lagging. It means, it is a series *R-L* circuit.

(*a*) $Z = V_m / I_m = 15/5$ = **3 Ω**

(*b*) $\cos\phi = \frac{R}{Z}$ or $\cos 30° = \frac{R}{3}$ or $R = 3 \times \frac{\sqrt{3}}{2}$ = **2.6 Ω**

(*c*) $\frac{X_L}{Z} = \sin\phi \qquad \therefore \quad X_L = Z\sin\phi = 3 \times \sin 30° = 1.5\ \Omega$

$X_L = \omega L \qquad \therefore \quad 1.5 = 314\,L, \quad L$ = **4.78 mH** (*d*) $I = I_m/\sqrt{2} = 5/\sqrt{2}$

$$P = I^2R = (5/\sqrt{2})^2 \times 2.6 = \mathbf{32.5\ W}$$

(*e*) power factor, p.f. = cos 30° = **0.866 lag**

Q. 15. An inductive load is connected in series with a non-inductive resistance of 8 ohms. The combination is connected across an a.c. supply of 100 volts, 50 Hz. A voltmeter connected across the non-inductive resistor and then across the inductive load gives the readings of 64 volts and 48 volts respectively. Calculate the following:

(*i*) impedance of the load (*ii*) impedance of the combination (*iii*) power absorbed by the load (*iv*) power absorbed by the resistor (*v*) total power taken from the supply (*vi*) power factor of load (*vii*) power factor of the whole circuit. (*A.M.I.E.*)

Ans. $I = 64/8 = 8$ A, $Z = 48/8 =$ **6 Ω**

(*i*) Combination $Z = 100/8 =$ **12.5 Ω**

(*ii*) From ΔBCD of Fig. 12(*b*),

$$BC^2 + CD^2 = 48^2$$

From ΔACD,

$$(64 + BC)^2 + CD^2 = 100^2$$

$\therefore$ $BC = 28.1$ V

$$CD = \sqrt{48^2 - 28.1^2}$$

$$= 38.9 \text{ V}$$

$$R_L = 28.1/8 = 3.5\ \Omega$$

$$X_L = 38.9 / 8 = 4.87\ \Omega$$

$$Z_L = \sqrt{3.5^2 + 4.87^2} = \mathbf{6\ \Omega}$$

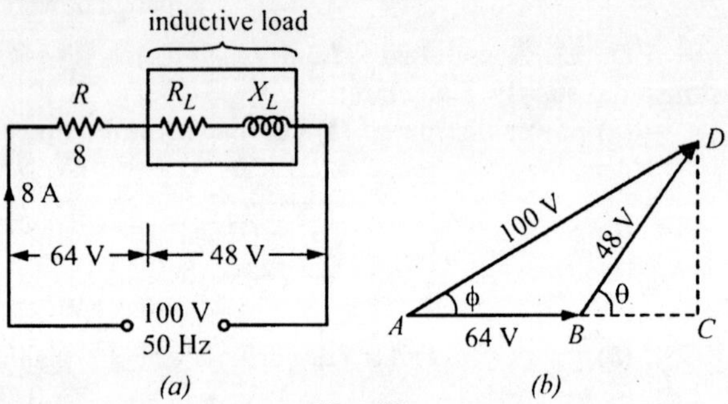

Fig. 12

(*iii*) $P_L = I^2 R_L = 8^2 \times 3.5 =$ **224 W**

(*iv*) $P = I^2 R = 8^2 \times 8 =$ **512 W**

(*v*) Total $P = 224 \times 512 =$ **736 W**

(*vi*) $\cos\theta = 28.1/48 =$ **0.585 (lag)**

(*vii*) $\cos\phi = AC/AD = 92.1/100 =$ **0.921**

Q. 16. A series circuit contains two pure elements and has the following applied voltage and current through it.

$$v = 200 \sin(500\,t + 10°) \text{ V}, \; i = 17.89 \sin(500\,t - 53.4°) \text{ A}$$

Find the two elements of the circuit.

Ans. In a two element series circuit, when current lags the voltage it means that it is an *R-L* circuit. As seen, current lags the voltage by (53.4° + 10°) = 63.4° as shown in Fig. 13.

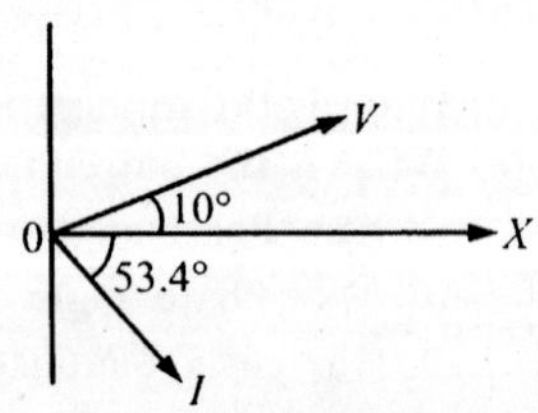

Fig. 13

Now, $\tan 63.4° = 2$

Since $\tan 63.4° = X_L/R = \omega L/R$

$\therefore$ $\omega L/R = 2$

or $\omega L = 2R$

Now, Z of the circuit $= V_m / I_m = 200/17.89 = 11.179\ \Omega$

Also, $Z = \sqrt{R^2 + (\omega L)^2}$ $\qquad \therefore\ 11.179 = \sqrt{R^2 + (2R)^2}$

or $R =$ **5 Ω**

Also, $\omega L = 2R$

or $L = 2R/\omega = 2 \times 5/500 =$ **0.02 H**

7.2 SERIES R-C CIRCUIT

Q. 17. Discuss the phase relationship between the different voltage drops in a series *R-C* circuit. Does the current lag or lead the applied ac voltage and by how much?

Ans. A *pure* resistance and *a pure* capacitor are shown connected in series in Fig. 14.

Let V = rms value of the applied voltage

I = rms value of the circuit current

$V_R = IR$ = voltage drop across R (in phase with I)

$V_C = I(-X_C) = -IX_C$ = drop over capacitor (lagging I by 90°)

It should be noted that X_C has been taken negative and hence, V_C is shown along negative direction of Y-axis because it lags behind the current by 90° as shown in Fig. 15. It means that current leads V_C by 90°. Here current has been taken as reference quantity because it is common to both R and C.

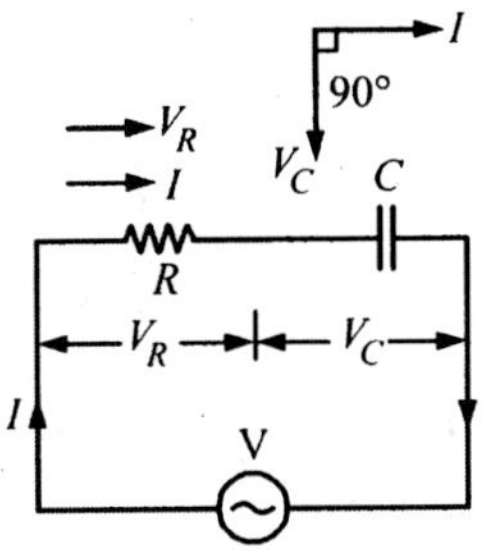

Fig. 14

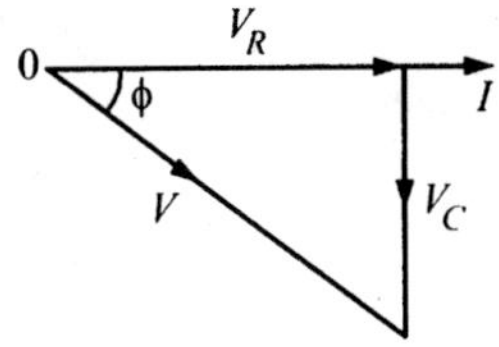

Fig. 15

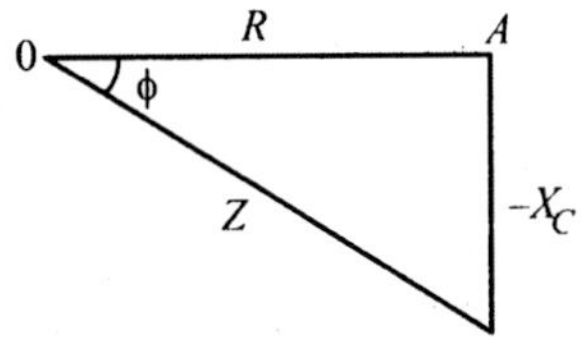

Fig. 16

As seen from the voltage triangle of Fig. 15

$$V = \sqrt{V_R^2 + V_C^2} = \sqrt{(IR)^2 + (-I\,X_C)^2}$$

$$I = \frac{V}{\sqrt{R^2 + X_C^2}} = \frac{V}{Z}$$

Obviously, the impedance of the R-C circuit is given by

$$Z = \sqrt{R^2 + X_C^2}$$

The impedance triangle is shown in Fig. 16.

It is seen from Fig. 15 that I leads V by angle ϕ such that $\tan\phi = X_C/R$.

If the equation of the applied ac voltage is

$$v = V_m \sin \omega t$$

then, the equaton of the circuit current is

$$i = I_m \sin(\omega t + \phi)$$

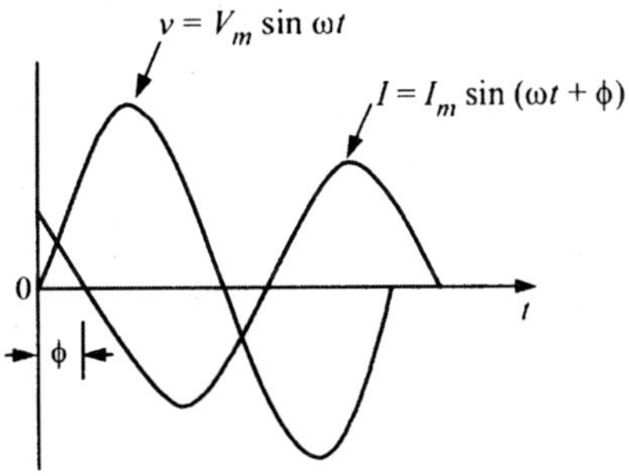

Fig. 17

As seen from Fig. 17, the current wave leads the voltage wave by ϕ.

Power consumed by the circuit is = $VI\cos\phi$. In fact, it is only the *resistor* which consumes power. A pure capacitor *does not consume any power*. Hence $P = I^2R$ watt.

Q. 18. A non-inductive resistor of 10 Ω is in series with a capacitor of 100 μF across a 250-V, 50-Hz ac supply. Determine the current taken by the capacitor and p.f. of the circuit. (*Basic Elect. Engg., VTU Karnataka, 2008*)

Ans. Let us first find the value of circuit impedance, Z.

$$Z = \sqrt{R^2 + X_C^2}$$

Now, $$X_C = \frac{1}{\omega C} = \frac{1}{2\pi f C} = \frac{1}{2\pi \times 50 \times 100 \times 10^{-6}} = \frac{10^4}{314} = 31.85\ \Omega$$

$$\therefore \quad Z = \sqrt{10^2 + 31.85^2} = 33.2\ \Omega$$

$$I = V/Z = 250/33.2 = \mathbf{7.53\ A}$$

It should be noted that it is the *rms value* of the current because 250 volts is the rms voltage.

$$\text{p.f. } \cos\phi = R/Z = 10/33.2 = \mathbf{0.3\ lead}$$

Q. 19. A two-element series circuit is connected across an ac source $v = 200\sqrt{2}$ sin (314t + 20°) volt. The current in the circuit is found to be $i = 10\sqrt{2}$ cos (314t – 25°) amperes. Determine the parameters of the circuit. (*Electromechanics, Allahabad Univ.*)

Ans. The given current is $i = 10\sqrt{2}\cos(314t - 25°)$. It can be written as

$$i = 10\sqrt{2}\sin(314t - 25° + 90°) \qquad \text{— } \because \sin(\theta + 90°) = \cos\theta$$

$$= 10\sqrt{2}\sin(314t + 65°)$$

It is seen that it leads the reference vector by 65° whereas voltage leads by 20°. It means that between them, current *leads* the voltage by 65° – 20° = 45° as shown in vector diagram of Fig. 18.

Now, $$V_{rms} = \frac{V_m}{\sqrt{2}} = \frac{200\sqrt{2}}{\sqrt{2}} = 200 \text{ volt}$$

$$I_{rms} = \frac{I_m}{\sqrt{2}} = \frac{10\sqrt{2}}{\sqrt{2}} = 10 \text{ A}$$

$$\therefore \quad Z = V/I = 200/10 = 20\ \Omega$$

Also, $$\text{p.f.} = \cos\phi = \cos 45° = \frac{1}{\sqrt{2}} \text{ (lead)}$$

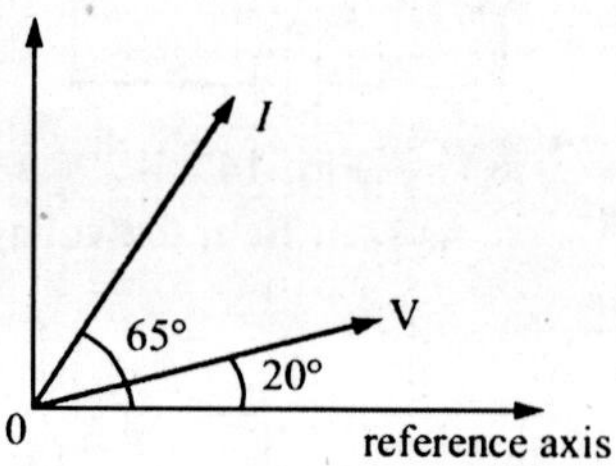

Fig. 18

$$R = Z\cos\phi = 20/\sqrt{2} = \mathbf{14.1\ \Omega}$$

Similarly, $$X_C = Z\sin\phi = 20/\sqrt{2} = 14.1\ \Omega$$

Now, $$X_C = 1/2\pi f C,\ C = 1/2\,\pi \times 50 \times 14.1 \text{ F} = \mathbf{226\ \mu F}$$

Hence, the given circuit is a series *R-C* circuit having a resistor of **14.1 Ω** and capacitor of capacitance **226 μF**.

Q. 20. A circuit dissipates 200 W at 0.7 power factor leading when connected to a 240-V, 50 Hz sinusoidal supply. Calculate:

(*i*) resistance (*ii*) impedance and (*iii*) capacitance of the circuit.

Ans. (*i*) Let us first find the value of current

$$P = VI\cos\phi$$

$$\therefore \quad I = \frac{P}{V.\cos\phi} = \frac{200}{240 \times 0.7} = 1.19 \text{ A}$$

Also, $$P = I^2R \quad \text{or} \quad R = P/I^2 = 200/1.19^2 = \mathbf{141\ \Omega}$$

(*ii*) $$\cos\phi = \frac{R}{Z} \quad \text{or} \quad Z = \frac{R}{\cos\phi} = \frac{141}{0.7} = \mathbf{201\ \Omega}$$

(*iii*) $$X_C = \sqrt{Z^2 - R^2} = \sqrt{(201)^2 - (141)^2} = 143\ \Omega$$

$$X_C = \frac{1}{2\pi f C} \text{ or } C = \frac{1}{2\pi f X_C} \text{ farad} = \frac{10^6}{2\pi \times 50 \times 143}\ \mu\text{F} = \mathbf{223\ \mu F}$$

Q. 21. A supply of 400 V, 50 Hz is applied to a series *R-C* circuit. Find the value of *C* if the power absorbed by the resistor be 500 W at 150 V. What is the energy stored in the capacitor? (*Basic Elect. Engg. R.G.P.V., Bhopal, 2005*)

Ans. Power = $V_R I$ ∴ $I = P/V_R = 500/150 = 3.3$ A

As seen from the vector diagram of Fig. 19(*b*).

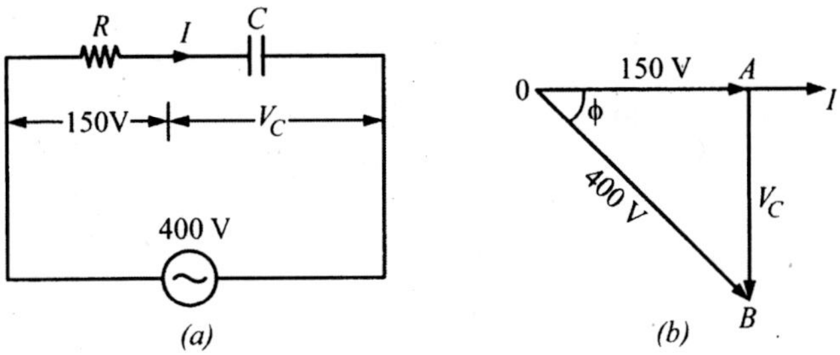

Fig. 19

$$V_C = \sqrt{400^2 - 150^2} = 371 \text{ V}$$

Now $V_C = I.X_C$ or $X_C = V_C/I = 371/3.3 = 112.4\ \Omega$

Now, $X_C = \dfrac{1}{2\pi fC}$ or $C = \dfrac{1}{2\pi f\, X_C}$ farad $= \dfrac{10^6}{314 \times 112.4} = \mathbf{28.3\ \mu F}$

Energy stored in the capacitor $= \dfrac{1}{2} CV_c^2 = \dfrac{1}{2} \times 28.3 \times 10^{-6} \times 371^2 = \mathbf{1.95\ J}$

Q. 22. The equations of voltage and current in a two-element series circuit are :

$$v(t) = 325.3 \sin(6.2\, kt + \pi/3) \text{ volt}$$
$$i(t) = 14.14 \sin(6.2\, kt + \pi/2) \text{ ampere}$$

(*i*) plot the power wave (*ii*) determine power factor and its nature (*iii*) determine the element values. (*Elect. Engg., Nagpur Univ., 2002*)

Ans. (*i*) The voltage, current and power waves are shown in Fig. 20.

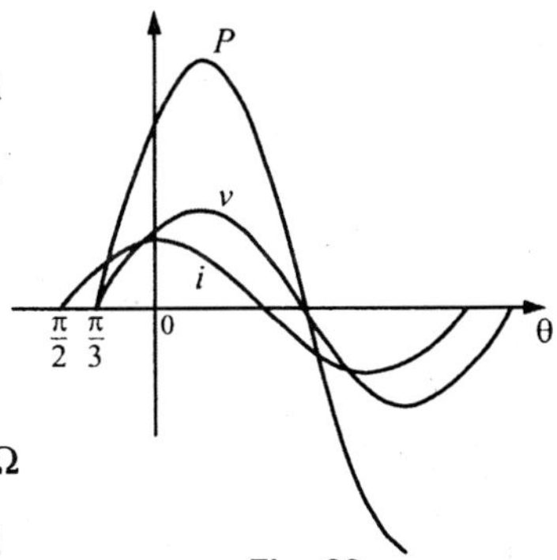

Fig. 20

(*ii*) As seen, the phase angle between the voltage and current is $\left(\dfrac{\pi}{2} - \dfrac{\pi}{3}\right) = (90° - 60°) = 30°$ with current leading.

∴ power factor = cos 30° = **0.866 (leading)**

(*iii*) $V_{rms} = \dfrac{V_m}{\sqrt{2}} = \dfrac{325.3}{\sqrt{2}}$ V

$$I_{rms} = \frac{14.14}{\sqrt{2}} \text{ A} \; ; \; Z = \frac{V_{rms}}{I_{rms}} = \frac{325.3}{14.14} = 23\ \Omega$$

$$R = Z \cos\phi = 23 \times 0.866 = \mathbf{20\ \Omega}$$

$$X_C = Z \sin\phi = 23 \times \sin 30° = 11.5\ \Omega$$

$$X_C = \frac{1}{\omega C} \text{ or } C = \frac{1}{\omega X_c} = \frac{1}{(6.2 \times 10^3) \times 11.5} \text{ farad}$$

$$= \frac{10^6}{6.2 \times 10^3 \times 11.5} = \mathbf{13.846\ \mu F}$$

Q. 23. A series *R-C* circuit dissipates 1 kW at 0.3 power factor leading when connected to a 240-V, 50-Hz sinusoidal supply. Calculate:

(*a*) the kVA (*b*) the current (*c*) the kVAR (*d*) the capacitance (*e*) the active and reactive components of current.

Ans. (*a*) kVA = kW/cos ϕ = 1/0.3 = **3.33 kVA**

(*b*) $$I = \frac{\text{kVA} \times 1000}{V} = \frac{3.33 \times 1000}{240} = \mathbf{13.9\ A}$$

(*c*) $$\text{kVAR} = \sqrt{(\text{kVA})^2 - (\text{kW})^2} = \sqrt{3.3^2 - 1^2} = \mathbf{3.2\ kVAR}$$

(*d*) $$Z = V/I = 240/13.9 = 17.3\ \Omega$$

$$\cos\phi = 0.3,\ \phi = \cos^{-1}(0.3) = 72.5°,\ \sin\phi = 0.954$$

$$X_C = Z \times \sin\phi = 17.3 \times 0.954 = 16.5\ \Omega$$

$$C = \frac{1}{2\pi f \times X_C}\ \text{farad} = \frac{10^6}{314 \times 16.5}\ \mu\text{F} = \mathbf{193\ \mu F}$$

(*e*) active component of I = I cos ϕ = 13.9 × 0.3 = **4.17 A**

reactive component of I = I sin ϕ = 13.9 × 0.954 = **13.2 A**

Q. 24. A 240-V, 50-Hz series circuit takes an rms current of 20 A. The maximum value of current occurs 1/900 second before the maximum value of voltage. Calculate:

(*a*) power factor (*b*) average power (*c*) *R* and *C* of the circuit.

Ans. (*a*) Let us convert the time interval of 1/900 second into degrees.

Now, time period = $\frac{1}{50}$ = 0.02 second. It means 0.02 second covers 360°.

$$\therefore \quad \frac{1}{900}\ \text{second} = \frac{360}{900 \times 0.02} = 20°$$

$$\therefore \quad \cos\phi = \cos 20° = \mathbf{0.94\ (leading)}$$

as shown in Fig. 21.

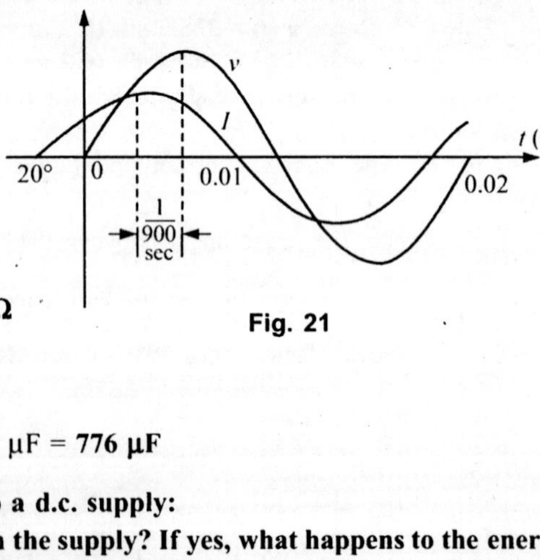

Fig. 21

(*b*) $P = VI \cos\phi$

= 240 × 20 × 0.94

= **4510 W**

(*c*) $Z = V/I$ = 240/20 = 12 Ω

$R = Z \cos\phi$ = 12 × 0.94 = **11.3 Ω**

$X_C = Z \sin\phi$ = 12 × sin 20°

= 12 × 0.342 = 4.1 Ω

$$C = \frac{1}{2\pi f X_C}\ \text{farad} = \frac{10^6}{314 \times 4.1}\ \mu\text{F} = \mathbf{776\ \mu F}$$

Q. 25. An *R-L* circuit is connected to a d.c. supply:

(*i*) Does the resistor take energy from the supply? If yes, what happens to the energy?

(*ii*) Does the inductance take energy from supply. If yes, what happens to the energy?

(*iii*) When the circuit is opened, what happens to the energy taken by resistor and inductor?

(*Basic Elect. Engg., Kerala Univ, 2004*)

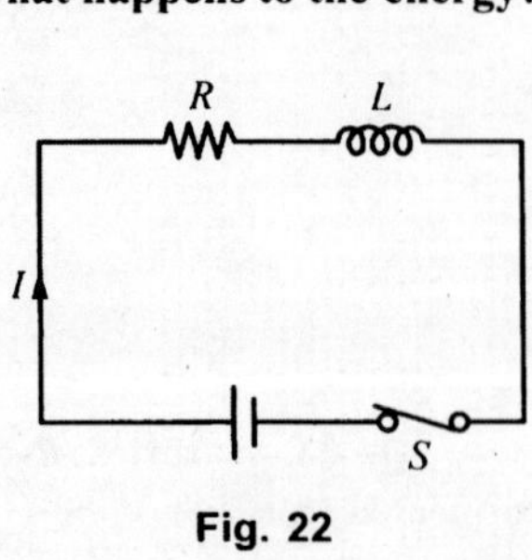

Fig. 22

Ans. (*i*) Yes, the resistor takes energy from the supply. This energy is converted into heat which raises its temperature.

(*ii*) Yes, an inductance also takes energy from the supply and stores it in the form of a magnetic field.

(*iii*) When dc supply is cut-off to the resistor, it no longer produces heat. The hot resistor cools down by radiating heat to its surroundings. Its resistance decreases to its previous value at room temperature.

When the circuit is opened (Fig. 22), the magnetic field of the inductor collapses rapidly. This rapid collapse of the magnetic field momentarily produces a very high voltage which causes an arc at the open switch.

Q. 26. A series circuit containing two pure elements has the following applied voltage and current

$$v = 200 \sin (2000t + 50°)\ V,\ i = 2 \cos (2000t + 13.2°)\ A$$

Find the two elements of the circuit.

Ans. Since sin (θ + 90°) = cos θ, hence the given current equation can be written as

$$i = 2 \sin (2000t + 90° + 13.2°) = 2 \sin (2000t + 103.2°).$$

Hence, current leads the voltage by

$$(103.2° - 50°) = 53.2° \quad \text{— Fig. 23}$$

It is obvious that the two elements of the circuit are R and C.

$$\tan 53.2° = 1.33 = X_C/R = 1/\omega CR$$

$$\therefore \quad 1/\omega C = 1.33\ R$$

$$Z = V_m/I_m = 100/2 = 50\ \Omega$$

Also,

$$Z = \sqrt{R^2 + (1/\omega C)^2}$$

$$50 = \sqrt{R^2 + (1.33R)^2}$$

$$\therefore \quad R = \mathbf{30\ \Omega}$$

$$C = 1/(1.33\omega R) = 1/(1.33 \times 2000 \times 30)\ F$$

$$= 12.5 \times 10^{-6}\ F = \mathbf{12.5\ \mu F}$$

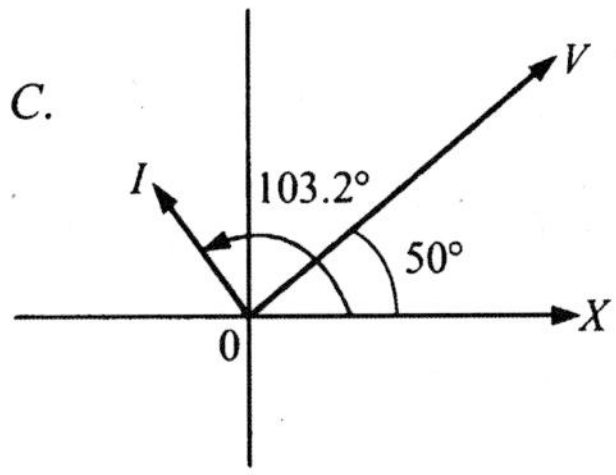

Fig. 23

7.3 SERIES R-L-C CIRCUIT

Q. 27. How are the voltage drops across R, X_L and X_C related to each other in a series R-L-C circuit when it is connected to a sinusoidal voltage source. How does this circuit behave when $X_L > X_C$ and when $X_C > X_L$?

Ans. Let the equation of the applied sinusoidal source be $v = V_m \sin \omega t$. The circuit is shown in Fig. 24. Since I is common to all the elements, its vector has been taken as reference vector.

Let $V_R = IR$ = voltage drop across R (in phase with I)

$V_L = I.X_L$ = voltage drop across L (ahead of I by 90°)

$V_C = I.X_C$ = voltage drop across C (behind I by 90°)

In voltage triangle of Fig. 25, OA represents V_R, AB represents inductive drop V_L and AC represents capacitive drop V_C. As seen, V_L and V_C are 180° out of phase with each other *i.e.* these are in direct opposition to each other. Of course, we are assuming that V_L is greater than V_C in magnitude.

Subtracting BD (= AC) from AB, we get the net *reactive* drop.

$$AD = (V_L - V_C) = I\ (X_L - X_C)$$

The rms applied voltage V as represented by OD is the vector sum of OA and AD.

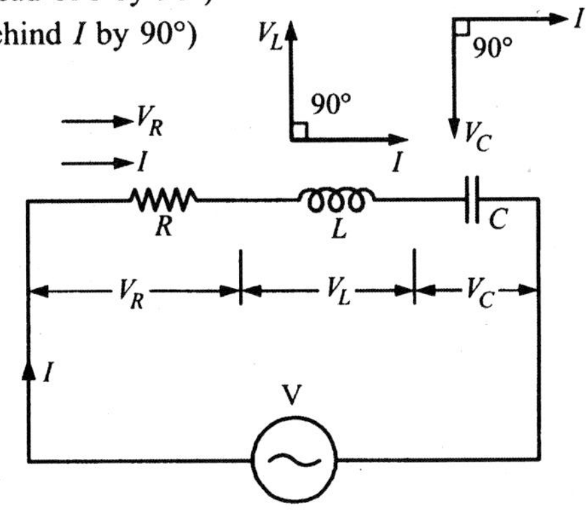

Fig. 24

$$OD = \sqrt{(OA)^2 + (AD)^2}$$

or $$V = \sqrt{(IR)^2 + (IX_L - IX_C)^2} = I\sqrt{R^2 + (X_L - X_C)^2}$$

or $$I = \frac{V}{\sqrt{R^2 + (X_L - X_C)^2}} = \frac{V}{\sqrt{R^2 + X^2}} = \frac{V}{Z}$$

The impedance, Z of the circuit is given by

$$Z = \sqrt{R^2 + (X_L - X_C)^2} = \sqrt{R^2 + X^2}$$

where X is the *net reactance.*

In the present case, net reactance is inductive because we have assumed $X_L > X_C$. If $X_C > X_L$, then net reactance is *capacitive.*

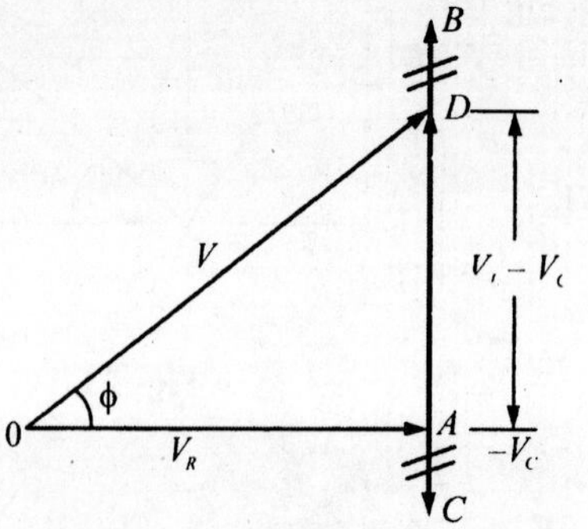

Fig. 25

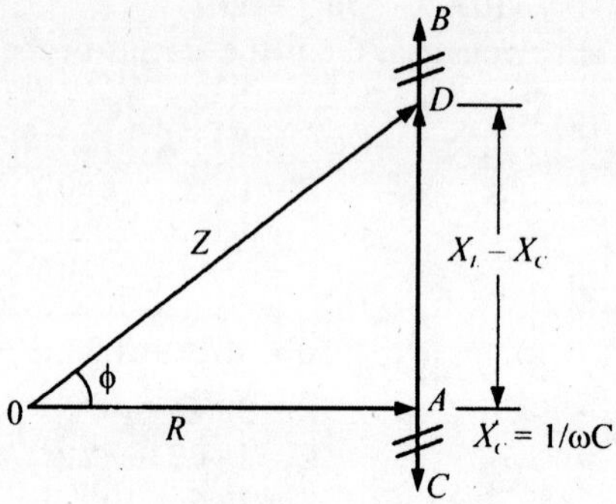

Fig. 26

Power factor :

As seen from impedance triangle of Fig. 26.

Power factor, $\cos \phi = \frac{R}{Z} = \frac{R}{\sqrt{R^2 + X^2}}$

Hence, it is seen that if the equation of the applied voltages is

$$v = V_m \sin \omega t$$

then equation of the resulting current in an R-L-C circuit is given by

$$i = I_m \sin (\omega t - \phi) \quad \text{– when } X_L > X_C$$

$$i = I_m \sin (\omega t + \phi) \quad \text{– when } X_C > X_L$$

As stated above, when $X_L > X_C$ so that $V_L > V_C$, then the circuit behaves like inductive circuit and has a lagging power factor.

If, on the other hand, $X_C > X_L$ so that $V_C > V_L$, the circuit behaves like a capacitive circuit with a leading power factor. However, it makes no difference to the amount of *power consumed by the circuit.*

Q. 28. A resistance of 12 Ω, an inductance of 0.15 H and a capacitance of 130 μF are connected in series across a 100 volt, 50 Hz supply. Calculate the impedance, current, phase angle and power factor. (*Basic Elect. Engg., R.G.P.V. Bhopal, 2008*)

Ans. $X_L = \omega L = 2\pi \times 50 \times 0.15 = 314 \times 0.15 = 47.1\ \Omega$ — *AB* in Fig. 27

$$X_C = \frac{1}{\omega C} = \frac{10^6}{2\pi \times 50 \times 130} = 24.5\ \Omega$$ — *AD* in Fig. 27

$$X = X_L - X_C = 47.1 - 24.5 = 22.6\ \Omega$$ — *AC* in Fig. 27

As seen, the net reactance is inductive as shown in Fig. 27.

(*i*) $Z = \sqrt{R^2 + X^2}$

$= \sqrt{12^2 + 22.6^2}$

$= \mathbf{25.6\ \Omega}$

(*ii*) current, $I = V/Z = 100/25.6$

$= \mathbf{3.9\ A}$

(*iii*) $\tan\phi = R/Z = 12/25.6 = 0.469$

$\phi = \tan^{-1}(0.469)$

$= \mathbf{25.1°}$

(*iv*) power factor $= \cos\phi = \cos 25.1°$

$=$ **0.9 (lagging)**

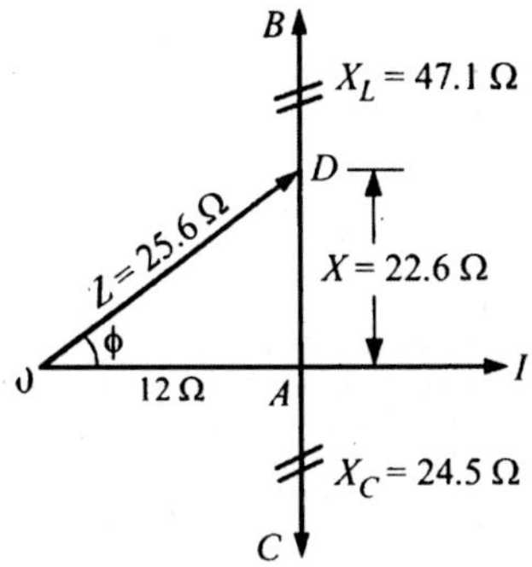

Fig. 27

Q. 29. A resistance of 20 Ω, an inductance of 0.2 H and a capacitance of 100 μF are connected in series across a 220 V, 50 Hz supply. Determine the following (*a*) impedance (*b*) current (*c*) voltages across *R*, *L* and *C* and (*d*) power factor.

(*Basic Elect. Engg., R.G.P.V. Bhopal 2007*)

Ans. $X_L = 2\pi \times 50 \times 0.2 = 62.8\ \Omega$

$$X_C = \frac{10^6}{314 \times 100} = 31.8\ \Omega$$

$X = X_L - X_C = 62.8 - 31.8 = 31\ \Omega$ (inductive)

(*a*) $Z = \sqrt{R^2 + X^2} = \sqrt{20^2 + 31^2} = \mathbf{36.9\ \Omega}$

(*b*) $I = V/Z = 220/36.9 = \mathbf{5.96\ A}$

$V_R = IR = 5.96 \times 20 = \mathbf{119.2\ V}$

(*c*) $V_L = I \,.\, X_L = 5.96 \times 62.8 = \mathbf{374.3\ V}$

$V_C = I \,.\, X_C = 5.96 \times 31.8 = \mathbf{189.5\ V}$

(*d*) Power factor $= \cos\phi = R/Z = 20/36.9 =$ **0.54 (lag)**

Q. 30. A coil having a resistance of 10 Ω and an inductance of 0.2 H is connected in series with a 100 × 10⁻⁶ F capacitor across a 230 V, 50 Hz supply. Calculate:

(*i*) the active and reactive components of the current

(*ii*) the voltage across the coil

Draw the phasor diagram. (*Basic. Elect. Engg., B.Tech. JNTU, Hyderabad 2007*)

Sol. $X_L = 2\pi \times 50 \times 0.2 = 62.8\ \Omega$

$X_C = 1/\omega C = 1/314 \times 100 \times 10^{-6} = 31.8\ \Omega$

$Z = \sqrt{R^2 + (X_L - X_C)^2} = \sqrt{10^2 + 31^2} = 32.6\ \Omega$

$\therefore$ $I = 230/32.6 = 7\ A$

As seen from Fig. 28(*b*), $\cos\phi = R/Z = 10/32.6 = 0.31$, $\sin\phi = 31/32.6 = 0.95$, $\phi = 72°$

(*i*) acitve component of current $= I\cos\phi = 7 \times 0.31 =$ **2.17 A**

reactive component of current $= I\sin\phi = 7 \times 0.95 =$ **6.65 A**

(*ii*) $Z_{coil} = \sqrt{10^2 + 62.8^2} = 63.6\ \Omega$

Coil p.f. angle θ is given by

$\tan\theta = X_L/R = 62.8/10 = 6.28$ $\therefore$ $\theta = \tan^{-1}(6.28) = 81°$

$V_{coil} = I\, Z_{coil} = 7 \times 63.6 = \mathbf{445.2\ V}$

$V_C = I\, X_C = 7 \times 31.8 = 222.6\ V$

The phasor diagram is shown in Fig. 28(*c*). Vector *OA* represents V_C which lags behind *I* by 90° (remember current *leads* the voltage in a capacitor). Vector *OB* represent V_{coil} which is 81° ahead of *I* (remember current lags behind in inductive coil). The applied voltage *V* is the vector sum of V_{coil} and V_C. It is represented by vector *OC*.

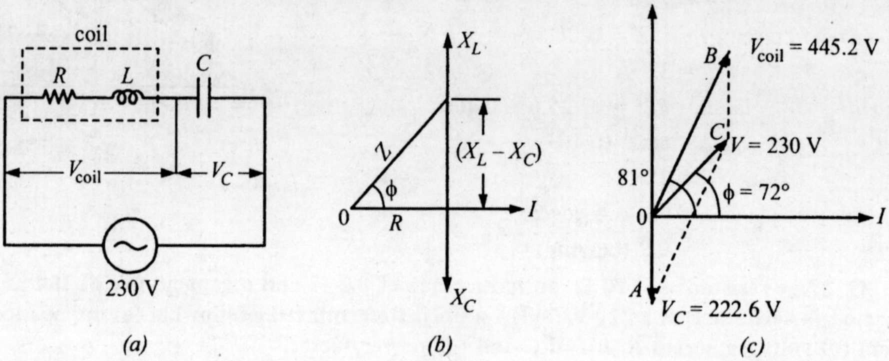

Fig. 28

If may also be noted that φ is the *circuit* power factor angle. But θ is the power factor angle of the *coil alone*.

Q. 31. A circuit takes a current of 3 A at a power factor of 0.6 lagging when connected to a 115 V, 50 Hz supply. Another one circuit takes a current of 5A at a power factor of 0.707 leading when connected to the same supply after the first circuit is removed. If the two circuits are connected in series across a 230 V, 50 Hz supply, calculate : (*a*) the current drawn from the source (*b*) the power consumed (*c*) the power factor of the circuit.

(*Basic Elect. Engg., West Bengal Tech. Univ. 2009*)

Ans. First circuit:

$$Z_1 = 115/3 = 38.3\ \Omega$$

$$\cos\phi_1 = 0.6,\ \sin\phi_1 = 0.8$$

$$R_1 = Z_1 \cos\phi_1 = 38.3 \times 0.6 = 23\ \Omega$$

$$X_L = Z_1 \sin\phi_1 = 38.3 \times 0.8 = 30.6\ \Omega$$

Second circuit :

$$Z_2 = 115/5 = 23\ \Omega$$

$$R_2 = Z_2 \cos\phi_2 = 23 \times 0.707 = 16.3\ \Omega$$

$$X_C = Z_2 \sin\phi_2 = 23 \times 0.707 = 16.3\ \Omega$$

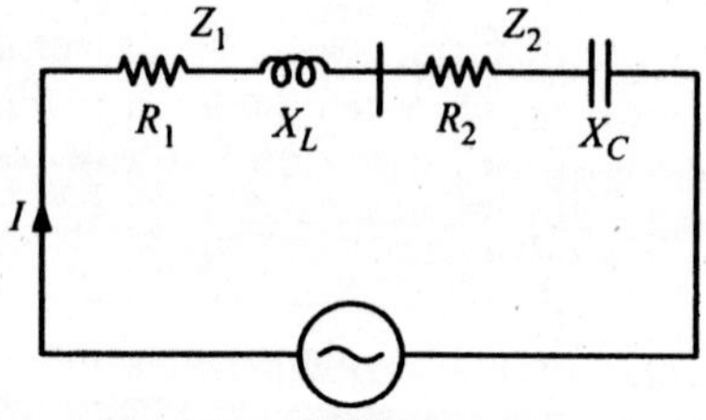

Fig. 29

Consider the circuit shown in Fig. 29 where the two circuits have been connected in series

$$R = R_1 + R_2 = 23 + 16.3 = 39.3\ \Omega$$

$$X = X_L - X_C = 30.6 - 16.3 = 14.3\ \Omega$$

$$Z = \sqrt{39.3^2 + 14.3^3} = 41.8\ \Omega \text{ (inductive)}$$

(*a*) $I = V/Z = 230/41.8 =$ **5.5 A**

(*b*) $P = I^2 R = 5.5^2 \times 39.3 =$ **1189 W**

(*c*) If φ is the phase angle of the *entire* circuit, then

power factor $= \cos\phi = R/Z = 39.3/41.8 =$ **0.07 (lag)**

Q. 32. A voltage of (100 + *j* 60) volts drives a current of (4 – *j*5) A through a series *R-L-C* circuit. Determine :

(*a*) the complex expression for impedance

(*b*) power consumed

(c) power factor

(d) draw the phasor diagram *(Basic Elect. Engg. B. Tech. JNTU, Hyderabad 2007)*

Sol. (a)
$$\mathbf{Z} = \frac{\mathbf{V}}{\mathbf{I}} = \frac{100 + j\,60}{4 - j5}$$

Rationlizing the above expression, we get

$$\mathbf{Z} = \frac{(100 + j60)\,(4 + j5)}{(4 - j5)\,(4 + j5)}$$

$$= \frac{400 + j\,500 + j240 - 300}{41} = \frac{100 + j\,740}{41} = \mathbf{2.44 + j18}$$

Since j-term is positive, hence impedance of 18 Ω is *inductive* [Fig. 30(*b*)].

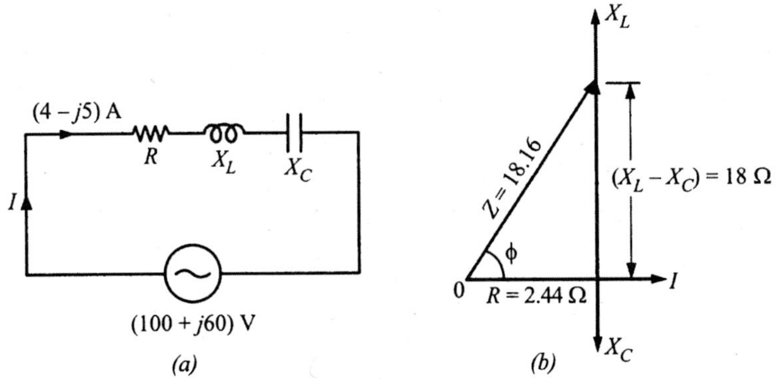

Fig. 30

(b) We will use the current conjugate to find the power consumed

$$P_{VA} = VI^* = (100 + j60)\,(4 + j5) = 100 + j740$$

Hence, real power is = **100 W**

Alternatively, $I = \sqrt{4^2 + 5^2} = 6.4\text{ A},\ R = 2.44\ \Omega$

Power consumed $= I^2 R = 6.4^2 \times 2.44 = 100$ W

(c) $\tan\phi = 18/2.44 = 7.377,\ \phi = 82.5°$

∴ power factor = cos 82.5° = **0.1276 (lag)**

(d) Now, $Z = 2.44 + j18 = 18.16\angle 82.5°$

It is obvious that it is an inductive circuit as is clear from Fig. 30(*b*).

Q. 33. A choke coil having a resistance of 10 Ω and inductance of 0.05 H is connected in series with a capacitor of 100 μF. The whole circuit is connected to 200-V, 50-Hz supply. Calculate :

(*i*) impedance (*ii*) current (*iii*) power factor (*iv*) power input (*v*) apparent and reactive power. *(Basic Elect. Engg., R.G.P.V, Bhopal, 2006)*

Ans. (*i*) Impedance :

$$X_L = \omega L = 2\pi f L = 314 \times 0.05 = 15.7\ \Omega$$

$$X_C = \frac{1}{2\pi f C} = \frac{1}{2\pi \times 50 \times 100 \times 10^{-6}} = 31.8\ \Omega$$

$$X = X_C - X_L = 31.8 - 15.7 = 16.1\ \Omega\ \text{(capacitive)}$$

$$Z = \sqrt{R^2 + X^2} = \sqrt{10^2 + 16.1^2} = \mathbf{19\ \Omega}$$

(ii) Current : $I = V/Z = 200/19 =$ **10.53 A (leading V)**

(iii) Power factor : $\cos\phi = R/Z = 10/19 =$ **0.53 (leading)**

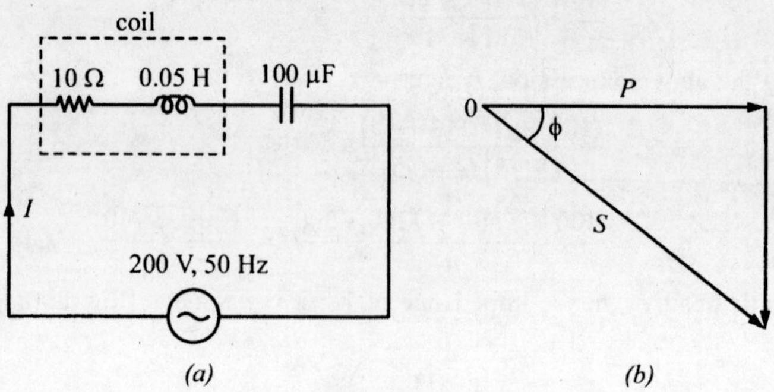

Fig. 31

(iv) Power input : $P = VI\cos\phi = 200 \times 10.53 \times 0.53 =$ **1116 Ω**

(v) Active power : First let us find $\sin\phi$.

$$\sin\phi = X/Z = 16.1/19 = 0.85$$

As seen from the power trinagle of Fig. 31(*b*).

$$\text{Apparent power, } S = VI = 200 \times 10.53 = 2106 \text{ VA} = \mathbf{2.106 \text{ kVA}}$$

$$\text{Reactive power } Q = S\sin\phi = VI\sin\phi = 200 \times 10.53 \times 0.85 = \mathbf{1791 \text{ VAR}}$$

$$= \mathbf{1.791 \text{ kVAR}}$$

Q. 34. A choke coil has a resistance of 2 Ω and an inductance of 5 H. A capacitor *C* is connected in series with the choke coil and the combination is fed from 230-V, 50 Hz source. What should be the value of *C* so that the voltage across choke coil is 250 V?

(*Basic Elect Engg., R.G.P.V, Bhopal 2006*)

Ans. It is obvious that circuit current is the same as produced by 250 V when applied across the coil

$$X_L = 2\pi \times 50 \times 5 = 1572\ \Omega$$

$$Z_{coil} = \sqrt{2^2 + 1572^2} \cong 1572\ \Omega$$

$$I = V/Z_{coil} = 250/1572 = 0.159 \text{ A} = 159 \text{ mA}$$

Let the required capacitor have the capacitance of C farad. Its capacitive reactance is X_C.

Assuming that $X_L > X_C$, the impedance of the *R-L-C* circuit is

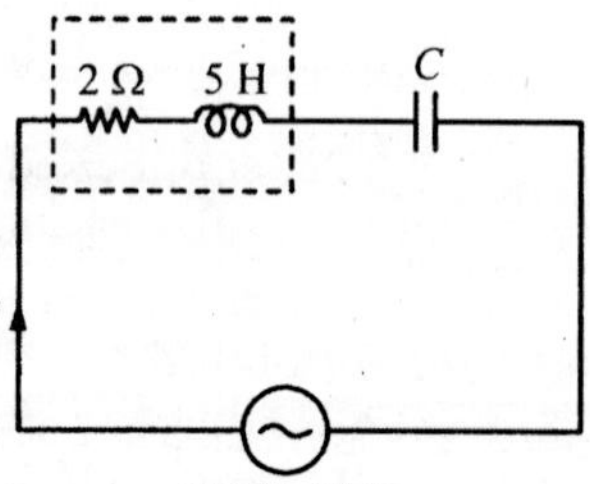

Fig. 32

$$Z = \sqrt{R^2 + (X_L \sim X_C)^2} = \sqrt{R^2 + X^2}$$

Since the value of circuit current is 159×10^{-3} A

$$\therefore \quad 159 \times 10^{-3} = \frac{230}{Z} \quad \text{or} \quad Z = \frac{230}{159 \times 10^{-3}} = 1446\ \Omega$$

$$\therefore \quad 1446 = \sqrt{R^2 + X^2} \quad \text{or} \quad 1446^2 = 4 + X^2$$

$$X^2 = 20{,}905\ \Omega, \qquad X = \sqrt{20{,}905} = 144.6\ \Omega$$

$$\therefore \quad 144.6 = X_L - X_C = 1572 - X_C \quad \text{or} \quad X_C = 1426\ \Omega$$

$$C = \frac{10^6}{314 \times 1426}\ \mu\text{F} = \mathbf{2.23\ \mu F}$$

Q. 35. A coil of p.f. 0.8 is connected in series with a 110 μF capacitor. The supply frequency is 50 Hz. The potential difference across the coil is found to be equal to the potential difference across the capacitor. Calculate the resistance and inductance of the coil. *(Elect. Engg, Nagapur Univ. 2000)*

Ans. $X_C = \frac{1}{2 \times \pi \times 50 \times 110 \times 10^{-6}} = \frac{10^6}{314 \times 110} = 29\ \Omega$

Now $V_L = V_C$ or $IZ_L = IZ_C$ or $Z_L = Z_C = 29\ \Omega$

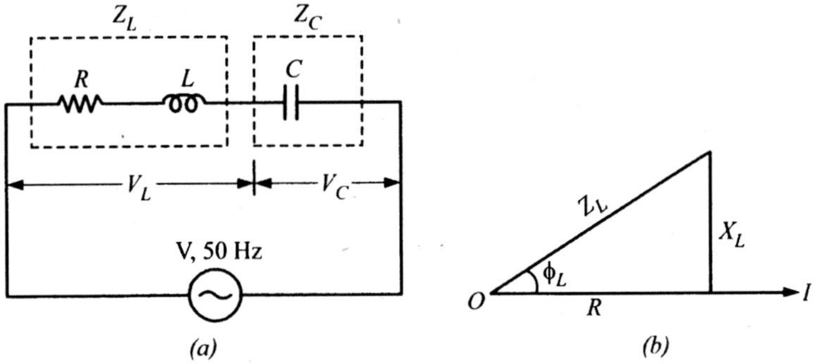

Fig. 33

The value of R can be found from the impedance triangle shown in Fig. 33(*b*).

$$\text{Coil p.f.} = \frac{R}{Z_L} \quad \text{or} \quad 0.8 = \frac{R}{29}, \qquad \therefore \quad R = \mathbf{23.1\ \Omega}$$

Obviously, $\sin \phi_L = \sqrt{1^2 - 0.8^2} = 0.6$

$$\sin \phi_L = \frac{X_L}{Z_L} \quad \text{or} \quad 0.6 = \frac{X_L}{29}, \quad X_L = 17.4\ \Omega$$

Now, $X_L = 2\pi f L$ or $17.4 = 2\pi \times 50 \times L$

$\therefore$ $L = 17.4/314 = 0.055\ \text{H} = \mathbf{55\ mH}$

Q. 36. An air-cored coil takes 3 A and dissipates 500 W when connected across a 240 V, 50 Hz sinusoidal supply. Calculate:

(*a*) the inductance of the coil.

(*b*) the two values of capacitance, either of which in series with the coil, will take a current of 4 A from the above supply.

(*c*) the power factor for each of the three conditions.

Ans. (*a*) $P = I^2 R \qquad \therefore \quad R = P/I^2 = 500/3^2 = 55.6\ \Omega$

$Z = V/I = 240/3 = 80\ \Omega$

$$X_L = \sqrt{Z^2 - R^2} = \sqrt{80^2 - 55.6^2} = 57.6\ \Omega$$

$$L = \frac{X_L}{2\pi f} = \frac{57.6}{314} = \mathbf{0.183\ H}$$

(*b*) $Z = V/I = 240/4 = 60\ \Omega$

Net $X = \sqrt{Z^2 - R^2} = \sqrt{60^2 - 55.6^2} = 22.7\ \Omega$

Now, two cases arise *i.e.* when X_C is less than X_L by 22.7 Ω and when X_C is greater than X_L by 22.7 Ω.

(*i*) When $X_C < X_L$, we have $X_C = 57.6 - 22.7 = 34.9\ \Omega$

Now, $X_C = \dfrac{1}{\omega C}$ or $C = \dfrac{1}{\omega . X_C} = \dfrac{10^6}{314 \times 34.9} = \mathbf{91.2\ \mu F}$

(*ii*) When $X_C > X_L$, we have $X_C = 57.6 + 222.7 = 80.3\ \Omega$

$$C = \frac{1}{\omega . X_C} = \frac{10^6}{314 \times 80.3} = \mathbf{39.6\ \mu F}$$

(*c*) (*i*) with no capcitor : $\cos\phi = R/Z = 55.6/80 =$ **0.695 (lagging)**

(*ii*) with $C = 91.2\ \mu F$: $\cos\phi = R/Z = 55.6/60 =$ **0.927 (lagging)**

(*iii*) with $C = 39.6\ \mu F$: $\cos\phi = R/Z = 55.6/60 =$ **0.927 (leading)**

Q. 37. A coil and a pure capacitor are connected in series across a 250-V, 50-Hz sinusoidal supply. The power dissipated is 3 kW at an overall power factor of 0.6 lagging. The voltage across the coil is 500 volts. Calculate:

(*a*) the total kVA (*b*) the current

(*c*) the kVAR of the capacitor (*d*) the capacitance

Ans. (*a*) Total kVA $= \dfrac{\text{kW}}{\cos\phi} = \dfrac{3}{0.6} =$ **5 kVA.**

(*b*) $I = \dfrac{\text{kVA} \times 1000}{V} = \dfrac{5 \times 1000}{250} =$ **20 A**

(*c*) Total kVAR = kVA × sin ϕ = 5 × 0.8 = 4 kVAR

kVA of coil $= \dfrac{V_{coil} \times I}{1000} = \dfrac{500 \times 20}{1000}$

= 10 kVA

kVAR of coil $= \sqrt{(\text{kVA of coil})^2 - \text{kW}^2}$

$= \sqrt{10^2 - 3^2} =$ **9.54 kVAR**

kVAR of capacitor = kVAR of coil – total kVAR

= 9.54 – 4 = **5.54 kVAR**

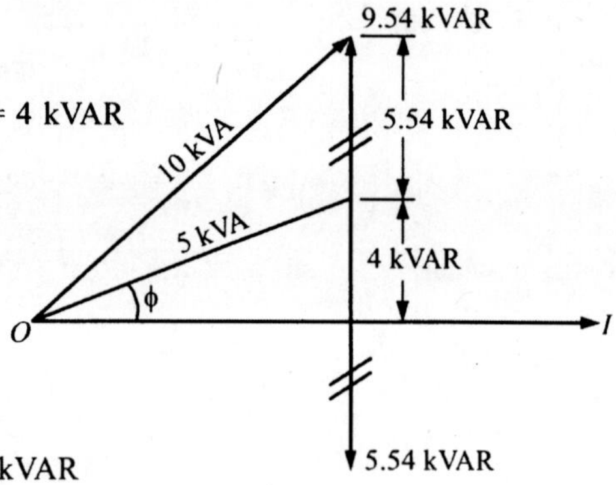

Fig. 34

The above facts have been shown in Fig. 34

(*d*) Voltage across capacitor $= \dfrac{(\text{kVAR of capacitor}) \times 1000}{I} = \dfrac{5540}{20} = 277$ volts

$\therefore$ $277 = I.X_C$ $\therefore X_C = 277/20 = 13.85\ \Omega$

Now, $C = \dfrac{1}{\omega . X_C} = \dfrac{10^6}{314 \times 13.85}\ \mu F =$ **230 μF**

Q. 38. A resistance of 20 Ω and inductance of 0.2H and capacitance of 100 μF are connected in series across 220 V, 50 Hz mains. Determine

(*i*) impedance (*ii*) current (*iii*) voltage across *R*, *L* and *C* (*iv*) power in watts (*v*) p.f and angle of lag. (*Basic Elect. Engg., R.G.P.V Bhopal, 2007*)

Ans. $X_L = \omega L = 2\pi f L = 2\pi \times 50 \times 0.2 = 63\ \Omega$

$$X_C = \frac{1}{\omega C} = \frac{1}{2\pi \times f \times C} = \frac{1}{2\pi \times 50 \times 100 \times 10^{-6}}$$

$$= \frac{10^4}{314} = 32\ \Omega$$

Net reactance, $X = X_L - X_C = 63 - 32 = 31\ \Omega$ (inductive) (*i*) $Z = \sqrt{R^2 + X^2} = \sqrt{20^2 + 31^2} = \mathbf{37\ \Omega}$

(*ii*) $I = V/Z = 220/37 = \mathbf{6\ A}$

(*iii*) $V_R = I.R = 6 \times 20 = \mathbf{120\ V}$ (in phase with I)

$V_L = I.X_L = 6 \times 63 = \mathbf{378\ V}$ (ahead of I by 90°)

$V_C = I.X_C = 6 \times 32 = \mathbf{192\ V}$ (behind I by 90°)

(*iv*) Power $= I^2R = 6^2 \times 20 = \mathbf{720\ W}$

(*v*) p.f. $= \cos\phi = R/Z = 20/37 = \mathbf{0.54\ (lag)}$

$\phi = \cos^{-1} 0.54 = \mathbf{57.32°}$

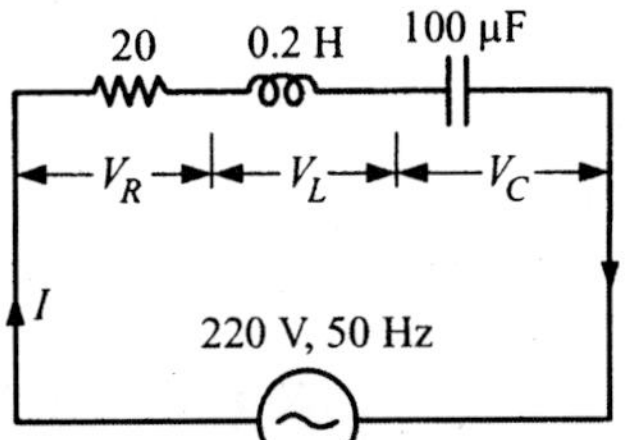

Fig. 35

Q. 39. For the series circuit shown in Fig. 36(*a*) and with current and voltages as indicated, calculate the values of *R*, *r* and *L* and frequency of the applied voltage and its magnitude. (*FY, Elect. Engg. Nagpur Univ., 2004*)

Ans. In vector diagram of Fig. 36(*b*), *OA* represents $V_R = 25$ V. The vector *OC* represents the drop over *R* and coil. The vector *AC* represents the drop over *coil only*. The point *C* is located by drawing two arcs from points *O* and *A* respectively. The angle between *I* and vector *AC* is found by measurement and equals 80°.

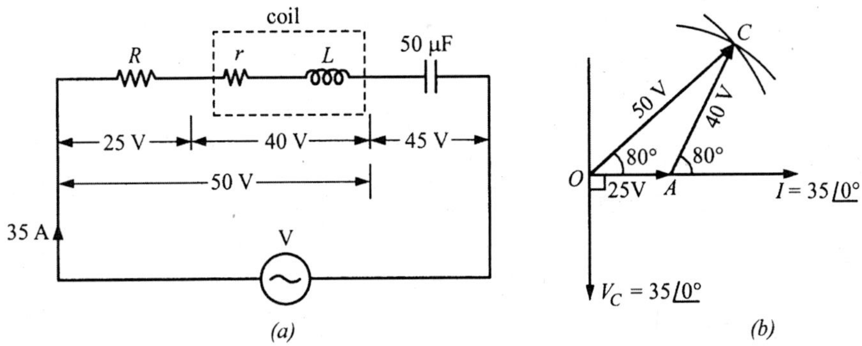

Fig. 36

$$V_C = I.\ X_C \quad \text{or} \quad X_C = V_C/I = 45/35 = 1.28\ \Omega$$

To find frequency, we note that

$$X_C = 1/2\ \pi f C \quad \text{or} \quad f = 1/2\ \pi\ CX_C$$

$\therefore$ $$f = 10^6/2\pi \times 50 \times 1.28 = 2490 \text{ Hz} = \mathbf{2.49\ kHz}$$

Now, $$R = V_R/I = 25/35 = \mathbf{0.714\ \Omega}$$

We have found that $\mathbf{V}_{coil} = 40\angle 80°$

$\therefore$ $$\mathbf{Z}_{\text{coil}} = 40\angle 80°/35\angle 0° = 1.14\angle 80°$$

or $$\mathbf{Z}_{\text{coil}} = 0.198 + j\ 1.12$$

$\therefore$ $$r = \mathbf{0.198\ \Omega} \text{ and } X_L = 1.12\ \Omega$$

Now, $$X_L = 2\ \pi f L \quad \text{or} \quad L = \frac{X_L}{2\pi f} = \frac{1.12}{2\pi \times 2490}$$

$$= \mathbf{7.16 \times 10^{-5}\ H}$$

The applied voltage is the vector sum of the three voltage drops V_R, V_{coil} and V_C

$\therefore$ $$\mathbf{V} = 25\angle 0° + 40\angle 80° + 45\angle -90°$$

$$= (25 + j0) + (6.8 + j\ 39.2) + (0 - j45)$$

$$= (31.8 - j5.8) = \mathbf{32.3\angle -10.7°}$$

7.4 SERIES RESONACE

Q. 40. When is a series R-L-C circuit said to be in resonance? Derive the value of resonant frequency of the circuit.

Ans. A series circuit is said to be in resonance when its *net reactance is zero.* [Fig. 37(*b*)] The frequency f_o at which this happens is known as the *resonant* frequency.

The net reactance of a series *R-L-C* circuit shown in Fig. 37(*a*) is

$$X = X_L - X_C \quad \text{and} \quad Z = \sqrt{R^2 + X^2}$$

At resonance, $X = 0$ or $X_L - X_C = 0$

$\therefore$ $X_L = X_C$ — Fig. 37(*b*)

or $\omega L = 1/\omega C$ or $\omega^2 = 1/LC$

$$\therefore \quad (2\pi f_0)^2 = \frac{1}{LC} \quad \text{or} \quad f_0 = \frac{1}{2\pi\sqrt{LC}}$$

It will be noted that f_o depends on the product LC but is *independent of circuit resistance.* Incidentally, the resonant angular frequency is given by

$$\omega_0 = \frac{1}{\sqrt{LC}}$$

Since under resonant conditions, X is zero, hence $Z = R$. It is the *minimum* possible value of impedance Z. Hence, circuit current is *maximum* and is given by

$$I_{max} = V/Z = V/R$$

The circuit behaves like a pure resistive circuit as shown in Fig. 37(*c*) and current is in phase with the applied voltage V as shown in Fig. 37(*b*). Obviously, the power factor of the circuit is *unity at resonance*

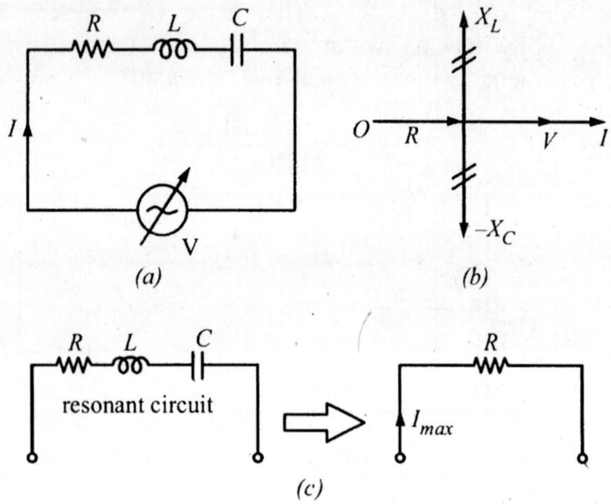

Fig. 37

As current is maximum, the voltage drops V_L and V_C are quite large but equal in magnitude. Since they are antiphase, they cancel each other out. For this reason, series resonance is sometimes called 'voltage' resonance.

Q. 41. Explain with the help of a suitable diagram, how different elements of a series *R-L-C* circuit vary when a sinusoidal voltage of varying frequency is applied across the circuit?

Ans. Suppose an *R-L-C* circuit is connected across a sinusoidal voltage of varying frequency. The changes which occur in different elements of the circuit when frequency is *changed* are shown in Fig. 38.

(*i*) **Resistance, *R*.** It is independent of *f.* Hence, it is shown by a straight horizontal line in Fig. 38.

(*ii*) **Inductive reactance, X_L**

$$X_L = 2\pi f L$$

It is directly proportional to *f.* Since it increases linearly with *f,* its graph is a straight line starting from zero.

(*iii*) **Capacitive reactance X_C**

$$X_C = 1/2\pi fC$$

Obviously, X_C is *inversely* proportional to f. Its graph is a rectangular hyperbola as shown in Fig. 38. It is drawn in the fourth quadrant because X_C is taken as negative.

(*iv*) **Net reactance, *X*.** Net reactance, $X = X_L \sim X_C$. Its graph is also a hyperbola (but not rectangular) and crosses the *X*-axis at point *A*. The value of frequency at *A* gives the resonant frequency of the circuit because here $X = 0$.

(*v*) **Impedance, *Z***

$$Z = \sqrt{R^2 + (X_L \sim X_C)^2}$$

$$= \sqrt{R^2 + X^2}$$

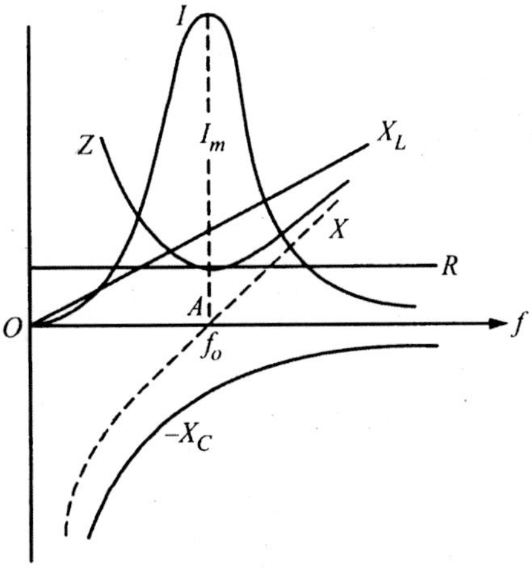

Fig. 38

At frequencies lower than f_o, *Z* is large. Since at lower frequencies $X_C > X_L$, *Z* is capacitive and p.f. is leading. At frequencies higher than f_o, *Z* is again large but is *inductive* because $X_L > X_C$ and p.f. is lagging. The *minimum* value of *Z* is *R* at f_o when $X_L = X_C$.

(*vi*) **Current, *I* :** It has low value on both sides of f_o because *Z* is again large. However, it has *maximum* value of $I_{max} = V/R$ at f_o.

Q. 42. Explain with a suitable diagram, the meaning of resonance curve for a series circuit. Which factor determines the shape of this curve?

Ans. The curve between current and frequency for a circuit is known as its *resonance curve*. The shape of such a curve depends on the circuit resistance as shown in Fig. 39. For smaller values of *R*, the resonance curve is sharply peaked whereas for larger values of *R*, the curve is flat. A series circuit having a sharply-peaked resonance curve has sharp *tuning* characteristics. It is said to possess high *selectivity* which means it is able to distinguish between the desired signal frequency to which it is resonant and other signals on nearby frequencies. Hence, it avoids adjacent channel interference.

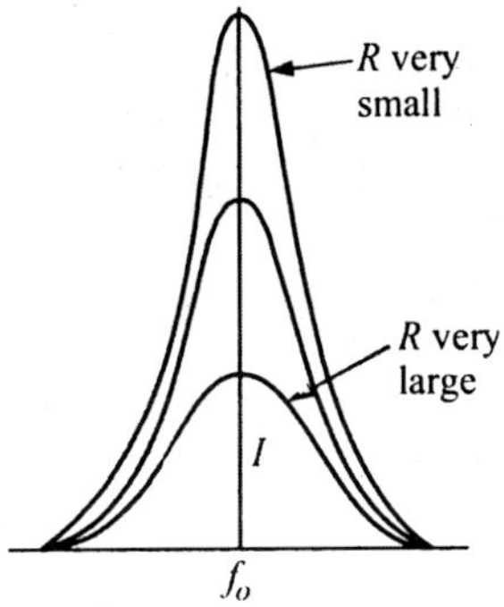

Fig. 39

Q. 43. How do you define the *Q*-factor of a series resonant circuit? Does it affect the bandwidth and selectivity of a resonant circuit?

Ans. The *Q*-factor of a series *R-L-C* circuit may be defined in the following two ways:

(*i*) It is given by the ratio of the inductive reactance of the coil to its resistance.

$$\therefore \quad Q = \frac{X_L}{R} \qquad \text{Since } X_C = X_L \qquad \therefore \quad Q = \frac{X_C}{R}$$

(*ii*) It is given by the voltage magnification produced in the circuit *at resonance.*

$$\text{Voltage magnification} = \frac{I_m X_L}{I_m R} = \frac{X_L}{R}$$

It is the same value as given above. Since *Q* is the ratio of like units, it has no unit itself. As found in *Q*. 40 earlier,

$$f_o = \frac{1}{2\pi\sqrt{LC}}$$

Substituting this value in the above expression, we get

$$Q = \frac{X_L}{R} = \frac{2\pi f_o L}{R} = \frac{2\pi L}{R} \times \frac{1}{2\pi\sqrt{LC}}$$

$$\therefore \qquad Q = \frac{1}{R}\sqrt{\frac{L}{C}} = \frac{\sqrt{L/C}}{R}$$

In a series resonant circuit, higher Q-factor means higher *voltage magnification* and higher *selectivity*. In other words, higher the value of Q, *shorter* the bandwidth of the series resonant circuit.

It is obvious that Q of an R-L-C series circuit can be increased by

(*i*) decreasing the value of circuit resistance and

(*ii*) increasing the L/C value.

Q. 44. What is meant by the resonance curve of a series *R-L-C* circuit? Is its shape affected by the circuit *Q*?

Ans. It is a curve which shows how circuit current varies with the frequency of the applied voltage. It is seen from Fig. 40(*a*) that as frequency of the applied voltage is increased, the magnitude of the the current increases. The current becomes maximum (= V/R) at resonant frequency f_o. Then it decreases towards zero at frequencies above. f_o. In other words, circuit current is maximum at the resonant frequency f_o and drops off on either side of this frequency.

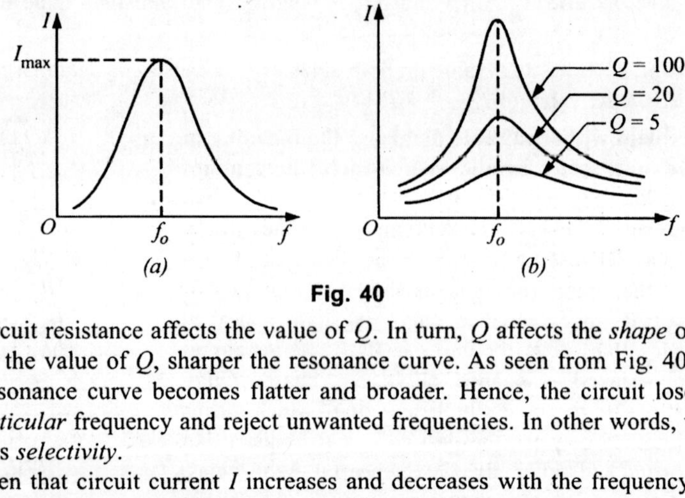

Fig. 40

The circuit resistance affects the value of Q. In turn, Q affects the *shape* of the resonance curve. Higher the value of Q, sharper the resonance curve. As seen from Fig. 40(*b*), at lower Q values, the resonance curve becomes flatter and broader. Hence, the circuit loses its ability to select one *particular* frequency and reject unwanted frequencies. In other words, with *low* Q, the circuit loses its *selectivity*.

It is seen that circuit current I increases and decreases with the frequency of the applied voltage. Similarly, the power developed in the circuit also increases or decreases with applied frequency. As frequency increases from zero to f_o, the current increases from zero to its maximum value I_{max} at f_o. Power has maximum value of $P_{max} = I_{max}^2 R$.

Q. 45. What is meant by the half-power points on the resonance curve of a series *R-L-C* circuit? Define bandwidth of a resonant circuit.

Ans. Half-power points are the points on the resonance curve where the power from the source is one-half the power delivered at the resonant frequency f_o (Fig. 41). This will happen only when currents at these points are $= I_{max}/\sqrt{2} = 0.707\ I_{max}$. The frequencies f_1 and f_2 corresponding to these two points A and B are called *half-power frequencies*. The two frequencies f_1 and f_2 where circuit current is 0.707 I_{max} and hence circuit power $P = 0.5\ P_{max}$. are also called lower and upper cut-off frequencies.

The bandwidth of a resonant circuit is given by the range of frequencies lying between f_1 and f_2.

$\therefore$ bandwidth, $BW = f_2 - f_1$

It can be proved that,

(i) $$BW \cong \frac{f_o}{Q}$$

(ii) $$f_o = \frac{f_1 + f_2}{2}$$

As seen, higher the value of Q, sharper the current response curve and hence, narrower the bandwidth of the resonant circuit.

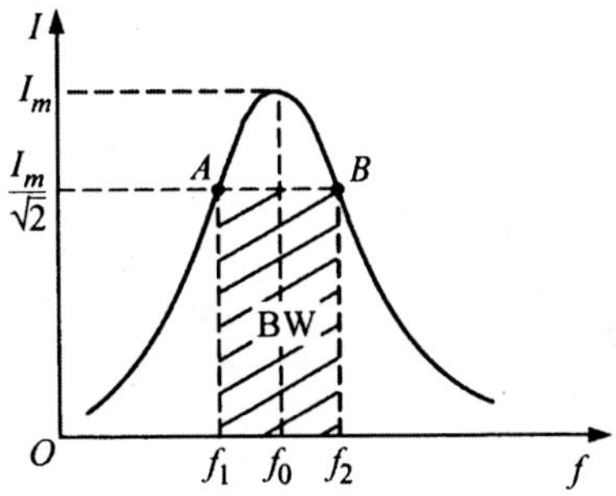

Fig. 41

Q. 46. A circuit having a resistance of 4 Ω and inductance of 0.5 H and a variable capacitance in series is connected across 100-V, 50-Hz supply. Calculate (*i*) the capacitance to give resonance and (*ii*) the voltage across inductance and the capacitance.

(*Elect. Engg.-I, Punjab Univ.*)

Ans. (*i*) Here, the resonance frequency = 50 Hz

Now $$f_o = \frac{1}{2\pi\sqrt{LC}} \quad \text{or} \quad f_o^2 = \frac{1}{4\pi^2 LC}$$

$$\therefore \quad C = \frac{1}{4\pi^2 L f_o^2} = \frac{10^6}{40 \times 0.5 \times 2500} = \mathbf{20\ \mu F}$$

(*ii*) At resonance,

$$I = V/R = 100/4 = 25\text{ A}$$
$$X_L = 2\pi f_o L = 2\pi \times 50 \times 0.5 = 157\ \Omega$$
$$V_L = IX_L = 25 \times 157 = \mathbf{3925\ V}$$

Since, $V_L = V_C$, hence $V_C = \mathbf{3925\ V}$

Q. 47. A coil having an inductance of 48 mH and resistance of 9.8 Ω is connected in series with a 25 μF capacitor across 230-volt, 50-Hz supply. Calculate the resonant frequency, current flowing at resonance and *Q*-factor of the circuit?

(*Basic Elect. Engg. Kerala Univ. 2005*)

Ans. $$f_o = \frac{1}{2\pi\sqrt{LC}} \quad \text{or} \quad f_0^2 = \frac{1}{4\pi^2 LC}$$

$$\therefore \quad f_0^2 = \frac{1}{4 \times 10 \times 48 \times 10^{-3} \times 25 \times 10^{-6}} = \frac{10^9}{40 \times 48 \times 25} = \frac{10^6}{48} = 20{,}833$$

$$f_o = \sqrt{20{,}833} = \mathbf{144\ Hz}$$
$$I_{max} = 230/9.8 = \mathbf{23.47\ A}$$
$$Q = X_L/R$$

Now, $X_L = 2\pi f_o L = 2\pi \times 50 \times 48 \times 10^{-3} = 15\ \Omega$

$\therefore$ $Q = X_L/R = 15/9.8 = \mathbf{1.53}$

Q. 48. A coil of resistance 24 Ω and inductance 0.316 H is connected in series with a 50 μF capacitor across a 240 volt variable frequency sinusoidal supply. Calculate the value of frequency at which resonance occurs and calculate the following for this condition.

(*a*) circuit impedance at resonance **(*b*) power factor of the circuit**

(*c*) current at resonance **(*d*) voltage across capacitor**

(*e*) voltage across the coil

Sketch the vector diagram.

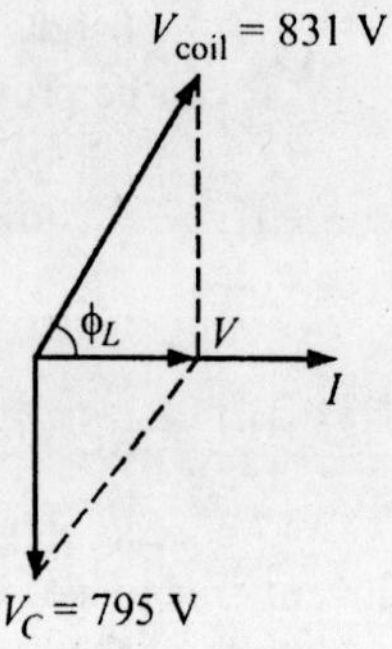

Fig. 42

Ans. $$f_o = \frac{1}{2\pi\sqrt{LC}} = \frac{1}{2\pi \times \sqrt{0.316 \times 50 \times 10^{-6}}} = \mathbf{40\ Hz}$$

(a) $Z = R = \mathbf{24\ \Omega}$

(b) $\cos\phi = R/Z = 24/24 = \mathbf{1}$

(c) $I = V/Z = 240/24 = \mathbf{10\ A}$

(d) $$X_C = \frac{1}{2\pi f_0 C} = \frac{10^6}{2\pi \times 40 \times 50} = 79.5\ \Omega$$

$$V_C = I.X_C = 10 \times 79.5 = \mathbf{795\ V}$$

(e) $X_L = X_C = 79.5\ \Omega$

$$Z \text{ of coil} = \sqrt{R^2 + X_L^2} = \sqrt{24^2 + 79.5^2} = 83.1\ \Omega$$

$$V \text{ across coil} = I \times (Z \text{ of coil}) = 10 \times 83.1 = \mathbf{831\ volt}$$

It should be noted that $V_{coil} > V_C$ because of coil resistance.

However, $V_L = V_C$

Vector diagram is shown in Fig. 42.

Q. 49. A pure capacitor is connected in series with a practical inductor (coil), the voltage source is of 10 volt at 10 kHz frequency. It was observed that a maximum current of 2A flows in the circuit when the value of capacitor is 1 μF. Find the parameters (*R* & *L*) of the coil. *(Elect. Engg. Nagpur Univ. , 2002)*

Ans. Obviously, the given circuit is an *R-L-C* circuit with *R* as the resistance of the inductor.

$$R = V/I_{max} = 10/2 = 5\ \Omega$$

At resonance, $X_L = X_C$ or $2\pi f_o L = 1/2\pi f_o C$

$$\therefore \quad L = \frac{1}{(2\pi f_o)^2 C} = \frac{1}{(2\pi \times 10{,}000)^2 \times 1 \times 10^{-6}} = \mathbf{0.25\ mH}$$

Q. 50. A series *R-L-C* circuit with $R = 10\ \Omega$, $L = (100/314)$ H and $C = (1/31400)$ F is excited from a 10 V, 50 Hz ($\omega = 314$ rad/sec) source. Determine the rms values of voltage across (*i*) resistance (*ii*) inductance (*iii*) capacitance.

Give explanation if any of the answers is more than the source voltage.

(Basic Elect. Engg. R.G.P.V., Bhopal 2004)

Ans. $$X_L = \omega L = 314 \times \frac{100}{314} = 100\ \Omega \qquad (\because 2\pi f = 2\pi \times 50 = 314)$$

$$X_C = \frac{1}{\omega C} = \frac{1 \times 31400}{314 \times 1} = 100\ \Omega$$

Since, $X_L = X_C = 100\ \Omega$, it means *R-L-C* circuit is in resonance.

$$I_{max} = V/Z = V/R = 10/10 = 1\ A$$

(*i*) $V_R = IR = 1 \times 10 = \mathbf{10\ volt}$

(*ii*) $V_L = IX_L = 1 \times 100 = \mathbf{100\ volt}$

(*iii*) $V_C = I.\ X_C = 1 \times 100 = \mathbf{100\ volt}$

It is seen that the voltages developed across X_L and X_C are 100/10 = 10 times greater than the supply voltage. It always happens when the circuit is in *resonance* because the circuit current is at its maximum value.

Incidentally, *Q* factor of the circuit is equal to the voltage magnification *i.e.* 10.

Q. 51. A 20Ω resistor is connected in series with an inductor, a capacitor and an ammeter across 25V variable frequency supply. When the frequency is 400 Hz, the current is at its maximum value of 0.5A and potential drop across capacitor is 150 V. Calculate :

(*i*) capacitance of the capacitor

(*ii*) resistance and inductance of the inductor.

Draw phasor diagram representing voltage across each element and current.

(*Basic Elect. Engg. C.S.V. Tech. Univ. Bhilai, 2007*)

Ans. Since circuit is in resonance, $X_L = X_C$

Now, $V_C = I.\,X_C \qquad \therefore \quad X_C = \dfrac{V_C}{I} = \dfrac{150}{0.5} = 300\ \Omega$

(*i*) $X_C = 1/2\pi f_o C$

$C = 1/2\pi f_o X_C = 1/2\ \pi \times 400 \times 300$

$= 1.326 \times 10^{-6}$ F = **1.326 μF**

(*ii*) Now, $X_L = 300$ or $I = X_L/2\pi f_o = 300/2\pi \times 400$

$L = \mathbf{0.119\ H}$

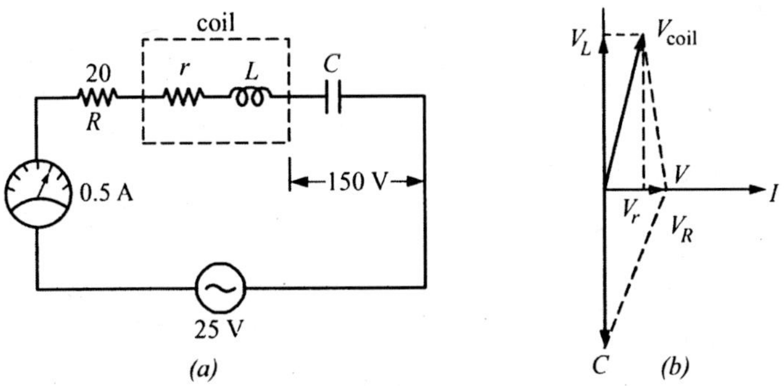

Fig. 43

Circuit impedance, $Z = V/I = 25/0.5 = 50\ \Omega$

Now $\qquad Z = 20 + r$ or $50 = 20 + r$, $r = \mathbf{30\ \Omega}$

$V_R = 0.5 \times 20 = 10$ V, $V_L = V_C = 150$ V, $V_r = 30 \times 0.5 = 15$ V

As seen from the vector diagram of Fig. 43(*b*),

V_{coil} = vector sum of V_L and V_r

V = vector sum of V_{coil}, V_R and V_C.

Q. 52.A circuit having a resistance of 5 Ω, an inductance of 0.4 H and a variable capacitance in series, is connected across a 110 V, 50 Hz supply. Calculate:

(*i*) the value of capacitance to give resonance

(*ii*) current

(*iii*) voltage across the inductance

(*iv*) voltage across the capacitance and

(*v*) *Q*-factor of the circuit (*Elect. Engg, Nagpur Univ. 2001*)

Ans. (*i*) At resonance, $X_L = X_C$. Since resonance occurs at the supply frequency of 50 Hz hence $f_o = 50$ Hz.

$$\therefore \quad 2\pi f_o L = \frac{1}{2\pi f_o C} \quad \text{or} \quad C = \frac{1}{(2\pi f_o)^2 L}$$

$$\therefore \quad C = \frac{1}{(2\pi \times 50)^2 \times 0.4} = \mathbf{25.3\ \mu F}$$

(*ii*) Current at resonance, $I_{max} = V/R = 110/5 =$ **22 A**

(*iii*) $V_L = I \cdot X_L = 22 \times 2\pi \times 50 \times 0.4 =$ **2764 V**

(*iv*) $V_C = V_L =$ **2764 V**

(*v*) $Q = \frac{1}{R}\sqrt{\frac{L}{C}} = \frac{1}{5}\sqrt{\frac{0.4}{25.3 \times 10^{-6}}} = \mathbf{25.14}$

Q. 53. A series resonant circuit has lower and upper half-power frequencies of 8 kHz and 12 kHz. Calculate the following :

(*a*) bandwidth (*b*) resonant frequency and (*c*) *Q*-factor.

Ans. (*a*) bandwidth, $BW = f_2 - f_1 = 12 \text{ kHz} - 8 \text{ kHz} =$ **4 kHz**

(*b*) $f_o = \frac{f_1 + f_2}{2} = \frac{12+8}{2} =$ **10 kHz**

(*c*) $BW = f_o/Q$ or $Q = f_o/BW = 10/4 = \mathbf{2.5}$

EXERCISE

R-L Circuits

1. A circuit takes a current of 8 A at 100 V, the current lagging by 30° behind the voltage. Calculate the values of equivalent resistance and reactance of the circuit.
   (*Elect. Engg. Nagpur Univ. 2003*)
   (10.82 Ω, 6.25 Ω inductive)
2. A series *RL* circuit has $R = 20\ \Omega$ and $L = 0.05$ H and is connected across 100 V, 50 Hz supply. Calculate circuit current and power consumed by the circuit.
   (3.39 A, 309 W)
3. A coil has an inductance of 0.2 H and a resistance of 12 Ω. If a voltage of 200 V at a frequency of 50 Hz is applied, calculate the energy stored in the magnetic field during the quarter period when the current is rising from zero to a maximum value. **(1.95J)**
4. A 60 Hz sinusoidal voltage $v = 100 \sin \omega t$ is applied to a series *RL* circuit. Assuming $R = 10\ \Omega$, $L = 0.01$.H, find the steady state current and relative phase angle. Also, compute the effective magnitude and phase of voltage drops across each circuit elements.
   (*Basic Elect. Engg. R. G. P. V Bhopal 2004*)
   ($I = 6.62\angle{-20.65°}$ A, $V_R = 66.2\angle{-20.65°}$ V, $V_L = 24.96\angle{69.35°}$ V)
5. A voltage of 260 V at 50 Hz applied to a coil produced a current of 2A. The application of 100 V (d.c.) was necessary to produce 2A of direct current through the coil. Calculate the inductance of the coil in millihenrys. (**Hint :** $Z = 130\ \Omega$, $R = 100/2 = 50\ \Omega$)
   (382.5 mH)
6. An inductor having a resistane of 10 Ω is connected to a 240 V, 50 Hz ac supply. The current flowing through the coil is found to be 12A.
   (*a*) Calculate Z, X_L and *L* of the inductor
   (*b*) Determine the phase angle between the current and the applied voltage
   (20 Ω, 17.32 Ω, 0.055 H, 60°)
7. Tow coils, *A* and *B*, are connected in series across a 250 V sinusoidal supply. Coil *A* has a resistance of 3 Ω and coil *B* has a reactance of 4 Ω. The overall power factor is 0.6 lagging and the supply current is 10 A.
   Calculate the reactance of coil *A* and the resistance of coil *B*. Also, calculate the kW, kVA and kVAR of the coil *A* only. **(15.2 Ω, 11.4 Ω, 0.3 kW, 1.55 kVA, 1.52 kVAR)**

8. A single-phase series circuit consists of 10 Ω resistance and 0.1 henry inductance. A voltage of 230 V, 50 Hz is applied across the circuit. What is the voltage across the resistance? (*Basic Elect. Engg. Kerala Univ. 2002*) **(69.7 V)**

R-C Circuits

9. Determine the circuit constants for the given ac voltage and current

$$v = 311 \sin (2500t + 170°) \quad \text{and} \quad i = 15.5 \sin (2500t - 145°).$$

Is the current lagging or leading? Draw vector and impedance diagrams.

(R = 14.14 Ω, C = 28.3 μF, leading by 45°)

10. A pure resistance of 8.31 Ω is connected in series with a capacitor having a reactance of 10 Ω when the supply voltage is v = 184 sin 942t. Calculate :
(*a*) frequency (*b*) capacitance (*c*) rms value of current, power factor.

(150 Hz ; 106 μF ; 10 A, 0.64 leading)

11. A capacitor of 80 μF takes a current of 1A when sinusoidal voltage of 250 V is applied across it. Find : (*a*) the frequency and (*b*) the resistance to be connected in series with the capacitor to reduce the current to 0.5 A at this frequency. **(7.98 Hz, 433 Ω)**
12. A capacitor having a capacitance of 20 μF is connected in series with a non-inductive resistance of 120 Ω across a 100 V, 50 Hz supply. Calculate (*a*) the current and (*b*) the phase difference between the supply voltage and current and (*c*) the power.

(0.501 A, 52.9°, 30.2 W)

13. A series *RC* circuit consists of 5 Ω resistor and 20 μF capacitor. If i (t) = 2 cos 5000t, find (*a*) expression for $v(t)$ (*b*) V_R and V_C and (*c*) power dissipated in the circuit.

[(*a*) 22.4 sin (5000t + 26.6°), (*b*) V_R = 7.07∠90°, V_C = 14.14∠0°, (*c*) 10 W]

14. A voltage of v (t) = 70.7 sin 2000t is applied across an *RC* circuit having R = 6 Ω. Find (*a*) value of C that limits the power consumed to 150 W and (*b*) expression for i (t).

[62.5 μF, 7.07 sin (2000t + 53.1°)]

15. A capacitor having a capacitance of 10 μF is connected in series with a non-inductive resistance of 120 Ω across a 100-V, 50-Hz supply. Calculate (*a*) the current (*b*) the phase difference between the current and the supply voltage and (*c*) the power.

(*Elect. Engg.-I, Punjab Univ.*)

(0.294 A, 69.3°, 10.4 W)

16. A voltage of 125 V at 60 Hz is applied across a non-inductive resistor connected in series with a capacitor. The current in the circuit is 2.2 A. The power loss in the resistor is 96.8 watts and that in the capacitor is negligible. Calculate the resistance and the capacitance. Draw the vector diagram. **(20 Ω, 50 μF)**

R-L-C Circuits

17. A circuit is made up of 10 Ω resistance, 12 mH inductance and 281.5 μF capacitance in series. The supply voltage is 100 V (constant). Calculate the value of the current when the supply frequency is (*a*) 50 Hz and (*b*) 150 Hz. **[8 A (leading), 8 A (lagging)]**
18. A series *R-L-C* circuit consists of R = 20 Ω, L = 95.6 mH and C = 318 μF. It is connected to a 500 V, 50 Hz supply. Calculate the current in the circuit and the power factor. **(24.3 A, 0.97 leading)**
19. A series circuit consists of a resistance of 4 Ω, a capacitance of 200 μF and an inductance of 50.7 mH. The supply voltage is 240 V at 50 Hz. Calculate (*a*) the capacitive reactance (*b*) the inductive reactance (*c*) the impedance and (*d*) the current.

(15.9 Ω, 15.9 Ω, 4 Ω, 60 A)

20. An air-cored coil took 0.8 kW and 2 kVAR when connected across a 240 V, 50 Hz sinusoidal supply. If it is now connected in series with a 100 μF capacitor across the same supply, calculate :
(*a*) the current (*b*) the power
(*c*) the power factor (*d*) the voltage across the coil
(*e*) the voltage across the capacitor **(19.7 A, 3880 W, 0.82 leading, 528 V, 629 V)**

21. A resistance *R* and inductance *L* = 0.01 H and a capacitance *C* are connected in series. When a voltage *v* = 400 (3000 *t* – 10°) volts is applied to the series combination, the current flowing is $10\sqrt{2}$ cos (3000*t* – 55°) ampere. Find *R* and *C*.
(*Elect. Circuits-I, Punjab Univ.*) **(20 Ω ; 33.3 μF)**

22. A circuit consists of 10 Ω resistance, 15 mH inductance and 281 μF capacitance in series. The supply voltage is 110 V. Calculate the current when supply frequency is 150 Hz. (*Basic Elect. Engg. Kerala Univ. 2004*) **(9.17 A lead)**

23. A coil of resistance 10 Ω and inductance 0.1 H is connected in series with a capacitor of 150 μF across a 230 V, 50 Hz supply. Determine :
(*i*) impedance (*ii*) circuit current
(*iii*) voltage drop across the coil (*iv*) power factor of the circuit
(*Basic Elect. Engg. RGPV, Bhopal 2006*)
(14.3 Ω, 16.1 A, 531 V, 0.7 lag)

24. A series *R-L-C* circuit consists of *R* = 10 Ω, *L* = 0.318 H and *C* = 63.6 μF and emf source *e* (*t*) = 100 sin 314 *t*. Calculate :
(*i*) expression for *i*(*t*) (*ii*) phase angle between voltage and current
(*iii*) power factor and (*iv*) active power consumed.
Draw the phase diagram. (*Elect Engg. Nagpur Univ. 2002*)
[*i*(*t*) = 1.97 sin (314 *t* – 78.64°), 78.64° lag, 0.197 lag, 19.4 W]

25. A coil of inductance 0.0805 H takes a current of 5 A when connected in series with a 50 μF loss-free capacitor across a 240 V, 50 Hz sinusoidal supply. Calculate :
(*a*) coil resistance (*b*) coil power factor
(*c*) overall p.f. (*d*) power
Sketch the vector diagram. **(28.8 Ω, 0.75 lagging, 0.6 leading, 720 W)**

Series Resonance

26. A circuit consists of a 150 μF capacitor in series with an inductor of resistance 10 Ω and inductance 48 mH. Calculate the frequency at which the current is maximum and determine the value of current for an ac supply of 250 V. **(59.5 Hz, 25 A)**

27. A resistor and a capacitor are connected in series with a variable inductor. When the circuit is connected to a 240 V, 50 Hz supply, the maximum current given by varying the inductance is 0.5 A. At this current, the voltage across the capacitor is 250 V. Calculate the value of (*a*) resistance (*b*) capacitance and (*c*) inductance. Neglect the resistance of the inductor. **(480 Ω, 6.36 μF, 1.59 H)**

28. What is the voltage magnification in a resonant circuit consisting of a coil of resistance 5 Ω and inductance 90 mH in series with a capacitor of 25 μF capacitance. **(12)**

29. A circuit has a non-inductive resistance of 100 Ω, an inductance of 1 H and a capacitance of 100 μF, all connected in series. Find (*a*) resonance frequency and (*b*) the current and power taken at resonance if the supply voltage is 230 V a.c.
(15.95 Hz, 2.3 A, 529 W)

30. An ac series circuit has a resistance of 10 Ω, an inductance of 0.2 H and a capacitance of 60 μF. Calculate : (*a*) resonant frequency (*b*) current (*c*) power at resonance. Given that the applied voltage is 200 V. **(46 Hz, 20 A, 4 kW)**

31. A large coil of inductance 1.405 H and resistance 40 Ω is connected in series with a capacitor of 20 μF. Calculate the frequency at which the circuit resonates.
If a voltage of 100 V is applied to the circuit at resonant condition, calculate the current drawn from the supply and the voltage across the coil and capacitor.
(*Elect. Engg., Nagpur Univ. 2000*)
(32 Hz, 2.5 A, $V_L = V_C = 663$ V, $V_R = 100$ V)

MULTIPLE CHOICE QUESTIONS

1. In a series *R-L* circuit, current
(*a*) is in phase with the voltage (*b*) lags the voltage
(*c*) leads the voltage (*d*) is anti-phase with the voltage.
2. The power developed in a pure reactance is .... power.
(*a*) true (*b*) radiated (*c*) reactive (*d*) apparent
3. The power factor of a pure resistive ac circuit is
(*a*) zero (*b*) unity (*c*) lagging (*d*) leading
4. The apparent power in a circuit is 50 VA and the reactive power is 40 VAR. The true power is
(*a*) 50 W (*b*) 10 W (*c*) 30 W (*d*) none of the above
5. An ac circuit has an apparent power of 50 VA. Its resistance is 60 Ω and true power is 30 W. The absolute value of the circuit impedance is
(*a*) 60 Ω (*b*) 50 Ω (*c*) 100 Ω (*d*) none of the above
6. Inductive reactance of a coil of inductance 0.5 H at 50 Hz is .... ohm.
(*a*) 15.7 (*b*) 157 (*c*) 50 (*d*) 25
7. Inductive reactance of an ac circuit increases with
(*a*) increase in frequency (*b*) decrease in resistance
(*c*) increase in resistance (*d*) decrease in frequency
8. Adding series resistance to a pure inductance
(*a*) reduces the power factor (*b*) increases the impedance
(*c*) increases the phase angle (*d*) changes the *j* value.
9. In a certain *RL* circuit, $V_R = 2$ V and $V_L = 3$ V. The magnitude of the total voltage is
(*a*) 2 V (*b*) 3 V (*c*) 5 V (*d*) 3.6 V
(*Basic Elect. Engg. VTU Karnataka, 2009*)
10. The hypotenuse of a kVA triangle represents
(*a*) true power (*b*) VARs (*c*) apparent power (*d*) active power
11. The power factor of an ac circuit is equal to
(*a*) apparent power divided by true power
(*b*) reactive power divided by apparent power
(*c*) reactive power divided by true power
(*d*) true power divided by apparent power
12. When dielectric medium of a capacitor is changed, keeping all other things the same
(*a*) value of X_C increases (*b*) value of X_C decreases
(*c*) value of X_C does not change (*d*) can't say without more data
13. As the frequency of the voltage applied across a capacitor is decreased, the value of X_C
(*a*) increases (*b*) decreases
(*b*) remains the same (*d*) depends on the current

14. When the frequency of the applied voltage in a series *R-C* circuit is increased, its capacitive reactance
(*a*) increases (*b*) decreases (*c*) becomes zero (*d*) remains same
(*Basic Elect. Engg., VTU Karnataka, 2009*)
15. In a series *RC* circuit, as the ratio X_C/R gets closer to zero, the phase angle
(*a*) gets closer to – 90° (*b*) gests closer to 0°
(*c*) stays the same (*d*) can't say
16. If in a series *RC* circuit, $X_C = R$, the circuit phase angle is
(*a*) 0° (*b*) – 45°
(*c*) – 90° (*d*) can't say from the given data
17. When supply frequency is increased, voltage across a capacitor in a series *RC* circuit
(*a*) decreases (*b*) increases (*c*) remains constant (*d*) becomes infinite
18. In a series *RC* circuit, the phase angle between voltage and current is
(*a*) between 0° and 90° leading (*b*) constant at 45°
(*c*) 90° all the time (*d*) 0°
19. The impedance of an ac circuit is
(*a*) voltage divided by current (*b*) resistance of the ac circuit
(*c*) vector sum of voltage drops (*d*) current divided by voltage.
20. In a series *R-L-C* circuit, if the difference $(X_L \sim X_C)$ increases
(*a*) impedance increases (*b*) impedance decreases
(*c*) impedance remains the same (*d*) resistance decreases
21. The phase angle of an *R-L-C* series circuit is leading if
(*a*) $X_C = 0$ (*b*) $R = 0$ (*c*) $X_C > X_L$ (*d*) $X_C < X_L$
22. Unit of reactive power is
(*a*) VA (*b*) watt (*c*) VAR (*d*) ohm
(*Basic Elect. Engg. RGPV, Bhopal 2008*)
23. A series *R-L-C* circuit becomes resonant when its
(*a*) $X_L > X_C$ (*b*) $X_C > X_L$ (*c*) $X_L = X_C$ (*d*) $X_L = R$
24. Unit of inductive reactance is
(*a*) henry (*b*) millihenry (*b*) weber (*d*) ohm
25. Which of the following elements must be present in a circuit in order to produce resonance?
(*a*) inductance (*b*) capacitance (*c*) resistance (*d*) both (*a*) and (*b*)
26. Which of the following *does not* influence the resonant frequency of a series *R-L-C* circuit?
(*a*) X_L (*b*) X_C (*c*) R (*d*) Z
27. In a series *R-L-C* circuit, the current is .... at resonance
(*a*) maximum (*b*) minimum (*c*) leading (*d*) lagging
28. When in resonance, a series *R-L-C* circuit behaves like a pure
(*a*) inductor (*b*) capacitor (*c*) resistor (*d*) none of the above
29. The *Q*-factor of a series *R-L-C* circuit is given by
(*a*) X_C/R (*b*) X_L/R (*c*) Z/R (*d*) both (*a*) and (*b*)
30. When the resistance of a resonant *R-L-C* circuit is increased, its
(*a*) *Q* is increased (*b*) bandwidth is increased
(*c*) resonant frequency is increased (*d*) power is increased

31. In a series *R-L-C* resonant circuit, when $X_L = X_C$, the circuit
(*a*) voltage leads the current (*b*) current leads the voltage
(*c*) impedance is minimum (*d*) impedance is maximum

32. It is possible to pass a specific signal frequency through a series resonant circuit by the proper selection of
(*a*) *C* and *R* (*b*) *C* and *L* (*c*) *R* and *L* (*d*) *R* and *Z*

33. A series *R-L-C* circuit would be in resonance when
(*a*) $X_L = 20\ \Omega$, $X_C = 10\ \Omega$ and $R = 15\ \Omega$
(*b*) $X_L = 15\ \Omega$, $X_C = 15\ \Omega$ and $R = 20\ \Omega$
(*c*) $X_L = 50\ \Omega$, $X_C = 30\ \Omega$ and $R = 25\ \Omega$
(*d*) $X_L = 100\ \Omega$, $X_C = 50\ \Omega$ and $R = 100\ \Omega$

34. For a resonant series circuit, $f_o = 10$ kHz and $Q = 10$. The bandwidth is
(*a*) 9 – 11 kHz (*b*) 1 – 10 kHz (*c*) 10 – 11 kHz (*d*) 9.5 – 10.5 kHz

35. When either *L* or *C* of a series circuit is increased, its resonance frequency
(*a*) increases (*b*) decreases
(*c*) remains the same (*d*) is determined by the resistor.

36. The resonant frequency of a series resonant circuit is determined by the product
(*a*) *LR* (*b*) *CR* (*c*) $L\sqrt{C}$ (*d*) *LC*

ANSWERS

1. *b*	2. *c*	3. *b*	4. *c*	5. *c*	6. *b*	7. *a*
8. *b*	9. *d*	10. *c*	11. *d*	12. *d*	13. *a*	14. *b*
15. *b*	16. *b*	17. *a*	18. *a*	19. *a*	20. *a*	21. *c*
22. *c*	23. *c*	24. *a*	25. *d*	26. *c*	27. *a*	28. *c*
29. *d*	30. *b*	31. *c*	32. *b*	33. *b*	34. *d*	35. *b*
36. *d*.						

CHAPTER 8
PARALLEL AC CIRCUITS

8.1 PARALLEL AC CIRCUITS

Q. 1 (*a*) Mention different methods available for solving parallel ac circuits. Give full details of the vector method for solving such circuits.

Ans. When impedance are joined in parallel, we can solve them by any one of the following three methods:

(*i*) **Vector method :** It is well-suited for solving *two-branched* parallel circuits

(*ii*) **Admittance method :** It is particularly suited for solving *series-parallel* circuits.

(*iii*) **Complex algebra :** It can solve *all types* of parallel circuits equally well. Vector method will be explained with the help of the parallel circuit shown in Fig. 1(*a*). Here, an *RL* circuit has been joined in parallel with an *R-C* circuit and both of them have been connected across an RMS supply of *V* volts. Even though voltage across the two branches is the same, their branch currents are different.

Students may please note that in parallel circuits unless stated otherwise voltage, vector is normally taken as the *reference vector* as shown in Fig. 1(*b*). It is so because voltage is common for all parallel elements.

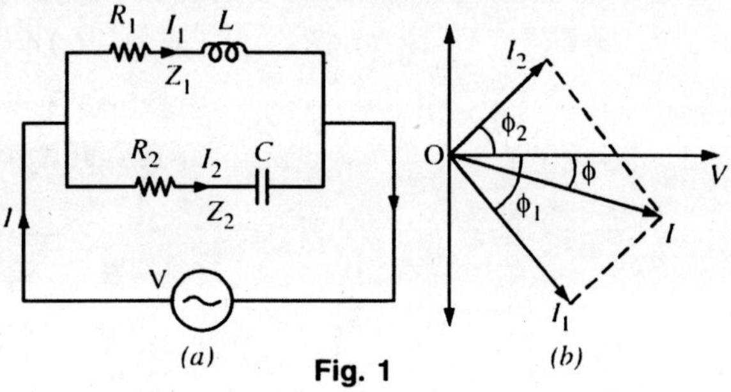

Fig. 1

For R-L Branch :

$$Z_1 = \sqrt{R_1^2 + X_L^2} \quad \text{and} \quad I_1 = V/Z_1$$
$$\cos\phi_1 = R_1/Z_1 \quad \text{or} \quad \phi_1 = \cos^{-1}(R_1/Z_1)$$

As seen from Fig. 1(*b*), I_1 *lags* behind the applied voltage *V* by ϕ_1.

For R-C branch :

$$Z_2 = \sqrt{R_2^2 + X_C^2} \quad \text{and} \quad I_2 = V/Z_2$$
$$\cos\phi = R_2/Z_2 \quad \text{or} \quad \phi_2 = \cos^{-1}(R_2/Z_2)$$

As seen, current, I_2 *leads* *V* by ϕ_2.

Circuit current, *I*

As seen from Fig. 2, the two branch currents have been split into their active and reactive components.

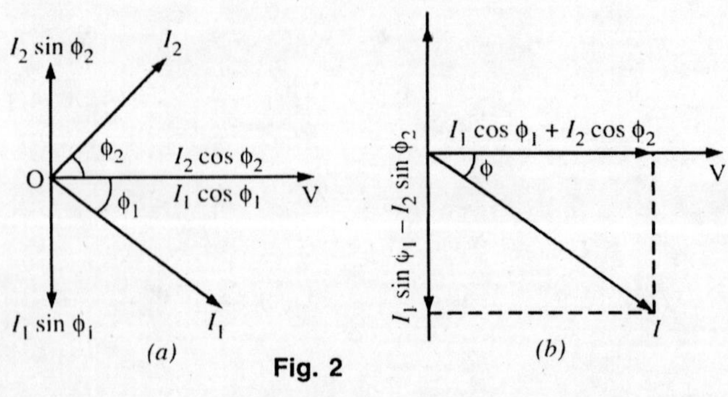

Fig. 2

Sum of the active components = $I_1 \cos \phi_1 + I_2 \cos \phi_2$

Sum of the reactive components = $I_1 \sin \phi_1 - I_2 \sin \phi_2$

These must equal the respective components of the total circuit current, I.

$$\therefore \quad I \cos \phi = I_1 \cos \phi_1 + I_2 \cos \phi_2$$

$$I \sin \phi = I_1 \sin\phi_1 - I_2 \sin \phi_2$$

Squaring both sides and adding them, we get

$$\therefore \quad I^2 (\sin^2\phi + \cos^2\phi) = (I_1 \cos\phi_1 + I_2 \cos\phi_2)^2 + (I_1 \sin \phi_1 - I_2 \sin\phi_2)^2$$

or
$$I = \sqrt{(I_1 \cos\phi_1 + I_2 \cos \phi_2)^2 + (I_1 \sin\phi_1 - I_2 \sin\phi_2)^2}$$

$$= \sqrt{(X\text{-comp.})^2 + (Y\text{-comp.})^2}$$

The power factor of the combined circuit is given by

$$\cos \phi = \frac{X\text{-comp.}}{I} = \frac{I_1 \cos\phi_1 + I_2 \cos \phi_2}{I}$$

Note : If $I_1 \sin \phi_1 > I_2 \sin \phi_2$, then circuit power factor would be *lagging*.

If $I_1 \sin \phi_1 < I_2 \sin \phi_2$, then circuit power factor would be *leading*.

Q. 2. A 10 μF capacitor is connected in parallel with a 200 Ω resistor across a 240-V, 50 Hz sinusoidal supply. Sketch the vector diagram and using the vector method, calculate: (*i*) supply current (*ii*) overall power factor (*iii*) total power dissipated.

Ans. (*i*) $X_C = \frac{1}{\omega C} = \frac{1}{2\pi f C}$

$$= \frac{10^6}{314 \times 10}$$

$$= 318 \ \Omega$$

$$I_C = V/X_C = 240/318$$

$$= 0.75 \text{ A}$$

$$I_R = V/R = 240/200$$

$$= 1.2 \text{ A}$$

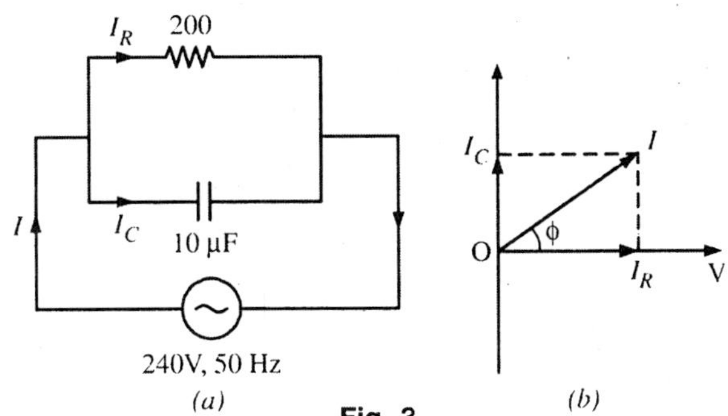

Fig. 3

The supply current is the vector sum of these two currents as shown in Fig. 3(*b*).

$$\therefore \quad I = \sqrt{I_R^2 + I_C^2} = \sqrt{0.75^2 + 1.2^2} = \mathbf{1.417 \text{ A}}$$

(*ii*) $$\cos \phi = \frac{I_R}{I} = \frac{1.2}{1.417} = \mathbf{0.85 \text{ leading}}$$

(*iii*) $$P = I_R^2 \times R = 1.2^2 \times 200 = \mathbf{288 \text{ W}}$$

or $$P = VI \cos \phi = 240 \times 1.417 \times 0.85 = 288 \text{ W}$$

Q. 3. A circuit having a resistance of 20 Ω and inductance of 0.07 H is connected in parallel with a series combination of 50 Ω resistance and 60 μF capacitance. Calculate the total current when the parallel combination is connected across 230 V, 50 Hz supply as shown in Fig. 4(*a*). (*Basic Elect. Engg., VTU Karnataka 2007*)

Ans. $$X_L = 2\pi \times 50 \times 0.07 = 314 \times 0.07 = 22 \ \Omega$$

$$Z_1 = \sqrt{R_1^2 + X_L^2} = \sqrt{20^2 + 22^2} = 29.7 \ \Omega$$

Now, $\cos\phi_1 = 20/29.7 = 0.673$, $\sin\phi_1 = 22/29.7 = 0.741$, $\phi_1 = 47.70°$

$\therefore$ $\mathbf{Z_1} = 29.7\angle 47.7°$

$$\mathbf{I}_1 = \frac{230\ \angle 0°}{29.7\ \angle 47.7°} = 7.74\ \angle{-47.7°}$$

It is seen that the current I_1 lags behind the voltage by 47.7° as shown in Fig. 4(*b*).

$X_C = 10^6/314 \times 60$
$= 53\ \Omega$

$\therefore\ \mathbf{Z_2} = 50 - j53$
$= 72.86\ \angle{-46.7°}$

$\cos\phi_2 = 50/72.86$
$= 0.686$

$\sin\phi_2 = 53/72.86$
$= 0.727,$

$\phi_2 = 46.7°$

$$\mathbf{I}_2 = \frac{230\ \angle 0°}{72.86\angle -46.7°}$$
$= 3.15\ \angle 46.7°$

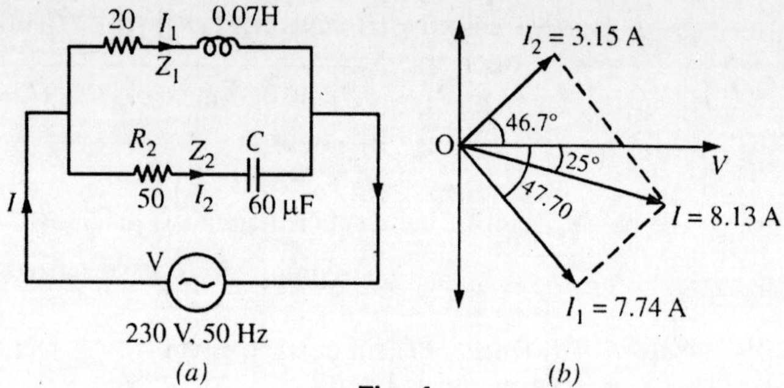

Fig. 4

It is seen from Fig. 4(*b*). that I_2 leads V by 46.7°

X-comp. of $I = I_1 \cos\phi_1 + I_2 \cos\phi_2$
$= 7.74 \times 0.673 + 3.15 \times 0.686$
$= 5.21 + 2.16 = 7.37\text{A}$

Y-comp. of $I = I_1 \sin\phi_1 - I_2 \sin\phi_2$
$= 7.74 \times 0.741 - 3.15 \times 0.727$
$= 5.73 - 2.29 = 3.44$ A

$\therefore$ $I = \sqrt{7.37^2 + 3.44^2} = \sqrt{54.3 + 11.8} = \mathbf{8.13\ A}$

Alternative solution

$\mathbf{Z_1} = 20 + j22 = 29.7\ \angle 47.7°$

$\mathbf{Z_2} = 50 - j53 = 72.86\ \angle{-46.7°}$

$\mathbf{Z_1} \cdot \mathbf{Z_2} = 29.7\ \angle 47.7° \times 72.86\ \angle{-46.7°} = 2165\ \angle 1°$

$\mathbf{Z_1} + \mathbf{Z_2} = 70 - j31 = 76.5\ \angle{-24°}$

$$\therefore\quad \mathbf{Z} = \frac{\mathbf{Z_1\ Z_2}}{\mathbf{Z_1 + Z_2}} = \frac{2165\angle 1°}{76.5\angle -24°} = 28.3\ \angle 25°$$

$$\therefore\quad I = \frac{230\ \angle 0°}{28.3\ \angle 25°} = \mathbf{8.13\ \angle{-25°}\ A}$$

Q. 4. For the circuit shown in Fig. 5, find

(*i*) total current
(*ii*) p.f. of the circuit
(*iii*) total power taken from the source
(*iv*) energy stored in inductor and capacitor (*Elect. Engg., Nagpur Univ. 2006*)

Ans. $X_L = \omega L = 2\pi. \times 50 \times 0.5 = 157\ \Omega$

$$X_C = \frac{1}{2\pi fC} = \frac{10^6}{2\pi \times 50 \times 20} = 159\ \Omega$$

$\mathbf{Z_1} = 90 + j\ 157 = 181\angle 60.1°$

$\mathbf{Z_2} = -\ j\ 159$ — **along –ve *Y*-axis**

Keeping in mind that in parallel circuits, usually the common voltage V is taken as the reference vector, we get $\mathbf{V} = 230 \angle 0°$.

(*i*) $\mathbf{I}_1 = \dfrac{\mathbf{V}}{\mathbf{Z}_1} = \dfrac{230 \angle 0°}{181 \angle 60.1°}$

$= 1.27 \angle -60.1°$

$= 0.63 - j1.1$

$\mathbf{I}_2 = \dfrac{\mathbf{V}}{\mathbf{Z}_2} = \dfrac{230 \angle 0°}{-j159}$

$= \dfrac{230 \angle 0°}{159 \angle -90°} = 1.445 \angle 90°$

$= 0 + j1.445$

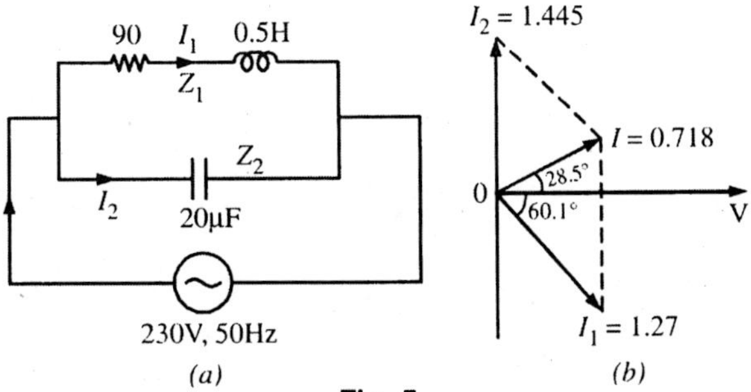

Fig. 5

$\therefore \quad I = I_1 + I_2 = 0.63 - j\,1.1 + 0 + j\,1.445$

$= 0.63 + 0.345 = \mathbf{0.718 \angle 28.5°}$ — Fig. 5(*b*)

(*ii*) circuit p.f. $= \cos\phi = \cos 28.5° = \mathbf{0.87}$ **(leading)**

(*iii*) total power $= VI\cos\phi = 230 \times 0.718 \times 0.87 = \mathbf{144\ W}$

or $P = I_1^2 R_1 = 1.27^2 \times 90 = 144$ W

(*iv*) energy stored in inductor $= \dfrac{1}{2} L\, I_1^2 = \dfrac{1}{2} \times 0.5 \times 1.2^2 = \mathbf{0.4\ J}$

energy stored in capacitor $= \dfrac{1}{2} CV^2 = \dfrac{1}{2} \times 20 \times 10^{-6} \times 230^2 = \mathbf{0.53\ J}$

Q. 5. The current in each branch of a two-branched parallel circuit is given as :

$$i_a = 7.07 \sin\left(314t - \frac{\pi}{4}\right) \quad ; \quad i_b = 21.2 \sin\left(314t + \frac{\pi}{3}\right)$$

and supply voltage is $v = 354 \sin 314t$. Calculate:

(*i*) total current in the same form

(*ii*) ohmic value of components in each branch. (*Elect. Engg., Nagpur Univ. 2004*)

Ans. Since i_a lags the voltage by $\pi/4$ radian or 45°, it means that circuit A is an R-L circuit. Since i_b leads V by 60°, it means branch B is an R-C circuit as shown in Fig. 6(*a*).

Moreover, *maximum* value of current in branch A is 7.07 A and in branch B is 21.2 A. The resultant current can be found vectorially with the help of vector diagram of Fig. 6(*b*).

X-comp. $= 21.2 \cos 60° + 7.07 \sin 45°$

$= 15.6$ A

Y-comp. $= 21.2 \sin 60° - 7.07 \sin 45°$

$= 13.36$ A

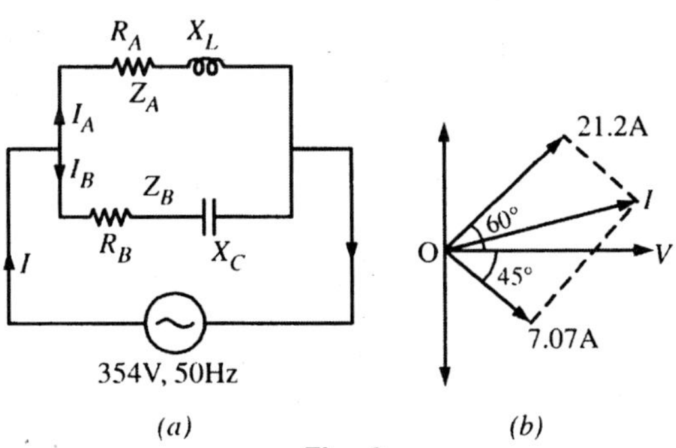

Fig. 6

Hence, *maximum* value of resultant current is

$I = \sqrt{15.6^2 + 13.36^2} = 20.54$ A

$\phi = \tan^{-1}(13.36/15.6) = \tan^{-1}(0.856) = 40.5°$ leading (*i*) Hence, the expression for the supply current I is

$$i = \mathbf{20.54 \sin (314t + 40.5°)}$$

(*ii*)
$$Z_A = \frac{V_m}{I_m} = \frac{354}{7.07} = 50\ \Omega,\ \cos \phi_A = \cos 45° = \frac{1}{\sqrt{2}}$$

$$\sin \phi_A = \sin 45° = \frac{1}{\sqrt{2}}$$

$$R_A = Z_A \cos \phi_A = 50 \times \frac{1}{\sqrt{2}} = \mathbf{35.35\ \Omega}$$

$$X_L = Z_A \sin\phi_A = 50 \times \frac{1}{\sqrt{2}} = \mathbf{35.35\ \Omega}$$

$$Z_B = 354/21.2 = 16.7\ \Omega,\ R_B = 16.7 \times \cos 60° = \mathbf{8.35\ \Omega}$$

$$X_C = 16.7 \times \sin 60° = \mathbf{14.46\ \Omega}$$

Q. 6. Two impedance Z_1 and Z_2 are connected in parallel. The first branch takes a leading current of 16 A and has a resistance of 5 Ω while the second branch takes a lagging current at power factor 0.8. The total power supplied is 5 kW, the applied voltage being (100 + *j* 200) V. Determine the complex expressions for branch currents and total current and for the circuit constants. (*Basic Elect. Engg. C.S.V. Tech Univ. Bhilai, 2007*)

Ans. It is worth reminding that while using complex algebra, positive *X*-axis is taken as the reference vector unless indicated otherwise

$$\mathbf{V} = 100 + j200 = 223.6 \angle 63.4°$$

$$|\mathbf{Z}_1| = V/I_1 = 223.6/16 = 13.98\ \Omega$$

Since it has been given that 16A is a leading current, hence $\mathbf{Z}_1$ must be a capacitive impedance.

Now, $\mathbf{Z}_1 = 5 - jX_C$ $\qquad \therefore\ 13.98 = \sqrt{5^2 + (-X_C)^2}$

or $13.98^2 = 25 + X_C^2$ $\qquad \therefore\ X_C = 13.1\ \Omega$

$$\mathbf{Z}_1 = 5 - j\ 13.1 = 14 \angle -69.1°$$

$$I_1 = \frac{\mathbf{V}}{\mathbf{Z}_1} = \frac{223.6 \angle 63.4°}{13.98 \angle -69.1°} = 16 \angle 132.5° = -10.8 + j11.8$$

It would be seen from Fig. 7(*b*) that I_1 leads the reference vector by 132.5° but leads the applied voltage by (132.5 – 63.4) = 69.1°.

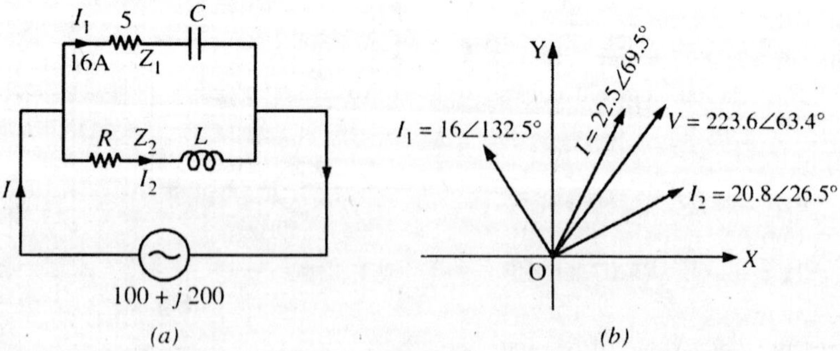

Fig. 7

Total power = 5 kW = 5000 W

Power in $Z_1 = I_1^2 R = 16^2 \times 5 = 1280$ W

$\therefore$ Power in $Z_2 = 5000 - 1280 = 3720$ W

$$\text{VA of second branch} = \frac{W}{\text{p.f.}} = \frac{3720}{0.8} = 4650 \text{ volt-amperes}$$

Now,
$$I_2 = \frac{VA}{A} = \frac{4650}{223.6} = 20.8 \text{ A}$$

$$Z_2 = \frac{V}{I_2} = \frac{223.6}{20.8} = 10.75\ \Omega$$

Since the p.f. of the second branch is 0.8 lagging

$$\therefore \quad \mathbf{Z}_2 = 10.75\,(0.8 + j\,0.6) = 10.75\angle 36.9° = 8.6 + j6.45$$

$$\mathbf{I}_2 = \frac{\mathbf{V}}{\mathbf{Z}_2} = \frac{223.6\angle 63.4°}{10.75\angle 36.9°} = 20.8\angle 26.5° = \mathbf{18.7 + j9.2}$$

It is seen that I_2 lags behind the applied voltage by (63.4° – 26.5°) = 36.9°

$$\boldsymbol{I} = \boldsymbol{I}_1 + \boldsymbol{I}_2 = (-10.8 + j\,11.8) + (18.7 + j9.3)$$
$$= 7.9 + j21.1 = \mathbf{22.5\angle 69.5°}$$

As seen, the supply current I leads the applied voltage by (69.5° – 63.4°) = 6.1° as shown in Fig. 7(*b*).

The total impedance of the circuit

$$= \frac{\mathbf{V}}{\mathbf{I}} = \frac{223.6\angle 63.4°}{22.5\angle 69.5°} = 9.94\angle -6.1° = 9.85 - j\,1.05.$$

Q. 7. A coil of inductance 80 mH and resistance 120 Ω is connected to a 230 V, 50 Hz single phase supply. In parallel with it is a series combination of 16 μF capacitor and a 40 Ω non-inductive resistor. Determine the power factor of combined circuit, the total power taken from the supply and current through each branch. (*Elect. Engg, Nagpur Univ. 2005*)

Ans.

$$X_L = 2\pi \times 50 \times 80 \times 10^{-3} = 25\ \Omega$$

$$\therefore \quad \mathbf{Z}_1 = R_1 + jX_L = 120 + j25$$
$$= 122.6\angle 11.82°$$

$$X_C = \frac{1}{\omega C} = \frac{1}{314 \times 16 \times 10^{-6}} = \frac{10^6}{314 \times 16} = 199\ \Omega$$

$$\therefore \quad \mathbf{Z}_2 = 40 - j199 = 203\angle -78.63°$$

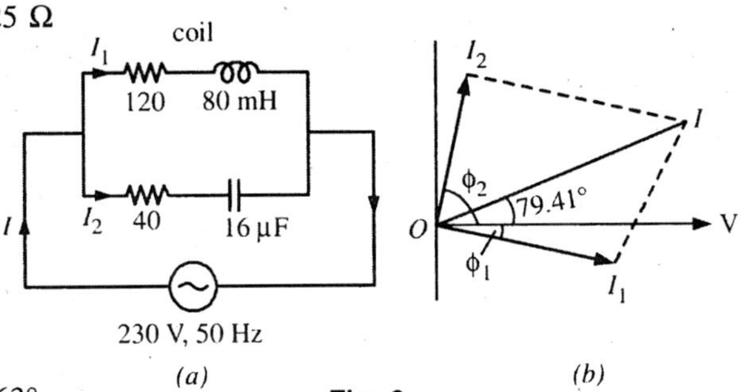

Fig. 8

The combined impedance of the parallel circuit

$$\mathbf{Z} = \mathbf{Z}_1 \parallel \mathbf{Z}_2 = \frac{\mathbf{Z}_1\,\mathbf{Z}_2}{\mathbf{Z}_1 + \mathbf{Z}_2}$$

$$= \frac{(122.6\angle 11.82°)\,(203\angle -78.63°)}{(120 + j\,25)\,(40 - j\,199)}$$

$$= 99.3 - j35 = 105.3\angle -19.41°$$

Power factor of the combined circuit is

$$\cos\phi = \frac{\text{combined resistance}}{\text{combined impedance}} = \frac{99.3}{105.3} = \mathbf{0.943\ (lead)}$$

circuit current,
$$\mathbf{I} = \frac{\mathbf{V}}{\mathbf{Z}} = \frac{230\angle 0°}{105.3\angle -19.41°}$$

$\therefore \quad \mathbf{I} = 2.18 \angle 19.41°$

Total power $= I^2 \times$ combined resistance

$= 2.18^2 \times 99.3 =$ **472 W**

$$I_1 = \frac{\mathbf{V_1}}{\mathbf{Z_1}} = \frac{230 \angle 0°}{122.6 \angle 11.82°} = \mathbf{1.87 \angle{-11.82°}} \quad \textbf{– lagging}$$

$$\mathbf{I_2} = \frac{\mathbf{V}}{\mathbf{Z_2}} = \frac{230 \angle 0°}{203 \angle -78.63°} = \mathbf{1.13 \angle 78.63°} \quad \textbf{– leading}$$

Q. 8. Two impedances given by $Z_1 = (10 + j\,5)$ and $Z_2 = (8 + j6)$ are joined in parallel across a voltage of $V = (200 + j\,0)$ volts. Calculate the circuit current, its phase and the branch currents. Draw the vector diagram. (*Eect. Circuits-I, Punjab Univ. 2003*)

Ans. We are given that

$V = (200 + j0) = 200\angle 0°$

$\mathbf{Z_1} = 10 + j\,5 = 11.2\angle 26.56°$

$\mathbf{Z_2} = 8 + j6 = 10\angle 36.87°$

$\mathbf{I_1} = \mathbf{V/Z_1} = 200\angle 0°/11.2\angle 26.56°$

$= 17.86 \angle{-26.56°}$

$\mathbf{I_2} = \mathbf{V/Z_2} = 200\angle 0°/10\angle 36.87°$

$= \mathbf{20\angle{-36.78°}}$

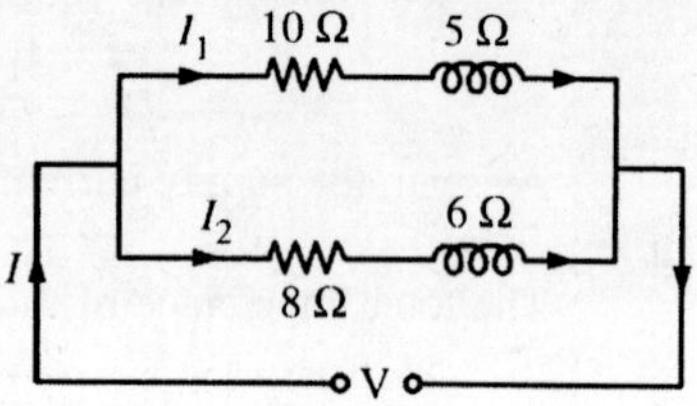

Fig. 9

As expected, both currents lag behind the applied voltage.

$$\mathbf{Z} = \frac{\mathbf{Z_1\,Z_2}}{\mathbf{Z_1 + Z_2}} = \frac{11.2 \angle 26.56° \times 10 \angle 36.87°}{18 + j11} = 5.3\angle 32°$$

$$\mathbf{I} = \frac{200 \angle 0°}{5.3 \angle 32°} = \mathbf{37.7 \angle{-32°}}$$

Power factor = cos (32°) = **0.848 lag**

Q. 9. If the voltmeter in Fig. 10 reads 60 V, find the reading of the ammeter.

(*Elect. Engg. Nagpur Univ. 2007*)

Ans. $I_2 = 60/4 = 15$ A.

Taking I_2 as the reference quantity,

$\mathbf{I_2} = 15\angle 0°$.

Obviously voltage across *R-C* branch is

$\mathbf{V} = 15\angle 0° \times (4 - j4) = 84.5\angle{-45°}$

Since the same voltage is applied across the first *R-L* branch,

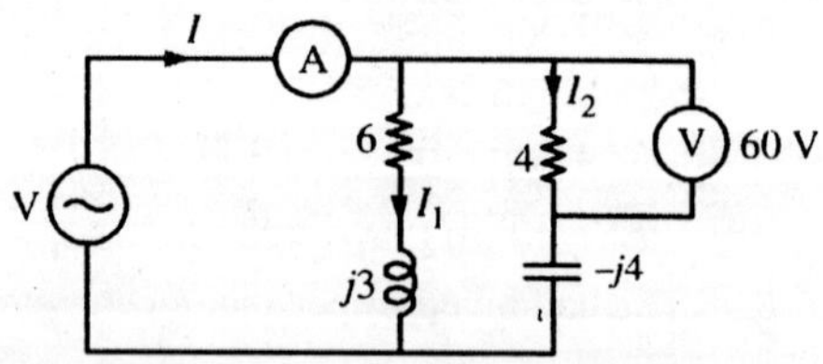

Fig. 10

$$\therefore \quad \mathbf{I_1} = \frac{84.5 \angle -45°}{6 + j3} = \frac{84.5 \angle -45°}{6.7 \angle 26.6°}$$

$= 12.6 \angle{-71.6°} = 4 - j12$

$\mathbf{I} = \mathbf{I_1} + \mathbf{I_2} = 15 + j0 + (4 - j12) = 19 - j12 = 22.47 \angle{-32.3°}$

Hence, the ammeter reads a value of **22.47 A.**

Q. 10. What is meant by admittance of a circuit? Find its value for a series *R-L* circuit.

Ans. Admittance of a circuit is defined as the reciprocal of its *impedance*. Its symbol is *Y*.

$$\mathbf{Y} = \frac{1}{\mathbf{Z}} \quad \text{or} \quad \mathbf{Y} = \frac{\text{rms amperes}}{\text{rms voltage}}$$

Its unit is siemens (S).

Consider the *R-L* circuit shown in Fig. 11. As seen from the impedance triangle of Fig. 11(*b*).

$$\mathbf{Z} = R + jX_L$$

As seen, X_L is positive and 90° ahead of *R*.

If *Y* is the admittance of the circuit, then

$$Y = \frac{1}{\mathbf{Z}} = \frac{1}{R + jX_L} = \frac{(R - jX_L)}{(R + jX_L)(R - jX_L)} = \frac{R - jX_L}{R^2 + X_L^2}$$

$$= \frac{R}{R^2 + X_L^2} - \frac{jX_L}{R^2 + X_L^2} = G - jB_L$$

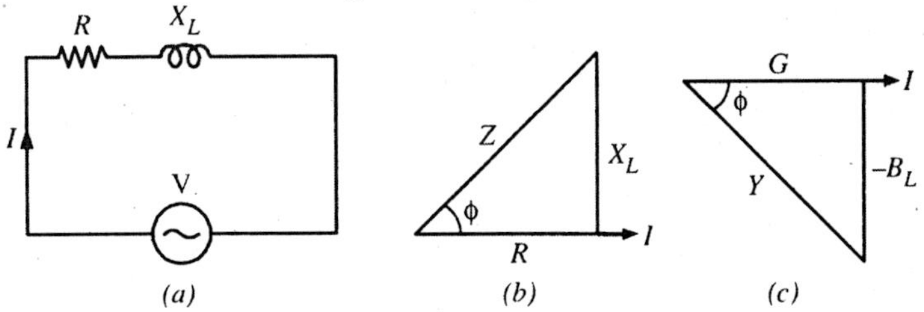

Fig. 11

The first part of the expression gives the *conductance* and the second part gives the inductive *susceptance* of the circuit.

As seen,

Conductance, $$G = \frac{R}{R^2 + X_L^2} = \frac{R}{Z^2}$$ – Fig. 11(*c*)

Susceptance, $$B_L = \frac{-X_L}{R^2 + X_L^2} = -\frac{X_L}{Z^2}$$ – Fig. 11(*c*)

Important notes :

(*i*) if the circuit has no reactance *i.e.* $X_L = 0$, then $G = 1/R$

(*ii*) if the circuit has no resistance *i.e.* $R = 0$ then $B_L = -1/X_L = -1/\omega L$

(*iii*) Also, $$\mathbf{B}_L = \frac{-j}{\omega L} = \frac{1}{\omega L}\angle{-90°}$$

(*iv*) It may be particularly noted that whereas X_L is *positive*, B_L is *negative*.

Q. 11. Derive an expression for the admittance of an R-C circuit shown in Fig. 12(*a*).

Ans. Let us consider the *R-C* circuit of Fig. 12(*a*) and its impedance and admittance triangles.

$$\mathbf{Z} = R - jX_C$$

$$\mathbf{Y} = \frac{1}{R - jX_C} = \frac{R + jX_C}{(R - jX_C)(R + jX_C)} = \frac{R + jX_C}{R^2 + X_C^2}$$

$$= \frac{R}{R^2 + X_C^2} + \frac{jX_C}{R^2 + X_C^2} = G + jB_C$$

As seen form above,

Conductance, $$G = \frac{R}{R^2 + X_C^2} = \frac{R}{Z^2}$$

Capacitive susceptance, $$B_C = \frac{X_C}{R^2 + X_C^2} = \frac{X_C}{Z^2}$$

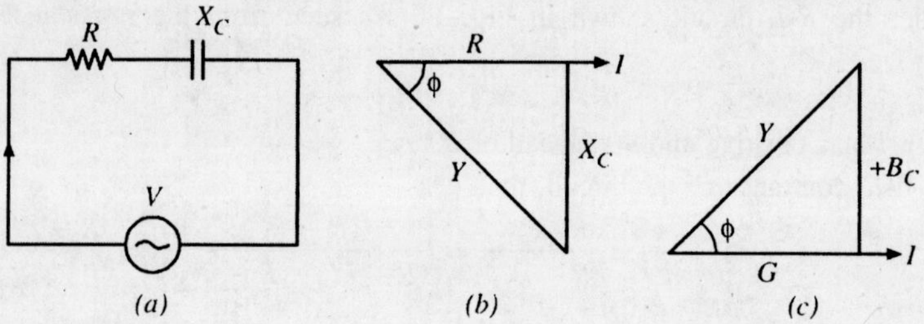

Fig. 12

Important notes :

(*i*) if the circuit has no reactance, *i.e.* $X_C = 0$, then

$$G = 1/R$$

(*ii*) if the circuit has no resistance *i.e.* $R = 0$, then $B_C = 1/X_C$

$$\therefore \quad B_C = \frac{1}{1/\omega C} = \omega C$$

(*iii*) Also,
$$\mathbf{B_C} = \frac{j}{X_C} = \frac{j}{1/\omega C} = j\omega C = \omega C \angle 90°$$

(*iv*) It may be particularly noted that whereas X_C is *negative,* B_C is *positive.*

Q. 12. Calculate the admittance of the two parallel circuits given in Fig. 13.

Ans. (*i*) Inductive reactance : The two-branch circuit with single elements of resistance and inductance is shown in Fig. 13(*a*). We will find Y_1 and Y_2 separately and then add them to find the admittance of the parallel circuit.

Since the upper branch contains only resistance but no reactance.

$$\therefore \quad \mathbf{Y_1} = \frac{1}{R}$$

and
$$\mathbf{Y_2} = -jB_L = -\frac{j}{X_L}$$

As mentioned above, admittances are additive.

$$\therefore \quad \mathbf{Y} = \mathbf{Y_1} + \mathbf{Y_2}$$

or
$$\mathbf{Y} = \frac{1}{R} - jB_L = \frac{1}{R} - \frac{j}{X_L}$$

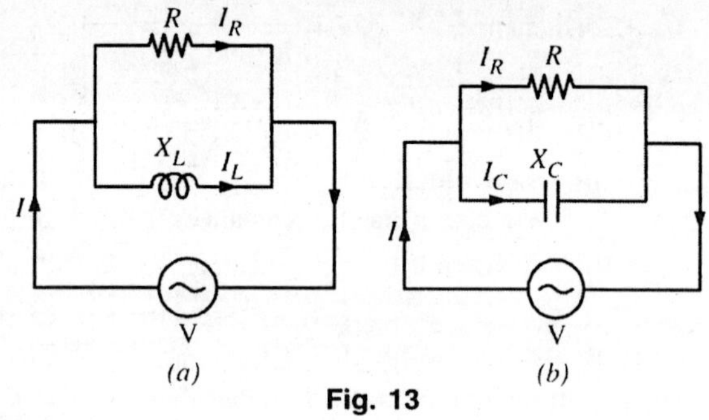

Fig. 13

(*ii*) **Capacitive reactance :** The circuit is shown in Fig. 13(*b*).

Here,
$$\mathbf{Y_1} = \frac{1}{R} \quad \text{and} \quad \mathbf{Y_2} = jB_C = j\omega C$$

$$\therefore \quad \mathbf{Y} = \frac{1}{R} + jB_C = \frac{1}{R} + j\omega C$$

Q. 13. Express in rectangular form the admittances of the circuits having the following impedances : (*i*) (6 + *j*8) and (*ii*) (8 – *j*6).

Ans. (*i*)
$$\mathbf{Y_1} = \frac{1}{6+j8} = \frac{6-j8}{6^2+8^2} = \frac{6-j8}{100} = 0.06 - j\,0.08 \text{ siemens}$$

$$\mathbf{Y_2} = \frac{1}{8-j6} = \frac{8+j6}{8^2+6^2} = \frac{8+j6}{100} = 0.08 + j\,0.06 \text{ siemens}$$

Q. 14. Find the admittance and current drawn by the circuit shown in Fig. 14.

Ans. The parallel circuit is shown in Fig. 14

$$\mathbf{Y}_1 = \frac{1}{12 + j16} = \frac{12 - j16}{12^2 + 16^2}$$

$$= 0.03 - j\,0.04 \text{ siemens}$$

$$\mathbf{Y}_2 = \frac{1}{10 - j\,20} = \frac{10 + j20}{10^2 + 20^2}$$

$$= 0.02 + j\,0.04 \text{ siemens}$$

$$\mathbf{Y} = \mathbf{Y}_1 + \mathbf{Y}_2 = \mathbf{0.05 \text{ siemens}}$$

$$\mathbf{I} = \mathbf{VY} = (200 + j0) \times 0.05 = \mathbf{10 \text{ A}}$$

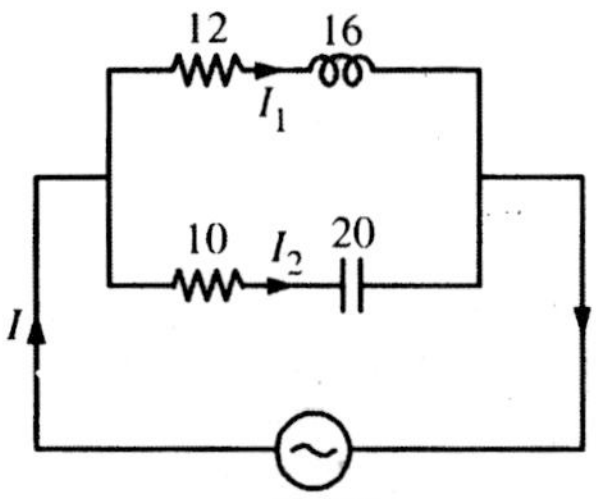

Fig. 14

Q. 15. A coil of resistance 15 Ω and inductance 0.05 H is connected in parallel with a non-inductive resistor of 20 Ω. Find

(*a*) the current in each branch

(*b*) the total current supplied and

(*c*) the phase angle of the combination when a voltage of 200 V at 50 Hz is applied. The circuit is shown in Fig. 15.

(*Basic Elect. Engg., West Bengal Tech. Univ,. 2005*)

Ans.

$$\mathbf{Z}_1 = 15 + j314 \times 0.05 = 15 + j\,15.7$$

$$\mathbf{Y}_1 = \frac{1}{\mathbf{Z}_1} = \frac{1}{15 + j\,15.7} = \frac{1}{21.7\,\angle 46.2°} = 0.046\angle{-46.2°}$$

$$I_1 = VY_1 = 200 \times 0.046\angle{-46.2°} = 9.2\angle{-46.2°} = 6.4 - j\,6.6$$

$$I_2 = \frac{200\angle 0°}{20\angle 0°} = 10 + j0$$

$$\therefore \quad I = I_1 + I_2 = 10 + (6.4 - j\,6.6) = 16.4 - j\,6.6 = 17.6\angle{-22°}$$

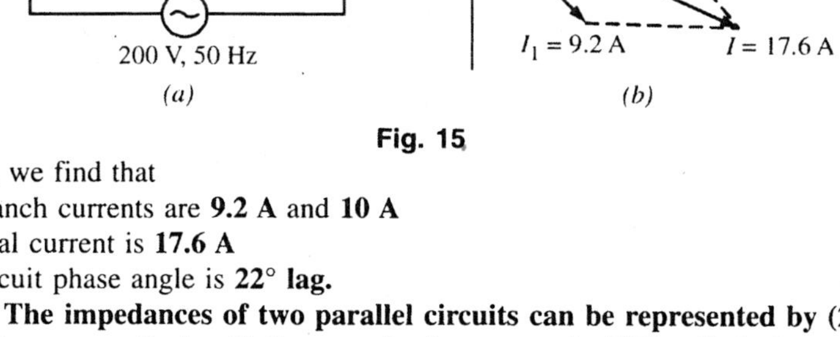

Fig. 15

Hence, we find that

(*a*) branch currents are **9.2 A** and **10 A**

(*b*) total current is **17.6 A**

(*c*) circuit phase angle is **22° lag.**

Q. 16. The impedances of two parallel circuits can be represented by (20 + *j*15) and (10 – *j*60) ohm respectively. If the supply frequency is 50Hz, find the resistance and inductance or capacitance of each circuit.

Also derive a complex expression for the admittance of the combined circuit and hence find the phase angle between the applied voltage and resultant current. State whether this current is leading or lagging with respect to the voltage

(*Applied Electricity, London Univ.*)

Ans. From the expression (20 + j15), it is obvious that the first branch consists of a 20 Ω resistor joined in series with a 15 Ω inductor (Fig. 16). Since the impedance of the second branch is (10 – j60), it shows that it consists of a 10 Ω resistor joined in series with a pure capacitor of 60Ω reactance. It is so because X_L is taken positive and X_C is taken as negative.

$$X_L = 2\pi \times 50 \times L$$
$$= 2\pi \times 50\ L$$
$$15 = 314\ L$$
or $$L = 0.0477\ \text{H} = \mathbf{47.7\ mH}$$

$$X_C = \frac{1}{\omega C} \quad \text{or} \quad C = \frac{10^6}{314 \times 60}$$

∴ $$C = \mathbf{53\ \mu F}$$

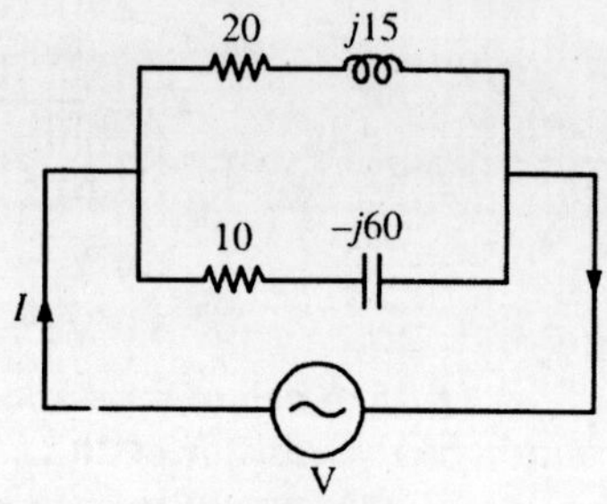

Fig. 16

$$\mathbf{Y}_1 = \frac{1}{20 + j15} = \frac{20 - j15}{20^2 + 15^2} = \frac{20 - j15}{625}$$
$$= 0.032 - j\ 0.024 \text{ siemens}$$

$$\mathbf{Y}_2 = \frac{1}{10 - j60} = \frac{10 + j60}{10^2 + 60^2} = \frac{10 + j60}{3700} = 0.0027 + j\ 0.0162$$

$$\mathbf{Y} = \mathbf{Y}_1 + \mathbf{Y}_2 = (0.032 - j\ 0.024) + (0.0027 + j\ 0.0162)$$
$$= 0.0347 + j\ 0.0078 \text{ siemens.}$$

$$\tan\phi = \frac{0.0078}{0.0347} = 0.2248 \qquad \therefore \phi = \tan^{-1}(0.2248) = \mathbf{12°40'\ lag}$$

Q. 17. Find the reading of the ammeter when the voltmeter across the 3 ohm resistor in the circuit of Fig. 17 reads 45 V. *(Elect. Engg. & Electronics, Bangalore Univ.)*

Ans. Obviously I_1 = 45/3 = 15 A.

If we take it as reference quantity, $\mathbf{I_1} = 3\ \angle 0°$

Now, $\mathbf{Z}_1 = 3 - j3 = 4.24\ \angle{-45°}$.

Hence, $\mathbf{V} = \mathbf{I}_1\ \mathbf{Z}_1 = 15\ \angle 0° \times 4.24\ \angle{-45°}$
$$= 63.6\ \angle{-45°}$$

$$\mathbf{I}_2 = \frac{\mathbf{V}}{\mathbf{Z_2}} = \frac{63.6\angle{-45°}}{5 + j2} = \frac{63.6\angle{-45°}}{5.4\angle 21.8°}$$
$$= 11.77\ \angle{-66.8°} = 4.64 - j10.8$$
$$\mathbf{I} = \mathbf{I}_1 + \mathbf{I}_2 = 19.64 - j10.8$$
$$= \mathbf{22.4\ \angle 28.8°}$$

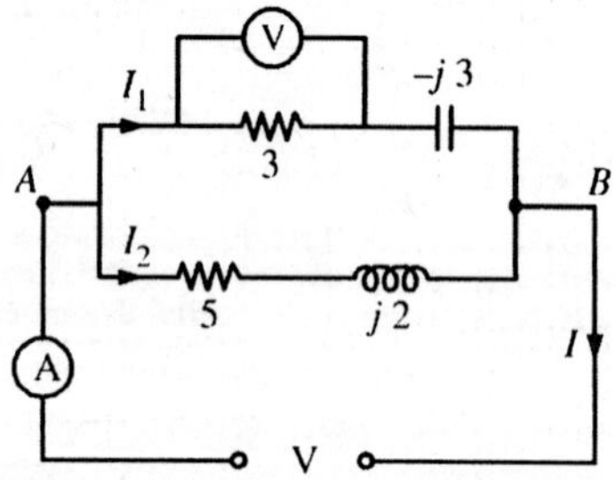

Fig. 17

8.2 SERIES PARALLEL CIRCUITS

Q. 18. In the series-parallel circuit of Fig. 18, calculate :
(*a*) current in the capacitor (*b*) current in the shunted resistor and (*c*) total current

Ans. The circuit is shown in Fig. 18.

$$\mathbf{Y_{AB}} = \frac{1}{100} + j\omega C = 0.01 + j\,314 \times 50 \times 10^{-6}$$
$$= 0.01 + j\,0.0157 = 0.0186\angle 57.5°$$

∴ $$\mathbf{Z_{AB}} = \frac{1}{\mathbf{Y_{AB}}} = \frac{1}{0.0186\ \angle 57.5°}$$
$$= 53.76\ \angle{-57.5°} = 28.89 - j\ 45.36$$

$$\mathbf{Z_{AC}} = 50 + (28.89 - j\,45.36)$$
$$= 78.89 - j\,45.36$$
$$= 91\angle -29.9^\circ\ \Omega$$

(*c*) $\mathbf{I} = 200\angle 0^\circ / 91\angle -29.9^\circ = 2.2\angle 29.9^\circ\text{ A}$

This I will divide at point A in the *direct ratio* of the admittances of the two branches of the parallel circuit.

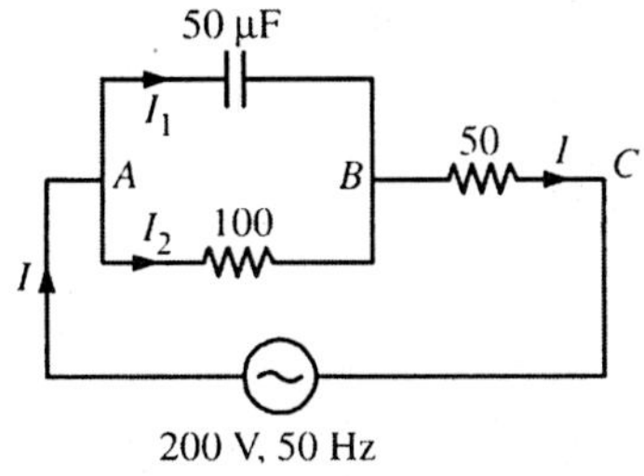

Fig. 18

(*a*) ∴ $$\mathbf{I_1} = \mathbf{I} \times \frac{\text{capacitive admittance}}{\mathbf{Y_{AB}}}$$

$$= 2.2\angle 29.9^\circ \times \frac{0.0157\angle 90^\circ}{0.0186\angle 57.5^\circ} = \mathbf{1.86\angle 62.4^\circ}$$

(*b*) $$\mathbf{I_2} = \mathbf{I} \times \frac{\text{resistive admittance}}{\mathbf{Y_{AB}}}$$

$$= 2.2\angle 29.9^\circ \times \frac{0.01\angle 0^\circ}{0.0186\angle 57.5^\circ} = \mathbf{1.18\angle -27.6^\circ}$$

Q. 19. In the series-parallel circuit shown in Fig. 19, calculate
(*a*) Z_{AC}
(*b*) I and branch currents I_1 and I_2
(*c*) Power taken by each impedance and the total power
(*d*) Overall power factor of the circuit.
(*e*) V_{AB} and V_{BC}.

Ans. (*a*) With reference to Fig. 19, we have

$$\mathbf{Z_{AB}} = \frac{\mathbf{Z_1 Z_2}}{\mathbf{Z_1 + Z_2}} = \frac{(8 + j4)(6 - j7.5)}{(8 + j4) + (6 - j\,7.5)}$$

$$= \frac{1218 - j\,231}{208}$$

$$= 5.86 - j\,1.1 = 5.96\angle -10.6^\circ$$

$$\mathbf{Z_{BC}} = 2.8 + j\,6.1 = 6.7\angle 65.3^\circ$$

$$\therefore\quad \mathbf{Z_{AC}} = \mathbf{Z_{AB}} + \mathbf{Z_{BC}}$$
$$= (5.86 - j1.1) + (2.8 + j6.1)$$
$$= 8.66 + j\,5 = \mathbf{10\angle 30^\circ}$$

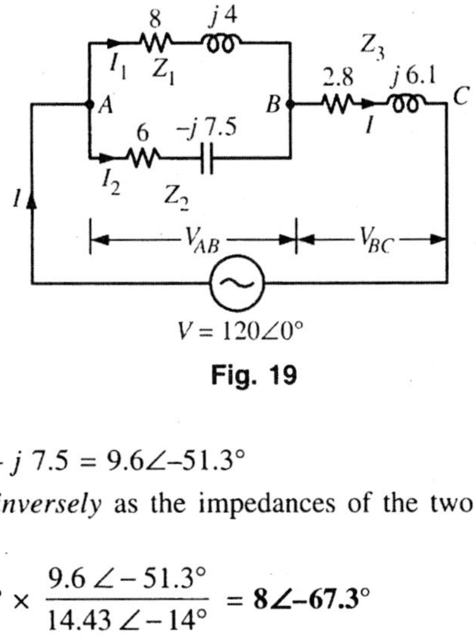

Fig. 19

(*b*) $$\mathbf{I} = \frac{\mathbf{V}}{\mathbf{Z_{AC}}} = \frac{120\angle 0^\circ}{10\angle 30^\circ} = 12\angle -30^\circ$$

$\mathbf{Z_1} = 8 + j4 = 8.94\angle 26.6^\circ$ and $\mathbf{Z_2} = 6 - j\,7.5 = 9.6\angle -51.3^\circ$

This I will be divided into two parts at point A *inversely* as the impedances of the two branches (as per current divider rule.)

$$\therefore\quad \mathbf{I_1} = \mathbf{I} \times \frac{\mathbf{Z_2}}{\mathbf{Z_1 + Z_2}} = 12\angle -30^\circ \times \frac{9.6\angle -51.3^\circ}{14.43\angle -14^\circ} = \mathbf{8\angle -67.3^\circ}$$

$$\mathbf{I_2} = \mathbf{I} \times \frac{\mathbf{Z_1}}{\mathbf{Z_1 + Z_2}} = 12\angle -30^\circ \times \frac{8.94\angle 26.6^\circ}{14.43\angle -14^\circ} = \mathbf{7.45\angle 10.6^\circ}$$

(*c*) $P_1 = 8^2 \times 8 = \mathbf{512\ W}$ $\qquad P_2 = 7.45^2 \times 6 = \mathbf{333\ W}$

$P_3 = 12^2 \times 2.8 = \mathbf{403\ W}$ $\qquad P = 333 + 512 + 403 = \mathbf{1248\ W}$

(*d*) Overall p.f. = W/VA = 1248/120 × 12 = **0.866 lag**

(*e*) $\mathbf{V_{AB}} = \mathbf{IZ_{AB}} = 12\angle -30^\circ \times 5.96\angle -10.6^\circ = \mathbf{71.5\angle -40.6^\circ}$

$\mathbf{V_{BC}} = \mathbf{IZ_{BC}} = 12\angle -30^\circ \times 6.7\angle 65.3^\circ = \mathbf{80.4\angle 35.3^\circ}$

Q. 20. In the circuit shown in Fig. 20, find I, V_1 and V_2. The values of impedance are in ohm. Represent them in phasor diagram. Take supply $V = 99\sqrt{2}$ volts (rms)

(*Basic Elect. Engg. CSV Tech. Univ. Bhilai, 2007*)

Ans. For solving this question, we need to find $\mathbf{Z_{AB}}$ and $\mathbf{Z_{BC}}$

Now, $\mathbf{Z_{AB}} = \mathbf{Z_1} = 6 - j6 = 8.48\angle{-45°}$

Also, $\mathbf{Z_{BC}} = \mathbf{Z_2} \parallel \mathbf{Z_3}$

Now, $\mathbf{Z_2} = 3 + j3 = 4.24\angle 45°$. $\mathbf{Z_3} = -j3 = 3\angle{-90°}$

$$\mathbf{Z_{BC}} = \frac{\mathbf{Z_2}\,\mathbf{Z_3}}{\mathbf{Z_2} + \mathbf{Z_3}} = \frac{(3+j3)\times(-j3)}{(3+j3)-j3} = \frac{9-j9}{3} = 3 - j3 = 4.24\angle{-45°}$$

$$\mathbf{Z_{AC}} = \mathbf{Z_{AB}} + \mathbf{Z_{BC}} = (6 - j6) + (3 - j3) = 9 - j9 = 12.72\angle{-45°}$$

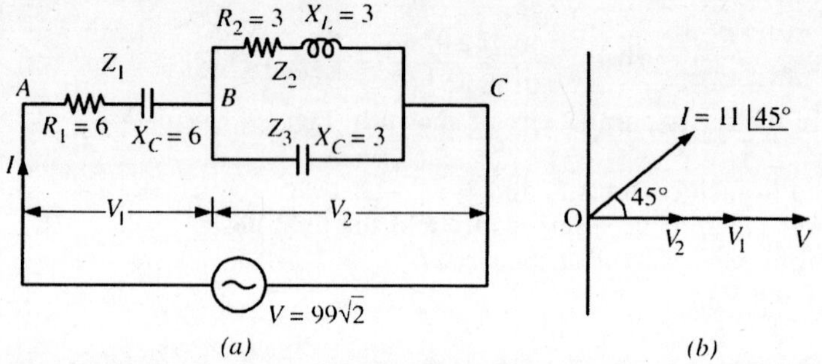

Fig. 20

$$\mathbf{V} = 99 \times \sqrt{2} = 140\angle 0° \qquad \text{— reference vector}$$

$$\mathbf{I} = \frac{\mathbf{V}}{\mathbf{Z_{AC}}} = \frac{140\angle 0°}{12.72\angle -45°} = \mathbf{11\angle 45°} \qquad \text{— Fig. 20(b)}$$

$$\mathbf{V_1} = \mathbf{IZ_1} = 11\angle 45° \times 8.48\angle{-45°} = \mathbf{93.28\angle 0°}$$

$$\mathbf{V_2} = \mathbf{IZ_{BC}} = 11\angle 45° \times 4.24\angle{-45°} = \mathbf{46.64\angle 0°}$$

Surprisingly, the three voltages are in-phase with each other as shown in Fig. 20(*b*).

Q. 21. When 220 volt ac supply is applied across terminals *AB* of the circuit shown in Fig. 21, the total power input is 3.25 kW and the current is 20A. Find the current through Z_3 and Z_2. Take V_{AC} as reference vector. (*Basic Elect. Engg. VTU Karnataka 2009*)

Ans. Let the *combined* resistance of the parallel circuit be *R* and the *combined* reactance be *X* as shown in Fig. 21(*b*). Then,

$$\mathbf{Z_{AC}} = (R + jX)$$

Also, $\mathbf{Z_{AB}} = \mathbf{Z_{AC}} + \mathbf{Z_{CB}} = (R + jX) + (5 + j10) = (R + 5) + j(X + 10)$

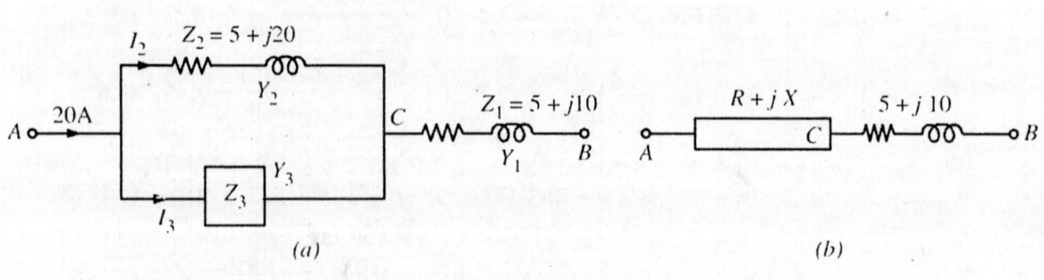

Fig. 21

It shows that the *combined resistance* of the entire series-parallel circuit is $(R + 5)$ and *combined reactance* is $(X + 10)$. Since we are given the power consumed by the entire circuit.

$\therefore \quad 20^2 \times (R + 5) = 3250, \quad$ hence $R = 3.125\ \Omega$

Now, we will find the value of X as under,

Since V and I are given,

$$Z_{AB} = 220/20 = 11\ \Omega$$

Also, $$Z_{AB} = \sqrt{(R+5)^2 + (X+10)^2}$$

or $$11^2 = (R + 5)^2 + (X + 10)^2 \qquad \therefore \quad X = -2.58\ \Omega$$

$$\therefore \quad \mathbf{Z_{AC}} = R + jX = 3.125 - j\,2.58 = 4.05\angle{-39.5°}$$

$$V_{AC} = 20.Z_{AC} = 20 \times 4.05 = 81\text{ V}$$

Taking this voltage as the reference vector, we have

$$\mathbf{V_{AC}} = 81\angle 0°$$

$$\mathbf{Y_{AC}} = \frac{1}{\mathbf{Z_{AC}}} = \frac{1}{3.125 - j2.58} = \frac{3.125 + j2.58}{16.4} = 0.190 + j0.157$$

$$\mathbf{Y_2} = \frac{1}{\mathbf{Z_2}} = \frac{1}{5 + j20} = \frac{5 - j20}{425} = 0.0485\angle{-76°} = 0.0117 - j0.047$$

$$\mathbf{Y_3} = \mathbf{Y_{AC}} - \mathbf{Y_Z} = (0.190 + j0.157) - (0.0117 - j0.047)$$

$$= 0.1783 + j0.204 = 0.271\angle 48.7°$$

$$\therefore \quad \mathbf{I_3} = \mathbf{V_{AC}}\,\mathbf{Y_3} = 81\angle 0° \times 0.271\angle 48.7° = \mathbf{22\angle 48.7°\ A}$$

Also, $$\mathbf{I_2} = \mathbf{V_{AC}}.\mathbf{Y_2} = 81\angle 0° \times 0.0485\angle{-76°} = \mathbf{3.9\angle{-76°}\ A}$$

Q. 22. In the circuit shown in Fig. 22, determine the voltage at a frequency of 50 Hz. to be applied across AB in order that the current in the circuit is 10 A. Draw the phasor diagram. *(Elect. Engg. & Electronies Bangalore Univ.)*

Ans. $X_{L1} = 2\pi \times 50 \times 0.05 = 15.71\ \Omega$; $X_{L2} = 2\pi \times 50 \times 0.02 = 6.28\ \Omega$,

$$X_C = 1/2\pi \times 50 \times 400 \times 10^{-6} = 7.95\ \Omega$$

$$\mathbf{Z_1} = R_1 + jX_{L1} = 10 + j\,15.71 = 18.6\angle 57°33'$$

$$\mathbf{Z_2} = R_2 + jX_{L2} = 5 + j6.28 = 8\angle 51°30'$$

$$\mathbf{Z_3} = R_3 - jX_C = 10 - j\,7.95 = 12.77\angle{-38°30'}$$

$$\mathbf{Z_{BC}} = \mathbf{Z_2} \parallel \mathbf{Z_3} = (5 + j6.28) \parallel (10 - j7.95)$$

$$= 6.42 + j2.25 = 6.8\angle 19°18'$$

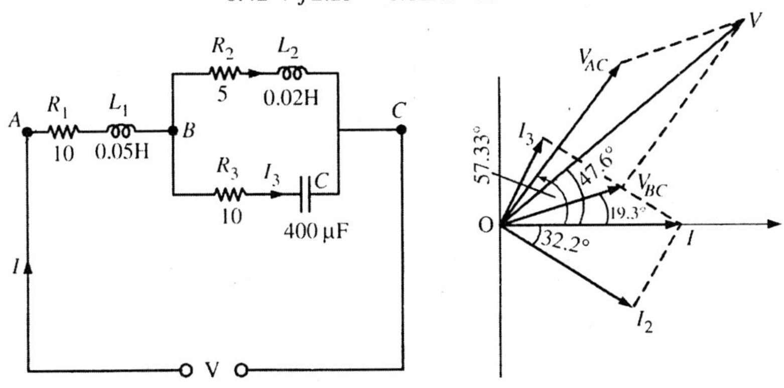

(a) (b)

Fig. 22

$$\mathbf{Z} = \mathbf{Z}_1 + \mathbf{Z}_{BC} = (10 + j15.71) + (6.42 + j2.25)$$
$$= 16.42 + j17.96 = 24.36\angle 47°36'$$

Let $\mathbf{I} = 10\angle 0°$; $\therefore$ $\mathbf{V} = \mathbf{IZ} = 10\angle 0° \times 24.36\angle 47°36 = 243.6\angle 47°36'$

$$\mathbf{V_{BC}} = \mathbf{IZ_{BC}} = 10\angle 0° \times 6.8\angle 19°18' = 68\angle 19°18'$$

$$\mathbf{I_2} = \frac{\mathbf{V_{BC}}}{\mathbf{Z_2}} = \frac{68\angle 19°18'}{8\angle 51°31'} = 8.5\angle -32°\ 12'$$

$$\mathbf{I_3} = \frac{\mathbf{V_{BC}}}{\mathbf{Z_3}} = \frac{68\angle 19°18'}{12.77\angle -38°30'} = 5.32\angle 57°48'$$

$$\mathbf{V_{AC}} = \mathbf{IZ_1} = 10\angle 0° \times 18.6\angle 57°33' = 186\angle 57°33'$$

The phasor diagram is shown in Fig. 22.

8.3 PARALLEL RESONANCE

Q. 23. Derive the condition for the resonance of a two-branch parallel L-C circuit. Does it continuously draw current from the supply source? If not, why?

Ans. The circuit is shown in Fig. 23(*a*). This *L-C* circuit is said to be in resonance when $X_L = X_C$. Surprisingly, it is the *same condition* as for series resonance. Moreover, the value of resonance frequency f_o is also the same. Since branch reactances are equal, the two currents are also equal in magnitude. However, I_L lags behind V by 90° whereas I_C leads V by 90° as shown in Fig. 23(*b*). Since the two currents are equal but opposite, they cancel each other out. It means that the current taken from the line after switching on the supply is zero.

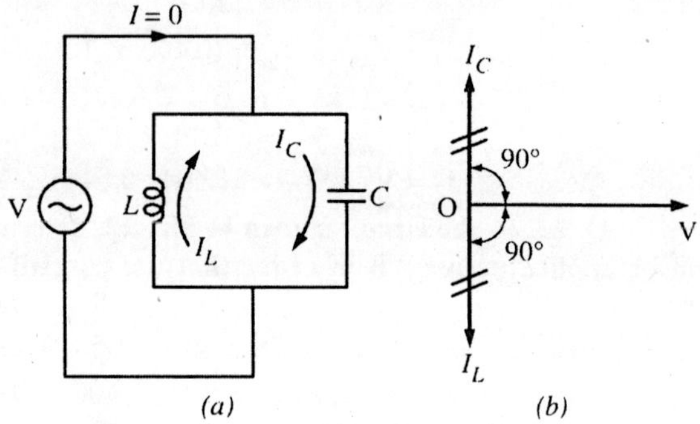

Fig. 23

The current *once* taken by the circuit *at the start* keeps circulating back and forth between *L* and *C* but no fresh current is taken from the source. In fact, the inductive and capacitive branches continuously exchange the energy of their magnetic and electric fields to circulate the current between themselves without taking any current *from the source*. When magnetic field of the coil collapses, it releases energy to charge the capacitor which produces an electric field. Half a cycle later, the electric field collapses and releases energy for producing magnetic field in the coil. This process keeps going on and on.

Since no source current is drawn by the circuit, it is said to offer *infinite* impedance to the supply source whose frequency equals f_o. Hence, this circuit can be used to block or reject any unwanted frequency in the input signal.

The circuit shown in Fig. 23(*a*) is also called a 'tank' circuit because it stores energy in its inductor and capacitor. When frequency is varied, the reading of the ac ammeter records zero at resonant frequency.

Q. 24. Under which condition is the circuit of Fig. 24(*a*) said to be in resonance with the supply frequency?

Ans. Consider the case of a *practical* coil connected in parallel with a pure capacitor and supplied by a sinusoidal source of variable frequency as shown in Fig. 24(*a*). Such a circuit is said to be in resonance when the reactive (*i.e.* wattless) component of the supply current *I* is zero. The frequency at which this happens is called the *resonance* frequency of the parallel circuit As seen from the vector diagram of Fig. 24(*b*),

$I_L = V/Z_L$ — lagging V by ϕ_L

$I_C = V/X_C$ — leading V by 90°

Reactive component of $I_L = I_L \sin \phi_L$ but I_C is totally reactive.

Net reactive component of the circuit current I is

$$= I_C - I_L \sin \phi_L$$

For resonance, reactive component of the current is zero.

$$\therefore \quad I_C - I_L \sin \phi_L = 0 \quad \text{or} \quad I_C = I_L \sin \phi_L \qquad ...(i)$$

Now, $$I_L = \frac{V}{Z_L}, \ \sin \phi_L = \frac{X_L}{Z_L}$$

and $$I_C = \frac{V}{X_C}$$

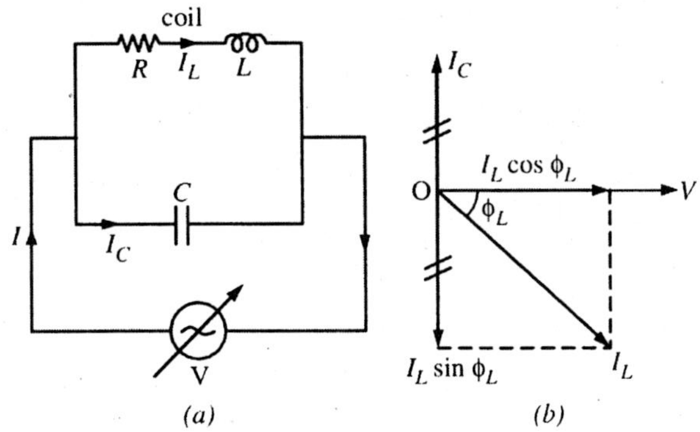

Fig. 24

Substituting these values in (*i*) above, we get

$$\frac{V}{X_C} = \frac{V}{Z_L} \times \frac{X_L}{Z_L}$$

or $$X_C \cdot X_L = Z_L^2$$

Now, $$X_L = \omega L$$

and $$X_C = 1/\omega C$$

$$\therefore \quad X_L . X_C = Z_L^2$$

or $$\frac{\omega L}{\omega C} = Z_L^2 \quad \text{or} \quad \frac{L}{C} = Z_L^2$$

$$\therefore \quad \frac{L}{C} = R^2 + X_L^2 = R^2 + (2\pi f_o L)^2$$

$$\therefore \quad (2\pi f_o L)^2 = \frac{L}{C} - R^2 \quad \text{or} \quad 2\pi f_o = \sqrt{\frac{1}{LC} - \frac{R^2}{L^2}}$$

$$\therefore \quad f_o = \frac{1}{2\pi}\sqrt{\frac{1}{LC} - \frac{R^2}{L^2}}$$

It may be noted that at resonance, current drawn by the resonant circuit is *minimum* I_{min} and is equal to the *in-phase or wattful* component of the current drawn by the circuit.

$$\therefore \quad I_{\min} = I_L \cos \phi_L = \frac{V}{Z_L} \times \frac{R}{Z_L} = \frac{VR}{Z_L^2}$$

As seen from above, $Z_L^2 = L/C$

$$\therefore \quad I_{\min} = \frac{VR}{Z_L^2} = \frac{VR}{L/C} = \frac{V}{L/CR}$$

The denominator L/CR is called the *dynamic* impedance of the resonant parallel circuit. It is *resistive* in nature because current is in phase with the voltage. This dynamic impedance L/CR represents the *maximum* impedance of the parallel circuit when in resonance. As shown in Fig. 25(*a*), the resonant parallel circuit is equivalent to a single-element circuit of Fig. 25(*b*). Since the resonant

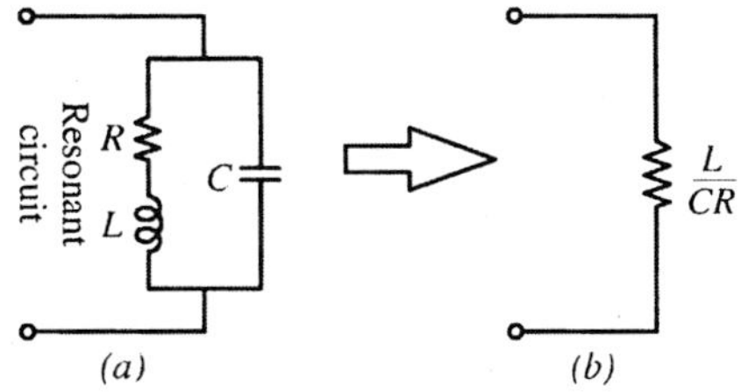

Fig. 25

circuit draws *minimum* current at a particular frequency, it is also known as *a rejector* circuit because it practically rejects that input signal frequency to which it resonates. Parallel resonance is also called '*current*' resonance because the current circulating *between* the two branches is many times greater than the minimum current taken by them from the supply source.

Q. 25. Compare the properties of series and parallel circuits at resonance in the form of a table.

Ans. Different properties shown by a series and a parallel resonant circuit are tabulated below:

	Item	Series *R-L-C* Circuit	Parallel *R-L-C* Circuit
1.	resonant frequency, f_o	$\frac{1}{2\pi\sqrt{LC}}$	$\frac{1}{2\pi}\sqrt{\left(\frac{1}{LC}-\frac{R^2}{L^2}\right)}$
2.	effective impedance	R	$\frac{L}{CR}$
3.	current at resonance	maximum = $\frac{V}{R}$	minimum = $\frac{V}{L/CR}$
4.	impedance at resonance	minimum	maximum
5.	p.f. at resonance	unity	unity
6.	it magnifies	voltage	current
7.	is also called	acceptor circuit	rejector circuit

Q. 26. An inductive circuit of resistance 2 ohm and inductance 0.01 H is connected to a 250-V, 50-Hz supply. What capacitance placed in parallel will produce resonance? Find the total current taken from the supply and the current in the branch circuits.

(*Elect. Engg., Kerala Univ.*)

Ans. As seen from Q. 24 at resonance $C = L/Z_L^2$

Now, $R = 2\ \Omega,\ X_L = 314 \times 0.01 = 3.14\ \Omega\ ;\ Z_L = \sqrt{2^2 + 3.14^2} = 3.74\ \Omega$

$C = 0.01/3.74^2 = 714 \times 10^{-6}$ F = **714 μF** ; $I_{RL} = 250/3.74 =$ **66.83 A**

$\tan \phi_L = 3.14/2 = 1.57\ ;\ \phi_L = \tan^{-1}(1.57) = 57.5°$

Hence, current in *R-L* branch lags the applied voltage by 57.5°

$$\therefore \quad I_C = \frac{V}{X_C} = \frac{V}{1/\omega C} = \omega VC = 250 \times 314 \times 714 \times 10^{-6} = \mathbf{56.1\ A}$$

This current leads the applied voltage by 90°.

Total current taken from the supply under resonant condition is

$$I = I_{RL} \cos \phi_L = 66.83 \cos 57.5° = 66.83 \times 0.5373 = \mathbf{35.9\ A} \left(\text{or } I = \frac{V}{L/CR}\right)$$

Q. 27. A coil of resistance 12 Ω and inductance 0.12 H is connected in parallel with a 60 μF capacitor to a 100 volt variable frequency supply. Calculate :

(*a*) the frequency at which the circuit will behave as a non-reactive resistor.

(*b*) the value of dynamic impedance at resonance.

Ans. (*a*) The circuit will behave as a non-reactive *i.e.* ohmic resistor at the resonant frequency given by

$$f_o = \frac{1}{2\pi}\sqrt{\frac{1}{LC}-\frac{R^2}{L^2}} = \frac{1}{2\pi}\sqrt{\frac{1}{0.12\times 60\times 10^{-6}}-\frac{12^2}{0.12^2}} = \mathbf{57.2\ Hz}$$

(*b*) dynamic impedance = $\frac{L}{CR} = \frac{0.12}{60\times10^{-6}\times12}$ = **167 Ω.**

Q. 28. A resonant circuit consists of a 4μF capacitor in parallel with an inductor of 0.25H having a resistance of 50Ω. Calculate the frequency of resonance and comment on the effect that the inductor resistance has upon it.

Ans. $$f_o = \frac{1}{2\pi}\sqrt{\frac{1}{LC}-\frac{R^2}{L^2}} = \frac{1}{2\pi}\sqrt{\frac{1}{0.25\times4\times10^{-6}}-\frac{50^2}{0.25^2}} = \mathbf{156\ Hz}$$

Higher value of resistance *reduces* the resonant frequency. It is obvious from the formula given above. When R is large, the value of the expression under the square root is decreased. It will obviously reduce f_o.

Q. 29. A coil having a resistance of 25 Ω and an inductance of 0.3 H is connected in parallel with a variable capacitor across a 200 V, 50 Hz supply. Calculate

(*a*) capacitance for which current drawn is minimum.

(*b*) the value of current.

Draw the vector diagram showing the supply voltage and the various currents.

Ans. The current drawn by the circuit would be minimum at resonant frequency.

(*a*) at resonance,

$Z_L^2 = L/C,\ C = L/Z_L^2$

Coil reactance,

$X_L = 314 \times 0.3$

$= 94.2\ \Omega$

$Z_L = \sqrt{25^2 + 94.2^2}$

$= 97.5\angle 75°$

∴ $C = L/Z_L^2 = 0.3/97.5^2$

$= 31.4 \times 10^{-6}$ F

$= \mathbf{31.4\ \mu F}$

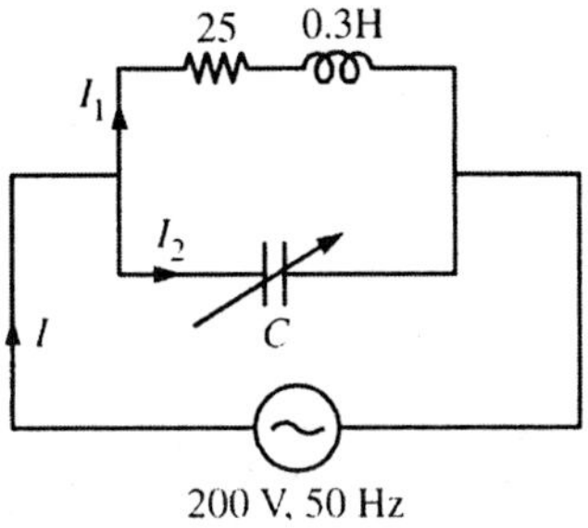

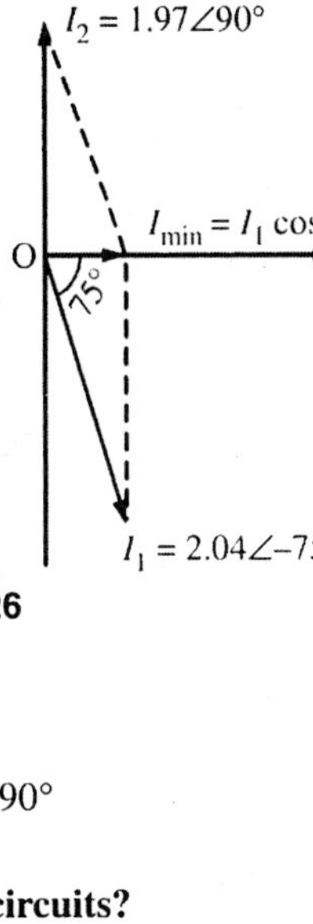

Fig. 26

(*b*) $I_{min} = \frac{VR}{Z_L^2} = \frac{200\times25}{97.5^2}$ = **0.525 A**

For drawing the vector diagram, we need the following:

$I_1 = V/Z_L = 200\angle 0°/97.5\angle 75° = 2.04\angle -75°$

$I_2 = jV\omega C = j \times 200 \times 314 \times 31.4 \times 10^{-6} = j1.97 = 1.97\angle 90°$

The vector diagram is shown in Fig. 26(*b*).

Q. 30. Briefly mention the practical applications of resonant circuits?

Ans. Phenomenon of resonance (whether series or parallel) finds many electronic applications. These resonant circuits make it possible for a radio receiver to receive or tune in a particular frequency. When we change stations on a radio set, we rotate the turning knob which operates a variable capacitor. Different resonant frequencies can be tuned in by varying the capacitance. In this way, it is possible to select one station's signal from among all the other signals present in the air at different frequencies. Hence, such circuits are used to *select* one particular frequency and reject others (Fig. 27) or *reject* a particular frequency and allow others to pass through (Fig. 28). These circuits are essential for operating radio transmitters, amateur radio and other communication equipments such as radar and telemetry.

Q. 31. Briefly explain why a resonant *R-L-C* circuit is called an acceptor circuit? In the same way why a parallel *R-L-C* circuit is called a rejector circuit?

Ans. Acceptor Circuit : If a radio or T.V. signal containing waves of many frequencies is applied to a resonant series *R-L-C* circuit (Fig. 27), the circuit allows only that particular frequency to pass through which is equal to its resonant frequency f_o. It does not allow to get through other frequencies of the input signal. That is why it is called an acceptor circuit.

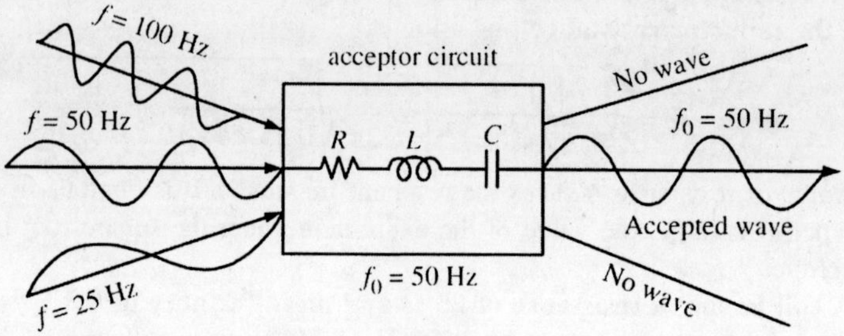

Fig. 27

As seen from Fig. 27, the input signal contains three waves of frequencies 25 Hz, 50 Hz and 100 Hz. As shown, only signal of 50 Hz frequency is accepted whereas all others are blocked. In other words, the series resonant circuit 'tunes' in 50Hz signal while rejecting others of frequencies lower or higher than its resonant frequency f_o.

Rejector Circuit : The parallel rejector circuit shown in Fig. 28 rejects or blocks that particular signal which is equal to its resonant frequency f_o but allows other signals to pass through. It is called rejector circuit because it rejects or blocks only that frequency which equals f_o (exactly opposite of what a resonant series circuit does.)

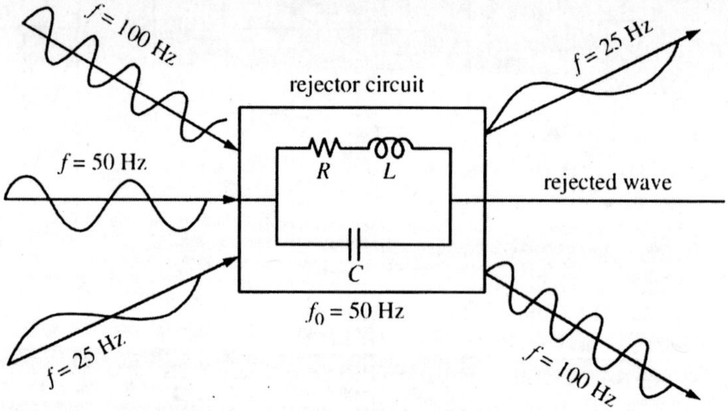

Fig. 28

As seen from Fig. 28, out of the three input signals only the signal of 50 Hz frequency has been rejected while others have been allowed to pass through. Such circuits are also necessary for rejecting any unwanted input signal.

Q. 32. A given parallel circuit has a Q of 200 and a resonant frequency of 600 kHz. Find the bandwidth and half-power frequencies, f_1 and f_2.

Ans. $BW = f_o/Q = 600/200 = \mathbf{3\ kHz}$

$$\therefore \quad f_1 = fo - \frac{BW}{2} = 600 - \frac{3}{2} = \mathbf{598.5\ kHz}$$

$$f_2 = f_o + \frac{BW}{2} = 600 + \frac{3}{2} = \mathbf{601.5\ kHz}$$

EXERCISE

Parallel Circuits

1. A coil having a resistance of 150 Ω and inductance 0.5 H is connected in parallel with a circuit containing a 30 μF capacitor and 100 Ω resistor across a 230 V, 50 Hz supply. Calculate :

   (*a*) current taken from the supply (*b*) power factor of the circuit

   Draw the vector diagram representing voltage and currents. **[1.84∠11°40′, 0.979 lead]**

2. A coil of resistance 5 Ω and inductance 0.0347 H is connected in parallel with a 20 Ω resistor across a 240 V, 50-Hz sinusoidal supply. Calculate:

   (*a*) supply current

   (*b*) overall power factor

   (*c*) total power consumed **[27.3 A, 0.75 lag, 4880 W]**

3. A parallel circuit having two branches is connected across a 240 V, 50 Hz sinusoidal supply. Branch-1 is a coil of resistance 4 Ω and inductance 0.0292 H. Branch-2 is a 356 μF capacitor connected in series with an 8 Ω resistor. Calculate:

   (*a*) supply current (*b*) overall power factor (*c*) total power consumed.

   Sketch the vector diagram. **[24 A, 0.96 lag, 5500 W]**

4. Two impedance $Z_1 = (8 + j6)$ and $Z_2 = (3 - j4)$ are in parallel. If the total current of the combination is 25 A, find the current taken and power consumed by each impedance.

   (*F.Y. Engg. Pune Univ.*)

   [I_1 = 11.175∠–63.4°, 990 W; 22.35∠26.57°, 1500 W]

5. Two impedance given by $Z_1 = (10 + j\,5)$ and $Z_2 = (8 + j6)$ are joined in parallel across a voltage of $V = (200 + j\,0)$ volts. Calculate the circuit current, its phase and the branch currents. Draw the vector diagram. (*Eect. Circuit-I, Punjab Univ. 2003*)

   [37.7∠–32°, 0.848 lag, I_1 = 17.86∠–26.56°, 20∠–36.78°]

6. Two impedance $(14 + j5)$ Ω and $(18 + j10)$ Ω are connected in parallel across a 220 V, 50 Hz supply. Determine

   (*a*) impedance of the entire circuit

   (*b*) total current, power and p.f

   (*b*) draw the phasor diagram for the above circuit. (*Elect. Engg. Nagpur Univ. 2004*)

   [8.66∠23.6°, 25.4∠–23.6°, 5122 W, 0.917 lag]

7. A coil having a resistance of 20 Ω and an inductance of 0.0382 H is connected in parallel with a circuit consisting of a 150 μF capacitor in series with a 10 Ω resistor. It is connected across a 240 V, 50 Hz supply. Calculate :

   (*i*) current in each branch (*ii*) total supply current (*iii*) p.f. (*vi*) power.

   [10.28 A, 10.2 A; 13.75 A; 0.96 lead; 3160 W]

Series-Parallel Circuits

8. A 100 Ω resistor shunted by a 0.4 H inductor, is in series with a capacitor *C*. A voltage of 250 V at 50 Hz is applied to the circuit. Find:

   (*a*) value of *C* to give unity power factor

   (*b*) the total current

   (*c*) current in the inductive branch.

   Draw the vector diagram **[64.9 μF, 4.07 A, 2.54 A]**

9. For the circuit shown in Fig. 29, calculate I_1, I_2 and I_3. The values marked on the inductance and capacitance give their reactances. *(Elect. Science-I Allahabad Univ.)*
(1.72∠–74.3°, 1.5∠10.3°, 1.3∠–41.8°)

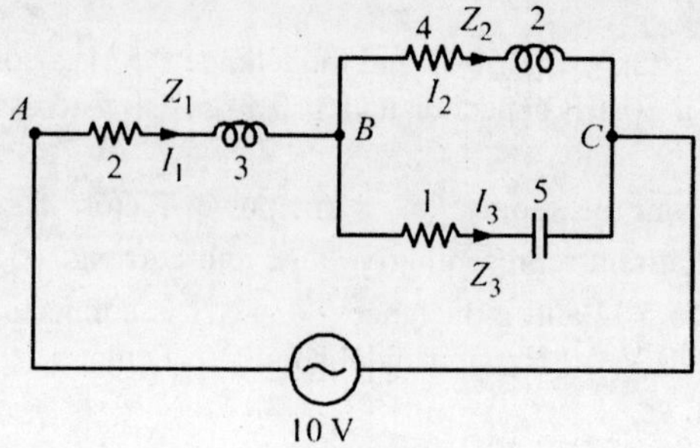

Fig. 29

10. An impedance of $(2 + j6)\ \Omega$ is connected in series with two impedances of $(10 + j4)\ \Omega$ and $(12 - j8)\ \Omega$ which are in parallel. Calculate the magnitude and power factor of the main current when the combined circuit is supplied at 200 volt. **[18.75 A, 0.839 lag]**

Parallel Resonance

11. A capacitor is connected in parallel with a coil having L = 5.52 mH and R = 10 Ω to a 100 V, 50 Hz supply. Calculate the value of the capacitance for which the current taken from the supply is in phase with voltage. *(Elect. Machines, A.M.I.E. Sec. B)* **[54.6 μF]**
12. A coil of resistance 10 Ω and inductance 0.159 H is connected across a 240 V, 50 Hz sinusoidal supply. Calculate the value of capacitance to be connected in parallel to cause resonance. **[61.2μF]**
13. A coil having a resistance of 25 Ω and inductance of 20 mH is connected in parallel with a variable capacitor.
(*a*) for what value of C will the circuit be resonant if 200 V, 400 Hz source is applied?
(*b*) what will be the line current under this condition? **[6.35 μF, 3.6 A]**
14. Calculate the resonant frequency of a two-branch parallel circuit for which the following particulars are given:
R_L = 30 Ω, L = 0.106 H; R_C = 30 Ω, C = 48 μF **[70.5 Hz]**

MULTIPLE CHOICE QUESTIONS

1. Susceptance and conductance of a circuit add to form its
(*a*) impedance (*b*) inductance (*c*) reactance (*d*) admittance
2. Capacitive susceptance of a circuit is
(*a*) positive and real (*b*) negative and real
(*c*) positive and imaginary (*d*) negative and imaginary
3. An RL circuit has R = 3 Ω and X_L = 4 Ω. Its conductance is .... siemens.
(*a*) 0.12 (*b*) 0.33 (*c*) 0.6 (*d*) 0.43
4. The capacitive susceptance of an R-C circuit is given by
(*a*) $-jX_C/Z^2$ (*b*) $-jX_C/Z$ (*c*) jX_C/Z^2 (*d*) jX_C
5. A circuit has an impedance of $(6 + j8)\ \Omega$. Its admittance is .... siemens.
(*a*) $0.08 + j0.06$ (*b*) $0.06 - j0.08$ (*c*) $0.06 + j0.08$ (*d*) $0.08 - j0.06$

6. A parallel circuit has an admittance of 0.06 siemens. When connected across a sinusoidal voltage of $(200 + j0)$, it will draw a line current of
   (*a*) 3 A (*b*) 12 A (*c*) 1.2 A (*d*) 120 A
7. The branch current in a parallel circuit is
   (*a*) directly proportional to its impedance
   (*b*) inversely proportional to its admittance
   (*c*) directly proportional to its admittance
   (*d*) depends on the ratio of the branch admittances.
8. In an *R-L-C* parallel circuit, the impedance at resonance is
   (*a*) minimum (*b*) maximum (*c*) infinity (*d*) zero
   (*Basic Elect. Engg. West Bengal Tech. Univ. 2003*)
9. When an *R-L* and *C* parallel circuit is in resonance, it draws
   (*a*) minimum current
   (*b*) maximum current
   (*c*) quadrature component of the line current
   (*d*) no current at all
10. As the resistance of a parallel circuit is decreased
    (*a*) its resonant frequency is increased (*b*) its resonant frequency is decreased
    (*c*) its dynamic impedance is decreased (*d*) it becomes an acceptor circuit.
11. A parallel resonant circuit is often used for
    (*a*) voltage magnification (*b*) decreasing circuit impedance
    (*c*) accepting a particular frequency (*d*) rejecting a particular frequency.
12. In an *R-L* and *C* parallel circuit, the capacitance required for producing resonance is given by
    (*a*) Z_L/L (*b*) Z_L^2/L (*c*) L/Z_L^2 (*d*) $L \times Z_L^2$

ANSWERS

1. *d* 2. *c* 3. *a* 4. *c* 5. *b* 6. *b* 7. *c*
8. *b* 9. *a* 10. *a* 11. *d* 12. *c*

9 CHAPTER POLYPHASE SYSTEMS

Q. 1. What are the main points of difference between single-phase, two-phase and three-phase systems of voltages and currents?

Ans. Single-phase System : A single-phase system is one which has *one* voltage source. It has only one *winding* and two *conductors* for supplying the load as shown in Fig. 1.

The single-phase system is suitable for most applications but is not so well-suited for power transmission and electric machines. In the case of power transmission, it requires very heavy copper conductors. The single-phase system provides *pulsating* power instead of constant power. Moreover, single-phase motors are not self-starting and operate at very low power factor.

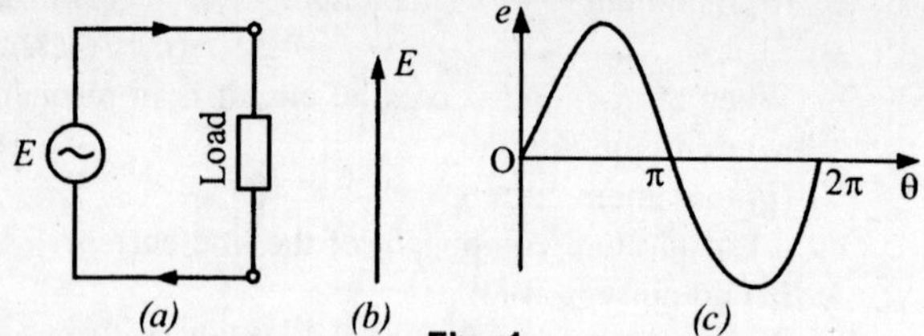

Fig. 1

Two-phase System : This system has *two* voltage sources E_1 and E_2 which are 90° out of phase with each other. For this purpose, it uses two similar windings as shown in Fig. 2.

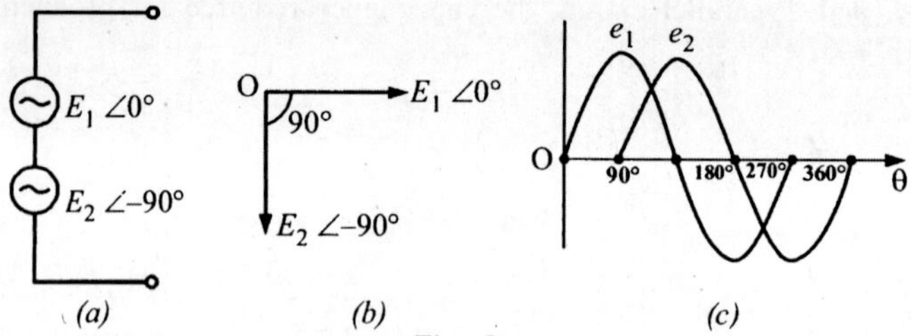

Fig. 2

This system is not employed these days because it has been superseded by three-phase system. However, it has recently become important for servomechanism.

Three-phase System : A three-phase system is one which has three voltage sources as shown in Fig. 3(*a*). It uses three windings which produce three voltages E_1, E_2 and E_3 that are equal in magnitude but have phase differences of 120° with each other as shown in Fig. 3(*b*). For CCW rotation, E_1 leads E_2 by 120° and leads E_3 by 240°.

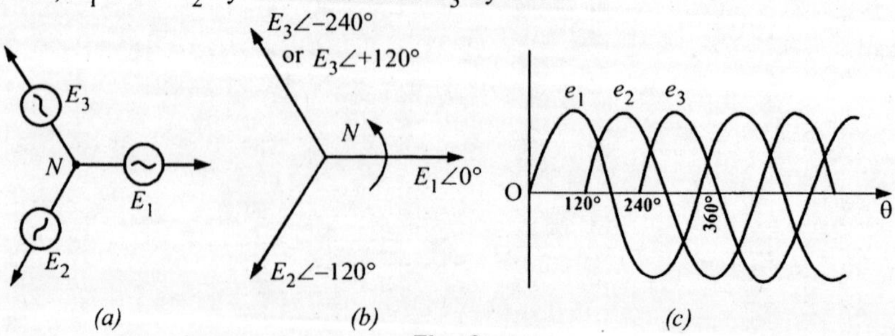

Fig. 3

Three-phase systems are the most common although, for certain special jobs, six-phase and twelve-phase systems are also used.

Q. 2. What is meant by a polyphase system of voltages and currents? Why has three-phase system become the most popular system throughout the world?

Ans. The word *poly* means many or several. Hence, polyphase system means a system which has two or more separate sources of voltages. The two-phase, three-phase, six-phase and twelve-phase systems etc. are examples of polyphase systems. However, all commercial energy throughout the world is generated, transmitted and distributed in the form of three-phase energy. The three voltages are sinusoidal and equal in magnitude but differ in phase by 120° as shown in Fig. 4. As seen, the three voltage waves do not reach their maximum value at the same time but with a time interval of $T/3$ seconds or one-third of a cycle or phase angle of 120°.

This fact makes the output power of a 3-ϕ system quite *smooth* instead of *pulsating* as is the case with single-phase system.

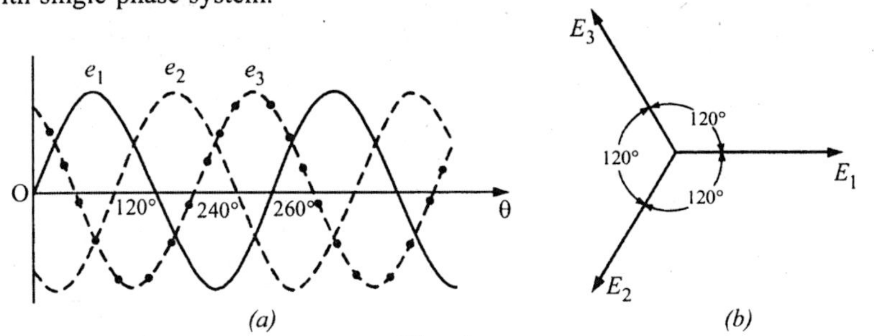

Fig. 4

For most applications, three-phase system is preferred to a single-phase system for the following reasons:

(*i*) The kW rating of 3-ϕ motors and the kVA rating of 3-ϕ generators for a *given physical size*, are larger. The capacity of a three-phase generator or an induction motor is about 150% of that of a 1-ϕ machine of comparable frame size.

(*ii*) The power delivered by a 1-ϕ system is pulsating while the *total power* supplied by a 3-ϕ system is *constant*. Hence, the operation of 3-ϕ generators and motors etc is less noisy and vibration-free. The constant power supply of the 3-ϕ system reduces eddy current and hysteresis losses in 3-ϕ equipment thereby increasing their efficiency.

(*iii*) Three-phase motors are self-starting whereas single-phase motors are not. They need additional apparatus for making them self-starting thereby increasing the cost.

(*iv*) Any 3-ϕ system can be converted (with the help of transformers) to a higher or lower 3-ϕ or 1-ϕ system. But a single-phase system cannot be converted to a true 3-ϕ system even with transformers.

(*v*) Power factor of a 1-ϕ motor is lower than that of a 3-ϕ motor of the same speed and output.

(*vi*) To transmit a given amount of power at a given voltage and at a given distance, 3-ϕ system uses only 75% of the copper required for a 1-ϕ system.

(*vii*) 3-ϕ currents can produce *rotating* magnetic fields when passed through stationary coils (as in induction and synchronous motors). Single-phase currents cannot do this.

Q. 3. Fully explain the exact meaning of the word 'phase' so often used in ac work. Give some examples of how the word 'phase' means different things in different contexts?

Ans. When two or more current waveforms are simultaneously present in a circuit, the angle difference between them is called their *phase* or *phase angle*. One wave leads or lags the other by a certain phase angle. However, the term *phase* means different things depending on the *context in which it is used.*

Given below are the examples showing different ways in which the word 'phase' is used:

(*i*) When it is said that current is out of phase with the voltage, it refers to the difference of angle between their vectors.

(*ii*) The three phases of a transmission line means the three conductors of that line.

(*iii*) The phase sequence means the *order* in which the voltage or current vectors follow each other.

(*iv*) The burned-out phase means the burned-out *winding* of a 3-ϕ machine.

(*v*) The 3-phase voltage means the *line voltage* of a 3-ϕ system.

(*vi*) Phase-to-ground fault means a short-circuit between a line or winding and ground.

(*vii*) The phase-to-phase fault means a short-circuit between two line conductors.

Q. 4. Explain the meaning and significance of the term 'phase-sequence' of a 3-ϕ system. How are different phases numbered in a 3-ϕ system? Can the phase sequence be reversed? If yes, explain the effects of phase reversal.

Ans. Apart from line voltage and frequency, the third important property of a 3-ϕ system is its *phase sequence*. The phase sequence of a 3-ϕ system means the *order* (or sequence) in which its three line voltages reach their peak or maximum values.

As seen from Fig. 4(*a*), e_1 is the first to reach its maximum value followed by e_2 120° later and by e_3 240° later. Hence, the order in which the three emfs achieve their maximum value is 1-2-3. It is called the phase order or *phase sequence* 1-2-3. Similarly, the voltage vectors of Fig. 4(*b*) follow the order E_1-E_2-E_3 if they rotate counterclockwise (CCW). It is called *positive* phase sequence. However, if vectors rotate clockwise, then E_1 would be followed by E_3 and then by E_2. Obviously, in this case, the phase sequence would be 1-3-2 or E_1-E_3-E_2 as shown in Fig. 5(*b*). It is called *negative* phase sequence.

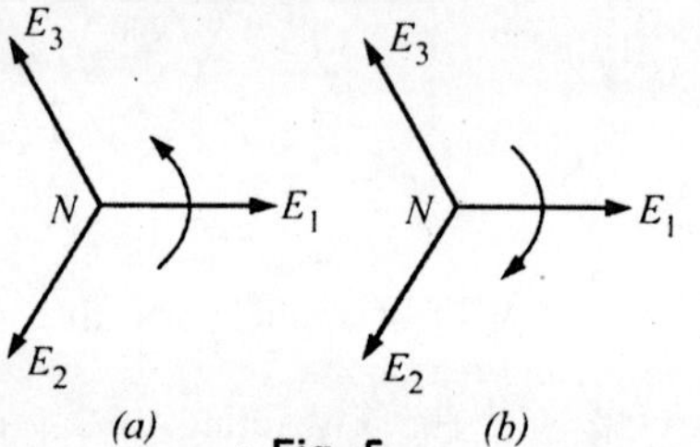

Fig. 5

Very often, the capital alphabets *A*, *B* and *C* are used to denote the phase sequence. Similarly, three different colours are also used to identify the three phases. The three colours used commercially are red, yellow and blue. In that case, the positive phase sequence is *RYB* and the negative phase sequence is *RBY*.

Since there are only two possible directions of vector rotation, there are only two possible phase sequences.

The phase sequence of a 3-ϕ circuit can be reversed by reversing the direction of rotation of the 3-ϕ generator or by interchanging the connections of any *two* of the three line wires supplying the 3-ϕ load. It should be noted that unless stated otherwise, the term phase sequence refers to the phase sequence of the 3-ϕ *voltages*.

Phase reversal effects :

1. The direction of rotation of a 3-ϕ motor depends on the phase sequence of the applied phase voltages. If phase sequence is reversed, the direction of rotation of the motor is also reversed.
2. When phase sequence is reversed, the readings of the two wattmeters used for measuring power in balanced 3-ϕ loads are also reversed.
3. In unbalanced systems, reversing voltage phase sequence changes the magnitude of certain branch currents.

Q. 5. Briefly discuss the generation of 3-ϕ voltages using simple sketch of a 3-ϕ generator.

Ans. The 3-ϕ generator (also called an alternator) has three 1-ϕ windings spaced in such a way that voltage induced in any one winding has a phase difference of 120° from the other two. In the simplified diagram shown in Fig. 6 the windings of each phase have been *lumped together* as one winding. Moreover, the rotor has been omitted for simplicity. The voltage waveforms generated in each phase winding are shown in Fig. 6(*b*). As seen, these waveforms are displaced by 120° from one another. For example, *wave* produced by winding *Y* lags behind wave produced by winding *R* by 120° whereas *wave B* lags by 240°.

Fig. 7(*a*) shows a cross-section of the stator and rotor of a simplified 3-ϕ alternator, with the rotor rotating in the counterclockwise (CCW) direction. When the rotor is in the position shown in Fig. 7(*a*), the voltage induced in winding R_1R is maximum. It is so because the rate of rotor flux cutting by the conductors of winding R_1R is *maximum*. Let us take this voltage as the reference voltage. As the rotor rotates CCW by 120°, its *N*-pole moves from coil end R_1 to coil end Y_1 where it induces maximum voltage in winding *Y*. Obviously, emf induced in winding *Y* reaches its maximum value 120° later than the maximum value in winding *R*. In the same way, maximum emf is induced in *B* winding 120° later than that in winding *Y* or 240° later than that in winding *R*.

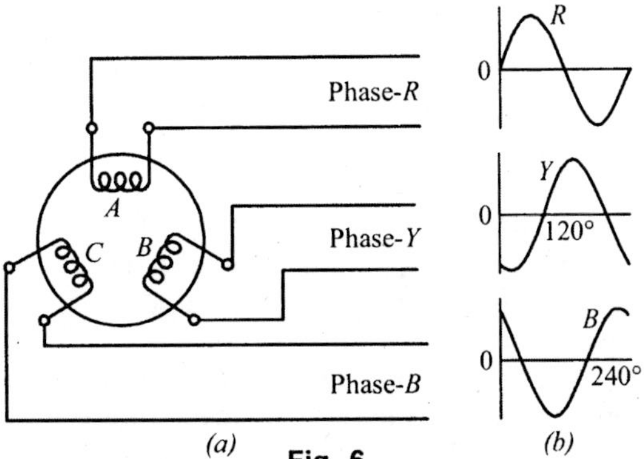

Fig. 6

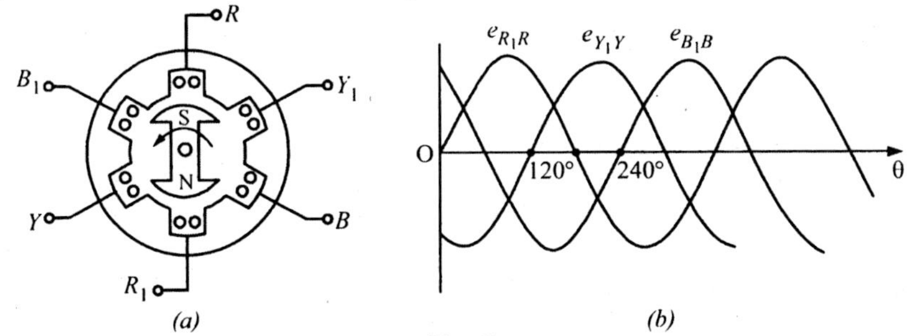

Fig. 7

The three emfs induced in the three windings are exactly similar though they are 120° out of phase with each other. Each voltage wave has a maximum value of $E_m = E_{R_1R} = E_{Y_1Y} = E_{B_1B}$. Of course, the rms value would be $= E_m/\sqrt{2}$.

The above points may be summarized as follows :

(*i*) Each of the three sinusoidal waveforms has a positive maximum value displaced by 120° from the other two.

(*ii*) For CCW rotation, the phase sequence of the generated voltages is *R-Y-B* (for CW rotation, the phase sequence would be *R-B-Y*).

(*iii*) When any phase voltage is positive maximum, the other two phase voltages are *negative* and each is *half of its maximum value* as seen from Fig. 7(*b*). Hence, the vector sum of the three phase voltages at any instant is always *zero*.

(*iv*) The instantaneous values of the three sinusoidal voltages can be represented in time domain as under :

$$e_R = E_m \sin \omega t$$
$$e_Y = E_m \sin (\omega t - 120°)$$
$$e_B = E_m \sin (\omega t - 240°)$$
or $$= E_m \sin (\omega t + 120°)$$

Q. 6. What is double-subscript notation? What is its necessity? Illustrate by suitable examples.

Ans. Double subscripts provide a convenient method for describing a quantity that has a *certain direction*. It is used to avoid any doubt about the *direction* of a current, emf or potential difference (p.d.). In the case of voltages, this notation tells us which voltage is positive at some node compared to another node. For example, V_{12} tells us that node 1 is more positive than node-2. If V_{12} = + 10 V, then V_{21} = –10 V so that $V_{12} = -V_{21}$.

When double-subscript notation is applied to currents, it indicates the *direction* of current flow between the two nodes. For example, I_{12} represents the direction of current flowing from node-1 to node-2. If I_{12} = 10 A, then I_{21} = –10 A. Similarly, voltage V_{ab} represents voltage between nodes *a* and *b*. The voltage V_{ba} is in the *opposite* direction. Incidentally, if we have 3-ϕ voltages of V_{ab}, V_{bc}, V_{ca}, then the order of the *first subscript* gives us the phase sequence of the source. In the present case, it would be *a-b-c*. In the case of voltages V_{ab}, V_{ca}, V_{bc}, the phase sequence would be *a-c-b*.

Q. 7. Why are the three armature windings of an ac generator interconnected? Mention the two general methods of interconnection.

Ans. A three-phase generator has *three* windings, hence there are *six* conductors which come out of it as shown in Fig. 8(*a*). If these conductors are not interconnected, then each transmission cable would contain six conductors. It will make the whole system complicated and costly. If the three windings are interconnected, there would be a big saving of copper.

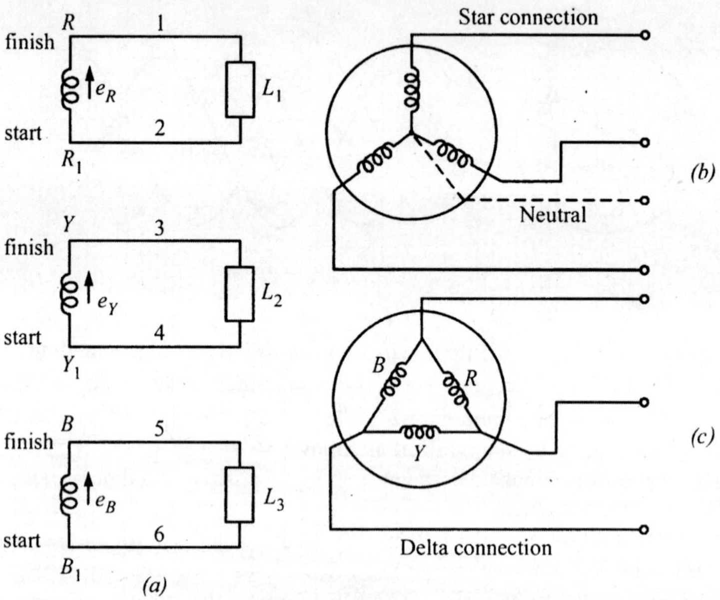

Fig. 8

The two common methods of interconnecting the three generator windings are :

(*a*) Star or Wye (Y) connection and — Fig. 8(*b*)

(*b*) Mesh or Delta (Δ) connection. — Fig. 8(*c*)

In star connection, the *similar* ends of the three generator windings are connected together. The voltage between any two lines is equal to the *vector difference* of the emfs induced in the two *generator windings* connected between them. However, line current is the *same* as the current of the generator winding to which the line is connected.

When the generator windings are connected end-to-end, we get delta connection. In this connection, line voltages are equal to phase voltages but line currents are the *vector difference* of the two respective phase currents.

Q. 8. What is a Star/Y-connection? What is meant by a three-phase, four wire system? What advantages does it possess?

Ans. In this method of interconnection, the three similar* ends R_1, Y_1, B_1 of the three windings are joined together at a common point as shown in Fig. 9(*a*). This point is known as *star* point or *neutral* point *N*. The three conductors meeting at *N* are replaced by a *single* conductor known as *neutral* conductor (or simply, neutral) as shown in Fig. 9(*b*). Such an interconnected system is called a 3-ϕ, four-wire system. If this 3-ϕ voltage system is applied across a *balanced load*, the neutral wire will carry three load currents which are exactly equal in magnitude but 120° out of phase with each other. Hence, their *vector sum* is zero *i.e.* $I_R + I_Y + I_B = 0$. In other words, $I_N = 0$.

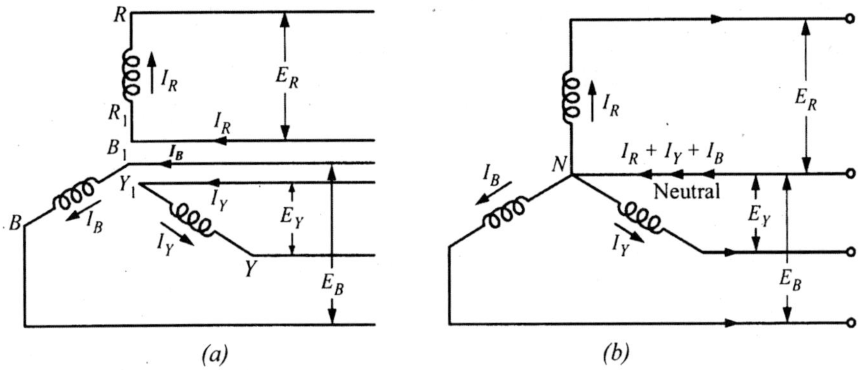

Fig. 9

In the case of balanced loads, the neutral may be omitted giving us 3-ϕ, 3-wire system. The 3-ϕ, four-wire systems are widely used to supply electric power to domestic as well as commercial and industrial users because it provides two different values of 3-ϕ voltage. The potential difference between any terminal (or line) and the neutral is called the *phase voltage* (E_{ph}). But the potential difference between any two lines gives the *line voltage* (V_L).

While considering the distribution of current in a 3-ϕ system, it is essential to keep the following points in mind:

(*i*) The arrows placed alongside the phase currents I_R, I_Y and I_B indicate the direction of current when they are *assumed to be positive* and not the *direction at a particular instant.* It is important to understand that at no instant will all the three currents flow in the *same* direction either inwards or outwards. It is so because the three currents have a phase differences of 120° with each other.

(*ii*) The current flowing outwards in one conductor is equal to the currents flowing inwards in the other *two conductors*. It means that each conductor, in turn, provides a *return path* for the currents of the other two conductors.

Q. 9. Derive the relationship between the phase and line voltages of a 3-ϕ, *Y*-connected generator. What is the phase difference between line voltages and line currents?

Ans. The voltage induced in each winding of the generator is called the *phase voltage* (E_{ph}) and current in each winding is called *phase current* (I_{ph}). However, the voltage available between any *pair* of terminals (or outers) is called *line voltage* (V_L) and the current flowing in each line is called the *line current* (I_L). As seen from Fig. 10(*a*), in *Y*-connection there are *two* phase windings between each pair of terminals. Since their *similar* ends have been joined together, their induced emfs are in *opposition* to each other. Hence, the line voltage between any two terminals is the *vector difference* of the two winding emfs. The vector diagram for induced phase voltages and currents in the *Y*-connection is shown in Fig. 10(*b*).

---

*As an aid to memory, the first letter **S** of Similar is the same as that of **Star.**

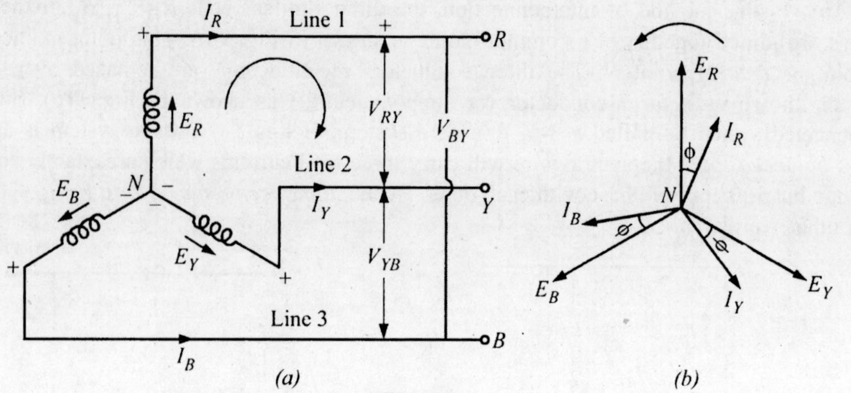

Fig. 10

Let $\quad E_R = E_Y = E_B = E_{ph}$

As seen from Fig. 10(*a*),

$$V_{RY} = \text{vector difference of } E_R \text{ and } E_Y$$

$$V_{YB} = \text{vector difference of } E_Y \text{ and } E_B$$

$$V_{BR} = \text{vector difference of } E_B \text{ and } E_R$$

The above would become obvious when we apply Kirchoff's Voltage Law to the loop shown in Fig. 10(*a*). If we do so, we get

$$-V_{RY} - E_Y + E_R = 0 \qquad V_{RY} = E_R - E_Y$$

(*i*) **Line voltages and phase voltages :** As shown in Fig. 11, E_Y has been reversed and combined with E_R to get the line voltage (V_{RY}). Obviously, the angle between E_R and ($-E_Y$) is 60°.

Let $\quad E_R = E_Y = E_B = E_{ph}$.

Then, $V_{RY} = 2 \times E_{ph} \times \cos(60°/2)$

$= 2 \times E_{ph} \times \cos 30°$

$= 2 \times E_{ph} \times \frac{\sqrt{3}}{2} = \sqrt{3}\, E_{ph}$

Similarly,

$V_{YB} = E_Y - E_B$ (vector difference)

$= \sqrt{3}\, E_{ph}$

$V_{BR} = E_B - E_R$ (vector difference)

$= \sqrt{3}\, E_{ph}$

Let, $\quad V_{RY} = V_{YB} = V_{BR}$ = line voltage, V_L

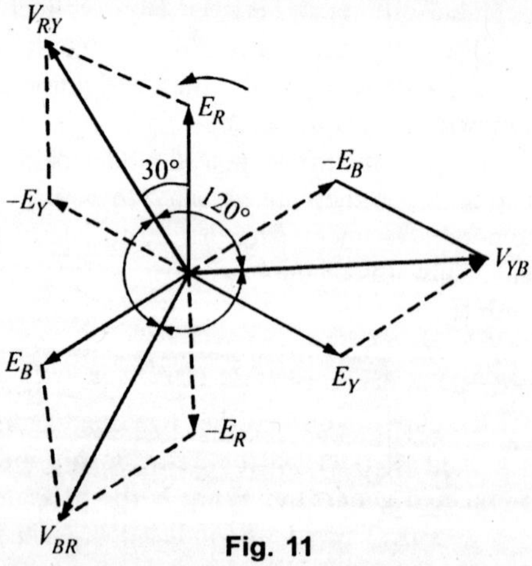

Fig. 11

It will be found from Fig. 11 that,

(*a*) line voltages are 120° apart

(*b*) line voltages are 30° ahead of their respective phase voltages. For example, V_{RY} is ahead of E_R by 30°.

(*c*) the angle between the *line* current and the corresponding *line* voltage is (30° + ϕ) with current lagging. For example I_R lags V_{RY} by (30° + ϕ) as shown in Fig. 15.

(*ii*) **Line current and phase current :** It is seen from Fig. 10(*a*) that each line is in *series* with a phase winding. Hence, each line current is the *same* as the current in the generator winding to which it is connected.

Current in line – 1 = I_R, current in line – 2 = I_Y, current in line – 3 = I_B.

Since, $I_R = I_Y = I_B = I_{ph}$ — the phase current

$\therefore$ $I_L = I_{ph}$

As seen from Fig. 15(*b*), this current is (30° + ϕ) behind its respective line voltage*.

(*iii*) **Power :** Total power in the circuit is the sum of the three 1-ϕ powers.

Hence, total power = 3 × phase power

or $P = 3 \times E_{ph} I_{ph} \cos \phi$

Now, $E_{ph} = V_L/\sqrt{3}$ and $I_{ph} = I_L$

Hence, in terms of line values, the above expression becomes,

$$P = 3 \times \frac{V_L}{\sqrt{3}} \times I_L \times \cos \phi$$

or $$P = \sqrt{3}\, V_L I_L \cos \phi$$

It should be noted that ϕ is the angle between the phase voltage and phase current and **NOT** between the line voltage and line current.

Q. 10 (*a*) What is the relation between the active, reactive and apparent power in a three-phase circuit?

(*b*) A 3-ϕ motor draws a line current of 10 A when connected to a 440 V line. If the power factor of the motor is 0.8, calculate

(*i*) total apparent power

(*ii*) total active power

(*iii*) total reactive power drawn by the motor

Ans. (*a*) As discussed earlier, the relationship between the active or real power P, reactive power Q and apparent power S is the same for balanced 3-ϕ circuit *i.e.*

$$S = \sqrt{P^2 + Q^2} \quad \text{— Fig. 12}$$

Also, $\cos \phi = P/S$

Here, S = total 3-ϕ apparent power (VA)

$= \sqrt{3}\, E_L I_L$

P = total 3-ϕ active/real power (W)

$= \sqrt{3}\, E_L I_L \cos \phi$

Q = total 3-ϕ reactive power (VAR)

$= \sqrt{3}\, E_L I_L \sin \phi$

ϕ = phase angle between *phase* voltage and *phase* current.

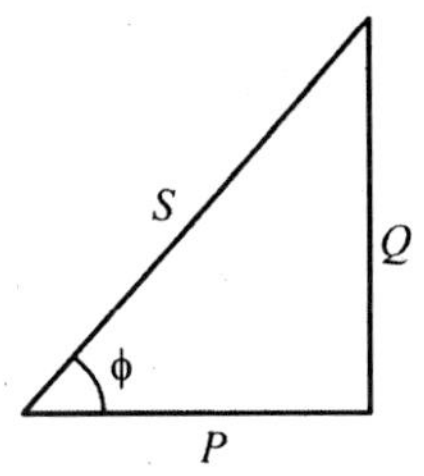

Fig. 12

(*b*) (*i*) $S = \sqrt{3}\, E_L I_L = 1.73 \times 440 \times 10 = 7620$ VA = **7.62 kVA**

(*ii*) $P = S \cos \phi = 7.62 \times 0.8 =$ **6.1 kW**

(*iii*) $Q = S \sin \phi = 7.62 \times 0.6 =$ **4.57 kVAR**

Q. 11. A balanced star-connected load of (8 + *j*6) ohm is connected across three-phase, 50 Hz, 440 V supply system.

Calculate :

(*i*) line current (*ii*) power absorbed (*iii*) reactive volt-amperes

(*Basic Elect. Engg. R.G.P.V. Bhopal 2008*)

Ans. The three-phase balanced star-connected load is shown in Fig. 13.

*The phase difference between V_{RY} and I_R is (30 + ϕ).

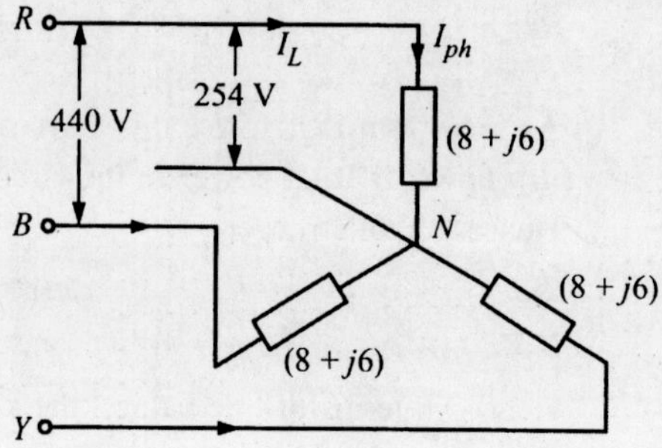

Fig. 13

Impedance per phase,

$$Z_{ph} = \sqrt{8^2 + 6^2} = 10\ \Omega$$

Voltage across each phase,

$$V_{ph} = 440/\sqrt{3} = 254\ \text{V}$$

$$\text{Phase current} = V_{ph}/Z_{ph} = 254/10 = 25.4\ \text{A}$$

Power factor, $\cos\phi = R_{ph}/Z_{ph} = 8/10 = 0.8$ (lag)

and $\sin\phi = 6/10 = 0.6$

(*i*) In star connection, line current equals phase current.

$\therefore$ $I_{ph} = I_L =$ **25.4 A**

(*ii*) Power absorbed is given by

$$P = \sqrt{3}\, V_L I_L \cos\phi = \sqrt{3} \times 440 \times 25.4 \times 0.8 = \mathbf{15{,}486\ W}$$

(*iii*) Reactive volt–amperes are given by

$$Q = \sqrt{3} \times V_L \times I_L \times \sin\phi_L$$

$$= \sqrt{3} \times 440 \times 25.4 \times 0.6 = \mathbf{11{,}614\ VAR}$$

Q. 12. A balanced 3-phase star-connected load of 150 kW takes a leading current of 100A with a line voltage of 1100 V, 50 Hz. Find the circuit constants of the load per phase?
(*Elect. Circuits. B.Tech, JNTU, Hyderabad 2006*)

Ans. Let us find the value of cos ϕ and then of sin ϕ which will be needed later.

$$P = \sqrt{3}\, V_L I_L \cos\phi$$

or $150 \times 10^3 = \sqrt{3} \times 1100 \times 100 \times \cos\phi$

$\therefore$ $\cos\phi = 0.787$ (lead)

Since, $\cos\phi = 0.787,\ \phi = 38.1°$

$\therefore$ $\sin\phi = 0.616$

$$I_{ph} = 100\ \text{A},\ V_{ph} = V_L/\sqrt{3} = 1100/\sqrt{3} = 635\ \text{V}$$

$$Z_{ph} = V_{ph}/I_{ph} = 635/100 = 6.35\ \Omega$$

As seen from the impedance triangle of Fig. 14(*b*).

$$R = Z_{ph}\cos\phi = 6.35 \times 0.787 = \mathbf{5\ \Omega}$$

$$X_C = Z_{ph}\sin\phi = 6.35 \times 0.616 = 4\ \Omega$$

Now $X_C = \dfrac{1}{\omega C}$

or $C = \dfrac{1}{\omega X_C} = \dfrac{1}{314 \times 4}$ farad

or $C = \dfrac{10^6}{314 \times 4} = \mathbf{796\ \mu F}$

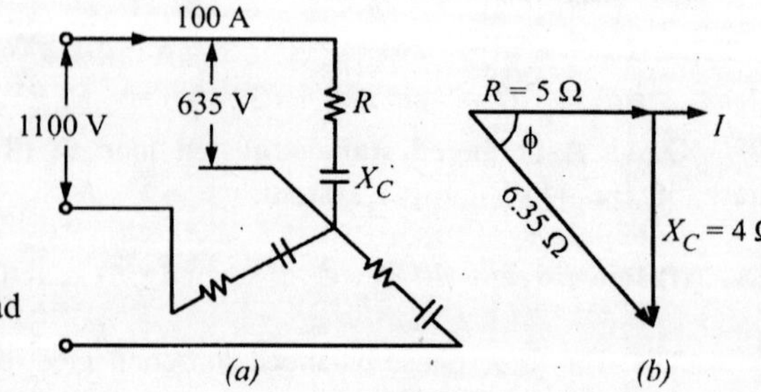

Fig. 14

Q. 13. A balanced star-connected load is supplied from a symmetrical three-phase, 400 V (line-to-line) supply. The current in each phase is 50 A and lags 30° behind the phase voltage. Find:

(*i*) phase impedance (*ii*) active and reactive powers drawn by the load (*iii*) draw phasor diagram showing phase and line voltages and line currents.

(*Elect. Engg. Nagpur Univ. 2004*)

Ans. We are given that, $V_L = 400$ V, $I_{ph} = 50\angle{-30°}$ A.

Since it is Y-connection, hence $I_L = I_{ph} = 50\angle{-30°}$ A

Also, $$V_{ph} = V_L/\sqrt{3} = 400/\sqrt{3} = 231 \text{ V}$$

(*i*) $$Z_{ph} = \frac{V_{ph}}{I_{ph}} = \frac{231\angle 0°}{50\angle -30°} = 4.62\angle 30° = \mathbf{4 + j\,2.31\ \Omega}$$

(*ii*) Active power, $$P = \sqrt{3}\, V_L I_L \cos\phi$$
$$= \sqrt{3} \times 400 \times 50 \times \cos 30° = 30{,}000 \text{ W} = \mathbf{30\ kW}$$

Reactive power, $$Q = \sqrt{3}\, V_L\, I_L \sin\phi$$
$$= \sqrt{3} \times 400 \times 50 \times \sin 30° = 17{,}320 \text{ VAR} = \mathbf{17.32\ kVAR}$$

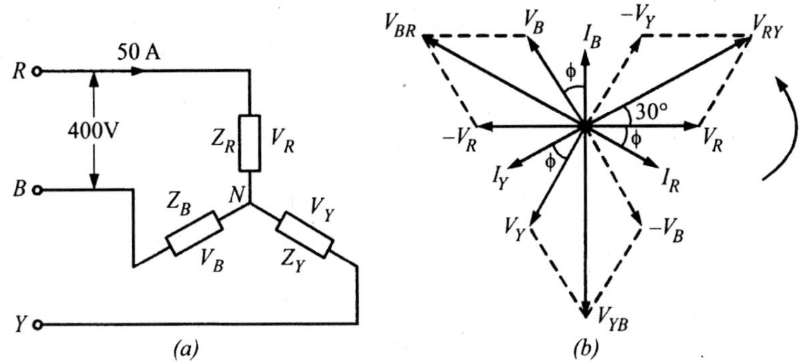

Fig. 15

Phase voltages :

$$\mathbf{V_R} = 231\angle 0°,\ \mathbf{V_Y} = 231\angle{-120°},\ \mathbf{V_B} = 231\angle 120°$$

Line currents :

In star-connection, line currents are equal to phase currents.

Hence, $$I_L = I_{ph}$$

$$\mathbf{I_R} = \frac{\mathbf{V_R}}{\mathbf{Z_R}} = \frac{231\angle 0°}{4.62\angle 30°} = \mathbf{50\angle{-30°}\ A}$$

$$\mathbf{I_Y} = \frac{\mathbf{V_Y}}{\mathbf{Z_Y}} = \frac{231\angle -120°}{4.62\angle 30°} = \mathbf{50\angle{-150°}\ A}$$

$$\mathbf{I_B} = \frac{\mathbf{V_B}}{\mathbf{Z_B}} = \frac{231\angle 120°}{4.62\angle 30°} = \mathbf{50\angle 90°}$$

Phasor diagram :

In Y-connection, line voltages are the *vector* difference of the respective phase voltages but line currents are equal to phase currents.

The line voltages have been constructed as shown in Fig. 15(*b*). The line currents or phase currents are equal in magnitude but have different phase angles of –30°, –150° and 90° with reference vector V_R as shown. But between *themselves*, they have a phase difference of 120°

Q. 14. A 3-phase, 230 V system supplies a total load of 2000 W at a line current of 6 A when three identical impedances are in star-connection across the line terminals of the system. Determine the resistance and reactance of each impedance.

(*Elect. Engg. Nagpur Univ. 2005*)

Ans. Since the three impedances are identical, power consumed by each is 2000/3 = 666.67 W.

$$I_{ph} = I_L = 6\text{A},\ V_{ph} = 230/\sqrt{3} = 132.8 \text{ V}$$

Now, power in each phase = $V_{ph} \cdot I_{ph} \cdot \cos\phi$

$\therefore$ $666.67 = 132.8 \times 6 \times \cos\phi$ $\quad\therefore \cos\phi = 0.836$

The value of sin ϕ will be 0.548

$$Z_{ph} = \frac{V_{ph}}{I_{ph}} = \frac{132.8}{6} = 22.13\ \Omega$$

$\therefore$ $R = Z_{ph}\cos\phi = 22.13 \times 0.836 = \mathbf{18.5\ \Omega}$

$X = Z_{ph}\sin\phi = 22.13 \times 0.548 = \mathbf{12.14\ \Omega}$

Q. 15. A balanced delta-connected load having an impedance of Z_L = (300 + j 210) ohm in each phase is supplied from 400 V, 3-ϕ supply through a three-phase line having an impedance of Z_s = (4 + j 8) ohm in each phase. Find the total power supplied to the load as well as the current and voltage in each phase of the load.

(*Basic. Elect. Engg. R.G.P.V. Bhopal, 2007*)

Ans. The equivalent Y-load of the given Δ-load is = $\mathbf{Z}_Y/3$ = (300 + j 210)/3 = (100 + j70) Ω as per Q. 22. The connection diagram is shown in Fig. 16.

$\mathbf{Z_{an}} = \mathbf{Z_S} + \mathbf{Z_L} = (4 + j8) + (100 + j70)$

$= 104 + j78 = 130\angle 36.9°$

$V_{an} = 400/\sqrt{3} = 231$ V

Let $\mathbf{V_{an}} = 231\angle 0°$

$\mathbf{I_{an}} = \mathbf{V_{an}}/\mathbf{Z_{an}}$

$= 231\angle 0°/130\angle 36.9°$

$= 1.78\angle -36.9°$

Now, $\mathbf{Z_{aa_1}} = Z_S = (4 + j\ 8)$

$= 8.94\angle 63.4°$

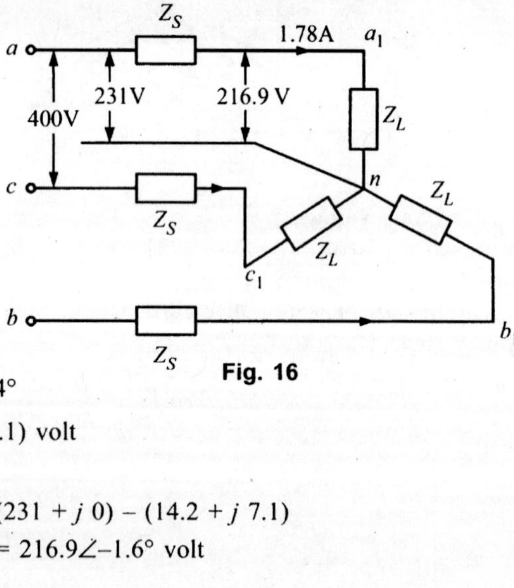

Fig. 16

Drop on $\mathbf{Z_S} = \mathbf{I_{aa_1}} \times \mathbf{Z_S} = \mathbf{I_{an}}\ \mathbf{Z_S}$

or $\mathbf{V_{aa_1}} = 1.78\angle -36.9° \times 8.94\angle 63.4°$

$= 15.9\angle 26.5° = (14.2 + j\ 7.1)$ volt

Voltage across Z_L is

$$\mathbf{V_{a_1n}} = \mathbf{V_{an}} - \mathbf{V_{aa_1}} = (231 + j\ 0) - (14.2 + j\ 7.1)$$
$$= (216.8 - j\ 7.1) = 216.9\angle -1.6° \text{ volt}$$

It would be seen from Fig. 16 that

Phase voltage at load end,

$$V_{a_1n} = \mathbf{216.9\ V}$$

Phase current at load end,

$$I_{an} = I_{a_1n} = \mathbf{1.78\ A}$$

Power supplied to the load = $3 \times (1.78)^2 \times 100$ = **951 W**

Q. 16. A manufacturing plant draws total of 415 kVA from a 2400 V, 3-ϕ line. The plant load consists of three identical star-connected impedances. If the plant power factor is 87.5% lagging, calculate the following :

(*i*) impedance of the plant per phase

(*ii*) phase angle between the line-to-neutral voltage and the line current.

(*iii*) complete phasor diagram for the plant.

Ans. (*i*) $V_L = 2400$ V, $V_{ph} = 2400/\sqrt{3} = 1390$ V

VA drawn by the plant = $\sqrt{3}\ V_L I_L$

$\therefore \quad 4{,}15{,}000 = \sqrt{3} \times 2400 \times I_L \qquad \therefore \quad I_L = 100$ A

In a *Y*-connection $I_{ph} = I_L = 100$ A

$Z_{ph} = V_{ph}/I_{ph} = 1390/100 = \mathbf{13.9\ \Omega}$

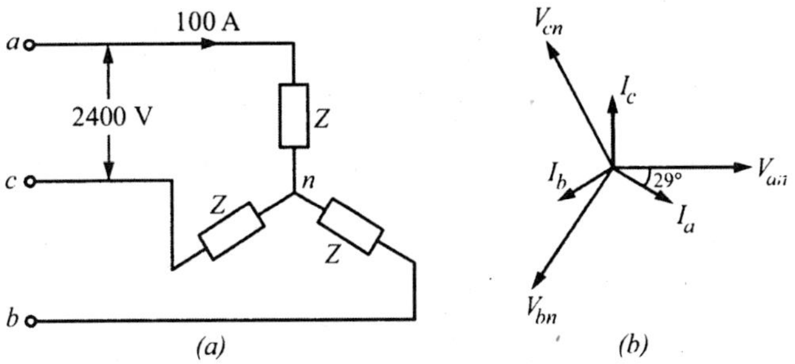

Fig. 17

(*ii*) The phase angle ϕ between V_{ph} and the corresponding I_{ph} can be found from the power factor of the plant.

$$\cos\phi = 0.875 \quad \text{or} \quad \phi = \mathbf{29°}$$

(*iii*) The phasor diagram is shown in Fig. 17(*b*).

Q. 17. Each phase of a star-connected load consists of a non-reactive resistance of 100 Ω in parallel with a capacitor of capacitance 31.8 μF.

Calculate the line current, the power absorbed, the total kVA and the power factor when connected to a 416 V, 3-phase, 50 Hz supply.

Ans. The circuit is shown in Fig. 18.

$$\mathbf{V_{ph}} = (416/\sqrt{3})\ \angle 0° = 240\ \angle 0° = (240 + j0)$$

Admittance of each phase is

$$\mathbf{Y_{ph}} = \frac{1}{R} + j\,\omega C = \frac{1}{100} + j\,2\pi \times 50 \times 31.8 \times 10^{-6}$$

$$= 0.01 + j\,0.01$$

$\therefore \quad \mathbf{I_{ph}} = \mathbf{V}_{ph} \cdot \mathbf{Y}_{ph}$

$= 240\ (0.01 + j\ 0.01)$

$= 2.4 + j\ 2.4 = 3.39\ \angle 45°$

Since $I_{ph} = I_L$ — for a star-connection

$\therefore \quad I_L = \mathbf{3.39\ A}$

Power factor = cos 45°

= **0.707 (leading)**

Now $\mathbf{V}_{ph} = (240 + j0)$

and $\mathbf{I}_{ph} = (2.4 + j2.4)$

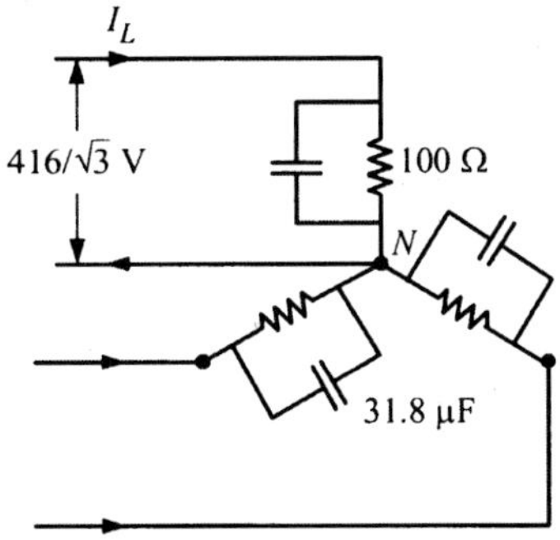

Fig. 18

$$\mathbf{P}_{VA} = (240 + j0)\ (2.4 - j2.4)$$
$$= 240 \times 2.4 - j2.4 \times 240 = 576 - j576$$
$$= 814.4 \angle{-45°} \text{ — per phase}$$

Now total power = 3 × 576 = 1728 W = **1.728 kW**

Total volt-amperes = 814.4 × 3 = 2443 VA

Kilovolt-amperes = **2.443 kVA**

Q. 18. Derive the numerical relationship between the line and phase currents for a 3-φ, Δ-connected generator.

Ans. In this form of interconnection, the dissimilar* ends of the three generator windings are joined together *i.e.* the *starting* end of one winding is joined to the *finishing* end of the other winding as shown in Fig. 19(*a*).

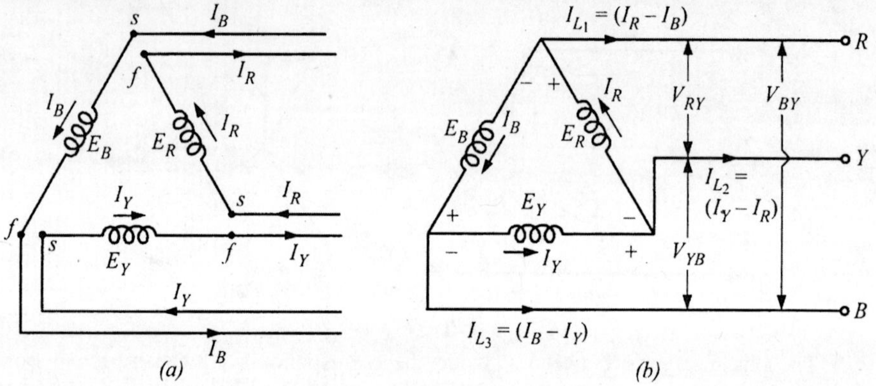

Fig. 19

Three leads are taken out from the three junctions as shown. The *outward direction* of flow is taken as positive.

***(i)* Line voltages and phase voltages :** As seen from Fig. 19(*b*), there is only one phase winding *completely* included between any pair of terminals.

Hence, line voltage is equal to the phase voltage of the winding connected between the two lines. The generator line voltages are

$$V_{RY} = E_R,\ V_{YB} = E_Y \text{ and } V_{BR} = E_B$$

These three voltages have a phase differences of 120° with each other.

Let $E_R = E_Y = E_B = E_{ph}$

Also, let $V_{RY} = V_{YB} = V_{BR} = \text{line voltage } V_L$

∴ $V_L = E_{ph}$

***(ii)* Line currents and phase currents :** As seen from Fig. 19(*b*), each line current is the *vector difference* of the two phase currents flowing through that line. Hence,

$$\left.\begin{aligned} I_{L_1} &= I_R - I_B \\ I_{L_2} &= I_Y - I_R \\ I_{L_3} &= I_B - I_Y \end{aligned}\right\} \text{vector difference}$$

As seen from Fig. 20, the generator line current I_{L_1} is found by combining I_R and $(-I_B)$. Angle between I_R and $(-I_B)$ is 60°. If $I_R = I_B = I_Y = \text{phase current } I_{ph}$, then

$$I_{L_1} = 2 \times I_{ph} \times \cos(60°/2) = 2 \times I_{ph} \times \sqrt{3}/2 = \sqrt{3}\, I_{ph}$$

*Remember that first letter **D** of **D**issimilar is the same as that of **D**elta.

Similarly, $I_{L_2} = I_Y - I_R = \sqrt{3}\, I_{ph}$

$I_{L_3} = I_B - I_Y = \sqrt{3}\, I_{ph}$

Since $I_{L_1} = I_{L_2} = I_{L_3} = I_L$

$\therefore \quad I_L = \sqrt{3}\, I_{ph}$

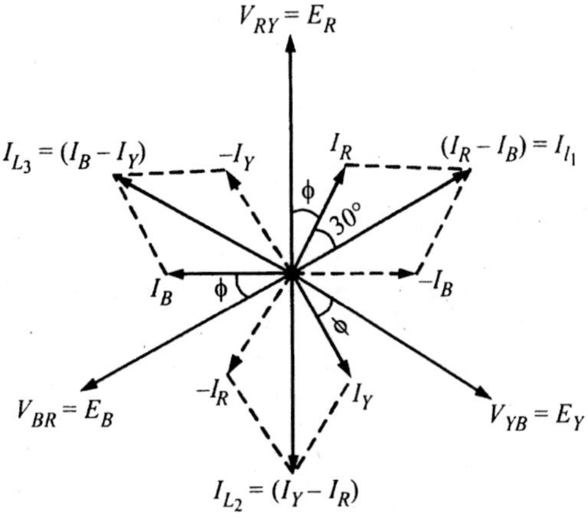

Fig. 20

From Fig. 20, it would be noted that

(*a*) line currents are 120° apart

(*b*) line currents are 30° behind the respective phase currents. For example, I_{L_1} lags behind I_R by 30°.

(*c*) the angle between the line current and the corresponding line voltage is (30° + ϕ) with the current lagging. For example, I_{L_1} lags behind V_{RY} by (30° + ϕ).

(*iii*) **Power :** Power in each phase $= V_{ph}\, I_{ph} \cos\phi$

Total power $= 3\, V_{ph}\, I_{ph} \cos\phi$

Now, $V_{ph} = V_L$ and $I_{ph} = I_L / \sqrt{3}$

Hence, in terms of *line values*, the total power becomes

$$P = 3 \times V_L \times \frac{I_L}{\sqrt{3}} \times \cos\phi = \sqrt{3}\ V_L\, I_L \cos\phi$$

where ϕ is the *phase* power factor angle. It is the same expression as for a star connection.

Q. 19. Give the summary of the voltage and current relationships in star-connected and delta-connected 3-ϕ circuits.

Ans. The relationships are summarized below :

Y–connection	**Δ-connection**
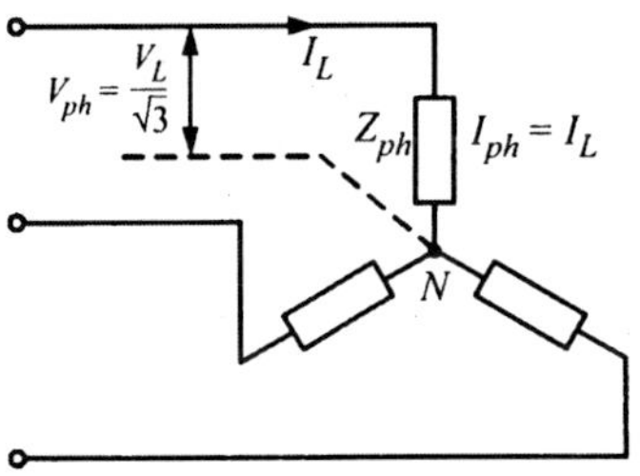 **Fig. 21**	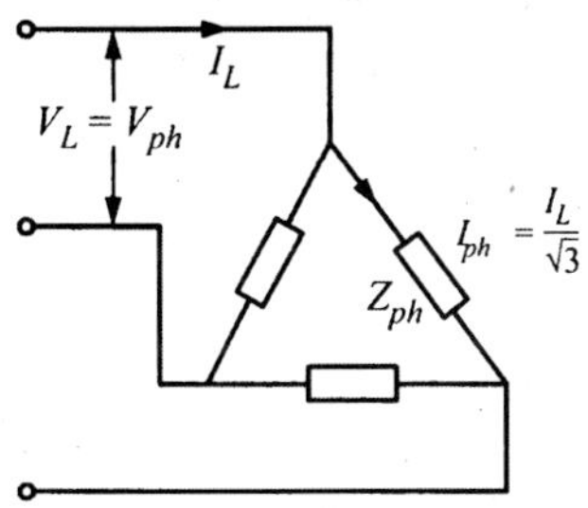 **Fig. 22**
1. The current in each load is equal to the line current *i.e.* $I_{ph} = I_L$	1. The current in each load is equal to the line current divided by $\sqrt{3}$. *i.e.* $I_{ph} = I_L / \sqrt{3}$
2. The voltage across each load is equal to the line voltage divided by $\sqrt{3}$. *i.e.* $V_{ph} = V_L / \sqrt{3}$	2. The voltage across each load is equal to the line voltage *i.e.* $V_{ph} = V_L$
3. The voltages across the loads are 120° out of phase with each other.	3. The voltages across the loads are 120° out of phase with each other.
4. Currents in the loads are 120° out of phase with each other	4. The currents in the loads are 120° out of phase with each other.

Q. 20. Compare the advantages and applications of star and delta connections.

Ans. (*a*) Star-connection :

(*i*) Three-phase generators are usually star-wound for the reason that phase voltage has to be only $1/\sqrt{3}$ of the line voltage, whereas for a delta winding, phase voltage has to be *equal* to the line voltage. It means in star-wound generator, number of coil turns required per phase is less than for a delta-connected generator.

(*ii*) *Star-connection* provides two voltages *i.e.* phase voltage and line voltage. Hence, lighting loads are connected across the three phases whereas power loads like 3-ϕ motors are connected across the lines as shown in Fig. 23.

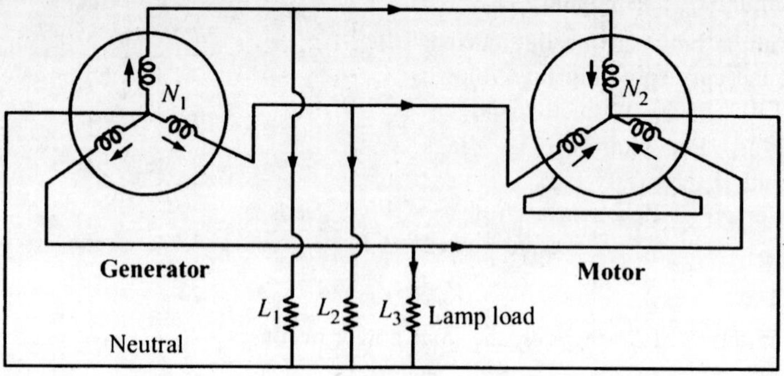

Fig. 23

(*iii*) Another advantage of *Y*-connection is that the neutral of the generator can be earthed. In that case, the p.d. between each line and earth is equal to phase voltage *i.e.* $V_L/\sqrt{3}$. Hence, if by fault, line conductor is earthed, the insulator will have to bear a voltage of $V_L/\sqrt{3} = 0.577\ V_L$ only. But in the case of a delta-connection, if any line conductor is earthed, the insulator will have to bear full line voltage V_L. It will increase the possibility of insulator breakdown.

(*b*) Delta-connection

(*i*) Transformers work better with delta-connection.

(*ii*) It is the only connection which is suitable for rotary converters.

(*iii*) This connection is mostly used for low-voltage 3-ϕ motors.

Q. 21. Discuss the voltage and current relationships when a balanced delta-connected load is connected across a delta-connected generator.

Ans. Fig. 24 shows a delta-connected 3-ϕ generator supplying a *balanced* 3-ϕ delta-connected load. The voltage across each load impedance is equal to the voltage induced in the

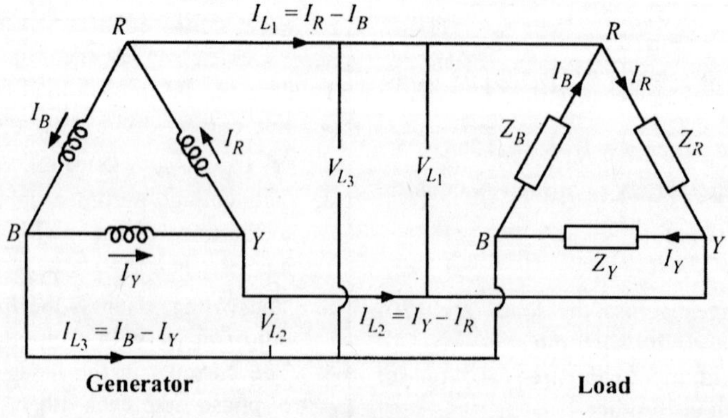

Fig. 24

generator phase winding. It is so because the three line voltages are equal to the respective phase voltages of the generator winding. However, each line carries a current which is the *vector difference* of the two currents induced in the two generator windings connected to that line.

As shown, I_{L_1} = vector difference of I_R and I_B because the line is connected to the phase windings R and B of the generator*. Same is the case with other line currents. The three line voltages are 120° apart. However, the phase difference between the line voltages and line currents is (30° + ϕ) where ϕ is the *phase* power factor angle.

Q. 22. Prove that for the same supply line voltages and currents, $Z_\Delta = 3Z_Y$

Ans. It may be proved that a 3-ϕ, Y-connected system having the line voltage of V_L and line current I_L may be replaced by a Δ-connected system in which phase voltage is V_L and phase current is $I_L/\sqrt{3}$.

Similarly, a balanced Y-connected load having branch impedances of $Z\angle\phi$ may be replaced by an equivalent Δ-connected load whose each branch impedance is 3 Z_Y. This equivalence is shown in Fig. 25.

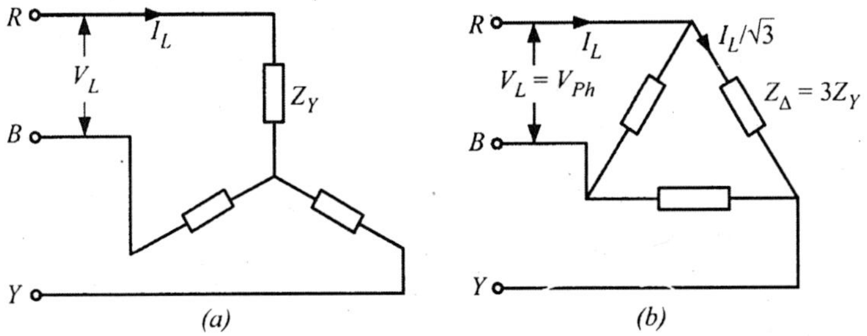

Fig. 25

For a balanced star-connected load, let

V_L = line voltage ; I_L = line current

Z_Y = impedance/phase ; $V_{ph} = V_L/\sqrt{3}$

$I_{ph} = I_L$; $Z_Y = V_{ph}/I_{ph}$

$= V_L/(\sqrt{3}\, I_L)$

In the equivalent Δ-connected system, we have

$$V_{ph} = V_L \;;\; I_{ph} = I_L/\sqrt{3}$$

$$Z_\Delta = V_{ph}/I_{ph} = V_L/(I_L/\sqrt{3})$$

$$= \sqrt{3}\, V_L/I_L = 3.V_L/\sqrt{3}\, I_L = 3Z_Y$$

$$\therefore \quad Z_\Delta\angle\phi = 3\, Z_Y\angle\phi$$

Q. 23. Show that for the same line voltages and same phase impedances, power drawn by a Δ-connected load is three times the power drawn by a Y-connected load.

Ans. Here we are given that $Z_Y = Z_\Delta = Z$ (say).

Star-connection :

$$V_{ph} = V_L/\sqrt{3} \quad ; \quad I_{ph} = V_{ph}/Z = V_L/\sqrt{3}\, Z$$

$$\therefore \quad P_Y = 3\, I_{ph}^2\, R_{ph} = 3 \times \left(\frac{V_L}{\sqrt{3}Z}\right)^2 \times R_{ph} = \frac{V_L^2}{Z^2} \times R_{ph}$$

*Applying KVL to point R in Fig. 24, we get $-I_{L_1} - I_B + I_R = 0$ or $I_{L_1} = I_R - I_B$

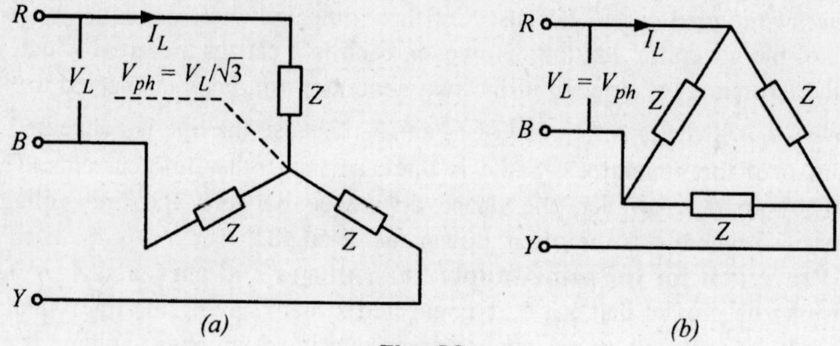

Fig. 26

Delta-connection :

$$V_{ph} = V_L, \quad I_{ph} = V_{ph}/Z = V_L/Z$$

$$\therefore \quad P_\Delta = 3\,I_{ph}^2\,R_{ph} = 3\left(\frac{V_L}{Z}\right)^2 R_{ph} = 3\,\frac{V_L^2}{Z^2}\,R_{ph}$$

As seen, $\quad P_\Delta = 3\,P_Y$ or $P_Y = \frac{1}{3}\,P_\Delta$

It is so because delta I_{ph} is 3 times star I_{ph} and $P \propto I_{ph}^2$

Q. 24. A 3-phase load has a resistance of 10 Ω in each phase and is connected in
(*i*) star and
(*ii*) delta against a 400 V, 3-phase supply. Compare the power consumed in both the cases. (*Elect. Circuit. B.Tech, JNTU, Hyderabad 2006*)

Ans. (*i*) Star Connection :

$$V_{ph} = V_L/\sqrt{3} = 400/\sqrt{3} = 231 \text{ V}$$

$$I_{ph} = V_{ph}/R_{ph} = 231/10 = 23.1 \text{ A}$$

$$P_Y = 3\,I_{ph}^2\,R_{ph} = 3 \times 23.1^2 \times 10 = \mathbf{16{,}000\ W}$$

(*ii*) **Delta Connection :**

In this case $\quad V_{ph} = V_L = 400$ V

$$\therefore \quad I_{ph} = V_{ph}/R_{ph} = 400/10 = 40 \text{ A}$$

$$P_\Delta = 3\,I_{ph}^2\,R_{ph} = 3 \times 40^2 \times 10 = \mathbf{48{,}000\ W}$$

Incidentally $\quad P_\Delta = 3P_Y$

Q. 25. A delta-connected load draws a current of 15 A at a lagging power factor of 0.85 from a 400 V, 50 Hz, 3-φ supply. Calculate :
(*i*) resistance and inductance of each phase (*ii*) power consumed.
(*Elect. Engg. Nagpur Univ 2003*)

Ans. $\quad V_{ph} = 400$ V, $I_{ph} = I_L/\sqrt{3} = 15/\sqrt{3} = 8.66$ A

$\cos\phi = 0.85 \quad \therefore\ \phi = \cos^{-1}(0.85) = 31.8°$

Let $\quad V_{ph} = 400\angle 0°,\ I_{ph} = I_L/\sqrt{3}$

$$= 15/\sqrt{3} = 8.66\angle -31.8°$$

(*i*) $\therefore \quad Z_{ph} = \dfrac{V_{ph}}{I_{ph}} = \dfrac{400\angle 0°}{8.6\angle -31.8°} = 46.5\angle 31.8°$

$$= 39.5 + j\,24.4$$

$\therefore \quad R = \mathbf{39.5\ \Omega}$, $X_L = 24.4\ \Omega$

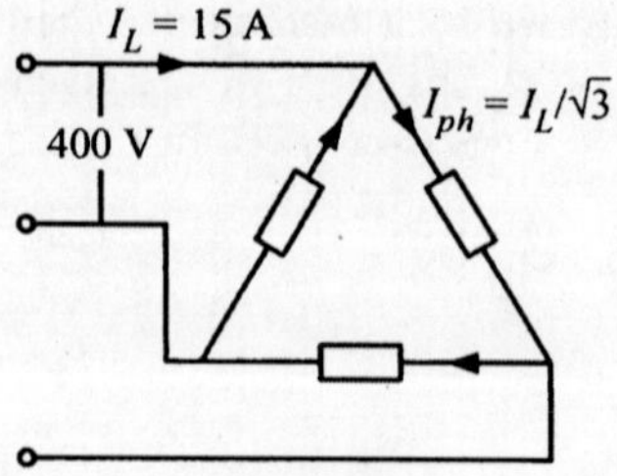

Fig. 27

$\therefore \quad L = 24.4/2\pi \times 50 = 0.077$ henry = **77 mH**

(*ii*) $\quad P = 3\, V_{ph}\, I_{ph} \cos\phi = 3 \times 400 \times 8.66 \times 0.85 = $ **8332 W**

Q. 26. Three impedances each of (3 – *j*4) Ω are connected in delta connection across a 3-phase, 230 V balanced supply. Calculate the line and phase currents in the Δ-connected load and the power delivered to the load? (*Elect Circuits. B.Tech, JNTU Hyderabad, 2006*)

Ans. Here, $V_L = V_{ph} = 230$ V

$Z_{ph} = 3 - j4 = 5\angle -53.1°$

$\therefore \quad I_{ph} = V_{ph}/Z_{ph} = 230/5 = \mathbf{46\ A}$

Also, $\quad I_L = \sqrt{3}\, I_{ph} = \mathbf{46 \times \sqrt{3}\ A}$

$P = 3\, I_{ph}^2\, R_{ph} = 3 \times 46^2 \times 3$

$= \mathbf{19{,}044\ W}$

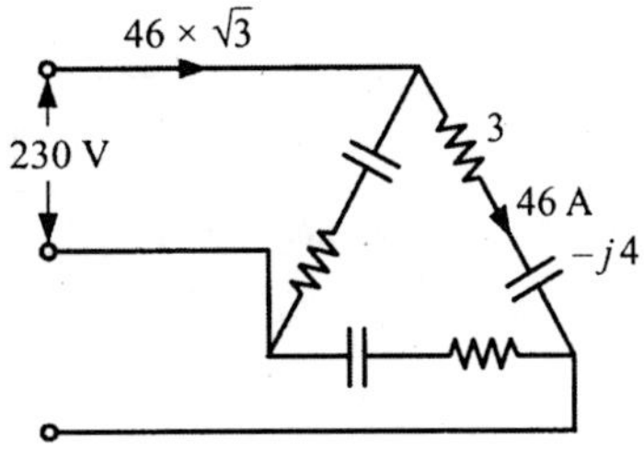

Fig. 27A

Alternatively, $\quad P = \sqrt{3}\, V_L I_L \cos\phi$

$= \sqrt{3} \times 230 \times 46 \times \sqrt{3} \times \cos 53.1°$

$= 19{,}044$ W

Q. 27. Three coils each of resistance 4 Ω and inductive reactance 3 Ω are connected in delta across 400 V, 50 Hz supply.

Find current in each coil, line current, active power and reactive power.

(*Basic Elect. Engg. Kerala Univ. 2005*)

Ans. The circuit is shown in Fig. 28.

$V_{ph} = V_L = 400$ V, $\quad Z_{ph} = 4 + j\,3 = 5\angle 37°$

$I_{ph} = V_{ph}/Z_{ph} = 400/5 = \mathbf{80\ A}$

$I_L = \sqrt{3}\, I_{ph} = \sqrt{3} \times 80 = \mathbf{138.6\ A}$

$\cos\phi = R_{ph}/Z_{ph} = 4/5 = 0.8$

$\sin\phi = X_L/Z_{ph} = 3/5 = 0.6$

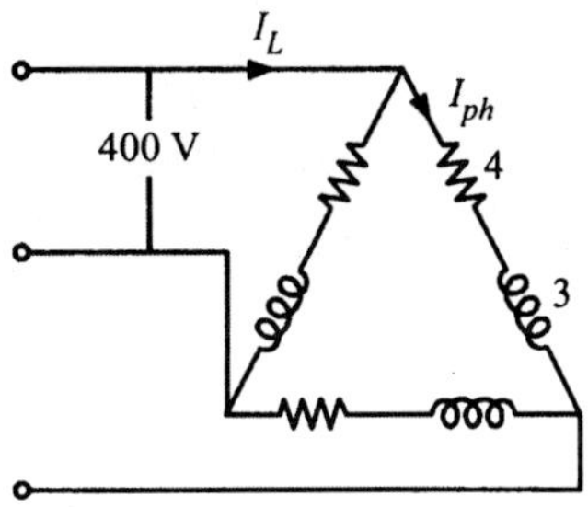

Fig. 28

Active or true power $= \sqrt{3}\, V_L\, I_L \cos\phi$

$= \sqrt{3} \times 400 \times (80 \times \sqrt{3}) \times 0.8$

$= 76{,}800$ W = **76.8 kW**

Reactive power, $\quad Q = \sqrt{3}\, V_L\, I_L \sin\phi$

$= \sqrt{3} \times 400 \times 80 \times \sqrt{3} \times 0.6$

$= 57{,}600$ VAR = **57.6 kVAR**

Q. 28. A 3-ϕ load of three equal impedances connected in delta across a balanced 400 V supply takes a line current of 10 A at a power factor of 0.7 lagging. Calculate from the first principles : (*i*) phase current (*ii*) total power (*iii*) total reactive kVA.

If the windings are connected in star, what will be the new values of the phase current and total power. (*Basic Elect. Engg. VTU Karnataka 2008*)

Ans. Delta-Connection :

In this connection $V_L = V_{ph}$ and $I_{ph} = I_L/\sqrt{3}$

$$Z_{ph} = \frac{V_{ph}}{I_{ph}} = \frac{400}{10/\sqrt{3}} = 40 \times \sqrt{3}\ \Omega$$

(*i*) $\therefore \quad I_{ph} = I_L/\sqrt{3} = \mathbf{10/\sqrt{3}\ A}$

(*ii*) $P = \sqrt{3}\, V_L\, I_L \cos\phi = \sqrt{3} \times 400 \times 10 \times 0.7 =$ **4,850 W**

(*iii*) $\cos\phi = 0.7$, hence $\sin\phi = 0.7$

total reactive power, $Q = \sqrt{3}\, V_L I_L \sin\phi$

$= \sqrt{3} \times 400 \times 10 \times 0.7 =$ **4,850 VAR**

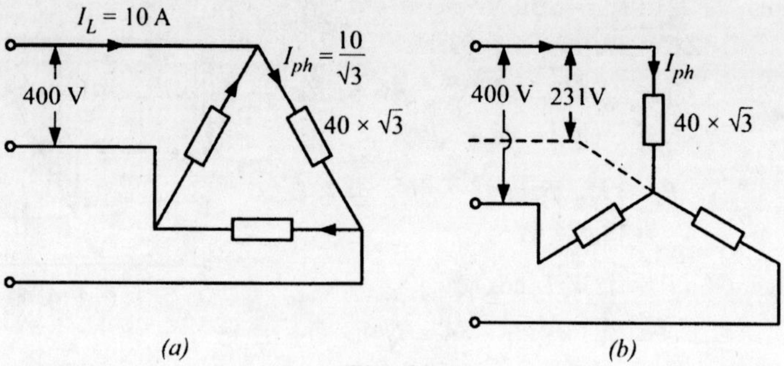

Fig. 29

The circuit is shown in Fig. 29(*a*)

Star Connection :

The same three impedances have been connected in star as shown in Fig. 29(*b*). In this connection

$$I_{ph} = I_L \quad \text{and} \quad V_{ph} = V_L/\sqrt{3}$$

$$\therefore \quad I_{ph} = \frac{V_{ph}}{Z_{ph}} = \frac{231}{40\sqrt{3}} = 3.33 \text{ A},\ I_L = 3.3 \text{ A}$$

$$\text{Total power} = \sqrt{3}\, V_L\, I_L \cos\phi = \sqrt{3} \times 400 \times 3.3 \times 0.7 = 1617 \text{ W}$$

It will be seen that $P_Y = \frac{1}{3} P_\Delta$.

Q. 29. Three impedance each of (15 – *j*20) Ω are connected in mesh across a 3-phase 400 V a.c. supply. Determine the phase current, line current, active power and reactive power drawn from the supply. (*Elect. Circuits-I, Punjab Univ. 2000*)

Ans. The mesh-connected impedances are shown in Fig. 30.

$V_{ph} = V_L = 400$ V

$I_{ph} = V_{ph} / Z_{ph}$,

$\mathbf{Z_{ph}} = 15 - j\,20$

$= 25 \angle -53.1°$

$\therefore \quad \mathbf{I_{ph}} = 400/25 \angle -53.1° = 16\angle -53.1°$ A

$I_L = 16\angle 3 = 27.7$ A, $\cos\phi = \cos 53.1° = 0.6$

Active power $= \sqrt{3} \times V_L \times I_L \times \cos\phi$

$= \sqrt{3} \times 400 \times 27.7 \times 0.6 =$ **11,520 W**

Reactive power $= \sqrt{3} \times V_L \times I_L \times \sin\phi$

$= \sqrt{3} \times 400 \times 27.7 \times 0.8 =$ **15,360 VAR**

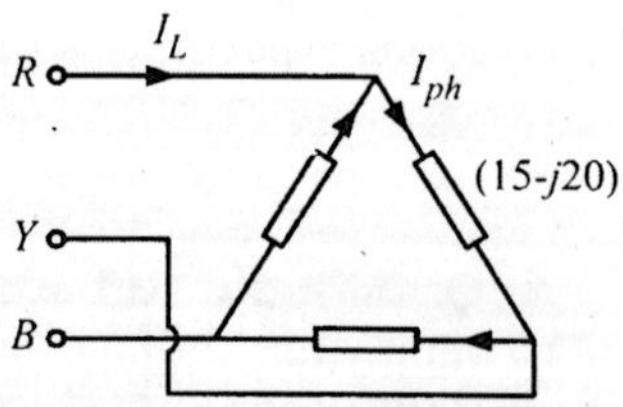

Fig. 30

Q. 30. Three similar resistors are connected in star across 400-V, 3-phase lines. The line current is 5 A. Calculate the value of each resistor. To what value should the line voltage be changed to obtain the same line current with the resistors delta-connected?

Ans. Star Connection :

$I_L = I_{ph} = 5$ A — (given)

$$V_{ph} = 400/\sqrt{3} = 231 \text{ V}$$

$$R_{ph} = 231/5 = \mathbf{46.2\ \Omega}$$

Delta Connection :

$$I_L = 5 \text{ A}$$

$$I_{ph} = 5/\sqrt{3} \text{ A}$$

$$R_{ph} = 46.2\ \Omega \qquad \text{— found above}$$

$$V_{ph} = I_{ph} \times R_{ph} = 5 \times 46.2/\sqrt{3} = \mathbf{133.3\ V}$$

Note : Voltage needed is 1/3rd the star value.

Q. 31. Three similar coils each having a resistance of 20 Ω and an inductance of 0.05H are connected in star to a 3-ϕ, 50 Hz supply with 400 V between the lines. Calculate power factor, total power absorbed and line current.

If the same coils are reconnected in delta across the same supply, what will be the power factor, total power absorbed and line current. (*Elect. Engg. Nagpur Univ. 2004*)

Ans. $X_L = 2\pi \times 50 \times 0.05 = 15.7\ \Omega$

$$Z_{ph} = \sqrt{20^2 + 15.7^2} = 25.4\ \Omega$$

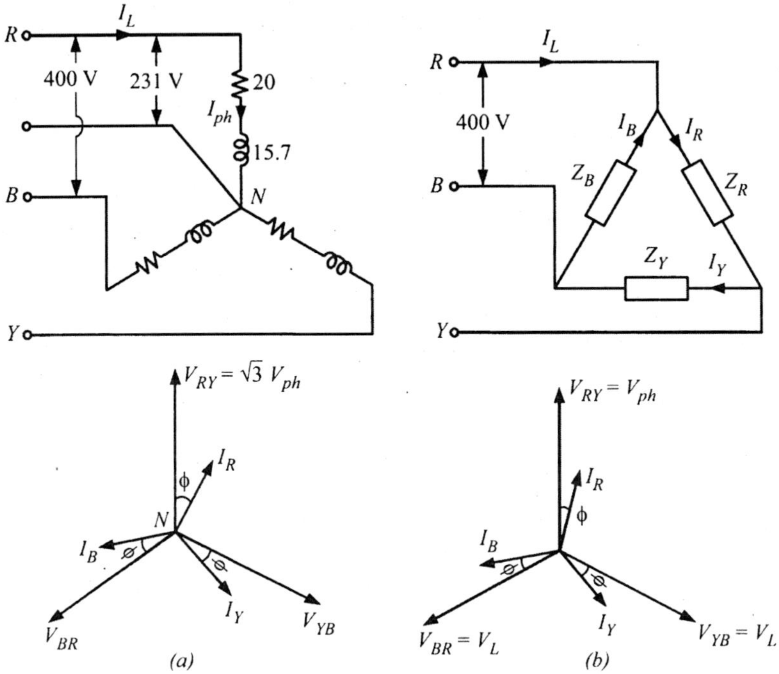

Fig. 31

(*a*) **Star-connection** [Fig. 31(*a*)]

$$V_{ph} = \frac{400}{\sqrt{3}} = 231 \text{ V}, \quad I_{ph} = \frac{V_{ph}}{Z_{ph}} = \frac{231}{25.4} = 9.09 \text{ A}$$

$$\therefore \quad I_L = I_{ph} = \mathbf{9.09\ A}$$

$$\text{power factor} = \cos\phi = R_{ph}/Z_{ph} = 20/25 = \mathbf{0.787}$$

$$P = \sqrt{3} \times 400 \times 9.09 \times 0.787 = \mathbf{4956\ W}$$

The circuit and the vector diagram are shown in Fig. 31(*a*).

(*b*) **Delta-connection** [Fig. 31(*b*)]

$$V_{ph} = V_L = 400 \text{ V}, \quad I_{ph} = V_{ph}/Z_{ph} = 400/25.4 = \mathbf{15.75\ A}$$

$$I_L = \sqrt{3}\ I_{ph} = \sqrt{3} \times 15.75 = \mathbf{27.27\ A}$$

Power factor of the coil is the same as before *i.e.* 0.787

$$P = \sqrt{3}\ V_L I_L \cos\phi$$

$$= \sqrt{3} \times 400 \times 27.27 \times 0.787 = \mathbf{14{,}868\ W}$$

It will be seen that $P_\Delta = 3\ P_Y$.

Q. 32. The currents in the *RY*, *YB* and *BR* branches of a delta-connected system with symmetrical voltages are : 25 A at p.f. 0.8 lag, 30 A at p.f. 0.7 lead and 20 A at unity power factor respectively. Determine the currents in each line. Phase sequence is *RYB*.

(*Basic Elect. Engg. R.G.P.V. Bhopal 2007*)

Ans. It may be noted that the Δ-connected load is *unbalanced*.

We are given that

$$\mathbf{I_{RY}} = 25\angle -\cos^{-1} 0.8 = 25\angle -37° = 20 - j\,15 \qquad \text{— lagging}$$

$$\mathbf{I_{YB}} = 30\angle \cos^{-1} 0.7 = 30\angle 45.5° = 21 + j\,21.42 \qquad \text{— leading}$$

$$\mathbf{I_{BR}} = 20\angle \cos^{-1} 1.0 = 20\angle 0° = 20 + j0 \qquad \text{— in phase}$$

Now, currents in the three lines are (Fig. 32)

$$I_{L_1} = \mathbf{I_{RY}} - \mathbf{I_{BR}}$$

$$= 20 - j\,15 - 20 - j\,0 = 0 - j\,15$$

$$= \mathbf{15\angle -90°}$$

$$\mathbf{I_{L_2}} = \mathbf{I_{YB}} - \mathbf{I_{RY}} = 21 + j\,21.42 - 20 + j\,15$$

$$= 1 + j\,36.42 = \mathbf{36.43\angle 88.14°}$$

$$\mathbf{I_{L_3}} = \mathbf{I_{BR}} - \mathbf{I_{YB}} = 20 + j0 - 21 - j\,21.42$$

$$= -1 - j\,21.42 = \mathbf{21.44\angle -92.3°}$$

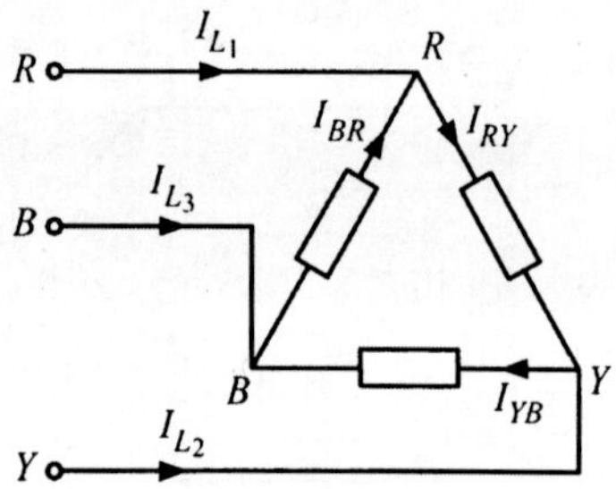

Fig. 32

Following points should be noted about *unbalanced* Δ-connected loads supplied from a *balanced* system of line voltages:

1. Since phase voltages in a delta are equal to line voltages, the *phase voltages are always balanced!*
2. the phase currents of the unbalanced Δ-loads are *unbalanced* and their vector sum is *not equal to zero.*
3. even though line currents are unbalanced, their *vector sum is zero.*

As seen from above, the vector sum of the three line currents

$$= I_{L_1} + I_{L_2} + I_{L_3}$$

$$= (0 - j\,15) + (1 + j\,36.42) + (-1 - j\,21.42) = 0$$

Q. 33. The delta-connected generator of Fig. 33 has the following voltages.

$$\mathbf{V_{RY} = 220\angle 0°, \quad V_{YB} = 220\angle -120° \text{ and } V_{BR} = 220\angle -240°}$$

The line current I_R is given as $10\angle 0°$. The load is balanced and delta-connected.

Find:

(*i*) impedance per phase (*ii*) current per phase and (*iii*) other line currents I_Y and I_B.

(*Elect. Engg. Nagpur 2000*)

Ans. In a balanced delta load (Fig. 33), the three line currents are equal in magnitude and have a phase difference of 120° with each other. We are given that $I_R = 10\angle 0°$, hence other line currents are $I_Y = 10\angle -120°$ and $I_B = 10\angle -240°$ or $10\angle 120°$.

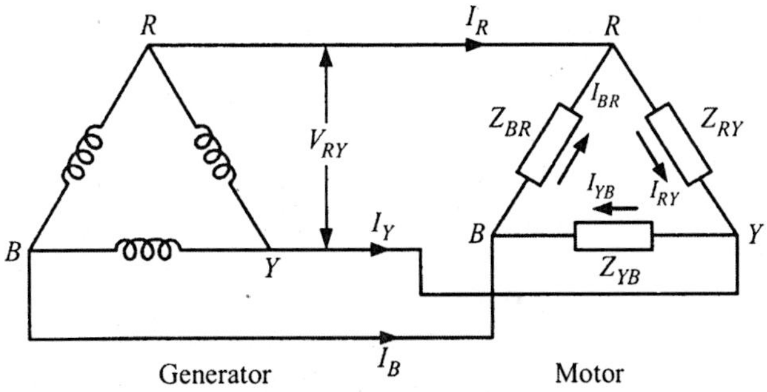

Fig. 33

(*ii*) Current per phase

$$I_{ph} = I_L/\sqrt{3} = I_R/\sqrt{3} = 10/\sqrt{3} = \mathbf{5.77\ A}$$

(*i*) impedance per phase

$$V_{ph} = V_L = V_{RY} = 220\text{V},\ I_{ph} = 5.77\text{ A}$$

$$\therefore \quad Z_{RY} = 220/5.77 = \mathbf{38.1\ \Omega}$$

(*iii*) other line currents are

$$\boldsymbol{I_Y = 10\angle{-120°},\ I_B = 10\angle{-240°}\ \text{or}\ 10\angle 120°}$$

Q. 34. A three-phase Δ-connected load consists of three similar coils, each of resistance 50 Ω and inductance 0.3 H. The supply is 415 V, 50 Hz. Calculate:

(*i*) the line currents (*ii*) the power factor (iii) total active and reactive powers when the load is Δ-connected. Draw the phasor diagram. (*Elect. Engg. Nagpur Univ. 2006*)

Ans. The circuit is shown in Fig. 34(*a*)

$$X_L = 2\pi fL = 314 \times 0.3 = 94.2\ \Omega$$

$$Z = R + jX_L = 50 + j\,94.2 = 106.7\angle 62°$$

$$V_{ph} = V_L = 415\text{ V},\ I_{ph} = \frac{V_{ph}}{Z_{ph}} = \frac{415}{106.7} = 3.89\text{ A}$$

(*i*) $\therefore$ $$I_L = \sqrt{3}\,I_{ph} = \sqrt{3} \times 3.89 = \mathbf{6.73\ A}$$

(*ii*) Power factor $$\cos\phi = \frac{R_{ph}}{Z_{ph}} = \frac{50}{106.7} = \mathbf{0.46},\ \phi = 62°$$

(*iii*) Total active power, $P = 3\ I_{ph}^2\ R_{ph} = 3 \times 3.89^2 \times 50 = \mathbf{2270\ W}$

Total reactive power, $Q = 3\ V_{ph}\ I_{ph} \sin\phi = 3 \times 415 \times 3.89 \times \sin 62° = \mathbf{4300\ VAR}$

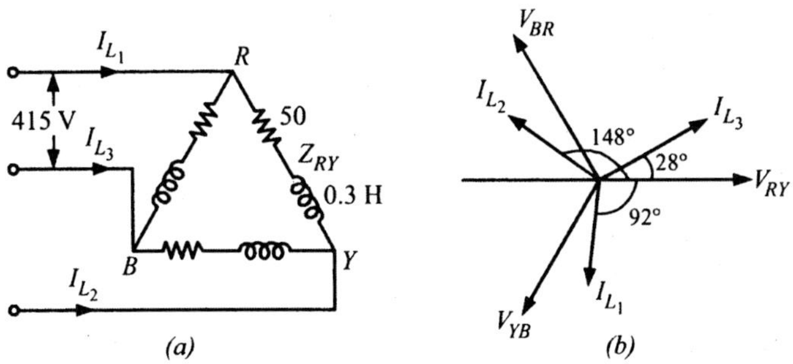

Fig. 34

Phase Currents :

$$\mathbf{I_{RY}} = \frac{\mathbf{V_{RY}}}{\mathbf{Z_{RY}}} = \frac{415\angle 0°}{106.7\angle 62°} = 3.89\angle{-62°}$$

$$\mathbf{I_{YB}} = \frac{\mathbf{V_{YB}}}{\mathbf{Z_{YB}}} = \frac{415\angle -120°}{106.7\angle 62°} = 3.89\angle -182°$$

$$\mathbf{I_{BR}} = \frac{\mathbf{V_{BR}}}{\mathbf{Z_{BR}}} = \frac{415\angle 120°}{106.7\angle 62°} = 3.89\angle 58°$$

These phase currents are shown in the vector diagram of Fig. 34(*b*).

Line Currents :

Line current, $\mathbf{I_{L_1}} = \mathbf{I_{RY}} - \mathbf{I_{BR}} = 3.89\angle -62° - 3.89\angle 58° = \mathbf{6.737\angle -92°}$

$\mathbf{I_{L_2}} = \mathbf{I_{YB}} - \mathbf{I_{RY}} = 3.89\angle -182° - 3.89\angle -62° = \mathbf{6.737\angle 148°}$

$\mathbf{I_{L_3}} = \mathbf{I_{BR}} - \mathbf{I_{YB}} = 3.89\angle 58° - 3.89\angle -182° = \mathbf{6.737\angle 28°}$

Q. 35. A delta-connected load has the following impedances; $Z_{RY} = j\,10\ \Omega$, $Z_{YB} = 10\angle 0°\ \Omega$ and $Z_{BR} = -j\,10\ \Omega$. If the load is connected across 100 volt balanced 3-φ supply, obtain the line currents. ***(Elect. Engg. R.G.P.V., Bhopal 2004)***

Ans. The unbalanced Δ-connected load circuit is shown in Fig. 35.

$\mathbf{V_{RY}} = 100\angle 0°$

$\mathbf{V_{YB}} = 100\angle -120°$

$\mathbf{V_{BR}} = 100\angle 120°$

Also, $\mathbf{Z_{RY}} = j\,10 = 10\angle 90°$

$\mathbf{Z_{YB}} = 10\angle 0°$

$\mathbf{Z_{BR}} = -j\,10 = 10\angle -90°$

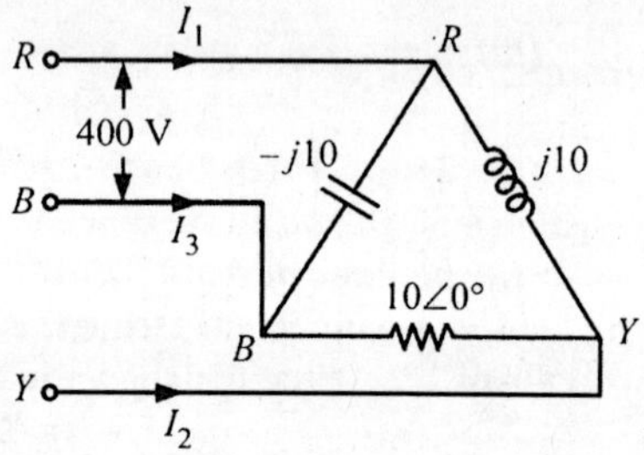

Fig. 35

$$\therefore \quad \mathbf{I_{RY}} = \frac{V_{RY}}{Z_{RY}} = \frac{100\angle 0°}{10\angle 90°} = 10\angle -90°$$

$$\mathbf{I_{YB}} = \frac{V_{YB}}{Z_{YB}} = \frac{100\angle -120°}{10\angle 0°} = 10\angle -120°$$

$$\mathbf{I_{BR}} = \frac{V_{BR}}{Z_{BR}} = \frac{100\angle 120°}{10\angle -90°} = 10\angle 210°$$

Line current, $\mathbf{I_{L_1}} = \mathbf{I_{RY}} - \mathbf{I_{BR}} = 10\angle -90° - 10\angle 210°$

$= -j\,10 - (-8.66 - j\,5) = 8.66 - j\,5 = \mathbf{10\angle -30°}$

Line current $\mathbf{I_{L_2}} = \mathbf{I_{YB}} - \mathbf{I_{RY}} = 10\angle -120° - 10\angle -90°$

$= -5 - j\,8.66 - (-j\,10) = -5 + j\,1.34 = \mathbf{5.18\angle 165°}$

Line current $\mathbf{I_{L_3}} = \mathbf{I_{BR}} - \mathbf{I_{YB}} = 10\angle 210° - 10\angle -120°$

$= -8.66 - j\,5 - (-5 - j\,8.66) = -3.66 + j\,3.66 = \mathbf{5.18\angle 135°}$

As expected, the vector sum of the *line currents is zero.*

Q. 36. A balanced load takes 20 kVA at a power factor of 0.8 lagging from a 250 V, 50 Hz, 3-φ circuit. Calculate the values of the components in the circuit if the load is (*i*) delta-connected (*ii*) star-connected.

Ans. (*i*) Delta-connection :

$20{,}000 = \sqrt{3} \times 250 \times I_L, \quad I_L = 80/\sqrt{3}\ \text{A}$

$I_{ph} = I_L/\sqrt{3} = 80/\sqrt{3} \times \sqrt{3} = 80/3\ \text{A}$

$$Z_{ph} = \frac{250}{80/3} = \frac{250 \times 3}{80} = \frac{75}{8}\ \Omega \text{ (Fig. 36)}$$

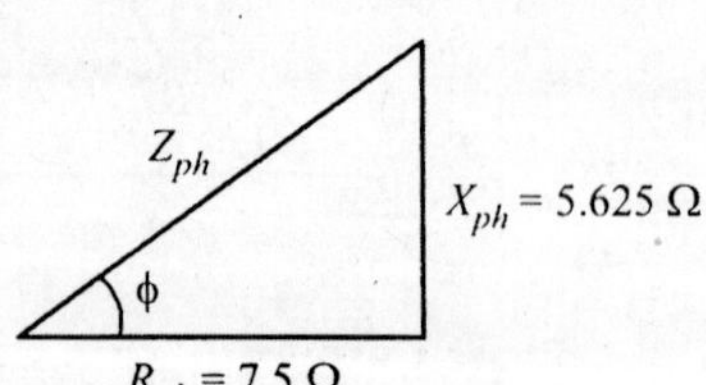

Fig. 36

$$R_{ph} = Z_{ph} \times \cos\phi = \frac{75}{8} \times 0.8 = \mathbf{7.5\ \Omega}$$

$$X_{ph} = Z_{ph} \sin \phi = \frac{75}{8} \times 0.6 = \mathbf{5.625\ \Omega}$$

Now, $$X_L = 2\pi f L \text{ or } 5.625 = 314 \times L \text{ or } L = 0.0179 \text{ H} = \mathbf{17.9\ mH}$$

***(ii)* Star-connection :** In *Y*-connection, the component values one-third of the delta values *i.e.* **2.5 Ω** and **5.97 mH.**

Q. 37. Three 40 Ω non-inductive resistances are connected *(i)* in star *(ii)* in meash across 200 V, 3-phase lines. Calculate the power taken from the mains in each case. If one of the two resistors is disconnected, what would be the power taken from the mains in each case? (*Elect. Engg. A.M.Ae.S.I. June 2001*)

Ans. *(i)* Star Connection :

$$V_{ph} = 200/\sqrt{3} \text{ V}$$

$$I_{ph} = 200/\sqrt{3} \times 40 = 5/\sqrt{3} \text{ A}$$

$$P = \sqrt{3}\, V_L\, I_L \cos \phi$$

$$= \sqrt{3} \times 200 \times \frac{5}{\sqrt{3}} \times 1 = \mathbf{1000\ W}$$

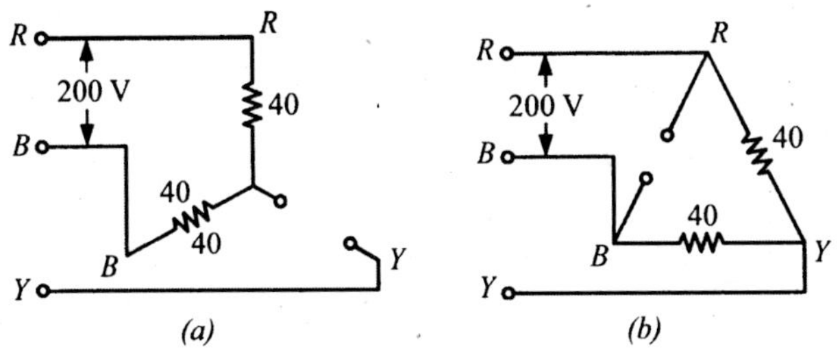

Fig. 37

(*ii*) **Delta Connection :**

$$V_{ph} = 200 \text{ V} ;\ R_{ph} = 40\ \Omega$$

$$I_{ph} = 200/40 = 5 \text{ A} ;\ I_L = 5 \times \sqrt{3} \text{ A}$$

$$P = \sqrt{3} \times 200 \times 5 \times \sqrt{3} \times 1 = \mathbf{3{,}000\ W}$$

When one of the resistors is disconnected

***(i)* Star Connection** [Fig. 37(*a*)]

The circuit no longer remains a 3-phase circuit but consists of two 40 Ω resistors in series across a 200 V supply. Current in lines *R* and *B* is = 200/80 = 2.5 A.

Power absorbed in both = 200 × 2.5 = **500 W**

Hence, by disconnecting one resistor, the power consumption is reduced by half.

***(ii)* Delta Connection** [Fig. 37(*b*)]

In this case, currents in *R* and *B* lines remain as usual 120° out of phase with each other.

Current in each phase = 200/40 = 5 A.

Power consumption in both = $2 \times 5^2 \times 40$ = **2,000 W** (or $P = 2 \times 5 \times 200 = 2000$ W)

In this case, when one resistor is disconnected, the power consumption is reduced by one-third.

Q. 38. What are the disadvantages of supplying low power factor load from a single-phase or three-phase generator?

Ans. The power drawn by a single-phase load (such as a motor) is given by

$$P = VI \cos \phi \quad \text{or} \quad W = VA \cos \phi$$

Similarly, the power drawn by a three-phase load is

$$P = \sqrt{3}\, V_L\, I_L \cos \phi$$

The capacity of ac generators (*i.e.* alternators) is always given in terms of kVA. As seen from above, the actual or real power is directly proportional to the *power factor* of the connected load. In other words, for the given kVA of the alternator, $P \propto \cos \phi$.

It means higher the power factor, *greater is the actual power delivered to the load.*

As an example, consider the case of a 3-ϕ generator rated to give 2000 A at a voltage of 500 V. Obviously, the rating of the generator is $2000 \times 500/1000 = 1000$ kVA.

If the load power factor is unity, then 1000 kVA are also 1000 kW. But if the power factor of the load is 0.5, then power supplied is = kVA $\cos \phi = 1000 \times 0.5 = 500$ kW only. As seen, the power delivered is halved though the generator is supplying its *full rated output* of 1000 kVA. The above fact may be put in an other way. We can say that for a *given power*, lower the load power factor, the *larger* must be the *size i.e.* kVA rating of the generator and, hence *greater* must be the cross-section of the conductors to transmit this power. It will obviously increase the cost of generation and transmission of the electrical energy. That is why the electric supply companies compel their customers to improve the power factor of their loads.

Moreover, kVA is directly proportional to current which itself is *inversely* proportional to the power factor for a *given power*. Hence, if power factor is low, then the current required for supplying the given amount of power is very high. This high current produces large voltage drops on the line conductors when the load is switched on. This drop decreases the voltage available *at the load.* The difference in the generator voltage and the actual load voltage is called *voltage regulation.* Such changes in voltage are not good from the users point of view.

To sum up the above, we can say that low power factor of the load leads to the following problems :

(*i*) large kVA relative to kW power.

(*ii*) poor voltage regulation *i.e.* larger voltage drops on the supply lines.

Q. 39. What are the different methods used for correcting (*i.e.* improving) the power factor of 1-ϕ or 3-ϕ loads?

Ans. Single-phase Load : The power factor of single-phase loads such as motors can be improved by connecting a suitable capacitor in parallel with the motor (Q. 40).

Three-phase Load : Following methods are available for improving the power factor of a 3-ϕ load :

(*i*) **Synchronous motors :** These motors when *running idle* draw *leading* kVARs from the 3-ϕ supply. These leading kVARs are used to reduce the lagging kVARs drawn by induction motor etc. thus improving their power factor.

(*ii*) **Static Capacitors :** These are used for improving the power factor of a *group* of 3-ϕ motors. They draw a current which leads the voltage by 90°. This leading current reduces the lagging current drawn by the motors.

(*iii*) **Phase advancers :** They are capacitors which are fitted with *individual* ac machines.

It may, however, be noted that the amount of power factor improvement depends on two factors. The saving in cable cost and I^2R losses are offset by cost of the power factor improvement equipment required. Hence, there is a point of balance at which the best saving can be obtained. Generally, the most desired power factor value ranges from 0.8 to 0.95.

Another point worth nothing is that even though the correction equipment changes the reactive power and the power factor, it, however, does not change the *real power* taken by the load.

Q. 40. A 200 V, 50 Hz single-phase ac motor draws a power of 10 kW at a power factor of 0.75 lagging. Calculate the change in current taken from the supply and the new power factor when a 250 μF capacitor is connected in parallel with the motor.

If the motor is supplied through a cable of 0.05 Ω resistance, calculate the power loss in the cable before and after connecting the capacitor.

Ans. The current I_m drawn by the motor of [Fig. 38(*a*)] can be found from the relation.

$$P_m = VI_m \cos\phi_m$$

$$\therefore \quad I_m = \frac{P_m}{V\cos\phi_m} = \frac{10{,}000}{200\times 0.75} = 66.7 \text{ A}$$

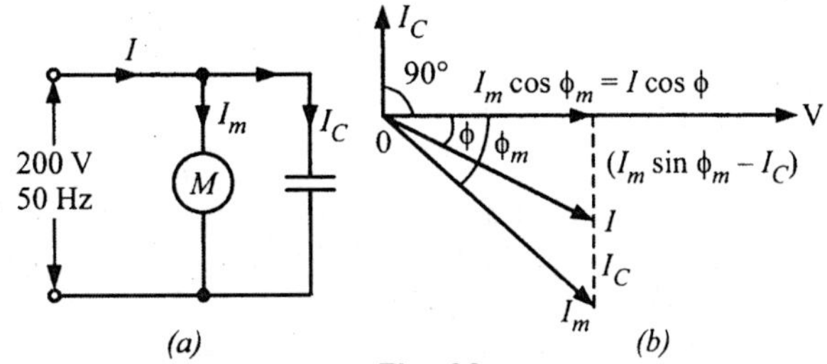

Fig. 38

This current lags the applied voltage by $\phi_m = \cos^{-1} 0.75 = 41.5°$

As seen from the vector diagram of Fig. 38(*b*)

$$X_C = \frac{1}{2\pi fC} \qquad \therefore \quad I_C = \frac{V}{X_C} = \frac{V}{1/2\pi fC}$$

$$I_C = 2\pi fC\,V$$

$$= 2\pi \times 50 \times (250 \times 10^{-6}) \times 200 = 15.8 \text{ A}$$

This current leads the applied voltage by 90° as shown in the vector diagram.

The new current I (Fig. 38) drawn by the motor is the vector sum of I_m and I_C.

$$I\cos\phi = I_m \cos\phi_m = 66.7 \times 0.75 = 50 \text{ A}$$

$$I\sin\phi = I_m \sin\phi_m - I_C = 66.7 \times 0.662 - 15.8$$

$$= 44.1 - 15.8 = 28.3 \text{ A}$$

$$\therefore \quad I = \sqrt{50^2 + 28.3^2} = 57.4 \text{ A}$$

As seen from the vector diagram, the new power factor is given by

$$\cos\phi = \frac{I\cos\phi}{I} = \frac{50}{57.4} = \mathbf{0.87\ lag}$$

Hence, the power factor has improved from 0.75 lag to 0.87 lag.

The current drawn from the supply has *decreased* from 66.7 A to 57.4 A *i.e.* a decrease of **9.3A.**

$$\text{Initial power loss} = I_m^2 R = 66.7^2 \times 0.05 = \mathbf{222.4\ W}$$

$$\text{Final power loss} = I^2 R = 57.4^2 \times 0.05 = \mathbf{164.7\ W}$$

Q. 41. A 50 MVA, 11 kV, 3-φ alternator supplies full load at a lagging power factor of 0.7. What would be the percentage increase in earning capacity if the load power factor is increased to 0.95?

Ans. The earning capacity of an alternator is proportional to the power (in MW or kW) supplied by it. MW supplied at 0.7 lagging p.f. = 50 × 0.7 = 35

MW supplied at 0.95 lagging p.f. = 50 × 0.95 = 47.5

∴ increase in MW supply = 12.5

$$\therefore \quad \text{percentage increase in earning capacity} = \frac{12.5}{35} \times 100 = \mathbf{35.7\%}$$

Q. 42. A 3-φ balanced load draws 10 kW power from a 400V, 3-φ, 50 Hz, 4-wire supply at 0.8 lagging power factor.

(*i*) Determine the line current

(*ii*) What would be the line current if power factor were raised to unity with power drawn remaining the same?

(*iii*) How can the power factor be improved to unity?

(*Basic Elect. Engg. R.G.P.V, Bhopal, 2004*)

Ans. (*i*) $P = \sqrt{3}\, V_L I_L, \cos\phi$

or $10 \times 10^3 = \sqrt{3} \times 400 \times I_L \times 0.8, \quad \therefore\ I_L = \mathbf{18.04\ A}$

(*ii*) If p.f. $= \cos\phi = 1$, then

$10 \times 10^3 = \sqrt{3} \times 400 \times I_L \times 1$ or $I_L = \mathbf{14.43\ A}$

It is seen that when power factor is improved, current drawn from the supply is *decreased* from 18.04 A to 14.43 A.

(*iii*) Please see *Q*. 39.

Q. 43. Describe the method of measuring power in a 3-φ circuit by using two wattmeters when balanced load with a lagging power factor is connected across it.

Ans. The two-wattmeter method can measure the power drawn by a 3-φ load whether *Y*-connected or Δ-connected *without regard to balance*. Consider a balanced *Y*-connected load having a lagging power factor as shown in Fig. 39(*a*).

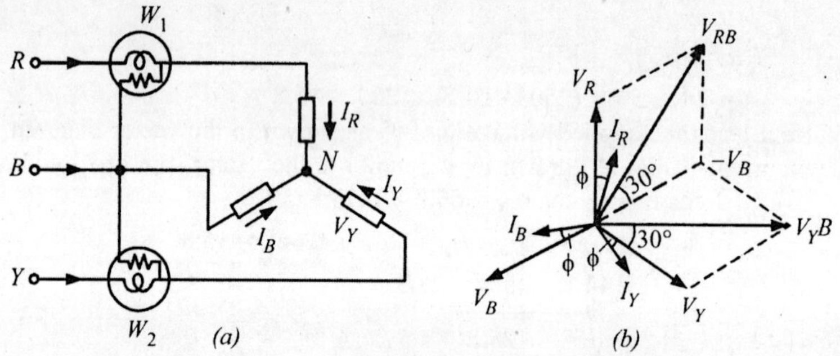

Fig. 39

The current coils of the wattmeters are inserted in *any two lines*, say, *R* and *Y* lines and their voltage coils are joined to the *B*-line.

Let V_R, V_Y and V_B be the rms values of the three phase voltages and I_R, I_Y and I_B the rms values of the phase currents. Assuming these voltages and currents to be sinusoidal, they have been represented by vectors in Fig. 39(*b*), with phase currents lagging behind their phase voltages by φ. As seen, current through wattmeters W_1 is I_R and p.d. across its voltage coil is V_{RB} which is the *vector difference* of V_R and V_B. It is seen that phase angle between V_{RB} and I_R is $(30° - \phi)$ for a *lagging* power factor.

$\therefore$ reading of $W_1 = V_{RB} I_R \cos(30° - \phi)$

Similarly, current through $W_2 = I_Y$

p.d. across $W_2 = V_{YB} = V_Y - V_B$ — vector difference

Phase angle between V_{YB} and I_Y is $(30° + \phi)$

$\therefore$ reading of $W_2 = V_{YB} I_Y \cos(30° + \phi)$

Since the load is balanced, hence

$$V_{RB} = V_{YB} = \text{line voltage, } V_L$$

$$I_R = I_Y = \text{line current, } I_L$$

$$\therefore \quad W_1 = V_L I_L \cos(30° - \phi)$$

$$\text{and} \quad W_2 = V_L I_L \cos(30° + \phi)$$

$$\therefore \quad W_1 + W_2 = V_L I_L \cos(30° - \phi) + V_L I_L \cos(30° + \phi)$$

$$= V_L I_L [\cos 30° \cos\phi + \sin 30° \sin\phi + \cos 30° \cos\phi - \sin 30° \sin\phi]$$

$$= V_L I_L (2 \cos 30° \cos\phi) = \sqrt{3}\, V_L I_L \cos\phi$$

= total power in the 3-phase load

Hence, we find that the sum of the two wattmeter readings gives the *total* power consumed by the 3-φ load. It should be noted that phase sequence *RYB* has been assumed. If we take the phase sequence *R-B-Y*, the readings of the two wattmeters will get *interchanged*.

Q. 44. Draw the vector diagram when the two-wattmeter method is used for measuring power in a 3-φ balanced load having a leading power factor. Are the two wattmeter readings interchanged with those for lagging power factor?

Ans. It will be shown that for a *leading* power factor load, the wattmeter readings get *interchanged*. It means what W_1 was reading on a *lagging* power factor is *now* being indicated by W_2 on a *leading* power factor and vice-versa. Fig. 40 shows the vector diagram for a 3-φ load having a *leading* power factor. As seen, I_R leads V_R, I_Y leads V_Y and I_B leads V_B by angle φ. A phase sequence of *RYB* and *CCW* rotation of vectors has been assumed.

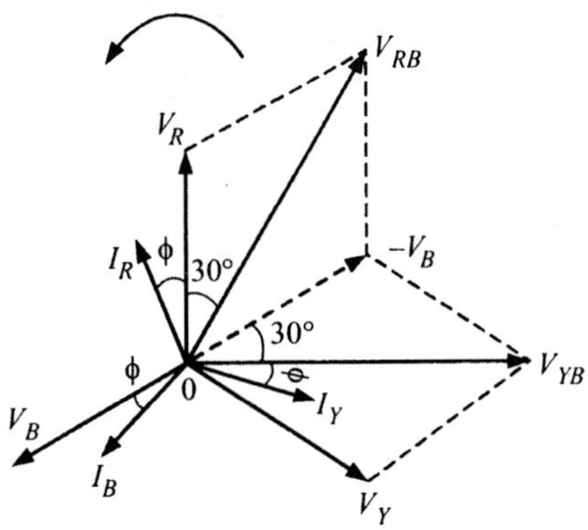

Fig. 40

It is seen that phase angle between V_{RB} and I_R is (30° + φ).

∴ reading of $W_1 = V_{RB}\, I_R \cos(30° + \phi)$

Current through $W_2 = I_Y$

p.d. across $W_2 = V_{YB}$

reading of $W_2 = V_{YB}\, I_Y \cos(30° - \phi)$

For a balanced load,

$$W_1 = V_L\, I_L \cos(30° + \phi)$$
$$W_2 = V_L\, I_L \cos(30° - \phi)$$

As seen, the two wattmeter readings have interchanged.

As shown in *Q*. 43,

$$W_1 + W_2 = V_L\, I_L \cos(30° + \phi) + V_L\, I_L \cos(30° - \phi)$$
$$= \sqrt{3}\, V_L\, I_L \cos\phi$$

Q. 45. How do the readings of the two wattmeters used for measuring power taken by a 3-φ load vary with the power factor of the load? For what lagging power factor will one of the wattmeters road zero?

Ans. The readings of the two wattmeters for a *lagging* power factor are

$$W_1 = V_L\, I_L \cos(30° - \phi)$$
$$W_2 = V_L\, I_L \cos(30° + \phi)$$

It is obvious that readings of the two wattmeters depend not only on the load but upon its *power factor* also.

Consider the following cases:

1. **When cos φ = 1**

It means φ = 0 *i.e.* the load is *purely resistive*. In this case,

$$W_1 = W_2 = V_L I_L \cos 30° = \frac{\sqrt{3}}{2} V_L I_L$$

As seen, both wattmeters indicate *equal and positive* readings. Each wattmeter reads one-half of the total power.

Also, $\quad W_1 + W_2 = 2 \times \dfrac{\sqrt{3}}{2} V_L I_L = \sqrt{3}\, V_L I_L$ = total power

2. **When cos φ = 0.866**

It means φ = 30°. In this case,

$$W_1 = V_L I_L \cos (30° - 30°) = V_L I_L \cos 0° = V_L I_L$$

$$W_2 = V_L I_L \cos (30° + 30°) = V_L I_L \cos 60° = \frac{1}{2} V_L I_L$$

Obviously, reading of W_1 is *twice* that of W_2.

It is seen that $W_1 + W_2 = V_L I_L + \frac{1}{2} V_L I_L = \frac{3}{2} V_L I_L$

$$P = \sqrt{3} V_L I_L \cos 30° = \sqrt{3} V_L I_L \frac{\sqrt{3}}{2} = \frac{3}{2} V_L I_L$$

$$\therefore \quad P = W_1 + W_2$$

3. **When cos ϕ = 0.5**

It means $\phi = \cos^{-1} 0.5 = 60°$.

In this case, $W_2 = V_L I_L \cos (30° + 60°) = V_L I_L \cos 90° = 0$

Hence, power is being measured by W_1 alone because $W_2 = 0$.

$$W_1 = V_L I_L \cos (30° - 60°) = V_L I_L \cos 30° = \frac{\sqrt{3}}{2} V_L I_L$$

$$P = \sqrt{3} V_L I_L \cos 60° = \frac{\sqrt{3}}{2} V_L I_L \quad \therefore P = W_1 + W_2$$

4. **When 0 < cos ϕ < 0.5**

It means angle ϕ lies between 60° and 90° *i.e.* $60° < \phi < 90°$

Whether ϕ = 60° or 90°, W_1 will read *positive.* However, reading of W_2 will be *negative* in both cases. This can be checked by putting these values of ϕ in the expression for W_2 reading given above. Hence, total power will be $(W_1 - W_2)$

5. **When cos ϕ = 0**

It means ϕ = 90° lagging

$$W_1 = V_L I_L \cos (30° - 90°) = V_L I_L \sin 30° = \frac{1}{2} V_L I_L$$

$$W_2 = V_L I_L \cos (30° + 90°) = - V_L I_L \sin 30° = - \frac{1}{2} V_L I_L$$

Since the two readings are *equal in magnitude but of opposite sign,*

$$\therefore \quad W_1 + W_2 = 0$$

As we know, ϕ = 90° only in the case of pure inductance or capacitance. These elements draw no power from the supply.

Q. 46. Prove that the power factor of a balanced 3-ϕ load can be determined from the readings of the two wattmeters.

Ans. If the three-phase load is balanced and the voltages and currents are sinusoidal, then for a *lagging* power factor

$$W_1 + W_2 = V_L I_L \cos (30° - \phi) + V_L I_L \cos (30° + \phi)$$

$$= \sqrt{3} V_L I_L \cos \phi \quad ...(i)$$

Similarly, $W_1 - W_2 = V_L I_L \cos (30° - \phi) - V_L I_L \cos (30° + \phi)$

$$= V_L I_L \sin \phi \quad ...(ii)$$

Dividing (*ii*) by (*i*), we have

$$\tan \phi = \sqrt{3} \frac{(W_1 - W_2)}{(W_1 + W_2)}$$

Knowning tan ϕ and hence ϕ from the above expression, the value of power factor cos ϕ can be found.

However, it is very important to keep in mind that if W_2 reading has been taken after *reversing* the voltage coil *i.e.* if W_2 is *negative*, then the above relation becomes

$$\tan \phi = \sqrt{3}\,\frac{W_1 - (-W_2)}{W_1 + (-W_2)} = \frac{\sqrt{3}\,(W_1 + W_2)}{W_1 - W_2}$$

However, an important point to remember is that in the above expression only *numerical values* of W_1 and W_2 should be substituted.

Q. 47. Discuss the variations in the readings of the two wattmeters when the 3-ϕ load has a leading power factor. Give the relation between tan ϕ and wattmeter readings.

Ans. As shown in Q. 44, the readings of the two wattmeters in the case of a balanced 3-ϕ load with a *leading* power factor or

$$W_1 = V_L I_L \cos (30° + \phi)$$

$$W_2 = V_L I_L \cos (30° - \phi)$$

As seen, the two readings have *interchanged* with those for a lagging power factor.

The variations in the readings due to change in leading power factor are as under :

1. **When cos ϕ = 1**

Here, $\phi = 0$. Hence, the two readings become

$$W_1 = W_2 = V_L I_L \cos 30° = \frac{\sqrt{3}}{2} V_L I_L$$

$$\therefore \qquad W_1 + W_2 = 2 \times \frac{\sqrt{3}}{2} V_L I_L = \sqrt{3}\, V_L I_L = \text{total power}$$

As seen, both wattmeters indicate *equal and positive* values. It happens in the case of pure resistive loads.

2. **When cos ϕ = 0.866**

Here, $\phi = 30°$. Hence, the two readings become

$$W_1 = V_L I_L \cos (30° + 30°) = V_L I_L \cos 60° = \frac{1}{2} V_L I_L$$

$$W_2 = V_L I_L \cos (30° - 30°) = V_L I_L \cos 0° = V_L I_L$$

Obviously, W_2 is twice that of W_1

It is opposite of that for a lagging power factor (Q. 45). The readings have interchanged.

3. **When cos ϕ = 0.5**

It means $\phi = 60°$

$$\therefore \qquad W_1 = V_L I_L \cos(30° + 60°) = V_L I_L \cos 90° = 0$$

It means that the power is being read by W_2 alone. Again for a lagging power factor, it is just the opposite.

4. **When 0 < cos ϕ < 0.5**

It means angle ϕ lies between 60° and 90°. Let $\phi = 75°$. Then

$$W_1 = V_L I_L \cos (30° + 75°) = V_L I_L \cos (90° + 15°)$$

$$= - V_L I_L \sin 15°$$

$$W_2 = V_L I_L \cos (30° - 75°) = V_L I_L \cos (-45°) = V_L I_L \cos 45°$$

It is seen that reading of W_1 is *negative* and that of W_2 is *positive*. Total power = $W_2 - W_1$. The results here are opposite to those for lagging power factor (*Q*. 45).

5. **When cos ϕ = 0**

It means $\phi = 90°$ *i.e.* the load is purely reactive

$$\therefore \qquad W_1 = V_L I_L \cos (30° + 90°) = - V_L I_L \sin 30° = -\frac{1}{2} V_L I_L$$

$$W_2 = V_L I_L \cos (30° - 90°) = V_L I_L \cos (-60°) = \frac{1}{2} V_L I_L$$

Since the two readings are *equal and opposite*, they cancel each other out.

$$W_1 + W_2 = 0$$

It is seen that W_1 is negative and W_2 is positive. It is just the opposite of that for a lagging power factor.

The relation between tan ϕ and the readings W_1 and W_2 for a *leading* load power factor is

$$\tan \phi = -\frac{\sqrt{3}(W_1 - W_2)}{W_1 + W_2} \qquad \text{— see Q. 48}$$

Another point worth remembering is that if W_1 is *positive but less than* W_2, then the load power factor is *leading*. If W_1 is positive and *greater* than W_2, then load power factor is *lagging*.

The above facts about the variation of wattmeter readings with the power factor of the three-phase load are summarized below.

Power Factor	**ϕ**	**Lagging** W_1	**Lagging** W_2	**Leading** W_1	**Leading** W_2
1.0	0°	+	+	+	+
		$W_1 = W_2$		$W_1 = W_2$	
0.866	30°	+	+	+	+
		$W_1 = 2W_2$		$W_2 = 2W_1$	
0.5	60°	+	0	0	+
0.259	75°	+	–	–	+
0.00	90°	+	–	–	+
		$W_1 = -W_2$		$W_2 = -W_1$	

Q. 48. Derive the relation between tan ϕ and the two wattmeter readings W_1 and W_2 for a balanced three-phase load having leading power factor.

Ans. The readings of the two wattmeters for a *leading* power factor load are

$$W_1 = V_L I_L \cos (30° + \phi)$$

$$W_2 = V_L I_L \cos (30° - \phi)$$

$$W_1 + W_2 = V_L I_L \cos (30° + \phi) + V_L I_L \cos (30° - \phi)$$

$$= \sqrt{3} V_L I_L \cos \phi \qquad \text{— (see Q. 44)}$$

$$W_1 - W_2 = V_L I_L \cos (30° + \phi) - V_L I_L \cos (30° - \phi)$$

$$= V_L I_L [(\cos 30° \cos\phi - \sin 30° \sin\phi) - (\cos 30° \cos\phi + \sin 30° \sin\phi)]$$

$$= V_L I_L (-2 \sin 30° \sin\phi) = -V_L I_L \sin\phi$$

$$\therefore \quad \frac{-V_L I_L \sin \phi}{\sqrt{3} V_L I_L \cos\phi} = \frac{W_1 - W_2}{W_1 + W_2}$$

or

$$\frac{-1}{\sqrt{3}} \tan \phi = \frac{W_1 - W_2}{W_1 + W_2}$$

$$\therefore \quad \tan \phi = -\sqrt{3} \frac{W_1 - W_2}{W_1 + W_2}$$

Q. 49. Show that the two wattmeter readings can be used to find the kVAR and kVA taken by a balanced 3-ϕ load of lagging p.f.

Ans. If in the two-wattmeter method of measuring 3-ϕ power, the two wattmeter readings are W_1 and W_2, then it can be proved that the term $\sqrt{3}(W_1 - W_2)$ measures the reactive power (kVAR) of the load whereas $(W_1 + W_2)$ gives the true power (kW) of the load as shown in the kVA triangle of Fig. 41.

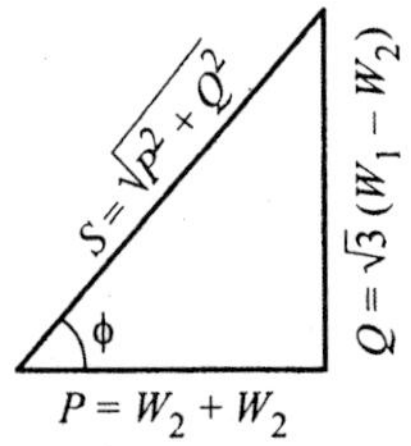

Fig. 41

It may be proved mathematically as follows:

$$\sqrt{3}(W_1 - W_2) = \sqrt{3}[V_L I_L \cos(30° - \phi) - V_L I_L \cos(30° + \phi)]$$
$$= \sqrt{3} V_L I_L (\cos 30° \cos\phi + \sin 30° \sin\phi - \cos 30° \cos\phi + \sin 30° \sin\phi)$$
$$= \sqrt{3} V_L I_L \sin\phi = \text{kVAR}$$

Q. 50. Show that in the two-wattmeter method of power measurement, the power consumed by a balanced 3-ϕ load with lagging power factor of 0.866 equals the sum of the two wattmeter readings.

Ans. As we know, power consumed by a 3-ϕ balanced load is given by

$$P = \sqrt{3} V_L I_L \cos\phi = \sqrt{3} V_L I_L \times 0.866$$
$$= \sqrt{3} V_L I_L \times \frac{\sqrt{3}}{2} = \frac{3}{2} V_L I_L$$

We are given that $\cos\phi = 0.866$ or $\phi = \cos^{-1} 0.866 = 30°$

As shown in Q. 45,

$$W_1 = V_L I_L \cos(30° - \phi) = V_L I_L \cos(30° - 30°)$$
$$= V_L I_L \cos 0° = V_L I_L \times 1 = V_L I_L$$
$$W_2 = V_L I_L \cos(30° + \phi) = V_L I_L \cos(30° + 30°)$$
$$= V_L I_L \cos 60° = \frac{1}{2} V_L I_L$$

$$\therefore \quad W_1 + W_2 = V_L I_L + \frac{1}{2} V_L I_L = \frac{3}{2} V_L I_L$$

Hence, it proves that $\boldsymbol{P = W_1 + W_2}$.

Q. 51. A balanced 3-ϕ , Y-connected load draws power from 440 V supply. The two wattmeters connected indicate W_1 = 5 kW and W_2 = 1.2 kW. Calculate power, power factor and current in the circuit. (*Basic Elect. Engg. VTU Karnataka, 2009*)

Ans. Total power $= W_1 + W_2 = 5 + 1.2 =$ **6.2 kW**

Now,
$$\tan\phi = \frac{\sqrt{3}(W_1 - W_2)}{W_1 + W_2} = \frac{\sqrt{3}(5 - 1.2)}{5 + 1.2} = \frac{\sqrt{3} \times 3.8}{6.2} = 1.0615$$

$\therefore \quad \phi = \tan^{-1} 1.0615 = 46.7°, \quad \therefore \cos\phi = \mathbf{0.686}$

Since, $W_1 > W_2$, it is a lagging power factor (refer to Q. 47)

Now, $\quad P = \sqrt{3} V_L I_L \cos\phi$

$\therefore \quad 6200 = \sqrt{3} \times 440 \times I_L \times 0.686$

$\therefore \quad I_L = \mathbf{11.86A}$

Q. 52. The measurement of power in a 3-ϕ, Y-connected load was done by using two-wattmeter method. In this experiment, the load supplied was 30 kW at 0.7 p.f. lagging. Find the readings of each wattmeter. For what power factor will one of the wattmeters read zero? (*Basic Elect. Engg.,CSV Tech. Univ. Bhilai 2007*)

Ans. Power factor = $\cos\phi = 0.7$ (lag) $\quad\therefore\quad \phi = \cos^{-1} 0.7,\ \phi = 45.5°$

$\therefore \quad \tan\phi = \tan 45.5° = 1.0176$

Now, $\quad \tan\phi = \dfrac{\sqrt{3}\,(W_1 - W_2)}{W_1 + W_2}$ or $1.0176 = \dfrac{\sqrt{3}\,(W_1 - W_2)}{30}$

$\therefore \quad W_1 - W_2 = \dfrac{1.0176 \times 30}{\sqrt{3}} = 17.62$ kW

Now, $\quad W_1 + W_2 = 30$ kW

$W_1 - W_2 = 17.62$ kW

$\therefore \quad W_1 = $ **23.81 kW** and $W_2 = $ **6.19 kW**

One wattmeter will read zero when $\cos\phi = 0.5$

Q. 53. The power flowing in a 3-ϕ , 3-wire balanced load is measured by the two-wattmeter method. The reading in W_1 = 750 W and W_2 = 1500 W. What is the power factor of the system. *(Basic Elect. Engg. VTU Karnataka, 2007)*

Ans. Since $W_1 < W_2$, it means the load has a *leading* power factor (refer to Q. 47)

$$\therefore \quad \tan\phi = -\frac{\sqrt{3}\,(W_1 - W_2)}{W_1 + W_2} = -\frac{\sqrt{3} \times (-750)}{2250} = 0.577$$

or $\quad \phi = \tan^{-1} 0.577 = 30° \therefore$ p.f. $= \cos 30° = $ **0.866 (lead)**

Q. 54. Two wattmeters connected to measure the input to a balanced, three-phase circuit indicate 1500 W and 500 W respectively. Find the power factor of the circuit (*i*) when both readings are positive and (*ii*) when the later is obtained after reversing the connection to the current coil of one instrument.

Ans. We are given that W_1 = 1500 W and W_2 = 500 W

(*i*) **When both readings are positive**

$$\tan\phi = \sqrt{3}\,\frac{W_1 - W_2}{W_1 + W_2} = \sqrt{3}\,\frac{(1500 - 500)}{(1500 + 500)}$$

$$= \sqrt{3}\,\frac{1000}{2000} = 0.866$$

$\therefore \quad \phi = \tan^{-1}(0.866) = 41°$, p.f. $= \cos\phi = \cos 41° = $ **0.754**

(*ii*) **When W_2 reading is reversed**

Now W_1 = 1500 W and W_2 = – 500 W

$$\therefore \quad \tan\phi = \sqrt{3}\,\frac{[W_1 - (-W_2)]}{W_1 + (-W_2)} = \sqrt{3}\,\frac{W_1 + W_2}{W_1 - W_2} \qquad \text{— Q. 46}$$

$$= \sqrt{3}\,\frac{(1500 + 500)}{(1500 - 500)} = 3.464$$

$\phi = \tan^{-1}(3.464) - 74°$, $\cos\phi = \cos 74° = $ **0.275**

Q. 55. Two wattmeters are used for measuring power in a three-phase balanced load. Find the power factor if (*i*) two readings are positive and equal and (*ii*) the two readings are equal and opposite.

Ans. (*i*) Here $\quad W_1 = W_2$

$$\tan\phi = \sqrt{3}\,\frac{W_1 - W_2}{W_1 + W_2} = \sqrt{3} \times 0 = 0 \therefore \phi = 0°$$

power factor $= \cos\phi = \cos 0° = $ **1** — (see Q. 45)

It means that the load is pure resistance.

(*ii*) When readings are equal and opposite *i.e.*

$$W_1 = -W_2$$

$$\tan \phi = \sqrt{3}\frac{W_1 - W_2}{W_1 + W_2} = \sqrt{3}\frac{-2W_2}{0} = \infty$$

$\therefore$ $\phi = 90°$, p.f. $= \cos 90° = \mathbf{0}$

It means that the load is purely reactive. (either a pure coil or a pure capacitor)

Q. 56. A 3-ϕ, 500-V motor load has a power factor of 0.4. Two wattmeters connected to measure the power show the input to be 30 kW. Find the reading on each wattmeter.

(*Basic Elect. Engg. CSV Tech. Univ. Bhilai, 2007*)

Ans. $\cos\phi = 0.4$, $\phi = \cos^{-1} 0.4 = 66.6°$

$\therefore$ $\tan 66.6° = 2.311$ and $W_1 + W_2 = 30$ kW ...(*i*)

Now, $\tan \phi = \dfrac{\sqrt{3}(W_1 - W_2)}{W_1 + W_2}$ or $2.311 = \dfrac{\sqrt{3}(W_1 - W_2)}{30}$

$\therefore$ $W_1 - W_2 = 40$ ...(*ii*)

From (*i*) and (*ii*), we get

$$W_1 = \mathbf{35\ kW} \text{ and } W_2 = \mathbf{-5\ kW}$$

It was shown earlier (Q. 45) that for a lagging p.f. $\cos \phi < 0.5$, W_2 reads negative.

Q. 57. Three similar choking coils each having resistance of 10 Ω and reactance 10 Ω are connected in star across a 440-V, 3-ϕ supply. Find line current and readings of each of the two wattmeters connected to measure power. (*Basic Elect. Engg. VTU Karnataka, 2007*)

Ans. $\mathbf{Z_{ph}} = R + jX_L = 10 + j10 = 14.14\angle 45°$

$\cos \phi = R/Z_{ph} = 10/14.14 = 0.707$ (lag)

$\therefore$ $\phi = \cos^{-1} 0.707 = 45°$

For a Y-load, $I_{ph} = \dfrac{V_{ph}}{Z_{ph}} = \dfrac{V_L/\sqrt{3}}{14.14} = \dfrac{440/\sqrt{3}}{14.14} = \dfrac{254}{14.14} = 17.96$ A

Now, $I_L = I_{ph} = \mathbf{17.96\ A}$

$$P = \sqrt{3}\, V_L I_L \cos \phi$$

$$= \sqrt{3} \times 440 \times 17.96 \times 0.707 = 9677 \text{ W}$$

$\therefore$ $W_1 + W_2 = 9677$ W — (i)

Now, $\tan\phi = \dfrac{\sqrt{3}(W_1 - W_2)}{(W_1 + W_2)}$ or $\tan 45° = \dfrac{\sqrt{3}(W_1 - W_2)}{9677}$

$\therefore$ $W_1 - W_2 = 5587$ W — (ii)

From (*i*) and (*ii*), we get

$$W_1 = \mathbf{7632\ W}, \quad W_2 = \mathbf{2045\ W}$$

Q. 58. A 3-ϕ, Y-connected balanced load of value $Z = 40\angle 60°$ Ω per phase is connected across 400 V, 3-ϕ balanced supply. Find the power drawn from the supply and the readings of the two wattmeters connected to measure the input power.

(*Basic Elect. Engg. CVS Tech Univ. Bhilai 2007*)

Ans. $V_{ph} = V_L/\sqrt{3} = 400/\sqrt{3} = 231$ V, $Z_{ph} = 40$ Ω

$\therefore$ $I_{ph} = V_{ph}/Z_{ph} = 231/40 = 5.775$ A

Since $\phi = 60°$, $\cos \phi = \cos 60° = 0.5$

$$R_{ph} = Z_{ph} \times \cos \phi = 40 \times 0.5 = 20\ \Omega$$

Total power $= 3\, I_{ph}^2\, R_{ph} = 3 \times 5.775^2 \times 20 = 2000$ W

$\therefore \quad W_1 + W_2 = 2000$ W

Now, $\tan\phi = \dfrac{\sqrt{3}(W_1 - W_2)}{(W_1 + W_2)}$ or $\tan 60° = \dfrac{\sqrt{3}(W_1 - W_2)}{2000}$

or $\sqrt{3} = \dfrac{\sqrt{3}(W_1 - W_2)}{2000} \quad \therefore \quad W_1 - W_2 = 2000$ W

Obviously, $W_2 = \mathbf{0}$ and $W_1 = \mathbf{2000\ W}$

Even otherwise, it is obvious that W_2 should be zero because

p.f. $= \cos 60° = 0.5$ — (see Q. 45)

Q. 59. Two wattmeters have been used to measure the power input to a 150 kW, 440-V, 3-ϕ, slip-ring induction motor running at full load. The wattmeter readings are 115 kW and 50 kW. Calculate :

(i) input of the motor (*ii*) power factor of the motor (*iii*) line current drawn by the motor (*iv*) efficiency of the motor. *(Basic Elect. Engg. R.G.P.V., Bhopal 2007)*

Ans. (*i*) Power input to the motor $= W_1 + W_2 = 115 + 50 = \mathbf{165\ kW}$

(*ii*) $\tan\phi = \dfrac{\sqrt{3}(W_1 - W_2)}{(W_1 + W_2)} = \dfrac{\sqrt{3}(115-50)}{115+50} = \dfrac{\sqrt{3}\times 65}{165} = 0.682$

$\phi = \tan^{-1} 0.682 = 34.3°$, p.f. $= \cos\phi = \cos 34.3° = \mathbf{0.826\ (lag)}$

(*iii*) Motor input $= \sqrt{3}\, V_L I_L \cos\phi$

or $165 \times 10^3 = \sqrt{3} \times 440 \times I_L \times 0.826, \quad I_L = \mathbf{262.1\ A}$

(*iv*) Motor output = 150 kW, Motor input = (150 + 115 + 50) = 165 kW

$\therefore$ motor efficiency $= \dfrac{\text{motor output}}{\text{motor input}} = \dfrac{150}{165} = 0.909 = \mathbf{90.9\%}$

Q. 60. The input power to a 3-ϕ motor was measured by two-wattmeter method. The readings of the two wattmeters are 5.2 kW and – 1.7 kW and the line voltage is 415 V. Calculate the total active power, power factor and the line current.

(Basic Elect. Engg. CSV Tech. Univ. Bhilai 2009)

Ans. Total active power = 5.2 – 1.7 = **3.5 kW**

$\tan\phi = \sqrt{3}\dfrac{5.2-(-1.7)}{5.2-1.7} = \sqrt{3}\times\dfrac{6.9}{3.5} = 3.41$

$\therefore \quad \phi = \tan^{-1} 3.41 = 73.6°, \ \cos 73.6° = \mathbf{0.281\ lag}$

Now, $P = \sqrt{3}\, V_L I_L \cos\phi$

or $3500 = \sqrt{3} \times 415 \times I_L \times 0.281, \quad \therefore I_L = \mathbf{17.32\ A}$

EXERCISE

Star-Connection

1. Three coils, each having a resistance of 40 Ω and reactance 30 Ω are connected in star across a 415 V, 3-ϕ, 50 Hz supply. Calculate:
(*i*) line current (*ii*) total kVA (*iii*) power factor (*iv*) total power
(4.8 A, 3.45 kVA, 0.8 lag, 2.76 kW)

2. Three similar coils each of impedance $Z = (8 + j\ 10)\ \Omega$ are connected in star and supplied from a 3-ϕ, 400 V, 50 Hz supply. Find the line current, power factor, power and total volt-amperes. (*Elect. Engg. Nagpur Univ. 2003*)
(18 A, 0.625 lag, 7815 W, 12.5 kVA)
3. Calculate the active and reactive components of the current in each phase of a star-connected 5,000-V, 3-ϕ generator supplying 3,000 kW at a power factor of 0.8.
(346 A, 260 A)
4. A star-connected load has impedance of $(6 + j\ 8)\ \Omega$ in each phase and is connected across a balanced 400 V, 3-ϕ supply. Obtain the line current, total real power and reactive power consumed by the load. (*Basic Elect. Engg. R.G.P.V, Bhopal 2003*)
(23.1 A, 9.6 kW, 16 kVA)
5. A 3-ϕ induction motor is equivalent to a balanced *Y*-connected load of 2.5 Ω resistance and 0.03 H inductance in series in each phase. Calculate the line current, power input, kVA and kVAR taken by the load from a 415-V, 50 Hz, 3-ϕ supply.
(24.6 A, 4.52 kW, 17.7 kVA, 17.1 kVAR)
6. A three-phase, *Y*-connected alternator has a line voltage of 11 kV. The output of the alternator is 12 MVA at a power factor of 0.85 lagging. Calculate (*i*) line-to-neutral voltage (*ii*) power output and (*iii*) line current. **(6.357 kV, 10.2 MW, 631 A)**
7. Three similar coils are connected in star and this load is supplied from a 3-ϕ, 450 V system. The line current is 17.32 A and power factor 0.8 lagging. Calculate :
(*i*) resistance, reactance and impedance of each coil.
(*ii*) total power in kilowatts **(12 Ω, 9 Ω, 15 Ω, 10.8 kW)**

Delta-Connection

8. Calculate the power absorbed by three similar choking coils each of resistance 150 Ω and inductance 1.275 H when delta-connected to a 440 V, 3-ϕ, 25 Hz supply. **(1390 W)**
9. A balanced, delta-connected load is supplied at a line voltage of 500 V. The line current is 17.3 A and the power factor 0.95. Calculate
(*i*) total input power and apparent power
(*ii*) value of each phase current. **(14.25 kW, 15 kVA, 10 A)**
10. A balanced 3-ϕ, 400-V system supplies a balanced delta-connected load. The current in each branch is $20\angle 40°\ \Omega$..
Find : (*i*) the line current and (*ii*) total power. **(34.64 A, 10.6 kW)**
11. Three similar coils each of resistance 20 Ω and inductance of 0.6 H are connected in (*i*) star (*ii*) delta to a 3-ϕ, 400 V, 50 Hz supply. Calculate the line currents and total active power absorbed in both cases. (*Elect. Engg. Nagpur. Univ 2001*)
(8.4 A, 4234 W, 25.2 A, 12,702 W)
12. Three inductive coils each with a resistance of 15 Ω and an inductance of 0.03 H are connected (*i*) in star and (*ii*) in delta to a 3-ϕ, 400 V, 50 Hz supply. Calculate for each of the above cases.
(*a*) phase current and line current (*b*) total power absorbed
(*Elect. Engg. Nagpur Univ. 2001*)
(13.05 A, 13.05 A, 7.664 kW, 22.6 A, 39.1 A, 22.95 kW)
13. A Δ-connected balanced 3-phase load is supplied from a 3-phase, 400 V supply. The line current is 20 A and the power taken by the load is 10,000 W. Find (*i*) impedance in each branch (*ii*) the line current, power factor and power consumed if the same load is connected in star. (*Elect. Machines, A.M.I.E. Sec. B., 2001*)
(34.64 Ω, 20/3 A, 0.7217, 3333 W)

14. Three similar resistors are connected in star across 400-V, 3-phase lines. The line current is 5 A. Calculate the value of each resistor. To what value should the line voltage be changed to obtain the same line current with the resistors delta-connected.

(46.2 Ω, 133.3 V)

15. A delta- connected load draws a line current of 20 A, at a lagging power factor of 0.8 from a 400 V, 3-ϕ, 50 Hz supply. Calculate:
    (*i*) resistance and inductance of each phase (*Elect. Engg. Nagpur 2005*)
    (*ii*) power consumed **(27.7 Ω, 66.18 mH, 11.088 kW)**

Two Wattmeter Method

16. Three choke coils each with a resistance of 10 Ω and reactance of 10 Ω are connected in star across a 3-ϕ, 400-V, 3-ϕ supply. Calculate line current and readings of the two wattmeters connected to measure the power. What is the total power drawn by the load?

(*Basic Elect. Engg. R.G.P.V, Bhopal 2005*)

(16.34 A, 6312.5 W, 1691.5 W, 8004 W)

17. The power readings of two wattmeters are +15 kW and -4 kW for a 3-ϕ balanced load. If the supply voltage is balanced 440 V, find the true power drawn by the load, power factor and line current. (*Basic Elect. Engg. R.G.P.V, Bhopal 2005*)

(11 kW, 0.317, 45.53 A)

18. Three similar coils each having resistance of 10 Ω and reactance of 8 Ω are connected in star across a 400 V, 3-ϕ supply. Determine (*i*) line current (*ii*) total power and (*iii*) readings of each of the two wattmeters connected to measure the power.

(*Basic Elect. Engg. VTU Karnataka, 2009*)

(18.04 A, 9736 W, W_1 = 7116 W, W_2 = 2620 W)

19. A three-phase 230 V load has a power factor of 0.7. Two wattmeters are used to measure power which show the input to be 10kW. Find the reading of each wattmeter.

(*Basic Elect. Engg. West Bengal Tech. Univ. 2009*)

(W_1 = 7.937 kW, W_2 = 2.063 kW)

20. Phase voltage and current of a star-connected inductive load is 150 V and 25 A. Power factor of load is 0.707 (lag). Assuming that the system is 3-wire and power is measured using two wattmeters, find the readings of wattmeters.

(*Elect. Instrument & Measurements, Nagpur Univ.*)

(6273 W, 1681 W)

21. In a balanced 3-phase 400-V circuit, the line current is 115.5 A. When power is measured by two wattmeter method, one meter reads 40 kW and the other zero. What is the power factor of the load? If the power factor were unity and the line current the same, what would be the reading of each wattmeter? **(0.5 lag, W_1 = W_2 = 40 kW)**
22. The input power to a three-phase motor was measured by two wattmeter method. The readings were 10.4 kW and – 3.4 kW and the voltage was 400 V. Calculate (*a*) the power factor (*b*) the line current. **(0.45 lag, 22.4 A)**
23. The power in a 3-phase circuit is measured by two wattmeters. If the total power is 100 kW and power factor is 0.66 leading, what will be the reading of each wattmeter? Give the connection diagram for the wattmeter circuit. For what p.f. will one of the wattmeter read zero? **(W_1 = 17.15 kW, W_2 = 82.85 kW, 0.5)**
24. Each phase of a 3-phase, Δ-connected load consists of an impedance $Z = 20\angle 60°$ ohm. The line voltage is 440 V at 50 Hz. Compute the power consumed by each phase impedance and the total power. What will be the readings of the two wattmeters connected? (*Elect. and Mech. Technology, Osmania Univ.*)

(4840 W, 14520 W, W_1 = 14520 W, W_2 = 0 W)

MULTIPLE CHOICE QUESTIONS

1. In a 3-ϕ system, the zero-going-positive peaks of the 3-phase emf waveforms follow each other at a phase angle of .... degrees.
(*a*) 90 (*b*) 120 (*c*) 240 (*d*) – 120
2. In a three-phase system, the emfs are .... apart.
(*a*) 30° (*b*) 60° (*c*) 90° (*d*) 120°
(*Basic Elect. Engg. VTU, Karnataka. Jan 2009*)
3. In a 3-ϕ, 4-wire balanced system, the neutral current is
(*a*) zero (*b*) dependent on load
(*c*) dependent on line voltage (*d*) determined by load power factor.
4. In a 3-ϕ balanced star-connection with *RYB* phase sequence, if phase voltage $V_R = 400\angle 0°$, then the value of V_B is
(*a*) $400\angle -120°$ (*b*) $400\angle 120°$ (*c*) $400\angle -240°$ (*d*) either (*b*) or (*c*)
5. In a 3-ϕ balanced star-Connection with *RYB* phase sequence, if phase voltage $V_R = 400\angle 30°$, then the value of V_Y is
(*a*) $400\angle 120°$ (*b*) $400\angle -90°$ (*c*) $400\angle -120°$ (*d*) $400\angle 150°$
6. In a delta-connected system, the relationship between the line current I_L and phase current I_{ph} is
(*a*) $I_L = I_{ph}$ (*b*) $I_L = I_{ph}/\sqrt{3}$ (*c*) $I_L = \sqrt{3}\, I_{ph}$ (*d*) $I_L = 3\, I_{ph}$.
(*Basic Elect. Engg. VTU, Karnataka 2009*)
7. The phase sequence of the voltages applied to a 3-ϕ load is determined by.
(*a*) generator phase sequence
(*b*) type of load
(*c*) the order in which 3-ϕ lines are connected
(*d*) power factor of the load
8. For a delta-connected circuit, the correct relationship is
(*a*) $V_L = V_{ph}$ (*b*) $I_{ph} \times \sqrt{3} = I_L$ (*c*) $V_L = I_L \times \sqrt{3}$ (*d*) (*a*) or (*b*)
(*Basic Elect. Engg. R.G.P.V. Bhopal, 2008*)
9. As compared to a single-phase system, one big advantage of 3-ϕ system is that power delivered by it is
(*a*) much higher (*b*) constant (*c*) pulsating (*d*) reactive
10. In a balanced *Y*-connected 3-ϕ load, the line voltage leads the phase voltage by
(*a*) 30° (*b*) 60° (*c*) 90° (*d*) 120°
11. In a balanced delta-connected 3-ϕ load, the line currents lag the phase currents by
(*a*) 60° (*b*) 30° (*c*) 90° (*d*) 120°
12. For the same line voltage and the same transmitted kVA for the same distance, the 3-ϕ system requires nearly .... % of the copper weight as compared to 1-ϕ system.
(*a*) 50 (*b*) 75 (*c*) 85 (*d*) 95
13. The constant power in a 3-ϕ system results in
(*a*) reduced eddy current loss (*b*) low hysteresis loss
(*c*) quiter operation of motors (*d*) all of the above
14. For a 3-ϕ balanced load, power factor is given by
(*a*) kVAR/kVA (*b*) kVR/kW (*c*) kW/kVA (*d*) kW/kVAR
15. The power taken by a 3-ϕ load is given by the expression
(*a*) $3\, V_L I_L \cos\phi$ (*b*) $\sqrt{3}\, V_L I_L \cos\phi$ (*c*) $3\, V_L I_L \sin\phi$ (*d*) $\sqrt{3}\, V_L I_L \sin\phi$
(*Basic Elect. Engg. VTU Karnataka 2009*)

16. A 3-ϕ, star-connected symmetrical load consumes *P* watts of power from a balanced supply. If the same load is connected in delta to the same supply, the power consumption will be
(*a*) *P* (*b*) $\sqrt{3}\,P$ (*c*) 3 *P* (*d*) *P*/3
(*Basic Elect. Engg R.G.P.V., Bhopal 2006*)
17. In 3-ϕ circuits, low load power factor has the disadvantage of
(*a*) large kVA as compared to true power
(*b*) poor voltage regulation
(*c*) higher current for a given power
(*d*) all of the above
18. The power-factor correction equipment does NOT change the .... of the load
(*a*) reactive power (*b*) power factor
(*c*) real power (*d*) current
19. For the same line voltages and same branch impedances, power drawn by a delta-connected load is .... times the power drawn by a star-connected load
(*a*) $\sqrt{3}$ (*b*) $\sqrt{2}$ (*c*) 3 (*d*) 2
20. For the same 3-ϕ supply line voltages and currents, the delta impedances equal .... times the star impedances
(*a*) 3 (*b*) $\sqrt{3}$ (*c*) 2 (*d*) $\sqrt{2}$

Note : For solving *Q*. 21 – 29, please consult table given in Q. 47.

21. While measuring power in a three-phase load by 2 WM method, the readings of the two wattmeters are equal and opposite, when
(*a*) p.f. is unity (*b*) load is balanced
(c) phase angle is between 60° and 30° (*d*) the load is purely inductive
(*Basic Elect. Engg. R.G.P.V Bhopal 2007*)
22. In the two-wattmeter method of power measurement in 3-ϕ circuits, the two wattmeters have equal and positive readings when power factor angle is
(*a*) 60° (*b*) 30° (*c*) 90° (*d*) 0°
23. In the two-wattmeter method of measuring power in 3-ϕ circuits, one of the wattmeters reads zero when power factor is
(*a*) unity (*b*) 0.866 (*c*) 0.5 (*d*) 0.4
24. In the two-wattmeter method of measuring power in 3-ϕ circuits, one wattmeter reading will be double of the other reading when power factor angle is
(*a*) 0° (*b*) 30° (*c*) 60° (*d*) 90°
25. In the two-wattmeter method of measuring 3-ϕ power, one of the two-wattmeters reads zero when power factor angle is
(*a*) 0° (*b*) 30° (*c*) 60° (*d*) 90°
26. In the two-wattmeter method of measuring 3-ϕ power, the two-wattmeters indicate equal and opposite readings when the load power factor angle is .... degrees
(*a*) 60 (*b*) 0 (*c*) 30 (*d*) 90
(*Basic Elect. Engg. VTU Karnataka 2009*)
27. In the two-wattmeter method of measuring power of a balanced 3-ϕ load with lagging power factor, if one-wattmeter reads twice the other, then load power factor angle is
(*a*) 30° (*b*) 45° (*c*) 60° (*d*) 90°
28. In the two-wattmeter method of measuring power of a balanced 3-ϕ load, if reading of W_2 is twice that of W_1, then load power factor angle is
(*a*) 30° lag (*b*) 30° lead (*c*) 60° lag (*d*) 60° lead.

29. In the two-wattmeter method of measuring power of a balanced 3-ϕ load, if W_1 reads zero, then power factor angle of the load is
(*a*) 30° lead (*b*) 60° lag (*c*) 60° lead (*d*) 90° lag.

30. In the two-wattmeter method of measuring power of a balanced 3-ϕ load, if W_1 and W_2 readings are positive but $W_1 = W_{2/2}$, then load power factor is
(*a*) 0.866 lead (*b*) 0.5 lag (*c*) 0.866 lag (*d*) 0.5 lead

ANSWERS

1. *b*	2. *d*	3. *a*	4. *d*	5. *b*	6. *c*	7. *c*
8. *d*	9. *b*	10. *a*	11. *b*	12. *b*	13. *d*	14. *c*
15. *b*	16. *c*	17. *d*	18. *c*	19. *c*	20. *a*	21. *d*
22. *d*	23. *c*	24. *b*	25. *c*	26. *d*	27. *a*	28. *b*
29. *c*	30. *a*					

CHAPTER 10

ELECTRICAL INSTRUMENTS AND MEASUREMENTS

Q. 1 (*a*) Distinguish between absolute and secondary instruments. Give one example of each.

Ans. All electrical instruments are of two types :

(*i*) absolute instruments and (*ii*) Secondary instruments

Absolute instrument is one which gives the value of the quantity to be measured in terms of the *constants* of the instrument and its *deflection* only. It does not need any comparison with or reference to any other instrument. The example of such an instrument is tangent galvanometer. It gives the value of current in terms of the tangent of deflection produced by the current and the radius and number of turns of wire used and the horizontal component (H) of Earth's magnetic field. Such instruments are used in laboratories only.

Secondary instrument is that which has to be pre-calibrated by comparison with an absolute instrument. Without calibration, its deflection has *no meaning*. Such instruments are generally used in everyday work.

Q. 2. Mention and briefly describe the different categories of secondary instruments.

Ans. There are three types of secondary instruments

(*i*) indicating instruments (*ii*) recording instruments (*iii*) integrating instruments

As the name shows, an indicating instrument is the one which *indicates* the value of the electrical quantity being measured at the *time of its measurement.* The value is *indicated* by a pointer moving over a pre – calibrated scale [Fig. 1(*a*)]. Examples are : ammeters, voltmeters and wattmeters etc.

Recording instruments are those which keep a *continuous* record of the quantity being measured over a certain *period of time.*

The instruments which store the data of the incoming and outgoing quantities (like voltage, current etc) at a power station are examples of a recording instrument. These instruments also provided a graph and chart of these quantities over a certain period of time [Fig. 1(*b*)].

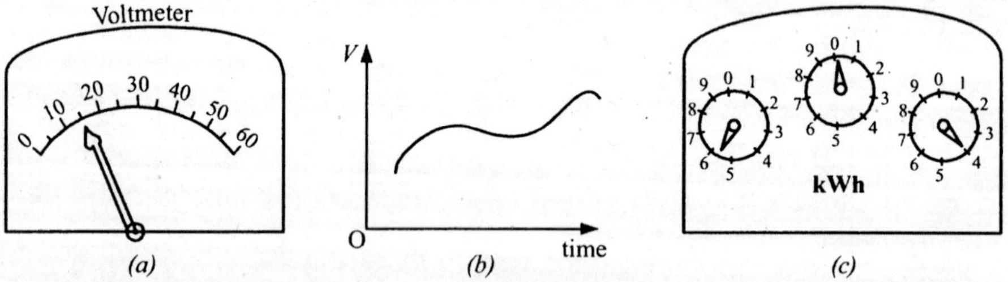

Fig. 1

Integrating instruments measure the amount of electrical quantity, then add it to the previous value and show the reading [Fig. 1(*c*)]. The domestic kWh meter is an example of such an instrument. It shows the amount of electrical energy consumed in a *certain period of time.*

Q. 3. State the essential requirements of indicating type instruments. Explain each of them in brief.

Ans. Indicating instruments are those which give the value of quantity at the time it is measured. Such instruments consist of a pointer which moves over a pre-calibrated scale. This pointer is attached to a moving system pivoted in jewelled bearings. The moving system of the indicating instruments is acted upon by the following three torques.

(*i*) a deflecting (or operating) torque

(*ii*) a controlling (or restoring) torque

(*iii*) a damping torque.

The deflection torque is produced by using any property of the electric current. This torque causes the moving system to move from its zero position.

The controlling torque *opposes* the deflecting torque. The pointer comes to rest at a position where the deflecting and controlling torques *are equal.* Due to the deflecting torque, different currents produce different deflections proportional to their magnitude. Without a deflecting torque, all currents whether large or small will produce the *same deflection.* The controlling torque for indicating instruments is obtained either by a spring or by gravity.

A damping torque is produced by a damping or stopping force which acts on the moving system only *when it is moving* and always *opposes* its motion. Such a force is necessary to bring the pointer to rest *quickly.* If there is no damping torque, then due to the inertia of the moving system, the pointer will keep moving to and fro about its final deflected position quite for some time before coming to rest.

Q. 4. Explain the two common methods of providing controlling torque in indicating type of measuring instruments. Compare the merits and demerits of the two methods.

Ans. The controlling torque is obtained either by a spring or by gravity.

(*a*) Spring Control : In this method, a fine spring (usually of phosphor bronze) is attached to the moving system of the instrument as shown in Fig. 2(*a*).

When the pointer is deflected, the spring is twisted in the *opposite direction*. This *twist* in the spring produces controlling torque which is directly proportional to the deflection of the moving system. The pointer comes to rest when the deflecting torque (T_d) and controlling torque (T_c) are *equal and opposite.*

For example, in a permanent-magnet moving-coil (PMMC) type of instruments, the deflecting torque is proportional to the current I passing through them.

$$\therefore \qquad T_d \propto I \quad \text{and} \quad T_c \propto \theta$$

Because in the final defelected position, $T_c = T_d$.

$$\therefore \qquad \theta \propto I$$

Since deflection is directly proportional to the current I, the spring-controlled instruments have *uniform* scales.

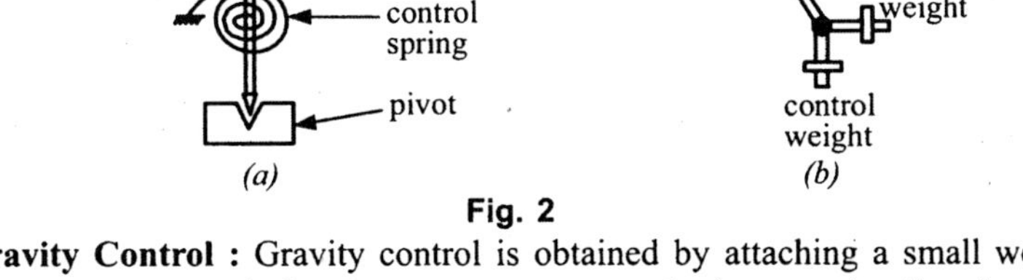

Fig. 2

(*b*) Gravity Control : Gravity control is obtained by attaching a small weight to some part of the moving system such that the two exert torques in the opposite directions as shown in Fig. 2(*b*).

The controlling or restoring torque is proportional to the sine of deflection angle *i.e.*

$$T_c \propto \sin\theta$$

Now $\quad T_d \propto I$. Since at the position of rest; $T_d = T_c \quad \therefore \; I \propto \sin\theta$

As compared to spring control, the disadvantages of gravity control are

(*i*) it gives an uneven scale

(*ii*) the instrument has to be kept vertical.

However, gravity control has the following advantages

(*i*) it is cheap

(*ii*) it is not affected by temperature

(*iii*) it does not spoil with time.

Q. 5. Why is damping torque provided in electrical instruments? What are the different methods of providing the damping torque? Describe them in brief.

Ans. A damping force is one which acts on the moving system of the instrument only *when it is moving* and always *opposes* its motion. Such a force is necessary to bring the pointer to rest *quickly*. If this force is not there, the pointer will oscillate about its final deflected position for quite sometime before coming to rest at the final deflected position. This damping force provides such a damping torque as to make the instrument dead-beat.

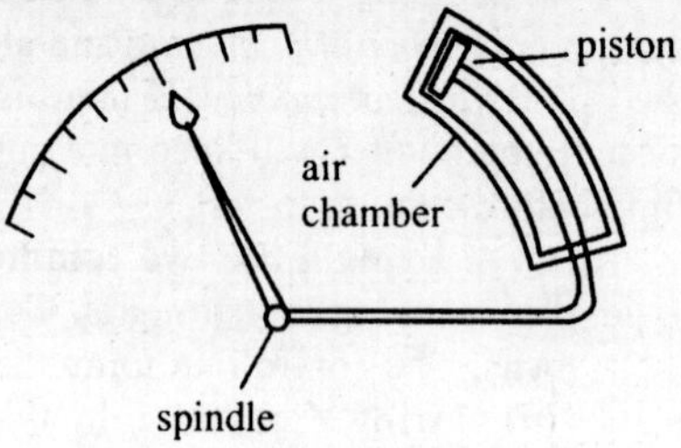

Fig. 3

The two most commonly used methods for producing damping torque are provided by

(*i*) air friction and (*ii*) eddy currents.

The two methods are discussed below :

Air-friction damping method is shown in Fig. 3. In this method, a light aluminimum piston attached to the moving system travels in a fixed air chamber closed at one end. Damping of the oscillations is due to the compression and suction action of the piston on the air enclosed in the chamber.

The two more commonly used air friction methods are shown in Fig. 4. In Fig. 4(*a*), a light aluminium vane is mounted on the spindle of the moving system

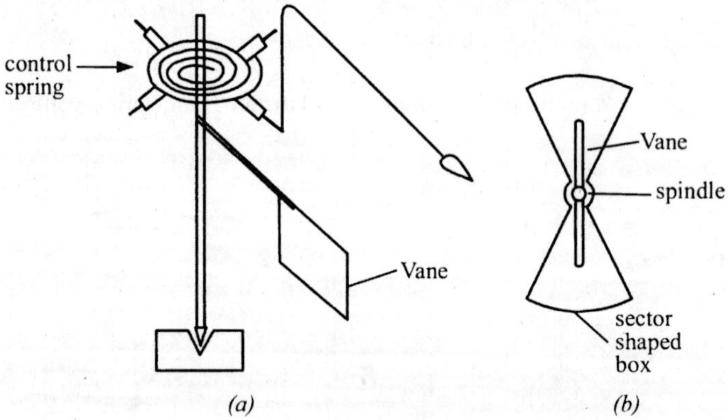

Fig. 4

As the pointer moves, the vane provides the necessary air friction. In the method shown in Fig. 4(*b*), the vanes, mounted on the spindle of the moving system, move in a closed sector-shaped box. Their to and for movement provides the necessary air-friction.

Eddy-current form of damping is the *most efficient*. Two forms of such a damping are shown in Fig. 5 and Fig. 6. As shown in Fig. 5, a thin disc of a conducting but *non-magnetic material* (copper or aluminium) is mounted on a spindle which carries the moving system and the pointer. The disc is so placed that when it rotates, its edge cuts the magnetic flux between the poles of a permanent magnet. Hence, eddy currents are induced in the disc which flow and so produce a damping force in such a direction as to oppose the very cause producing them (Lenz's

Law). Since the cause producing them is the rotation of the disc, the eddy-currents oppose the rotation of the disc and the moving system as a whole. Hence, the pointer is quickly brought to rest at the final deflected position.

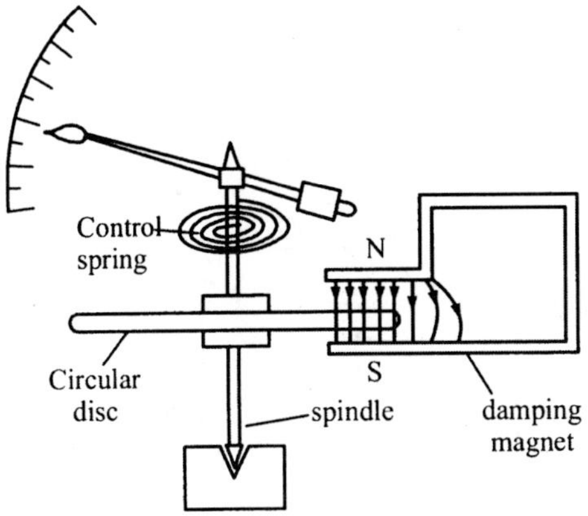

Fig. 5

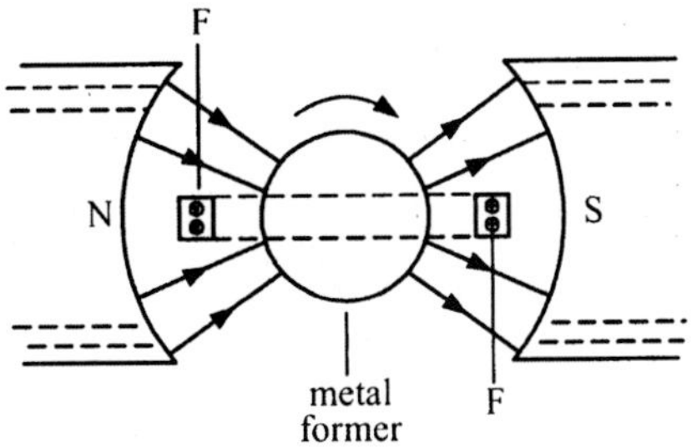

Fig. 6

In Fig. 6, the coil of the instrument is wound on a thin light aluminium former in which eddy-currents are produced when the coil moves in the field of a permanent magnet. The induced eddy currents provide the damping force.

Q. 6. Why are the moving–iron instruments called so? Name the two types of such instruments. Can they be used for ac measurement?

Ans. Such instruments are called moving iron instruments because they use *soft iron pieces* which become strongly magnetized when placed in a magnetic field. They are of two types (*i*) **attraction** type and (*ii*) **repulsion** type. In the attraction type, the soft iron piece is *attracted* towards a magnetic field. But in the repulsion type, the two soft iron pieces *repel each other* when magnetized. Such instruments are used as ammeters and voltmeters. Their current range can be increased by using a shunt and voltage range by using a series resistance as shown in Fig. 13 and 14 respectively.

Both attraction type and repulsion type moving-iron instruments can be used for ac and dc measurements.

Q. 7. Describe the construction and operation of a moving-iron attraction type instrument. Can such instruments be used both for ac and dc? Justify your answer.

Ans. (*i*) Basic principle : Attraction type moving-iron instrument is based on the fact that a soft iron piece *is always attracted by a magnetic field.*

(*ii*) Construction : As shown in Fig. 7, the instrument consists of a working coil *C*. This coil carries the line current to be measured (as ammeter) or a current proportional to the voltage to be measured (as voltmeter). A soft iron disc *D* is attached to a spindle and is placed near the coil *C*. The spindle also carries a pointer which moves over a calibrated scale. Usually, a spring is used to produce controlling torque. Damping is by air frication.

(*iii*) Operation : When current is passed through the working coil, it produces a magnetic field as shown. Since the field is strongest at the *centre* of the coil, the oval- shaped iron disc *D* moves towards the centre of the coil. The pointer also moves over the scale. The deflection of the pointer is proportional to the current carried by the coil *C*. Whatever the direction of current through the coil, the iron disc is so magnetized that it is always pulled inside the coil. Hence, such instruments can be used for both direct as well as alternating currents.

(*iv*) Deflecting torque : The force (*F*) pulling the iron disc towards centre of the magnetic field depends on (*a*) Strength of the field H produced by the coil and

(*b*) Pole strength (m) developed by the disc.

$\therefore \quad F \propto mH$

Now, H depends on the current I and m depends on H and hence upon I,

$\therefore \quad F \propto mH \propto H^2 \propto I^2$

For a spring control, $T_c \propto$ deflection of the disc, θ

At the rest position of the pointer

$$T_c = T_d \quad \text{or} \quad \theta \propto I^2$$

Obviously, the scale of such instruments is *uneven.*

As said above, these instruments can be used both for dc as well as ac currents or voltages. It is due to the fact that when coil current reverses its direction, the polarity of the magnetized iron disc as well as the direction of the coil magnetic field reverse *at the same time* Hence, the iron disc is *always* attracted by the coil field whatever the direction of current through the coil.

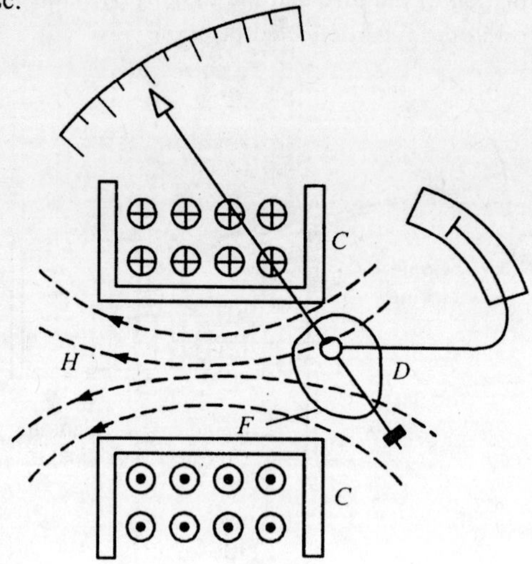

Fig. 7

Q. 8. Describe the construction and working of a repulsion type moving-iron instrument. Show that the deflecting torque of such instruments depends on the square of the current being measured.

Ans. (*i*) Basic principle : If two iron pieces are placed in a given magnetic field, they *repel* each other. If one of them is made movable, the repulsive force on it pushes it to one side. This movement of the iron pieces is used to measure the value of the current or voltage which produces the magnetic field.

(*ii*) Construction : As shown in Fig. 8, the instrument consists of a fixed coil C which carries a current proportional to the voltage or the circuit current to be measured. There are two iron bars inside the coil. One of them is fixed and the other is *movable.* The movable iron bar carries the pointer which moves over a calibrated scale. The controlling torque is produced by a spring. Damping is by air-friction.

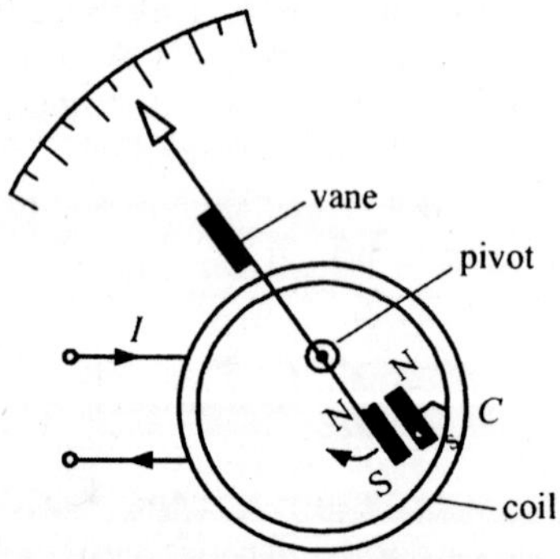

Fig. 8

(*iii*) Operation : When current to be measured is passed through the fixed coil, it sets up its own magnetic field. The field magnetises the two iron rods *similarly i.e.* the rods have the same polarity. Hence, they repel each other. The pointer is deflected against the controlling torque of a spring. The force of repulsion is proportional to the square of the current passing through the coil C. Moreover, whatever the direction of the current through the coil, the two rods get magnetized *similarly* and, hence, repel each other.

(*iv*) Deflecting torque : If m_1 and m_2 are the pole strengths of the two rods, the repulsive force between them is proportional to their product ($m_1 \times m_2$)

$\therefore \quad T_d \propto m_1 \, m_2$

Since pole strengths are proportional to the strength of the magnetic field (H), hence

$\therefore \quad T_d \propto H^2$

But H is itself proportional to the current (I) passing through the coil

$\therefore \qquad T_d \propto I^2$

For spring control, $T_c \propto \theta$. hence $\theta \propto I^2$

Obviously, scales of such instruments are *uneven.* Since the polarity of both iron rods reverses at the same time, the instruments can be used both for dc as well as ac current and voltage measurements.

Q. 9. Mention the two types of moving-coil instruments which are used frequently. Can they be used on dc as well as ac supply?

Ans. Following two moving-coil instruments are frequently used for current or voltage measurements.

(*i*) Permanent-magnet moving-coil (PMMC) instruments.

(*ii*) Dynamometer instruments.

PMMC instruments consist of a permanent magnet with a rectangular coil placed inside its field. It can be used only for dc currents.

The dynamometer instruments do not use a permanent magnet but, instead, use electromagnetic field produced by a fixed coil. There is a moving coil placed inside the electromagnetic field. These instruments can be used both for *dc and ac* measurements.

Q. 10. Describe the construction and working of a permanent-magnet moving- coil (PMMC) instrument. Find an expression for its deflecting torque. Give its advantages and disadvantages.

Ans. (*i*) Basic Principle : This instrument is based on the principle that when a current-carrying conductor is placed in a magnetic field, it is acted upon by a force which tries to push it out of the field. The movement of the conductor is used to measure either the current or the voltage

(*ii*) Construction : In Fig. 9, N and S are the pole pieces of a powerful permanent magnet. Between the magnetic poles is fixed a soft-iron cylinder which supports a rectangular coil of many turns on a light aluminum frame. The controlling torque is provided by the two fine springs which also lead the current in and out of the moving coil.

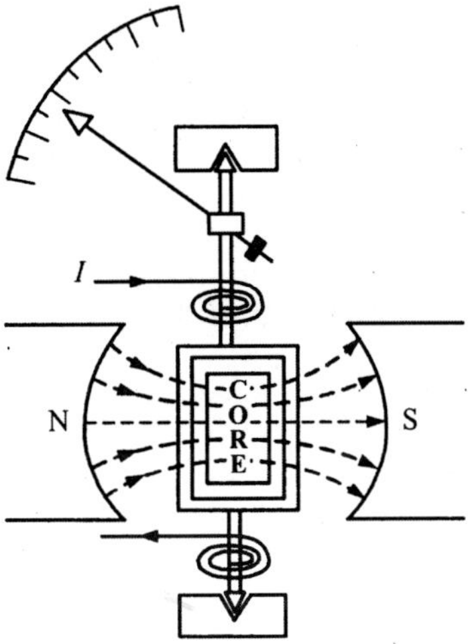

Fig. 9

(*iii*) Operation : When current is passed through the moving coil, its both *vertical* sides *ab* and *dc* experience force F as shown in Fig. 10. Hence, the coil is deflected. The pointer attached to the spindle is also deflected and moves over a calibrated scale. The deflection of the pointer is directly proportional to the current I being measured. Damping torque is produced by the eddy- currents induced in the aluminium former when it rotates in the magnetic field of the permanent magnetic

(*iv*) Deflecting torque : When current is passed through the coil, two forces F, F act on its both vertical sides and produce deflection torque as shown in Fig. 10.

Let $\qquad B$ = flux density of the field

l = length of the coil

b = breadth of the coil

N = number of turns of the coil

If I is the current through the coil,

then $\qquad F = NBIl$

Now, $\qquad T_d$ = force × perpendicular distance

$= NBIl \times b = NBI\,(l \times b) = NBIA = kI$ — where A is the area of the coil

As seen, $T_d \propto I$

For spring control $T_c \propto \theta$

At the final deflected position,

$$T_d = T_c \qquad \therefore \; \theta \propto I$$

Hence, such instruments have uniform scales. Damping is by eddy-currents induced in the metal frame over which the coil is wound.

The PMMC instruments have the following advantages and disadvantages.

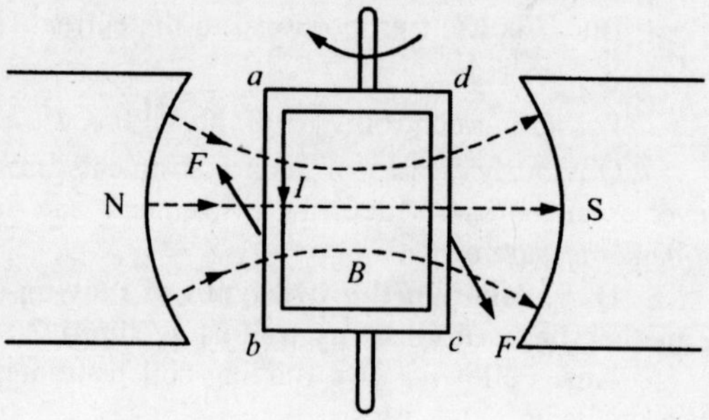

Fig. 10

Advantages :

1. they have low power consumption
2. their scales are uniform
3. their range can be easily extended
4. there is no hystersis loss
5. they have very effective eddy-current damping

Disadvantages :

1. they are costlier as compared to moving-iron instruments.
2. some errors are set in due to ageing of control springs and permanent magnet.

Q. 11. Briefly describe the construction and operation of a dynamometer type moving-coil instrument. Does it have a uniform scale?

Ans. (*i*) Basic Principle : In such instruments, no *permanent magnet* is used. Instead, a *current –carrying* coil is placed in the electromagnetic field of a *fixed coil.* The two electromagnetic fields produced by the fixed coil and the moving coil *react with each other* and produce rotation of the moving coil.

(*ii*) Construction : As shown in Fig. 11, the fixed coil is made up of two section C_1 and C_2. A moving coil M is placed in the space between C_1 and C_2. The coil M is attached to a spindle which carries a pointer. When the coil M rotates, the pointer moves over a calibrated scale. Two springs are used to give the required controlling torque. Damping is by air friction.

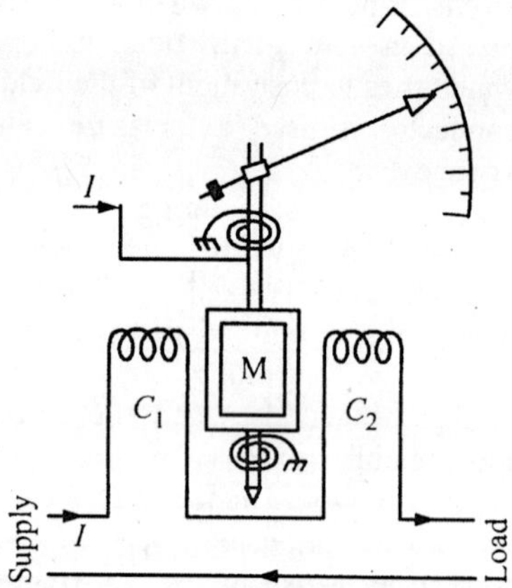

Fig. 11

(*iii*) Operation : Both the fixed coil and the moving coil carry either line current (as ammeter) or current proportional to voltage (as voltmeter). Hence, two electromagnetic fields are produced which react with each other. The electromagnetic force acts on the moving coil and makes it rotate. The attached pointer also moves over the scale.

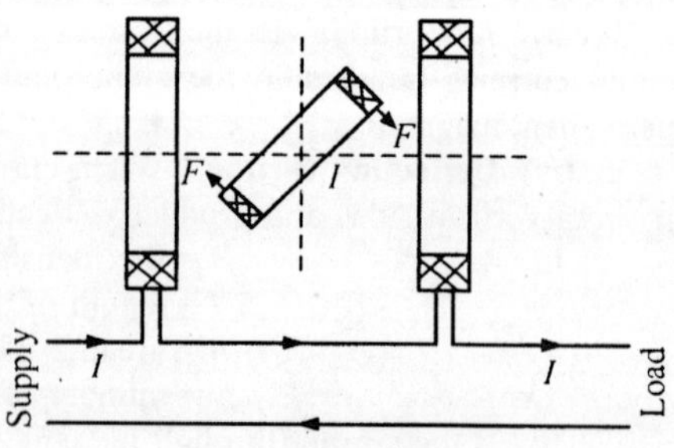

Fig. 12

(*iv*) Deflecting torque : Since the fixed and moving coils are connected *in series*, they carry the *same* current I. It may be the line current (as ammeter) or current proportional to the voltage (as voltmeter). The flux density produced by the fixed coil $B \propto I$ or $B = KI$. Let the moving coil be rectangular of size $l \times b$. Let it have N turns. Then, the force on its each side is

$$F = NBIl$$

The deflecting torque acting on the moving coil is

$$T_d = F \times b = NBIl \times b = NBIA$$

Substituting the value of $B = KI$ from above, we get

$$T_d = KNI^2 A \quad \text{or} \quad T_d \propto I^2$$

The controlling torque is given by $T_c \propto \theta$

In the final deflected position $\quad T_d = T_c \quad$ or $\quad \theta \propto I^2$

Since $\theta \propto I^2$ (and not I), hence scale is uneven.

Q. 12. What is meant by 'shunt' and 'multiplier'? How are these useful for extending the range of instruments?

Ans. By 'shunt' is meant a *low* resistance connected in *parallel* with the instrument. It extends its range as *ammeter*. Lower the value of shunt resistance, greater the extension of range.

A 'multiplier' is a *high* resistance connected in *series* with the instrument. It extends or multiplies its range *as a voltmeter.* Greater the value of series resistor, greater the range extension.

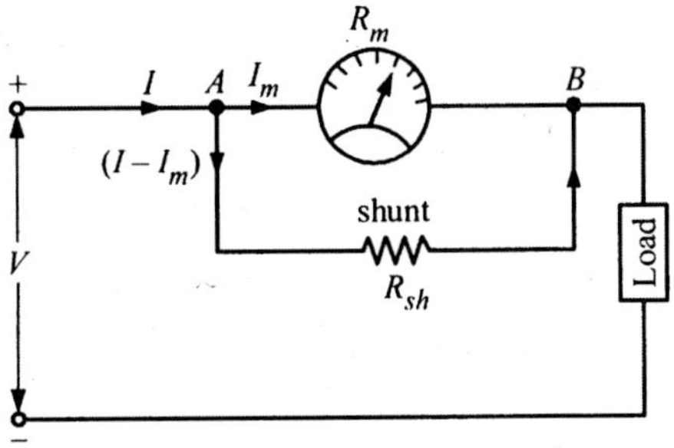

Fig. 13

1. Rage Extension as Ammeter : As shown in Fig. 13, a *low* shunt resistance R_{sh} has been connected in *parallel* with the instrument for increasing its range as an ammeter. The shunted instrument is connected in the supply line.

Let $\quad R_m$ = resistance of the meter

$\quad I_m$ = full – scale deflection (FSD) current of the meter

$\quad I$ = maximum current to be measured.

Since in Fig. 13, voltage across points A and B is the same

$$\therefore \quad I_m R_m = (I - I_m) R_{sh} \qquad \therefore \quad R_{sh} = \frac{I_m R_m}{(I - I_m)}$$

This is the value of the shunt resistance required for increasing the range of the meter from I_m to I. The range has been extended (I/I_m) times.

2. Range Extension as Voltmeter : The range of the instrument as voltmeter can be extended or *multiplied* by using a *high* resistance connected in series with the instrument as shown in Fig. 14. The meter along with R_{se} has been connected across the supply lines. Suppose the range has to be extended from V_m to V. Then, obviously, the extra voltage of $(V - V_m)$ has to be dropped across R_{se}.

Now $$V = V_m + I_m R_{se}$$

or $$I_m R_{se} = V - V_m = V - I_m R_m$$

$$\therefore \quad R_{se} = \left(\frac{V}{I_m} - R_m\right)$$

The amount by which voltage range has been multiplied *i.e. voltage multiplication* is given by

$$M = \frac{V}{V_m}$$

As seen from Fig. 14

$$V - V_m = I_m R_{se}$$

Dividing both sides by V_m, we get

$$\frac{V}{V_m} - \frac{V_m}{V_m} = \frac{I_m R_{se}}{V_m}$$

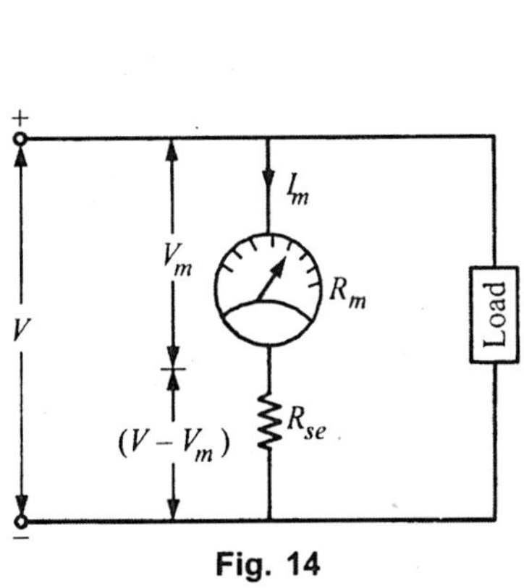

Fig. 14

or $\dfrac{V}{V_m} - 1 = \dfrac{I_m R_{se}}{I_m R_m}$ or $\dfrac{V}{V_m} = 1 + \dfrac{R_{se}}{R_m}$ $\therefore$ $M = \left(1 + \dfrac{R_{se}}{R_m}\right)$

Obviously, greater the value of R_{se}, greater is the increase in voltage range.

Q. 13. Full-scale deflection of a moving-iron meter is obtained when carrying a current of 100 mA. Show how it can be used to measure voltages upto 150 V and currents up to 20 A. Given that resistance of the instrument is 20 Ω.

(*Elect. Engg. BE 1st. Yr. Nagpur Univ. 2005*)

Sol. We are given that full-scale deflection current of the meter is I_m = 100 mA and meter resistances R_m = 20 Ω.

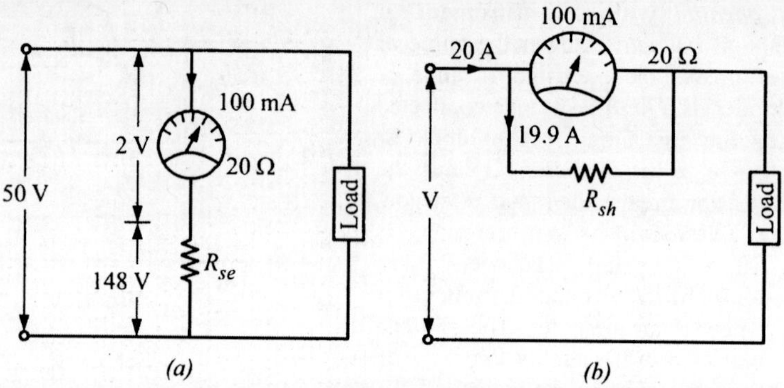

Fig. 15

(a) For measuring 150 V, series resistance R_{se} is connected in series with the meter. The meter is connected *across* the supply lines as shown in Fig. 15(*a*). As seen, voltage drop across the meter itself is = 20 × 100 × 10^{-3} = 2 V. Hence, the balance voltage of (150 – 2) = 148 V has to be dropped across R_{se}.

$$\therefore \quad 100 \times 10^{-3} \times R_{se} = 148 \text{ or } R_{se} = \mathbf{1480\ \Omega}$$

(b) For measuring a current of 20 A, a low resistance shunt R_{sh} is connected in parallel with the meter as shown in Fig. 15(*b*).

$$R_{sh} = \frac{I_m R_m}{I - I_m} = \frac{100 \times 10^{-3} \times 20}{(20 - 100 \times 10^{-3})} = \mathbf{1.005\ \Omega}$$

Q. 14. A moving-coil instrument of resistance 5 Ω requires a potential difference of 75 mV to give full-scale deflection. Calculate

(i) The value of shunt resistance needed to enable the instrument to work as an ammeter and to give a full-scale deflection at 30 A.

(ii) The value of the series resistance to allow the instrument to work as a voltmeter with a full-scale reading of 250 V. (*Elect. Engg. BE. Ist Yr, Nagpur Univ, 2006*)

Sol. (i) Let the required value of the shunt resistance be R_{sh}.

Now. $R_m = 5\ \Omega$ and $V_m = 75$ mV = 75 × 10^{-3} = 0.075 V

$\therefore$ $I_m = V_m/R_m = 0.075/5 = 0.015$ A, I = 30 A

$$\therefore \quad R_{sh} = \frac{I_m R_m}{I - I_m} = \frac{V_m}{I - I_m} = \frac{0.075}{30 - 0.015} = \mathbf{0.0025\ \Omega}$$

(ii) Let R_{se} be the series resistance connected with the meter

$$\therefore \quad R_{se} = \left(\frac{V}{I_m} - R_m\right) = \frac{250}{0.015} - 5 = \mathbf{16{,}660\ \Omega}$$

Q. 15. A moving-coil instrument has a resistance of 10 Ω and gives full-scale deflection when carrying a current of 50 mA. Show how it can be adopted to measure (*i*) voltages up to 750 V and (*ii*) currents up to 100 A.

(*Basic Elect. Engg. C.S.V. Tech Univ. Bhilai, 2007*)

Sol. We are given that

$$I_m = 50 \text{ mA} = 50 \times 10^{-3} \text{ A} = 0.05 \text{ A}, R_m = 10\ \Omega$$
$$V = 750 \text{ V}, I = 100 \text{ A}$$

(*i*)
$$R_{se} = \left(\frac{V}{I_m} - R_m\right) = \frac{750}{0.05} - 10 = \mathbf{14{,}990\ \Omega}$$

(*ii*)
$$R_{sh} = \frac{I_m R_m}{I - I_m} = \frac{0.05 \times 10}{100 - 0.05} = \mathbf{0.005\ \Omega}$$

Q. 16. A PMMC instrument has FSD current of 50 mA and 2 Ω resistance. How the instrument can be converted to:

(*i*) 0–5 A range ammeter?

(*ii*) 0–100 V range voltmeter? (*Elect. Engg. B.E. 1st Yr. Nagpur Univ. 2005*)

Sol. We are given that

$$I_m = 50 \text{ mA} = 0.05 \text{ A}, R_m = 2\ \Omega, I = 5 \text{ A}$$
$$V = 100 \text{ V}$$

(*i*) **0–5 A range ammeter**

$$R_{sh} = \frac{I_m R_m}{I - I_m} = \frac{0.05 \times 2}{5 - 0.05} = \mathbf{0.02\ \Omega}$$

(*ii*) **0–100 V range voltmeter**

$$R_{se} = \left(\frac{V}{I_m} - R_m\right) = \frac{100}{0.05} - 2 = \mathbf{1998\ \Omega}$$

Q. 17. Briefly explain the induction method of producing eddy-current deflecting torque in induction instruments.

Ans. All induction instruments employ the deflecting torque produced by the reaction between the flux of an ac magnet and the eddy-currents induced by this flux in a metal disc. Consider a thin aluminium disc D free to rotate about its axis as shown in Fig. 16(*a*) Two *ac magnets* M_1 and M_2 produce *alternating* fluxes Φ_1 and Φ_2 which cut this disc. The part of the disc immediately below M_1 is linked by Φ_1 and produces an alternating emf. e_1. This e_1 circulates the eddy-current i_1 which, as shown in Fig. 16(*b*), also passes under M_2. Similarly, Φ_2 produces an *emf* e_2 which induces the eddy-current i_2. This eddy-current i_2 also flows under the pole M_1.

The portion of the disc which is cut by flux Φ_1 and carries current i_2 experiences a force F_1 along the direction as shown. Obviously, $F_1 \propto \Phi_1 i_2$. Similarly, $F_2 \propto \Phi_2 i_1$

$$\therefore \quad F_1 \propto \Phi_1 i_2 = K \Phi_1 i_2 \text{ and } F_2 \propto \Phi_2 i_1 = K\Phi_2 i_1$$

If r is the radius at which these forces act, then the *net* torque acting on the disc is equal to the *difference* of the two torques.

$$\therefore \quad T_d = r(K\Phi_1 i_2 - K\Phi_2 i_1) = K'(\Phi_1 i_2 - \Phi_2 i_1)$$

On further simplifying the above expression (the actual derivation has been omitted), we get

$$T_d = k\,\omega\,\Phi_{1m}\,\Phi_{2m} \sin \alpha$$

Here, ω is the angular frequency of the supply, Φ_{1m} and Φ_{2m} are the maximum values of the two fluxes and α the phase angle between the two fluxes.

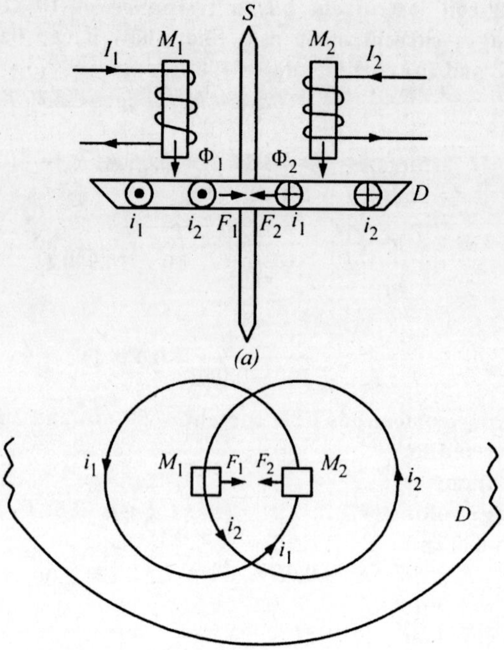

(b)

Fig. 16

Q. 18. Explain the construction and working of a dynamometer wattmeter with its advantages and disadvantages.

Ans. (*i*) Basic Principle : In such instruments, *no permanent magnet* is used. The operating field is produced by a fixed coil. The moving coil is placed in the electromagnetic field of the fixed coil. The reaction between the two magnetic fields produces the rotating force on the moving coil.

(*ii*) Construction : As shown in Fig. 17, the fixed coil is made up of two sections C_1 and C_2 kept parallel to each other. The moving coil M is placed in the electromagnetic field of the fixed coil. The moving coil is attached to a spindle which carries a pointer.

When coil M rotates, the pointer moves over a calibrated scale. Springs are used to produce controlling torque. Damping is by air friction.

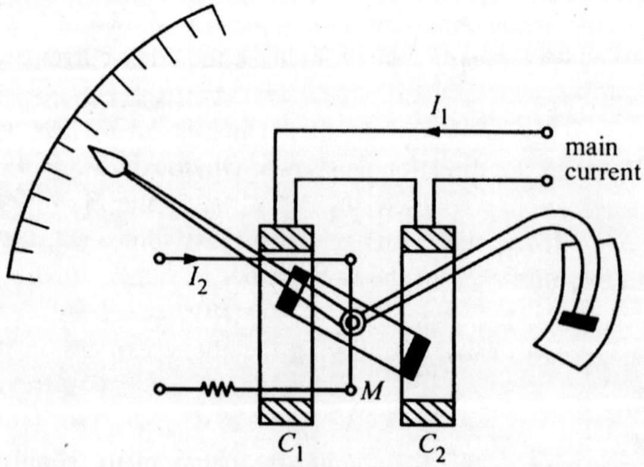

Fig. 17

(*iii*) Operation : The fixed coil carries the main circuit current I_1. However, the moving coil M carries current I_2 which is proportional to the supply voltage V. Hence, the deflection of the pointer depends on voltage V and I_1 which means circuit power.

(*iv*) Deflecting torque : The flux density B produced by the fixed coil is proportional to the current carried by it

i.e. $B \propto I_1$.

Similarly, $I_2 \propto V$

$\therefore\ T_d \propto BI_2 \propto VI_1$

$\propto$ power — on dc circuits

$$T_d \propto VI_1 \cos \phi \propto \text{power} \qquad \text{— on ac circuits}$$

For spring control, $T_c \propto \theta$

At the final *deflected position,*

$$T_d = T_c \quad \text{or} \quad \theta \propto \text{power}$$

The scales of induction wattmeters are uniform.

Advantages

1. they are highly accurate.
2. they are equally accurate on dc as well as ac circuits

Disadvantages

1. at low power factors, the inductance of voltage coil causes some error.

Q. 19. Discuss the construction and working of an induction wattmeter. Give its advantages and disadvantages.

Ans. (*i*) Basic Principle : This wattmeter has two *ac magnets*, one excited by line current and the other by the current proportional to the voltage *V*. The two alternating fluxes and the eddy currents produced by them react with each other. This reaction is the basic working principle of this wattmeter.

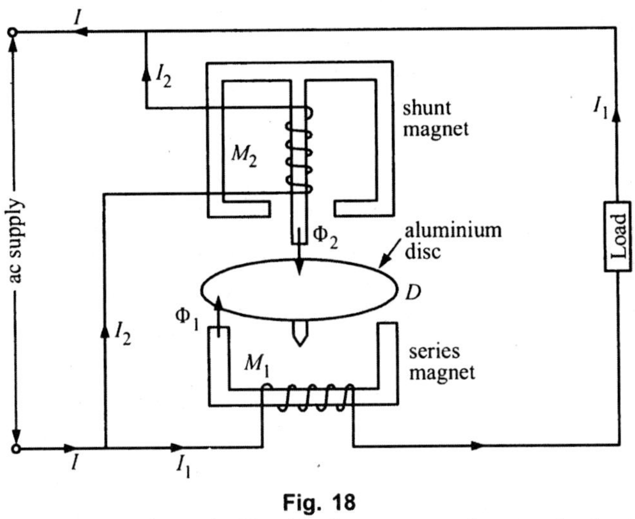

Fig. 18

(*ii*) Construction : As shown in Fig. 18, the wattmeter has two laminated electromagnets M_1 and M_2. M_1 is connected in series with the circuit and carries line current I_1. It is called *series* magnet. M_2 is connected *across* supply lines. It carries current I_2 which is proportional to the supply voltage *V*. It is called *shunt* magnet.

A thin aluminum disc *D* cuts the fluxes of both magnets. Hence, eddy currents are induced in it. The instrument is spring controlled by the two springs (not shown in the figure) fitted to the spindle of the moving system which also carries the pointer.

(*iii*) Operation : Since aluminium disc cuts the alternating fluxes of series and shunt electromagnets, two eddy currents are induced in the disc. The deflecting torque is produced due to the interaction of these eddy-currents with the two alternating fluxes Φ_1, and Φ_2. A pointer attached to the spindle moves over the calibrated scale.

(*iv*) Deflecting torque : The deflecting torque depends on the two currents I_1 and I_2 so that $T_d \propto I_1 I_2$. Since $I_2 \propto V$, $T_d \propto VI_1$. If ϕ is the phase angle between *V* and I_1, then

$$T_d \propto VI_1 \cos \phi \propto \text{power}$$

Since the instrument is spring-controlled, $T_c \propto \theta$.

In the final deflected position, $T_c = T_d$

$\therefore$ $\theta \propto$ power

Hence, the scales of such wattmeters are uniform i.e. equally spaced.

Advantages

1. they have fairly long scales (extending over 300°)
2. they are free from the effects of stray fields.
3. they have good damping
4. they are almost free from frequency errors.

The only disadvantage of these instruments is temprature errors which are due to the resistance of the eddy-current paths.

Q. 20. Explain the construction and working of a single-phase energy meter.

Ans. (*i*) Basic Principle : The basic principle of these meters is just the same as that of *induction wattmeters.* The reaction between the alternating fluxes of two electromagnets and eddy currents induced by them in a pivoted aluminium disc forms the basis of their working.

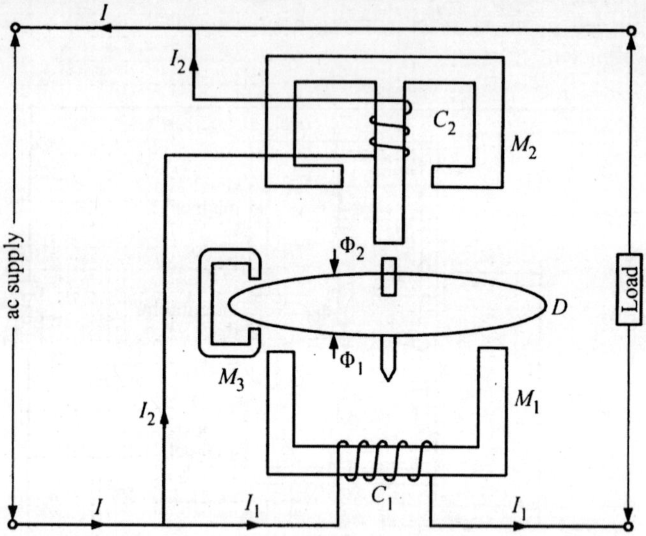

Fig. 19

(*ii*) Construction : The energy meter is an *integrating instrument* and has two electromagnets M_1 and M_2. M_1 is called the *series* magnet because it is connected in series with the circuit and carries line current. M_2 is called *shunt* magnet. The current I_2 carried by it is proportional to the supply voltage V. Both of them induce eddy currents in the pivoted aluminium disc D. These eddy currents react with the two magnetic fluxes and produce a rotating torque on the disc. Hence, the disc starts rotating continuously at a constant speed. Permanent magnet M_3 called braking magnet is used to produce the braking torque on the disc. It is this torque which makes the disc rotate continuously at a constant speed. The electric energy consumed by the customer is recorded by a register mechanism either of pointer type or cyclometer type.

(*iii*) Operation : The driving torque is provided by the *reaction* between the two magnetic fluxes Φ_1, Φ_2 and eddy currents induced in the aluminium disc D. The driving torque so produced is balanced by the braking torque produced by the braking magnet M_3.

The rotating torque $T_R \propto I_1 I_2 \propto VI_1$ because $I_2 \propto V$

If cos ϕ is the power factor of the load,

then $T_R \propto VI_1 \cos \phi$

Braking torque $T_B \propto N$ where N is the rotating speed of the disc.

The disc achieves a *steady* speed N when the two torques are equal.

i.e. $T_R = T_B$ or $N \propto$ power W

Hence, in a given period of time, the total number of revolutions $\int_0^t N.dt$ is proportional to $\int_0^t W.dt$ *i.e.* the electric energy consumed.

or $N \times t = W \times t$

or number of revolutions = energy consumed.

EXERCISE

1. A PMMC instrument has a resistance of 7.5 Ω and requires 15mA for full-scale deflection. Calculate the resistance to be connected
   (*i*) to read 2 A and (*ii*) to read 30 V. (*Elect Engg. B.E Ist Yr, Nagpur Univ, 2000*)
   (0.0566 Ω, 1992.5 Ω)
2. A moving-coil instrument gives full-scale deflection with 10 mA and has a resistance of 50 Ω. Calculate the resistance required to be connected in series/parallel with the instrument in order that it may be used as
   (*i*) 0–5 A ammeter and (*ii*) 0–200 V voltmeter. (*Elect Engg. B.E Ist Yr, Nagpur Univ*)
   (0.1 Ω, 19,950 Ω)
3. A PMMC instrument has 1 Ω resistance and gives full-scale deflection with 100 mA. Calculate the resistance required to be connected in series or parallel with the coil to measure
   (*i*) 10 A and (*ii*) 300 V. (*Elect Engg. B.E Ist Yr, Nagpur Univ, 2001*)
   [0.0101 Ω, 2999 Ω]
4. A permanent-magnet moving-coil instrument has 1 Ω resistance and gives full-scale deflection with 50 mA. Calculate the resistance required to be connected in series or parallel with coil to measure.
   (*i*) 5 A (*ii*) 50 V. (*Elect Engg. B.E Ist Yr, Nagpur Univ*)
   (0.01 Ω, 999 Ω)

MULTIPLE CHOICE QUESTIONS

1. Integrating meters are used for the measurement of
   (*a*) current (*b*) voltage (*c*) power (*d*) energy
   (*Basic Elect. Engg. V.T.U. Karnatka 2009*)
2. In the measuring instruments under equilibrium condition, the relations between controlling torque T_c and deflecting torque T_d is
   (*a*) $T_c = T_d$ (*b*) $T_c > T_d$ (*c*) $T_c < T_d$ (*d*) none of these
3. Which is NOT essential for the working of an indicating instrument
   (*a*) deflecting torque (*b*) braking torque
   (*c*) damping torque (*d*) controlling torque
4. The main function of the damping torque in an indicating electrical instrument is to
   (*a*) bring the pointer to rest quickly (*b*) prevent sudden movement of the pointer
   (*c*) make pointer deflection gradual (*d*) provide friction.
5. In a dynamometer wattmeter, the fixed coil is
   (*a*) current coil (*b*) voltage coil
   (*c*) current or voltage coil (*d*) none of the above
   (*Basic Elect. Engg. V.TU. Karnatka 2009*)

6. In elctrodynamic instruments, the damping is always
   (*a*) pneumatic (*b*) electromagnetic
   (*c*) by fluid friction (*d*) by braking magnet
7. In all induction instruments, deflection torque is produced due to the reaction between
   (*a*) two eddy currents (*b*) two alternating fluxes
   (*c*) flux and eddy current (*d*) voltage and current
8. In an induction type energy meter
   (*a*) there is no braking magnet (*b*) two d.c.magnets are used
   (*c*) no shading bands are used (*d*) disc revolves continuously
9. In an energy meter, constant speed of rotation of the disc is provided by
   (*a*) shunt magnet (*b*) series magnet
   (*c*) braking magnet (*d*) none of the above
10. Which of the following instrument has uniform scale
    (*a*) PMMC instrument (*b*) moving-iron instrument
    (*c*) electrodynamic instrument (*d*) none of the above
11. An induction type energy meter differs from the induction type wattmeter in that it has
    (*a*) only one ac magnet (*b*) no eddy currents
    (*c*) no rotating disc (*d*) a continuously rotating disc
12. A single-phase energy meter is built up of
    (*a*) three electromagnets
    (*b*) two electromagnets and one permanent magnet
    (*c*) two permanent magnets and one electromagnet
    (*d*) three permanent magnets

ANSWERS

1. *d*	2. *a*	3. *b*	4. *a*	5. *a*	6. *a*	7. *c*
8. *d*	9. *c*	10. *a*	11. *d*	12. *b*.		

CHAPTER 11

DC GENERATOR

Q. 1. Sketch an elementary ac generator and explain how it produces an ac voltage. What is the essential difference between an elementary ac generator and a dc generator?

Ans. Construction : A simple or an elementary ac generator consists of two parts (*i*) a loop of wire and (*ii*) a magnetic field.

The loop of wire is so placed that it can rotate in a uniform magnetic field. When the loop of wire cuts the magnetic flux, an induced emf is induced in it. This emf is taken out of the loop with the help of sliding contacts and is applied to the external load.

The magnetic field is provided by a pair of North and South magnetic poles (Fig. 1). The loop of wire that rotates within the magnetic field is called *armature*. The ends of armature are connected to two rings called *sliprings* that rotate along with the armature. Two carbon brushes ride up against the sliprings to connect the armature to the external circuit.

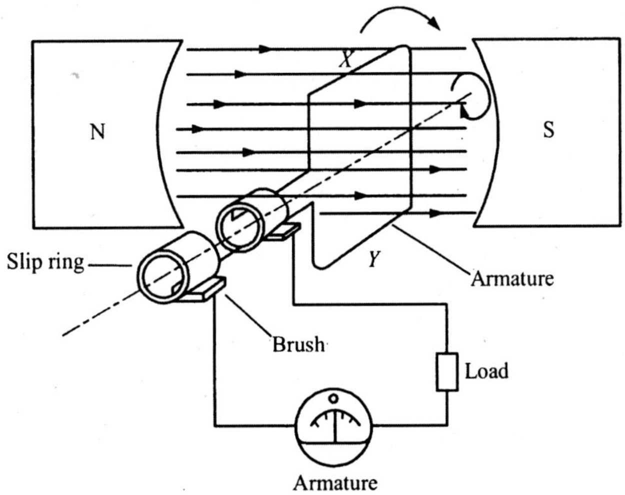

Fig. 1

Operation : To begin with, let the loop be in the vertical position inside the magnetic field. Now suppose it starts rotating in the clockwise direction. As the sides of the loop cut through the magnetic field, an emf is induced in them which causes a current to flow through the loop, sliprings, brushes, ammeter and the load — all connected in series.

The magnitude of the induced emf and current depends on the *instantaneous* position of the loop in the magnetic field.

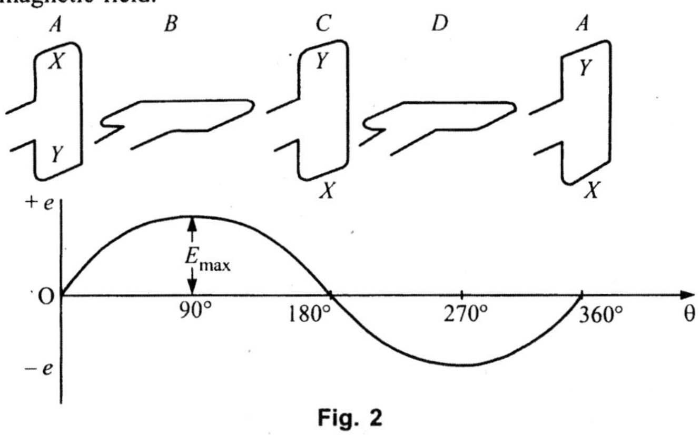

Fig. 2

The initial position A of the loop in Fig. 2 is taken as 0° position. Since in this position, the loop is perpendicular to the field, the conductors X and Y move *parallel* to the lines of the magnetic flux. Since there is no *cutting* of magnetic flux, so no emf is induced in the conductors. As a result., no current flows in the ammeter which reads zero. (Fig. 1)

As the loop rotates from position A to position B, the conductors cut through more and more lines of flux till at position B (90°), they cut *maximum* amount of flux. Hence, between 0° and 90°, the induced emf in the conductors increases from zero to a maximum value. Another point worth noting is that from 0° to 90°, the conductor X cuts the flux in the *downward* direction whereas conductor Y cuts the flux in the *upward* direction. Hence, the induced emfs in the two conductors are in series-aiding and the resultant voltage across the brushes is the *sum* of the two induced emfs. Hence, the ammeter deflects more to the right side.

As the loop moves from position B (90°) to position C (180°), the conductors cut lesser amount of flux until at point C, they cut *no flux* because they move parallel to the lines of flux. Hence, no emf is induced in the conductors and the ammeter reads zero again.

As the loop rotates beyond 180° back to the initial position A, the direction of flux cutting *reverses*. Now, conductor X moves *upwards* through the field and the conductor Y moves *downwards*. As a result, the polarity of the induced emf is reversed. Hence, from position C through D and back to position A, the polarity of the induced emf is reversed. The voltage waveform for one complete revolution is shown in Fig. 2. It may be noted that the voltage induced in the loop reverses its polarity everytime the loop goes through half revolution (0°–180°).

For converting induced ac voltage to dc voltage, a commutator is needed instead of two sliprings. For this purpose, one of the two sliprings is removed and the other is cut into two segments along its axis as shown in Fig. 3. The ends of the loop are then connected, one to each of the two segments of the slipring.

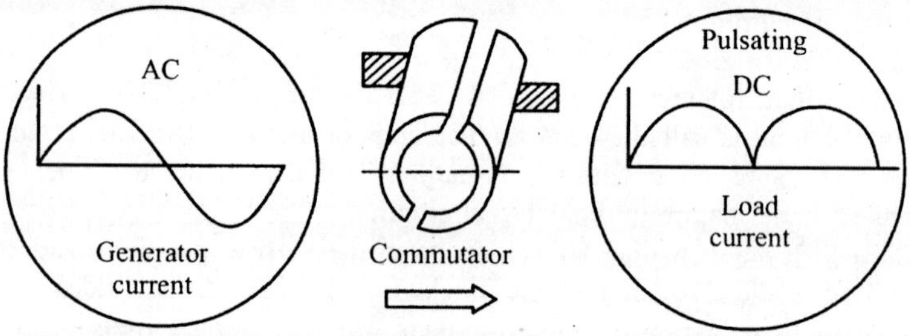

Fig. 3

Q. 2. Describe the basic construction and working of an elementary dc generator. How can the output voltage of such a generator be increased?

Ans. Construction

A simple or an elementary dc generator consists of (*i*) a magnetic field (*ii*) a rotating coil (*iii*) a commutator with two brushes as shown in Fig. 4. The brushes B_1 and B_2 are placed opposite each other on the two segments of the commutator. The generator voltage is supplied to the external load through the two brushes.

Operation : Let the armature start rotating in the clockwise direction from its initial position A (0°). At position A, the loop is perpendicular to the lines of the magnetic field (Fig. 4).

In this position, the loop wires X and Y move parallel to the field and do not cut any flux. Hence, no emf is induced in the loop conductors and so there is no current flow.

The moment the loop moves slightly beyond position A in the clockwise direction, the brush B_1 comes in contact with commutator segment S_1 and the brush B_2 comes in contact with segment S_2. As the loop rotates clockwise from position A to position B, the induced emf starts increasing from zero to its *maximum* value at B position (90°) as shown in Fig. 5. The induced current is also maximum at B position. As the loop continues rotating from B position to C position, the induced emf decreases till it becomes zero once again at C position (180°)

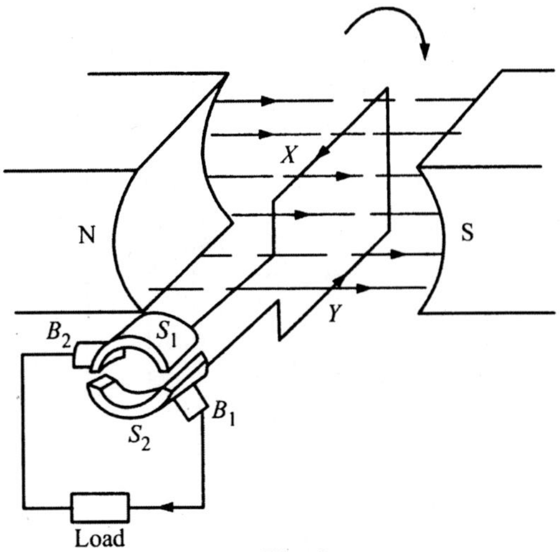

Fig. 4

Fig. 5 shows how the generator voltage various from 0° to 180°. The direction of the current in conductor X can be found by using Fleming's Right Hand rule and is as shown in Fig. 4. as well as in Fig. 5. The current is collected by brush B_1 and supplied to be external load. It may be noted that in position C, brush B_1 is sliding away from segment S_1 to S_2 and, at the same time, brush B_2 is moving away from segment S_2 to S_1. In this way, B_1 is always in contact with the conductor moving downwards and B_2 is always in contact with the conductor moving upwards. Since the downward moving conductor has the current flow *towards* the brush, hence B_1 is the *positive* terminal and B_2 is the *negative* terminal of the dc generator. It can be easily verified with the help of Fleming's Right Hand Rule.

As the loop continuous to rotate from C position (180°) through D position (270°) and back to A position (360° or 0°), B_1 is now connected to conductor Y which is moving *down* and B_2 is connected to conductor X which is moving *up*. As a result, the same polarity voltage waveform is generated across the brushes from 180° to 360° as was generated from 0° to 180°. Hence, current flows in the *same direction* through the external load even though it reverses its direction every half-cycle *within the loop itself.* The output voltage has the same polarity at all times though its value varies. Hence, it is called *pulsating* dc voltage (Fig. 5).

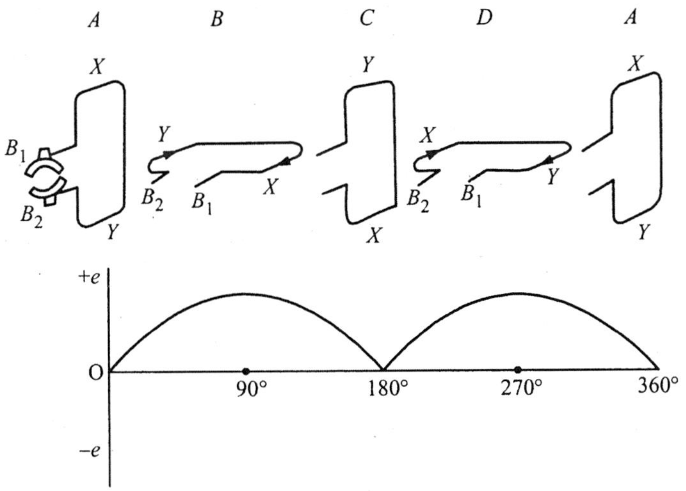

Fig. 5

This pulsating dc voltage is not smooth enough to operate many dc appliances and equipment. To make this output voltage smoother, more coils of wire are added to the armature, the two coils being positioned at right angles to each other. The commutator is divided into four segments with opposite segments connected to the ends of a coil. The pulsating output voltages of coil *A* and *B* are shown in Fig. 7. The output voltage can be increased and also made smoother by *increasing* the number of coils.

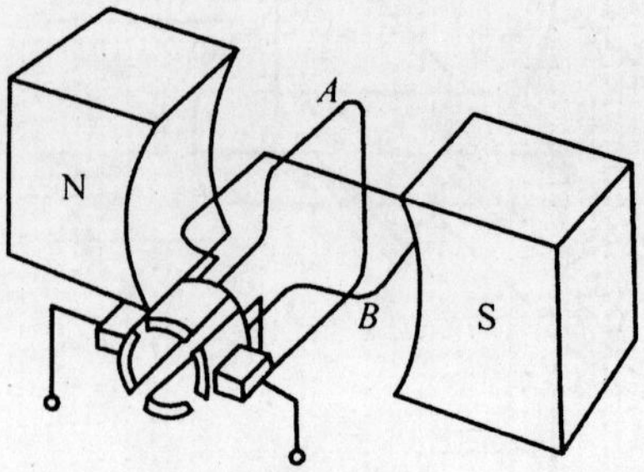

Fig. 6

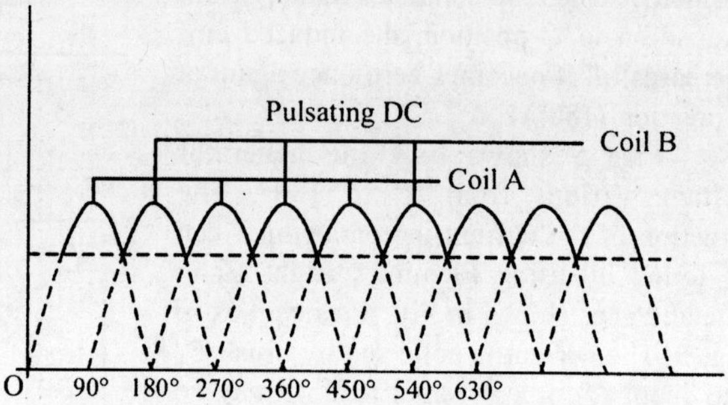

Fig. 7

Q. 3. Give the constructional details of a practical dc generator and briefly describe its working

Ans. DC generator design and its construction depend on its size, type and manufacturer. But the major parts of all dc generators are as under :

(*i*) magnetic frame or yoke
(*ii*) field poles and pole shoes
(*iii*) field coils or windings
(*iv*) armature assembly
(*v*) brushes and bearings

Of these, the yoke, the field poles, the armature and the air gap between the field poles and the armature form the *magnetic* circuit whereas the rest form the *electrical* circuit as shown in Fig. 8. Further details of other components are given below:

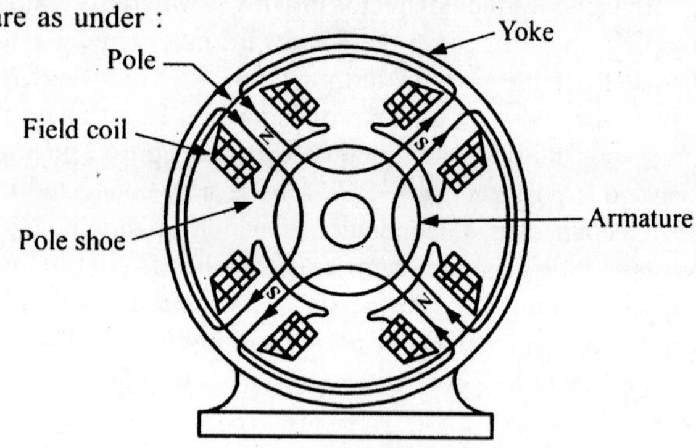

Fig. 8

(*i*) **Yoke :** The yoke or outer frame is of cylindrical shape and serves three purposes:

(*a*) it supports magnetic poles and the field coils which produce the magnetic flux.
(*b*) it carries the magnetic flux that passes from the field poles to the armature.
(*c*) it protects the whole machine.

(*ii*) **Field poles and pole shoes :** These are made of highly magnetic alloys. They become strong electromagnets when current is passed through the field coils.

(*iii*) **Field coils :** These coils consist of copper wire. When direct current is passed through them, they electromagnetise the field poles which produce strong magnetic field around the armaturc. When the armature revolves, its conductors cut the magnetic flux and, hence, produce emf in the armature winding.

(*iv*) **Armature assembly :** The armature assembly consists of a shaft, armature core, armature winding and a commutator (Fig. 10). The

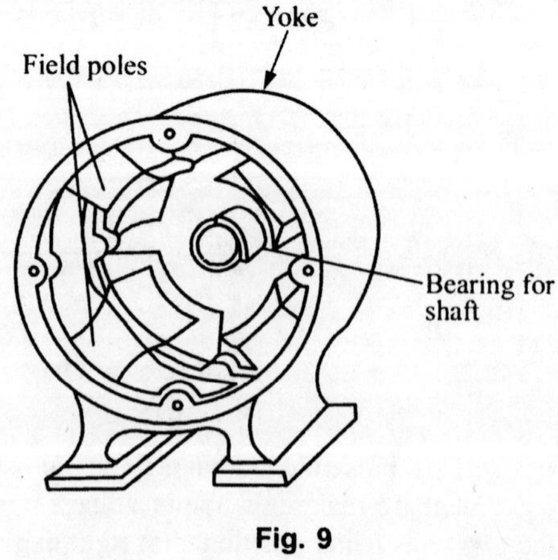

Fig. 9

armature core is laminated to reduce eddy current loss and has slots to receive the armature conductors. The commutator is made up of copper segments (or bars) which are insulated from one another and from the shaft. The shaft supports the entire armature assembly and rotates in the end bearings.

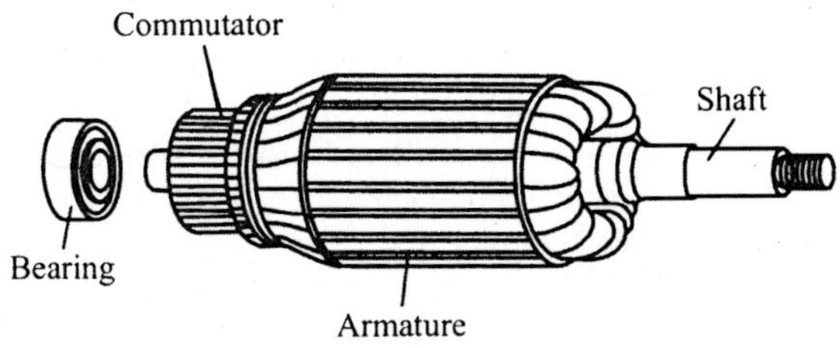

Fig. 10

There is a small air-gap between the armature and the field poles to prevent rubbing between the armature and the field poles during rotation.

(*v*) **Brushes and bearings :** The brushes ride on the commutator bars and carry the generated voltage to the external load. They are usually made of high-grade of carbon or a carbon-copper mixture and are held in place by brush holders. They have the shape of a rectangular block. (Fig. 11)

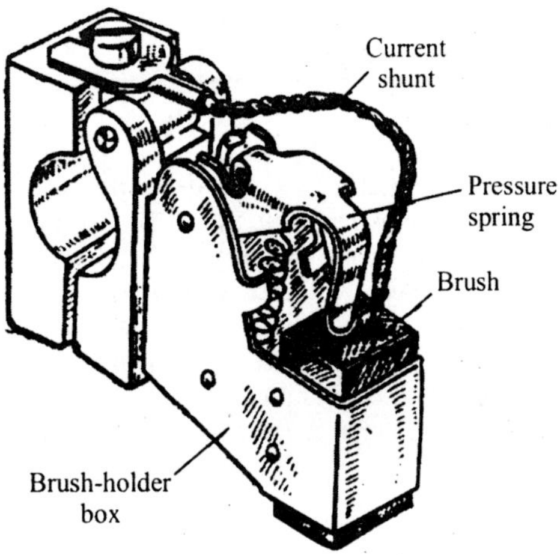

Fig. 11

Working : A dc generator is a machine which converts *mechanical energy into electrical energy* when driven by a prime mover. The prime mover may be a water turbine or a steam turbine, diesel engine or an electric motor etc. The prime mover rotates the armature which carries armature windings. These armature windings cut the magnetic flux produced by the field poles. It induces emf in the armature winding which is taken out through the brushes and applied across the external load.

The value of the generated voltage depends on

(*i*) the number of armature conductors.

(*ii*) armature speed.

(*iii*) strength of the magnetic field.

Q. 4. What is the function of armature winding in a dc generator? Distinguish between a lap and a wave armature winding with regards to the number of parallel paths.

Ans. The armature winding is, in fact, the *heart* of the generator; *it is where voltage is generated.* When the armature rotates in the magnetic field of the poles, the armature conductors cut the magnetic flux and, hence, produce induced emf. This induced emf is taken out via brushes to the external load.

There are two basic types of windings which are used in drum-type armature and are **known** as wave windings and lap winding. They differ from each other mainly in the way their coil ends are connected to the commutator bars.

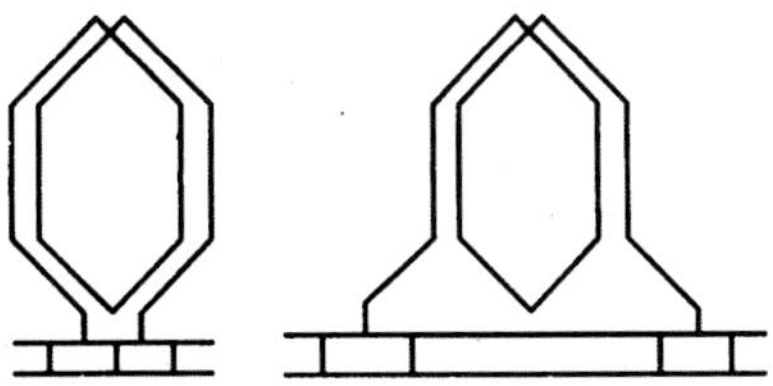

Lap winding (*a*)   Wave winding (*b*)

Fig. 12

In a lap winding, the coil ends are connected to *adjacent* commutator bars [Fig. 12(*a*)]. In wave windings, the coil ends are connected to the commutator bars under the *alternate* poles as shown in Fig. 12(*b*). Another important difference between the two armature windings is about the number of parallel paths they provide through the armature winding.

(*i*) **Wave winding :** It electrically divides the armature conductors into two parallel paths between the positive and negative brushes whatever may be the number of generator poles. As shown in Fig 13, when the armature current enters the negative brush, it finds two parallel paths of equal resistance available for going to the positive brush. Hence, it divides *equally* into two parts. If Z is the total number of armature conductors, then each path consists of $Z/2$ conductors connected in *series*. Each path carries a current of $I_a/2$ where I_a is the total armature current.

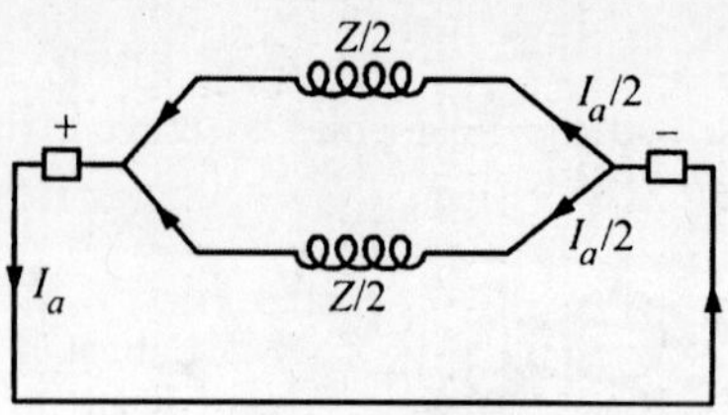

Wave Winding

Fig. 13

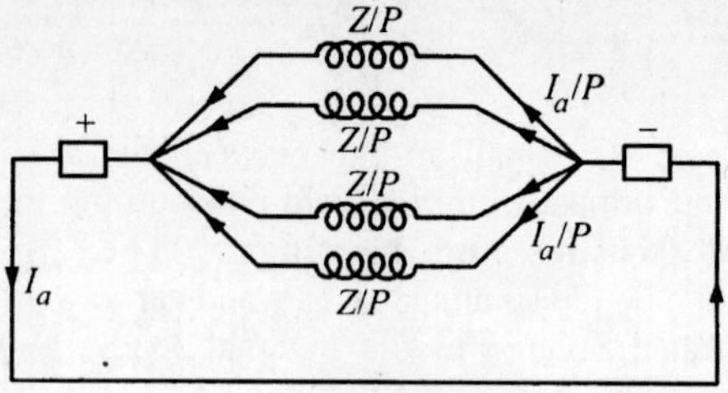

Lap Winding

Fig. 14

(*ii*) **Lap winding :** In this case, armature conductors are divided into as many parallel paths as the *number of poles* of the generator. If there are P poles and Z armature conductors, then there are P parallel paths, each consisting of Z/P conductors connected in series between the positive and negative brushes.

Fig. 14 shows the case of a 4-pole machine. As the armature current enters the negative brush, it has four parallel paths available for going to the positive brush. Each path has $Z/4$ conductors and carries a current of $I_a/4$.

Q. 5. Derive the expression for the armature resistance of a dc generator.

Ans. Let,
l = length of each armature conductor
S = its cross-section
A = No. of parallel paths
= 2 — for wave-winding
= P — for lap-winding
R = resistance of the whole winding

Then, $$R = \frac{\rho l}{S} \times Z$$

Resistance of each parallel path $\frac{\rho l Z}{SA}$.

It is the resistance of the whole armature winding. There are A such paths in parallel, hence equivalent resistance (called armature resistance) is

$$R_a = \frac{1}{A}\,\frac{\rho l Z}{SA} \quad \text{or} \quad R_a = \frac{\rho l Z}{SA^2}.$$

Q. 6. What is meant by the excitation of a dc generator? Differentiate between the separately-excited and self-excited dc generators.

Ans. Excitation of a generator means establishing its magnetic field. It can be done by passing a dc current through the field coils placed around the magnetic poles of the generator. The poles become magnetised and produce strong magnetic field in which armature rotates :

When the exciting current is taken from an outside source such as a battery, it is called a *separately-excited* generator. As shown in Fig. 15(*a*), the battery circuit is electrically *separate* from the armature circuit.

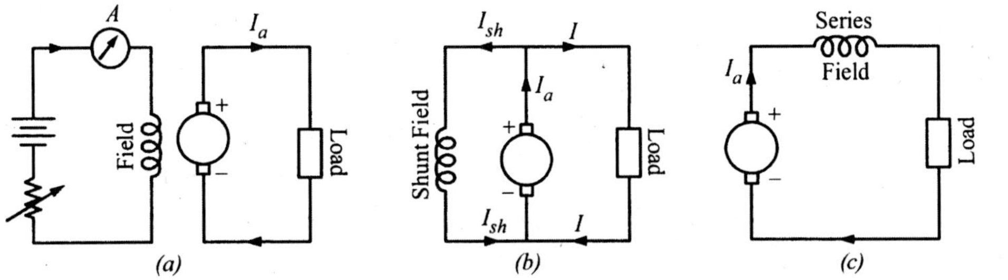

Fig. 15

In the case of a self-excited generator, part of the armature current is used for producing the magnetic field. That is why they are called self-excited. The exciting circuit is electrically *connected to the armature circuit* as shown in Fig. 15(*b*) and (*c*). In these generators, some *residual magnetism* is always present in the field poles. When the armature rotates, some emf and, hence, some induced current is produced which is partly or fully passed through the field-coils. Hence, magnetic flux of the generator is increased which increases the generator emf. Soon after, the generator builds-up to its full voltage.

Q. 7. Draw the circuit diagram of the different self-excited d.c. generators in common use.

Ans. There are *three* types of *self-excited* d.c generators depending on the way in which field windings are connected to their armatures.

(*i*) **Shunt Wound Generator :** Its field winding is connected in *parallel* with its armature circuit and has the full generator voltage applied across it as shown in Fig. 16.

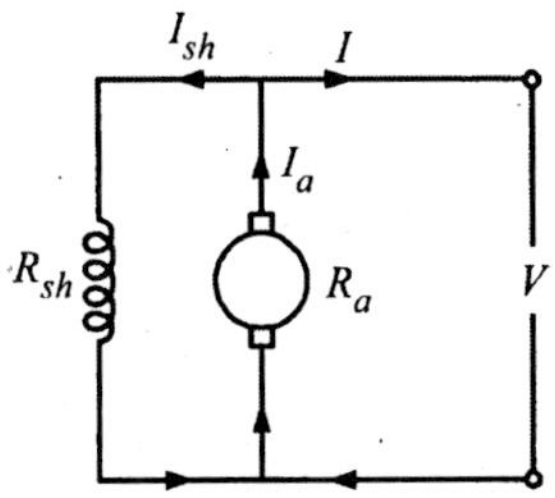

Fig. 16

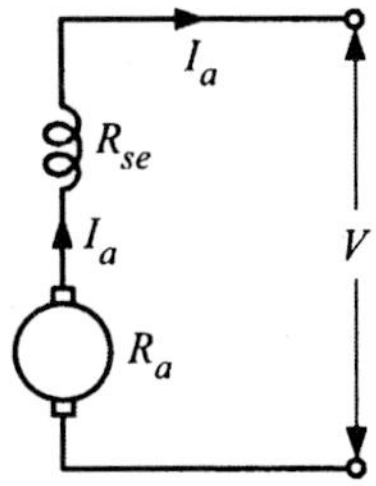

Fig. 17

(*ii*) **Series Wound Generator :** In this case, the field winding is connected in series with the armature. The series coil carries full armature current I_a as shown in Fig. 17.

(*iii*) **Compound Wound Generator :** It is a combination of a series winding and a shunt winding. It can be either a short-shunt generator or a long-shunt generator as shown in Fig. 18 and Fig. 19 respectively.

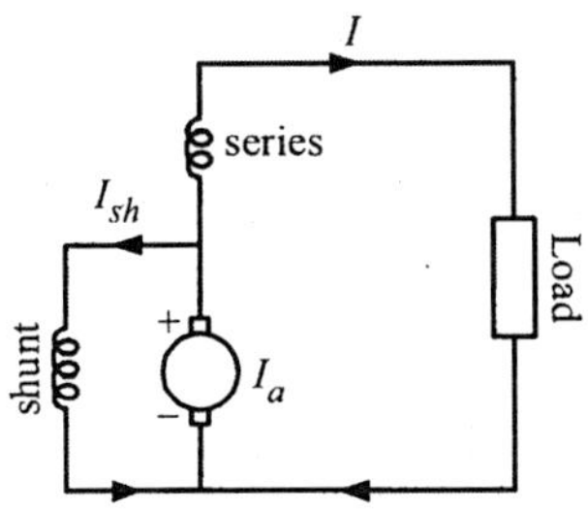

Fig. 18

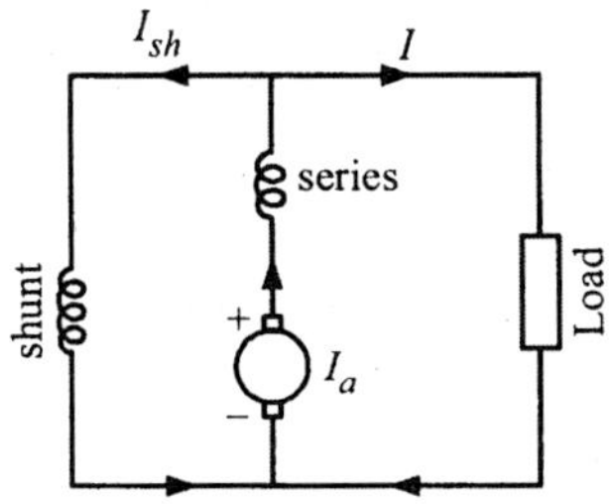

Fig. 19

Various types of dc generators and their sub-divisions are show in Fig. 20 below.

Q. 8. A 4-pole, long shunt, lap-wound generator supplies 25 kW at a terminal voltage of 500 V. The armature resistance is 0.03 Ω, series field resistance is 0.04 Ω and shunt field resistance is 200 Ω. The brush drop may be taken as 1 V. Determine the emf generated.

(*B.Tech, Elect Engg. JNTU, Hyderabad 2007*)

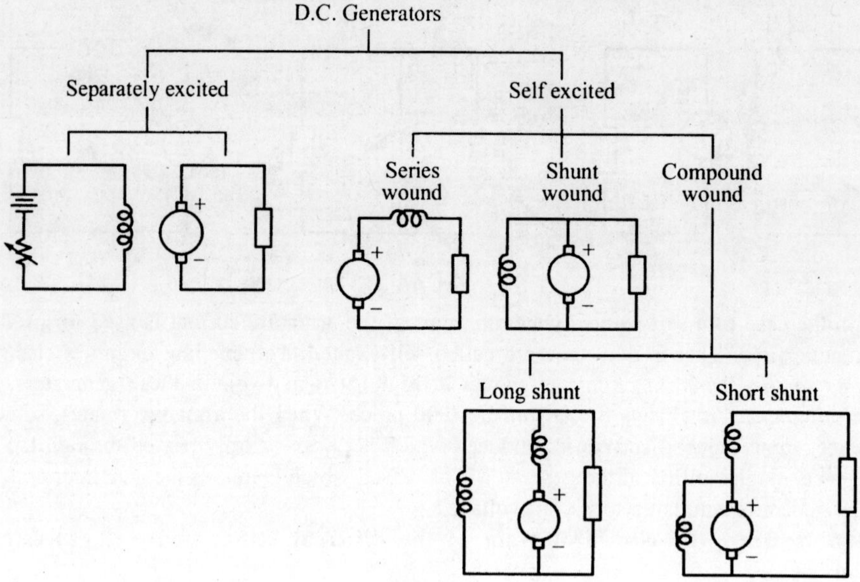

Fig. 20

Ans. As shown in Fig. 21

$$I = 25{,}000/500 = 50\text{ A}$$

$$I_{sh} = 500/200 = 2.5\text{ A}$$

$$\therefore \quad I_a = I + I_{sh} = 50 + 2.5 = 52.5\text{ A}$$

Series field drop $= 52.5 \times 0.04 = 2.1\text{V}$

Armature drop $= 52.5 \times 0.03 = 1.575\text{ V}$

Brush drop $= 2 \times 1 = 2\text{ V}$

Generated emf, $E_g = 500 + 2.1 + 1.575 + 2$

$= \mathbf{505.67\ V}$

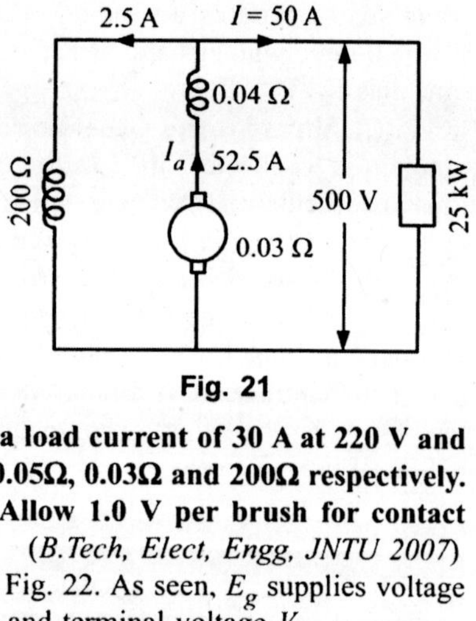

Fig. 21

Q. 9. A short shunt compound generator delivers a load current of 30 A at 220 V and has armature, series field and shunt field resistances of 0.05Ω, 0.03Ω and 200Ω respectively. Calculate the induced emf and the armature current. Allow 1.0 V per brush for contact drop. *(B.Tech, Elect, Engg, JNTU 2007)*

Ans. The short-shunt generator diagram is shown in Fig. 22. As seen, E_g supplies voltage drop across R_a ($I_a R_a$), drop across series resistance ($I R_{se}$) and terminal voltage V.

Now, $E_g - I_a R_a - I R_{se} - V = 0$

$\therefore \quad E_g - I_a R_a = V + I R_{se}$

Since $\quad E_g - I_a R_a = V_a \quad$ ...(*i*)

Hence from (*i*) above

$$V_a = V + I R_{se}$$

$\therefore \quad V_a = 220 + 30 \times 0.03 = 220.9\text{ V}$

Shunt field current $= 220.9/200 = 1.10\text{ A}$

Armature current $I_a = 30 + 1.10 = \mathbf{31.1\ A}$

$\therefore \quad E_g = I_a R_a + I R_{se} + V$

$= 31.1 \times 0.05 + 30 \times 0.03 + 220$

$= 1.55 + 0.9 + 220 = \mathbf{222.45\ V}$

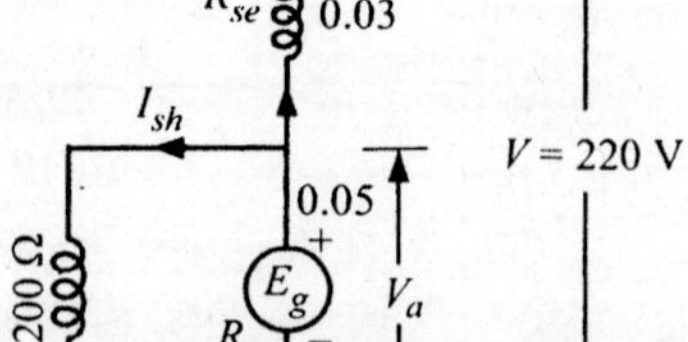

Fig. 22

Q. 10. A short-shunt cumulative compound d.c. generator supplies 7.5 kW at 230 V. The shunt field, series field and armature resistance are 100, 0.3 and 0.4 ohm respectively. Calculate the induced e.m.f. and the load resistance.

Ans. Generator circuit is shown in Fig. 23.

Load current, $I = 7.5 \times 1000/230 = 32.6$ A

Voltage drop in series winding

$= 0.3 \times 32.6 = 9.78$ V

Voltage across shunt winding

$= 230 + 9.78 = 239.78$ V

$I_{sh} = 239.78/100 = 2.4$ A

$I_a = 32.6 + 2.4 = 35$ A

$I_a R_a = 35 \times 0.4 = 14$ V

Induced e.m.f. $= 230 + 9.78 + 14$

$= \mathbf{253.78\ V}$

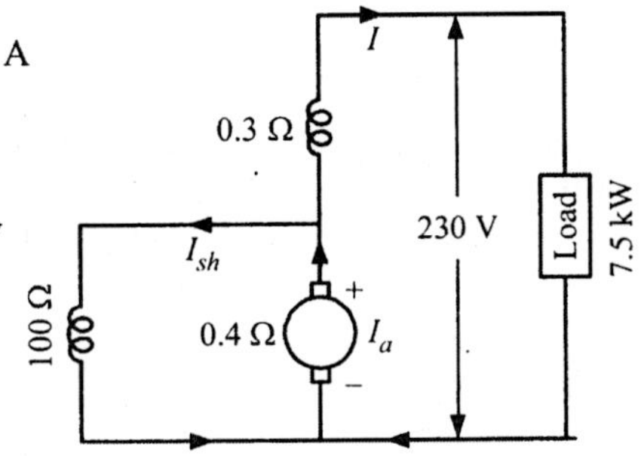

Fig. 23

If R is load resistance, then,

$W = V^2/R$

$\therefore$ $R = V^2/W = 230^2/7.5 \times 1000 = \mathbf{7.05\ \Omega}$

Q. 11. A 4-pole dc generator is delivering 20 A to a load of 10 Ω. If the armature resistance is 0.5 Ω and the shunt field resistance is 50 Ω, calculate the induced emf and the efficiency of the machine. *(B.E. First Year Nagpur Univ. 2009)*

Ans. Generator circuit is shown in Fig. 24

load voltage $V = I \times R_L = 20 \times 10 = 200$ V

$I_{sh} = V/R_{sh} = 200/50 = 4$ A

$\therefore$ $I_a = I + I_{sh} = 20 + 4 = 24$ A

generated emf $E_g = V + I_a R_a = 200 + 24 \times 0.5$

$= \mathbf{212\ V}$

output power $P_o = VI = 200 \times 20 = 4000$ W

Input power $P_i =$ output power + armature copper loss + shunt copper loss

$= 4000 + I_a^2 R_a + I_{sh}^2 R_{sh}$

$= 4000 + 24^2 \times 0.5 + 4^2 \times 50 = 5088$ W

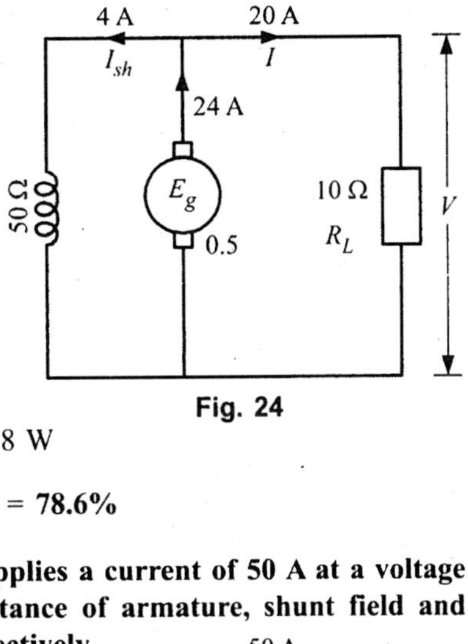

Fig. 24

efficiency $\eta = \dfrac{P_o}{P_i} \times 100 = \dfrac{4000}{5088} \times 100 = 0.786 = \mathbf{78.6\%}$

Q. 12. A short-shunt compound dc generator supplies a current of 50 A at a voltage of 200 V. Calculate the generated voltage if the resistance of armature, shunt field and series field windings are 0.04 Ω, 50 Ω and 0.02 Ω respectively. *(B.E. First Year Nagpur Univ. 2006)*

Ans. The generator circuit is shown in Fig. 25.

$E_g = I_a R_a + I\, R_{se} + V$

or $E_g - I_a R_a = I\, R_{se} + V$

Now, $E_g - I_a R_a = V_a$ — voltage across armature

$\therefore$ $V_a = I\, R_{se} + V$

$= 50 \times 0.02 + 200$

$= 201$ V

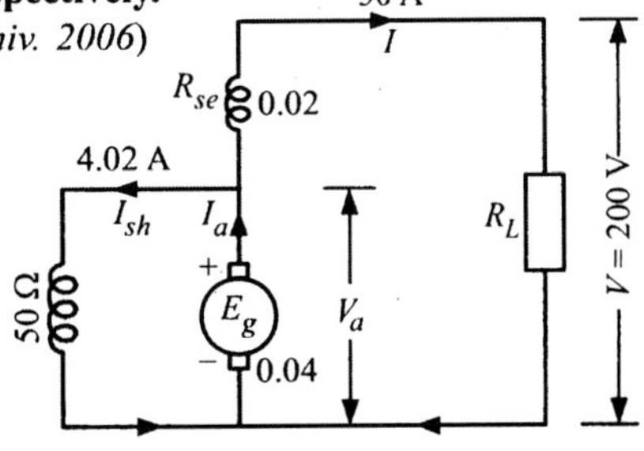

Fig. 25

$$I_{sh} = 201/50 = 4.02 \text{ A}$$

$\therefore$ $$I_a = I + I_{sh} = 50 + 4.02 = 54.02 \text{ A}$$

Now, $$E_g = V_a + I_a R_a = 201 + 54.02 \times 0.04 = \mathbf{203.16\ V}$$

Q. 13. A 4-pole shunt generator has a lap-wound armature with resistance of 0.1 Ω and field circuit resistance of 50 Ω. The generator supplies sixty 100 V, 40 W lamps. Find the current in each armature conductor and the generator emf. The brush contact drop is 1 volt/brush. (*Elect. Engg. Nagpur Univ. 2004*)

Ans. Current drawn by each lamp

$= W/V = 40/100$

$= 0.4$ A

Current drawn by all 60 lamps

$= 60 \times 0.4 = 24$ A

$$I_{sh} = V/R_{sh} = 100/50 = 2 \text{ A}$$

$$I_a = I + I_{sh} = 24 + 2 = 26 \text{ A}$$

generator voltage is given by

$$E_g = V + I_a R_a + \text{brush drop}$$

$= 100 + 26 \times 0.1 + 2 \times 1$

$= \mathbf{104.6\ V}$

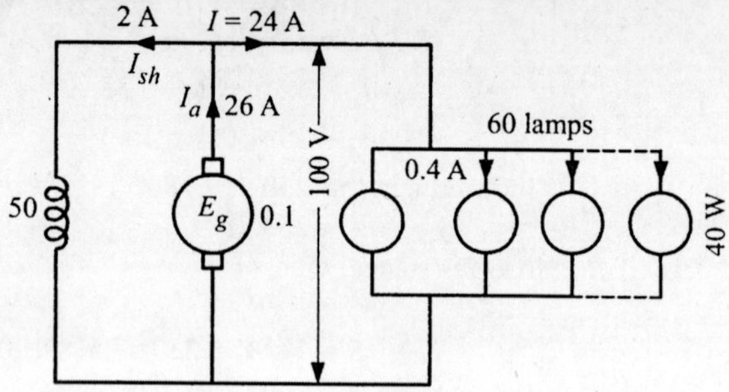

Fig. 26

Since there are 4 parallel paths in the armature, hence current in each armature conductor is = 26/4 = **6.5A**

Q. 14. Deduce the emf equation of a dc generator. Everything else being the same, which generator produces more emf?

Ans. Let Φ = flux/pole in webers

Z = total number of armature conductors

= No. of slots × conductors/slot

P = No. of poles

A = No. of parallel paths in armature

N = armature rotation in r.p.m.

E_g = e.m.f. induced *in any parallel path* in the armature

= emf available across the brushes

Generated e.m.f. E_g = e.m.f. generated in any one of the armature parallel paths.

Average e.m.f. generated/conductor

$$= N\frac{d\Phi}{dt} \text{ volt}$$

$$= \frac{d\Phi}{dt} \text{ volt} \qquad (\because N = 1)$$

Now, flux cut/conductor in one revolution = ΦP weber

$\therefore$ $$d\Phi = \Phi \times P$$

No. of revolutions/second = $N/60$

$\therefore$ time for one revolution = $60/N$ second

$\therefore$ $$dt = 60/N \text{ second}$$

$\therefore$ flux cut per conductor/second

$$= N\frac{d\Phi}{dt} = \frac{\Phi P}{60/N} = \frac{\Phi PN}{60} \text{ Wb/s}$$

$\therefore$ e.m.f. generated/conductor $= \dfrac{\Phi PN}{60}$ volt

For a wave-wound generator

No. of parallel paths = 2

No. of conductors (in series) in one path = Z/2

$$\therefore \quad \text{E.M.F. generated/path} = \frac{\Phi PN}{60} \times \frac{Z}{2} = \frac{\Phi ZPN}{120} \text{ volt}$$

For a lap-wound generator

No. of parallel paths = P

No. of conductors (in series) in one path = Z/P

$$\therefore \quad \text{e.m.f. generated/path} = \frac{\Phi PN}{60} \times \frac{Z}{P} = \frac{\Phi ZN}{60} \text{ volt}$$

In general, generated e.m.f. is

$$E_g = \frac{\Phi ZN}{60} \times \left(\frac{P}{A}\right) \text{ volt}$$

where $A = 2$ — for wave-winding

$= P$ — for lap-winding

Since for a lap-wound generator, $A = P$, hence generated emf is given by

$$E_g = \frac{\Phi ZN}{60} \times \left(\frac{P}{A}\right) = \frac{\Phi ZN}{60}\left(\frac{P}{A}\right) = \frac{\Phi ZN}{60}$$

For a wave-wound generator, $A = 2$

$$\therefore \quad E_g = \frac{\Phi ZN}{60}\left(\frac{P}{2}\right) = \frac{\Phi ZPN}{120}$$

Obviously, the emf generated by a wave-wound generator is greater than that produced by a lap-wound generator. It is so because value of (P/2) is always more than unity in dc generators.

Q. 15. What is meant by the magnetic losses of a dc generator? Where do they take place? How can they be minimised?

Ans. When the armature core of a dc generator rotates in the magnetic field of the poles, some losses take place continuously in its core. These losses are known as iron losses (or core losses). Iron losses consist of (*i*) hysteresis loss and (*ii*) eddy current loss.

(*i*) **Hysteresis Loss (W_h) :** This loss is due to the fact that magnetism of the armature core reverses as it rotates between the North and South poles of the generator. As it passes under North and South poles one after the other, it attains South and North polarity respectively. This loss is reduced by making armature core of special silicon steels which have low hysteresis coefficient but high electrical resistivity.

(*ii*) **Eddy Current Loss :** When armature core rotates, it cuts the pole flux. Hence, an emf is induced in the *body* of the core according to Faraday's laws of electromagnetic induction. This emf is quite small. But it sets up a large current in the body of the core due to its very *small resistance*. This current is known as *eddy current*. The power loss due to the flow of this current is known as eddy current loss. This loss would be large if solid iron core were used. It will heat up the core. To avoid this, iron core is built up of thin laminations which are insulated from each other. Thinner the laminations, lesser the loss.

The above two losses together are known as *magnetic losses*.

Q. 16. Enumerate the different power losses taking place in a dc generator. What is meant by stray losses?

Ans. Various losses taking place in a dc generator are given below :

(*a*) **Copper Losses (or I^2R losses) :** These consist of the following and are due to the resistance of copper conductors.

(*i*) **armature copper loss** = $I_a^2 R_a$

Here, R_a is the resistance of armature, interpoles and series field windings which are made of copper and I_a is armature current.

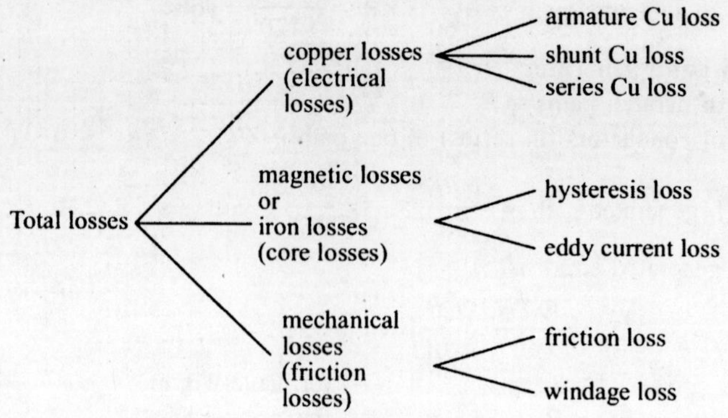

Fig. 27

(*ii*) **field copper loss :** It is the loss taking place in the field windings

(*iii*) The loss due to brush contact resistance.

(*b*) **Magnetic or iron Losses :** These losses consist of

(*i*) hysteresis loss.

(*ii*) eddy current loss.

These losses are almost constant for shunt and compound-wound generators.

(*c*) **Mechanical or Friction Losses :** These losses consist of

(*i*) friction loss at bearings and commutator

(*ii*) air-friction or windage loss of rotating armature.

The total losses taking place in a dc generator are summed up in Fig. 27.

The Cu losses are also called electrical losses.

Usually, magnetic and mechanical losses are together known as *stray losses*.

Q. 17. What are the different power stages in a dc generator? How do you define its mechanical, electrical and commercial efficiency? How are these efficiencies related to each other?

Ans. As we know, a dc generator is driven by a *prime mover*. The output of the prime mover is the input of the dc generator. When iron and mechanical losses are subtracted from it, the remaining equals the electric power developed in the armature and is equal to $E_g I_a$. From this power, copper losses are subtracted. The balance appears as the output power of the generator and is equal to *VI*. Various power stages for a dc generator are shown in Fig. 28. The three generator efficiencies are as under.

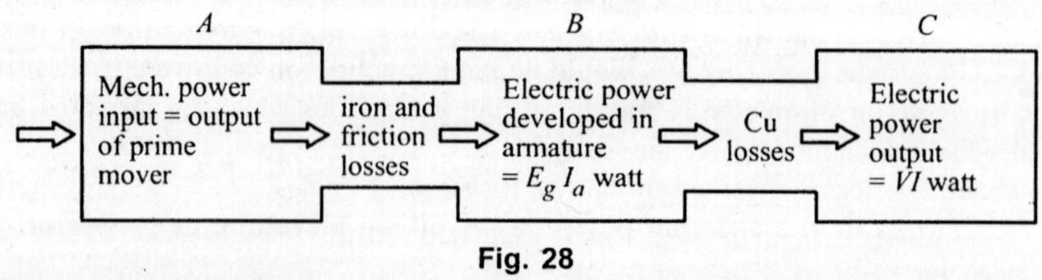

Fig. 28

(*i*) ***Mechanical Efficiency***

$$\eta_m = \frac{B}{A} = \frac{\text{electrical power produced in armature}}{\text{mechanical power input}}$$

(2) ***Electrical Efficiency***

$$\eta_e = \frac{C}{B} = \frac{\text{power supplied to load}}{\text{power produced in armature}} = \frac{VI}{E_g I_a}$$

(3) ***Overall or Commercial Efficiency***

$$\eta_c = \frac{C}{A} = \frac{\text{power supplied to load}}{\text{mechanical power input}}$$

Now $$\eta_c = \frac{C}{A} = \frac{C}{B} \times \frac{B}{A} = \eta_e \times \eta_m$$

Q. 18. Briefly describe the energy conversion process in a dc generator under load.

Ans. When a dc generator is under load, some important flux and current relationships take place which produce mechanical-electrical energy conversion process. Consider the case of a 2-pole generator that is driven counterclockwise (Fig. 29.). The direction of the induced current in the armature is as shown by dots and arrows. The current flows *into* the page, *away* from the reader in the conductors lying momentarily under a *S* pole (as per Right-hand-rule). However, current under the *N* pole flows *out* of the page *towards* the reader. As seen, these conductors lie in the magnetic field produced by field poles of the generator. Hence, current-carring armature conductors are acted upon by a force acting clockwise (Left-hand-rule). This force produces a torque that acts *opposite* to the direction in which the generator is being driven. To keep the armature rotating, a torque must be applied to the shaft to overcome this opposing electromagnetic torque. In this way, input mechanical power is converted into electrical power which is then supplied to generator load. This is how the energy conversion takes place in a dc generator.

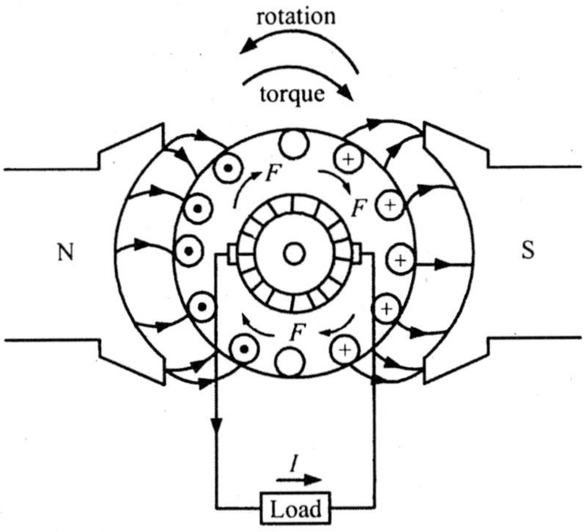

Fig. 29

Q. 19. An 8-pole lap-wound dc generator has 960 conductors, a flux of 40 milliwebers and is driven at 400 rpm. Find induced emf.

(*B.Tech Basic Elect. Engg. JNTU, Hyderabad 2007*)

Ans. We are given that

$$P = 8,\ A = 8,\ Z = 960,\ \Phi = 40 \text{ mWb} = 0.04 \text{ Wb},\ N = 400 \text{ rpm}$$

$$\therefore \quad E_g = \frac{\Phi Z N}{60} \times \left(\frac{P}{A}\right)$$

$$= \frac{0.04 \times 960 \times 400}{60} \times \left(\frac{8}{8}\right) = \mathbf{256\ V}$$

Q. 20. An 8-pole wave-connected dc generator has 900 armature conductors and a flux per pole of 0.04 Wb. At what speed it must be driven to generator 500 V?

(*Basic Elect. Engg. R.G.P.V. Bhopal Dec 2007*)

Ans. $$E_g = \frac{\Phi Z N}{60} \times \left(\frac{P}{A}\right) \text{volt}$$

Now, $$\Phi = 0.04 \text{ Wb},\ Z = 900,\ P = 8,\ E_g = 500 \text{ V}$$

Substituting the given values in the above equation, we get

$$500 = \frac{0.04 \times 900 \times N}{60} \times \left(\frac{8}{2}\right), \quad \therefore N = \mathbf{208.33\ r.p.m.}$$

Q. 21. The armature of a 6-pole dc generator has a wave winding containing 650 conductors.

Calculate the generated emf when flux per pole is 0.055 Wb and speed is 300 rpm. Calculate the speed at which the armature must be driven to generate an emf of 550 V if the flux is reduced to 0.05 Wb. (*Basic Elect. Engg. CSV Tech. Univ. Bhilai, 2006*)

Ans.
$$E_g = \frac{\Phi ZN}{60} \times \left(\frac{P}{A}\right) \text{ volt}$$

$$E_g = \frac{0.055 \times 650 \times 300}{60} \times \left(\frac{6}{2}\right) = \mathbf{536.25\ V}$$

Substituting the given values in the above equation, we get

$$550 = \frac{0.05 \times 650 \times N}{60} \times \left(\frac{6}{2}\right) \quad \therefore N = \mathbf{338.5\ r.p.m.}$$

Q. 22. An 8-pole dc generator has per pole flux of 40 mWb and winding is connected in lap with 960 conductors. Calculate the generated EMF on open circuit when it runs at 400 rpm. If the armature is wave wound, at what speed must the machine be driven to generate the same voltage. (*B. Tech. Elect. Machines JNTU, Hyderabad 2008*)

Ans. We are given the following data:

$$\Phi = 40 \times 10^{-3} \text{ Wb}, Z = 960, N = 400 \text{ rpm}, P = 8, \ A = P = 8$$

Now,
$$E_g = \frac{\Phi ZN}{60} \times \left(\frac{P}{A}\right)$$

$$= \frac{(40 \times 10^{-3}) \times 960 \times 400}{60} \times \left(\frac{8}{8}\right) = \mathbf{256\ V}$$

When armature is wave-wound :

Now, $A = 2$. We have to find the speed for generating the same voltage of 256 V.

From the above equation we get,

$$N = \frac{60\, E_g}{\Phi Z} \times \left(\frac{A}{P}\right)$$

$$= \frac{60 \times 256}{40 \times 10^{-3} \times 960} \times \left(\frac{2}{8}\right) = \mathbf{100\ rpm}$$

Q. 23. A 4-pole dc shunt generator with a shunt field resistance of 100 ohm and an armature resistance of 1.0 ohm has 378 wave-connected conductors in its armature. The flux per pole is 0.02 Wb. If a load resistance of 10 ohm is connected across the armature terminals and the generator is driven at 1000 r.p.m., calculate the power absorbed by the load. (*Elect. Engg. Nagpur Univ. 2005*)

Ans.
$$E_g = \frac{\Phi ZN}{60} \times \left(\frac{P}{A}\right) \text{ volt}$$

$$= \frac{0.02 \times 378 \times 1000}{60} \times \left(\frac{4}{2}\right) = 252 \text{ V}$$

$I = V/R_L = V/10$ A

$I_{sh} = V/R_{sh} = V/100$ A

$$I_a = I + I_{sh} = \frac{V}{10} + \frac{V}{100} = \frac{11V}{100} \text{ A}$$

$$E_g = V + I_a R_a \text{ or } 252 = \frac{11V}{100} \times 1 + V = \frac{111V}{100}$$

$$\therefore \quad V = \frac{252 \times 100}{111} = 277 \text{ V}$$

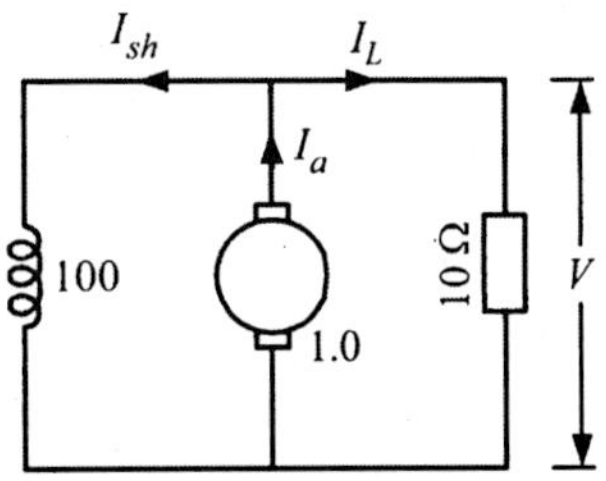

Fig. 30

$I = V/R_L = 277/10 = 22.7$ A

Power absorbed by the load $= V \times I_L = 227 \times 22.7$

$= 5153.5$ W $=$ **5.15 kW.**

Q. 24. A separately-excited generator when running at 1200 rpm supplies 200 A at 125 V to a circuit of constant resistance. What will be the current when the speed is dropped to 1000 rpm if the field current is unaltered? Armature resistance is 0.04 Ω. Total drop at brushes is 2 V. Ignore change in armature reaction.

(B.Tech. Elect. Technology JNTU, Hyderabad 2007)

Ans. Load resistance, $R = 125/200 = 0.625$ Ω as shown in Fig. 31.

$E_{g_1} = 125 + 200 \times 0.04 + 2 \times 1$

$= 135$ V, $N_1 = 1200$ rpm

Since $E_g \propto N$, hence

$E_{g_2} = E_{g_1} \times N_2/N_1$

$= 135 \times 1000/1200 = 112.5$ V

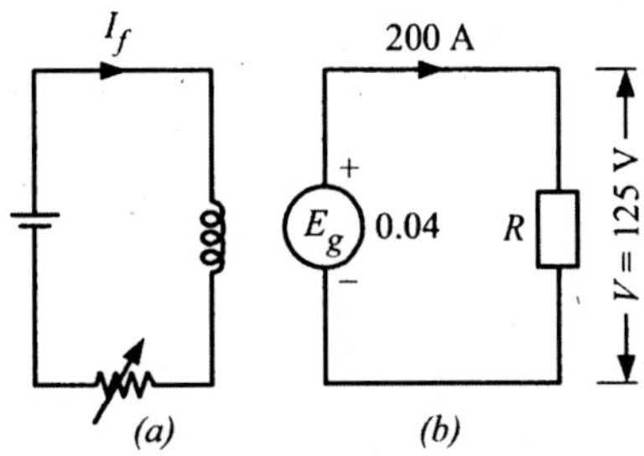

Fig. 31

If I is the new generator current, then generator terminal voltage is given by

$V = 112.5 - 0.04\, I - 2 \times 1$

$= 110.5 - 0.04\, I$

$$I = \frac{V}{R} = \frac{110.5 - 0.04\, I}{0.625} \text{ or } I = \mathbf{166.2\ A}$$

Q. 25. The armature of a four-pole, lap-wound shunt generator has 120 slots with 4 conductors per slot. The flux per pole is 0.05 Wb. The armature resistance is 0.05 Ω and the shunt field resistance is 50 Ω. Find the speed of the machine when supplying 45 A at a terminal voltage of 250 V. *(Elect. Engg. Nagpur Univ. 2006)*

Ans. No. of slots = 120, conductors/slot = 4 $\therefore$ $Z = 120 \times 4 = 480$

$R_a = 0.05$ Ω, $R_{sh} = 50$ Ω, $\Phi = 0.05$ Wb, $V = 250$ V

$$I_a = I + I_{sh} = I + \frac{V}{R_{sh}} = 45 + \frac{250}{50} = 50 \text{ A}$$

$$E_g = V + I_a R_a = 250 + (50 \times 0.05) = 252.5 \text{ V}$$

Now,
$$E_g = \frac{\Phi Z N}{60} \times \left(\frac{P}{A}\right)$$

or
$$252.5 = \frac{0.05 \times 480 \times N}{60} \times \left(\frac{4}{4}\right) \quad \therefore N = \mathbf{631\ r.p.m.}$$

Q. 26. A shunt generator delivers 195 A at terminal voltage of 250 V. The armature resistance and shunt field resistance are 0.02 Ω and 50 Ω respectively. The iron and friction losses equal 950 watt.

Find: (*i*) EMF generated (*ii*) Cu losses (*iii*) output of the prime mover (*iv*) commercial, mechanical and electrical efficiencies. (*Elect. Engg. Nagpur Univ. 2006*)

Ans. (*i*) $I = 195$ A, $I_{sh} = 250/50 = 5$ A, $I_a = 195 + 5 = 200$ A

armature voltage drop $= I_a R_a = 200 \times 0.02 = 4$ V

$\therefore$ generated e.m.f., $E_g = 250 + 4 =$ **254 V**

(*ii*) armature Cu loss $= I_a^2 R_a = 200^2 \times 0.02 = 800$ W

Shunt Cu loss $= V.\ I_{sh} = 250 \times 5 = 1250$ W

$\therefore$ total Cu loss $= 1250 + 800 =$ **2050 W**

(*iii*) stray losses (*i.e.* iron and friction losses) = 950 W

total losses = 2050 + 950 = 3000 W

generator output $= VI = 250 \times 195 = 48{,}750$ W

generator input = 48,750 + 3000 = 51,750W

$\therefore$ output of prime mover = **51,750 W**

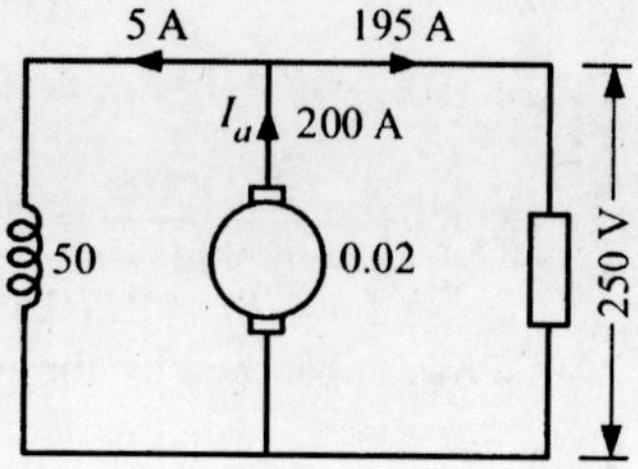

Fig. 32

(*iv*) generator input = 51,750 W, stray losses = 950 W

electrical power produced in armature = 51,750 – 950 = 50,800 W

Mechanical eff. $\eta_m = \dfrac{\text{electrical power produced in armature}}{\text{output of prime mover}}$

$= \dfrac{50{,}800}{51{,}750} \times 100 =$ **98.2%**

Electrical eff. $\eta_e = \dfrac{\text{power supplied to load}}{\text{power produced in armature}}$

$= \dfrac{48{,}750}{50{,}800} \times 100 =$ **95.9%**

Commercial eff. $\eta_c = \dfrac{\text{power supplied to load}}{\text{prime mover input}}$

$= \dfrac{48{,}750}{51{,}750} \times 100 =$ **94.2%**

Alternatively, $\eta_c = \eta_m \times \eta_e = 0.982 \times 0.959 = 0.942$ or **94.2%**

Q. 27. A 400 V, 50-kW, long-shunt, compound generator delivers its full-load output at the rated voltage. The resistances of the armature, series field and shunt field are 0.08, 0.05 and 200 Ω respectively. Find :

(*a*) the load current.

(*b*) the shunt field current and armature current.

(*c*) the volt drop across the series field windings.

(*d*) the p.d. across the armature.

(*e*) the e.m.f. generated in the armature.

Ans. (*a*) $I = \dfrac{\text{output power}}{\text{terminal voltage}}$

$= \dfrac{50{,}000}{400} =$ **125 A**

(*b*) $I_{sh} = 400/200 =$ **2 A** — Fig. 33

$I_a = I_{sh} + I = 2 + 125 =$ **127 A**

(*c*) Voltage drop across series field windings

$= 127 \times 0.05 =$ **6.35 V**

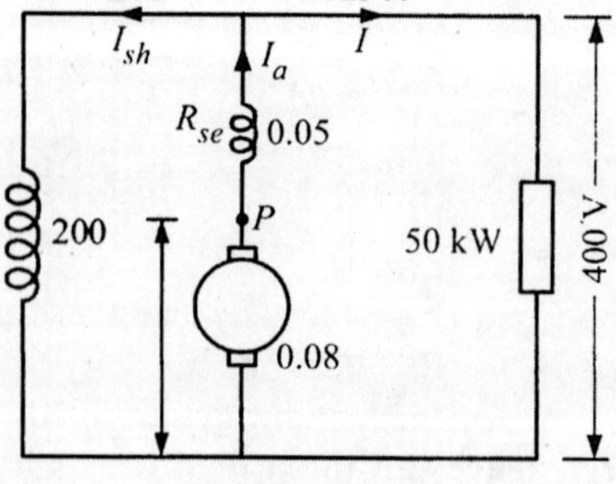

Fig. 33

(*d*) the potential difference across the armature *i.e.* voltage of point *P*.

$= V + I_a R_{se} = 400 + 6.35 =$ **406.35 V**

(*e*) E_g = p.d. across armature + voltage drop in armature

$= 406.35 + (127 \times 0.08) =$ **416.51 V**

or $E_g = V + I_a R_a + I_a R_{se} = 400 + 127(0.08 + 0.05) = 416.51$ V

Q. 28. A 250-volt d.c. machine has armature and field resistances of 0.06 and 100 respectively. Determine the total armature power developed when working

(*i*) as a generator delivering 25 kW output and

(*ii*) as a motor taking 25 kW input.

Ans. (*i*) as a generator (Fig. 34)

Load current $= I = 25{,}000/250 = 100$ A

$I_{sh} = V/R_{sh} = 250/100 = 2.5$ A

$I_a = I + I_{sh} = 100 + 2.5 = 102.5$ A

$E_g = V + I_a R_a = 250 + (102.5 \times 0.06) = 256.15$ V

power developed in armature

$= E_g I_a = 256.15 \times 102.5 = 26{,}255$ W = **26.255 kW**

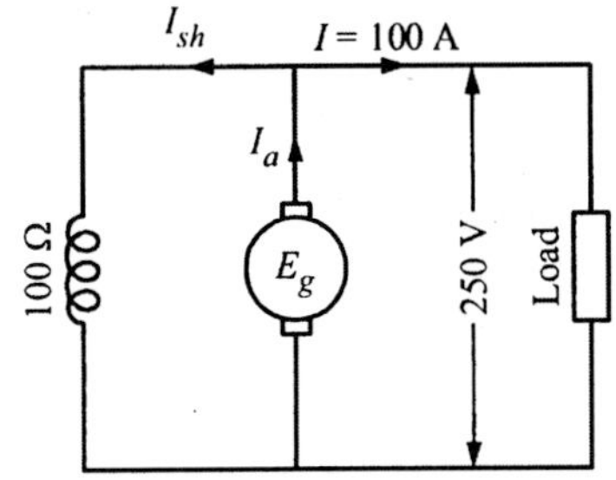

Fig. 34

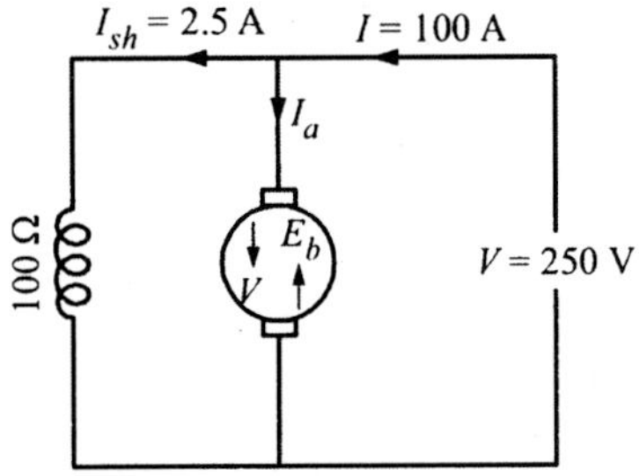

Fig. 35

(*ii*) **as a motor** (Fig. 35)

$I_a = I - I_{sh} = 100 - 2.5 = 97.5$ A

Now, $V = E_b + I_a R_a$

Back e.m.f., $E_b = V - I_a R_a = 250 - (97.5 \times 0.06) = 244.15$ V

power developed in armature is $= E_b I_a = 244.15 \times 97.5 = 23{,}804$ W = **23.804 kW**

Q. 29. A 4-pole d.c. generator is delivering 20 A to a load of 10 Ω. If the armature resistance is 0.5 Ω and the shunt field resistance is 50 Ω, calculate the induced emf and the efficiency of the machine. Allow a drop of 1 V per brush.

(*Elect. Technology-I, Osmania Univ.*)

Ans. Terminal voltage $= 20 \times 10 = 200$ V

$I_{sh} = 200/50 = 4$ A

$I_a = 20 + 4 = 24$ A

$I_a R_a = 24 \times 0.5 = 12$ V

Brush drop $= 2 \times 1 = 2$ V

$\therefore$ $E_g = 200 + 12 + 2$

$=$ **214 V**, as in Fig. 36.

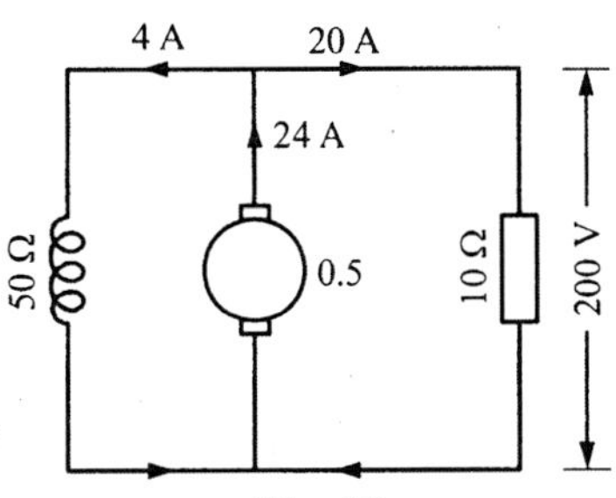

Fig. 36

Since iron and friction losses are not given, only electrical efficiency of the machine can be found out.

Total power generated in the armature

$$= 214 \times 24 = 5{,}136 \text{ W}$$

$$\text{useful output} = 200 \times 20 = 4{,}000 \text{ W}$$

$$\eta_e = 4{,}000/5{,}136 = 0.779 \quad \text{or} \quad \mathbf{77.9\%}.$$

Q. 30. A separately-excited generator with constant excitation is connected to a constant load. When the speed is 1000 rpm, it delivers 120 A at 500 V. At what speed will the current be reduced to 60 A? Armature resistance is 0.1 Ω, contact drop/brush is 1 V. Armature reaction may be ignored. (*B. Tech. Elect. Machine JNTU, Hyderabad 2008*)

Ans. With reference to Fig. 37(*a*), we have

armature *circuit* drop $= I_a R_a$ + brush drop

$$= 120 \times 0.1 + 2 \times 1 = 14 \text{ V}$$

$$\therefore \quad E_{g1} = V + \text{arm circuit drop}$$

$$= 500 + 14 = 514\ V,\ N_1 = 1000 \text{ rpm}$$

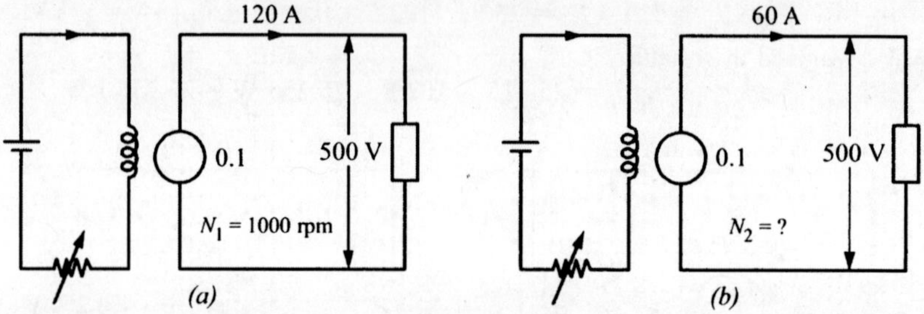

Fig. 37

As seen from Fig. 37(*b*)

armature *circuit* drop $= 60 \times 0.1 + 2 \times 1 = 8$ V

$$\therefore \quad E_{g2} = 500 + 8 = 508 \text{ V},\ N_2 = ?$$

Since field excitation remains constant, hence

$$\frac{E_{g2}}{E_{g1}} = \frac{N_2}{N_1} \quad \text{or} \quad \frac{508}{514} = \frac{N_2}{1000} \quad \therefore N_2 = \mathbf{988.3 \text{ rpm}}$$

Q. 31. A 4-pole 500 V, 25 kW, long-shunt compound generator delivers full load at the rated voltage. Calculate the e.m.f. generated if the armature resistance is 0.03 Ω, series field resistance is 0.04 Ω and the shunt field resistance is 200 Ω. Take contact drop per brush as 0.9 V. (*Elect. Engg., Nagpur Univ. 2004*)

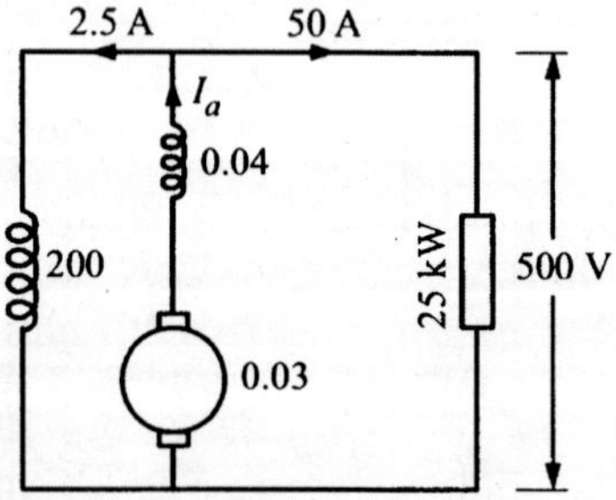

Fig. 38

Ans. Load power $= V \times I$

or $$I = 25 \times 10^3/500 = 50 \text{ A}$$

$$I_{sh} = V/R_{sh} = 500/200 = 2.5 \text{ A}$$

$$I_a = I + I_{sh} = 50 + 2.5 = 52.5 \text{ A}$$

$$\text{Brush drop} = 2 \times 0.9 = 1.8 \text{ V}$$

$$E_g = V + \text{brush drop} + I_a (R_a + R_{se})$$

$$= 500 + 1.8 + 52.5\,(0.03 + 0.04) = \mathbf{505.47 \text{ V}}$$

Q. 32. In a long-shunt compound generator, the terminal voltage is 230 V when generator delivers 150 A.

Determine :

(*i*) induced emf (*ii*) total power generated and (*iii*) distribution of this power.

Given that shunt field, series field, diverter and armature resistances are 92 Ω, 0.015 Ω, 0.03 Ω and 0.032 Ω respectively.

Ans. As seen, series field resistance and diverter resistance are in parallel.

Hence, their combined resistance is

$$= 0.031 \parallel 0.015 = 0.03 \times 0.015/0.045 = 0.01\ \Omega$$

Total armature circuit resistance is

$$R_t = 0.032 + 0.01 = 0.042\ \Omega$$

$$I_{sh} = 230/92 = 2.5\ \text{A},\ I_a = 150 + 2.5 = 152.5\ \text{A}$$

Voltage drop on armature *circuit* resistance is

$$I_a R_t = 152.5 \times 0.042 = 6.4\ \text{V}$$

(*i*) Voltage induced in the armature.

$$E_g = V + I_a R_t = 230 + 6.4$$

$$= \mathbf{236.4\ V}$$

(*ii*) Total power generated in the armature.

$$= E_g \cdot I_a = 236.4 \times 152.5$$

$$= \mathbf{36{,}051\ W}$$

(*iii*) Power lost in the armature *itself.*

$$= I_a^2 R_a = 152.5^2 \times 0.032 = \mathbf{744\ W}$$

Power lost in series field and diverter.

$$= 152.5^2 \times 0.01 = \mathbf{232\ W}$$

Power dissipated in shunt winding

$$= V \cdot I_{sh} = 230 \times 2.5 = \mathbf{575\ W}$$

Power delivered to the load.

$$= 230 \times 150 = \mathbf{34{,}500\ W}$$

Total power distributed

$$= 744 + 232 + 575 + 34{,}500 = \mathbf{36{,}050\ W}$$

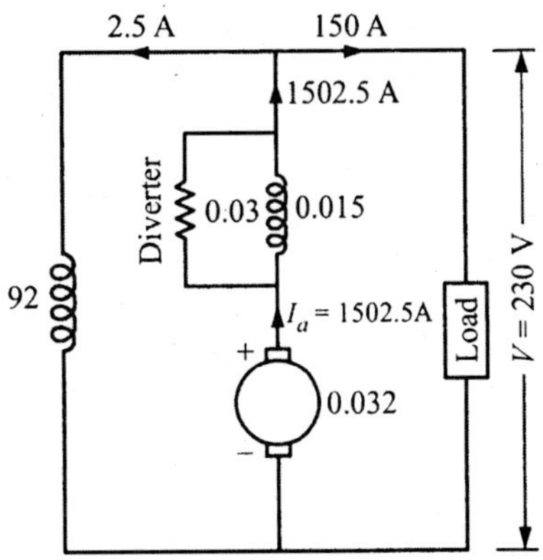

Fig. 39

Hence, we find that power distributed is equal to the power generated.

Q. 33. A separately-excited dc generator when running at 1000 rpm supplies 50 A at 250 V. Find how much current it will deliver when the speed falls to 800 rpm. Take armature resistance as 0.01 Ω and brush drop of 1 V/brush.

(*Basic Elect Engg VTU, Karnataka 2009*)

Ans. The emf generated by a dc generator is given by

$$E_g = \frac{\Phi Z N}{60}\left(\frac{P}{A}\right) \quad \text{or} \quad E_g \propto N$$

Total armature *circuit* drop

$$= 50 \times 0.01 + 2 \times 1 = 2.01\ \text{V}$$

Now, $E_{g1} = V + I_a R_a + \text{brush drop}$

$$= 250 + 50 \times 0.01 + 2 = 252.5\ \text{V}$$

∴ $252.5 \propto 1000$, and $E_{g2} \propto 800$

∴ $$\frac{E_{g2}}{E_{g1}} = \frac{800}{1000}$$

∴ $E_{g2} = 252.5 \times 800/1000 = 202\ \text{V}$

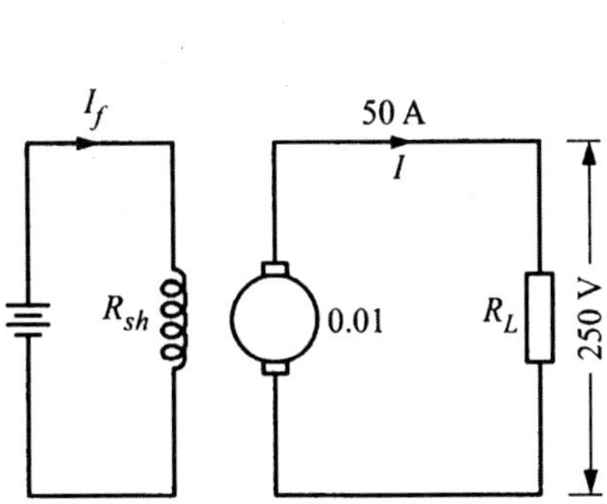

Fig. 40

Now $I R_L = 250$ ar $50 \times R_L = 250$ ∴ $R_L = 5\ \Omega$

The value of load current at 800 rpm can be found as follows.

$202 = I \times (5 + 2.01)$ ∴ $I = \mathbf{28.8\ A}$

Incidentally, output voltage = 5 × 28.8 = 144 V

Q. 34. A dc shunt generator has a full-load output of 10 kW at a terminal voltage of 240 V. The armature and shunt field winding resistances are 0.6 Ω and 160 Ω respectively. The sum of the mechanical and core losses is 500 W. Calculate the power required in kW at the driving shaft at full load and the corresponding value of efficiency.

(*Basic Elect Engg R.G.P.V Bhopal 2005*)

Ans. The power flow from input of the generator to its output is as under :

Mechanical power input from the prime mover

$$W_i = \text{copper losses } (W_{cu}) + \text{constant loss } (W_c) + \text{electrical power output}$$

Also, constant loss, W_c = mechanical loss + core loss = 500 W

$$W_{cu} = \text{armature Cu loss} + \text{shunt field loss} = I_a^2 R_a + I_{sh}^2 R_{sh}$$

Load current, $I = 10 \times 10^3/240 = 41.66$ A

Shunt field current,

$$I_{sh} = V/R_{sh} = 240/160 = 1.5 \text{ A}$$

Armature current,

$$I_a = I_{sh} + I = 1.5 + 41.66$$
$$= 43.16 \text{ A}$$

$$\therefore \quad W_{cu} = (43.16^2 \times 0.6) + (1.5^2 \times 160)$$
$$= 1478 \text{ W}$$

$$\therefore \quad W_i = 1478 + 500 + 10 \times 10^3$$
$$= 11{,}978 \text{ W} = \mathbf{11.978 \text{ kW}}$$

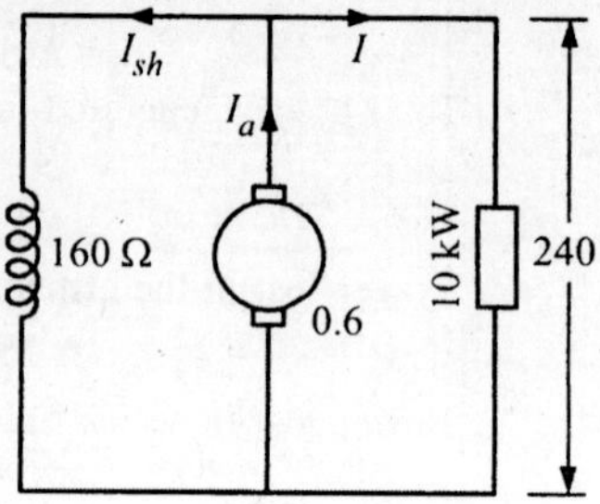

Fig. 41

$$\therefore \text{ generator efficiency} = \frac{\text{output power}}{\text{input power}} \times 100 = \frac{10 \text{ kW}}{11.978 \text{ kW}} \times 100 = \mathbf{83.48\ \%}$$

Q. 35. Briefly discuss the three most important characteristics of a d.c generator. Why are they important?

Which one of these characteristics is almost the same for all d.c. generators?

Ans. The three most important characteristics of a d.c. generator are as under:

1. Open-Circuit Characteristic (O.C.C) : It is also known as *magnetic characteristic* or *no-load characteristic*. It shows the value of the no-load generated emf E_o for various field currents (I_f) at a given *fixed* speed. In fact, it shows how E_o increases with exciting current I_f. Basically, it is simply the magnetisation curve for the material of the electromagnets used in the generator. Its shape is almost the same for all dc generators whether separately-excited or self-excited as shown in Fig. 42. It is so because they have similar material for their electromagnets.

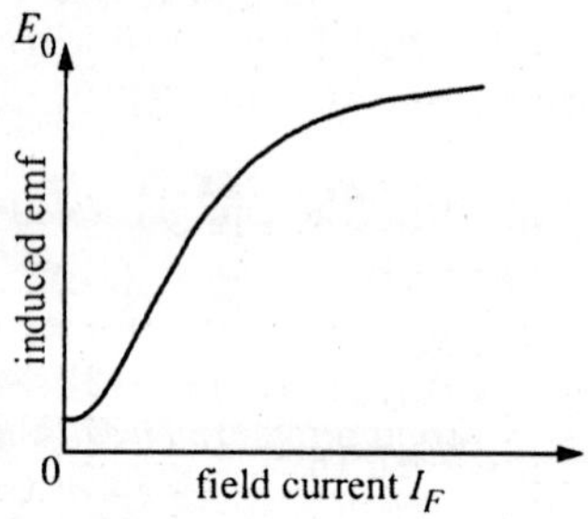

Fig. 42

The O.C.C. is helpful in finding the following :

(*i*) the voltage to which the generator will build up when run at a *particular* speed.

(*ii*) the critical resistance of the generator for a given speed and

(*iii*) the critical speed N_c of the generator.

2. Internal or Total Characteristic (E/I_a) : It shows how the emf E *actually* induced in the armature increases with armature current I_a. This characteristic is of use to a design engineer.

3. External Characteristic (V/I) : It is also known as performance characteristic or voltage-regulation curve. It shows how terminal voltage V varies with load current I. It takes into account the voltage drop due to the armature circuit resistance. The different values of V are obtained by subtracting I_aR_a drop from the values of E_g.

This characteristic is helpful in selecting a dc generator for a particular job.

Q. 36. Describe a simple procedure for obtaining the open-circuit characteristic (O.C.C) of a dc shunt generator.

Ans. The procedure for finding O.C.C of a dc generator is as under :

First, the field winding of the generator (whether shunt or series-wound) is disconnected from the machine. Then, it is connected to a battery which supplies the field or exciting current I_f as shown in Fig. 43(*a*). The value of I_f is changed with the help of shunt rheostat and is read from the ammeter *A*. The generator is run at a constant speed, say, *N*, and the value of the no-load armature emf E_o is read from the voltmeter connected across the generator terminals. Current I_f is increased in suitable steps and the corresponding values of E_o are noted down. When we plot the values of I_f and E_o, we get the open-circuit characteristic shown in Fig. 43(*b*).

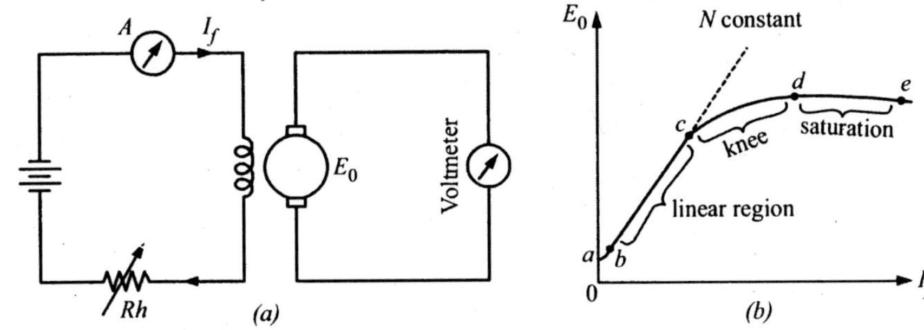

Fig. 43

The main three features of O.C.C are as under:

1. *Oa* is the emf due to residual magnetism present in the field electromagnets. This emf is produced even when I_f is zero.
2. For the portion *bc*, E_o increases linearly with I_f. It is due to the fact that at low flux densities, the magnetic flux is directly proportional to the exciting current I_f. At such densities, iron is still unsaturated.
3. The portion *cd* indicates that the magnetic poles have started getting saturated. It means that now a *greter* increase in I_f is required for producing a given increase in E_o.
4. The portion *de* shows that magnetic circuit is getting more saturated. Hence, further increase in I_f gives only a very small increase in E_o.

Q. 37. Explain the terms "critical field resistance" and "critical speed" as applied to self-excited dc generators.

Ans. Fig. 44(*b*) shows the open-circuit characteristic of the shunt generator of Fig. 44(*a*). Line *OA* represents the shunt resistaance R_{sh}. The maximum voltage developed by the generator for the given shunt resistance and given speed is represented by point *B*.

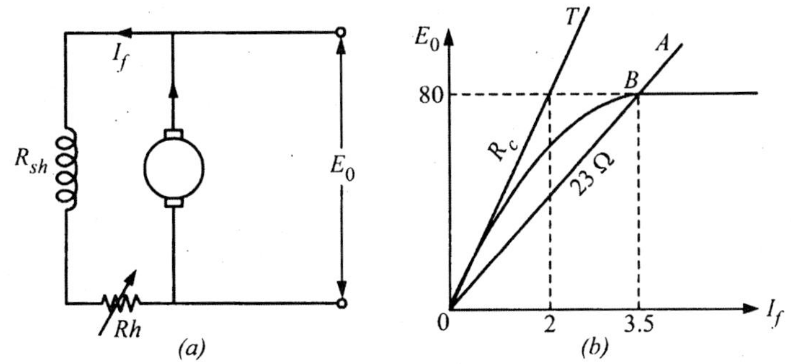

Fig. 44

Since point B represents a voltage of 80 V and a field current of $I_f = 3.5$ A, hence $R_{sh} = 80/3.5 = 23\ \Omega$. If now, field resistance is increased with the help of rheostat, then I_f is decreased. As a result, pole flux is decreased. Consequently, generator voltage is decreased. If the value of R_{sh} is so increased that the resistance line OT does not cut the O.C.C. at all, then the generator will not build up its voltage. The value of shunt resistance whose line is tangent to the O.C.C is called *critical field resistance* R_c for a given generator speed.

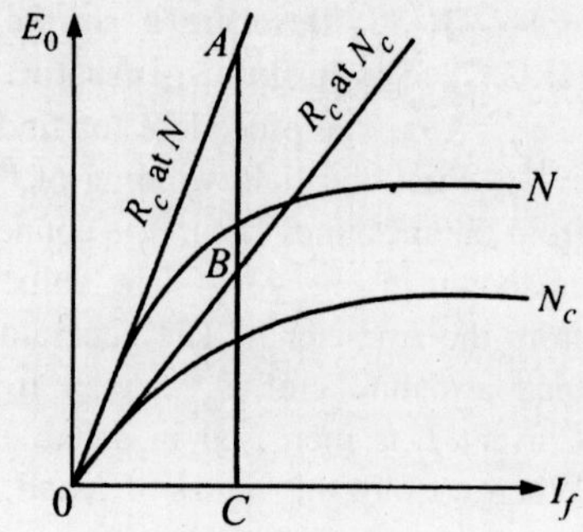

Fig. 45

Fig. 45 shows the O.C.C of a given generator at two speeds N and N_C. Resistance line OA represents critical resistance of the generator at a speed of N. As seen, the given shunt resistance line OB becomes tangent to the O.C.C drawn at a *speed lower than N*. The speed at which the *given* shunt field resistance becomes critical resistance of a generator is called its critical speed. The value of the *critical speed* can be found from Fig. 45

$$\frac{BC}{AC} = \frac{N_c}{N} \quad \text{or} \quad N_c = \frac{BC}{AC} \times N$$

Q. 38. Fully explain the voltage build-up of a shunt generator.

Ans. Before loading a shunt generator, it is allowed to build up its full voltage on open circuit. For this purpose, the generator depends on the residual magnetism of its field poles. When the generator is started, a small emf is induced in the armature. It produces a small current I_f in the field circuit which strengthens the pole flux. As a result, the generator voltage E_o is increased because $E_o \propto \Phi$. The increased emf, in turn, produces larger I_f which means more field flux. This build-up process goes on till the generator reaches its operating voltage. After this, switch S in Fig. 46(*a*) is closed and the generator is employed for supplying the load current.

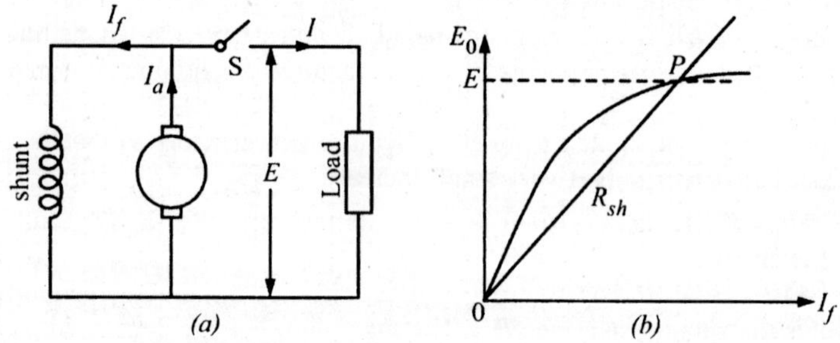

Fig. 46

It should be noted that generator voltage may fail to build up for various reasons including

(*i*) lack of residual magnetism (*ii*) wrong shunt field connection (*iii*) too low generator speed (*iv*) too much field resistance (*v*) dirty brushes or commutator (*vi*) an open- or short-circuited field circuit or loose brush connections.

Q. 39. What are the conditions necessary for the voltage build-up of a d.c shunt generator?

Ans. There are *four* conditions necessary for the voltage build-up of a shunt generator.

1. There must by some residual magnetism in the field poles.
2. The generator speed should be higher than critical speed N_C.
3. The shunt field resistance must be less than the critical resistance (which can be found from the O.C.C of the generator).

4. For the given direction of generator rotation, the shunt field coils should be correctly connected to the armature. It means coils should be so connected that the induced current *strengthens* the emf E_r produced initially due to residual magnetism.

Q. 40. The O.C.C. of a dc shunt generator driven at rated speed is as follows:

Field current I_f in amperes	0.5	1.0	1.5	2.0	2.5	3.0	3.5
Induced voltage in volts	**60**	**120**	**138**	**145**	**149**	**151**	**152**

If resistance of the field circuit is adjusted to 53 Ω, calculate the open-circuit voltage and load current when the terminal voltage is 100 V. Neglect armature reaction and assume an armature resistance of 0.1 Ω.. (*Elect Engg, Nagpur Univ. 2000*)

Ans. Let us first draw the shunt resistance line. We know that $R_{sh} = E_o/I_f$ or $E_o = I_f \cdot R_{sh} = 2 \times 53 = 106$ V. The point *B* in Fig. 47 represents (2 A, 106 V). The line *OB* is the shunt resistance line of 53 Ω. It cuts the O.C.C at point *A*. A line drawn from point *A* parallel to the base cuts the voltage axis at 150 V. Hence, the open circuit voltage with R_{sh} = 53 Ω is **150 V**.

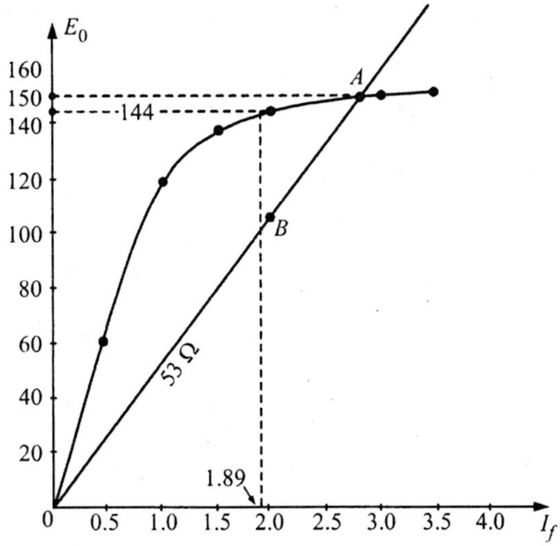

Fig. 47

Now when terminal voltage V = 100 V, the shunt current $I_{sh} = I_f = V/R_{sh}$ = 100/53 = 1.89A. As seen from the O.C.C, value of E_o for I_f = 1.89 A is 144 V.

Now, $E_o = V + I_a R_a$

∴ $I_a R_a = E_o - V$

$= 144 - 100 = 44$ V

or $I_a \times 0.1 = 44$

∴ $I_a = 44/0.1 =$ **440 A**

Q. 41. The open-circuit characteristic of a separately-excited d.c generator driven at 1000 rpm is as follows:

Field current (A)	0.2	0.4	0.6	0.8	1.0	1.2	1.4	1.6
E.M.F. (Volts)	**30**	**55**	**75**	**90**	**100**	**110**	**115**	**120**

If the machine is connected as shunt generator and driven at 1000 rpm and has a field resistance of 100 Ω, find (*a*) open-circuit voltage and exciting current (*b*) the critical resistance and (*c*) resistance to induce 115 volts on open circuit.

(*Elect Technology, Punjab Univ. 2000*)

Ans. The O.C.C has been plotted from the given data as shown in Fig. 48. The shunt resistance line *OA* is drawn with the help of point (1 A, 100 V)

(*a*) Point *D* represents *OC* voltage of **100 V** for exciting or field current of **1.0 A**.

(*b*) Line *OT* is tangent to the initial part of the O.C.C. It represents critical resistance R_c. It is seen from point *C* that R_c = 90/0.6 = **150 Ω.**

(*c*) As seen, 115 V corresponds to a field current of 1.4A. Hence, line *OB* represents a shunt resistance R_{sh} = 115/1.4 = **82.1 Ω.**

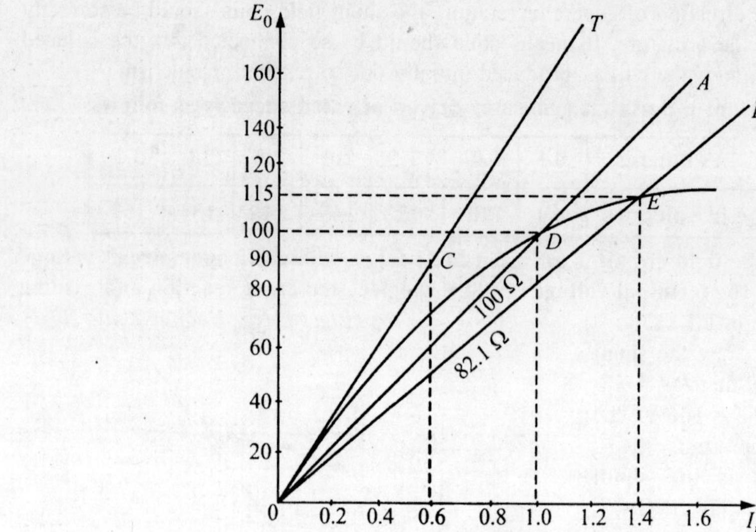

Fig. 48

Q. 42. What is meant by the term armature reaction in a dc generator? How is its effect neutralised?

Ans. The effect of armature field on the main field of a d.c. generator is called *armature reaction.* When a generator is loaded, the current in the armature conductors produces a magnetic field of *its own.* This armature field reacts with the magnetic field of the main poles. This distorts and weakens the *resultant field* in the air gap. This effect produced by the armature magnetic field on the pole field is known as armature reaction.

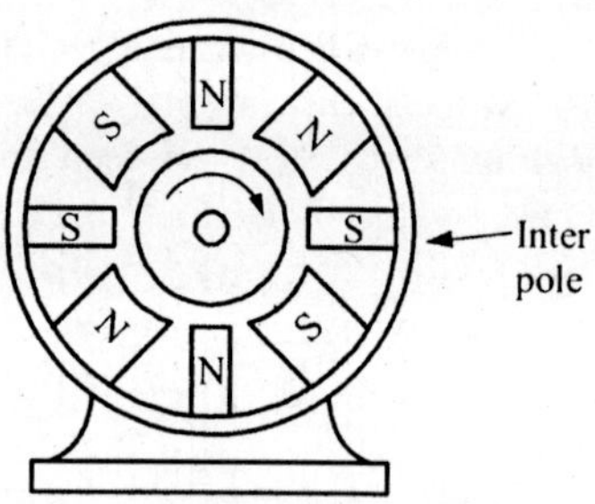

Fig. 49

The distorting effect produces sparking at the brushes. It also decreases the generated voltage to some extent. These effects of armature reaction are minimized by adding interpoles (or commutating poles) to the field system of the generator. Interpoles are small field poles fixed midway between the main poles as shown in Fig. 49. Their polarity is the *same* as that of the main pole *further ahead* according to the direction of rotation.

Q. 43. Briefly describe the external characteristic of a separately-excited d.c generator. What causes its voltage to drop when it is loaded?

Ans. A separately-excited dc generator draws its exciting curent from an independent d.c. source such as a storage battery as shown in Fig. 50(*a*). A field rheostat is connected in series with field resistance for changing the exciting current I_f. The generator is first excited on no load so that it develops its full-circuit voltage E_o. Then, the load is switched on. As seen from Fig. 50(*b*), when the load current I is increased, the terminal voltage is *decreased.*

There are two reasons for this decrease in terminal voltage.

(*i*) The first reason is armature reaction due to which the resultant field flux is reduced, thereby decreasing the generated emf E_o.

(*ii*) The second reason is voltage drop due to armature circuit resistance.

Due to these two reasons, the terminal voltage V is reduced as shown as Fig. 50(*b*). Unlike a self-excited generator, a separately-excited generator operates in a stable condition with any field exciting current. Hence, a large number of output voltages can be obtained.

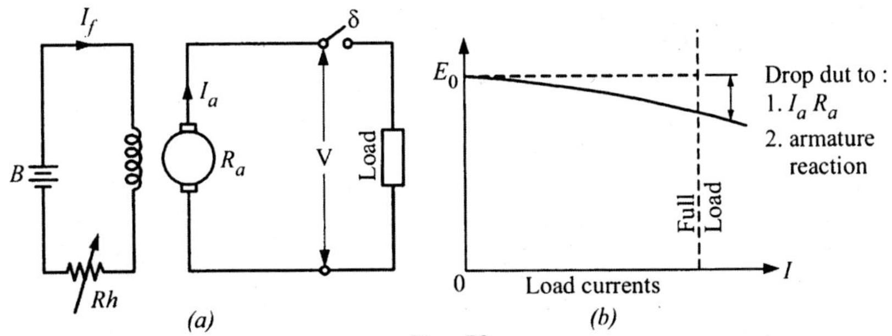

Fig. 50

Q. 44. With the help of the load characteristic of a shunt generator, explain why its terminal voltage decreases as its load current is increased.

Ans. A self-excited generator depends on *residual magnetism* of its field poles for its operation. First, the generator is excited to its full open-circuit voltage E_o and then the load is switched on by closing the switch *S*. (Fig. 51)

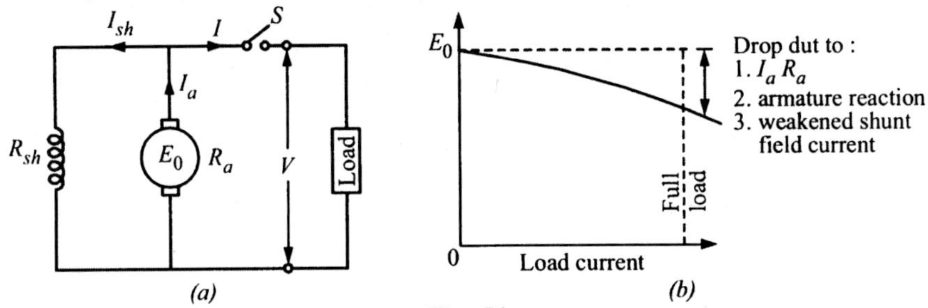

Fig. 51

As the load on the generator is increased, its terminal voltage *V* decreases due to the following three reasons :

(*i*) $I_a R_a$ drop in the armature (*ii*) armature reaction (*iii*) weakened shunt field current.

When load curret *I* is increased, I_a is also increased. The $I_a R_a$ drop increases thereby decreasing the terminal voltage. Also, as I_a increases, armature field strengnth also increases which distorts and weakens the pole flux. Hence, the output voltage is decreased. Now, shunt field resistance is connected across the terminal voltage. Hence, shunt current is reduced which further reduces the main field flux. Due to these reasons, the terminal voltage *V* is decreased on load.

Q. 45. What is meant by the voltage regulation of a generator? Give a formula for finding its value.

Ans. The voltage regulation of a generator means the *change in its terminal voltage* with changes in load current when it is run at a constant speed. If the change in voltage between no load and full load is small, the generator is said to have good regulation. If the change in voltage is large, it has poor regulation. Voltage regulation is defined as the percentage change in terminal voltage from no load to full load with respect to the full voltage.

$$\therefore \qquad \% \text{ voltage regulation} = \frac{V_{NL} - V_{FL}}{V_{FL}}$$

where V_{NL} is the no-load terminal voltage and V_{FL} is full-load terminal voltage.

It is important to know how the terminal voltage of a generator changes with the load in order to determine whether the given generator is suitable for a particular job. For example, a generator used for supplying lighting load should have a very low voltage regulation. In other words, its terminal voltage should be very nearly the same at full load as at no load.

Q. 46. What is a compound generator? How does its terminal voltage vary with load?

Ans. A generator with both a series and a shunt field winding is called a compound generator as shown in Fig. 18 and 19. The current of the shunt field winding is almost constant. However, current through the series winding varies and depends on the load current. At no load, the shunt winding provides all the field flux because there is no current in the series field winding (Fig. 18). As load is increased on the generator, the series field adds more flux to the shunt field. The terminal voltage of the loaded generator depends on the relative *strengths* of the two fields.

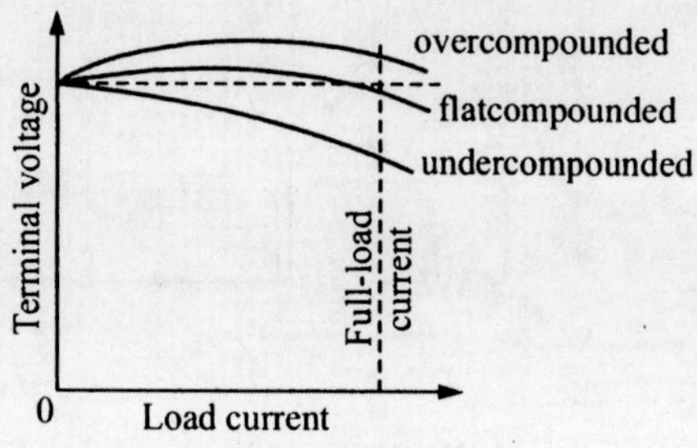

Fig. 52

(*i*) If the full-load terminal voltage V_{FL} is equal to the no-load terminal voltage V_{NL}, the generator is said to be flat-compounded.

(*ii*) If series flux is large enough to compensate for the armature reaction and $I_a R_a$ drop, the terminal voltage is higher than at no load. In that case, the generator is called overcompounded.

(*iii*) However, if $V_{FL} < V_{NL}$, then the generator is said to be under-compounded. The load characteristics of the three types of compound generator are shown in Fig. 52.

EXERCISE

1. A shunt generator has an induced voltage of 127 V and its terminal voltage is 120 V on full load. Find the armature current if the field resistance is 15 Ω and armature resistance is 0.02 Ω. Ignore armature reaction and also brush drop.
   (*Basic Elect Engg. BE. VTU Karnatka 2009*)
   (350 A)
2. Calculate the useful flux per pole in a four-pole generator with 722 armature conductors generating 500 V at 1000 r.p.m. when the armature is (*i*) lap-connected (*ii*) wave-connected. **(0.0416 Wb, 0.0208 Wb)**
3. A 4-pole, wave-wound armature has 39 slots with 18 conductors per slot. The useful flux per pole is 0.02 weber. Calculate the speed at which the armature must be rotated in order to generate 440 V. **(940 r.p.m.)**
4. A six-pole generator has a flux per pole of 0.08 Wb. The armature is lap-wound with 288 conductors and the e.m.f. generated is 480 V. Find the speed of the armature.
   (1250 r.p.m.)
5. A four-pole lap-wound shunt generator delivers 200 A at terminal voltage of 250 volts. It has field and armature resistances of 50 and 0.05 Ω respectively. Neglecting brush drop, determine :
   (*i*) armature current (*ii*) current in each conductor
   (*iii*) emf induced (*iv*) power developed (*Elect. Engg. Nagpur Univ. 2003*)
   (205 A; 51.25 A; 260.25 V; 50 kW)
6. A 4-pole dc shunt generator with lap-connected armature supplies a load of 100 A at 200 V. The armature resistance is 0.1 ohm and the shunt field resistance is 80 ohm. Find : (*i*) total armature current (*ii*) current per armature path and (*iii*) E.M.F. generated.
   (*Basic Elect Engg R.G.P.V. Bhopal 2002*)
   (102.5 A, 25.625 A, 210.25 A)

7. A six-pole lap-wound armature has 840 conductors and flux per pole of 0.018 Wb. Calculate the e.m.f. generated when the machine is running at 600 r.p.m.

(*Basic Elect Engg R.G.P.V. Bhopal 2005*)

(151.2 V)

8. An 8-pole wave-connected dc generator has 900 armature conductors and flux per pole of 0.04 Wb. At what speed it must be driven to generate 500 V?

(*Basic Elect Engg. R.G.P.V, Bhaopal 2007*)

(208.32 rpm)

MULTIPLE CHOICE QUESTIONS

1. A dc generator
   (*a*) changes electrical energy to mechanical energy
   (*b*) changes mechanical energy to electrical energy
   (*c*) is always self-excited
   (*d*) is always separately-excited
2. The armature of a d.c generator is made of
   (*a*) iron laminations (*b*) solid iron (*c*) copper (*d*) steel alloys
3. Armature coils of D.C. machines are usually made of
   (*a*) mica (*b*) copper (*c*) brass (*d*) graphite
   (*Basic Elect Engg, R.G.P.V. 2007*)
4. Which of the following is **not** part of a d.c. machine?
   (*a*) armature (*b*) commutator (*c*) field wdg (*d*) damper wdg
   (*Basic Elect Engg. R.G.P.V Bhopal 2007*)
5. The armature conductors of a 6-pole, lap-wound d.c generator are divided into .... parallel paths.
   (*a*) 2 (*b*) 3 (*c*) 6 (*d*) 4
   (*Basic Elect. Engg. West Bengal Tech Univ 2009*)
6. A separately-excited dc generator has the field connected
   (*a*) across the armature (*b*) in series with the armature
   (*c*) to an external circuit (*d*) none of these
7. The emf generated by a given d.c. generator depends upon
   (*a*) flux only (*b*) speed only (*c*) flux and speed (*d*) terminal voltage
   (*Basic Elect Engg. VTU Karnataka 2009*)
8. Commutating poles are
   (*a*) fastened to the centre of the commutator
   (*b*) located midway between the main poles
   (*c*) secondary poles induced by cross-magnetising the armature
   (*d*) used to regulate the voltage of the armature
9. The function of brushes in a dc generator is to
   (*a*) carry the current to the external circuit
   (*b*) prevent sparking
   (*c*) keep the commutator clean.
   (*d*) convert ac into dc

10. The voltage of a shunt generator is built up by
    (*a*) permanent magnetism (*b*) proper operation of the field rheostat
    (*c*) residual magnetism (*d*) increasing the speed
11. In a long-shunt compound-wound generator, the shunt field is connected in parallel with
    (*a*) armature
    (*b*) series field
    (*c*) parallel combination of armature and series field.
    (*d*) series combination of armature and series field.
12. If the flux/pole of a d.c generator is halved but its speed is doubled, its generator emf will
    (*a*) be halved (*b*) remain the same (*c*) be doubled (*d*) be quadrupled
13. Stray losses in a d.c. generator consist of .... losses.
    (*a*) magnetic and mechanical (*b*) magnetic and electrical
    (*c*) electrical and mechanical (*d*) copper and iron.
14. A d.c. generator may fail to build up its voltage due to
    (*a*) lack of residual magnetism (*b*) wrong shunt field connection
    (*c*) too low generator speed (*d*) all of the above
15. The voltage of a separately-excited dc generator may be increased by
    (*a*) increasing the speed of armature rotation
    (*b*) decreasing the magnetic flux.
    (*c*) both (*a*) and (*b*).
    (*d*) neither (*a*) nor (*b*).
16. To raise generator voltage, the
    (*a*) field current should be increased (*b*) field current should be decreased
    (*c*) speed should be decreased (*d*) brushes should be shifted forward.
17. Cutting resistance out of a shunt field circuit
    (*a*) cuts down the magnetic flux (*b*) decreases the terminal voltage
    (*c*) increases the load (*d*) increases the terminal output voltage
18. The number of parallel paths in wave-connected winding of a dc machine is
    (*a*) equal to the number of poles (*b*) half the number of poles
    (*c*) two (*d*) four
19. Voltage regulation of a generator is
    (*a*) the same thing as voltage control (*b*) the increase in output voltage
    (*c*) the decrease in output voltage (*d*) change in the output voltage
20. Which type of generator may be used for loads quite distant from the generator.?
    (*a*) overcompounded (*b*) undercompounded
    (*c*) flat-compounded (*d*) differential compounded.

ANSWERS

1. *b*	2. *d*	3. *b*	4. *d*	5. *c*	6. *c*	7. *c*
8. *b*	9. *a*	10. *c*	11. *d*	12. *b*	13. *a*	14. *d*
15. *a*	16. *a*	17. *d*	18. *c*	19. *d*	20. *a*	

CHAPTER 12
DC MOTOR

Q. 1. Is there any fundamental difference between a dc generator and a dc motor? List out operating differences between the two.

Ans. So for as *construction* is concerned, there is no basic difference between a dc generator and a dc motor. In fact, the same machine could be used either as a generator or as a motor. Each consists of magnetic poles, a rotating armature, a commutator, brushes and brush holders etc. However, a dc generator converts mechanical energy into electrical energy. But a motor converts *electrical* energy into *mechanical* energy.

Some of the operating differences between a dc motor and a dc generator are as under:

1. When a dc generator operates, it is driven by a prime mover. The rotation of its armature through a magnetic field generates the voltage in armature conductors.
   When a dc motor is in operation, it is *fed* by an electric current from a dc voltage source. The motor input current produces *two magnetic fields*, one by the field poles and the other by the rotating armature. The two fields react with each other and develop torque that produces mechanical rotation.
2. The load on a generator is electrical in nature whereas that on the motor is mechanical.
3. The voltage of a generator changes with load. Generally, there is a drop in voltage when load increases. However, for compound generators, the voltage may fall, rise or even remain constant as the load changes.
   In the case of a motor, it is the *speed* of rotation which changes as the load varies. When the load is increased, the speed of a shunt motors drops *slightly,* that of a compound motor drops *considerably* and that of a series motor drops *drastically.*
4. The voltage of a generator can be adjusted by either changing the speed or by changing the strength of the magnetic field or both.
   The speed of rotation of a dc motor can be changed by varying either the strength of the magnetic field or the voltage applied across the armature terminals or both.
5. Generators can be operated in parallel with each other to supply power to a common load. Motors, on the other hand, usually operate as single independent units to drive their own loads.
6. Generators are *always* started without electrical load *i.e.* on open circuit. They first build up their terminal voltage and then are switched on to supply line current. However, motors may or may not have a mechanical load when they are started.

Q. 2. Briefly describe the construction of a dc motor. How do you classify such motors?

Ans. DC motors look just like dc generators. In fact, it is difficult to identify them by their looks only. A motor has the same two main parts as a generator.

(*i*) a stationary frame or yoke supporting field magnets and

(*ii*) an armature assembly consisting of armature core, armature winding, commutator and brushes etc. as shown in Fig. 1.

The field windings magnetise the field poles to produce a strong magnetic field.

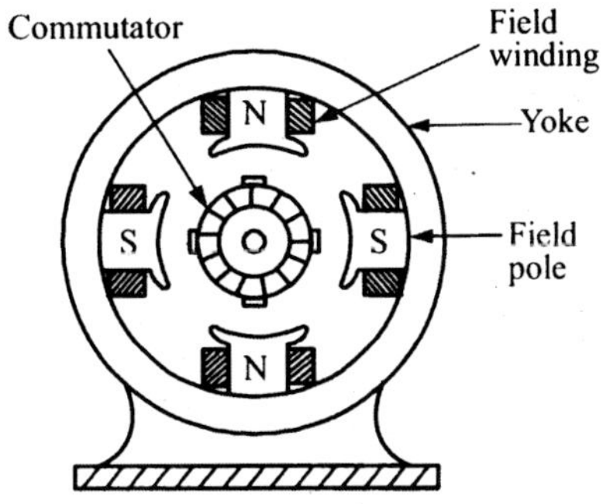

Fig. 1

The armature core is made up of thin mild-steel stampings. The armature winding consists of machine-wound copper coils placed in parallel slots on the surface of the armature core. The brushes are usually made of carbon and they press on the commutator segments. They carry current from the dc supply to the armature conductors.

D.C. motors are classified as per the windings used. Like generators, they are of three types:

(*i*) shunt – wound motors (*ii*) series – wound motors (*iii*) compound – wound motors

In shunt motors, the shunt winding and armature are connected in *parallel* across the dc supply as shown in Fig. 2(*a*). Because of this, the shunt field current I_{sh} is only a *small portion* of the line current.

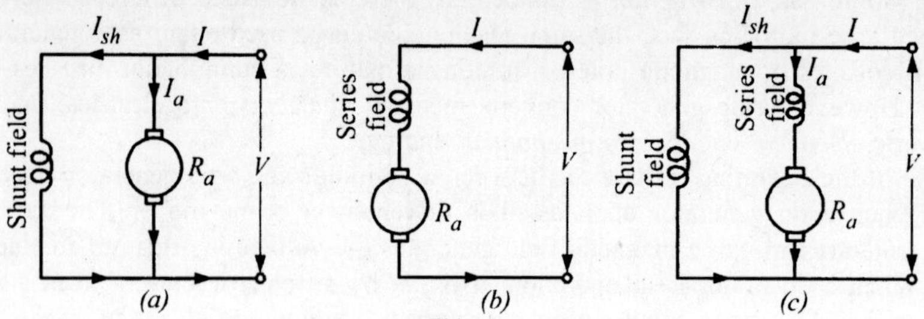

Fig. 2

In series motors, the field coils and armature are connected in series [Fig. 2(*b*)]. Hence, the *entire* line current I passes through the series field coils.

The compound motor has both the series and shunt windings. [Fig. 2(*c*)]. As seen, armature current I_a passes through the series winding.

Q. 3. Briefly explain the principle and operation of a d.c motor.

Ans. When a dc motor is connected to a dc voltage source, direct current flows through the armature conductors. This flow of current produces the armature field. Now, there are *two* magnetic fields in the air-gap present between the field shoes and the armature core. These two fields *react* with each other to rotate the armature.

The commutator plays a very important role in keeping the armature rotating in the *same* direction. It reverses the direction of flow of armature current at regular intervals so that armature field is always repelled by the pole field. Since the armature is continuously repelled by the field poles, it keeps rotating in the *same* direction thus producing the armature torque.

Q. 4. Why is the generated emf in the motor armature called counter emf or back emf? What is its value?

Ans. When a dc motor is switched on, its armature rotates in the magnetic field provided by the stationary field poles. Hence, armature conductors cut the pole flux. As per the laws of electromagnetic induction, an emf is induced in them. The direction of this emf can be found by Fleming's Right-hand Rule. As shown in Fig. 3, its direction is *opposite* to that of the applied voltage V. Because of its opposite direction, it is called counter emf. (E_c) or back emf (E_b). The equivalent circuit of a motor is as shown. Obviously, the applied voltage V has to drive I_a *against* the opposition of E_b. The power required to overcome this opposition is $E_b I_a$ watts.

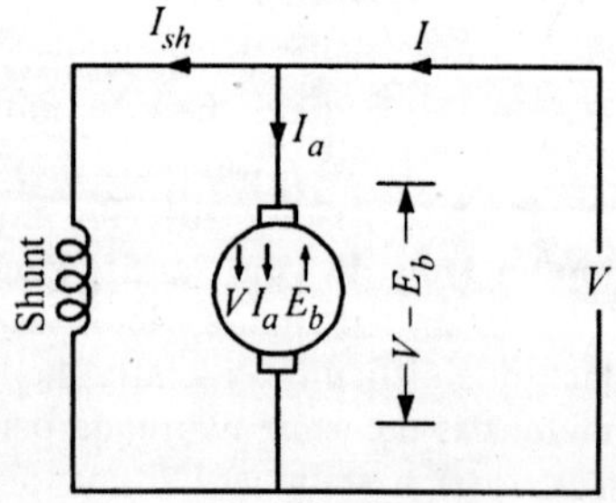

Fig. 3

It should be noted that ($V - E_b$) is the actual voltage which drives the armature current I_a. The value of the back emf. E_b is the same as that of the induced emf E_g in a dc generator.

$$\therefore \qquad E_b = \frac{\Phi ZN}{60}\left(\frac{P}{A}\right) \text{ volt}$$

Among other factors, E_b depends on the motor speed. If speed is high, E_b is large and I_a is small. If the speed is less, then E_b is less, hence, I_a is large and develops more torque. We find that E_b acts like a *governor.* It makes the motor self-regulating so that it draws just that much current as is required by the load.

Q. 5. When a dc motor is in operation, why is the armature current not equal to the line voltage divided by the armature resistance? Give the equation for armature current.

Ans. When a dc motor is in operation, there are *two* voltages present in its armature (Fig. 3). One is the applied voltage V which is *constant.* The other is the induced *emf* called back *emf* E_b which is *variable.* Its value is zero when the motor is stationary and is maximum when the motor runs at its rated speed. Hence, the net voltage which produces armature current is $(V - E_b)$*. If R_a is the armature resistance, then

$$I_a = \frac{V - E_b}{R_a}$$

Hence, armature current is not given by $I_a = V/R_a$. As seen, the value of I_a depends on the value of E_b. As E_b changes with the speed of the motor, I_a also changes to meet the load requirement.

It will be seen from above that $I_a = V/R_a$ only when $E_b = 0$ *i.e.* when motor is *stationary*, not otherwise.

Q. 6. Can the value of E_b become equal to the value of the applied motor voltage V? If not, explain why.

Ans. The back *emf* E_b can *never* be equal to the voltage V applied across the armature terminals. This E_b must *always* be less than V. In fact, E_b is about 80-95 percent of the applied voltage V. The net voltage across the armature is $(V - E_b)$. Its value has to be positive because some armature current is essential for rotating the armature and for meeting the rotational losses in the motor.

Value of E_b depends on the speed of the motor. Its value is zero when motor is stationary *i.e.* when $N = 0$. Its value is maximum when N is maximum. As seen form the equation $I_a = (V - E_b)/R_a$. If $E_b = V$, then $I_a = 0$ When I_a is zero, then armature cannot rotate.

Q. 7. Give the voltage equation of a dc motor. Derive an expression for the mechanical power developed by a dc motor.

Ans. As shown in Fig. 4, V is the dc voltage applied across the armature terminals. This voltage is used to

(*i*) overcome the back emf E_b and

(*ii*) supply the armature ohmic drop $I_a R_a$ (neglecting the brush dorp)

$\therefore \quad V = E_b + I_a R_a$

It is known as voltage equation of a dc motor.

Now, multiplying both sides by I_a, we get

$$VI_a = E_b I_a + I_a^2 R_a$$

where, VI_a = electrical input to the armature

$E_b I_a$ = electrical equivalent of the mechanical power P_m developed in the armature.

$I_a^2 R_a$ = Cu loss in the armature

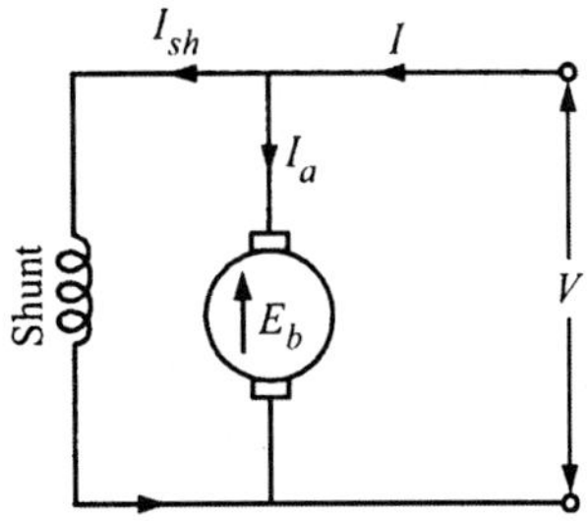

Fig. 4

Hence, out of the armature input, some is wasted as I^2R loss and the rest is converted into mechanical power developed by the armature.

*We have neglected the brush voltage drop. Hence, the supply voltage V has been taken as the voltage across the armature. Actually, voltage across the armature $V_a = V$ – brush drop.

Hence, $\quad I_a = (V_a - E_b)/R_a$.

Q. 8. Prove that power developed by a dc motor is maximum when its back emf E_b is equal to half of its applied voltage V.

Ans. The mechanical power developed by the armature of a dc motor is

$$P_m = VI_a - I_a^2 R_a$$

Differentiating both sides with respect to I_a, we get

$$\frac{dP_m}{dI_a} = V - 2I_a R_a$$

Equating dP_m/dI_a to zero, we get

$$V - 2I_a R_a = 0 \quad \text{or} \quad I_a R_a = V/2$$

Now, motor voltage equation is $V = E_b + I_a R_a$

Substituting the value of I_aR_a, we get

$$V = E_b + \frac{V}{2} \quad \text{or} \quad E_b = \frac{V}{2}$$

As seen from above, mechanical power developed by a motor is maximum when its back emf equals half the applied voltage.

In practice, this condition is not realised because the current equired would be much beyond the normal current of the motor. Morerover, half the input would be wasted in the form of heat thus decreasing motor efficiency.

Q. 9. A 220 V dc shunt motor has armature resistance of 0.2 Ω and brush contact drop of 2 V. Calculate the armature current which will flow when

(*i*) back emf is 215 V

(*ii*) back emf drops to 212 V on increasing motor load

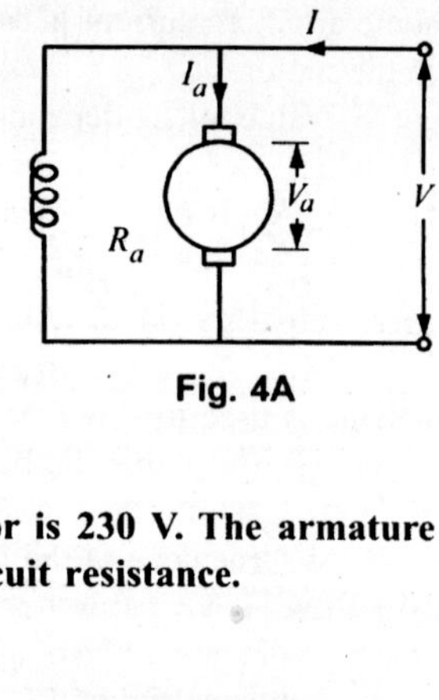

Fig. 4A

Ans. (*i*) $I_a = \dfrac{V_a - E_b}{R_a}$

where V_a = voltage across the armature

= 220 – brush drop

= 220 – 2 = 218 V

$\therefore$ $I_a = \dfrac{218 - 215}{0.2}$ = **15 A**

(*ii*) $I_a = \dfrac{218 - 212}{0.2}$ = **30 A**

Q. 10. The voltage applied to the armature of a d.c. motor is 230 V. The armature current is 12 A and the back emf 226.4 V. Find the armature circuit resistance.

Ans. The motor voltage equation is

$$V = E_b + I_aR_a.$$

Putting the given values, we get

$$230 = 226.4 + 12 R_a \quad \therefore \; R_a = \mathbf{0.3\ \Omega}$$

Q. 11 A shunt motor of armature circuit resistance 0.25Ω is connected to a 230 – V d.c. supply. The input line current is 8A and the shunt field current 0.8 A. Calculate the emf generated in the armature.

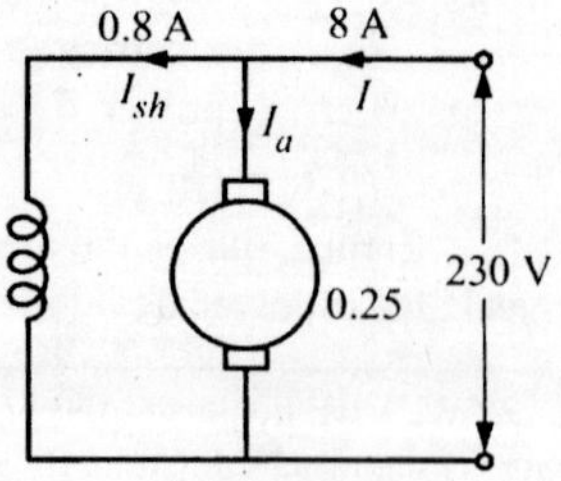

Fig. 5

Ans. As seen, from Fig. 5,

$$I_a = I - I_{sh} = 8 - 0.8 = 7.2 \text{ A}$$

Now, $V = E_b + I_a R_a$

or $230 = E_b + 7.2 \times 0.25$

$\therefore$ $E_b = 230 - 7.2 \times 0.25 =$ **228.2 V**

Q. 12. When a current of 18 A flows through the armature of a d.c. shunt motor, the back emf is 248 V. The armature resistance is 0.15 Ω and the total brush voltage drop 2.3 V. What is the supply voltage?

Ans. The *complete* motor voltage equation is

$$V = E_b + I_a R_a + \text{brush drop}$$

Putting the given values, we get

$$248 = E_b + (18 \times 0.15) + 2.3 \qquad \therefore \; V = \mathbf{253\ V}$$

Q. 13. A 4-pole, wave-wound dc motor has 48 armature slots each containg 4 conductors. The flux per pole is 58.7 mWb and armature resistance is 0.2 Ω. When connected to a 400 V supply, the motor runs at a spped of 1050 rpm. Calculate the value of the back emf developed by the motor and the armature current.

Ans. The back emf is given by

$$E_b = \frac{\Phi ZN}{60}\left(\frac{P}{A}\right)$$

$$= \frac{(58.7 \times 10^{-3}) \times (48 \times 4) \times 1050}{60} \times \frac{4}{2} = 395 \text{ V}$$

Now, $V = E_b + I_a R_a \qquad \therefore \; I_a = \dfrac{V - E_b}{R_a}$

or $I_a = \dfrac{400 - 395}{0.2} = \mathbf{25\ A.}$

Q. 14. A long- shunt compound motor takes a line current of 52 A at a supply voltage of 440 V. The shunt field resistance is 220 Ω, the armature circuit resistance 0.16 Ω and the series field resistance 0.04 Ω. Find (*a*) the armature current (*b*) the back emf of the armature (*c*) power developed by the motor.

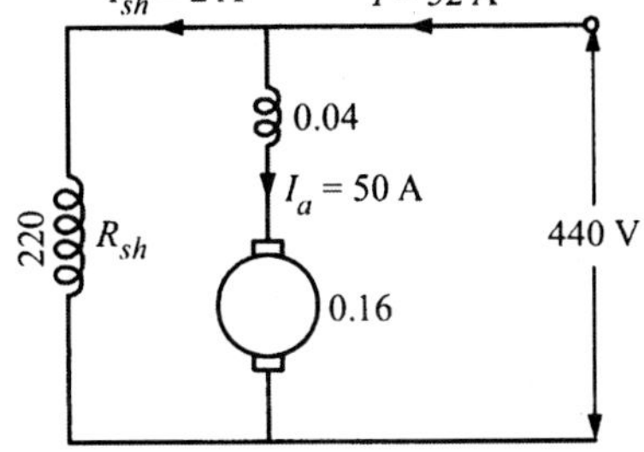

Fig. 6

Ans. (*a*) As seen from Fig. 6,

$$I_{sh} = V/R_{sh} = 440/220 = 2 \text{ A}$$

$$\therefore \quad I_a = I - I_{sh} = 52 - 2 = \mathbf{50\ A}$$

(*b*) Armature *circuit* resistance

$$= 0.04 + 0.16 = 0.2\ \Omega$$

$$E_b = V - I_a \times \text{armature } \textit{circuit} \text{ resistance}$$

$$= 440 - 50 \times 0.2 = \mathbf{430\ V}$$

(*c*) Power developed by armature

$$P_m = E_b I_a = 430 \times 50$$

$$= 21{,}500 \text{ W} = \mathbf{21.5\ kW}$$

Q. 15. A 25-kW, 250 V, d.c. shunt generator has armature and field resistances of 0.06 Ω and 100 Ω respectively. Determine the total armature power developed when working (*i*) as a generator delivering 25 kW output and (*ii*) as a motor taking 25 kW input.

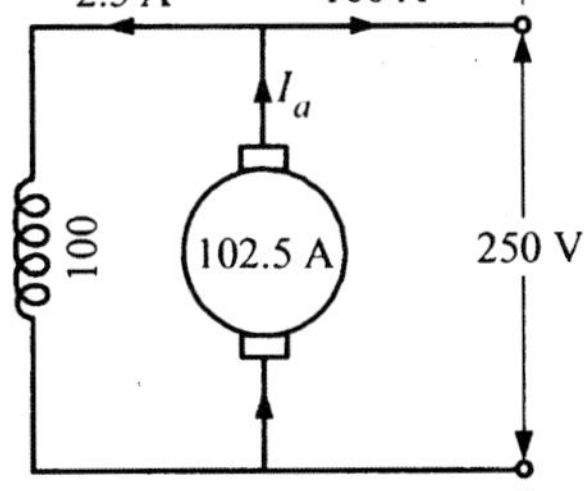

Fig. 7

Sol. As generator (Fig. 7)

$$\text{Output current} = 25{,}000/250 = 100 \text{ A}$$

$$I_{sh} = 250/100 = 2.5 \text{ A}$$

$$I_a = 102.5 \text{ A}$$

Generated e.m.f. $E_g = 250 + I_a R_a = 250 + 6.15 = 256.15$ V

Power developed in armature

$$= E_g\, I_a = \frac{256.15 \times 102.5}{1000} = \mathbf{26.25\ kW}$$

As Motor (Fig. 8)

Motor input current $= 100$ A

$I_{sh} = 2.5$ A

$I_a = 100 - 2.5 = 97.5$ A

$E_b = 250 - (97.5 \times 0.06)$

$= 250 - 5.85 = 244.15$ V

Power developed in armature $= E_b I_a$ watt

$$= \frac{244.15 \times 97.5}{1000} = \mathbf{23.8\ kW}$$

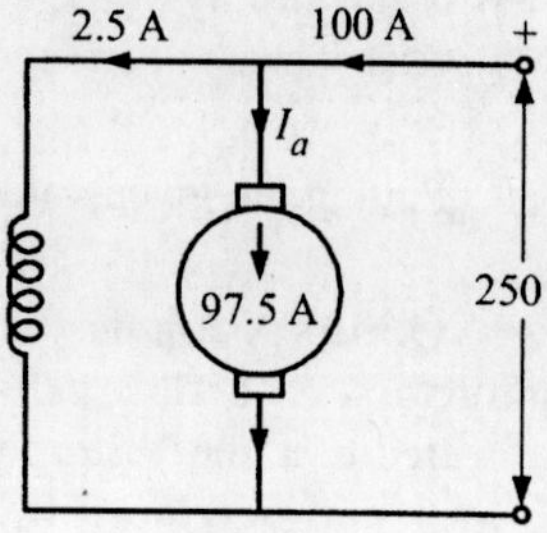

Fig. 8

Q. 16. Define the term torque giving its units. Find the value of power produced by a torque *T* newton-metre acting on a pulley of radius *R* metre and revolving at a speed of *N* revolution/second.

Ans. By the term *torque* is meant the turning or *twisting* moment of a force about an axis. It is measured by the *product* of the force and the radius at which this force acts.

Consider a pulley of radius R metres acted upon by a *tangential* force of F newton which causes it to rotate at N rev/second (Fig. 9)

Then, torque $T = F \times R$ newton-metre

Work done by this force in *one* revolution

$= \text{Force} \times \text{distance} = F \times 2\pi R$ joules

Work done per second

$W = F \times 2\pi R \times N$ joules/ second

$= (F \times R) \times 2\pi N$ joules/second

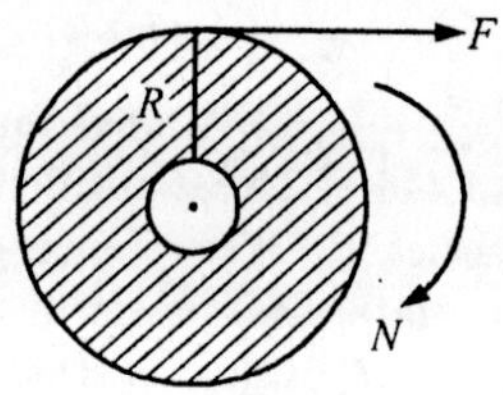

Fig. 9

Now, $2\pi N =$ angular velocity ω in radian/second

Also $F \times R =$ torque, T

$\therefore$ Work done/ second $= T \times \omega$ joules/second

If T is the torque in N-m and ω the angular velocity in radian/second, then

Power developed, $P = T \times \omega$ watt

$= T \times 2\pi N$ watt

Q. 17. Derive a general expression for the gross armature torque developed by a dc motor. How is this expression modified for (*a*) a shunt motor and (*b*) a series motor?

Ans. Let T_a be the *gross* torque developed by the armature of a motor running at N r.p.s. *i.e.* with an angular velocity of $\omega = 2\pi N$ rad/second. If T_a is in *N-m* then,

$$\text{Power developed} = T_a \times \omega = T_a \times 2\pi N \text{ watt} \quad ...(i)$$

$$= T_a \times \frac{2\pi N}{60} = \frac{NT_a}{60/2\pi} = \frac{NT_a}{9.55} \text{ watt}$$

Now, the electrical power converted into mechanical power in the armature is

$$= E_b\, I_a \text{ watt} \quad ...(ii)$$

Equating (*i*) and (*ii*), we get

$$T_a \times 2\pi N = E_b\, I_a \quad ...(iii)$$

Now, $E_b = \Phi ZN \times (P/A)$ volt — N in rps

$\therefore$ $T_a \times 2\pi N = \Phi ZN\,(P/A).\, I_a$

or $$T_a = \frac{1}{2\pi} \Phi Z I_a \left(\frac{P}{A}\right) \text{ N-m} = 0.159\ \Phi Z I_a\ (P/A) \text{ N-m}$$

From the above, we find that $T_a \propto \Phi I_a$.

(a) Torque for a series Motor :

In the case of a series motor, Φ is directly proportional to I_a (before magnetic saturation) because series winding carries full armature current

$$\therefore \qquad T_a \propto \Phi I_a \quad \text{or} \quad T_a \propto I_a^2$$

As seen, armature torque varies as the square of the armature current.

(b) Torque For shunt Motor :

For shunt motors, Φ is practically constant,

hence, $$T_a \propto \Phi I_a \quad \text{or} \quad T_a \propto I_a$$

As seen, torque varies linearly with the armature current.

Note : It can be seen from Eqn. (*iii*) that

$$T_a \times 2\pi N = E_b I_a \qquad \text{— } N \text{ in rps}$$

$$\therefore \qquad T_a = \frac{1}{2\pi} \cdot \frac{E_b I_a}{N} \text{ N-m}$$

$$= 0.159 \frac{E_b I_a}{N} \text{ N-m}$$

If N is in rpm, then

$$T_a \times \frac{2\pi N}{60} = E_b I_a$$

$$\therefore \qquad T_a = \frac{E_b I_b}{2\pi N/60} = \frac{60}{2\pi} \frac{E_b I_a}{N} = 9.55 \frac{E_b I_a}{N} \text{ N-m}$$

The above torque is also called the gross torque T_g.

Q. 18. What is meant by shaft torque of a dc motor? Why is it less than the gross armature torque?

Ans. The *whole* of the armature *gross* torque is not available for doing useful work because some part of it is used for supplying iron, friction and windage losses in the motor.

The torque which is *actually* available for doing useful work is known as *shaft torque* T_{sh}. It is so called because it is available *at the shaft.*

If T_{sh} is the shaft torque of a dc motor running at N rps, then

$$T_{sh} \times 2\pi N = \text{motor output}$$

$$\therefore \qquad T_{sh} = \frac{\text{motor output}}{2\pi N} \text{ N-m}$$

If N is in rpm, then

$$T_{sh} = \frac{\text{motor output}}{2\pi N/60} = \frac{60}{2\pi} \times \frac{\text{motor output}}{N}$$

$$= 9.55 \frac{\text{motor output}}{N} \text{ N-m}$$

The difference $T_a - T_{sh}$ is known as lost torque. Shaft torque is less than the armature torque because of the friction and windage losses in the motor

Q. 19. A 250 V, 4-pole wave-wound dc series motor has 782 conductors on its armature. It has armature and series field resistance of 0.75 Ω. The motor takes a current of 40A. Determine its speed and gross torque developed if it has a flux per pole of 25 mWb.

(*Basic Eect. Engg. Anna. Univ. Combatore 2009*)

Ans. $$N = \frac{E_b}{\Phi}\left(\frac{60A}{ZP}\right) \text{ rpm}$$ — see Q. 4

$$E_b = V - I_a R_t = 250 - 40 \times 2.75 = 220 \text{ V}$$

$$\therefore \quad N = \frac{220}{25\times 10^{-3}}\left(\frac{60\times 2}{782\times 4}\right) = \mathbf{337.6 \text{ rpm}}$$

gross torque T_a or $T_g = 9.55 \dfrac{E_b I_a}{N} = 9.55 \dfrac{220\times 40}{337.6} = \mathbf{249 \text{ N-m}}$

Q. 20. A dc motor has 6 poles, flux per pole of 0.05 Wb with lap-wound armature of 600 conductors. Motor speed is 500 r.p.m. Determine the applied voltage and back e.m.f. given:

armature resistance is 0.25 Ω, armature current 40 amperes.

Also, determine the torque developed by the motor in N-m.

(Elements of Elect. Engg. Punjab Univ.)

Sol. $$E_b = \frac{\Phi Z N}{60}\left(\frac{P}{A}\right) = \frac{0.05\times 600\times 500}{60}$$

$$= \mathbf{250 \text{ V}}$$

Arm. drop $= I_a R_a = 40 \times 0.25 = 10$ V

∴ applied voltage,

$$V = E_b + I_a R_a = 250 + 10 = \mathbf{260 \text{ V}}$$

$$T_a = 0.159\, E_b I_a/N \quad \text{— } N \text{ in rps}$$

$$\therefore \quad T_a = 0.159 \times 250 \times 40/(500/60) = \mathbf{1908 \text{ N-m}}$$

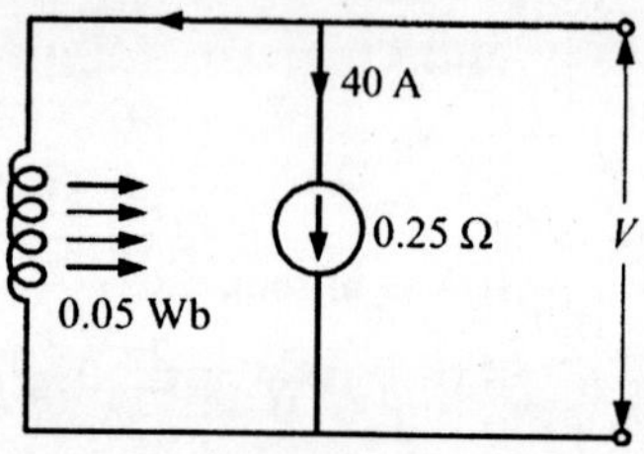

Fig. 10

Q. 21. The armature of a 6-pole, 6-circuit d.c. shunt motor takes 300 A at a speed of 400 revolutions per minute. The flux per pole is 75 × 10^{-3} Wb. The number of armature turns is 500. The torque lost in friction, windage and iron losses can be assumed at 2.5%.

Calculate :

(*i*) torque developed by the armature

(*ii*) shaft torque

(*iii*) shaft power in kW. *(Basic Elect Engg. RGPV Bhopal 2006)*

Ans. Since armature has a coil of 500 turns , the number of conductors Z = 500 × 2 1000. It is a lap connected motor because the number of armature circuits equals the No. of poles

(*i*) Now, $$T_a = 0.159\, \Phi\, Z\, I_a\, (P/A) \text{ N-m}$$

$$= 0.159 \times (75 \times 10^{-3}) \times 1000 \times 300 \times (6/6) = \mathbf{3{,}577.5 \text{ N-m}}$$

(*ii*) Lost torque due to friction, windage and iron losses is

$$= 2.5\% \text{ of } T_a = 2.5 \times 3{,}577.5/100 = 89.44 \text{ N-m}$$

$$\therefore \quad T_{sh} = 3{,}577.5 - 89.44 = \mathbf{3{,}488 \text{ N-m}}$$

(*iii*) $$\text{shaft power} = T_{sh} \times \left(\frac{2\pi N}{60}\right) = \frac{3488\times 2\pi\times 400}{60} = 146{,}000\text{W} = \mathbf{146 \text{ kW}}$$

Q. 22. Derive the fundamental equation for the speed of a dc motor. What are its different forms in the case of (*i*) a series motor and (*ii*) a shunt motot?

Sol. The value of back emf E_b of a dc motor is

$$E_b = \Phi Z N \times \left(\frac{P}{A}\right) \text{ volt} \quad \text{— } N \text{ is rps}$$

$$\therefore \qquad N = \frac{E_b}{\Phi}\left(\frac{A}{ZP}\right) \text{ rev/second} = k\,\frac{E_b}{\Phi}$$

It shows that speed is directly proportional to back emf E_b and inversely to the flux Φ.

$$\text{or} \qquad N \propto \frac{E_b}{\Phi}$$

(*a*) For series Motor

Let, N_1 = speed in the 1st case

I_{a1} = armature current in the 1st case

Φ_1 = Flux/ pole in the 1st case

N_2, I_{a2}, Φ_2 = corresponding values in the second case

$$\text{Now,} \qquad N_1 \propto \frac{E_{b1}}{\Phi_1} \qquad \text{— where } E_{b1} = V - I_{a1} R_a$$

$$N_2 \propto \frac{E_{b2}}{\Phi_2} \qquad \text{— where } E_{b2} = V - I_{a2} R_a$$

$$\therefore \qquad \frac{N_2}{N_1} = \frac{E_{b2}}{E_{b1}} \times \frac{\Phi_1}{\Phi_2}$$

Prior to *saturation** of the magnetic poles, $\Phi \propto I_a$

$$\therefore \qquad \frac{N_2}{N_1} = \frac{E_{b2}}{E_{b1}} \times \frac{I_{a1}}{I_{a2}}$$

(*b*) For shunt motor

In this case, the same equation applies

$$\text{i.e.} \qquad \frac{N_2}{N_1} = \frac{E_{b2}}{E_{b1}} \times \frac{\Phi_1}{\Phi_2}$$

Since in the case of shunt motors, $\Phi_1 = \Phi_2$

$$\therefore \qquad \frac{N_2}{N_1} = \frac{E_{b2}}{E_{b1}}$$

Q. 23. Define speed regulation of a dc motor. Why does it not apply to a series motor?

Ans. By speed regulation is meant the change in the speed of a motor when its load is changed. By change in speed here is meant the change which occurs due to the *basic properties of the motor itself.* It does not mean the change in speed brought about with the help of rheostats or other controlling devices. This type of change in speed is called *speed control.*

The speed regulation of a motor is defined as the change in speed of a motor when its load is reduced from full load to no load expressed as the percent of the full load speed

$$\therefore \qquad \%\text{ speed regulation} = \frac{\text{N.L. speed — F. L. speed}}{\text{F. L. speed}} \times 100 = \frac{N_0 - N}{N} \times 100$$

Speed regulation has no meaning for a series motor. It is so because it is not possible to measure the speed of series motor at no load. When started on no load, the motor builds up to a very dangerously high speed and develops huge centrifugal force which may destroy the motor itself.

*Magnetic saturation means that flux does not change with current

Q. 24. A 250 V shunt motor runs at 1000 rpm at no load and takes 8 A. The total armature and shunt field resistances are 0.2 Ω and 250 Ω respectively. Calculate the speed when loaded and taking 50 A. Assume the flux to be constant.

(*Elect Engg. B.E.1st Yr., Nagpur Univ. 2009*)

Ans. We will use the relation

$$\frac{N_2}{N_1} = \frac{E_{b2}}{E_{b1}} \qquad \text{— since } \Phi \text{ is constant}$$

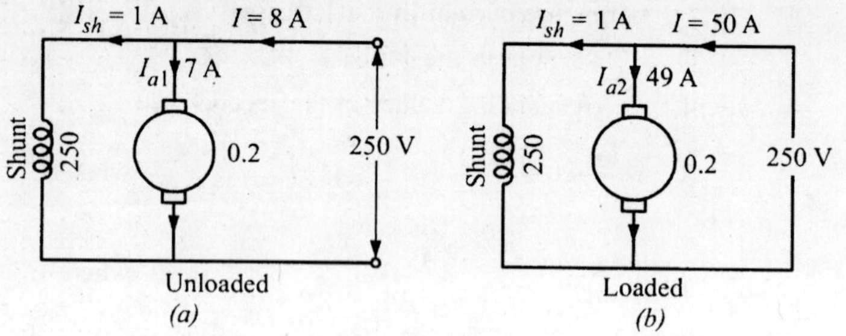

Fig. 11

At no-load [Fig. 11(*a*)]

$$I_{a1} = 8 - 1 = 7 \text{ A}$$

$$E_{b1} = V - I_{a1} R_a = 250 - (7 \times 0.2) = 248.6 \text{ V}$$

When loaded [Fig. 11(*b*)]

$$I_{a2} = 50 - 1 = 49 \text{ A}$$

$$E_{b2} = V - I_{a2} R_a = 250 - (49 \times 0.2) = 240.2 \text{ V}$$

$$\therefore \quad \frac{N_2}{1000} = \frac{240.2}{248.6} = \mathbf{966 \text{ rpm}}$$

Q. 25. A dc series motor draws a line current of 100 A from the mains while running at 1000 rpm. Its armature resistance is 0.15 Ω and field resistance is 0.1 Ω. Assuming that the flux corresponding to a current of 25 A is 40% of that corresponding to 100 A, find the speed of the motor when it is drawing 25 A from 230 V supply.

(*B. Tech. Elect Technology JNTU, Hyderabad 2007*)

Ans. The motor speed equation is

$$\frac{N_2}{N_1} = \frac{E_{b2}}{E_{b1}} \times \frac{\Phi_1}{\Phi_2}$$

Now, $\Phi_2 = 0.4\ \Phi_1 \qquad \therefore \quad \dfrac{\Phi_1}{\Phi_2} = \dfrac{1}{0.4}$

$$E_{b1} = 230 - 100\ (0.15 + 0.1) = 205 \text{ V}$$

$$E_{b2} = 230 - 25\ (0.15 + 0.1) = 223.75 \text{ V}$$

$$\therefore \quad \frac{N_2}{1000} = \frac{223.75}{205} \times \frac{1}{0.4} \qquad \therefore \quad N_2 = \mathbf{2729 \text{ rpm}}$$

Q. 26. A 250 V shunt motor on no load runs at 1000 rpm and takes 5 A. The total armature and shunt field resistances are 0.2 Ω and 250 Ω respectively. Calculate the speed when loaded and taking a current of 50 A, if the armature reaction weakens the field by 3%.

(*Basic Elect Engg. C.S.V. Tech. Univ., Bhilai 2007*)

Ans. We will use the relation $\dfrac{N_2}{N_1} = \dfrac{E_{b2}}{E_{b1}} \times \dfrac{\Phi_1}{\Phi_2}$

Now, $I_{a1} = 5 - 1 = 4$ A, $E_{b1} = 250 - 4 \times 0.2 = 249.2$ V, $N_1 = 1000$ rpm

$I_{a2} = 50 - 1 = 49$ A, $E_{b2} = 250 - 49 \times 0.2 = 240.2$ V, $N_2 = ?$

Also, $\Phi_2 = 97\%$ of $\Phi_1 = 0.97\ \Phi_1$

$$\therefore \quad \frac{N_2}{1000} = \frac{240.2}{249.2} \times \frac{\Phi_1}{0.97\Phi_1} \qquad \therefore \quad N_2 = \mathbf{964\ rpm}$$

Q. 27. A dc series motor is running with a speed of 1000 rpm while taking a current of 22 A from the supply. If the load is changed such that the current drawn by the motor is increased to 55 A, calculate the speed of the motor on new load. The armature and series winding resistances are 0.3 Ω and 0.4 Ω respectively. Assume voltage as 250 V.

(*Basic Elect Engg. VTU Karnatka 2009*)

Ans. We will use the relation.

$$\frac{N_2}{N_1} = \frac{E_{b2}}{E_{b1}} \times \frac{I_{a1}}{I_{a2}} \qquad \because \quad \Phi \propto I_a \text{ for a series motor}$$

The Fig. 12. shows the motor circuit

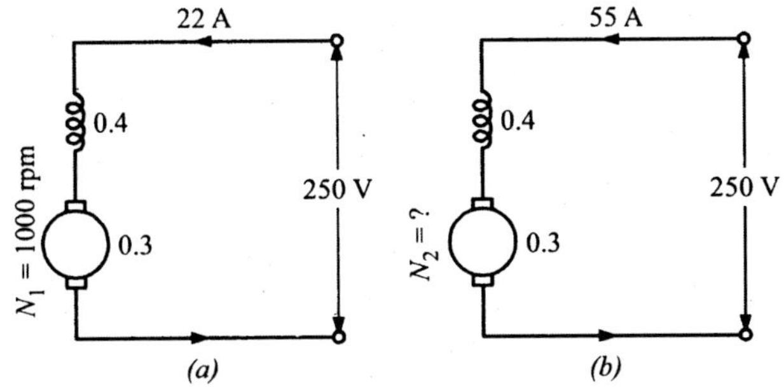

Fig. 12

First Case $I_{a1} = 22$ A, $E_{b1} = 250 - 22\ (0.3 + 0.4) = 234.6$ V, $N_1 = 1000$ rpm

Second Case $I_{a2} = 55$ A, $E_{b2} = 250 - 55\ (0.3 + 0.4) = 211.5$ V, $N_2 = ?$

$$\therefore \quad \frac{N_2}{1000} = \frac{211.5}{234.6} \times \frac{22}{55} \qquad \therefore \quad N_2 = \mathbf{360.6\ rpm}$$

As seen, motor speed reduces from 1000 rpm to 360.6 rpm when load is increased.

Q. 28. A 230 V dc shunt motor has an armature resistance of 0.5 Ω and field resistance of 115 Ω. At no load, speed is 1200 rpm and the armature current 2.5 A. On operation at rated load, the speed drops to 1120 rpm. Determine the line current and power input when the motor delivers rated load. (*Elect. Engg Nagpur Univ. 2005*)

Ans. $N_1 = 1200$ rpm ; $E_{b1} = 230 - (2.5 \times 0.5)$

$= 288.75$ V

$N_2 = 1120$ rpm; $E_{b2} = 230 - 0.5\ I_{a2}$

Since in a shunt motor, Φ remains costant

$$\therefore \quad \frac{N_2}{N_1} = \frac{E_{b2}}{E_{b1}}$$

or
$$\frac{1120}{1200} = \frac{230 - 0.5\ I_{a2}}{228.75} \qquad \therefore \quad I_{a2} = 33 \text{ A}$$

Line current drawn by the motor (Fig. 13) is

$= I_{a2} + I_{sh} = 33 + (230/115) = \mathbf{35\ A}$

Power input at rated load $= 230 \times 35 = \mathbf{8050\ W}$

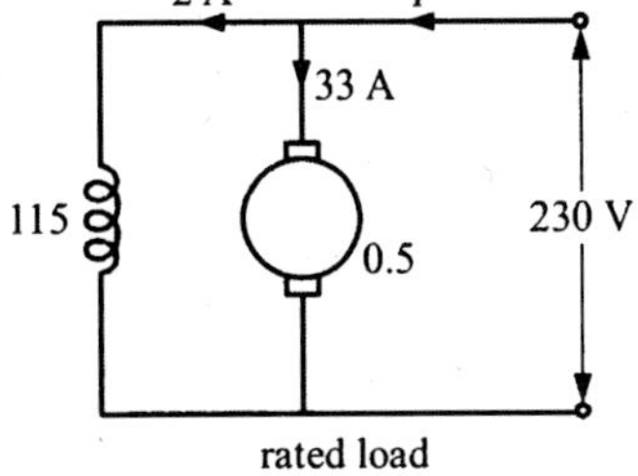

Fig. 13

Q. 29. A 250 V dc shunt motor has armature resistance of 0.25 Ω, on load it takes armature current of 50 A and runs at 750 rpm. If the flux of the motor is reduced by 10% without changing the load torque, find the new speed of the motor

(*Basic Elect. Engg. Nagpur Univ. 2006*)

Ans. We will use the relation

$$\frac{N_2}{N_1} = \frac{E_{b2}}{E_{b1}} \times \frac{\Phi_1}{\Phi_2}$$

The motor armature torque, $T_a \propto \Phi\, I_a$

$$\therefore \quad T_{a1} \propto \Phi_1\, I_{a1} \quad \text{and} \quad T_{a2} \propto \Phi_2\, I_{a2}$$

Since torque has not been changed though the flux has been reduced by 10%

$$\therefore \quad Ta_1 = T_{a2} \quad \text{or} \quad \Phi_1\, I_{a1} = \Phi_2\, I_{a2}$$

Since $\quad \Phi_2 = 0.9\, \Phi_1, \qquad \therefore \quad \Phi_1 \times I_{a1} = (0.9\Phi_1) \times I_{a2}$

or $\quad \Phi_1 \times 50 = (0.9\, \Phi_1) \times I_{a2} \qquad \therefore \quad I_{a2} = 55.6$ A

$$E_{b1} = 250 - (50 \times 0.25) = 237.5 \text{ V}$$

$$E_{b2} = 250 - (55.6 \times 0.25) = 236.1 \text{ V}$$

$$\therefore \quad \frac{N_2}{750} = \frac{236.1}{237.25} \times \frac{\Phi_1}{0.9\Phi_1} \qquad \therefore \quad N_2 = \mathbf{829.3 \text{ rpm}}$$

Q. 30. A 460 V dc series motor runs at 1000 rpm taking a current of 40 A. Calculate the speed and percentage change in torque if the load is reduced so that the motor is taking 30 A. Total resistance of the armature and field circuits is 0.8 Ω. Assume flux is proportional to the field current. (*B. Tech., Elect. Engg, JNTU, Hyderabad 2007*)

Ans. Armature torque $T_a \propto \Phi\, I_a \propto I_a^2 \qquad -(\because \Phi \propto I_a)$

$$\therefore \quad T_{a1} \propto 40^2 \text{ and } T_{a2} \propto 30^2$$

$$\therefore \quad \frac{T_{a2}}{T_{a1}} = \frac{9}{16}$$

$\therefore$ percentage change in torque is

$$= \frac{T_{a1} - T_{a2}}{T_{a1}} \times 100 = \frac{7}{16} \times 100 = \mathbf{43.75\%}$$

Now, $\quad E_{b1} = 460 - (40 \times 0.8) = 428$ V

$$E_{b2} = 460 - (30 \times 0.8) = 436 \text{ V}$$

$$\therefore \quad \frac{N_2}{N_1} = \frac{E_{b2}}{E_{b1}} \times \frac{I_{a1}}{I_{a2}} \qquad \therefore \quad \frac{N_2}{1000} = \frac{436}{428} \times \frac{40}{30} = \mathbf{1358 \text{ rpm}}$$

Q. 31. A 250 volt series motor has an armature resistance of 0.08 Ω and field resistance of 0.02 Ω and produces full-load torque when running at 500 r.p.m taking a current of 40 A.

Calculate : (*a*) armature current (*b*) the speed when producing full-load torque. Neglect effect of armature reaction and saturation.

Sol. (*a*) Since saturation is neglected

$$T \propto \Phi\, I_a \propto I_{a2}$$

$$T_{a1} \propto I_{a1}^2 \propto 40^2 \quad \text{and} \quad T_{a2} \propto I_{a2}^2$$

Now $\quad T_{a2}/T_{a1} = I_{a2}^2/40^2 \quad \text{or} \quad I_{a2}^2/40^2 = 1/2$

or $\quad I_{a2} = 40/\sqrt{2} = 28.28$ A

(b)
$$\frac{N_2}{N_1} = \frac{E_{b2}}{E_{b1}} \times \frac{I_{a1}}{I_{a2}}$$
$$E_{b1} = 250 - (40 \times 01) = 246 \text{ V}$$
$$E_{b2} = 250 - (28.28 \times 0.1) = 247.2 \text{ V}$$
$$\therefore \quad \frac{N_2}{500} = \frac{247.2}{246} \times \frac{40}{28.28} \qquad \therefore \quad N_2 = \mathbf{711 \text{ r.p.m.}}$$

Q. 32. A 220 V, dc shunt motor runs at 500 rpm when the armature current is 50A. Calculate the speed if the torque is doubled. Given that R_a = 0.2 Ω.

(*Elect Engg. Nagpur Univ. 2003*)

Ans. Since in a shunt motor, flux Φ remains almost constant, hence $T_a \propto I_a$

$$\therefore \quad T_{a1} \propto I_{a1} \propto 50 \text{ A}, T_{a2} \propto I_{a2}$$

Now, it is given that $\frac{T_{a2}}{T_{a1}} = 2$

or $\frac{I_{a2}}{I_{a1}} = 2 \qquad \therefore \quad I_{a2} = 2 \times 50 = 100 \text{ A}$

$$E_{b1} = V - I_{a1} R_a = 220 - 50 \times 0.2 = 210 \text{ V}$$
$$E_{b2} = 220 - 100 \times 0.2 = 200 \text{ V}$$

Now, $\frac{N_2}{N_1} = \frac{E_{b2}}{E_{b1}} \qquad \therefore \quad N_2 = N_1 \left(\frac{E_{b2}}{E_{b1}}\right)$

or $N_2 = 500 \times (200/210) = \mathbf{476 \text{ rpm}}$

Q. 33. A 220 V shunt motor has an armature circuit resistance of 0.2 Ω and field circuit resistance of 110 Ω. At no load, the motor takes 5 A and runs at 1500 rpm. If the motor draws 52 A at rated voltage and rated load, calculate the motor speed and its rated shaft torque in N-m. The rotational losses at no load and full load are the same. Neglect armature reaction. (*B. Tech. Elect. Technology JNTU, Hyderabad 2007*)

Ans.
$$I_{sh} = 220/110 = 2 \text{ A}, I_{a1} = 5 - 2 = 3 \text{ A}$$
$$I_{a2} = 52 - 2 = 50 \text{ A}$$
$$E_{b1} = 220 - 3 \times 0.2 = 219.4 \text{ V}$$
$$E_{b2} = 220 - 50 \times 0.2 = 210 \text{ V}$$

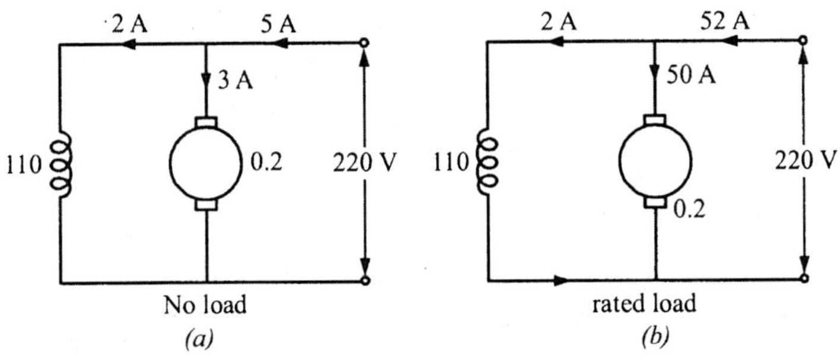

Fig. 14

Now, $\frac{N_2}{N_1} = \frac{E_{b2}}{E_{b1}}$ — because Φ is constant

$$\therefore \quad \frac{N_2}{1500} = \frac{210}{219.4}, \qquad N_2 = \mathbf{1436 \text{ rpm}}$$

For calculating the shaft torque, T_{sh}, we will find the motor output when it draws a line current of 52 A as shown in Fig. 14(*b*). However, we will first use the no-load data for finding the constant losses of the motor.

No-load motor input = 220 × 5 = 1100 W, Arm Cu loss = $3^2 \times 0.2 = 2$ W

Constant losses of the motor = 1100 – 2 = 1098 W

When loaded, arm Cu loss = $50^2 \times 0.2 = 500$ W

∴ Total motor loss = 1098 + 500 = 1598 W

Motor output on load = 220 × 52 = 11,440 W

Output, P_{out} = 11,440 – 1598 = 9,842 W

T_{sh} = 9.55 × output/*N* = 9.55 × 9842/1436

= **65.5N-m**

Q. 34. What are the three important characteristic curves of a dc motor? Give the motor equation which helps to draw these curves.

Ans. The characteristic curves of a motor represent graphically the relation between its two quantities as explained below.

1. Torque and armature current *i.e.* T_a/I_a *characteristic* : It is also known as *electrical* characteristic. It shows how the torque of a motor varies with its armature current.

2. Speed and armature current *i.e.* N/I_a *characteristic* : It shown how speed and armature current of a motor are related to each other.

3. Speed and torque *i.e.* N/T_a *characteristic* : It is also known as *mechanical* characteristic. It can be found form (1) and (2) above.

While drawing these curves, following motor equations will be helpful

$$(i)\ T_a \propto \Phi I_a \qquad (ii)\ I_a = \frac{V - E_b}{R_b} \quad \text{and} \quad (iii)\ N \propto \frac{E_b}{\Phi}$$

Q. 35. Draw the three important characteristics of a d.c. shunt motor and comment on them.

Ans. The three important characteristics of a dc shunt motor are shown in Fig. 15.

1. T_a/I_a characteristic : Motor torque $T_a \propto \Phi I_a$. Since I_{sh} remains almost constant, hence Φ is constant. Therefore, $T_a \propto I_a$. The characteristic is shown in Fig. 15(*a*). It is almost a straight line through the origin. Obviously, the starting torque of the motor is *low* because I_a is low at the start.

Shunt Motor

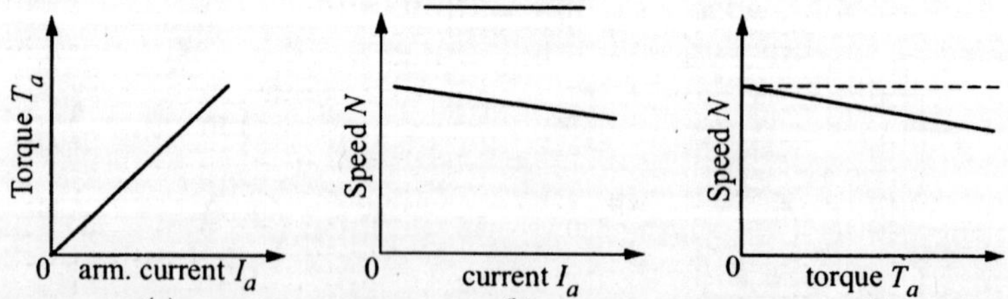

(*a*) (*b*) (*c*)

Fig. 15

2. N/I_a Characteristic : The speed of a motor is given by $N \propto E_b/\Phi$. Since Φ is constant, $N \propto E_b$. However, value of E_b decreases from no-load to full load by about 6-10 percent only. Therefore speed of a shunt motor remains almost constant [Fig. 15(*b*)]. Hence, shunt motor is considered a *constant-speed motor.* As seen, the speed of a shunt motor does not change much from no load to full load. Hence, such motors are well-suited for loads which are totally and suddenly thrown off without any fear of very high speed resulting. Due to this reason, shunt motors are suitable for drawing machine tools, lathes, wood-working machines and centrifugal pumps etc.

3. N/T_a characteristic : In a shunt motor, $T_a \propto I_a$. Hence, N/T_a characteristic shown in 15(*c*) is similar to the N/I_a characteristic.

Q. 36. Sketch and explain the shape of the torque-current, speed-current and speed-torque curves of a series motor. State the factors which influence their shapes.

Ans. The three important characteristics of a dc series motor are shown in Fig. 16.

1. T_a/I_a characteristic : In general, armature motor torque $T_a \propto \Phi\, I_a$. In the case of a series motor, $\Phi \propto I_a$ till field magnets become saturated.

Hence, $$T_a \propto \Phi\, I_a \propto I_a^2.$$

After saturation, Φ becomes constant. Hence, $T_a \propto I_a$. The T_a/I_a curve shown in Fig. 16(*a*) is a parabola *before* saturation and a straight line *after* saturation. That is why a series motor has a *very large* starting torque. Hence, this motor is suitable for traction work where a heavy starting torque is required.

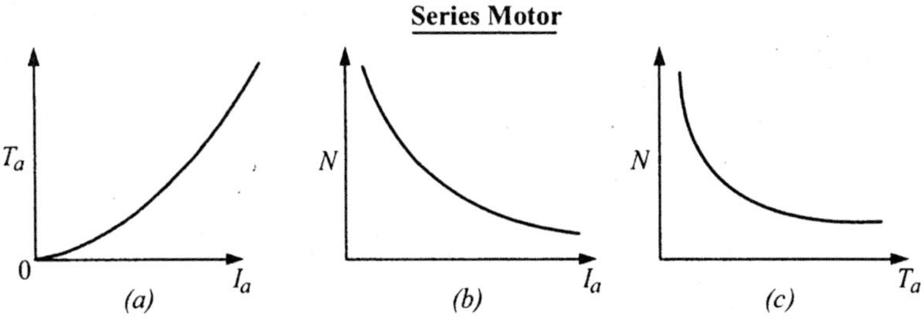

Fig. 16

2. N/I_a characteristic : The speed of a motor is given by $N \propto E_b/\Phi$. Usually changes in E_b are small and may be neglected. Flux Φ increase as I_a increases so that $N \propto 1/\Phi \propto 1/I_a$. Hence, speed of the series motor varies *inversely* as armature current as shown in Fig. 16(*b*).

When load is heavy, I_a is large. Hence, speed is low. But when load becomes very small, I_a also becomes very small so that speed becomes *dangerously high*. Hence, a series motor should never be started without some mechanical (not belt – driven) load on it, otherwise it may destroy itself. It should be noted that a series motor is a *variable speed motor.*

3. N_a/T_a or mechanical characteristic : It can be found from (1) and (2) above and is shown in Fig. 16(*c*). As seen, when speed is high, E_b is large so that I_a becomes small. As a result, torque becomes small. On the other hand, when speed is low, E_b is small which means I_a is large. Hence, torque is large at low speeds.

Q. 37. Briefly describe the characteristics of a cumulative compound motor.

Ans. In such motors, the series field helps the shunt field. It combines the features of series and shunt motors. In compound motors, increase in load decreases not only the speed but also greatly increases the developed torque. Their starting toque is also very large.

Hence, a cumulative compound motor is a fairly constant-speed motor with excellent pulling power for heavy loads and good starting torque. Such motors are used where series characteristics are required and also where load is likely to be *removed totally* such as some types of coal-cutting machines or for driving heavy deep cuts quite often and in rolling mills where there are sudden temporary loads.

Q. 38. Draw the torque–armature current curves of shunt, series and cumulative compound dc motors and compare their performance.

Ans. Fig. 17. shows the T_d/I_a curves for the three motors. For a shunt motor, torque T_a depends only on the armature current I_a. Hence, its graph is a *straight line.* For the cumulative compound motor, the series field increases with load thereby aiding the shunt field. Its graph is somewhat curved as shown. In a series motor, *before* magnetic saturation, T_a depends on I_a^2 which is a *parabola.* But after saturation, the graph becomes a straight line at heavy loads since the flux remains constant.

It should be noted that between no load and full load, the shunt motor develops the *greatest* torque while the series motor develops the *least* torque. The torque developed by the compound motor falls between these two.

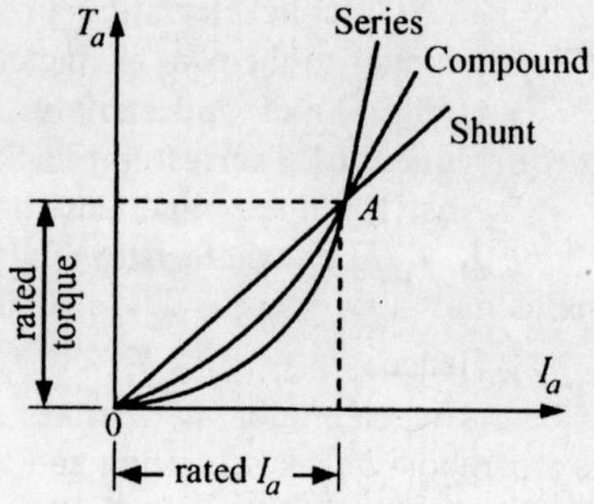

Fig. 17

However, at overloads, the picture becomes different. It is seen that the curves for the series and compound motors become *straight* line after point *A* due to magnetic *saturation.* Now, the overload torque of a series motor is much higher than that of a shunt motor. As usual, the compound motor falls between the other two.

For currents *greater* than the rated current, the series motor develops the highest torque. It means that the starting torque of a series motor is the *highest* which makes it suitable for traction work.

Q. 39. Summarise the applications of different dc motors.

Ans. The different characteristics of dc motors and their applications are summarised below.

Type of Motor	*Characteristics*	*Applications*
Shnut	Approximately constant speed Adjustable speed Medium starting torque	For driving constant-speed line shafting Lathes Wood planers Centrifugal pumps Machine tools Blowers and fans Reciprocating pumps
Series	Variable speed Adjustable varying speed High starting torque	For traction drive *i.e.* Electric locomotives Rapid transit systems Trolly cars etc. Crances and hoists Conveyors Compressors Steel mills
Cumulative compound	Variable speed Adjustable varying speed High starging torque	For intermittent high torque loads For shears and punches Elevators Conveyors Heavy machines tools Rolling mills Large presses

Q. 40. Draw and explain the power flow diagram of a dc motor. Define its various efficiencies.

Ans. The various stages of energy conversion in a dc motor and also the various losses taking place in it are shown in the power flow diagram of Fig. 18. As seen, $T_a = 9.55\ E_b\ I_a/N$ and shaft torque $T_{sh} = 9.55\ P_{out}/N$

Different efficiencies of the motor are given below:

Electrical efficiency, $\eta_e = \dfrac{B}{A}$

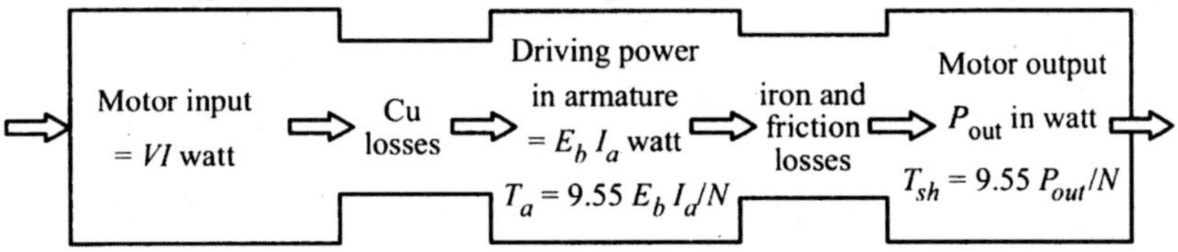

Fig. 18

Mechanical efficiency, $\eta_m = \dfrac{C}{B}$

Overall or commercial efficiency. $\eta_c = \dfrac{C}{A}$

Q. 41. Describe the Swinburne test for finding the no-load losses of a dc shunt motor. Mention its advantages and disadvantages.

Ans. This test is particularly suited for shunt and compound motors whose flux remains almost constant. After knowing the no-load losses, the efficiency of the motor can be found *at any given load.*

For this purpose, the motor is run on no load at its rated voltage. The speed can be adjusted with the help of shunt rheostat as shown in Fig. 19. The ammeter A_1 reads the no-load input current I_o. The ammeter A_2 reads the shunt current I_{sh}. The no-load armature current $I_{ao} = (I_o - I_{sh})$. It is seen from Fig. 19 that

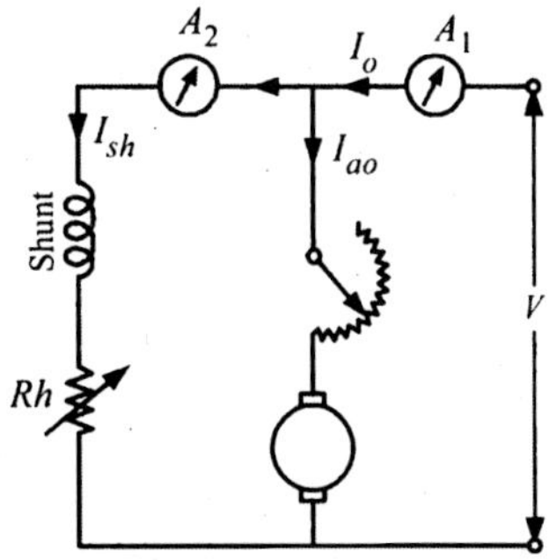

Fig. 19

No-load motor input = VI_o

Power input to armature = $VI_{ao} = V(I_o - I_{sh})$

Power input to the shunt winding = VI_{sh}

No-load power input to armature supplies the following losses:

(*i*) iron or core losses (*ii*) friction loss

(*ii*) windage loss (*iv*) armature Cu loss $I_{ao}^2\ R_a$

Here, R_a stands for the total resistance of the armature *circuit* (and not armature resistance alone).

If we subtract armature circuit Cu loss form the no-load losses, we get constant losses (W_c). Knowing W_c, we can determine the efficiency of the motor at any other load.

Advantages of this test are that it is convenient and economical. Also, motor efficiency can be pre-determined *at any other load.*

However it does not take into account the change in the iron losses from no load to full load. Also, it does not consider rise in temperature and effect of commutation.

Q. 42. A dc series motor takes 40 A at 220 V and runs at 800 rpm. If the armature and field resistances are 0.2 Ω and 0.1 Ω respectively and friction losses are 0.5 kW, find the torque developed in the armature. What will be the output of the motor?

(*Elect Engg. B.E. 1st Yr. Nagpur Univ. 2007*)

Ans. $T_a = 9.55\ \dfrac{E_b I_a}{N}$ N-m — N in rpm

$E_b = V - I_a(R_a + R_{se}) = 220 - 40\ (0.2 + 0.1) = 208$ V

$\therefore$ $T_a = 9.55 \times 208 \times 40/800 =$ **99.3 N-m**

Cu loss in armature *circuit* = $40^2\ (0.2 + 0.1) = 480$ W

Iron and friction losses = 500 W
Total loss = 500 + 480 = 980 W
Motor input = 220 × 40 = 8800 W
Motor output = 8800 – 980 = 7820 W = **7.82 kW**

Q. 43. A dc shunt motor takes 22 A at 220 V and has an armature resistance of 0.5 Ω and shunt field resistance of 110 Ω. If total mechanical and magnetic losses together are 160 W, calculate

***(i)* field Cu loss**	***(ii)* armature Cu loss**
***(iii)* total loss**	***(iv)* motor input**
***(v)* motor output**	***(vi)* motor efficiency**

Ans. The motor circuit connections are shown in Fig. 19A.

As seen $I = 22$ A and $I_{sh} = 220/110 = 2$ A

(*i*) Field Cu loss = $I_{sh}^2 \; R_{sh} = 2^2 \times 110 =$ **440 W**

(*ii*) Arm Cu loss $I_a^2 \; R_a = 20^2 \times 0.5 =$ **200 W**

(*iii*) Total loss = Cu loss + (magnetic and mechanical losses)

= (440 + 200) + 160 = **800 W**

(*iv*) Motor input = VI = 220 × 22 = **4848 W**

(*v*) Motor output = motor input – losses

= 4840 – 800 = **4040 W**

(*vi*) Motor efficiency, η = output/input = 4040/4840

= 0.83 or **83%**

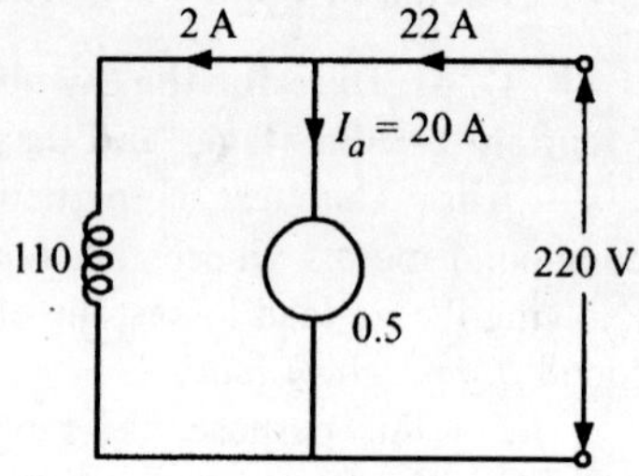

Fig. 19A

Q. 44. A dc series motor is supplied at 110 V and the input current is 20 A. The armature resistance is 0.35 Ω and the field resistance 0.15 Ω. The iron and friction losses amount to 200 W. Allowing a total voltage drop at the brushes of 2 V, calculate the efficiency of the motor.

Ans. Fig. 20 shows the motor circuit

Loss in armature = $I_a^2 R_a = 20^2 \times 0.35 = 140$ W
Loss in series windings = $20^2 \times 0.15 = 60$ W
Loss due to brushes = 2 × 20 = 40 W
Iron and friction losses = 200 W

Total loss = 140 + 60 + 40 + 200 = 440 W
Motor input = 110 × 20 = 2200 W
Motor output = 2200 – 440 = 1760 W

$$\text{efficiency} = \frac{\text{output}}{\text{input}} = \frac{1760}{2200} = 0.8 \text{ or } \mathbf{80\%}$$

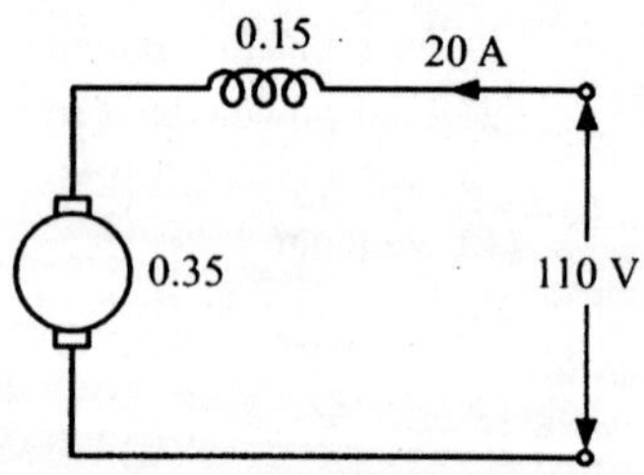

Fig. 20

Q. 45. A 660 V dc series motor runs at 550 rpm and takes 50 A when on full load. The armature resistance is 0.25 Ω, the field resistance is 0.15 Ω and the sum of the iron, friction and windage losses is 2.1 kW. Calculate the full-load output power, shaft torque and efficiency.

Ans. Motor input = 660 × 50 = 33,000 W = 33 kW

Armature circuit Cu loss = $50^2 \times (0.25 + 0.15)$ = 1000 W = 1 kW

Iron, friction and windage losses = 2.1 kW

Total loss = 2.1 + 1.0 = 3.1 kW

(*a*) Full-load motor output = 33 – 3.1 = **29.9 kW**

(*b*) $T_{sh} = 9.55 \times \text{output}/N = 9.55 \times 29.9 \times 10^3 \times 550 =$ **519 N-m**

(*c*) $$\eta = \frac{\text{output}}{\text{input}} \times 100 = \frac{29.9}{33} \times 100 = \mathbf{90.6\%}$$

Q. 46. A 440 V dc shunt motor takes a current of 3 A at no load. The armature resistance including brushes is 0.3 Ω and the field current is 1 A. Calculate the output and efficiency when the input current is 20 A. (*B. Tech. Elect. Machine JNTU, Hyderabad 2008*)

Ans. Let us first find the constant losses of the motor from its no-load test. The constant losses (W_c) of a shunt motor consist of (*i*) stray losses and (*ii*) field Cu loss.

Now, $I_a = 3 - 1 = 2$ A

No-load input = armature Cu loss + constant losses

Or $= I_a^2 R_a + W_c$

$\therefore$ W_c = no-load input $- I_a^2 R_a$

$= 440 \times 3 - 2^2 \times 0.3 = 1320 - 1.2 = 1318.8$ W

When loaded $I_a = 20 - 1 = 19$ A, $I_a^2 R_a = 19^2 \times 0.3 = 108.3$ W

Total loss = 108.3 + 1318.8 = 1427 W

Input = 440 × 20 = 8800 W

Output = 8800 − 1427 = 7373 W

$\therefore$ $\eta = 7373 \times 100/8800 = 0.838$ or **83.8%**

Q. 47. A 500 V dc shunt motor takes a current of 5 A on no-load. The resistances of the armature and field circuit are 0.2 Ω and 300 Ω respectively. Find the efficiency when loaded and taking a current of 125 A and percentage change of speed.

(*B. Tech. Elect. Machines JNTU Hyderabad 2008*)

Ans. This question is similar to Q. 46.

No-load test

Shunt current = 500/300 = 5/3 A, $I_a = 5 - (5/3) = 10/3$ A

Arm. Cu loss $= I_a^2 R_a = (5/3)^2 \times 0.2 \cong 0.5$ W

No-load input = 500 × 5 = 2500 W

Constant losses = no-load input − arm Cu loss

= 2500 − 0.5 = 2499.5 W

When loaded

$I_a = 125 - (5/3) = 123.33$ A

arm Cu loss $= 123.33^2 \times 0.2 = 3042$ W

motor input = 500 × 125 = 62,500 W

Output = 62,500 − 3042 = 59,458 W

$$\therefore \quad \eta = \frac{59,458}{62,500} \times 100 = 0.95 \text{ or } \mathbf{95\%}$$

Motor speed

$$\text{Now,} \quad \frac{N_2}{N_1} = \frac{E_{b2}}{E_{b1}} \times \frac{\Phi_1}{\Phi_2}$$

Since in a shunt motor flux remain the same,

$$\therefore \quad \frac{N_2}{N_1} = \frac{E_{b2}}{E_{b1}} \text{ or } \frac{N_1}{N_2} = \frac{E_{b1}}{E_{b2}}$$

Now, $E_{b1} = 500 - 5 \times 0.2 = 499$ V

$E_{b2} = 500 - 125 \times 0.2 = 475$ V

$$\therefore \quad \frac{N_1}{N_2} = \frac{499}{475}$$

$$\therefore \quad \frac{N_1 - N_2}{N_1} = \frac{499 - 475}{499} = 0.048 \text{ or } \mathbf{4.8\%}$$

Q. 48. A 460 V, dc series motor runs at 500 rpm and takes a current of 40 A. The total resistance of the armature and field circuit is 0.8 Ω. Calculate the speed and percentage change in torque when the load is reduced so that the motor takes 30 A. Assume the flux to be proportional to the field current.

Ans. We will use the relation

$$\frac{N_2}{N_1} = \frac{E_{b2}}{E_{b1}} \times \frac{I_{a1}}{I_{a2}}$$

$$E_{b1} = 460 - 40 \times 0.8 = 456.8 \text{ V}$$

$$E_{b2} = 460 - 30 \times 0.8 = 457.6 \text{ V}$$

$$\therefore \quad \frac{N_2}{500} = \frac{457.6}{456.8} \times \frac{40}{30} \quad \therefore N_2 = \mathbf{668\ rpm}$$

Also, $T_{a1} \propto 40^2 \propto 1600$ and $T_{a2} \propto 30^2 \propto 900$

∴ change in torque = 1600 – 900 = 700 N-m

$$\therefore \text{ Percentage change in torque} = \frac{700}{1600} \times 100 = \mathbf{43.75\%}$$

Q. 49. Mention the factors which control the speed of a dc motor. Explain the flux control method of varying the speed of a shunt motor. Does it increase or decrease the motor speed?

Ans. The speed of a dc motor is given by the relation

$$N = \frac{V - I_a R_a}{\Phi}\left(\frac{60A}{Zp}\right) = k\,\frac{E_b}{\Phi} \qquad \text{where } R_a = \text{armature } \textit{circuit resistance}$$

As seen, the motor speed can be controlled by varying

(*i*) flux per pole of the motor-called flux control method

(*ii*) resistance R_a of the armature *circuit* called rheostatic control method

Flux control Method : Since E_b does not change much, hence $N \propto 1/\Phi$. By decreasing the flux, the speed can be increased and vice versa. The flux can be changed by changing I_{sh} with the help of shunt field rheostat (Fig. 21). Since I_{sh} is small, shunt rheostat I^2R loss is small. That is why this method is very efficient. This method can be used to either increase or decrease the motor speed.

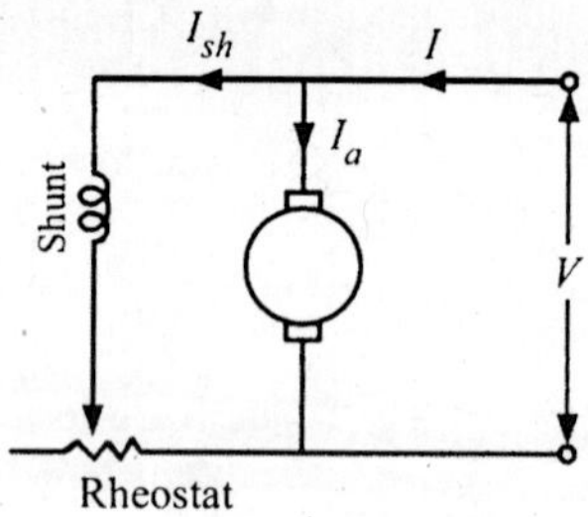

Fig. 21

Q. 50. A 500 V dc shunt motor runs at its normal speed of 300 rpm when the armature current is 250 A. The resistance of armature is 0.16 Ω. Calculate the speed when an additional resistance is inserted in the field circuit reducing the shunt field to 80% of its normal value and the armature current is 100 A.

(*Elect Engg. B.E. 1st Yr. Nagpur Univ. 2005*)

Ans.

$$E_{b1} = V - I_{a1}R_a = 500 - 250 \times 0.12 = 470 \text{ V}$$

$$E_{b2} = V - I_{a2}R_a = 500 - 100 \times 0.12 = 488 \text{ V}$$

$$\Phi_2 = 0.8\,\Phi_1 \text{ or } \Phi_1/\Phi_2 = 1/0.8$$

$$\therefore \quad \frac{N_2}{300} = \frac{488}{470} \times \frac{1}{0.8} \qquad \therefore \quad N_2 = \mathbf{389\ rpm}$$

Q. 51. A 230 V dc shunt motor runs at 800 rpm and takes armature current of 50 A. Find the resistance to be added to the field circuit to increase speed to 1000 rpm at an armature current of 80 A. Assume flux proportional to field current. Armature resistance = 0.15 Ω and field winding resistance = 250 Ω. (*Elect. Engg. B.E. 1st Yr. Nagpur Univ. 2006*)

Ans. The circuit diagram is shown in Fig 22.

$$E_{b1} = V - I_{a1} R_a = 230 - (50 \times 0.15) = 222.5 \text{ V}$$
$$E_{b2} = V - I_{a2}R_a = 230 - (80 \times 0.15) = 218 \text{ V}$$
$$I_{sh1} = V/R_{sh1} = 230/250 = 0.92 \text{ A}$$
$$I_{sh2} = V/R_{sh2} = 230/R_{sh2}$$

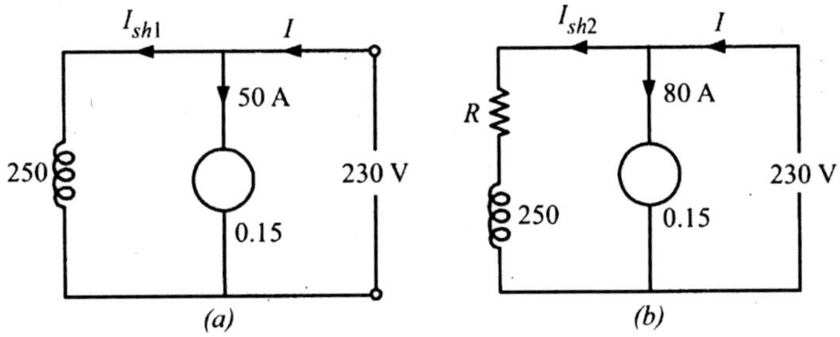

Fig. 22

Also, $\Phi_1 \propto I_{sh1}$ and $\Phi_2 \propto I_{sh2}$

Now, $$\frac{N_2}{N_1} = \frac{E_{b1}}{E_{b2}} \times \frac{\Phi_1}{\Phi_2} = \frac{E_{b1}}{E_{b2}} \times \frac{I_{sh1}}{I_{sh2}}$$

Or $$\frac{1000}{800} = \frac{218}{222.5} \times \frac{0.92}{230/R_{sh2}} \qquad \therefore \quad R_{sh2} = 318.95 \ \Omega$$

Additional resistance R required is

$$= R_{sh2} - R_{sh1} = 318.95 - 250 = \mathbf{68.95 \ \Omega}$$

Q. 52. Briefly describe the armature or rheostatic control method of changing the speed of a dc motor. Is it used for increasing or decreasing the motor speed?

Ans. This method is used when speeds below the no-load speed are required. In this method, voltage across the armature is decreased by adding controller resistance in series with the armature as shown in Fig 23(a). As controller resistance is increased, voltage across the armature is decreased, thereby decreasing the armature speed.

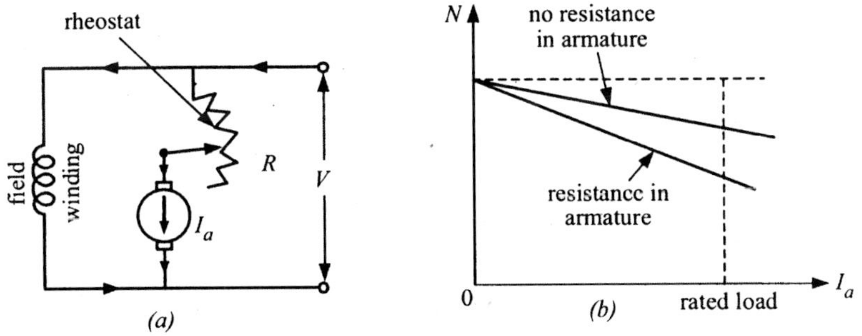

Fig. 23

If load torque remains constant, motor speed is proportional to the voltage across the armature. From the speed versus armature current characteristic of Fig. 23(b), it is seen that greater the controller resistance, greater is the fall in speed.

Let I_{a1} = armature current in the first case

I_{a2} = armature current in the second case

N_1, N_2 = corresponding speeds

V = supply voltage

If load is of *constant torque,* then $I_{a1} = I_{a2}$

Hence $\quad N_1 \propto (V - I_{a1} R_a) \propto E_{b1}$

Let some controller resistance of value R be added to the armature circuit. In that case, armature resistance becomes $(R + R_a) = R_t$

Hence $\quad N_2 \propto (V - I_{a2} R_t) \propto E_{b2} \quad \therefore \quad \frac{N_2}{N_1} = \frac{E_{b2}}{E_{b1}}$

This method is very wasteful and unsuitable for loads which change rapidly.

Q. 53. A dc series motor runs at 500 rpm on 220 V mains drawing a current of 50 A. The resistance of the machine is 0.15Ω. Calculate the value of resistance which when connected in series with the machine will reduce the speed to 350 rpm, the load torque being then half of the previous value. Take flux proportional to current

(*B. Tech. Elect. Engg. JNTU, Hyderabad 2006*)

Ans. $\quad \frac{N_2}{N_1} = \frac{E_{b2}}{E_{b1}} \times \frac{\Phi_1}{\Phi_2} = \frac{E_{b2}}{E_{b1}} \times \frac{I_{a1}}{I_{a2}} \qquad -(\because \Phi \propto I_a)$

Since field is unsaturated,

$$T_a \propto \Phi I_a \propto I_a^2$$

$\therefore \quad T_1 \propto I_{a1}^2$ and $T_2 \propto I_{a2}^2$ or $\frac{1}{2} = \left(\frac{I_{a2}}{I_{a1}}\right)^2$

$\therefore \quad I_{a2} = I_{a1}/\sqrt{2} = 50/\sqrt{2} = 35.36$ A

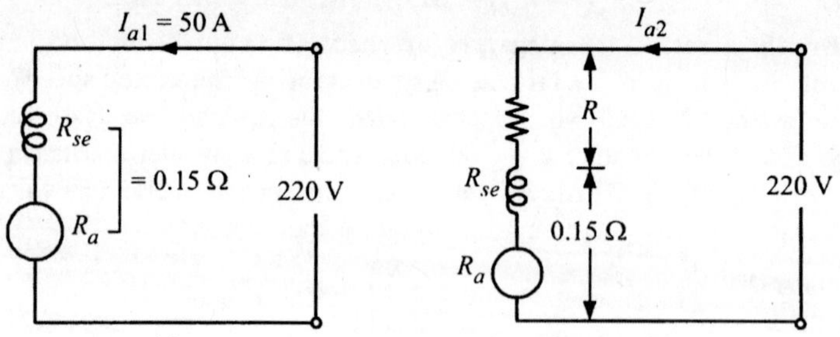

Fig. 24

Let R be the resistance added to the motor as shown in Fig. 24(*b*).

$\therefore$ total resistance, $\quad R_t = (0.15 + R)$

Now $\quad E_{b1} = 220 - 50 \times 0.15 = 212.5$ V

$E_{b2} = 220 - 35.36 \times R_t$ volt

$N_1 = 500$ rpm, $N_2 = 350$ rpm

$$\therefore \quad \frac{350}{500} = \frac{220 - 35.36 \times R_t}{212.5} \times \frac{50}{35.36}$$

$R_t = 3.2\ \Omega$

$\therefore \quad 0.15 + R = 3.2 \quad \therefore \quad R = \mathbf{3.05\ \Omega}$

EXERCISE

1. The following particulars refer to a 4-pole dc motor :
   Number of armature conductors, 520 wave-wound.
   Flux per pole, 0.036 Wb.
   Armature current 12.5 A.
   Find the gross torque developed by armature. **[74.25 N-m]**
2. A 4-pole motor has its armature lap-wound with 840 conductors. The speed is 1,000 rev/min. The supply voltage is 425 V dc and the armature current 25 A. Armature circuit resistance, 0.2 Ω. Calculate
   (*a*) The back e.m.f.
   (*b*) The useful flux per pole.
   (*c*) The gross torque **[420 V, 0.03 Wb, 99.5 N-m]**
3. The armature of a 240 V, 4-pole, d.c shunt motor is wave-wound with 888 conductors and the useful flux per pole is 0.015 Wb.
   Calculate the gross torque when the armature current is 25 A. **[106 N-m]**
4. A dc series motor connected across a 460 V supply runs at 500 rev/min when the current is 40 A. The total resistance of the armature and field circuit is 0.6 Ω. Calculate the torque of the armature in newton-metres. **[333 N-m]**

MULTIPLE CHOICE QUESTIONS

1. In a dc motor, the role of commutator is to
   (*a*) reduce armature reaction (*b*) convert dc to ac
   (*c*) rotate the armature (*d*) strengthen pole field
   (*Basic Elect. Engg. RGPV, Bhopal, 2006*)
2. Although back emf E_b opposes the applied voltage, it
   (*a*) welcomes flow of armature current (*b*) increases motor torque
   (*c*) decreases motor speed (*d*) cannot equal the applied voltage
3. The E_b/V ratio of a dc motor is an indication of its
   (*a*) efficiency (*b*) speed regulation (*c*) starting torque (*d*) running torque
4. Induced back emf in a dc motor acts as .... regulator
   (*a*) currrent (*b*) voltage (*c*) efficiency (*d*) torque
5. As load is applied to a dc shunt motor, its
   (*a*) field current increases (*b*) counter emf increases
   (*c*) armature current increases (*d*) torque developed decreases
6. When a dc motor runs at high speed
   (*a*) its E_b is large (*b*) its armature current is small
   (*c*) it develops less torque (*d*) all of the above
7. The armature current drawn by a dc motor is given by $I_a = V/R_a$ only when the motor
   (*a*) is stationary (*b*) runs at half load (*c*) runs at full load (*d*) none of the above
8. The armature power developed by a dc motor is given by
   (*a*) VI_a (*b*) E_bI_a (*c*) $I_a^2\ R_a$ (*d*) all of the above
9. Condition for maximum torque developed by a dc motor is
   (*a*) $E_b = V$ (*b*) $E_b = V/2$ (*c*) $E_b < V$ (*d*) $E_b > V$
10. As load is applied to a dc series motor, the
    (*a*) series field current decreases (*b*) armature current decreases
    (*c*) counter emf increases (*d*) voltage across armature increases

11. In a cumulative compound motor, an increase in load
(*a*) decreases the speed (*b*) greatly increases the torque
(*c*) increases the shunt field current (*d*) both (*a*) and (*b*)
12. Which of the following motor develops the highest torque between no load and full load?
(*a*) series motor (*b*) shunt motor
(*c*) comulative compound motor (*d*) differential compound motor
13. A drawback of a dc series motor is its
(*a*) low starting torque (*b*) constant speed
(*c*) speed runaway (*d*) high starting torque
14. In a cumulative compound dc motor, an increase in load
(*a*) decreases speed (*b*) increases armature torque
(*c*) both (*a*) and (*b*) (*d*) none of the above
15. The armature torque developed by a dc shunt motor
(*a*) varries directly with armature current
(*b*) directly depends on applied voltage
(*c*) is independent of speed
(*d*) decreases with increasing load.
16. Which of the following dc motor is well-suited for operating line shafts?
(*a*) shunt motor (*b*) series motor (*c*) self-excited motor (*d*) compound motor
17. Because of its high starting torque, a series motor is used for
(*a*) machine tools (*b*) centrifugal pumps (*c*) electric trains (*d*) large presses
18. Cumulative compound dc motors are best suited for
(*a*) traction work (*b*) intermittent high torque loads
(*c*) centrifugal pumps (*d*) cranes and hoists

ANSWERS

1. *b*	2. *d*	3. *b*	4. *a*	5. *c*	6. *d*	7. *a*
8. *b*	9. *b*	10. *d*	11. *d*	12. *b*	13. *c*	14. *c*
15. *a*	16. *a*	17. *c*	18. *b*.			

13 CHAPTER SINGLE-PHASE TRANSFORMERS

Q. 1. What is a transformer? Briefly describe its working principle.

Ans. A transformer, in its simple form, consists of two windings on a laminated iron core (Fig. 1). The winding connected to the source voltage is called the *primary* winding and the other connected to the load is called the *secondary* winding. It transfers electric *power* form one circuit to another at the *same* frequency. When an ac voltage is applied to the primary, current flows through the winding which creates an alternating magnetic flux in the iron core. This flux cuts the secondary winding and induces voltage in it.

A transformer has no moving or rotating parts like some other electric machines.

Working Principle

The working principle of a transformer is based on *mutual induction* between the primary and secondary circuits linked together by a common magnetic flux. As shown in Fig. 1, the primary and secondary windings are electrically *separate* but magnetically linked through a path of *low reluctance* provided by the iron core. When primary winding is energised by an ac source, an alternating flux is set up in the core. This flux is linked up with the secondary winding in which it produces *mutually induced* emf as per Faraday's Laws of Electromagnetic Induction. If secondary winding is closed, a current flows in it and so electric power is transferred *inductively* from the primary to the secondary. Following points about the working of the transformer should be kept in mind:

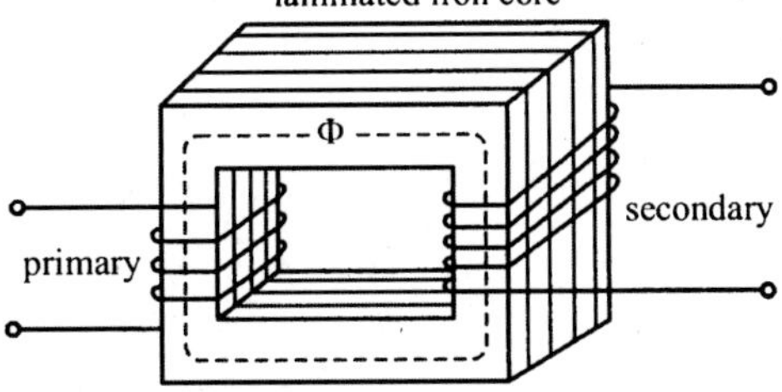

Fig. 1

1. There is no *electrical* connection between its primary and secondary windings.
2. It transfers electric power from one winding to another.
3. It transfers the electrical power by *induction* and not by *conduction.*
4. It does so without change of frequency.
5. It obeys the laws of electromagnetic induction.
6. Its two circuits are linked by a common alternating flux.

Q. 2. What are the two types of constructions generally used in transformers? Compare the two types.

Ans. Construction wise, transformers are of two types depending on how the primary and secondary windings are placed around the iron core. The two types are known as

(*i*) Core type and (*ii*) Shell type

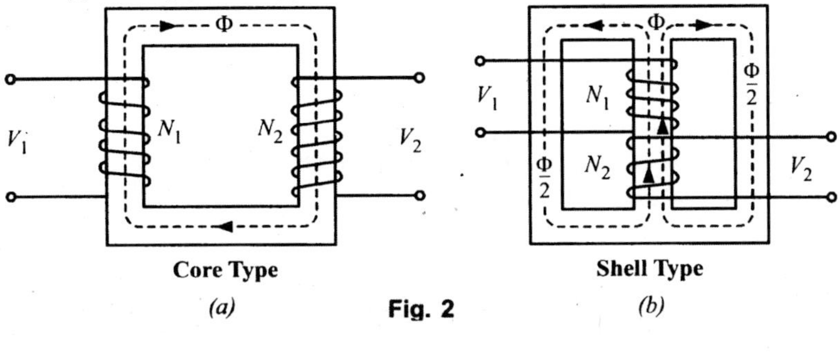

Fig. 2

As shown in Fig. 2(*a*), in core-type transformer, the two *windings surround* the core. In shell-type transformer, the *core surrounds* the two windings as shown in Fig. 2(*b*).

As seen, in core type transformers, the primary and secondary winding are located on the *opposite* limbs of the core. In shell type transformers, the two windings are located on the *same central limb*.

Comparison of core and shell type transformers is given below:

Sl. No.	*Core type*	*Shell type*
1.	The windings surround the core	The core surrounds the two windings.
2.	There is a single magnetic circuit	There is a double magnetic circuit.
3.	The core has two limbs	The core has three limbs.
4.	Simple cylindrical coils are used.	Multilayer disc type coils are used.
5.	It is preferred for low-voltage transformers.	It is preferred for high-voltage transformers.

Q. 3. What are the voltage and current ratios of a transformer? What is the phase relationship between the primary and secondary voltages?

Ans. Fig. 3(*a*) shows the schematic diagram of a transformer. The relationship between V_1 and V_2 depends on the number of turns in the primary (N_1) and the number of turns in the secondary (N_2). It can be expressed as

$$\frac{V_2}{V_1} = \frac{N_2}{N_1} = K$$

where K is known as voltage *transformation ratio.* In an ideal transformer,

(*i*) if $N_2 > N_1$ *i.e.* $K > 1$, then transformer is called *step-up* transformer.

(*ii*) if $N_2 < N_1$ *i.e.* $K < 1$, then transformer is called *step- down* transformer.

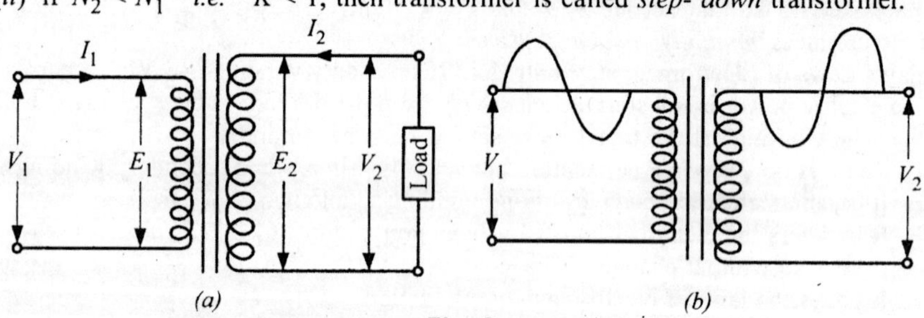

Fig. 3

Also, input VA = output VA or $V_1 I_1 = V_2 I_2$

or $$\frac{I_2}{I_1} = \frac{V_1}{V_2} = \frac{1}{K}$$

Hence, currents are in the *inverse* ratio of voltage transformation ratio (K). It means that if voltage is *stepped up,* the current is *stepped down* so as to keep the two VAs equal.

The value of secondary voltage V_2 depends on the turn ratio K. But it will be *out of phase* with the primary voltage V_1 as shown in Fig. 3(*b*). So will be the secondary current I_2 if load is connected across the secondary winding.

Q. 4. The number of turns on the primary and secondary windings of a single-phase transformer are 350 and 35 respectively. If the primary is connected to a 2.2 kV, 50 Hz supply, determine the secondary voltage. (*B. Tech. Elect. Tech. JNTU, Hyderabad 2007*)

Ans. Transformation ratio $K = N_2/N_1 = 35/350 = 1/10$

$$\therefore \quad V_2 = K \times V_1 = \frac{1}{10} \times 2200 = \mathbf{220\ V}$$

Q. 5. What is an ideal transformer? Draw its phasor diagram.

Ans. A transformer is said to be *ideal* if it has the following properties:

1. The resistance of its primary and secondary windings is zero. It means that it has no Cu losses. In other words, the two coils are *purely inductive coils.*
2. It has no iron loss in the core *i.e.* there are no eddy-current and hysteresis losses.
3. There is no leakage flux *i.e.* all the primary flux is linked with the secondary winding. In other words, the primary flux remains *within the core.*
4. Permeability of the core is so high that only negligible current is required to establish flux in it. It is so because high permeability means low reluctivity of the core.

Consider the *ideal* transformer shown in Fig 4(a). Its secondary is open but its primary is connected to an alternating voltage V_1. Since there is no output, primary draws current only for producing magnetic flux Φ. It is called *magnetizing* current I_m. It is small in magnitude but lags behind V_1 by 90° because primary coil is purely *inductive.* Current I_m is in phase with Φ. The alternating flux produces *self-induced* emf E_1 in the primary which is *equal to* but in *opposition* to V_1.

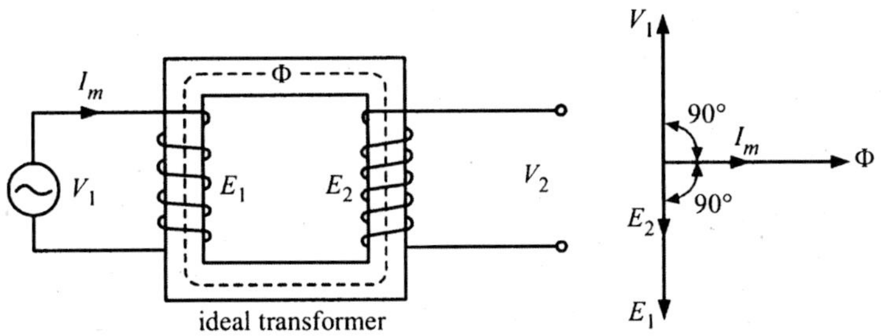

(a) (b)

Fig. 4

Similarly, *mutually-inducted* emf E_2 is produced in the secondary which is also in *opposition* to V_1. Fig. 4(b) shows the phasor diagram of an *ideal* transformer when on *no load.* As seen, I_m and Φ are in phase with each other but I_m lags behind V_1 by 90°. Obviously, there is no *real* power input in the primary but only wattless input $V_1 I_m$.

Q. 6. Derive the emf equation of an ideal single-phase transformer.

Ans. Let N_1 = No. of turns in primary

N_2 = No. of turns in secondary

Φ_m = maximum flux in the core = $B_m \times A$

f = frequency of ac input in hertz (Hz)

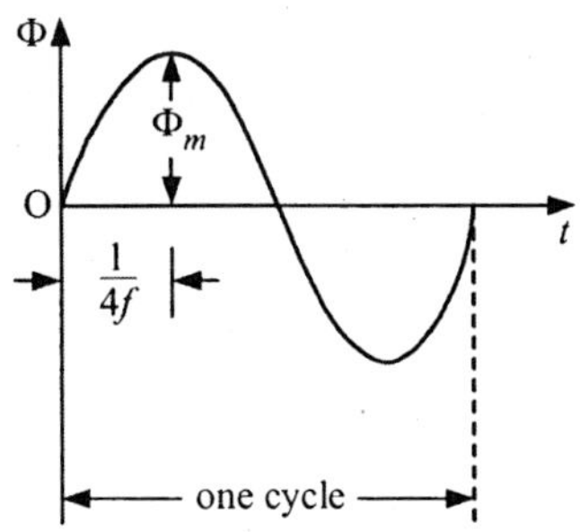

Fig. 5

As shown in Fig. 5, the core flux increases form its zero value to maximum value Φ_m in one quarter of a cycle *i.e.* in $1/4f$ second.

∴ average rate of change of flux

$$= \frac{\Phi_m}{1/4f} = 4f\,\Phi_m \text{ Wb/s}$$

Now, rate of change of flux per turn means induced emf in volts (remember, $e = N\,\Phi/dt$)

∴ average emf induced/ turn = $4f\Phi_m$ volt

Since flux Φ varies sinusoidally, the rms value of induced emf is obtained by multiplying the average value with form factor.

$$\text{Form factor} = \frac{\text{rms. value}}{\text{average value}} = 1.11$$

$\therefore$ rms value of emf/turn = $1.11 \times 4f\,\Phi_m$ volt

$= 4.44f\,\Phi_m$ volt

Now, rms value of induced emf in the whole of primary winding.

= (induced emf/turn) × No. of primary turns

$\therefore\ E_1 = 4.44\,f N_1 \Phi_m$

$= 4.44 f N_1 B_m A$

Similarly, rms value of induced emf in the secondary

$E_2 = 4.44\,f N_2\,\Phi_m$

$= 4.44 f N_2\,B_m$ A

In an ideal transformer on no load

$$V_1 = E_1 \quad \text{and} \quad E_2 = V_2$$

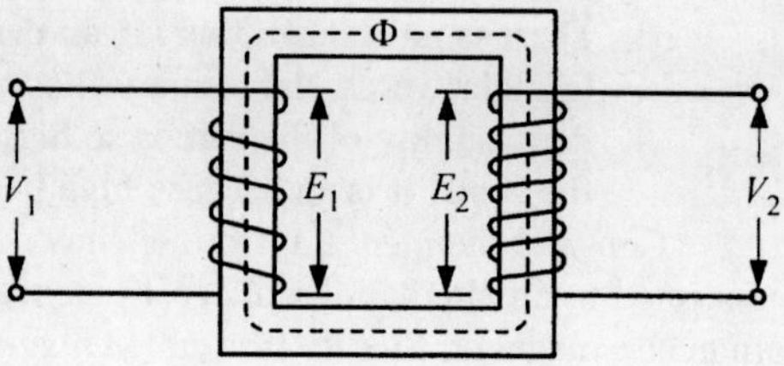

Fig. 6

where V_2 is terminal voltage (Fig. 6) If K is the voltage transformation ratio, then

$$V_2 = KV_1 \quad \text{and} \quad I_2 = I_1/K$$

Note : Though values of V_1 and V_2 are different, the value of induced *emf per turn is the same* both in the primary and secondary windings.

Q. 7. Describe the working of a transformer having losses but no magnetic leakage and draw its phasor diagram when on no load.

Ans. Actual transformers have Cu and iron losses as well as magnetic leakage which produces reactive voltage drops. However, we will first consider a transformer which *has losses* but no magnetic leakage. When such a transformer is on no load [Fig. 7(*a*)] it will draw some primary power for supplying (*i*) core losses and (*ii*) a very small amount of Cu loss in the primary. Since secondary is open, there is no Cu loss in it. The primary draws no-load current of I_0 which lags behind V_1 by an angle $\phi_0 < 90°$. It is due to the fact that primary has some resistance and hence some Cu loss whereas core has iron loss. No-load primary input power.

$$W_0 = V_1\,I_0 \cos \phi_0$$

where $\cos \phi_0$ is primary power factor *under no-load* condition. The phasor diagram is shown in Fig. 7(*b*).

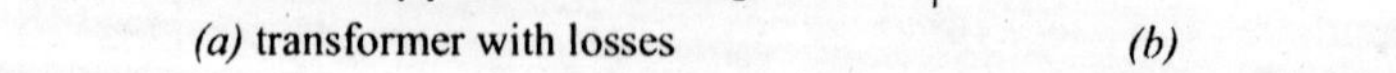

(*a*) transformer with losses (*b*)

Fig. 7

As seen from Fig. 7(*b*), the primary current I_0 has two compoents.

(*i*) active or iron loss component I_w which supplies iron loss plus a small amount of primary Cu loss.

As seen, $I_w = I_0 \cos \phi_0$

(*ii*) The magnetising component I_m whose function is to *maintain* the alternating flux Φ in the core. It lags behind V_1 by 90° and is watteless.

As seen, $I_m = I_0 \sin \phi_0$

Obviously, $I_0 = \sqrt{I_m^2 + I_w^2}$

Q. 8. Explain in detail how primary current changes when load current is set up in the secondary winding of a transformer.

Ans. In Fig. 8(*a*) is shown a transformer when on no load. The primary draws a current of I_0 which is used for (*i*) mostly supplying power to maintain flux in the core and (*ii*) partly for supplying very small amount of iron loss in the core. Suppose, now, load is put on the secondary winding and it draws a current of I_2 as shown in Fig. 8(*b*). Obviously, this current has to come from the primary side. The question is: Will primary current change by the same amount I_2? The answer is that primary current *does not* change by the *same* amount as I_2. However, it *does* change but by a different amount. It is so because a transformer works on the principle of *induction* and not *conduction*. The changes in the primary and secondary voltages and currents are governed by the voltage transformation ratio K.

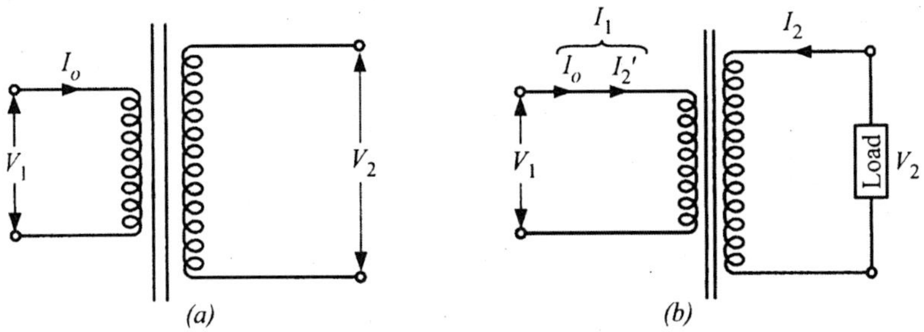

Fig. 8

As we know, $V_2 = KV_1$ and $I_1 = KI_2$. It means that for supplying I_2, primary current will have to change by KI_2. We denote this *additional primary current* needed for supplying I_2 as I_2'. It equals KI_2. Hence, $I_2' = KI_2$. In fact, any quantity with a *dash over it* belongs to the *other* winding. For example I_2' is in the *primary winding* and I_1' is in the *secondary winding*.

When on load, the primary current I_1 consists of two components I_0 and I_2'. Hence, I_1 is the *vector sum* of I_0 and I_2' as shown in Fig. 8(*b*).

Q. 9. Discuss the working of a transformer having losses but no magnetic leakage when on load. Draw its phasor diagram.

Ans. When the secondary of such a transformer is loaded, secondary current I_2 is set up. It produces a flux Φ_2 which *apposes* the primary flux Φ thereby weakening it. As a result, primary draws an additional current I_2' whose flux Φ_2' cancels Φ_2. Hence, the *net flux* in the core *remains the same* Φ as under no load. [Fig. 9(*a*)].

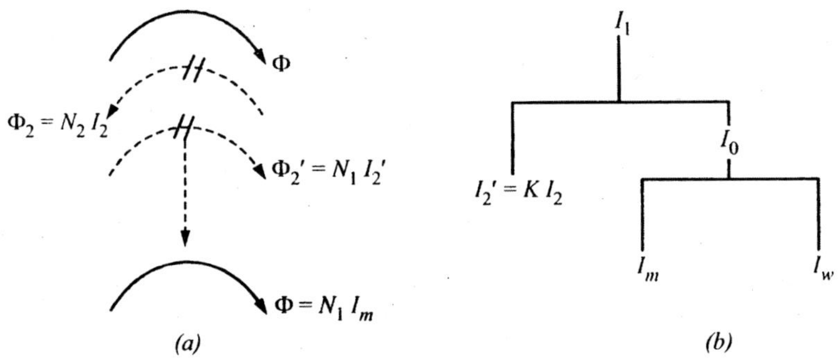

Fig. 9

Since, $\Phi_2 = \Phi_2'$ $\therefore$ $N_2 I_2 = N_1 I_2'$

$$\therefore \quad I_2' = \frac{N_2}{N_1} \times I_2 = KI_2$$

Hence, when such a transformer is *on load,* the primary winding has two currents in it.

(*i*) Current I_0 which it draws when on no load

(*ii*) I_2' which it draws when on load

Current I_2' is *anti-phase* with I_2 and K times in magnitude. The primary current on load is the vector sum of I_0 and I_2' as shown in Fig. 9(*b*).

Q. 10 Derive the emf equation of a 1-phase transformer and calculate the emf/turn if the maximum flux is 0.15 Wb at a frequency of 50 Hz.

(*B.Tech. Elect. Engg, JNTU, Hyderabad 2007*)

Ans. Before sloving this question, it should be recalled that emf/turn is the *same* both in the primary and secondary windings of a transformer. Hence, we may use the emf equation for either E_1 or E_2. Considering the primary winding we have

$$E_1 = 4.44\, f N_1\, \Phi_m$$

$$\therefore \quad E_1/N_1 = 4.44\, f\, \Phi_m = 4.44 \times 50 \times 0.15 = \mathbf{33\ V/turn}$$

Q. 11. A 6,600/440 V, single-phase 600 kVA transformer has 1200 primary turns. Find (*i*) transformation ratio (*ii*) secondary turns (*iii*) voltage per turn (*iv*) secondary current when it supplies a load of 400 kW at 0.8 power factor lagging.

(*Basic Elect. Engg., RGPV, Bhoal, 2007*)

Ans. (*i*) $K = V_2/V_1 = 440/6{,}600 = \mathbf{1/15}$.

(*ii*) $N_2 = KN_1 = 1200 \times (1/15) = \mathbf{80}$

(*iii*) Primary and secondary voltages may be different, but voltage per turn in each winding is the same

$\therefore$ Voltage per turn $= 6{,}600/1200 = \mathbf{5.5\ V}$

(*iv*) Output power $= 600 \times 0.8 = 480$ kW $= 480{,}000$ W

$\therefore$ secondary current $= 480{,}000/440 = \mathbf{1091\ A}$

Q. 12. The no-load ratio of a 50 Hz, single-phase transformer is 6,000/250 V. Estimate the number of turns in each winding if the maximum flux is 0.06 Wb in the core

(*Electircal Engg. AMAe.S.I.*)

Ans. Using the transformer emf equation, we get

$$E_1 = 4.44\, f N_1\, \Phi_m$$

$\therefore$ $6{,}000 = 4.44 \times 50 \times N_1 \times 0.06$ $\therefore$ $N_1 = \mathbf{450}$

Similarly, $250 = 4.44 \times 50 \times N_2 \times 0.06$ $\therefore$ $N_2 = \mathbf{19}$

Alternatively, $K = V_2/V_1 = 250/6000 = 1/24$

$\therefore$ $N_2 = KN_1 = (1/24) \times 450 = 19$

Q. 13. A 2000 kVA, 3300/240 volt, 50 Hz single-phase transformer has 80 turns on the secondary winding. Assuming an ideal transformer, calculate (*i*) primary and secondary currents on full-load (*ii*) the maximum value of flux and (*iii*) the number of primary turns.

(*Basic Elect. Engg. Kerala Univ, 2004*)

Ans. A 3300/240 V transformer is one whose normal primary and secondary voltages are 3300 V and 240 V respectively.

$$K = V_2/V_1 = 240/3300 = 12/165$$

(*iii*) $\dfrac{N_2}{N_1} = \dfrac{80}{N_1} = \dfrac{12}{165}$, $N_1 = \mathbf{1100}$

(*i*) $I_2 = 2{,}000{,}000/\ 240 = \mathbf{8333\ A}$

$$I_1 = KI_2 = (12/165) \times 8333 = \mathbf{606\ A}$$

(*ii*) $$3000 = 4.44 \times 50 \times 1100 \times \Phi_m$$

$\therefore$ $$\Phi_m = \mathbf{13.5\ mWb}$$

Q. 14. The maximum flux density in the core of a 250/3000 volt 50 Hz single phase transformer is 1.2 Wb/m^2. If the emf per turn is 8 V, determine primary and secondary turns and area of the core. (*B. Tech, Basic Elect. Engg. JNTU, Byderabad 2007*)

Ans. $$E_1 = 4.44\, f N_1\, B_m \times A$$

or $$E_1 / N_1 = 4.44\, f B_m\, A$$

$\therefore$ $$8 = 4.44 \times 50 \times 1.2 \times A \quad \text{or} \quad A = \mathbf{0.03\ m^2}$$

EMF per turn is the same both for primary as well as secondary windings.

$\therefore$ $$N_1 = E_1/8 = 250/8 = \mathbf{32}$$

$$N_2 = E_2/8 = 3000/8 = \mathbf{375}$$

Q. 15. (*a*) A 2,200/200 V transformer draws a no-load primary current of 0.6 A and absorbs 400 watts. Find the magnetising and iron- loss currents.

(*b*) A 2,200/250 V transformer takes 0.5 A at a p.f. of 0.3 on open circuit. Find magnetising and working components of no-load primary current.

Ans. (*a*) $$\text{Iron-loss current} = \frac{\text{no-load input in watts}}{\text{primary voltage}}$$

$$= 400/2200 = \mathbf{0.182\ A}$$

Now $$I_0^2 = I_w^2 + I_m^2$$

Magnetising component

$\therefore$ $$I_m = \sqrt{(0.6^2 - 0.182^2)} = \mathbf{0.572\ A}$$

The two components are shown in Fig. 10 (*a*)

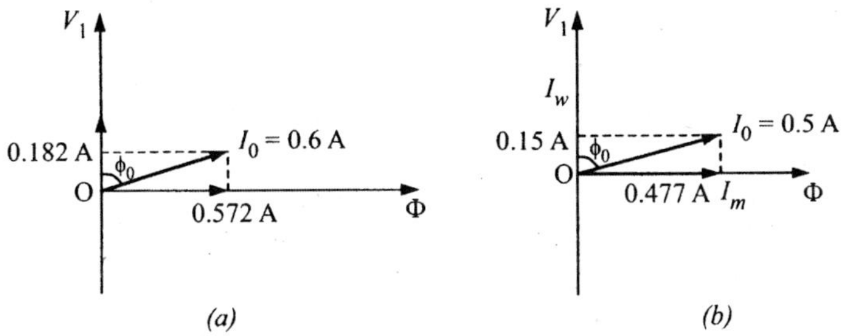

Fig. 10

(*b*) $$I_0 = 0.5\ \text{A}\ ;\ \cos\phi_0 = 0.3$$

$$I_w = I_0 \cos\phi_0 = 0.5 \times 0.3 = \mathbf{0.15\ A}$$

$$I_m = \sqrt{(0.5^2 - 0.15^2)} = \mathbf{0.477\ A}$$

The two components are shown in Fig. 10(*b*).

Q. 16. The no-load current of a transformer is 5A at 0.25 power factor when supplied at 235 V, 50 Hz. The number of turns on the primary winding is 200. Calculate:

(*a*) the maximum value of flux in the core

(*b*) the core loss

(*c*) the magnetising component. (*Elect. Machines-I, Nagpur Univ*)

Ans. (*a*) $$E_1 = 4.44 f N_1\, \Phi_m$$

$$235 = 4.44 \times 50 \times 200\ \Phi_m$$

$\therefore$ $\Phi_m = \mathbf{5.29\ mWb}$

(*b*) Primary no-load input represents the core loss.

$\therefore$ Core loss $= V_1\, I_0 \cos \phi_0$

$= 235 \times 5 \times 0.25 = \mathbf{294\ W}$

(*c*) Magnetising component is (Fig. 11)

$$I_m = I_0 \sin \phi$$

Now $\sin \phi_0 = \sqrt{1 - \cos \phi_0} = \sqrt{15/16}$

$= 0.9682$

$I_m = 5 \times 0.9682 = \mathbf{4.84\ A}$

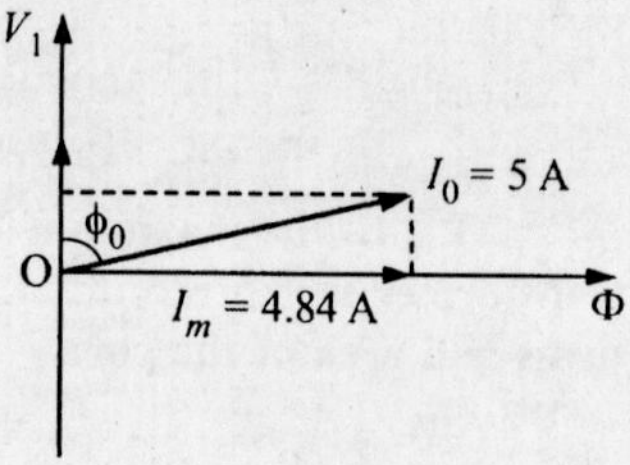

Fig. 11

Q. 17. A single-phase transformer of 3300/220 V, 50 Hz takes a no-load current of 0.8 A and 500 W. Calculate:

(*i*) the active and magnetizing currents

(*ii*) the power factor. (*Basic Elect. Engg. RGPV. Bhopal 2006*)

Ans. (*i*) Iron loss current

$$I_w = \frac{\text{no-load input in watt}}{\text{primary voltage}}$$

$\therefore$ $I_w = \dfrac{500}{3300} = \mathbf{0.151\ A}$

Now $I_0^2 = I_w^2 + I_m^2$

where I_m is magnetizing current,

$\therefore$ $I_m = \sqrt{(0.8)^2 - (0.151)^2} = \mathbf{0.786\ A}$

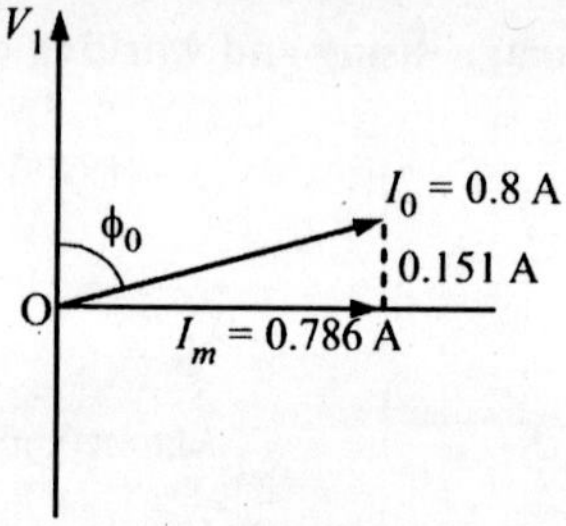

Fig. 12

(*ii*) We know that,

$I_w = I_0 \cos \phi_0$ or $\dfrac{0.151}{0.8} = \cos \phi_0$; $\therefore$ $\cos \phi_0 = \mathbf{0.189}$

Q. 18. A single-phase, 50 Hz transformer has 80 turns on the primary winding and 280 turns on the secondrary winding. The voltage applied across the primary winding is 240 volts. Calculate:

(*i*) the maximum flux density in the core

(*ii*) induced emf in the secondary winding.

The net cross-sectional area of the core can be taken as 200 cm².

(*Basic Elect. Engg. RGPV. Bhopal 2008*)

Ans. (*i*) $E_1 = 4.44\, f\, N_1\, B_m\, A$

or $240 = 4.44 \times 50 \times B_m \times 200 \times 10^{-4} \times 80$

$\therefore$ $B_m = \mathbf{0.676\ Wb/m^2}$

(*ii*) $\dfrac{E_2}{E_1} = \dfrac{N_2}{N_1}$

$\therefore$ $E_2 = E_1 \times \dfrac{N_2}{N_1} = 240 \times \dfrac{280}{80} = \mathbf{840\ V}$

Q. 19. The primary winding of a transformer is connected to a 240V, 50Hz supply. The secondary winding has 1500 turns. If the maximum value of the core flux is 0.00207 Wb, determine (*i*) the secondary induced emf (*ii*) number of turns in the primary (*iii*) cross-sectional area of the core if the flux density has maximum value of 0.465 Tesla.

(*Basic Elect. Engg. VTU karnatka 2009*)

Ans. $E_2 = 4.44\, f N_2\, \Phi_m$

$= 4.44 \times 50 \times 1500 \times 0.00207$

$= \mathbf{689.3\ V}$

(*ii*) $\dfrac{E_1}{E_2} = \dfrac{N_1}{N_2}$ or $\dfrac{240}{689.3} = \dfrac{N_1}{1500}$

$\therefore\ N_1 = \mathbf{522}$

(*iii*) $\Phi_m = B_m \times A$

$\therefore\ A = \Phi_m / B_m$

$\therefore\ A = 0.00207/0.465 = 0.0445\ m^2$

$= 0.0445 \times 10^4\ cm^2 = \mathbf{445\ cm^2}$

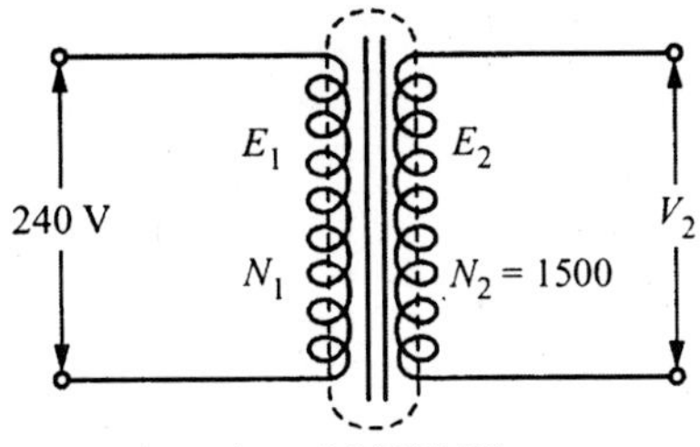

Fig. 13

Q. 20. A transformer has a primary winding of 800 turns and secondary winding of 200 turns. When the load current of the secondary is 80 A at 0.8 power factor lagging, the primary current is 25 A at 0.707 power factor lagging. Determine the no-load current of the transformer and its phase with respect to the primary voltage.

(*Basic Elect. Engg. RGPV. Bhopal 2006*)

Ans. The vector diagram is shown in Fig. 14.

For drawing the vector diagram, following points should be kept in mind.

(*i*) Current I_2' has been drawn in a direction directly opposite to that of I_2. Strictly speaking. it should be written as $(- I_2')$ but, in practice, the negative sign is dropped.

(*ii*) Phase angle of I_2' with V_1 equals ϕ_2.

$\cos \phi_2 = 0.8,\ \phi_2 = \cos^{-1}(0.8) = 36.9°$

$\phi_1 = \cos^{-1}(0.707) = 45°$

$K = \dfrac{N_2}{N_1} = \dfrac{200}{800} = \dfrac{1}{4}$

$I_2' = KI_2 = \dfrac{1}{4} \times 80 = 20\ A$

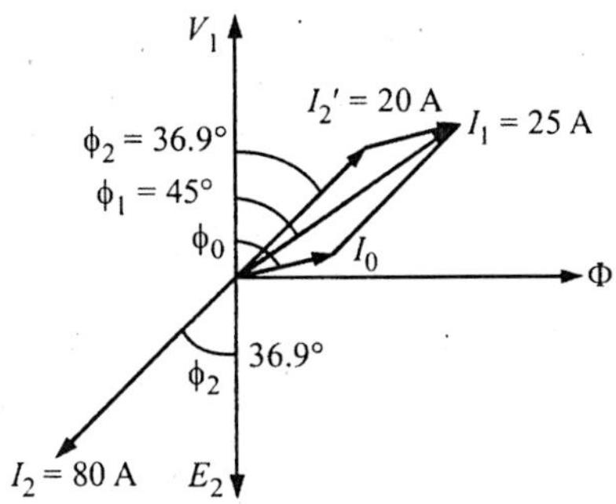

Fig. 14

As seen, I_1 is the vector sum of I_0 and I_2'. Here, I_0 is no-load primary current and I_2' the secondary current referred to primary.

Using the *X*-and *Y*-components of the three currents involved, we get

$I_0 \cos \phi_0 + 20 \cos 36.9° = 25 \cos 45°$

or $I \cos \phi_0 = 25 \times 0.707 - 20 \times 0.8 = 1.675$ A ...(*i*)

Similarly, $I_0 \sin \phi_0 + 20 \sin 36.9° = 25 \sin 45°$

or $I_0 \sin \phi_0 = 25 \times 0.707 - 20 \times 0.6 = 5.675$ A ...(*ii*)

Dividing (*ii*) by (*i*)

we get, $\tan \phi_0 = \sin \phi_0 / \cos \phi_0 = 5.675/\ 1.675 = 3.388$

$\therefore\ \phi_0 = \tan^{-1}(3.338) = \mathbf{73.55°}$,

Also $I_0 = \sqrt{(I_0 \sin\phi_0)^2 + (I_0 \cos\phi_0)^2} = \sqrt{1.675^2 + 5.675^2} = \mathbf{5.91\ A}$

Q. 21. A 20 kVA transformer has 400 turns on the primary and 40 turns on the secondary winding. The primary is connected to 400 volt, 50 Hz supply. Find the full-load primary and secondary currents, secondary emf and the maximum flux in the core. Neglect leakage drop and no-load current. (*Basic Elect. Engg. West Bengal Tech. Univ., 2007*)

Ans. $K = \dfrac{N_2}{N_1} = \dfrac{40}{400} = \dfrac{1}{10}$

$V_2 = KV_1 = (1/10) \times 400 = \mathbf{40\ V}$

Full-load $I_2 = 20{,}000/40 = \mathbf{500\ A}$

Full-load $I_1 = KI_2 = (1/10) \times 500 = \mathbf{50\ A}$

Now, $E_1 = 4.44\, f N_1\, \Phi_m$

or $400 = 4.44 \times 50 \times 400 \times \Phi_m$

$\therefore$ $\Phi_m = 0.0045\ \text{Wb} = \mathbf{4.5\ mWb}$

Q. 22. The primary winding of a 50 Hz single-phase transformer has 480 turns and is fed from 6400 V supply. The secondary winding has 20 turns. Find the peak value of flux in the core and the secondary voltage. (*B. Tech. Basic Elect Engg. JNTU, Hyderabad 2007*)

Ans. $E_1 = 4.44\, f N_1 \Phi_m$

$\therefore$ $6400 = 4.44 \times 50 \times 480 \times \Phi_m$ $\quad \therefore \ \Phi_m = \mathbf{0.06\ Wb}$

Now, $\dfrac{E_2}{E_1} = \dfrac{N_2}{N_1}$ or $\dfrac{E_2}{6400} = \dfrac{20}{480}$

$\therefore$ $E_2 = \dfrac{20}{480} \times 6400 = \mathbf{267\ V}$

Q. 23. A transformer with winding resistances but no magnetic leakage is loaded with resistive, inductive and capacitive loads. Draw its vector diagrams for the three cases.

Ans. In this case, we will take into account the primary resistance R_1 and secondary resistance R_2 while drawing the phasor diagrams. These resistances will produce voltage drops in both windings. As a result, E_1 would be *less* than V_1 but E_2 would be *greater* than V_2.

While drawing phasor diagrams, students should remember the following points:

(*i*) The flux vector is drawn along positive *X*-axis.

(*ii*) Primary voltages are shown along positive *Y*-axis whereas secondary voltages are drawn along negative *Y*-axis. It means vector for V_1 would be drawn vertically up whereas vector for $\mathbf{V_2}$ would be drawn vertically down

(*iii*) Since E_1 and E_2 are in *opposition* to V_1, they are both drawn along negative *Y*-axis.

(*iv*) However, since E_1 belongs to primary side, it would be *reversed* so that it becomes $(-E_1)$ and is combined with $I_1\, R_1$ to give V_1.

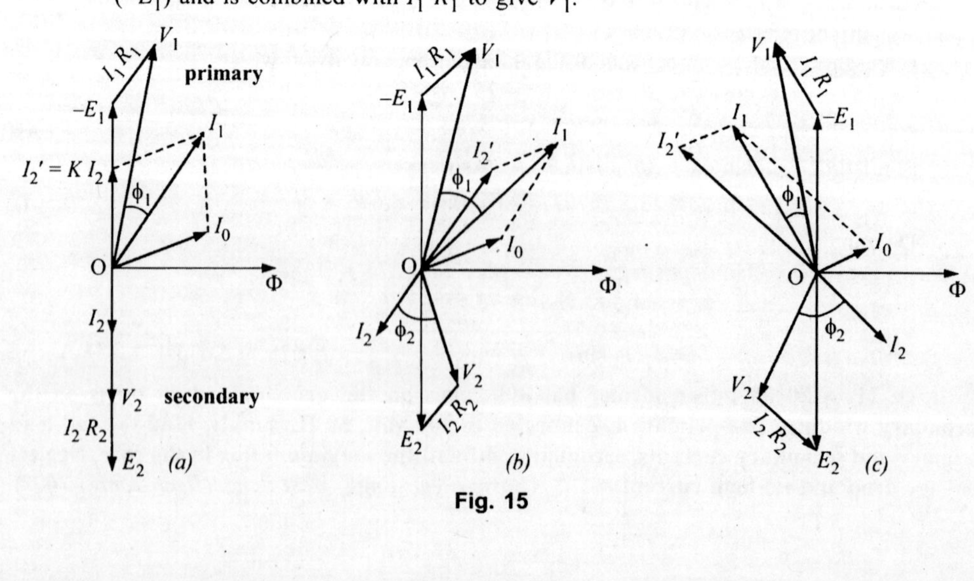

Fig. 15

It would be seen from Fig. 15(*a*) that

$$V_1 = (-E_1) + I_1R_1$$

Also $$V_2 = E_2 - I_2R_2$$

The voltage vector diagram for resistive, inductive and capacitive loads are shown in Fig. 15 (*a*), 15(*b*) and 15(*c*) respectively. As seen, I_2' is antiphase with I_2 and K times in magnitude. Primary current I_1 is the vector sum of I_0 and I_2'. Vector for voltage drop I_1 R_1 has been drawn parallel to I_1. Vector sum of $(-E_1)$ and I_1 R_1 gives V_1.

It is interesting to note from Fig. 15(*a*) that secondary load power factor is unity but primary power factor is less than unity and lagging. In Fig 15(*b*), both primary and secondary load factors are less than *unity* and *lagging*. However, in Fig. 15(*c*), both secondary and primary power factors are less than unity but *leading*.

Q. 24. Derive an expression for the primary resistance of a transformer as referred to the secondary and vice-versa. Explain the usefulness of such referred values.

Ans. Fig. 16. shows a transformer whose primary and secondary windings have resistances of R_1 and R_2 respectively. The resistances have been shown *external* to the windings.

It can be shown that the resistance of the primary winding can be transferred to (or referred to) the secondary winding and vice-versa. It will be proved that R_2 in the secondary can be replaced by R_2/K^2 in the primary winding. This referred resistance is denoted by R_2'. The primary resistance as referred to secondary winding is $R_1' = K^2R_1$.

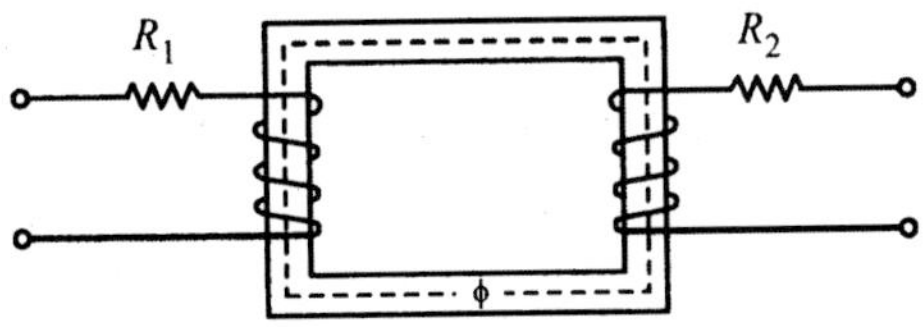

Fig. 16

Proof

The copper loss in secondary is I_2^2 R_2. This loss is supplied by the primary which draws a current of I_1. If R_2' is the *equivalent* resistance in primary which would cause the *same* loss as R_2 in secondary, then

$$I_1^2\ R_2' = I_2^2\ R_2 \quad \text{or} \quad R_2' = (I_2/I_1)^2\ R_2$$

or $$R_2' = R_2/K^2$$

Similarly, equivalent primary resistance as referred to secondary is

$$R_1' = K^2\ R_1$$

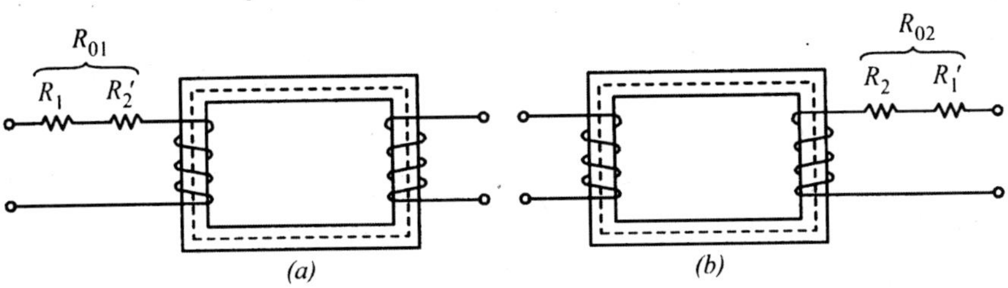

Fig. 17

In Fig. 17(*a*), secondary resistance has been transferred to the primary side leaving the secondary winding without resistance. Hence, the *equivalent* or *effective* resistance of the *transformer* as referred to *primary* is

$$R_1 + R_2' = R_1 + R_2/K^2$$

It is denoted by R_{01}

$$\therefore \quad R_{01} = R_1 + R_2' = R_1 + R_2/K^2 \qquad \text{— Fig. 17(a)}$$

Similarly, the equivalent resistance of the transformer as referred to *secondary* is

$$R_{02} = R_2 + K^2 R_1 \quad \text{— Fig. 17}(b)$$

The following points should be noted:

1. While transferring R_1 to secondary, *muliply* it by K^2.
2. While transfering R_2 to the primary, *divide* it by K^2.

The advantage of shifting the two resistances to one winding is that it makes calculations very simple because one has to work in one winding only.

Q. 25. What is meant by magnetic leakage in a transformer? How is it accounted for? Does it help in transferring power from one winding to the other?

Ans. In an *actual* transformer, all the flux produced by the primary *does not* link with the secondary. Part of it *i.e.* Φ_{L1} is linked with the primary winding only as shown in Fig. 18(*a*) It is called primary *leakage flux* and is produced by primary current I_1. It induces an emf e_{L1} in primary but *none in secondary*.

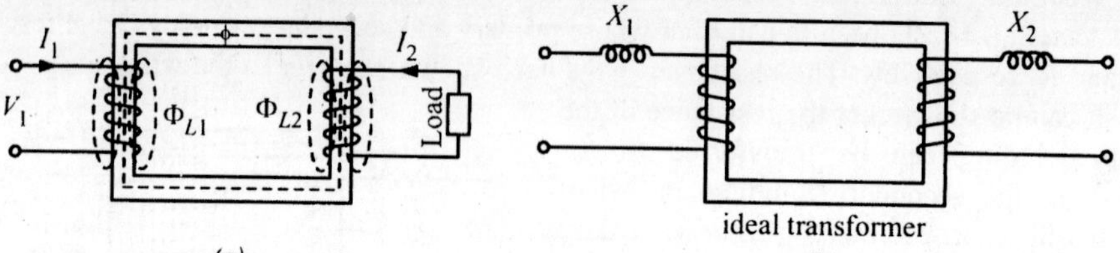

(*a*) (*b*)

Fig. 18

Similarly, leakage flux Φ_{L2} is linked with the *secondary only* and is produced by the secondary current I_2. It induces an emf e_{L2} in the secondary but *none in the primary*.

The *leakage* reactances of the primary and secondary windings are given by

$$X_1 = e_{L1}/I_1 \quad \text{and} \quad X_2 = e_{L2}/I_2$$

These can be represented by two inductive coils each connected in series with the two windings as shown in Fig. 18(*b*). The induced emfs in the two coils are equal to those due leakage fluxes in the actual transformer. Since leakage fluxes have been accounted for by the two coils of reactances X_1 and X_2 connected *externally*, the transformer becomes an *ideal* transformer.

The primary voltage V_1 of an *actual* transformer has to provide the resistive drop $I_1 R_1$ and inductive drop $I_1 X_1$ and the remaining will appear as E_1. Similarly, E_2 will have to supply $I_2 R_2$ and $I_2 X_2$ drops and the remaining will apear as V_2.

Since leakage fluxes are linked only with *one winding*, they do not help in transferring any power form one winding to the other.

Q. 26. Give diagramatic representation of an actual transformer having winding resistances and reactances. Derive an expression for total treansformer impedance as referred to primary and to secondary.

Ans. As shown in Fig. 19, an *actual* transformer has primary resistance of R_1 and reactance of X_1. Similarly, the secondary has resistance of R_2 and reactance of X_2. These are actually *internal* to the two windings. However, these can be taken out of the windings and connected externally as shown in Fig. 19 thus making the transformer *ideal*.

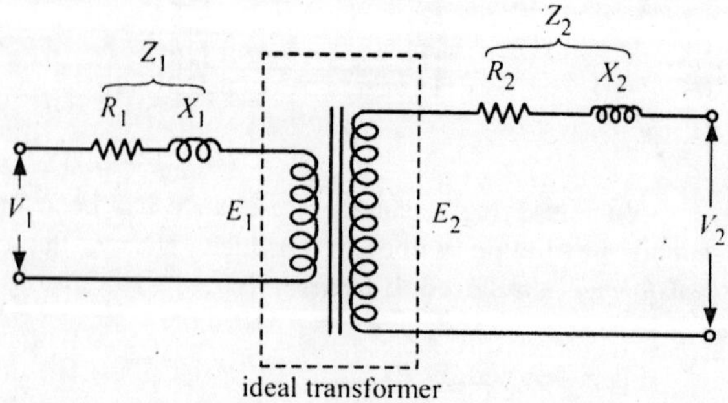

Fig. 19

The primary impedance is $\mathbf{Z_1} = (R_1 + j\,X_1)$. Similarly, secondary impedance is $\mathbf{Z_2} = (R_2 + jX_2)$. Obviously there would be some voltage drop over these impedances. As a result, we have

$$\mathbf{E_1} = \mathbf{V_1} - \mathbf{I_1}\,\mathbf{Z_1} \quad \text{and} \quad \mathbf{V_2} = \mathbf{E_2} - \mathbf{I_2 Z_2}$$

In Fig. 20(*a*), the secondary resistance and reactance have been transferred to the primary. In Fig. 20(*b*), the primary resistance and reactance have been transferred to the secondary.

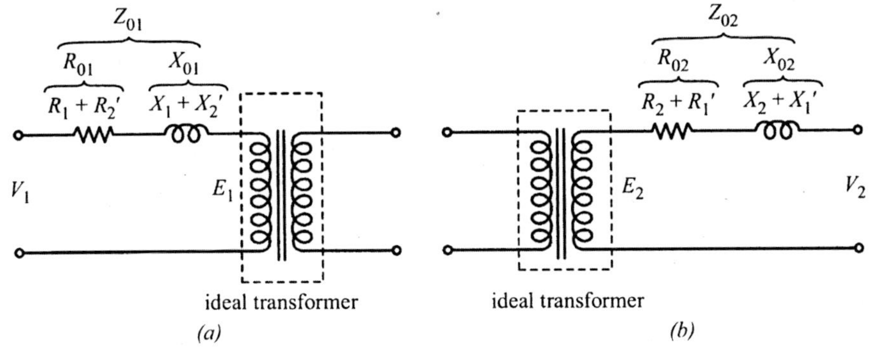

Fig. 20

In Fig. 20(*a*), total transformer resistance as referred to primary is R_{01} and total reactance is X_{01}. Hence, the transformer impedance referred to primary is

$$Z_{01} = \sqrt{R_{01}^2 + X_{01}^2}$$

Similarly, R_{02} is transformer resistance and X_{02} its reactance both referred to secondary. Hence, transformer impedance referred to secondary is given by

$$Z_{02} = \sqrt{R_{02}^2 + X_{02}^2}$$

Q. 27. A 25 kVA, 2200/220 V, 50 Hz single-phase transformer has the following resistances and leakage reactances :

$$R_1 = 0.8\ \Omega \qquad R_2 = 0.009\ \Omega$$
$$X_1 = 3.2\ \Omega \qquad X_2 = 0.03\ \Omega$$

Calculate :

(*i*) equivalent resistance referred to primary side
(*ii*) equivalent reactance referred to primary side
(*iii*) equivalent resistance referred to secondary side.
(*iv*) equivalent reactance referred to secondary side.

(*Basic Elect. Engg. RGPV, Bhopal 2008*)

Ans. Here, $K = V_2/V_1 = 220/2200 = 1/10$

(*i*) $R_{01} = R_1 + R_2 K^2 = 0.8 + 0.009/(1/10)^2 = \mathbf{1.7\ \Omega}$

(*ii*) $X_{01} = X_1 + X_2/K^2 = 3.2 + 0.03/(1/10)^2 = \mathbf{6.2\ \Omega}$

(*iii*) $R_{02} = R_2 + K^2 R_1 = 0.009 + 0.8 \times (1/10)^2 = \mathbf{0.017\ \Omega}$

(*iv*) $X_{02} = X_2 + K^2 X_1 = 0.03 + 3.2 \times (1/10)^2 = \mathbf{0.062\ \Omega}$

Q. 28. A 40 kVA transformer with a voltage ratio of 2000/250 volt has a primary resistance of 1.15 Ω and a secondary resistance of 0.0155 Ω. Calculate:

(*i*) the total resistance referred to the primary
(*ii*) the total resistance drop on full load
(*iii*) the total copper loss on full load. (*B. Tech. Elect. Engg. JNTU, Hyderabad 2007*)

Ans. $K = V_2/V_1 = 250/2000 = 1/8$

Full-load $I_2 = 40{,}000/\ 250 = 160$ A

$$R_1 = 1.15\Omega,\ R_2 = 0.0155\ \Omega$$

(*i*) $R_{01} = R_1 + R_2/K^2 = 1.15 + 0.0155/(1/8)^2 = \mathbf{2.142\ \Omega}$

(*ii*) $R_{02} = R_2 + K^2R_1 = 0.0155 + (1/8)^2 \times 1.15 = \mathbf{0.0335\ \Omega}$

$\therefore$ total resistance drop $= I_2R_{02} = 160 \times 0.0335 = \mathbf{5.36\ V}$

(*iii*) Total Cu loss on full-load $= I_2^2\ R_{02} = 160^2 \times 0.0335 = \mathbf{857.6\ W}$

Q. 29. Give the expression for the approximate total voltage drop in a transformer as referred to its secondary. Does it depend on load power factor?

Ans. When a transformer is loaded, there is a voltage drop of $I_1\ Z_1$ in primary and $I_2\ Z_2$ in secondary. The approximate expression for the *total* voltage drop in a transformer as referred to secondary is given by

$$\text{secondary drop} = I_2\ (R_{02}\ \cos\ \phi \pm X_{02}\ \sin\ \phi)$$

Similarly, approximate voltage drop as referred to primary is

$$\text{primary drop} = I_1\ (R_{01}\ \cos\ \phi \pm X_{01}\ \sin\ \phi)$$

Here, it has been assumed that $\phi_1 = \phi_2 = \phi$

The positive sign is used for a *lagging* power factor and negative sign for a *leading* power factor.

It is clear from the above expressions that voltage drop depends on the power factor.

When power factor is lagging, V_2 will *decrease* from its no-load value. Hence, voltage regulation of the transformer will be *positive.* However, when the power factor is leading, V_2 will *increase* from its no-load value. Hence, voltage regualtion of the transformer would be *negative.*

Q. 30. A 230/460 V transformer has a primary resistance of 0.2 Ω and a reactance of 0.5 Ω and the corresponding values for the secondary are 0.75 Ω and 1.8 Ω respectively. Find the secondary terminal voltage when supplying (*a*) 10A at 0.8 power factor lagging (*b*) 10A at p.f. 0.8 leading. [*Elect. Machines- I Allahabad Univ.*]

Ans. Here, $K = 460/230 = 2$

$$R_{02} = 0.75 + (2^2 \times 0.2) = 1.55\ \Omega$$

Similarly, $X_{02} = 1.8 + (2^2 \times 0.5) = 3.8\ \Omega$

$$I_2 = 10\ \text{A}\ ;\ \cos\ \phi = 0.8,\ \sin\ \phi = 0.6$$

(*a*) voltage drop $= I_2\ (R_{02}\ \cos\ \phi + X_{02}\ \sin\ \phi)$

$$= 10\ (1.55 \times 0.8 + 3.8 \times 0.6) = 35.2\ \text{V}$$

$$V_2 = 460 - 35.2 = \mathbf{424.8\ V}$$

(*b*) Voltage drop $= 10\ (1.55 \times 0.8 - 3.8 \times 0.6) = -\ 10.4\ \text{V}$

$$V_2 = 460 - (-\ 10.4) = \mathbf{470.4\ V}$$

Note : We have assumed $\phi_2 = \phi_1 = \phi$.

Q. 31. Define voltage regulation of a transformer. Does it depend on the load power factor?

Ans. When a transformer is loaded, its secondary terminal voltage V_2 changes provided the applied primary voltage V_1 *is held constant.* For a lagging power factor, V_2 *decreases* but for a leading power factor, V_2 *increases.* This change in secondary voltage from no load to full load is called the *voltage regualtion* of a transformer. It is expressed as a percentage of the secondary no-load voltage.

$$\therefore \quad \text{percentage regulation} = \frac{\text{sec. volt (N.L.)} - \text{sec. volt (F. L)}}{\text{sec. volt (N. L.)}} \times 100$$

Let ${}_0V_2$ = secondary terminal voltage on no load

V_2 = secondary terminal voltage on full load

$$\text{\% age regulation} = \frac{{}_0V_2 - V_2}{{}_0V_2} \times 100$$

Since $\quad {}_0V_2 - V_2 = I_2\,(R_{o2}\cos\phi \pm X_{o2}\sin\phi)$

$$\therefore \quad \text{\% age regn.} = \frac{I_2\,(R_{02}\cos\phi \pm X_{02}\sin\phi)}{{}_0V_2}$$

In the above expression, +ve sign is taken for a *lagging* power factor and –ve sign for a *leading* power factor.

Q. 32. A 75 kVA transformer has 500 turns in the primary and 100 turns in the secondary. The primary and secondary resistances are 0.4 Ω and 0.02 Ω respectively and corresponding leakage reactances are 1.5 Ω and 0.045 Ω respectively. The supply voltage is 6600 volt.

Calculate:

(*i*) equivalent impedance referred to the primary

(*ii*) equivalent impedance referred to the secondary

(*iii*) the voltage regualtion at power of 0.8 lagging.

(*Basic Elect. Engg. West Bengal Tech Univ. 2009*)

Ans. $\quad K = \dfrac{N_2}{N_1} = \dfrac{100}{500} = \dfrac{1}{5}$

$$\frac{V_2}{V_1} = \frac{1}{5} \text{ or } \frac{V_2}{6600} = \frac{1}{5} \qquad \therefore \quad V_2 = 1320 \text{ V}$$

Full-load $\quad I_2 = 75{,}000/1320 = 56.8\text{A}$

(*i*) $\quad R_{o1} = R_1 + R_2/K^2 = 0.4 + 0.02/(1/5)^2 = 0.9\ \Omega$

$$X_{o1} = X_1 + X_2/K^2 = 1.5 + 0.045/(1/5)^2 = 2.625\ \Omega$$

$$Z_{o1} = \sqrt{R_{o1}^2 + X_{o1}^2} = \sqrt{0.9^2 + 2.625^2} = \mathbf{2.775\ \Omega}$$

(*ii*) $\quad R_{o2} = R_2 + K^2R_1 = 0.02 + (1/5)^2 \times 0.4 = 0.036\ \Omega$

$$X_{o2} = X_2 + K^2X_1 = 0.045 + (1/5)^2 \times 1.5 = 0.105\ \Omega$$

$$Z_{o2} = \sqrt{R_{o2}^2 + X_{o2}^2} = \sqrt{0.036^2 + 0.105^2} = \mathbf{0.111\ \Omega}$$

$$(\text{or } Z_{o2} = K^2 Z_{o1} = (1/5)^2 \times 2.775 = 0.111\ \Omega)$$

(*iii*) $\quad \text{Voltage drop} = I_2\,(R_{o2}\cos\phi + X_{o2}\sin\phi)$

$$= 56.8\,(0.036 \times 0.8 + 0.105 \times 0.6) = 5.2 \text{ V}$$

$$\therefore \quad \text{\% age regulation} = \frac{5.2}{1320} \times 100 = \mathbf{0.4\%}$$

Q. 33. What is the purpose of performing open-circuit and short-ciruit tests on a transformer?

Ans. These transformer tests are very economical and convenient because they provide transformer parameters without *actually loading it.*

The open-circuit or no-load test provides the following data:

1. iron loss or core loss under no-load condition. It is the same *at all loads.* That is why it is called constant loss.
2. No-load primary current I_o which is used for finding X_o and R_o.
3. Transformation ratio K of the transformer.

The short-circuit test determines the following:

1. equivalant impedance, Z_{o1} or Z_{o2}, leakage reactance X_{o1} or X_{o2} and R_{o1} or R_{o2}.

2. Cu loss at *full load*. This loss can be found at any other load. It is ∝ I^2
3. Since X_{o1} and X_{o2} become known, voltage drop in the transformer can be found.

The above data can be used to find the efficiency as well as the voltage regualtion of a transformer at any load.

Q. 34. With the help of a neat circuit diagram, explain the open-circuit or no-load test of a transformer.

Ans. In this test, usually *high-voltage* (H.V.) winding is *kept open* and the other *low-voltage* (L.V.) winding is connectd to its normal voltage supply. A wattmeter *W*, voltmeter *V* and ammeter *A* are connected in the low voltage winding *i.e.* primary winding in the present case (Fig. 21). The secondary winding is kept open in order to avoid Cu losses.

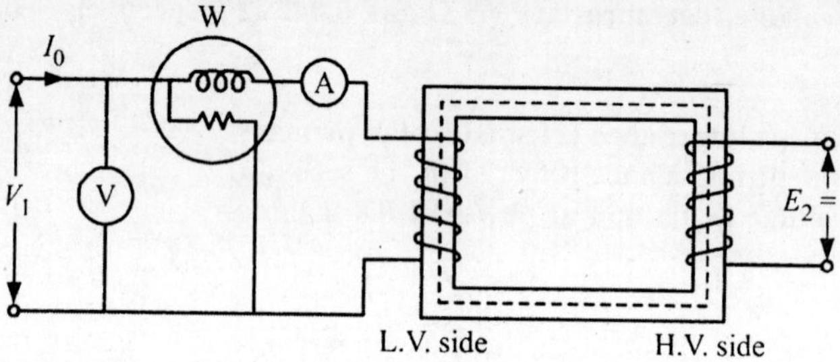

Fig. 21

Since primary is conneted to its normal voltage, normal flux will be set up in the core. Hence, *normal iron losses* will occur which are recorded by the wattmeter. Since primary no-load current I_o is very small, Cu loss is negligible in primary and nil in secondary (being open).The wattmeter reading gives only iron loss at no-load which is the same *for all loads*.

If we connect a *very high* resistance voltmeter across the secondary, it will read the secondary voltage V_2. From this, we can find out the transformation ratio *K*.

Q. 35. Explain the short-circuit test on a single-phase transformer.

Ans. In this test, usually *low-voltage* winding is short-circuited by a thick conductor. A very small voltage is applied across the high-voltage winding and is gradually increased till full-load currents flow in both windings.

Since voltage applied to the H.V. winding is 5 to 10 percent of its normal value, flux produced is *very small*. Hence, iron or core losses are *negligible*. The wattmeter reading gives the *full-load* Cu loss for the transformer.

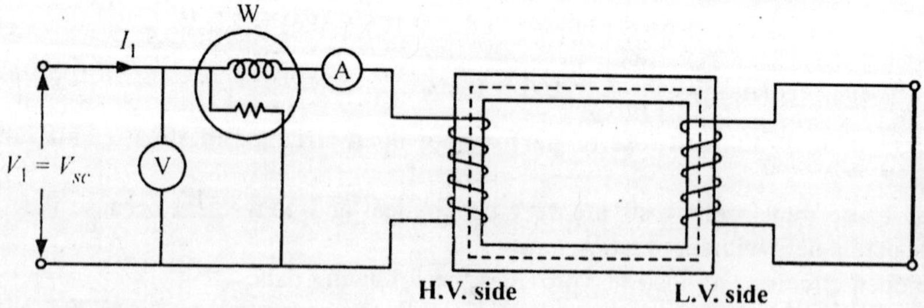

Fig. 22

If V_{sc} is the voltage applied to the H.V. winding for producing full-load currents in both windings, then

$$Z_{o1} = V_{sc}/I_1$$

Also, $$W = I_1^2 R_{o1}$$

$\therefore \quad R_{o1} = W/I_1^2$ and $X_{o1} = \sqrt{(Z_{o1}^2 - R_{o1}^2)}$

Note : If SC test is carried out at a load either less or greater than full load, then Cu loss has to be found for full load.

Q. 36. What are the losses that take place in a transformer? On what factors do they depend? How are they minimized?

Ans. In a static transformer, there are no friction or windage losses. The only two losses occurring in a transformer are as under:

1. Core or iron loss : It includes hysteresis loss and eddy-current loss. Since flux in a transformer core remains *almost constant* for all loads, the core loss is *almost the same at all loads.*

Hysteresis loss $\quad W_h \propto B_{max}^{1.6} f$

Eddy current loss $\quad W_e \propto B_{max}^2 f^2 t^2$

As seen, hysteresis loss depends on the maximum flux density in the core and supply frequency. The eddy current loss depends on maximum flux density in the core, supply frequency and thickness (t) of the laminations used for the core.

Hysteresis loss is minimized by using silicon steel for the core. The eddy-current loss is minimized by using very thin laminations for constructing the core. The iron loss is found from the no-load test on a transformer.

Since it remains the same at all loads, it is called *constant loss.*

2. Cu loss : This loss is due to the resistance of the two transformer windings. Total Cu loss $= I_1^2 R_1 + I_2^2 R_2 = I_1^2 R_{o1} = I_2^2 R_{o2}$. As seen, Cu loss is $\propto I^2$ or (kVA)^2. It means that Cu loss at *half load is one-fourth* of that at full load. This loss can be found from the short-circuit test on a transformer. It is called *variable* loss because it varies with load

Q. 37. Define the efficiency of a transformer. How can it be found? Does it depend on load power factor?

Ans. The efficiency of a transformer at a *particular load and power fator* is defined as the output divided by the input.

$$\therefore \quad \text{efficiency} = \frac{\text{output}}{\text{input}}$$

A transformer has very high efficiency. Its output and input are almost of the same size. Hence their ratio cannot be found accurately. Better method is to measure transformer losses *separately* and then find the efficiency of the transformer.

$$\therefore \quad \text{efficiency, } \eta = \frac{\text{output}}{\text{input}} = \frac{\text{output}}{\text{output + losses}}$$

$$= \frac{\text{input} - \text{losses}}{\text{input}} = 1 - \frac{\text{losses}}{\text{input}}$$

The efficiency depends on load power factor and has maximum value at a power factor of *unity.*

As said above, efficiency can be found from the iron loss given by no-load test and Cu loss from the short-circuit test on the transformer.

Q. 38. Derive the condition for the maximum efficiency of a single-phase transformer.

Ans. $\quad$ Cu loss $= I_1^2 R_{o1}$ or $I_2^2 R_{o2} = W_{cu}$

Iron loss = hysteresis loss + eddy current loss

$= W_h + W_e = W_i$

Considering primary side, we have

$$\text{Primary input} = V_1 I_1 \cos \phi_1$$

$$\therefore \quad \eta = \frac{V_1 I_1 \cos\phi_1 - \text{losses}}{V_1 I_1 \cos\phi_1} = \frac{V_1 I_1 \cos\phi_1 - I_1^2 R_{o1} - W_i}{V_1 I_1 \cos\phi_1}$$

$$= 1 - \frac{I_1 R_{o1}}{V_1 \cos\phi_1} - \frac{W_i}{V_1 I_1 \cos\phi_1}$$

Differentiating both sides with respect to I_1, we get

$$\frac{d\eta}{dI_1} = 0 - \frac{R_{o1}}{V_1 \cos\phi_1} + \frac{W_i}{V_1 I_1 \cos\phi_1}$$

For η to be maximum, $\frac{d\eta}{dI_1} = 0$. Hence, the above equation becomes

$$\frac{R_{o1}}{V_1 \cos\phi_1} = \frac{W_i}{V_1 I_1^2 \cos\phi_1} \quad \text{or} \quad W_i = I_1^2 R_{o1} \text{ (or } I_2^2 R_{o2})$$

or Cu loss = iron loss

or variable loss = constant loss

The output current I_2 corresponding to maximum efficiency can be found from

$$I_2^2 R_{o2} = W_i \quad \text{or} \quad I_2 = \sqrt{\frac{W_i}{R_{o2}}}$$

It is this value of the load current I_2 which will make the Cu loss = the iron loss.

Note : If full-load Cu loss and iron loss of a transformer are known, the load at which it will operate at maximum efficiency can be found easily. If x is the load at which maximum efficiency occurs, then its value is given by

$$x = \text{F.L. kVA} \times \sqrt{\frac{\text{iron loss}}{\text{full-load Cu loss}}}$$

$$= \text{F.L. kVA} \times \sqrt{\frac{W_i}{W_{cu}}}$$

The value of η_{max} can be found by taking total loss = 2 W_i.

Q. 39. Calculate the effective resistance and leakage reactance of a transformer in terms of primary from the following data on test with the secondary terminals short-circuited : Applied voltage, 60 V; current, 100 A; power input, 1.2 kW.

(B. Tech, Elect. Engg. JNTU, Hyderabad 2007)

Ans. The transformer inpedence as referred to primary is given by

$$Z_{o1} = V_{sc}/I_1 = 60/100 = 0.6\ \Omega$$

Also, $\quad W_{cu} = I_1^2 R_{o1}$ or $1200 = 100^2 \times R_{o1}$

$$\therefore \quad R_{o1} = 1200/100^2 = \mathbf{0.12\ \Omega}$$

The transformer leakage reactance X_{o1} is given by

$$X_{o1} = \sqrt{(Z_{o1}^2 - R_{01}^2)} = \sqrt{(0.6^2 - 0.12^2)} = \mathbf{0.588\ \Omega}$$

Q. 40. The following readings were obtained for O.C. and S.C. tests on 8 kVA, 400/120 V, 50 Hz transformer:

O.C. Test : 120 V, 4 A, 75 W — L.V. Side

S.C. Test : 9.5 V, 20 A, 110 W — H.V. Side

Calculate :

(*i*) Voltage regulation and efficiency for 0.8 lagging power factor at full load.

(*ii*) The efficiency at half full- load and 0.8 power factor lagging load.

(*Basic Elect. Engg. RGPV, Bhopal, 2007*)

Ans. (*i*) It should be noted that during the S.C. test, instruments have been placed in the primary *i.e.* high voltage (HV) side as shown in Fig. 23(*b*).

$\therefore \qquad Z_{o1} = 9.5/20 = 0.475\ \Omega$

Now, $\qquad I_1^2 R_{o1} = W_{cu},\ R_{o1} = 110/20^2 = 0.275\ \Omega$

$$\therefore \qquad X_{o1} = \sqrt{Z_{o1}^2 - R_{o1}^2} = \sqrt{0.475^2 - 0.275^2} = 0.15\ \Omega$$

Full-load primary current $I_1 = 8000/400 = 20$ A

$$\text{Voltage drop} = I_1\,(R_{o1}\cos\phi + X_{o1}\sin\phi)$$
$$= 20\,(0.275 \times 0.8 + 0.15 \times 0.6) = 6.2\text{ V}$$

$$\%\text{ regn.} = \frac{6.2}{400} \times 100 = \mathbf{1.55}$$

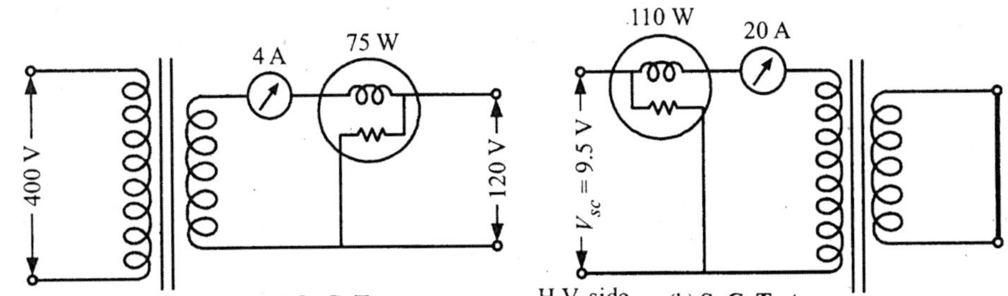

(*a*) **O. C. Test** H.V. side (*b*) **S. C. Test**

Fig. 23

F.L. Cu loss from S.C. test = 110 W

iron loss from O.C. test = 75 W

It may be noted that above Cu loss is the full-load Cu loss corresponding to full load primary current $I_1 = 20$ A

Full-load losses = 110 + 75 = 185 W

Full-load output = 8,000 × 0.8 = 6400 W

Full-load input = 6400 + 185 = 6585 W

$$\text{Full load efficiency} = \frac{\text{output}}{\text{input}} = \frac{6400}{6585} = 0.9719 \quad \text{or} \quad \mathbf{97.19\%}$$

(*ii*) Half-load

Cu loss = $(1/2)^2 \times 110 = 27.5$ W

Iron loss = 75 W, total loss = 27.5 + 75 = 102.5 W

output = 4,000 × 0.8 = 3200 W

input = 3200 + 102.5 = 3302.5 W

$$\therefore \qquad \eta = \frac{3200}{3302.5} \times 100 = \mathbf{97\%}$$

Q. 41. A 5 kVA, 200/400 V, 50 Hz single-phase transformer gave the following results:

O.C. test : 200 V, 0.7 A, 60 W on L.V. side

S.C. test : 22 V, 10 A, 120 W on H.V. side

Determine the efficiency of transformer at full load having a load power factor of 0.8 lagging. (*Elect. Engg. Nagpur Univ.*)

Ans. Iron loss = 60 W — from O.C. test

The S.C. test on the secondary side was carried out with secondary current $I_2 = 10$ A. It is *not the full load* secondary current.

Full load secondary current = 5,000/400 = 12.5 A

We are given that at 10 A, the Cu loss is 120 W.

Let us find the Cu loss at full load with $I_2 = 12.5$ A.

$$\text{Full-load Cu loss} = (12.5/10)^2 \times 120 = 187.5 \text{ W}$$

$$\text{Total loss} = 187.5 + 60 = 247.5 \text{ W}$$

$$\text{Full-load output} = 5,000 \times 0.8 = 4000 \text{ W}$$

$$\text{Full-load input} = 4,000 + 247.5 = 4247.5 \text{ W}$$

$$\therefore \quad \eta = \frac{4,000}{4247.5} \times 100 = \mathbf{94.17\%}$$

Q. 42. A 4 kVA, 200/400 volt 50 Hz 1-phase transformer gives the following test results :

O.C. Test (On primary side) – 200 volt, 0.8A, 50 W

S.C. Test (On secondary side) – 17.5 volt, 9 A, 50 W

Calculate the full-load efficiency at 0.8 power factor lagging.

(*Basic Elect. Engg. RGPV, Bhopal, 2006*)

Ans. (*i*) It may be noted that O.C. test has been conducted on the primary side at its rated voltage of 200V. [Fig. 24(*a*)].

(*ii*) Short-circuit test has been performed on the secondary side with a voltage much lower than rated voltage of 400V. Please also note that this test has *not been performed on full-load* [Fig. 24(*b*)].

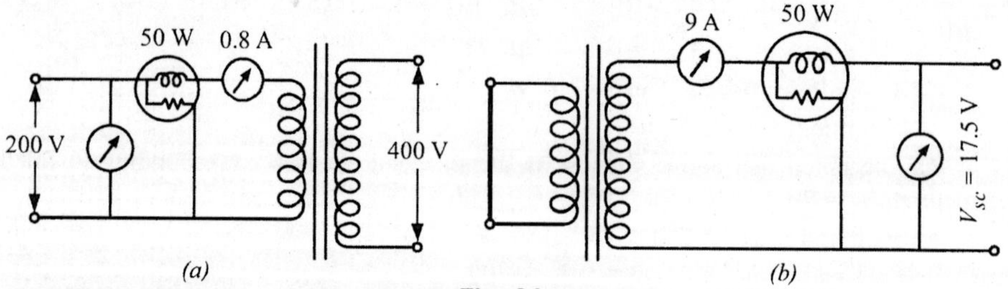

Fig. 24

Full-load secondary current, = 4,000/400 = 10 A

We are given that Cu loss with $I_2 = 9$ A is 50 W.

∴ full-load Cu loss with 10 A = 50 × $(10/9)^2$ = 62 W

Total full-load loss = $W_{cu} + W_i$ = 62 + 50 = 112 W

Full-load output = 4,000 × 0.8 = 3200 W

Full-load input = 3200 + 112 = 3312 W

$$\eta = \frac{3200}{3312} \times 100 = \mathbf{96.6\ \%}$$

Q. 43. A 150 kVA transformer has an iron loss of 1400 watts and full-load copper loss of 1600 watts. Find the efficiency of the transformer at 50% full load (*i*) u p.f (*ii*) 0.8 p.f. lagging. (*B. Tech. Elect Tech. JNTU, Hyderabad 2007*)

Ans. Half full load = 150/2 = 75 kVA

(*i*) Half load unity p.f.

Iron loss = 1400 W — it is the same at all loads

$$\text{Cu loss} = (1/2)^2 \times 1600 = 400 \text{ W} \qquad \text{— it is} \propto \text{kVA}^2$$

$$\text{Total loss} = 1400 + 400 = 1800 \text{ W} = 1.8 \text{ kW}$$

$$\text{Output} = 75 \times 1 = 75 \text{ kW}$$

$$\text{Input} = 75 + 1.8 = 76.8 \text{ kW}$$

$$\eta = \frac{75}{76.8} \times 100 = \mathbf{97.6\%}$$

(*ii*) **Half load 0.8 p.f.**

$$\text{Output} = 75 \times 0.8 = 60 \text{ kW}$$

$$\text{Input} = 60 + 1.8 = 61.8 \text{ kW}$$

$$\eta = \frac{60}{61.8} \times 100 = \mathbf{97.08\%}$$

Q. 44. A 5 kVA, 400/200 V, 50 Hz single-phase transformer on test gives the following results with instruments connected on H.V. side.

O.C. test :	**400 V**	**2 A**	**120 W**
S.C. test :	**20 V**	**20 A**	**80 W**

Find out the percentage regulation at full load 0.8 lagging p.f. and efficiency.

(*Basic Elect Engg. CSV. Tech, Univ, Bhilai 2009*)

Ans. As shown in Fig. 25, the instruments have been connected on the H.V. side *i.e.* primary side for both test.

$$Z_{o1} = \frac{V_{sc}}{I_{sc}} = \frac{20}{20} = 1 \ \Omega$$

$$I_{sc}^2 R_{o1} = 80 \text{ W}, \ R_{o1} = \frac{80}{20^2} = 0.2 \ \Omega$$

$$X_{o1} = \sqrt{Z_{o1}^2 - R_{o1}^2} = \sqrt{1^2 - (0.2)^2} = 0.774 \ \Omega$$

Full load $\quad I_2 = 5000/200 = 25 \text{ A}, \ K = 200/400 = 1/2$

$\therefore \quad I_1 = KI_2 = (1/2) \times 25 = 12.5 \text{ A}$

$$\text{Voltage drop} = I_1 (R_{o1} \cos\phi + X_{o1} \sin\phi) = 12.5 (0.2 \times 0.8 + 0.774 \times 0.6) = 7.8 \text{ V}$$

$$\% \text{ regulation} = \frac{7.8}{200} \times 100 = \mathbf{3.9\%}$$

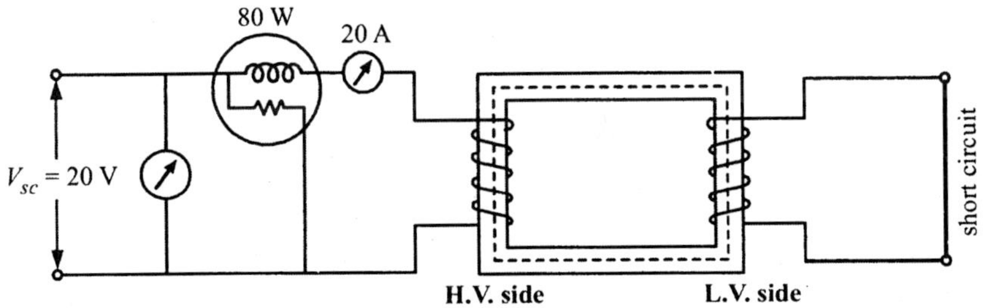

Fig. 25

The S.C. test has been carried out with a primary current of 20 A. It is more than the full-load primary current of 12.5 A.

Hence, full-load Cu loss is $= 80 \times (12.5/20)^2 = 31$ W

Iron loss from O.C. test = 120 W

$\therefore$ Total loss at full load = 120 + 31 = 151 W = 0.151 kW

Full load output at 0.8 p.f. = 5 × 0.8 = 4 kW

Full load input = 4 + 0.151 = 4.151 kW

$$\eta = \frac{4}{4.151} = 0.964 \quad \text{or} \quad \mathbf{96.4\%}$$

Q. 45. In a 50 kVA, 1100/220 V transformer, the iron and copper losses at full load are 350 W and 425 W respectively. Calculate the efficiency at:

(*i*) full load with unity power factor.

(*ii*) half load with unity power factor. (*Basic Elect. Engg. RGPV, Bhopal 2008*)

Ans. (*i*) Full load unity p.f.

$$\text{Total loss} = 350 + 425 = 775 \text{ W}$$

$$\text{F.L. output at u.p.f} = 50 \times 1 = 50 \text{ kW}$$

$$\text{F.L. input at u.p.f} = 50 + 0.775 = 50.775 \text{ W}$$

$$\therefore \quad \eta = \frac{50}{50.775} \times 100 = \mathbf{98.47\%}$$

(*ii*) Half load u.p.f

$$\text{Cu loss at half load} = (1/2)^2 \times 425 = 106 \text{ W}$$

$$\text{Total loss} = 350 + 106 = 456 \text{ W}$$

$$\text{half-load output at u.p.f.} = 25 \times 1 = 25 \text{ kW}$$

$$\text{Half-load input at u.p.f} = 25 + 0.456 = 25.456 \text{ kW}$$

$$\therefore \quad \eta = \frac{25}{25.456} \times 100 = \mathbf{98.2\%}$$

Q. 46. A 100 kVA, 50 Hz, 440/11000 V, single-phase transformer has an efficiency of 98.5% when supplying full-load current at 0.8 p.f lagging and an efficiency of 99% when supplying half full-load current at unity power factor. Find the core losses and Cu losses corresponding to full-load current. (*Basic Elect. Engg. CSV. Tech Univ,, Bhilai 2007*)

Ans. Full load, 0.8 p.f.

$$\text{Output} = 100 \times 0.8 = 80 \text{ kW}$$

$$\text{Input} = 80/0.985 = 81.2 \text{ kW}$$

$$\text{Total loss} = 81.2 - 80 = 1.2 \text{ kW}$$

Let x = iron loss — it remains constant at all loads

y = F.L Cu loss — it is $\propto$ (kVA)^2

$\therefore \quad x + y = 1.2$ kW ...(*i*)

Half load, u.p.f.

$$\text{Output} = 50 \times 1 = 50 \text{ kW}$$

$$\text{Input} = 50/0.99 = 50.5 \text{ kW}$$

$$\text{Total losses} = 50.5 - 50 = 0.5 \text{ kW}$$

The iron loss would remain the same but Cu loss would become $\left(\frac{1}{2}\right)^2 y = y/4$

$\therefore \quad x + y/4 = 0.5$ kW ...(*ii*)

Subtracting (*ii*) from (*i*), we get

$$\frac{3y}{4} = 0.7 \quad \text{or} \quad y = \mathbf{0.93 \text{ kW}}$$

$$\therefore \quad x = 1.2 - y = 1.2 - 0.93 = \mathbf{0.27 \text{ kW}}$$

Q. 47. A 600 kVA, single–phase transformer has an efficiency of 92% both at full load and half full load u.p.f. Determine its efficiency at 75% full load, 0.9 p.f.

(*Basic Elect. Engg. VTU, Karnatka 2009*)

Ans. The fact that efficiency is the same *i.e.* 92% at full load and half load will help us to find the iron and Cu losses.

At full-load :

$$\text{Output} = 600 \times 1 = 600 \text{ kW}$$

$$\text{input} = 600/0.92 = 652.2 \text{ kW}$$

$$\text{Total losses} = 652.2 - 600 = 52.2 \text{ kW}$$

Let x = iron loss — it remains constant at all loads

y = F. L. Cu loss — it is $\propto$ (kVA)^2

$\therefore$ $x + y = 52.2$ ...(*i*)

At half-load :

Output $300 \times 1 = 300$ kW ; input = 300/0.92 = 326.1 kW

$\therefore$ losses = 326.1 − 300 = 26.1 kW

Since Cu loss becomes one-fourth of its F.L. value, hence

$$x + y/4 = 26.1$$

Solving for x and y, we get

$$x = 17.4 \text{ kW} \quad \text{and} \quad y = 34.8 \text{ kW}$$

At 75% full-load 0.9 p.f :

$$\text{Cu loss} = (3/4)^2 \times 34.8 = 19.57 \text{ kW}$$

$$\text{Total loss} = 17.4 + 19.57 = 36.97 \text{ kW}$$

$$\text{Output} = (600 \times 0.75) \times 1 = 450 \text{ kW}$$

$$\text{input} = 450 + 36.97 = 486.97 \text{ kW}$$

$$\eta = \frac{450}{486.97} \times 100 = \mathbf{92.4\%}$$

Q. 48. A 25 kVA transformer has an efficiency of 94% at full load unity p.f. and at half full-load 0.9 p.f. Determine the iron loss and full load Cu loss.

(*Basic. Elect Engg. VTU, Karnatka, 2009*)

Ans. Full-load :

Output at full-load u.pf. = 25×1 = 25 kW

Input at full-load u.p.f. = 25/0.94 = 26.596 kW

Total full-load losses = 26.596 − 25 = 1.596 kW

Let x = iron loss — it remain constant at all loads.

y = Cu loss — it is $\propto$ (kVA)^2

$\therefore$ $x + y = 1.596$ kW ...(*i*)

Half-load :

Output at 0.9 p.f. = $12.5 \times 0.9 = 11.25$ kW

Input at 0.9 p.f. = 11.25/0.94 = 11.968 kW

Total loss = 11.968 − 11.25 = 0.718 kW

Since Cu loss becomes $(1/2)^2$ *i.e.* one-fourth of its F.L. value,

hence $x + y/4 = 0.718$ ...(*ii*)

Solving for x any y, we get,

$$x = \mathbf{0.426 \text{ kW}}, y = \mathbf{1.17 \text{ kW}}$$

Q. 49. A single-phase 25 kVA, 1000/2000 V, 50 Hz transformer has maximum efficiency of 98% at full load u.p.f. Determine its efficiency at:

(*i*) 3/4 full load u.p.f.

(*ii*) 1/2 full load 0.8 p.f.

(*iii*) 1.25 full load 0.9 p.f. (*Basic Elect Engg. VTU, Karnataka 2008*)

Ans. Output at full load u.p.f. = $25{,}000 \times 1 = 25{,}000$ W

Input = 25,000/0.98 = 25,510 W

Total losses = 25,510 − 25,000 = 510 W

Now, at maximum efficiency, Cu loss equals iron loss

$\therefore$ iron loss = 510/2 = 255 W

Full load Cu loss = 255 W

(*i*) 3/4 full load u.p.f.

$$W_{cu} = \left(\frac{3}{4}\right)^2 \times 255 = 143.4 \text{ W} ; W_i = 255 \text{ W}$$

$$\text{Total loss} = 255 + 143.4 = 398.4 \text{ W}$$
$$\text{Output at u.p.f.} = (3/4) \times 25{,}000 \times 1 = 18{,}750 \text{ W}$$
$$\text{Input} = 18{,}750 + 398.4 = 19{,}148.4 \text{ W}$$
$$\therefore \quad \eta = \frac{18{,}750}{19{,}148.4} \times 100 = \mathbf{97.2\%}$$

(*ii*) 1/2 full load 0.8 p.f.

$$W_{cu} = (1/2)^2 \times 255 = 63.7 \text{ W}, W_i = 255 \text{ W}$$
$$\text{Total loss} = 255 + 63.7 = 318.7 \text{ W}$$
$$\text{Output at u.p.f.} = (25{,}000/2) \times 0.8 = 10{,}000 \text{ W}$$
$$\text{Input} = 10{,}000 + 318.7 = 10{,}318.7 \text{ W}$$
$$\eta = \frac{10{,}000}{10{,}318.7} \times 100 = \mathbf{96.9\%}$$

(*iii*) 1.25 full load 0.9 p.f.

$$W_{cu} = 1.25^2 \times 255 = 398 \text{ W}, W_i = 255 \text{ W}$$
$$\text{Total loss} = 255 + 398.4 = 653 \text{ W}$$
$$\text{Output} = (1.25 \times 25{,}000) \times 0.9 = 28{,}125 \text{ W}$$
$$\text{Input} = 28{,}125 + 653 = 28{,}778 \text{ W}$$
$$\eta = \frac{28{,}125}{28{,}778} \times 100 = \mathbf{97.7\%}$$

Q. 50. A 1 kVA single-phase transformer has core loss of 15 W and full load Cu loss of 20 W. Calculate the efficiency at

(*i*) full load, 0.9 p.f lag

(*ii*) half full-load unity p.f.

(*iii*) 3/4 full-load, 0.707 power factor, lag/lead.

(Basic Elect. Engg VTU, Karnataka 2007)

Ans. (*i*) Full-load 0.9 p.f. lag

$$\text{Output} = 1000 \times 0.9 = 900 \text{ W}$$
$$\text{Total losses} = 15 + 20 = 35 \text{ W}$$
$$\text{Input} = 900 + 35 = 935 \text{ W}$$
$$\eta = \frac{900}{935} \times 100 = \mathbf{96.25\%}$$

(*ii*) Half-load unity p.f.

$$\text{Output} = (1000/2) \times 1 = 500 \text{ W}$$
$$\text{Cu loss at half load} = (1/2)^2 \times 20 = 5 \text{ W}$$
$$\text{Total loss} = 15 + 5 = 20 \text{ W}$$
$$\text{Input} = 500 + 20 = 520 \text{ W}$$
$$\eta = \frac{500}{520} \times 100 = \mathbf{96.16\%}$$

(*iii*) 3/4 full-load 0.707 p.f.

$$\text{Output} = (3/4) \times 1000 \times 0.707 = 530.25 \text{ W}$$
$$\text{Cu loss} = (3/4)^2 \times 20 = 11.25 \text{ W}$$
$$\text{Total loss} = 15 + 11.25 = 26.25 \text{ W}$$
$$\text{Input} = 530.25 + 26.25 = 556.5 \text{ W}$$
$$\eta = \frac{530.25}{556.5} \times 100 = \mathbf{95.28\%}$$

Q. 51. In a 25 kVA, 3300/230 V, single-phase transformer, the iron and full-load copper losses are respectively 350 and 400 W Calculate (*i*) the efficiency at half-load 0.8 p.f. and (*ii*) the load at which the efficiecy is maximum.

Ans. (*i*) Half-laod 0.8 p.f.

$$\text{Cu loss at half load} = ((1/2)^2 \times \text{F.L. loss}$$
$$= 400/4 = 100\text{W} = 0.1\text{kW}$$
$$\text{Output} = (25/2) \times 0.8 = 10\text{kW}$$
$$\text{Input} = 10 + 0.1 = 10.1 \text{ kW}$$
$$\therefore \quad \eta = \frac{10 \times 100}{10.1} = \mathbf{99.02\%}$$

(*ii*) Load for maximum efficiency

$$= \text{Full load kVA} \times \sqrt{\frac{\text{iron loss}}{\text{F.L. Cu loss}}}$$
$$= 25 \times \sqrt{350/400} = \mathbf{23.4 \text{ kVA}}$$

Q. 52. A 150 kVA single-phase transformer working at unity power factor has an efficiency of 96 percents both at 70 percent of full load and at full load. Find the load at which maximum efficency occurs and also the value of maximum efficiency.

(*Elect. Enng. RGPV, Bhopal 2003*)

Ans. The fact that efficiency is the same i.e. 96% at both full load and 70% of full load will help us to find the iron and Cu losses.

Full load upf

$$\text{Output} = 150 \times 1 = 150 \text{ kW}$$
$$\text{Input} = 150/0.96 = 156.25 \text{ kW}$$
$$\text{Total loss} = 156.25 - 150 = 6.25 \text{ kW}$$

Let, x = iron loss — it remains constant at all loads

y = F.L. Cu loss — it is $\propto (\text{kVA})^2 \propto I^2$

$\therefore \quad x + y = 6.25$ kW ...(*i*)

70% full-load upf

$$\text{Output} = (150 \times 0.7) \times 1 = 105 \text{ kW}$$
$$\text{Input} = 105/0.96 = 109.37 \text{ kW}$$
$$\text{Total loss} = 109.37 - 105 = 4.37 \text{ kW}$$

Cu loss at 70% full load = $(0.7)^2 \times y = 0.49\, y$

$\therefore \quad x + 0.49\, y = 4.37$ kW ...(*ii*)

Subtracting (*ii*) form (*i*), we get

$$0.51\, y = 1.88 \text{ or } y = 3.68 \text{ kW}$$
$$\therefore \quad x = 6.25 - 3.68 = 2.57 \text{ kW}$$
$$\therefore \quad \text{iron loss} = 2.57 \text{ kW, Cu loss} = 3.68 \text{ kW}$$

kVA load corresponding to maximum efficiency is

$$= \text{full load kVA} \times \sqrt{\frac{\text{iron loss}}{\text{full-load Cu loss}}}$$
$$= 150 \times \sqrt{\frac{2.57}{3.68}} = \mathbf{125.35 \text{ kVA}}$$

At this load of 125.35 kVA, Cu loss equals iron loss. Hence, total loss is = $2 \times W_i = 2 \times 2.57 = 5.14$ kW. Because the transformer is working at u.p.f

$$\therefore \quad \text{Output} = 125.35 \times 1 = 125.35 \text{ kW}$$
$$\text{Input} = 125.35 + 5.14 = 130.49 \text{ kW}$$
$$\therefore \quad \eta_{max} = \frac{125.35}{130.49} \times 100 = \mathbf{96.1\%}$$

Q. 53. A 40 kVA single-phase transformer has core loss of 450 W and full-load copper loss of 850 W. If the power factor of the load is 0.8, calculate:

(*i*) full-load efficiency

(*ii*) maximum efficiency at u.p.f.

(*iii*) load for maximum efficiency. (*Basic Elect. Engg. VTU, Karnatka 2007*)

Ans. Total losses = 450 + 850 = 1300 W = 1.3 kW ; p.f. = 0.8

(*i*) Full-load output at 0.8 p.f. = 40 × 0.8 = 32 kW

Full-load input = 32 + 1.3 = 33.3 kW

$$\text{F.L. } \eta = \frac{32}{33.3} \times 100 = \mathbf{96.0\%}$$

(*ii*) Output at full load u.p.f. = 40 × 1 = 40 kW

Input = 40 + 1.3 = 41.3 kW

$$\eta = \frac{40}{41.3} \times 100 = \mathbf{96.8\%}$$

(*iii*) The load for maximum efficiency is

$$= 40 \times \sqrt{\frac{450}{850}} = \mathbf{29.1\ kVA}$$

EXERCISE

1. A 200/50 V, 50 Hz single-phase transformer is connected to a 200 V, 50 Hz supply with secondary winding open. Primary winding has 400 turns.
   (*i*) What is the value of maximum flux through the core if the primary winding has 400 turns?
   (*ii*) What is the peak value of flux if primary voltage is 200V, 25Hz?
   (*Basic Elect, Engg. RGPV, Bhopal 2002*)
   [2.25 mWb, 4.5 mWb]
2. A 230/115 volt, single-phase transformer is supplying a load of 5 A at power factor 0.866 lagging. The no-load current is 0.2 A at power factor 0.208 lagging. Calculate the primary current and primary power factor. (*Elect. Engg. Nagpur Univ 2003*)
   [2.638 A, 0.836 lag]
3. A 230/115-V single phase transformer is supplying a load of 5 A at power factor 0.866 lagging. The no-load current is 0.2 A at power factor 0.208 lagging. Calculate the primary current. Also draw the phasor diagram.
   (*Basic Elect. Engg. VTU Karnatka 2007*)
   [2.638 A]
4. A 3,300/300 V single-phase transformer gives 0.6 A and 60 W as ammeter and wattmeter readings when rated input voltage is given to low-voltage winding and high-voltage winding is kept open. Find
   (*i*) p. f. of no-load current.
   (*ii*) magnetising component of input current.
   (*iii*) iron loss component of input current. (*Elect. Engg. Nagpur Univ, 2005*)
   [0.02, 0.599 A, 0.012 A]
5. The required no-load voltage ratio in a single-phase 50 Hz core type transformer is 6600/500. Find the number of turns in each winding if the flux is to be 0.06 Wb.
   (*Basic Elect. Engg. CSV Tech Univ Bhilai 2007*)
   [495, 38]

6. A 5kVA, 250/500V, 50 Hz, single-phase transformer gave the following test data :

   No load test : 250 V, 0.75 A, 60 W – L.V. side

   S.C. test : 9 V, 6 A, 22 W – H.V. side

   Calculate :

   (*i*) The efficiency at full load unity p.f

   (*ii*) The corresponding terminal voltage on full load at 0.8 p.f. lagging

   (*Elect. Engg. Nagpur Univ, 2004*)

   [97.65%, 487 V]

7. The efficiency of a 400 kVA single-phase transformer is 98.77% when delivering full-laod at 0.8 p.f. and 99.13% at half load and unity power factor. Calculate:

   (*i*) iron losses

   (*ii*) the full load Cu losses. (*Basic Elect. Engg. RGPV, Bhopal, 2006*)

   [1007 W, 3973 W]

8. The following test data is obtained on 5 kVA, 220/440 volt, single-phase transformer:

   O.C. test : 220 V, 2 A, 100 W – L.V. side

   S.C. test : 40 V, 11.4 A, 200 W – H.V. side

   Determine :

   (*i*) % efficiency at full load 0.9 p.f lag.

   (*ii*) % efficiency at half load unity p.f. (*Elect. Engg. Nagpur, Univ, 2004*)

   [93.77%, 94.4%]

9. A 200/400 V, 10 kVA, 50 Hz single-phase transformer has at full-load a Cu loss of 120 W. If it has an efficiency of 98% at full-load unity power factor, determine the iron losses.

   What would be the efficiency of the transformer at half load 0.8 power factor lagging?

   [84 W; 97%]

10. A single-phase transformer working at unity power factor has an efficiency of 92% at both half-load and at full-load of 92% at both half-load and at full-load of 500kW. Determine the efficiency at 80% of full-load. (*Elect. Engg. AM.Ae S.I. 2002*)

    [92.4%]

11. A 600 kVA, single-phase transformer has an efficiency of 92% both at full-load and half-load at unity power factor. Determine its efficiency at 60% of full-load at 0.8 power factor lag. **[85.9%]**

12. A 25 kVA, 2200/220 V single-phase transformer, has a primary resistance of 1.0 Ω and a secondary resistance of 0.01 Ω. Find the equivalent secondary resistance and the full-load efficiency at 0.8 p.f if the iron loss of the transformer is 0.8% of the full-load Cu loss. (*Elect. Technology, Utkal Univ*)

    [0.02 W, 97.7%]

13. In a 25 kVA, 2000/200 V transformer, the iron and copper losses are 350 and 400 W respectively. Calculate the efficiency on upf at.

    (*i*) full load and

    (*ii*) half load (*B.Tech. Elect. Engg. JNTU, 2007*)

    [97.09%, 96.52%]

14. A 11,000/230 V, 150 kVA, single-phase, 50 Hz transformer has core loss of 1.4 kW and F.L. Cu loss of 1.6 kW. Determine

    (*i*) the kVA load for maximum efficiency and value of maximum efficiency at unity p.f

    (*ii*) the efficiency at half F.L. 0.8 p.f leading. (*Elect. Engg. Nagpur Univ.*)

    [98.2%, 97%]

MULTIPLE CHOICE QUESTIONS

1. A transformer basically consists of
(*a*) primary winding (*b*) secondary winding
(*c*) core (*d*) all of the above
2. A transformer transforms
(*a*) current (*b*) voltage (*c*) power (*d*) energy
(Basic Elect. Engg.R.G.P.V. Bhopal 2007)
3. In an ordinary transformer which of the following does **NOT** change?
(*a*) voltage (*b*) current (*c*) frequency (*d*) all of the abve
(Basic Elect. Engg. RGPV Bhopal 2006)
4. Transformer core is laminated in order to
(*a*) reduce eddy-current loss (*b*) reduce hystersis loss
(*c*) reduce weight of steel (*d*) improve cooling
(Basic Elect. Engg. R.G.P.V Bhopal 2007)
5. In no-load phasor diagram of a transformer, the core loss component of the primary current remains in phase with
(*a*) no-load current (*b*) primary supply voltage
(*c*) core flux (*d*) primary induced voltage
6. The magnetising current drawn by the primary of an ideal transformer on no load
(*a*) maintains the alternating flux through the core
(*b*) lags the primary voltage by 90°
(*c*) is wattless
(*d*) all of the above
7. The main purpose of performing open-circuit test on a transformer is to measure its
(*a*) Cu loss (*b*) core loss
(*c*) total loss (*d*) insulation resistance
(Basic Elect. Engg. RGPV Bhopal 2007)
8. When the secondary load current of a 400/200 V transformer changes by 10 A, the primary current changes by
(*a*) 10 A (*b*) 20 A (*c*) 5 A (*d*) 2.5 A
9. Open-circuit and short circuit tests on a transformer give
(*a*) windage losses (*b*) friction losses
(*c*) iron and Cu losses respectively (*d*) Cu and iron losses respectively
10. The main purpose of using core in a transformer is to
(*a*) decrease iron losses (*b*) prevent eddy current loss
(*c*) eliminate magnetic hysteresis (*d*) offer low reluctance path to magnetic flux
11. A step-up transformer increases
(*a*) voltage (*b*) current (*c*) power (*d*) frequency
12. The number of secondary turns in a 250/400V transformer having 1000 primary turns is
(*a*) 1600 (*b*) 250 (*c*) 400 (*d*) 1250
13. The equivalent primary resistance of a transformer having $K = 5$ and $R_1 = 0.1\ \Omega$ as referred to secondary is .... ohm
(*a*) 0.5 (*b*) 0.02 (*c*) 0.004 (*d*) 2.5
14. A two-winding transformer operates at maximum efficiency when its
(*a*) hysteresis loss equals eddy current loss
(*b*) Cu loss equal iron loss
(*c*) primary resistance equals secondary resistance
(*d*) voltage regulation is minimum.

15. If Cu loss of a trasformer at full-load is 800 W, its value at half load is .... watt.
(*a*) 400 (*b*) 1600 (*c*) 200 (*d*) 600
16. A transformer has negative voltage regulation when its load p.f is
(*a*) zero (*b*) unity (*c*) lagging (*d*) leading
17. The efficiency of a given transformer is maximum when
(*a*) it runs at half full load (*b*) it runs at full load
(*c*) its Cu loss equals iron loss (*d*) it runs at 0.8 power factor
18. The maximum efficiency of a 100 kVA transformer having iron loss of 900 W and full load Cu loss of 1600W ocuurs at .... kVA
(*a*) 56.3 (*b*) 75 (*c*) 133.3 (*d*) 177.7
19. The short circuit test of a transformer provides the value of its
(*a*) equivalent impedance (*b*) full-load Cu loss
(*c*) voltage drop (*d*) all of the above
20. A 200/400 V single-phase transformer draws a primary current of 25 A at 0.8 p.f lag. The secondary kVA developed by it is
(*a*) 5 kVA (*b*) 4 kVA (*c*) 10 kVA (*d*) 8 kVA

ANSWERS

1. *d*	2. *c*	3. *c*	4. *a*	5. *b*	6. *d*	7. *b*
8. *c*	9. *c*	10. *d*	11. *a*	12. *a*	13. *d*	14. *b*
15. *c*	16. *d*	17. *c*	18. *b*	19. *d*	20. *a*	

14
CHAPTER

THREE-PHASE INDUCTION MOTOR

Q. 1. Why is the 3-phase induction motor most commonly used ac motor in industry? Does it have any disadvantages?

Ans. The 3-phase induction motor (Fig. 1) is most frequently used for industrial drives because it has the following advantages.

1. It is very simple and has *extremely* rugged (almost unbreakable) construction.
2. Its cost is low.
3. It is very reliable.
4. It is easy to maintain.
5. It is self-starting.
6. It runs at *almost constant speed* from no-load to full load.
7. It is very efficient in converting electrical energy to mechanical energy.
8. It can be manufactured with characteristics to suit most industrial applications.

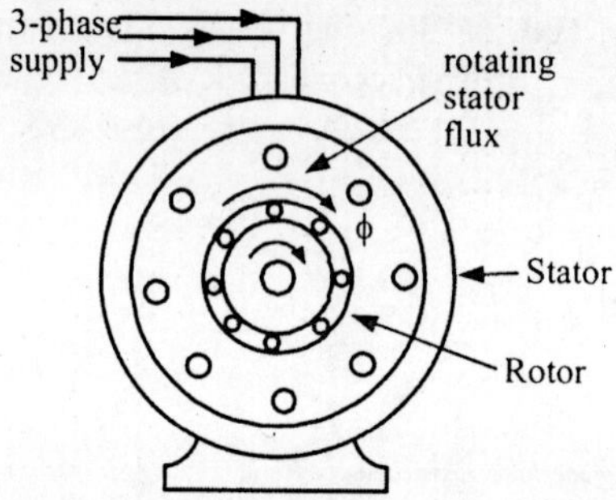

Fig. 1

A few disadvantages of this motor are as under:

1. Its speed is not easily controlled.
2. Its starting current is 5 to 8 times the full-load current.
3. It runs at low lagging power factor at light loads.
4. Its starting torque is *less* than that of a dc shunt motor.

How ever, for most applications, these disadvantages do not matter much.

Q. 2. With neat sketches, explain the constructional details of squirral-cage and phase-wound induction motors.

Ans. A 3-phase induction motor consists of two main parts:

(*a*) a stationary part called **stator** and

(*b*) a rotating part called **rotor.**

The rotor is mounted on bearings and is separated from the stator by an *air gap.*

(*a*) Stator : The stator consists of three parts [Fig. 2(*a*)].

(*i*) frame (*ii*) stator core and (*iii*) stator winding.

The function of the frame is to support the core and its stator windings.

The stator core is built up of thin laminations of high grade silicon steel and has parallel slots to house stator windings. The *six* terminals of the stator windings are brought out. They may be star-connected or delta-connected externally. The windings are usually delta connected for normal running.

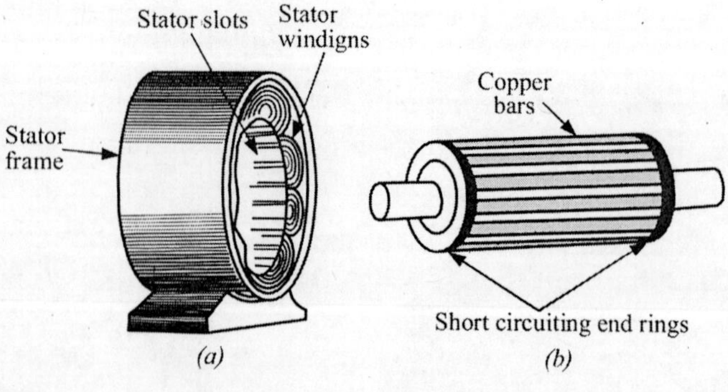

Fig. 2

(*b*) **Rotor :** There are two types of rotors used in an induction motor.

(*i*) squirrel-cage rotor (*ii*) phase-wound or wound rotor

The squirrel-cage rotor is shown in Fig. 2(*b*). This rotor consists of a cylindrical core having parallel slots. Each slot contains one copper or aluminium bar. The rotor bars are *electrically welded* to two copper end-rings.

As seen, the rotor bars are *short-circuited* upon themselves. Hence, it is not possible to *add any external resistance* in series with the rotor circuit for starting the motor. The squirrel-cage rotor adjusts to any number of stator poles.

The 3-phase motor using this type of rotor is known as *squirrel-cage* induction motor.

The phase-wound rotor has 3-phase windings placed in the slots of its rotor core.

The rotor is wound for as *many poles as the number of stator poles*.

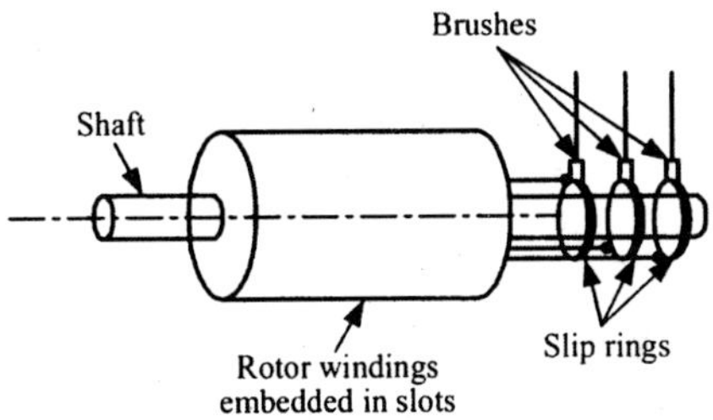

Fig. 3

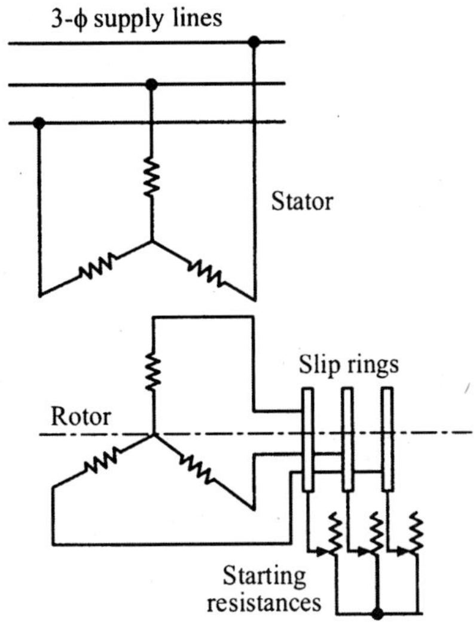

Fig. 4

The three phases of the wound rotor are connected in star internally. The other three winding terminals are brought out and connected to three slip-rings mounted on the shaft with brushes resting on them as shown in Fig. 3. These three brushes are further connected to a star-connected rheostat (Fig. 4). Hence, we can *add resistance in the rotor circuit* for starting the motor and for changing its speed/torque/current characteristics. When the motor runs normally, the slip-rings are automatically short-circuited. The brushes are then automatically lifted from the slip-rings. Hence, under normal running conditions, the wound rotor is *short-circuited on itself* just like the squirrel-cage rotor.

Q. 3. With neat sketches, explain how a rotating magnetic field is produced in the 3-phase stator winding of an induction motor. Which two factors determine the speed of its rotation?

Ans. The working of an induction motor depends on a *rotating* magnetic field which is established in the *air-gap* of the motor by the three stator currents.

When the stator winding of the motor is connected to a 3-phase supply [Fig. 5(*a*)], curents flow in each phase of the winding. These currents are out of phase with each other by 120° as shown in Fig. 5(*b*). The current *entering* the phase will be taken as *positive* and shown by a cross which means it moves *away* from the reader. The current *leaving* the phase will be taken as *negative* and shown by a *dot* which means it is moving *towards* the reader. The motor of Fig. 5(*a*) is a 2-pole motor.

In vector diagrams, the different currents will be taken *positive* when their directions *agree* with the arrows shown in Fig. 5(*b*). If the directions of the currents are *opposite* to the directions of the arrows, they will be taken as *negative.*

Each phase of the winding produces a magnetic field of its own. The three magnetic fields are 120° out of phase with each other. Hence, they combine to produce **one** resultant field which acts on the rotor. It can be shown that the resultant magnetic field *shifts its position around* the stator at a *constant* speed. At the end of one cycle of the applied ac current, the resultant field shifts 360° or one revolution.

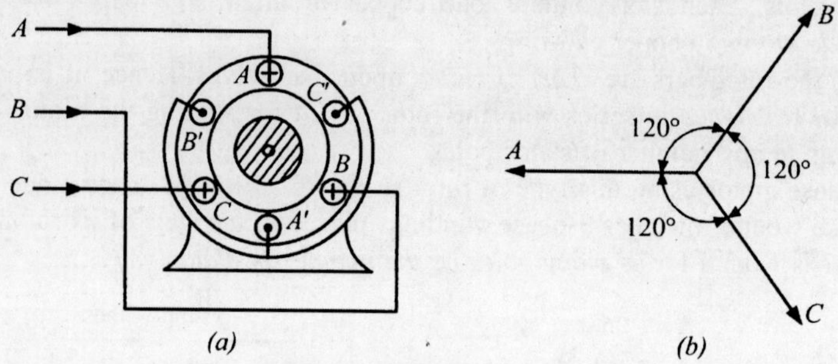

Fig. 5

The manner in which a rotating magnetic field is produced can be found by considering the directions of current flow through the three phases at different positions. We will consider four positions every one quarter of a cycle *i.e.* at intervals of 90° (Fig. 6).

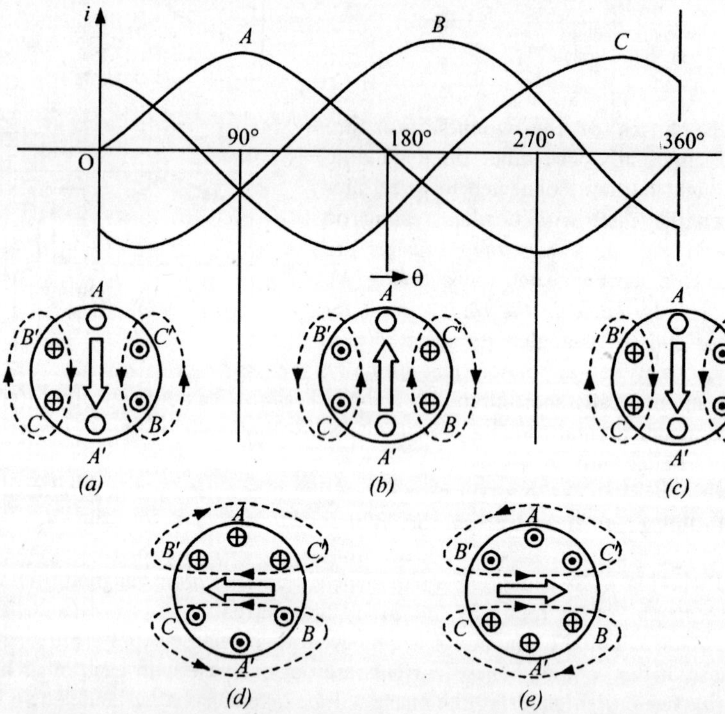

Fig. 6

Consider the position when $\theta = 0$. At this position, value of current wave A is zero, C is positive and B is negative. Since current C is positive, its direction has been shown by a cross at the *entering* point C and a dot at the *leaving* point C' in Fig. 6(*a*). Similarly, B has been shown by a dot (leaving point) and B' (entering point) by a cross in Fig. 6(*a*).

The directions of the two magnetic fields produced by the three current waves are shown by the *dotted* lines in Fig. 6(*a*). This can be confirmed by the Fleming's Right-hand rule. The direction of the *combined* magnetic field is shown by the arrow.

At 90° position, current A is positive but B and C are both negative. Their current directions, as found by right-hand rule, are as shown in Fig. 6(b). As seen, the combined magnetic field has turned clockwise through 90°. Similarly at position 180°, the resultant field has turned further clockwise by another 90°. By position 360°, the resultant field has rotated through *one full revolution* or one cycle of the stator currents.

It is obvious that the *combined* magnetic field produced by the 3-phase currents in the stator windings rotates around the stator at a *constant* speed. The speed of the resultant field is determined by two factors (i) frequency f of the ac currents and (ii) the *number* of stator poles P. The value of the speed is given by $N = 120f/P$. It is called the synchronous speed N_s.

Q. 4. What is the maximum value of the resultant flux in the rotating magnetic field of a 3-phase induction motor? Does it remain constant throughout?

Ans. When the three-phase windings of an induction motor are connected to a three-phase supply, they produce a resultant magnetic flux Φ_m which rotates in space as if *actual magnetic poles* were rotating mechanically.

Fig. 7(a) shows the three current waves as well as their flux waves. The positive directions of the currents and their fluxes are shown in Fig. 7(b). The resultant flux at any instant is given by the *vector sum* of the individual fluxes Φ_1, Φ_2 and Φ_3 due to the three phases. The directions of the fluxes will be taken as *positive* when they *agree with arrows of* Fig. 7(b). Otherwise, they will be taken as negative. Let us find the value of resultant flux Φ at four positions shown in Fig. 7.

(i) When $\theta = 0°$: At this position, $\Phi_1 = 0$. Flux Φ_2 is negative since its direction is opposite to the vector OII of Fig. 7(b). Its value is given by

$$\Phi_2 = -\Phi_m \sin 60 = -\frac{\sqrt{3}}{2}\Phi_m$$

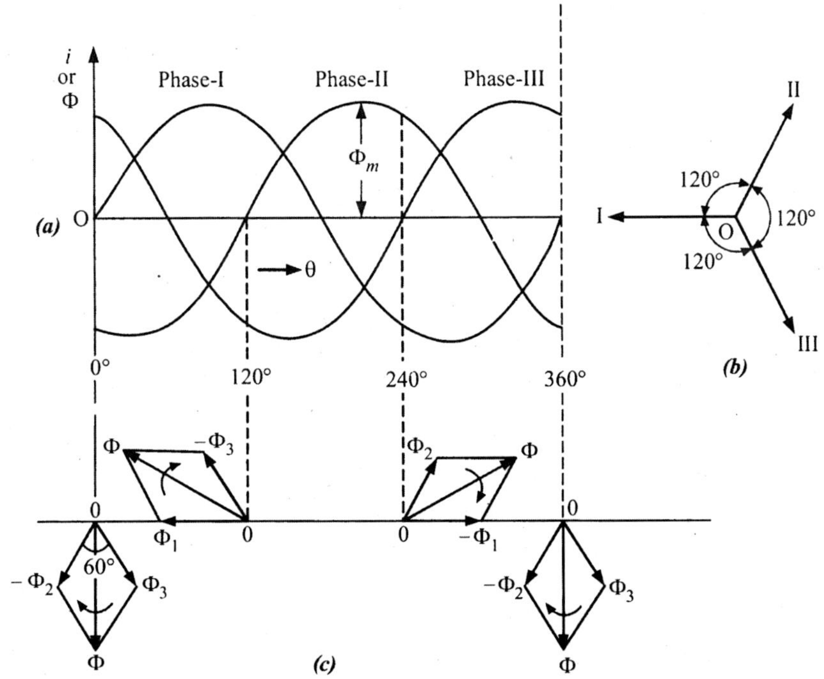

Fig. 7

However, Φ_3 is positive. Its value is given by $\Phi_3 = \frac{\sqrt{3}}{2}\Phi_m$

Here, Φ_m stands for the maximum value of the flux due to *each wave.*

The resultant flux is given by the vector sum of $(-\Phi_2)$ and Φ_3

$$\therefore \quad \Phi = 2 \times \frac{\sqrt{3}}{2}\Phi_m \times \cos\frac{60°}{2} = \sqrt{3} \times \frac{\sqrt{3}}{2}\Phi_m = \frac{3}{2}\Phi_m = 1.5\,\Phi_m.$$

***(ii)* When θ = 120° :** As seen from Fig. 7(*a*), Φ_1 is positive since it is along the direction OI of Fig. 7(*b*). Flux Φ_3 is negative since it is opposite to the direction OIII of Fig. 7(*b*). However, Φ_2 is zero.

The resultant flux is again $\frac{3}{2}\Phi_m$ but it has rotated clockwise through an angle of 120°.

***(iii)* When θ = 240° :** As seen from Fig. 7(*a*), Φ_2 is positive but Φ_1 is negative and Φ_3 is zero. The resultant field is given by

$$\therefore \quad \Phi = 2 \times \frac{\sqrt{3}}{2}\Phi_m \times \cos\frac{60°}{2} = \frac{3}{2}\Phi_m = 1.5\Phi_m$$

It is obvious from Fig 7(*c*) that resultant flux has the same value but has further rotated clockwise through an angle. of 120°.

***(iv)* When θ = 360° :** In this case, Φ_1 is zero, Φ_2 is negative and Φ_3 is positive as shown in Fig 7(*c*).

As before,

$$\Phi_2 = -\frac{\sqrt{3}}{2}\Phi_m \quad \text{and} \quad \Phi_3 = \frac{\sqrt{3}}{2}\Phi_m.$$

$$\therefore \quad \Phi = \frac{3}{2}\Phi_m = 1.5\,\Phi_m$$

Their resultant is again $\frac{3}{2}\Phi_m$ but has further rotated clockwise by an angle of 120° or by a total angle of 360°.

It will be seen from above that

1. The resultant stator flux has a *cosntant* value of 1.5 times the maximum value of the flux due to *each phase.*
2. The resultant flux rotates around the stator at synchronous speed given by $N_s = 120f/P$.

Q. 5. Can the direction of the revolving magnetic field in an induction motor be reversed? Explain how?

Ans. The direction of rotation of an induction motor can be reversed by *reversing the phase sequence* of its 3-phase supply. The supply phase sequence can be reversed by interchanging any of the *two* supply lines (Fig. 8) As a result, the direction of rotation of the magnetic field will be reversed. So will be the direction of rotation of the rotor because it *has to rotate* in the same direction as the rotating stator field. In this way, the direction of rotation of the induction motor can be reversed.

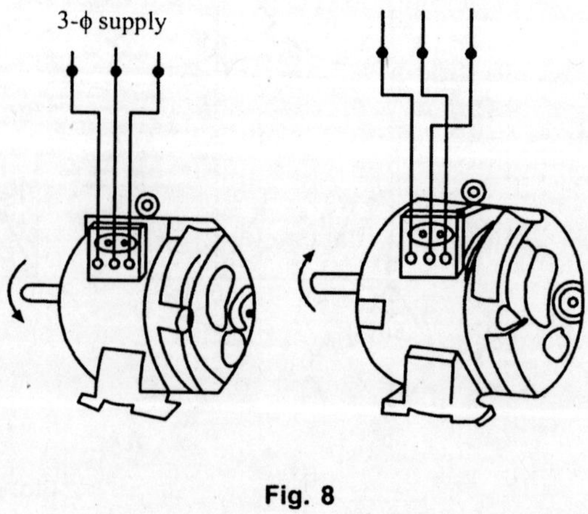

Fig. 8

Q. 6. Does the magnetic field produced by the stator windings of an induction motor actually revolve around the stator? If not, what else happens?

Ans. When three-phase currents flow in the stator windings of an induction motor, they produce three alternating fluxes which are out of phase with each other by 120°. They combine to produce a resultant magnetic flux Φ which has a constant value of 1.5 Φ_m where Φ_m is the maximum flux due to each phase. However, the *point* on the stator where this combination occurs *is not fixed* because the phase of the three current waves keeps changing with time. Suppose the combination of the three fluxes takes place first at point *A* (Fig. 9). Then after sometime, it happens at point *B* and then at *C* and so on. Hence, the *seat of resultant flux* Φ keeps shifting around the stator windings synchronously. Obviously, there is no actual revolution of the fluxes *i.e.* the three magnetic fluxes themselves do not move around the stator. It is only the *seat* of the resultant flux which keeps on shifting synchronously around the stator.

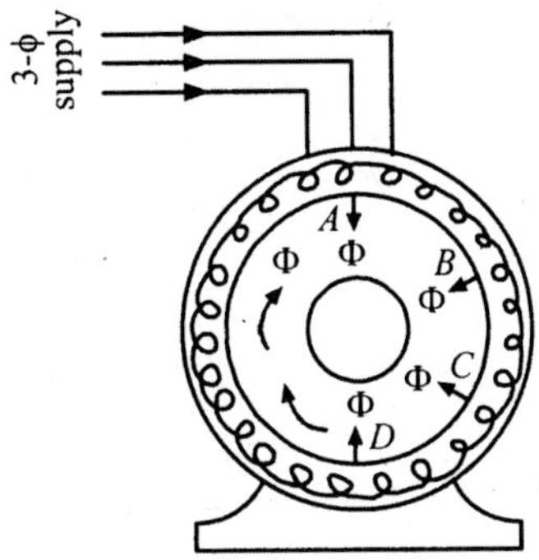

Fig. 9

Q. 7. Explain the principle of opreation of a 3-phase induction motor. Why is it known by the name of "induction" motor.

Ans. When the 3-phase stator windings of an induction motor are fed by a 3-phase supply, a magnetic field of constant magnitude but rotating at synchronous speed is set up. The flux lines pass through the air gap and are cut by the rotor conductors which are, as yet, *stationary.*

Since magnetic flux is rotating and the rotor is stationary, hence, there is *relative speed* between the two. According to Faraday's laws of electromagnetic induction, an emf is induced in the rotor conductors. The frequency of the induced emf is the *same as* the supply frequency when *rotor is stationary.*The magnitude of the induced emf is proportional to the relative speed between the rotating flux and the rotor conductors. Since the rotor conductors form a closed circuit, current is produced in then. According to Lenz's law, the induced rotor current *opposes* the very cause which produces it.

In this case, the cause is the *relative speed* between the rotating flux of the stator and the stationary rotor conductors. Hence, to reduce the relative speed, the rotor starts running in the **same** direction as the stator flux and tries to catch up with it though it does not succeed.

Simpler Explanation : When ac is applied to the stator windings, a rotating magnetic field is generated. This rotating field cuts the rotor cunductors and hence induces currents in them. These rotor currents produce their own rotating magnetic field. Thus, we have *two* rotating magnetic fields in the air gap which rotate at synchronous speed with *respect to the stator* but are stationary with respect to *each other.* The two magnetic fields react with each other and force the rotor to rotate in the direction of the rotating stator field (in a way, the stator field *drags* the rotor field along with itself).

It should be noted that electric power is transferred from stator to rotor by *induction* and not by *conduction.* Hence the name 'induction' motor.

Q. 8. Define motor slip and the role it plays in the operation of an induction motor.

Ans. The rotor of an induction motor *cannot* run with the same speed as the rotating magnetic field of the stator. If it were to run at the *same speed* as that of the rotating magnetic field, it would *not* cut any magnetic flux because there would be *no relative speed between the two*. As a result, there would be no rotor emf, no rotor current and hence, no torque to maintain rotation of the rotor. That is why the rotor runs at a speed which is **always** less than the speed of the rotating stator field. The difference in speed depends on the motor load (Fig. 10).

The differernce between the synchronous speed N_s and the *actual* rotor speed N is called **slip**. It is usually expressed as a percentage of the synchronous speed. In fact, the term slip beautifully describes the way in which the rotor 'slips black' from synchronism.

$$\% \text{ slip} = \frac{N_S - N}{N_S} \times 100$$

Sometimes, the speed difference ($N_s - N$) is called the *slip speed.*

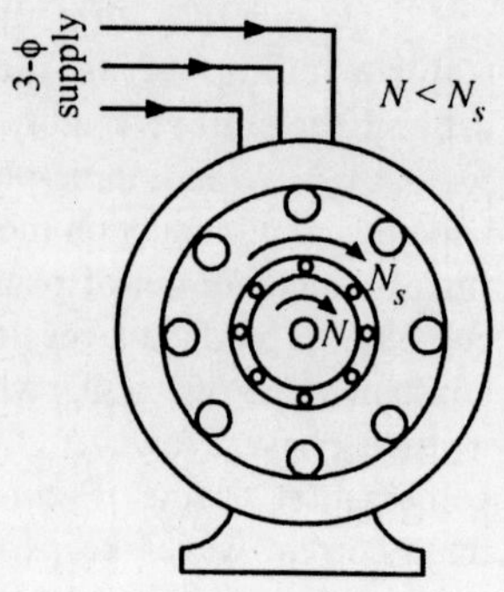

Fig. 10

The vlaue of slip for an induction motor varies from 1% to 5% from no load to full load As seen from above,

$$s = \frac{N_S - N}{N_S} \quad \text{or} \quad N = (1 - s)\, N_S$$

This gives the speed of the rotor in terms of the synchronous speed.

When induction motor runs at no load, it requires less rotor torque which means less rotor current. Now, rotor current will be less when rotor speed is high *i.e.* when slip is *small.* When motor load is increased, rotor torque has to increase to carry this load. The rotor torque will increase when rotor current increases. And rotor current will increase only when rotor *slows down* further i.e. when slip *increases.*

As seen, slip adjusts itself to meet the load rquirements of the motor. If motor load is less, slip is less and if load is more, slip is more. In this way, the slip plays a very important role in the operation of an induction motor.

Q. 9. Derive an expresion for the frequency of rotor currents in an induction motor. What is its value when rotor is locked?

Ans. When the motor is stationary, the frequency of rotor current is the *same* as the supply frequency f. But when the rotor starts rotating, then the frequency depends upon the relative speed or slip-speed. Let at any slip speed ($N_s - N$), the frequency of the rotor current be f'. Then

$$N_S - N = \frac{120 f'}{P}$$

Also, $$N_S = \frac{120 f}{P}$$

Dividing one by the other, we get

$$\frac{f'}{f} = \frac{N_S - N}{N_S} = s \qquad \therefore \quad f' = sf$$

Rotor frequency is important because as it varies, the rotor reactance also varies, thus affecting both *starting* and *running characteristics* of the motor. When rotor is locked, $f' = f$.

Q. 10. What is meant by the absolute and relative speed of a moving object? Do they have one or more than one value?

Ans. Absolute speed of an object is its speed with respect to a given stationary object like open space or Earth. The relative speed of an object is its speed as measured with respect to another *moving* object. As this speed is measured with respect to some other object moving with different speeds, hence the relative speed of the object has *different* values. It means that the relative speed of an object varies with the speed of the reference object. Hence, relative speed has no *fixed* value like the absolute speed of an object. In short, absolute speed of an object has only one value whereas relative speed has *many values.*

To illustrate the subject further, consider two trains A and B running in the *same* direction with speeds of 80 km/hr and 60 km/hr respectively as shown in Fig. 11. It is obvious that the train speeds given above are the *absolute* speeds of the two trains because they are with respect to the *stationary* Earth or *open space.*

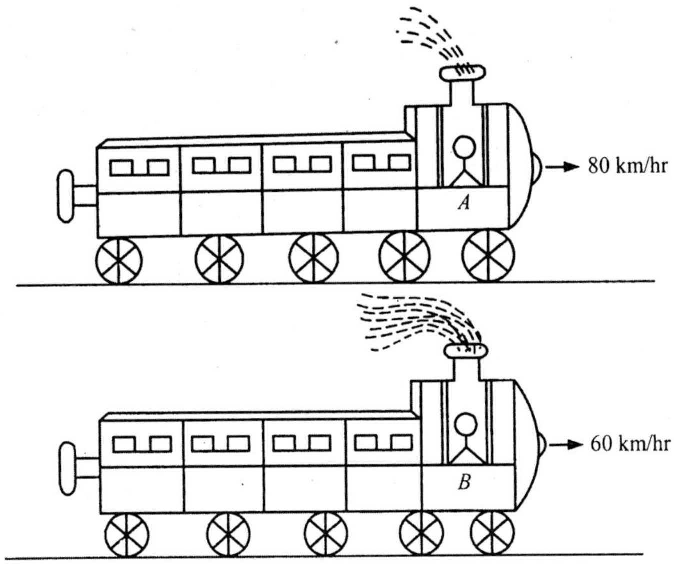

Fig. 11

Let us now calculate the different speeds of the driver of train *A*.

1. Since he is moving with the same speed as train *A*, his *absolute* speed is 80 km/hr.
2. His speed with respect to train *A* is *zero* because he is moving with the *same* speed as train *A*.
3. His speed with respect to train *B* is (80 – 60) = 20 km/hr in the same direction.

It is seen from above that absolute speed of driver *A* is only one whereas his relative speeds are many.

Q. 11. If a 6-pole induction motor supplied from a 3-phase, 50Hz supply has a rotor frequency of 2.3 Hz, Calculate:

(*i*) the percentage slip (*ii*) the speed of the motor.

(*Basic Elect. Engg. VTU Karnatka 2008*)

Ans. (*i*) $f' = sf$ or $2.3 = s \times 50$ $\therefore$ $s = 0.046$ or **4.6%**

(*ii*) $N_s = 120f/P = 120 \times 50/6 = 1000$ rpm

Now, rotor speed,

$$N = (1 - s)\, N_s = (1 - 0.046) \times 1000 = \mathbf{954\ rpm}$$

Q. 12. A 12-pole, 3-phase induction motor is fed from 100 Hz supply. If the frequency of rotor emf at full load is 5 Hz, find the full-load speed and % slip.

(*B. Tech Elect. Technology JNTU, Hydrabad 2007*)

Ans. $N_s = 120 \times 100/12 = 1000$ rpm

$f' = sf$ or $5 = s \times 100$ $\therefore$ $s = 0.05$ or **5%**

Full-load speed $N = (1 - s) \times N_s = (1 - 0.05) \times 1000 = \mathbf{950\ rpm}$

Q. 13. A 3-Φ, 4-pole, delta-connected induction motor is supplied from 3-ϕ, 430 V, 50 Hz source. It runs at a speed of 1425 rpm. Find the value of slip and frequency of rotor current. (*Elect. Engg. Nagpur Univ. 2005*)

Ans. We are given $P = 4, f = 50$ Hz, $N = 1425$ rpm

$N_s = 120f/P = 120 \times 50/4 = 1500$ rpm

$$\therefore \quad s = \frac{N_s - N}{N} = \frac{1500 - 1425}{1500} = \mathbf{0.05}$$

Frequency of rotor current, $f' = s \times f = 0.05 \times 50 = \mathbf{2.5\ Hz.}$

Q. 14. An 8-pole alternator running at 750 rpm supplies power to a 6-pole, 3-ϕ induction motor. The induction motor has a full-load slip of 3 percent. Find the full-load speed of the induction motor and the frequency of its rotor e.m.f.

(*Basic Elect. Engg. RGPV Bhopal 2005*)

Ans. We will first find the frquency of the 3-ϕ currents produced by the alternator. Its value is given by

$$f = \frac{PN}{120} = \frac{8 \times 750}{120} = 50 \text{ Hz}$$

This gives the frequency of the 3-ϕ currents supplied to the induction motor

Now, $N_s = 120 \times f/P = 120 \times 50/6 = 1000$ rpm

The motor (or rotor) speed is given by

$$N = N_s\,(1 - s) = 1000\,(1 - 0.03) = \mathbf{970\ rpm}$$

Frequency of rotor emf is

$$f' = s \times \text{supply frequency} \quad \therefore \quad f' = 0.03 \times 50 = \mathbf{1.5\ Hz}$$

Q. 15. Explain how a rotating magnetic field is produced in a 3-ϕ induction motor. A 3-ϕ, 6-pole, 50 Hz induction motor has a slip of 1% at no load and 3% at full load. Calculate :

(*i*) synchronous speed

(*ii*) no-load speed

(*iii*) full-load speed

(*iv*) frequency of rotor current at full load.

(*v*) frequency of rotor current at standstill

(*Basic Elect. Engg. West Bengal, Tech Univ 2007*)

Ans. (*i*) $N_S = 120 \times f/P = 120 \times 50/6 =$ **1000 rpm**

(*ii*) No-load slip = 1% = 0.01

$\therefore$ No load speed $= N_S\,(1 - s) = 1000 \times (1 - 0.01) =$ **990 rpm**

(*iii*) Full-load slip = 3% = 0.03

Full-load speed $= N_s\,(1 - s) = 1000 \times (1 - 0.03) =$ **970 rpm**

(*iv*) At full load, slip $s = 0.03$

$\therefore$ frequency of rotor current,

$$f' = 0.03 \times 50 = \mathbf{1.5\ Hz}$$

(*v*) At sandstill *i.e.* when rotor is stationary, $N = 0$

$$\therefore \quad s = \frac{N_S - N}{N_S} = \frac{N_S - 0}{N_S} = 1$$

$\therefore$ frequency of rotor current at standstill is

$$f' = s \times f = 1 \times 50 = \mathbf{50\ Hz}$$

Q. 16. A 3-ϕ, IM operates from a supply whose frequency is 50 Hz and rotates at a speed of 1485 rpm at no load and 1350 rpm at full load.

Calculate :

(*a*) the speed at which the magnetic field of stator is rotating

(*b*) the slip at no load and at full load

(*c*) the frequency of the rotor current at no load and full load

(*d*) the frequency of rotor current at standstill.

(*Elect. Mach-II, JNTU, Hydrabad 2007*)

Ans. (*a*) Let us find the number of poles by using the relation

$$f = PN/120$$

or $\quad 50 = P \times 1485/120, \quad \therefore \quad P = 4.04$

Since P cannot be a fraction, $P = 4$

$$N_s = 120f/P = 120 \times 50/4 = \mathbf{1500\ rpm.}$$

(*b*) slip at no load = $(N_s - N)/N_s$ = (1500 – 1485)/1500
= 1/100 or **1%**

slip at full load = (1500 – 1350)/1500 = 1/10 or **10%**

(*c*) Frequency of rotor currents at no load is

$$f' = (1/100) \times 500 = \mathbf{0.5\ Hz}$$

Frequency of rotor current at full load = (1/10) × 50 = **5 Hz**

(*d*) At standstill, $s = 1$

∴ frequency of currents = $sf = 1 \times 50 =$ **50 Hz**

Q. 17. A 3-φ induction motor runs at 290 rpm at full load with 50 Hz supply. Find the number of poles of the motor and slip. (*Basic Elect. Engg. C.S.V Tech. Univ. Bhilai 2007*)

Ans. It is obvious that the synchronous speed of the motor is 300 rpm.

Now, $N_s = 120f/P$ or $300 = 120 \times 50/P$ ∴ $P = \mathbf{20}$

$$\therefore \quad s = \frac{N_S - N}{N_S} = \frac{300 - 290}{300} = 0.33 \text{ or } \mathbf{3.3\%}$$

Q. 18. A 3-φ induction motor has 6 poles. It is connected to a 3-φ, 400 V, 50 Hz supply. The motor is operating at 5% slip. What is its speed?

(*Basic Elect. Engg. Kerala Univ. 2004*)

Ans. $N_s = 120f/P = 120 \times 50/6 = 1000$ rpm.

motor speed, $N = N_s(1 - s) = 1000\,(1 - 0.05) =$ **950 rpm**

Q. 19. A 3-phase, 4 pole, 50 Hz induction motor has a full-load speed of 1440 rpm. For this motor, calculate the following

(*i*) full-load slip and rotor frequency

(*ii*) speed of stator field with respect to

(*a*) stator structure and

(*b*) rotor structure and

(*iii*) speed of rotor field with respect to

(*a*) rotor structure

(*b*) stator structure and

(*c*) stator field.

Ans. (*i*) $N_s = 120f/P = 120 \times 50/4 =$ **1500 rpm**

$$\text{slip, } s = \frac{N_s - N}{N_s} = \frac{1500 - 1440}{1500} = 0.04 \text{ or } \mathbf{4\%}$$

$$f' = sf = 0.04 \times 50 = \mathbf{2\ Hz}$$

(*ii*) (*a*) Stator field revolves synchronously with respect to the stator structure which is stationary. Hence, its speed is **1500** rpm.

(*b*) The rotor is revolving with a speed of 1440 rpm in the *same* direction as the stator field. Hence, speed of stator field with respect to the rotating rotor is (1500 – 1440) = **60 rpm.**

The above is the relative speed of the revolving stator field and not its absolute speed.

(*iii*) (*a*) The frequency of rotor current is $f' = sf$. Hence, speed of the *rotor field* with respect to rotor is

$$= 120 \times (sf)/P = s \times 120f/P = \mathbf{sN_s}$$

(*b*) The speed of rotor field with respect to rortor is sN_s. The speed of rotor itself with respect to stator structure is $(N_s - N)$. Hence, the speed of rotor magnetic field with respect to stator structure is

$$= sN_s + (N_s - N) = N_s(1 + s) - N$$

$$= N_s + sN_s - N = \mathbf{N_s} \qquad \text{— } \because sN_s = N$$

It means the two magnetic fields are rotating synchronously. In fact, they are *stationary* with respect to each other.

Q. 20. A 4-pole, 3-ϕ induction motor operates form a supply whose frequency is 50Hz. Calculate:

(*a*) the speed at which magnetic field of the stator is rotating
(*b*) the speed of the rotor when the slip is 0.04
(*c*) the frequency of the rotor currents when the slip is 0.03
(*d*) the frequency of the rotor currents at standstill

(*Elect. Engg. BE Ist Yr. Nagpur Univ. 2006*)

Ans. (*a*) The stator magnetic field is rotating at the synchronous speed given by

$$N_s = 120f/P = 120 \times 50/4 = \mathbf{1500\ rpm.}$$

(*b*) Rotor speed is given by

$$N = N_s\,(1 - s) = 1500\,(1 - 0.04) = \mathbf{1440\ rpm}$$

(*c*) $f' = s \times f = 0.03 \times 50 = \mathbf{1.5\ Hz}$

(*d*) At standstill, $N = 0$. Hence, $s = 1$

$\therefore \quad f' = s \times f = 1 \times 50 = \mathbf{50\ Hz}$

As seen, at standstill rotor frequency f' equals supply frequency, f.

Q. 21. It the emf in the stator of an 8-pole induction motor has a frequency of 50 Hz and that in the rotor 1.5 Hz, at what speed is the motor running and what is the slip?

(*Basic Elect. Engg. VTU Karnatka 2009*)

Ans. We are given that $f = 50$ Hz and $f' = 1.5$ Hz

Now, $\quad f' = s \times f \quad$ or $\quad 1.5 = s \times 50 \quad \therefore \quad s = \mathbf{0.03}$

$$N_s = 120 \times f/P = 120 \times 50/8 = 750 \text{ rpm}$$

Motor speed, $\quad N = N_s\,(1 - s) = 750\,(1 - 0.03) = \mathbf{727.5\ rpm}$

Q. 22. A 6-pole induction motor is supplied by a 10-pole alternator which is driven at 600 rpm, If the motor is running at 970 rpm, determine the precentage slip.

(*Basic Elect. Engg V.T.U Karnatka 2008*)

Ans. Frequency of the 3-ϕ supply produced by the alternator is

$$f = PN/120 = 10 \times 600/120 = 50 \text{ Hz}$$

Synchronous speed of the motor is given by

$$N_s = 120f/P = 120 \times 50/6 = 1000 \text{ rpm}$$

Now, $\quad N = N_s\,(1 - s) \quad \therefore \quad 970 = 1000\,(1 - s)$

$\therefore \quad s = 1 - 0.97 = 0.03 \quad$ or $\quad \mathbf{3\%}$

Q. 23. On what three factors does the torque of an induction motor depend? What is the relation between the three?

Ans. The three factors on whcih the torque of an induction motor depends are :

(*i*) flux per stator pole; Φ
(*ii*) rotor current, I_2 and
(*iii*) power factor of the rotor, $\cos\phi_2$

Here, I_2 is the rotor current at *standstill i.e.* when motor is stationary. Also, ϕ_2 is the phase angle between *rotor* emf and *rotor current.*

The rotor torque is proportional to the product of the three quantities given above.

$\therefore \quad T \propto \Phi\, I_2 \cos\phi_2 \quad$ or $\quad T = K\Phi I_2 \cos\phi_2$

If E_2 be the rotor emf at *standstill*, then $E_2 \propto \Phi$

$\therefore \quad T \propto E_2\, I_2 \cos\phi_2$

or $\quad T = k\, E_2\, I_2 \cos\phi_2 \quad$ — where k is another constant.

Q. 24. Derive an expression for the starting (or standstill) torque of an induction motor.

Ans. While deriving the expression, it should be kept in mind that the rotor of an induction motor has resistance as well as reactance. Hence, it possesses *impedance.*

Let, $\quad I_2$ = rotor current/phase at standstill

R_2 = rotor resistance/phase

X_2 = rotor reactance/phase at *standstill*

$Z_2 = \sqrt{R_2^2 + X_2^2}$

= rotor impedance/phase at *standstill*

Then, $I_2 = E_2/Z_2 = E_2 \Big/ \sqrt{(R_2^2 + X_2^2)}$

$$\cos \phi_2 = \frac{R_2}{Z_2} = \frac{R_2}{\sqrt{R_2^2 + X_2^2}} \quad \text{— Fig. 12}$$

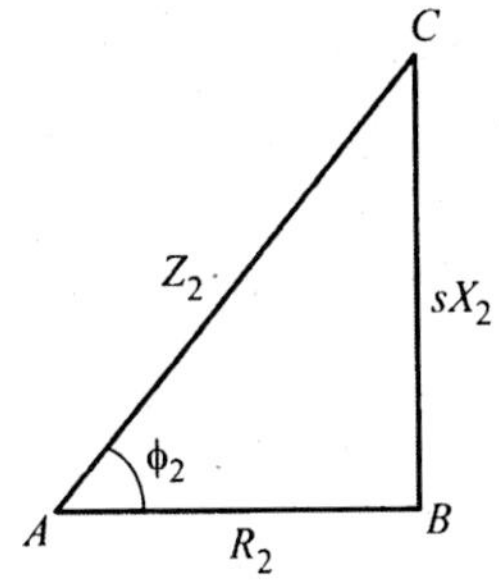

Fig. 12

Standstill or starting torque is

$$T_{st} = K\, E_2\, I_2 \cos \phi_2$$

$$= KE_2 \frac{E_2}{\sqrt{R_2^2 + X_2^2}} \times \frac{R_2}{\sqrt{R_2^2 + X_2^2}}$$

$$= \frac{KE_2^2\, R_2}{R_2^2 + X_2^2} = \frac{KE_2^2\, R_2^2}{Z_2^2}$$

If supply voltage V is constant, then flux Φ and hence E_2 both are constant.

$$\therefore \qquad T_{st} = k \frac{R_2}{R_2^2 + X_2^2} = k \frac{R_2}{Z_2^2}$$

where k is some other constant.

Q. 25. Explain why the starting torque of a squirrel-cage (SC) motor is much less as compared to that of a slip-ring (SR) motor.

Ans. The resistance (R_2) of squirrel-cage rotor is *constant* and is *small* as compared to its reactance (X_2) which is *very large* especially at the start. It is so because at standstill, the frequency of the rotor current is the *highest* and is equal to the supply frequency. As a result, the starting current I_2 of the rotor is very large but has a very small *active* component and a very large *reactive* component. The result is that the starting torque per ampere of the squirrel-cage motor is very poor. Hence, such motors are not useful where the motor has to start against heavy loads.

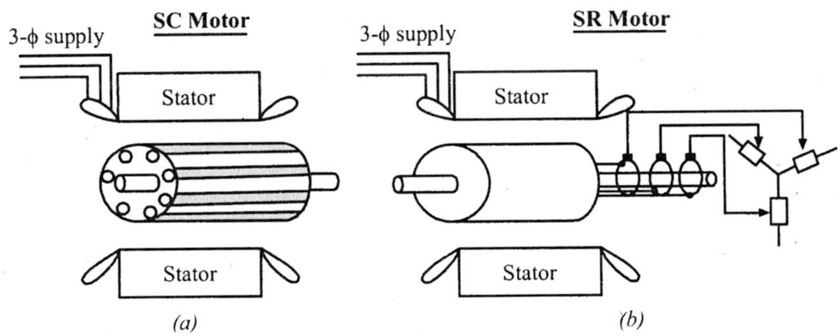

Fig. 13

The statrting torque of a slip-ring motor can be *increased* by improving its *power factor*. The power factor is increased by *adding* external resistance in the rotor circuit from the star-connected rheostat (Fig. 13). This rheostat resistance is gradually cut out as the motor speeds up. Of course, adding external resistance increases the rotor impedance and hence reduces the rotor current I_2. To start with, the effect of power factor improvement is more than the current-decreasing effect of rotor impedance. Hence, starting torque of the motor is increased at the *start.* But later on, the effect of reduced I_2 decreases the torque.

Q. 26. Derive the condition for maximum starting torque of an induction motor.

Ans. It can be proved that starting torque of an induction motor is maximum when its *rotor resistance is equal to the rotor reactance*. The starting torque of an induction motor is given by

$$T_{st} = \frac{k\,R_2}{R_2^2 + X_2^2} \qquad \therefore \quad \frac{dT_{st}}{dR_2} = k\left[\frac{1}{R_2^2 + X_2^2} - \frac{R_2(2R_2)}{\left(R_2^2 + X_2^2\right)^2}\right]$$

Now, for starting torque to be maximum, $dT_{st}/dR_2 = 0$

$$\therefore \qquad k\left[\frac{1}{R_2^2 + X_2^2} - \frac{R_2(2R_2)}{\left(R_2^2 + X_2^2\right)^2}\right] = 0$$

or $\qquad R_2^2 + X_2^2 = 2R_2^2 \quad \text{or} \quad R_2 = X_2$

Q. 27. How do the values of rotor emf, rotor current and rotor reactance of an induction motor get changed under running conditions as compared to standstill condition?

Ans. Let $\qquad E_2$ = *standstill* rotor induced emf/phase

X_2 = *standstill* rotor reactance/phase

f_2 = frequency of rotor current at *standstill*

When rotor is stationary *i.e.* $s = 1$, frequency of rotor emf is the same as the stator supply frequency. The value of rotor induced emf at standstill is *maximium* because *slip speed* is maximum.

When rotor starts running, the slip speed is *decreased.* Hence, the value of rotor emf is also decreased. Obviously, for a slip s, the rotor induced emf will be *s times its value at standstill.*

Hence, under running conditions

$$E_r = sE_2$$

The frequency of the rotor emf will also become

$$f_r = sf_2$$

Because rotor emf frequency has decreased, the rotor reactance is also decreased

$$\therefore \qquad X_r = sX_2$$

It should be noted that E_r and X_r are rotor emf and rotor reactance under *running conditions.*

Q. 28. Derive an expression for the torque developmed by an induction motor under running conditions. When does it achieve its maximum value?

Ans. The running torque of an induction motor is given by

$$T_r \propto E_r\, I_r \cos\phi_2$$

or $$T_r \propto \Phi\, I_r \cos\phi_2 \qquad ...(i) \qquad (\because E_r \propto \Phi)$$

Here, E_r is the rotor emf per phase and I_r is the rotor current per phase under *running* conditions.

Now, $$E_r = s\,E_2$$

$$\therefore \qquad I_r = \frac{E_r}{Z_r} = \frac{sE_2}{\sqrt{\left[R_2^2 + (sX_2)^2\right]}}$$

$$\cos \phi_2 = \frac{R_2}{\sqrt{\left[R_2^2 + (sX_2)^2\right]}} \quad \text{— Fig. 14}$$

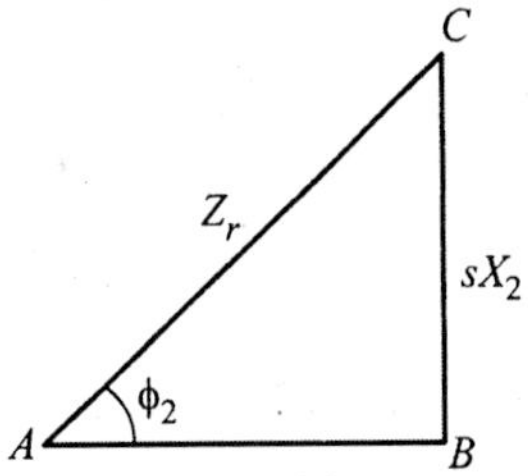

Fig. 14

Substituting the above values in Eqn. (*i*) we have

$$T_r \propto \frac{\Phi\, sE_2R_2}{R_2^2 + (sX_2)^2} = \frac{K\Phi .s.\, E_2R_2}{R_2^2 + (sX_2)^2}$$

Since $\Phi_2 \propto E_2$, we have

$$T_r = \frac{k\, s\, E_2^2\, R_2}{R_2^2 + (sX_2)^2}$$

It can be proved that for a 3-ϕ induction motor, $k, = 3/2\pi N_s$

$$\therefore \qquad T_r = \frac{3}{2\pi N_s} \cdot \frac{sE_2^2\, R_2}{R_2^2 + (sX_2)^2}$$

Obviously, at standstill when $s = 1$

$$T_{st} = \frac{k\, E_2^2\, R_2}{R_2^2 + X_2^2}$$

The *running* torque achieves its maximum value when $R_2 = s\, X_2$ or $s = R_2/X_2$

Q. 29. Discuss the relationship between the torque and slip of an induction motor. Does the maximum torque depend on the rotor resistance per phase?

Ans. As shown in Q. 28, the running torque of an induction motor is given by

$$T_r = \frac{K\Phi\, sE_2R_2}{R_2^2 + (sX_2)^2}$$

The torque/slip curves of Fig. 15 have been derived from this equation.

It is seen that when $s = o$, $T_r = o$. Hence, the curve starts from point O.

At normal speeds very close to synchronous speed, slip is small so that the term (sX_2) becomes very small. Hence, it becomes negligible with respect to R_2.

$$\therefore \qquad T_r \propto \frac{s}{R_2}$$

or $T_r \propto s$ because R_2 is constant.

It means that for small values of slip, the torque/slip curve is almost a straight line as shown. As slip increases when motor load is increased, the troque also increases and becomes, maximum when $s = R_2/X_2$. This torque is known as 'pull out' torque.

As the slip further increases *i.e.* motor speed decreases with further increase in motor laod, value of sX_2 increases. Hence, R_2 becomes negligible as compared to (sX_2).

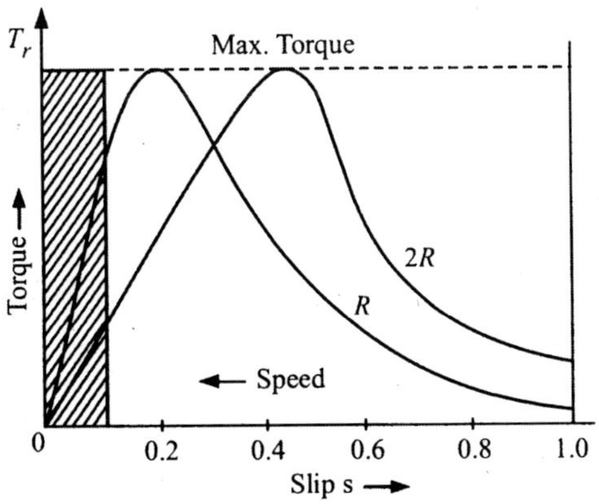

Fig. 15

$$\therefore \qquad T_r \propto \frac{s}{(sX_2)^2} \propto \frac{1}{s}$$

Hence, torque/slip curve becomes a rectangular hyperbola or T_r varies inversely as slip. It is seen that beyond the point of maximum torque, any further increase in motor load results in *decrease* of motor torque. As a result, the motor slows down and finally stops. The normal operation of the motor lies between the values of $s = 0$ and that corresponding to maximum torque. The operating range is shown by the shaded area in Fig. 15.

The maximum torque does not depend on the value of R_2 but the *exact location* of T_{max} is dependent on it as seen from Fig, 15. Greater the value of R_2, greater is the value of slip at which T_{max} occurs.

Q. 30. Explain why squirrel cage induction motor is considered a constant-speed motor?

Ans. In an induction motor, slip is essential to induce rotor currents required for developing motor torque. At no load, only a small torque is needed just to supply mechanical losses of the motor. Hence, motor slip is very small (about 1.5 percent or so). But when motor load is increased the rotor slip has also to increase in order to increase the rotor currents. The increased rotor currents will then produce higher torque required by the increased motor laod.

The short-circuited rotor bars of a squirrel-cage motor have low resistance and also low reactance under running conditions. Hence, rotor impedance is also very low. It means that only a *small* increase in rotor emf is required to produce a *large* increase in rotor currents. Therefore from no load to full load, only a small decrease in motor speed is required for producing large increase in rotor currents and hence in motor torque. That is why the speed regulation of a squirrel-cage motor is very small and, hence, it is called a constant-speed motor

Q. 31. Fully explain the different stages of power flow in an induction motor. Derive an expression for the gross torque and shaft torque developed by the rotor.

Ans. The losses taking place in an induction motor are as follows :

A. Stator losses :

(*i*) **Stator Cu loss :** It depends on the stator currents and the resistance of the stator winding.

(*ii*) **Stator Iron Loss :** It consists of eddy current loss and hystersis loss. These losses depend on the supply frequency and flux density in the iron core. Iron loss is almost *constant.*

B. Rotor Iron Loss : It is negligible because frequency of rotor currents under normal running conditions is always small.

C. Rotor Copper Loss : It is equal to $3\ I_2^2\ R_2$

D. Mechanical Losses : These consist of friction and windage losses.

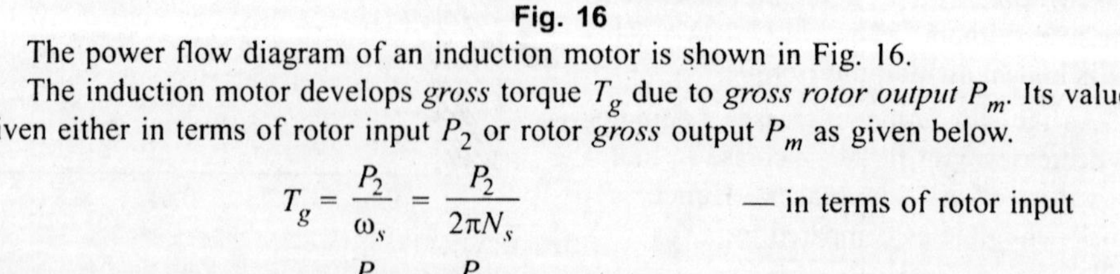

Fig. 16

The power flow diagram of an induction motor is shown in Fig. 16.

The induction motor develops *gross* torque T_g due to *gross rotor output* P_m. Its value can be given either in terms of rotor input P_2 or rotor *gross* output P_m as given below.

$$T_g = \frac{P_2}{\omega_s} = \frac{P_2}{2\pi N_s} \quad \text{— in terms of rotor input}$$

$$= \frac{P_m}{\omega} = \frac{P_m}{2\pi N} \quad \text{— in terms of rotor output}$$

The shaft torque T_{sh} is due to actual gross output power P_{out} which is less than P_m because of rotor friction and windage losses.

$$T_{sh} = P_{out}/2\pi N$$

In the above expressions, N and N_s are in rps. If they are in rpm, then the above equations become

$$T_g = \frac{P_2}{2\pi N_s / 60} = \frac{60}{2\pi} \cdot \frac{P_2}{N_s} = 9.55 \frac{P_2}{N_s} \text{ N-m}$$

$$= \frac{P_m}{2\pi N / 60} = \frac{60}{2\pi} \cdot \frac{P_m}{N} = 9.55 \frac{P_m}{N} \text{ N-m}$$

$$T_{sh} = \frac{P_{out}}{2\pi N / 60} = \frac{60}{2\pi} \cdot \frac{P_{out}}{N} = 9.55 \frac{P_{out}}{N} \text{ N-m}$$

Q. 32. Deduce the relationship between (*i*) rotor Cu loss and rotor input (*ii*) rortor gross output and rotor input and (*iii*) rotor Cu loss and rotor gross output

Ans. As seen from power flow diagram of Fig. 16.

$$T_g = \frac{P_m}{2\pi N} = \frac{\text{rotor gross output}}{2\pi N}$$

Also, $$T_g = \frac{P_2}{2\pi N_s} = \frac{\text{rotor input}}{2\pi N_s}$$

From the above two equations, we get

$$\text{rotor input} = T_g \times 2\pi N_s \quad ...(i)$$

$$\text{rotor gross outpur} = T_g \times 2\pi N \quad ...(ii)$$

As seen from Fig 16, the difference between rotor gross output and rotor input is equal to the rotor Cu loss. Hence, from Eqn. (*i*) and (*ii*), we get

$$\therefore \quad \text{rotor Cu loss} = \text{rotor input} - \text{rotor gross output}$$

$$= T_g \times 2\pi (N_s - N) \quad ...(iii)$$

Dividing Eqn. (*iii*) by (*i*) we get

$$\frac{\text{rotor Cu loss}}{\text{rotor input}} = \frac{N_s - N}{N_s} = s \quad ...(iv)$$

$$\therefore \quad \text{rotor Cu loss} = s \times \text{rotor input}$$

(*ii*) Again, as seen from Fig. 16.

$$\text{rotor gross output} = \text{rotor input} - \text{rotor Cu loss}$$

Putting the value of rotor Cu loss from above, we get,

$$\therefore \quad \text{rotor gross output} = \text{rotor input} - s \times \text{rotor input}$$

$$= (1 - s) \text{ rotor input}$$

$$\therefore \quad \text{rotor gross output} = (1 - s) \text{ rotor input}$$

or $$\frac{\text{rotor gross output}}{\text{rotor input}} = (1 - s) \quad ...(v)$$

(*iii*) Dividing Eqn. (*iv*) and (*v*), we get

$$\frac{\text{rotor Cu loss}}{\text{rotor gross output}} = \frac{s}{1 - s}$$

Q. 33. The power input to the rotor of a 440 V, 50 Hz, 3-phase, 12-pole induction motor is 75 W. The rotor emf has a frequency of 2 Hz.

Calulate :

(*a*) slip

(*b*) rotor speed

(*c*) rotor copper loss

(*d*) mechanical power developed. (*B.Tech Elect. Technology JNTU, Hydrabad 2007*)

Ans. (*a*) $f' = sf$ or $2 = s \times 50$ or $s = 0.04$ or **4%**

(*b*) $N_s = 120 \times 50/12 = 500$ rpm

$N = (1 - s) \times N_s = (1 - 0.04) \times 500 =$ **480 rpm.**

(*c*) rotor Cu loss = $s \times$ rotor input = $0.04 \times 75 =$ **3 kW**

(*d*) P_m = rotor input – rotor Cu loss = $75 - 3 =$ **72 kW**

Q. 34. A 3-ϕ, 50 Hz, 4-pole induction motor has a slip of 4%. Calculate : (*i*) speed of the motor (*ii*) frequency of rotor emf (*iii*) if the rotor has resistance of 1 Ω and stanstill reactance of 4 Ω, calculate p.f. at standstill and at a speed of 1400 rpm.

(*Elect. Engg. Nagpur Univ, 2004*)

Ans. (*i*) $N_s = 120f/P = 120 \times 50/6 = 1500$ rpm, $s = 4\% = 0.04$

Now, $$s = \frac{N_s - N}{N_s} \quad \text{or} \quad 0.04 = \frac{1500 - N}{1500} \quad \therefore \quad N = \mathbf{1440\ rpm}$$

(*ii*) $f' = s \times f = 0.04 \times 50 =$ **2 Hz**

(*iii*) The rotor impedance = $\sqrt{R_2^2 + (sX_2)^2}$ where X_2 is the rotor reactance *at standstill*

(*a*) Standstill Power Factor : At standstill *i.e.* when motor is stationary, $s = 1$

∴ standstill reactance = $sX_2 = 1 \times 4 = 4\ \Omega$

$$\therefore \quad Z_2 = \sqrt{1^2 + 4^2} = 4.123\ \Omega$$

$$\therefore \quad \text{p.f at standstill} = \cos\phi_2 = R_2/Z_2 = 1/4.123 = \mathbf{0.243\ (lag)}$$

At N = 1400 rpm

$$s = \frac{N_s - N}{N_s} = \frac{1500 - 1400}{1500} = 0.067$$

$$sX_2 = 0.067 \times 4 = 0.267\ \Omega$$

$$\therefore \quad Z_2 = \sqrt{1^2 + 0.267^2} = 1.035\ \Omega$$

$$\therefore \quad \text{p.f.} = \cos\phi = R_2/Z_2 = 1/1.035 = \mathbf{0.966\ (lag)}$$

It will be seen that power factor has increased from no load to full load by *almost 4 times.*

Q. 35. A 3-ϕ, slip-ring induction motor with a star-connected rotor has a slip of 4%. The rotor resistance per phase is 0.25 Ω and the standstill reactance per phase is 2Ω. The emf between slip rings at standstill is 104 V.

Calculate the rotor current and power factor when the slip-rings are short-circuited.

Ans. The circuit is shown in Fig. 17.

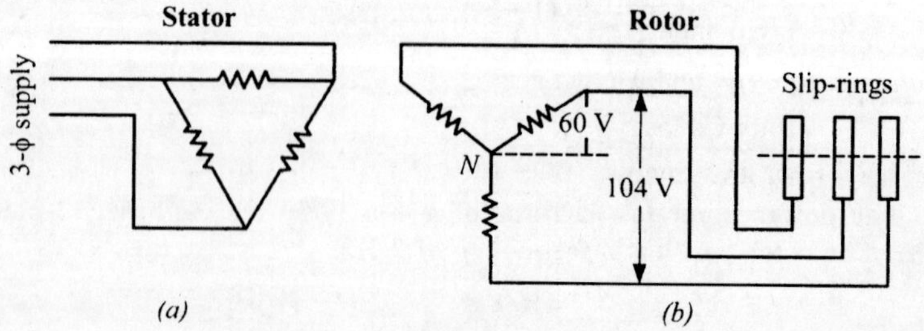

Fig. 17

Rotor voltage/phase at standstill

$$E_2 = 140/\sqrt{3} = 60\ \text{V}$$

Rotor phase voltage under runnings conditions = sE_2 = 0.04 × 60 = 2.4 V

$$Z_2 = \sqrt{R_2^2 + (sX_2)^2} = \sqrt{0.25^2 + (0.04 \times 2)^2} = 0.262\ \Omega$$

Rotor current/phase, I_2 = 2.4/0.262 = **9.15 A**

$$\text{p.f.} = R_2/Z_2 = 0.25/0.262 = \mathbf{0.95}\ \text{(lag)}$$

Q. 36. A 3-φ slip-ring induction motor has rotor resistance and standstill reactance per phase of 0.02 Ω and 0.1 Ω respectively. What should be the value of the external resistance per phase to be added to the rotor circuit for getting maximum torque at starting?

Ans. Let r be the required external resistance per rotor phase. Then, the rotor resistance becomes $R_2 = (0.02 + r)$. Also, $X_2 = 0.1\ \Omega$. The condition for getting maximum torque is

$$s = R_2/X_2$$

Now at start, $s = 1$. Hence, condition becomes

$$R_2 = X_2 \quad \text{or} \quad (0.02 + r) = 0.1 \quad \text{or} \quad s = \mathbf{0.08\ \Omega}$$

Q. 37. A 7.5 kW, 230 V, 3-φ, 50 Hz, 6-pole squirrel-cage induction motor operates at a full-load slip of 4% when rated voltage frequency is applied. Determine.

(*i*) full-load speed

(*ii*) full–load shaft torque in N-m

(*iii*) frequency of rotor current under this condition

(*iv*) speed of rotation of the stator mmf.

Ans. Let us first find the speed of rotation of the stator mmf which is the same as *synchronous speed.*

(*iv*) $N_s = 120\ f/P = 120 \times 50/6 = \mathbf{1000\ rpm}$

(*i*) $N = N_s\ (1 - s) = 1000\ (1 - 0.04) = \mathbf{960\ rpm}$

$$T_{sh} = 9.55\ \frac{P_{out}}{N} = 9.55 \times \frac{7.5 \times 10^3}{960} = \mathbf{74.6\ N\text{-}m}$$

(*iii*) $f' = s \times f = 0.04 \times 50 = \mathbf{2\ Hz}$

Q. 38. A 6-pole, 3-phase induction motor runs at a speed of 960 r.p.m. and the shaft torque is 135.7 N-m. Calculate the rotor Cu loss if the friction and windage losses amount to 150 watts. The frequency of supply is 50 Hz.

Ans.

$$\text{Output} = 2\pi \times N \times T_{sh}$$

$$= 2\pi \times (960/60) \times 135.7 = 13{,}642\ \text{W}$$

$$\text{Gross rotor output} = 13{,}642 + 150 = 13{,}792\ \text{W}$$

$$N_s = 120 \times 50/6 = 1000\ \text{r.p.m.}$$

$$s = (1000 - 960)/1000 = 0.04$$

Now

$$\frac{\text{rotor Cu loss}}{\text{rotor gross output}} = \frac{s}{1 - s}$$

∴

$$\text{rotor Cu loss} = \frac{13{,}792 \times 0.04}{1 - 0.04} = \mathbf{575\ W}$$

Q. 39. A 4-pole, 3-φ induction motor develops gross power of 7.5 kW at a speed of 1440 rpm when supplied from a 3-φ, 400 V, 50 Hz mains. If the power factor is 0.8, calculate for this load.

(*a*) slip (*b*) gross torque (*c*) rotor Cu loss/phase

Ans. (*a*)

$$N_s = 120f/P = 120 \times 50/4 = 1500\ \text{rpm}$$

∴

$$s = \frac{N_s - N}{N_s} = \frac{1500 \times 1440}{1500} = \mathbf{0.04}$$

(*b*) $$T_g = 9.55 \times \frac{P_m}{N} = 9.55 \times \frac{7500}{1440} = \mathbf{49.7\ N\text{-}m}$$

(*c*) $$\text{Rotor Cu loss} = \frac{s}{1-s} \times \text{rotor gross output}$$

$$= \frac{0.04}{1-0.04} \times 7500 = 312.5\ \text{W}$$

$$\text{Rotor Cu loss/phase} = 312.5/3 = \mathbf{104\ W}$$

EXERCISE

1. The frequency of stator emf is 50 Hz for an 8-pole induction motor. If the rotor frequency is 2.5 Hz, calculate the slip and the actual speed of rotor.
   (*Elect. Mach II, JNTU, Hydrabad 2007*)
   [5%, 1440 rpm]
2. A 6-pole induction motor is fed by three-phase 50 Hz supply and is running with a full load slip of 3%. Find the full load speed of induction motor and also the frequency of rotor emf. (*Basic Elect Engg. JNTU, Hydrabad 2007*)
   [970 rpm, 1.5 Hz]
3. The voltage appied to the stator of a 3-φ, 6-pole induction motor has a frequency of 50 Hz. The frequency of the induced emf in the rotor is 2 Hz. What is the slip and at what speed is the motor running? **[0.04, 960 rpm]**
4. A 3-φ, 400 V, 50 Hz induction motor has a speed of 950 rpm on full load. The motor has 6 poles. Calculate :
   (*i*) How many complete alternations will the rotor voltage make per second?
   (*ii*) speed of rotor field with respect to rotor core. **[2.5 Hz, 50 rpm]**
5. A 6-pole, 373-kW, 3-φ, 440 V, 50 Hz induction motor runs at a speed of 950 rpm on full load. Calculate the slip of the motor and frequency of rotor emf. **[0.05, 2.5Hz]**
6. In the case of an 8-pole induction motor, the supply frequency was 50 Hz and the shaft speed was 735 rpm. What were the magnitudes of the following :
   (*i*) synchronous speed
   (*ii*) slip speed
   (*iii*) percentage slip [750 rpm, 15 rpm, 2%]
7. A 400 V, 3-φ, 8-pole, 50 Hz induction motor when connected to supply runs at 730 rpm on no load. Find :
   (*i*) slip
   (*ii*) frequency of rotor currents **[0.0266, 1.333 Hz]**
8. A 400 V, 60 Hz, 6-pole, 3-φ induction motor runs at 1140 rpm. Calculate :
   (*i*) the slip
   (*ii*) synchronous speed
   (*iii*) frequency of rotor. **[5%, 1200 rpm, 3 Hz]**
9. An 8-pole alternator runs at 750 rpm and supplies power to a 6-pole induction motor which has at full-load a slip of 3%.
   Find the full-load speed of induction motor and the frequency of its rotor emf.
   [1000 rpm, 1.5 Hz]
10. A 4-pole, 3-φ, 275 kW, 440 V, 50 Hz induction motor has a speed of 1460 rpm on full load. Calculate the slip and the speed of the rotating magnetic field.
    (*Basic Elect. Engg. West Bengal, Tech Univ, 2003*)
    [0.0267, 1500 rpm]

11. The voltage applied to the stator of a 3-ϕ, 6-pole induction motor has a frequency of 50 Hz. The frequency of the EMF induced in the rotor is 2Hz. Calculate the slip and speed at which motor is running. (*Basic Elect. Engg. C.S.V., Tech Univ, Bhilai 2009*)
[0.04, 960 rpm]
12. A 3-ϕ, 4-pole, 50Hz induction motor is running at 1455 rpm. Find the slip speed and slip. [*Basic Elect. Engg. R.G.P.V, Bhopal, 2007*]
(45 rpm, 0.03)
13. An induction motor is supplied at a frequency of 50Hz. When the motor is fully loaded, the slip is 4.4%. Find the frequency of the rotor currents. **[2.2 Hz]**
14. A 12-pole, 50 Hz induction motor runs at 475 rpm. Calculate:
(*a*) the synchronous speed
(*b*) the slip speed and percentage slip
(*c*) the frequency of the rotor currents. **[500 rpm, 25 rpm, 5%, 2.5 Hz]**

MULTIPLE CHOICE QUESTIONS

1. The two main parts of an induction motor are
(*a*) stationary stator (*b*) revolving rotor
(*c*) slip-rings (*d*) both (*a*) and (*b*)
2. The rotor bars of a squirrel-cage induction motor are
(*a*) made of iron (*b*) slightly shorter than the rotor
(*c*) not connected to end-rings (*d*) short-circuited by end-rings
3. The air gap between the motor and stator of an induction motor is kept very small so as to obtain
(*a*) maximum field strength (*b*) maximum reluctance
(*c*) minimum air friction (*d*) minimum field strength
4. In a 3-ϕ induction motor, rotor currents are produced by
(*a*) conduction (*b*) direct connection (*c*) mutual induction (*d*) none of the above
5. The rotor speed of an induction motor
(*a*) is independent of motor load (*b*) is always less than synchronous speed
(*c*) may exceed synchronous speed (*d*) none of the above
6. The slip of an induction motor normally varies between
(*a*) 0 – 1 % (*b*) 1 – 5 % (*c*) 2 – 7 % (*d*) 5 – 10%
7. An induction motor is so-called because its operation depends on the phenomenon of
(*a*) self-induction (*b*) mutual induction
(*c*) eddy current (*d*) hysteresis
8. When an induction motor is at standstill, its
(*a*) stator current is zero
(*b*) stator frequency is zero
(*c*) rotor frequency equals supply frequency
(*d*) stator field is stationary
9. In an induction motor, there is no electrical connection to the
(*a*) stator (*b*) brushes (*c*) slip-rings (*d*) rotor
10. An induction motor operates on the basis of interaction between
(*a*) stator and rotor fields (*b*) two currents flowing in opposite directions
(*c*) two similar magnetic poles (*d*) none of the above
11. The two rotating fields produced by the stator and rotor of an induction motor
(*a*) rotate at the same speed
(*b*) are stationary with respect to each other

(*c*) rotate in opposite directions
(*d*) have the same speed with respect to rotor

12. In an induction motor, the frequency of rotor currents
(*a*) is the same at all rotor speeds
(*b*) equals supply frequency at standstill
(*c*) is greater than the supply frequency
(*d*) is never less than line frequency
13. The stator flux of an induction motor is
(*a*) stationary (*b*) alternating (*c*) rotating (*d*) reversing
14. The torque developed by an induction motor depends on
(*a*) flux per stator pole (*b*) rotor current
(*c*) rotor power factor (*d*) all of the above
15. Stator flux of an induction motor
(*a*) rotates at constant speed (*b*) rotates at varying speed
(*c*) has varying magnitude (*d*) alternates periodically
16. In an induction motor, the speed of stator rotating field
(*a*) depends on motor load (*b*) is independent of motor load
(*c*) is determined by rotor frequency (*d*) depends on the line frequency
17. When load on an induction motor is increased
(*a*) rotor current is increased (*b*) rotor torque is increased
(*c*) rotor slip is increased (*d*) all of the above
18. The no-load speed of an induction motor depends on
(*a*) the supply frequency (*b*) the number of its poles
(*c*) the maximum flux/phase (*d*) both (*a*) and (*b*)
19. The frequency of rotor current in a 6-pole, 50 Hz, 3-ϕ induction motor running at 950 rpm is .... Hz
(*a*) 2.5 (*b*) 1.5 (*c*) 5.0 (*d*) 0.05
20. Wound-rotor construction is used in induction motors to
(*a*) control motor speed (*b*) provide large starting torque
(*c*) minimise mechanical losses (*d*) both (*a*) and (*b*)
21. An induction motor differs from the synchronous motor in that its rotor
(*a*) carries no current (*b*) is not connected to any source of power
(*c*) is powered by magnetic induction (*d*) both (*b*) and (*c*)
22. The shaft torque developed by a 9.5 kW induction motor running at 950 rpm is
(*a*) 95.5 N-m (*b*) 955 N-m (*c*) 9.55 N-m (*d*) none of the above

ANSWERS

1. *d*	2. *d*	3. *a*	4. *c*	5. *b*	6. *b*	7. *b*
8. *c*	9. *d*	10. *a*	11. *b*	12. *b*	13. *c*	14. *d*
15. *a*	16. *d*	17. *d*	18. *d*	19. *a*	20. *d*	21. *d*
22. *a*.						

CHAPTER 15 SINGLE-PHASE INDUCTION MOTOR

Q. 1. Why in its elementary form, the single-phase induction motor is not self-starting? Describe various methods of starting a single-phase induction motor.

Ans. An *elementary* single-phase induction motor has *one* simple winding in its stator and a squirrel-cage rotor (Fig. 1). When this single winding is excited with single-phase current, it *does not* produce revolving magnetic field. A single-phase current produces only a *pulsating* or *alternating* flux. A rotating flux can be produced by either a 2-phase or 3-phase stator winding when supplied from a 2-phase or 3-phase ac supply respectively. The alternating flux of single-phase induction motor cannot produce a *starting* torque. Hence, the motor does not run.

The alternating flux of the single-phase motor induces currents in the rotor bars which produce their own flux. The stator and rotor fluxes react with each other and produce a starting torque during the first half cycle in one direction, say, in clockwise direction. In the next half cycle, the directions of the currents reverse. Hence, the directions of the two fluxes also reverse. Now, they produce a starting torque in counter-clockwise direction. It means that over a *full cycle* of the ac supply, net starting torque produceed by the motor will be zero. That is why a simple single-phase induction motor is not self-starting. Something has to be added to make it self-starting. For making the single-phase motor self-starting, we have to *temporarily* convert it into a 2-phase motor. For this purpose, we have to connect a second winding (called starting winding) to the same supply. It will convert the single-phase motor into a two-phase motor which will then produce a *revolving* flux around its stator. The two windings are connected in parallel and are spaced 90 electrical degrees apart [Fig. 2(*a*)]. As seen, two currents are drawn from the same ac supply and have a phase difference between them. The two fluxes produced by these currents combine to give a resultant revolving flux which develops starting torque. Now, the motor is called *split- phase* induction motor because there is a phase difference between running current and starting current.

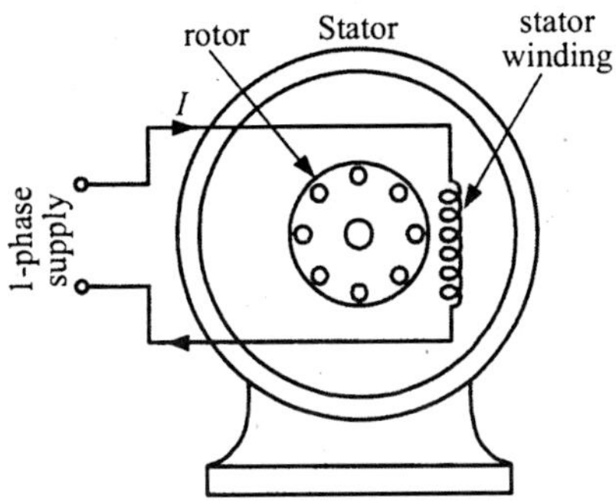

Fig. 1

Another way to make the motor self-starting is to connect a capacitor in the starting winding. Still another way is to use shaded-pole principle. Both these methods produce a rotating flux.

Q. 2. Describe the construction and working of a split-phase induction motor. Give its various applications.

Ans. (*a*) Construction : A split-phase motor is a single-phase induction motor that has *two* windings on the stator. One is called the *main or running* winding and the other is called *starting winding*. The two windings are connected in parallel and are 90° apart. The starting winding has *high resistance* whereas main winding has *high reactance* for creating phase difference between their currents. The motor behaves like a two-phase motor for starting purposes. A centrifugal switch S is connected in series with the starting winding and is located outside the motor. Its function is to automatically disconnect the starting winding from the supply when the motor reaches 75 to 80 percent of its synchronous speed.

(*b*) Working principle : The circuit connections of the motor are shown in Fig. 2(*a*). The starting winding has very high resistance but main winding has high reactance. Hence, starting current I_s *lags* behind the supply voltage by a small angle but the main winding current I_m lags by a large angle as shown in Fig. 2(*b*). The phase difference between the two currents is about 30°. These two currents produce a revolving flux and hence, make the motor self-starting. The starting torque of the motor is given by

$$T_{st} = k\, I_m\, I_s \sin \alpha$$

The angle α is known as the *torque angle*. It is the angle between I_m and I_s.

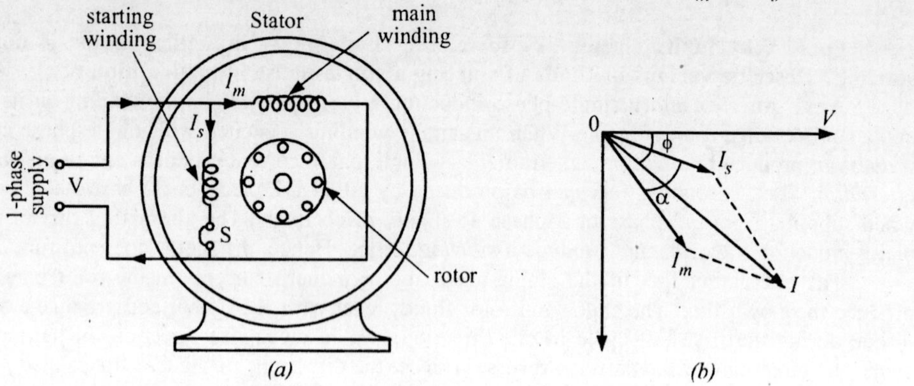

Fig. 2

(c) Applications

The split-phase motor is widely used for such applications as :

(*i*) washing machines
(*ii*) oil burners
(*iii*) blowers
(*iv*) business machines
(*v*) wood working tools
(*vi*) bottle washers
(*vii*) churns
(*viii*) buffing machines
(*ix*) grinders
(*x*) machine tools

Such motors are generally available in the 50-250 W range.

Q. 3. A 220 V, 50 Hz, 7.5 kW, single-phase induction motor operates at an efficiency of 85% and a power factor of 0.9 lagging. Find the capacitance of a capacitor to be connected in parallel with the motor so as to raise the line power factor to unity.

Ans. The capacitor *C* has been connected in parallel with the motor as shown in Fig. 3(*a*). The capacitor will draw current I_c which will be leading the supply voltage by 90° as shown in the phasor diagram of Fig. 3(*b*). This leading current will neutralize the lagging component of the current I_m drawn by the motor. Let us first find the value of motor current.

$$\text{motor input} = \text{output/efficiency}$$
$$= 7.5 \times 10^3/0.85 = 8{,}823 \text{ W}$$

Also, motor input $= V\, I_m \cos \phi$

$\therefore \quad VI_m \cos \phi = 8{,}823$

$\therefore \quad 220 \times I_m \times 0.9 = 8823 \qquad \therefore \quad I_m = 44.6 \text{ A}$

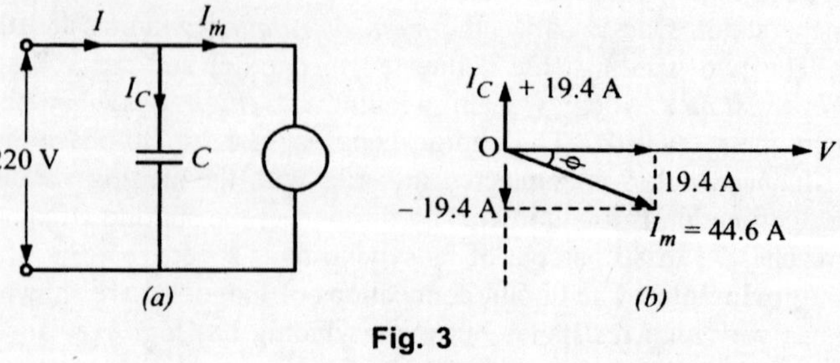

Fig. 3

The motor draws a current of 44.6 A lagging behind the supply voltage by $\phi = \cos^{-1}(0.9) = 25°50'$. Value of sin 25°50′= 0.4352. The lagging reactive current drawn by the motor is

$$I_m \sin 25°50' = 44.6 \times 0.4352 = 19.4 \text{ A}$$

It also represents the value of I_c which will raise the p.f. to unity.

Now $\qquad I_c \times X_c = 220$ or $19.4 \times X_c = 220$

or $\qquad X_c = 220/19.4 = 11.34\ \Omega$

Also $\qquad X_c = \dfrac{1}{\omega C} = \dfrac{1}{314C}$ or $11.34 = \dfrac{1}{314C}$

or $\qquad C = \dfrac{1}{11.34 \times 314}\ \text{F} = \dfrac{10^6}{11.34 \times 314}\ \mu\text{F} = \mathbf{281\ \mu F}$

Q. 4. Define a capacitor-start induction-run motor. Also, describe the construction and working of such a motor.

Ans. A capacitor-start motor is almost like a split-phase induction motor. The only difference is that, in this motor, a capacitor is added in series with the starting winding for *starting purpose.*

(*a*) Construction : Like split-phase motor, a capacitor-start motor also has two windings in the stator and a squirrel-cage rotor. The two stator windings are called *main winding* and *starting winding.* They are connected in parallel across the same single-phase supply and are spaced 90 electrical degrees apart. The starting winding has a capacitor and a centrifugal switch connected in series with it as shown in Fig. 4(*a*). The capacitor is mounted on top of the stator frame. The switch automatically disconnects the starting winding and the capacitor from the supply at nearly 75 to 80 percent of the synchronous speed.

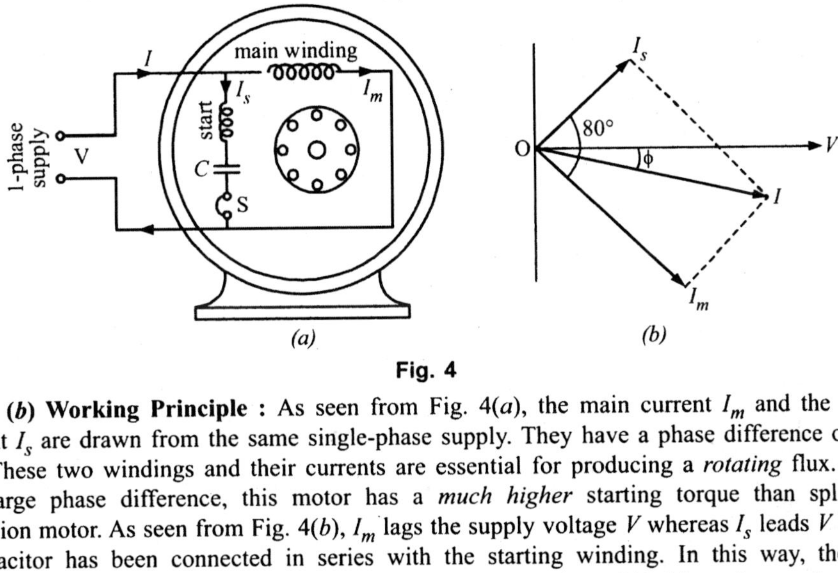

Fig. 4

(*b*) Working Principle : As seen from Fig. 4(*a*), the main current I_m and the starting current I_s are drawn from the same single-phase supply. They have a phase difference of about 80°. These two windings and their currents are essential for producing a *rotating* flux. Due to this large phase difference, this motor has a *much higher* starting torque than split-phase induction motor. As seen from Fig. 4(*b*), I_m lags the supply voltage V whereas I_s leads V because a capacitor has been connected in series with the starting winding. In this way, the phase difference between the two currents has been increased to about 80°. Now, the starting torque depends on the sine of the phase difference between the two winding currents. Hence, the starting torque of a capacitor-start motor is almost double that of the split-phase motor.

(c) Applications : The capacitor-start motors are very popular for heavy-duty general-purpose applications requiring high starting torques. These are generally used for such applications as compressors, jet pumps, farm and home work shop tools, swimming pool pumps and conveyors etc.

Q. 5. Differentiate between capacitor-start motor and two-value capacitor motor.

Ans. A capacitor-start motor is an induction motor which uses one capacitor only for starting purposes. That is why the easy-to-remember name 'capacitor-start' motor.

A two-value capacitor motor starts with one capacitor in series with the starting winding and runs with a different capacitor as shown in Fig. 5. The starting capacitor C_1 is cut out when the motor reaches near-synchronous speed. However, the starting winding and C_2 remain connected in the circuit *at all times*.

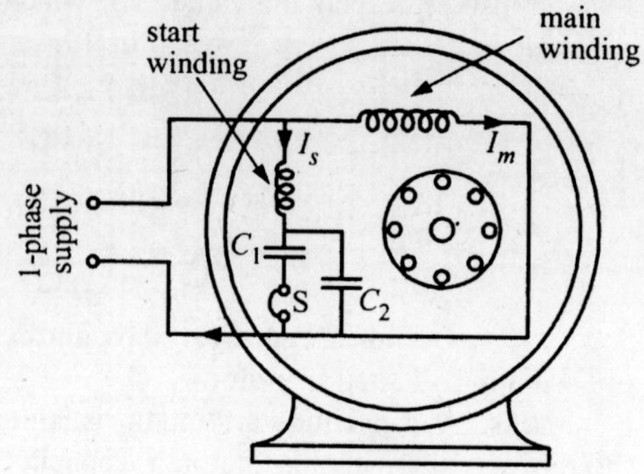

Fig. 5

As usual, two-value capacitor motor has two stator windings connected to the common ac supply and a squirrel-cage rotor. The two capacitances are added together, hence this motor has a high starting torque. Capacitor C_2 is called the *running* capacitor.

Such motors find applications in refrigerators, compressors, stokers, ceiling fans and blowers etc. In smaller sizes, they often compete with shaded-pole motors. Generally, they are not suitable for belted applications or for any other continous-duty application requiring large starting torque.

Q. 6. Describe the construction and working of a shaded-pole induction motor. Disscuss its advantages and disadvantages.

Ans. (*a*) Construction : As shown in Fig. 6, this motor has both a stator and a rotor. It has salient (*i.e.* projecting) poles which are different form any other poles. Each pole has a slot cut nearly one-third distance from one edge. Around the smaller part of the pole is placed a short-circuited copper coil known as *shading coil*. This part of the pole is known as shaded pole and the other as the unshaded pole. The rotor is a squirrel-cage type rotor

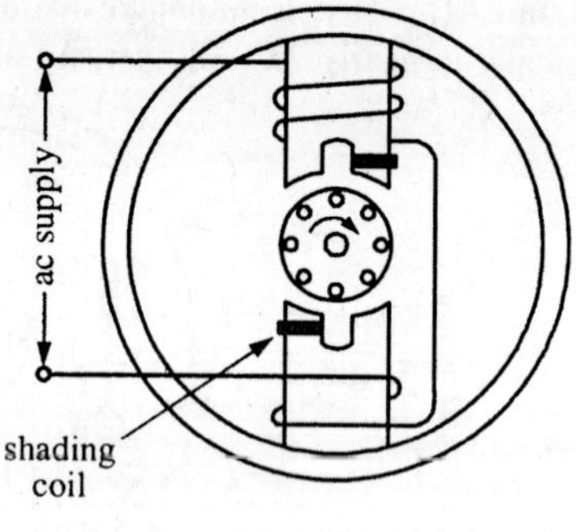

Fig. 6

(*b*) Working Principle : When an ac current is passed through the pole field winding, a magnetic field is set up. It is not a rotating field but one that sweeps *across the pole faces*. This field sweeps across the pole faces from the unshaded to the shaded portion. The *shifting* of this magnetic field is equivalent to the *actual* movement of the poles. Hence, the rotor starts rotating in the direction of this shift *i.e.* from the unshaded part of the stator pole to its shaded part.

(*c*) Applications : Shaded-pole motors are used widely for home applications such as fans of all kinds, slide projectors, humidifiers, small business machines such as photocopying machines, vending machines and advertising displays etc.

(*d*) Advantages : Such motors are built in very small sizes (5-50 W) but are simple in construction and are extremely rugged, reliable and cheap. They do not need any commutator, switches, brushes and slip-rings etc. They are available with gears to obtain almost any speed, even less than *one revolution per month*!

(*e*) Disadvantages

Some of their disadvantages are :

(*i*) low starting torque

(*ii*) very little overload capacity

(*iii*) very low efficiency from 5% to 35%

(*iv*) low power factor

Q. 7. A 40 W, 1550 rpm shaded-pole motor has a maximum torque of 0.3 N-m and a starting torque of 0.1 N-m. What percentage are these torques with respect to the full-load torque?

Ans. Full-load torque $= 9.55\ P_{out}/N$ — N in rpm

$= 9.55 \times 40/1550 = 0.246$ N-m

Percentage of max. torque $= \dfrac{0.3}{0.246} \times 100 = \mathbf{122\%}$

Percentage of starting torque $= \dfrac{0.1}{0.246} \times 100 = \mathbf{40\%}$

Q. 8. What is a universal motor? Why is it called so? Describe the construction and operation of a universal motor.

Ans. A universal motor is one which can be used with ac as well as dc supply at almost the same speed and output. Since its performance remains the same, hence it is called universal.

(*a*) Construction : It is a small *dc series motor* that is modified to run on ac supply. The powers of such motors vary from 50 W to 200 W and full speed varies from 6000 rpm to 8000 rpm. The basic construction of a small universal motor is shown in Fig. 7. The entire magnetic circuit is laminated to reduce eddy-current losses.

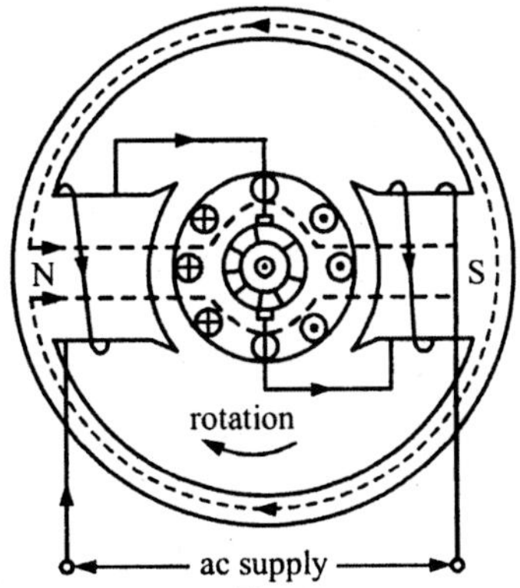

Fig. 7

(*b*) Operation : Since field winding and the armature winding are connected in series, the same current passes through both when motor is connected to either dc or ac supply. The magnetic fluxes of the series field and armature produced by this current react with each other and, hence, produce rotation. Because the series field magnetic flux and armature current reverse at the same time, the torque always acts in the same direction. Since universal motors are series-wound, they have *high* starting torque and high speed. They run at dangerously high speed at no load. That is why they are built into the device they drive.

(*c*) Applications : Such motors are commonly used for the following devices

(*i*) sewing machines
(*ii*) vacuum cleaners
(*iii*) food mixers and blenders
(*iv*) hair driers
(*v*) electric shavers
(*vi*) saws
(*vii*) projectors
(*viii*) portable power tools like drills etc.

Q. 9. Define a single-phase reluctance motor. Describe its construction, working principle and various applications.

Ans. It is a single-phase *synchronous* motor which does not require dc excitation for its rotor poles.

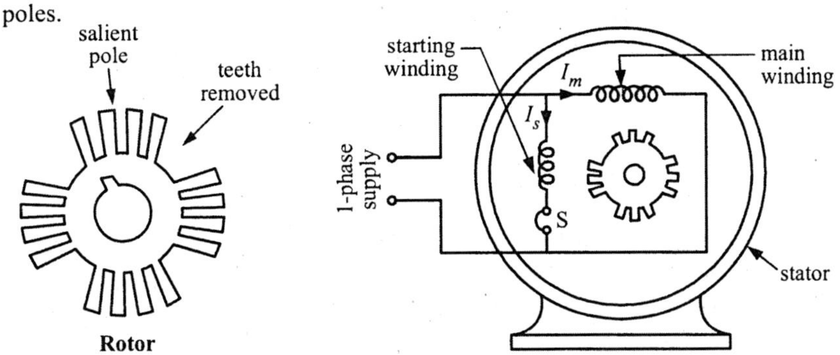

Fig. 8

(*a*) Construction : It consists of a stator which carries both the main and starting windings for producing a synchronously-rotating magnetic flux. There is a centrifugal switch for cutting out the starting winding. The rotor is squirrel-cage type but of unsymmetrical construction as shown in Fig. 8(*a*). By removing some of the teeth of a normal squirrel-cage rotor, salient poles are produced. These salient poles provide low-reluctance path to the stator flux since they offer low reluctance by becoming strongly-magnetised.

(*b*) Working Principle : When single-phase supply is switched on, the reluctance motor starts and accelerates like a squirrel-cage induction motor. But as it reaches 80-85 percent of its full-load speed, starting winding is cut off and the motor gets in step with the revolving stator flux thus running as a synchronous motor. This pull-in action takes place due to the attraction between the revolving stator poles and the magnetised low-reluctance rotor poles.

(*c*) Applications : As compared to an equivalent induction motor, such motors have poorer torque, power factor and efficiency. However, despite these drawbacks, such motors are widely used for constant-speed applications such as in

(*i*) recording instruments (*ii*) timing devices
(*iii*) control apparatus (*iv*) singnalling devices
(*v*) phonograph turntables (*vi*) control apparatus etc.

Q. 10. Describe the construction and working of a hysteresis synchronous motor. Mention same of its unique properties. What are its applications?

Ans. It is a single-phase synchronous motor having a non-salient pole rotor which has *no dc excitation.*

In a hysteresis motor, torque is produced by the hysteresis losses induced in the rotor.

(*a*) Construction : It consists of a stator and a rotor.

(*i*) ***Stator*** **:** For utilizing the split-phase principle, the stator has two windings which remain connected to the single-phase supply both at starting as well as during running of the motor. Hence, a revolving magnetic field is produced which induces hysteresis losses in the rotor.

(*ii*) ***Rotor*** **:** As shown in Fig. 9, the rotor consists of an outer ring and cross-bars of hard steel which have large hysteresis loss. The rotor is a cylinder having no salient poles.

(*b*) Working Principle : When stator is connected to the single-phase ac supply, a revolving field is produced. This field moves across the rotor and induces current in the cross-bars which magnetises them permanently. Hence, the motor starts as an *induction motor*. As the motor speeds up, hysteresis losses occur in the rotor ring which produce large driving torque. The motor quickly comes up to the synchronous speed.

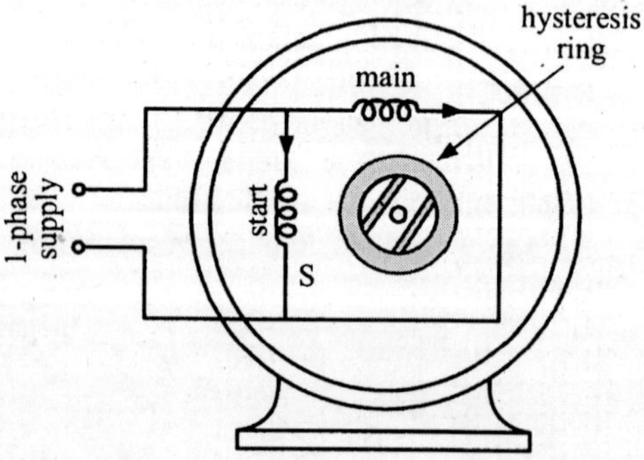

Fig. 9

(*c*) Unique Properties : The unique properties of a hysteresis synchronous motor are as under :

(*i*) since the rotor has no salient poles or any windings, it is extremely quiet in operation.
(*ii*) it produces no mechanical and magnetic vibrations.
(*iii*) torque exerted on the rotor *is always the same*. Hence, its accelerating torque is *constant*.

***(d)* Applications** : The hysteresis-synchronous motor is ideally suited for driving the following :

1. high-fidelity tape decks
2. phono turn-tables
3. electric clocks
4. precision audio equipment
5. various indicating devices.

MULTIPLE CHOICE QUESTIONS

1. An elementary single-phase induction motor has no starting torque because it has
   (*a*) a wound rotor (*b*) no starting winding
   (*c*) no rotating magnetic field (*d*) rotor without damping winding
2. The essential parts of a split-phase induction motor are
   (*a*) main winding (*b*) starting winding
   (*c*) squirrel-cage rotor (*d*) all of the above
3. The main winding of a split-phase induction motor has
   (*a*) more resistance than reactance (*b*) more reactance than resistance
   (*c*) less number of turns (*d*) very thin wire
4. A capacitor-start single-phase induction motor is one that uses
   (*a*) a capacitor for starting purposes only
   (*b*) starting winding and a capacitor continuously
   (*c*) one capacitor for starting and another for running
   (*d*) main and starting windings in series
5. The starting torque of a capacitor-start motor is
   (*a*) lower than that of split-phase motor
   (*b*) higher than that of split-phase motor
   (*c*) greater than that of a two-value capacitor motor
   (*d*) greater than that of a permanent-split motor
6. Use of a capacitor in a single-phase induction motor not only makes it self-starting but also increases
   (*a*) phase angle between main and starting currents
   (*b*) the starting torque
   (*c*) the running torque
   (*d*) both (*a*) and (*b*)
7. A two-value capacitor motor has two
   (*a*) running capacitors (*b*) stator windings
   (*c*) squirrel cage rotors (*d*) ac supply sources
8. Which of the following motor can have speeds as low as one revolution/month?
   (*a*) shaded-pole (*b*) reluctance (*c*) hysteresis (*d*) universal
9. In a shaded-pole motor working on split-phase principle
   (*a*) there is a rotating field (*b*) there is only sweeping field
   (*c*) non-salient poles are used (*d*) there is no shading coil
10. Which of the following single-phase motor is a synchronous motor without salient poles and dc excitation.
    (*a*) shaded-pole motor (*b*) hysteresis-synchronous motor
    (*c*) reluctance motor (*d*) capacitor-start motor

11. The single-phase motor that can be used both on ac and dc supply is called a .... motor
(*a*) unique (*b*) universal (*c*) reluctance (*d*) hysteresis
12. The rotating torque in a hysteresis-synchronous motor is produced by
(*a*) mutual attraction between stator and rotor poles
(*b*) permanent magnetism of the rotor bars
(*c*) hysteresis losses in the ring
(*d*) reaction between stator and rotor fluxes
13. Which of the following single-phase motor is quietest in operation?
(*a*) hysteresis-synchronous motor (*b*) split-phase induction motor
(*c*) two-value capacitor motor (*d*) reluctance motor
14. Which of the following single-phase motor starts as an induction motor but operates normally as a synchronous motor
(*a*) shaded-pole motor (*b*) capacitor-start motor
(*c*) hysteresis-synchronous motor (*d*) reluctance motor
15. A single-phase, salient-pole machine which runs as a synchronous motor without field excitation is called
(*a*) hysteresis motor (*b*) universal motor
(*c*) shaded-pole motor (*d*) reluctance motor

ANSWERS

1. *c* 2. *d* 3. *b* 4. *a* 5. *b* 6. *d* 7. *b*
8. *a* 9. *b* 10. *b* 11. *b* 12. *c* 13. *a* 14. *d*
15. *d*.

16 CHAPTER SYNCRONOUS GENERATOR

Q. 1. Bring out the similarities and differences between an ac generator and a dc generator.

Ans. Similarities :

1. Both types of generators are based on Faraday's laws of electromagnetic induction.
2. In both generators, action depends on the *relative* motion between conductors and a magnetic field.
3. Both require a stator and a rotor.
4. Both generators produce *alternating* emfs. In dc generators, ac voltage is converted to dc voltage with the help of commutator and brushes. However, ac generators directly deliver ac electric power to their loads.

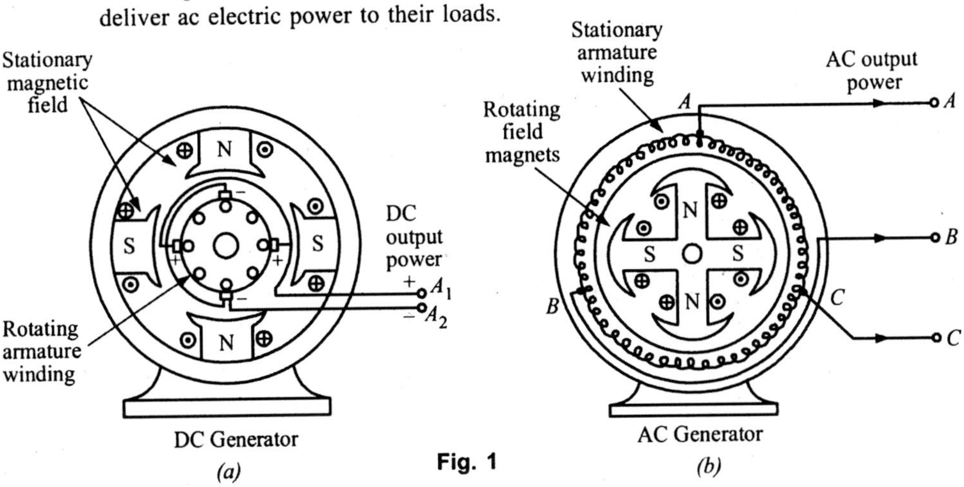

Fig. 1

Differences :

1. Unlike dc generators, ac generators always run at a *constant* speed. That is why they are called *synchronous* generators.
2. DC generators have a stationary field and a revolving armature. But ac generators employ a stationary armature and a revolving field (Fig. 1)
3. Synchronous generators are built in much larger sizes than dc generators.

Q. 2. Enumerate the advantages of having stationary armature and rotating field system in large size alternators.

Ans. The stationary-armature and revolving-field construction has the following advantages:

1. The armature winding of a synchronous generator is rated for very high voltages and currents. Hence, it is much larger and more complex than its field winding. Therefore, it can be better secured in the stationary stator from electrical point of view.
2. The armature windings experience less vibrations and centrifugal forces when housed in a stationary frame.
3. It is easier to insulate and protect the high-voltage armature winding used in synchronous generators when it is stationary.
4. The armature winding is easier to cool when stationary than when revolving.
5. The low-voltage field coils of a synchronous generator carry much smaller current as compared to its armature current. Hence, they are suitable for high-speed operations.

Q. 3. Name the two types of synchronous generator constructions.

Ans. There are two types of synchronous generators. They have the same stator construction but differ in the type of rotor used by them. The two types of rotors are :

(*i*) **salient pole rotor** and (*ii*) **smooth cylindrical rotor**.

The salient pole rotors are used only by low-speed generators such as those driven by diesel engines or water turbines. These rotors have projecting field poles as shown in Fig. 1(*b*). Because of their low speeds, they require a large number of poles. The salient pole structure is simpler and cheaper to manufacture than a cylindrical rotor structure.

The cylindrical rotors are used only in high-speed generators driven by steam turbines. Hnece, they are known as *turboalternators* or *turbogenerators*. They usually have two or four poles (Fig. 3).

Q. 4. With necessary sketches, distinguish between salient pole and cylindrical pole type of synchronous generators.

Ans. A synchronous generator consists of two main parts.

(*i*) **Stator** – it carries 3-phase armature windings as shown in Fig. 2 and Fig. 3.

(*ii*) **Rotor** – it provides rotating magnetic field structure.

***(a)* Stator :** Electrically speaking, the stator of a synchronous generator is similar to that of a 3-phase induction motor. It consists of a silicon-iron frame which supports the laminated core having slots on its inner side for housing the 3-phase winding. The winding is always star-connected and its neutral is earthed.

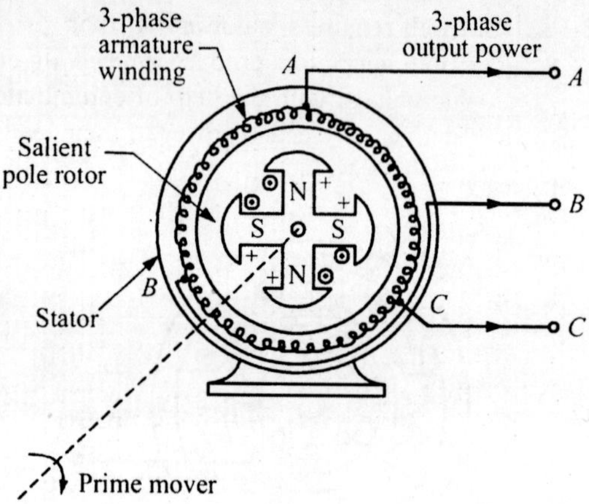

Fig. 2

***(b)* Rotor :** There are two types of rotors suited for low-speed and high-speed operations as discussed below :

1. **Salient-pole rotor :** It is like a flywheel which has a large number of alternate North and South poles bolted to it as shown on Fig. 2. The magnetic wheel is made of cost iron or steel of good magnetic quality.

These magnetic poles are energised or excited by a small dc generator mounted on the shaft of the alternator itself.

Such rotors are used in low and medium-speed alternators (50 – 300 rpm) which have large diameters and short axial lengths.

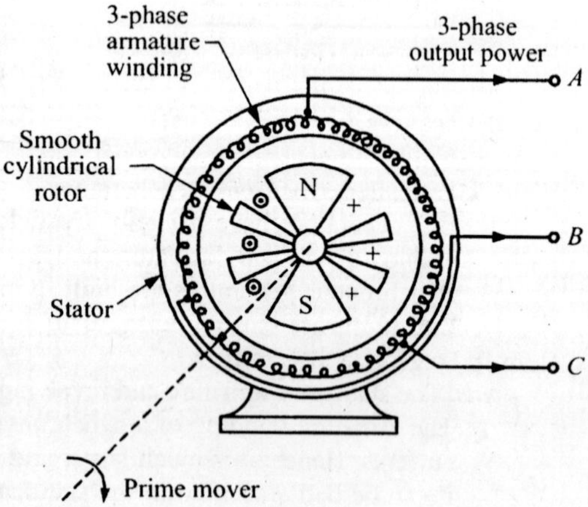

Fig. 3

2. **Smooth cylindrical rotor :** It consists of a smooth solid forged-steel cylinder having a number of slots milled out along its outer surface for housing magnetising field coils as shown in Fig. 3. Two or four regions are left unslotted for creating non-projecting poles. Such rotors are used in steam turbine-driven alternators which run at very high speeds (upto 3600 rpm). Such rotors have small diameters and very long axial lengths.

Q. 5. What determines the frequency of the generated voltage in a synchronous generator? How are these quantities related?

Ans. The frequency of the generated emf depends on (*i*) the number of field poles, *P* and (*ii*) on the generator speed, *N*. In a given armature coil, one complete cycle of emf is generated when a pair of rotor poles moves across it. Hence, number of cycles generated in one revolution of the rotor will be equal to the number of pairs of poles on the rotor i.e. *P*/2. If rotor speed *N* is in rpm, then number of revolutions per second is *N*/60. The frequency of the generated emf is given by

$$f = \frac{P}{2} \times \frac{N}{60} = \frac{PN}{120} \text{ Hz}$$

The most common power frequency used in india is 50 Hz although for railway electrification and traction work, low frequencies of 25 and $16\frac{2}{3}$ Hz are used.

Q. 6. Why are ac generators commonly called synchronous generators?

Ans. Unlike a dc generator, the synchronous generator has to be driven at a constant speed. The reason is that the frequency of the generated voltage has to match with that of the electrical network supplied by the generator. The speed of the generator which produces this matching frequency is known as synchronous speed. Since the speed of the ac generator has to be matched or synchronized with the electrical frequency of the network, hence ac generators are known as synchronous generators.

Q. 7. How is the rotor field system of a synchronous generator excited? What is brushless excitation?

Ans. The rotor poles of a synchronous generator are not permanent magnets but are electromagnets. They have to be supplied with dc current to excite (i.e energise or electromagnetise) them. There are several excitation systems which use rectifiers to provide dc excitation current for the generator. In one system, power is taken from the ac generator terminals, is rectified and then supplied to the rotor field system by means of slip-rings and brushes.

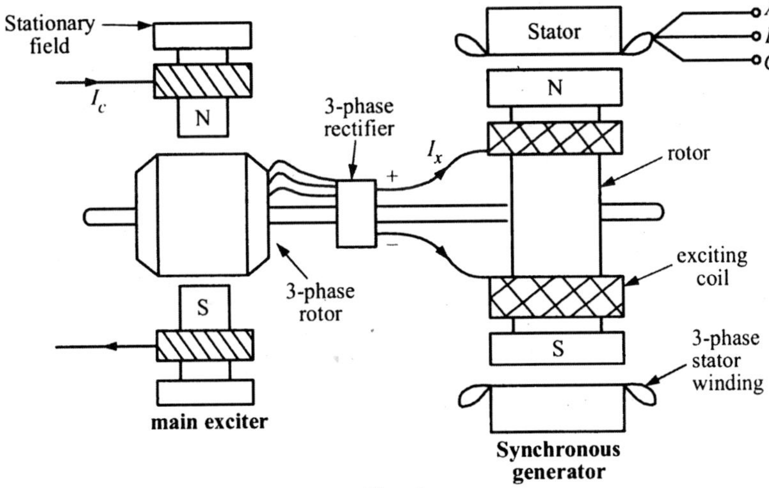

Fig. 4

In the brushless system of excitation, a small 3-phase generator mounted on the shaft of the main generator itself is used as an exciter. The output of the exciter is rectified and fed directly to the rotating field poles of the synchronous generator (Fig. 4). The brushless excitation system has no commutator, slip-rings or brushes which makes the system very simple and reliable.

Q. 8. Briefly discuss the process of voltage generation in a synchronous generator. How can the value of generated emf be controlled?

Ans. First of all, the ac generator is brought up to its proper speed by its prime mover. Next, its field poles are excited by dc current produced by a small dc generator mounted on the shaft of the alternator itself. As the field poles revolve, their flux is cut by the stationary armature winding. Hence, an induced emf is produced in the armature winding. It is an alternating emf because rotor poles of *opposite* polarity pass by a given stator conductor. This generated emf appears at the stator winding terminals.

The value of the generated emf depends on the rate at which flux lines are cut. In other words, it depends on the field strength and speed of the rotor. Since rotor speed is constant, the generated emf depends on the field strength. The field strength, in turn, depends on the dc exciting current supplied by the exciter. It means that the amount of generated emf can be controlled by adjusting the amount of *field excitation* supplied to the ac generator.

Power factor at which an ac generator operates is determined by the type of load it supplies. It could be lagging or leading or unity power factor.

Q. 9. How do the armature windings of a synchronous generator differ from those used in dc generators? Define the pitch factor and distribution factor of such a winding.

Ans. The stationary armature windings in alternators are different from those used in dc generators. The dc generators have *closed* circuit windings but alternator windings are *open* in the sense that there is no closed path for the armature current in the winding itself. Each end of the star-connected winding is joined to the neutral point and the other is brought out.

Pitch Factor (k_c)

The armature coils can have either *full pitch* or *fractional pitch* which is less than full pitch. The distance between two sides of a coil is called its *pitch*. When this distance is exactly equal to the distance between the centres of two adjacent poles (180°), the coil is said to be full-pitch and the winding a full-pitch winding (Fig. 5). Under this condition, the generated emfs in the two sides of the coil are *exactly in phase* with each other. Hence, voltage induced in the coil is the *arithmetic* sum of the two emfs.

However, if the distance between the two sides of a coil is less than 180°, the coil is said to have fractional pitch and the winding is called fractional-pitch winding [Fig. 5(b)]. Usually, fractional-pitch windings are used to make the generated emf sinusoidal.

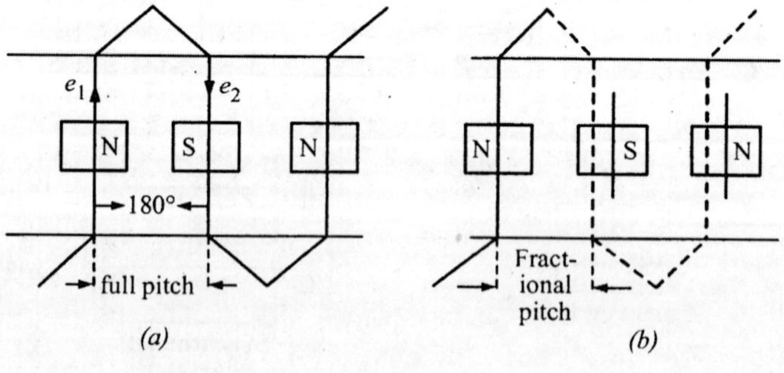

Fig. 5

When fractional-pitch windings are used, the emfs in the two sides of a coil are *not in phase with each other*. Hence, the resultant coil emf is less than it would be for a full-pitch winding.

The ratio of the emf generated in the fractional pitch coil to that generated in the full-pitch coil is called the *pitch factor* (k_c). Its value is always less than unity.

Distribution Factor (k_d)

Commercial windings used in alternators are not *concentrated* windings but *distributed* windings. In concentrated windings, all the conductors of each phase are included in *one* slot under each pole. In distributed windings, conductors of each phase group are distributed in *several* slots under the poles *i.e.* they are not concentrated in single slots under the poles. The result is that the emfs induced in coil sides of a polar group are *not in phase* with each other. Hence, value of the resultant emf is *reduced.* The ratio of the value of emf obtained with distributed winding to its value with concentrated winding is called the *distribution factor* or winding factor. Its value is always less than unity.

Q. 10. Derive an expression for the emf generated by a synchronous generator.

Ans. Let Z = No. of conductors (or coil sides) in series/phase

$= 2T$ — where T is the No. of *turns*/phase

P = No. of poles

Φ = flux/pole

N = rotor speed in rpm

Now, the average value of induced emf in a conductor is

$$= \frac{\text{total flux cut in one revolution}}{\text{time taken for one revolution}}$$

Since there are P poles, total flux cut by conductors in one revolution, $d\Phi = \Phi P$ weber.

Time taken for one revolution, $dt = 60/N$ second

$$\therefore \quad \text{average emf/conductor} = \frac{d\Phi}{dt} = \frac{\Phi P}{60/N} = \frac{\Phi NP}{60}$$

Alternator frequency, $f = PN/120$ or $N = 120f/P$

Putting this value of N in the above equation, we get

$$\text{average emf/conductor} = \frac{\Phi P}{60} \times \frac{120f}{P} = 2f\Phi \text{ volt}$$

Since there are Z conductors in series/phase

$$\therefore \quad \text{average emf/phase} = 2f\Phi Z \text{ volt} = 4f\Phi T \text{ volt} \qquad (\because Z = 2T)$$

Now, RMS value = form factor × average value

= 1.1 × average value

Hence, RMS value of the generated emf/phase is

$$E_{ph} = 1.11 \times 4f\Phi T = 4.44\, f\Phi T \text{ volt}$$

This would have been the *actual* value of the induced voltage if all the coils in each phase of the armature winding were (*i*) full-pitched and (*ii*) concentrated in one slot. But that is not the case. Hence, the actual voltage available is reduced in the ratio of these two factors.

$$\therefore \text{ emf/phase,} \quad E_{ph} = 4.44\, k_c\, k_d\, f\Phi T \text{ volt}$$

Since the armature windings are usually star-connected, the line voltage is $\sqrt{3}$ times the phase voltage. *i.e.* $E_L = \sqrt{3}\ E_{ph}$

Q. 11. A 6-pole, 3-ϕ, star-connected alternator has 90 slots and 8 conductors per slot and rotates at 1000 rpm. The flux per pole is 50 mWb. Find the induced emf across its lines. Take the winding factor as 0.97. (*Basic Elect. Engg. VTU, Karnatka 2007*)

Ans. Total No. of conductors = 90 × 8 = 720

conductors/phase, $Z = 720/3 = 240$, T/phase $= 240/2 = 120$

$f = PN/120 = 6 \times 1000/120 = 50$ Hz, $\Phi = 50$ mWb $= 50 \times 10^{-3}$ Wb

Winding factors, $k_d = 0.97$, $k_c = 1$ (assumed)

Now, $E_{ph} = 4.44\, k_c\, k_d\, f\Phi T$

$= 4.44 \times 1 \times 0.97 \times 50 \times (50 \times 10^{-3}) \times 120 = 1292$ V

Line voltage, $E_L = \sqrt{3}\, E_{ph} = \sqrt{3} \times 1292 =$ **2240 V**

Q. 12. A 4-pole, 1500 rpm, star-connected alternator has 9 slots/pole and 8 conductors per slot. Determine the flux per pole to give a terminal voltage of 3300 V. Take the winding factor as unity. *(Basic Elect. Engg. VTU, Karnatka 2008)*

Ans. $E_L = 3300$ V, $E_{ph} = 3300/\sqrt{3} = 1905$ V

$f = PN/120 = 4 \times 1500/120 = 50$ Hz

The number of slots/pole is 9. Total number of slots for four poles = 9 × 4 = 36

Total No. of conductors = 36 × 8 = 288

∴ turns per phase, $T = 288/2 = 144$

Since their values are not given, both k_c and k_d will be taken as unity

∴ $E_{ph} = 4.44\ k_c\ k_d\ f\Phi T$ volt

or $1905 = 4.44 \times 1 \times 1 \times 50 \times \Phi \times 144$ ∴ $\mathbf{\Phi = 0.06 Wb}$

Q. 13. A 3-ϕ, star-connected alternator on open circuit is required to generat a line voltage of 3600 V, 50 Hz when driven at 500 rpm. The stator has 3 slots/pole/phase and 10 conductors/slot. Calculate :

(*a*) the number of poles and (*b*) useful flux/pole.

Assume all the conductors/phase to be connected in series and the coils to be full-pitch. Winding factor, $k_d = 0.96$.

Ans. (*a*) $50 = P \times 500/120$, ∴ $P = \mathbf{12}$

(*b*) No. of slots/phase = 12 × 3 = 36

No. of conductors/phase, $Z = 36 \times 10 = 360$

No. of turns/phase, $T = Z/2 = 360/2 = 180$

$E_{ph} = E_L/\sqrt{3} = 3600/1.732 = 2080$V

Now, $E_{ph} = 4.44\ k_c\ k_d\ f\Phi T$ volt

∴ $2080 = 4.44 \times 0.96 \times 50 \times \Phi \times 180$ ∴ $\mathbf{\Phi = 0.054\ Wb}$

Q. 14. A 2-pole, 3-ϕ alternator running at 3000 rpm has 42 slots with 2 conductors/slot. Calculate the flux/pole required to generate a line voltage of 2300 V. Assume k_d = 0.952 and k_c = 0.956. *(Basic Elect. Engg. VTU, Karnatka 2008)*

Ans. $f = PN/120 = 2 \times 3000/120 = 50$ Hz

Total No. of conductors = 42 × 2 = 84

No. of conductors/phase = 84/3 = 28

No. of turns/phase, $T = 28/2 = 14$

$E_{ph} = E_L/\sqrt{3} = 2300/1.732 = 1328$ V

Now, $E_{ph} = 4.44\ k_c\ k_d\ f\Phi T$ volt

∴ $1328 = 4.44 \times 0.956 \times 0.952 \times 50 \times \Phi \times 14$ ∴ $\mathbf{\Phi = 0.47\ Wb}$

Q. 15. A 3-ϕ, 50 Hz, 16-pole alternator with star-connected winding has 144 slots with 10 conductors/slot. The flux per pole 24.8 mWb is sinusoidally distributed, the coils are full-pitched. Find (*i*) speed and (*ii*) the line emf. Assume winding factor k_d = 0.96. *(Basic Elect. Engg. VTU, Karnatka 2009)*

Ans. (*i*) $f = PN/120$ or $50 = 16 \times N/120$, $N = $ **375 rpm**

Total No. of conductors = 144 × 10 = 1440

No. of conductors/phase, $Z = 1440/3 = 480$

No. of turns/phase, $T = 480/2 = 240$

$k_d = 0.96$, $k_c = 1$ (given), $\Phi = 26.8 \times 10^{-3}$ Wb

(*ii*) $E_{ph} = 4.44 \times 1 \times 0.96 \times (24.8 \times 10^{-3}) \times 50 \times 240 = 1268$ V

$E_L = \sqrt{3} \times E_{ph} = 1.732 \times 1268 = \mathbf{2200\ V}$

Q. 16. Discuss the three factors responsible for the change in the generator terminal voltage upon loading.

Ans. As load on an alternator is varied, its terminal voltage V also varies as in dc generators. This change in V is due to the following reasons :

(*i*) voltage drop I_aR_a due to armature resistance, R_a

(*ii*) voltage drop I_aX_L due to armature reactance, X_L

(*iii*) voltage drop due to armature reaction.

To account for drop due to armature reaction, we assume an imaginary reactance X_a in the armature winding. The vector sum of X_L and X_a is called *synchronous reactance* X_s.

Hence, we can say that an alternator has (*i*) resistance R_a and (*ii*) synchronous reactance X_s. Their *vector* sum gives synchronous impedance Z_s.

$$\therefore \qquad Z_s = \sqrt{R_a^2 + X_s^2}$$

When an alternator is on load, there is voltage drop due to both R_a and X_s or due to Z_s alone. If V is the terminal voltage/phase on load and E is generated emf/phase on no load, then

$$\boldsymbol{E = V + I_a\,(R_a + jX_s) = V + I_a\,Z_s}$$

Q. 17. Draw and explain the significance of phasor diagrams of a loaded alternator. Is it possible for the terminal voltage to be greater than the generated emf?

Ans. As load is added to an alternator operating at a constant speed with fixed excitation, its terminal voltage changes. The amount of change depends on the power factor of the load.

Let E = no-load emf/phase

V = terminal voltage/phase

I_a = armature current/phase

ϕ = load p.f. angle

Z_s = armature synchronous impedance/phase

***(a)* Unity load p.f. :** In Fig. 6(*a*), V is taken as the reference quantity. Current I_a is in phase with V. Similarly, voltage drop $I_a\,R_a$ is in phase with I_a whereas voltage drop $I_a\,X_s$ is at right angles to it, their vector sum giving $I_a\,Z_s$. When $I_a\,Z_s$ is combined with V, we get E. It is seen that $V < E$.

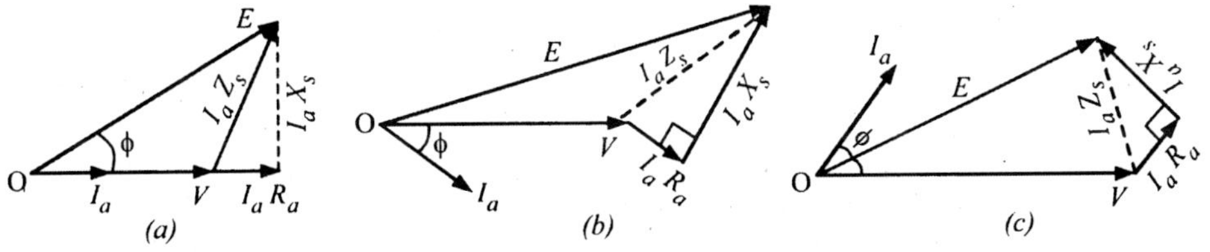

Fig. 6

***(b)* Lagging load p.f. :** In this case, I_a lags behind V by an angle ϕ. As usual, $I_a\,R_a$ is in phase with I_a whereas $I_a\,X_s$ is at right angles to it. When $I_a\,Z_s$ is combined with V, we get E as shown in Fig. 6(*b*). Again, $V < E$.

(c) Leading load p.f. : Such a case is shown in Fig. 6(*c*). Here, I_a leads V by an angle ϕ. As usual, E is the vector sum of V and $I_a\,Z_s$.

It is interesting to note that in this case, the generated voltage E is *less than* the terminal voltage V. The reason for this is obvious from Fig. 6(*c*).

Q. 18. Define voltage regulation of an alternator.

Ans. The terminal voltage V of an alternator varies with *load and its power factor.* Voltage regulation of an alternator is defined as the *rise* in voltage when load is reduced from full-rated value to zero, speed and field current remaining constant. It is normally expressed as a percentage of full-load voltage.

$\therefore$ % voltage regulation = $\dfrac{E - V}{V} \times 100$

Since for a leading power factor, $E < V$, hence regulation is *negative*.

Q. 19. A 50 kVA, 220 V, 50 Hz alternator has an armature resistance of 0.18 Ω and leakage reactance of 0.75 Ω per phase. Calculate the induced emf per phase when the generator is delivering rated current at a load power factor of unity.

Ans. Rated current

$$I_a = 50{,}000/220 = 227.3 \text{ A}$$

$$\text{resistance drop} = I_a\, R_a = 227.3 \times 0.018 = 4.1 \text{ V}$$

$$\text{reactance drop} = I_a \times X_s = 227.3 \times 0.075 = 17 \text{ V}$$

Hence, the induced emf in the generator per phase is given by

$$E = \sqrt{(V + I_a R_a)^2 + (I_a X_s)^2}$$

$$= \sqrt{(220 + 4.1)^2 + 17^2}$$

$$= \sqrt{50{,}221 + 289} = \mathbf{224.7\ V}$$

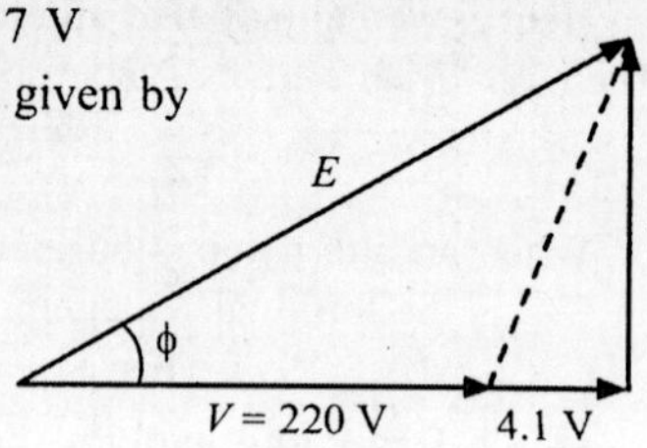

Fig. 7

As shown in Fig. 7. $E > V$

Q. 20. A 500 kVA, 1100 V, 50 Hz. Y-connected, 3-ϕ alternator has armature resistance/phase of 1.0 Ω and synchronous reactance/phase of 1.5 Ω. Find its voltage regulation for (*a*) unity p.f. (*b*) 0.9 lagging p.f. and (*c*) 0.8 leading p.f. Also, calculate voltage regulation in each case. (*Elect. Machines, Allahabad Univ.*)

Ans. $\sqrt{3}\, V_L\, I_L = 500{,}000$

$$I_L = 500{,}000/\sqrt{3} \times 1100 = 262 \text{ A}$$

armature current, $I_a = 262$ A

$$I_a\, R_a = 262 \times 0.1 = 26.2 \text{ V}$$

$$I_a\, X_s = 262 \times 1.5 = \mathbf{393\ V}$$

voltage/phase, $V = V_L/\sqrt{3} = 1100/\sqrt{3} = \mathbf{635\ V}$

(*a*) Unity p.f.

As shown in Fig. 8(*a*)

$$E = \sqrt{(V + I_a R_a)^2 + (I_a X_S)^2}$$

$$= \sqrt{(635 + 26.2)^2 + (393)^2} = \mathbf{769.2\ V}$$

$$\% \text{ regn.} = \frac{769.2 - 635}{635} \times 100 = \mathbf{21.1\%}$$

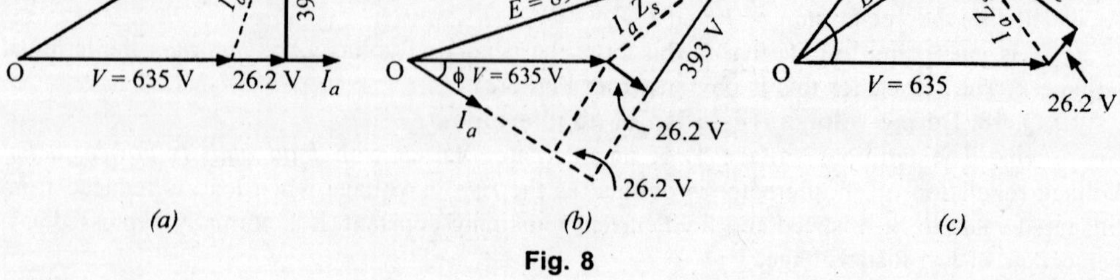

Fig. 8

(*b*) Lagging p.f.

This case is shown in Fig. 8(*b*)

$$\cos\phi = 0.9,\ \phi = 25.83°;\ \sin\phi = \sin 25.83° = 0.436$$

As seen from Fig. 8(*b*)

$$E = \sqrt{(V\cos\phi + I_a R_a)^2 + (V\sin\phi + I_a X_S)^2}$$

$$= \sqrt{(635 \times 0.9 + 26.2)^2 + (635 \times 0.436 + 393)^2} = 898 \text{ V}$$

$$\% \text{ regn.} = \frac{898 - 635}{635} \times 100 = \mathbf{41.4\%}$$

(*c*) Leading p.f.

As seen from Fig. 8(*c*).

$$E = \sqrt{(V\cos\phi + I_a R_a)^2 + (V\sin\phi - I_a X_S)^2}$$

$$= \sqrt{(635 \times 0.8 + 26.2)^2 + (635 \times 0.6 - 393)^2} = \mathbf{534\ V}$$

$$\% \text{ regn.} = \frac{534 - 635}{635} \times 100 = -\mathbf{15.9\%}$$

EXERCISE

1. A diesel-driven synchronous generator is operated at 200 rpm and has a frequency rating of 60 Hz. How many poles does it have? **(36)**
2. A 12-pole, 500 rpm star-connected alternator has 48 slots with 15 conductors per slot. The flux per pole is 0.02 Wb and is distributed sinusoidally. The winding factor is 0.97. Calculate the line emf. (*Basic Elect. Engg. VTU Karnataka, 2009*) **(896 V)**
3. The stator of a 3-ϕ, 8-pole, 750 rpm alternator has 72 slots each containing 10 conductors. Calculate the rms value of the emf per phase if the flux per pole is 0.1 Wb sinusoidally distributed. Assume full-pitch coils and a distribution factor of 0.96. **(2560 V)**
4. A 3-ϕ, star-connected alternator running at 900 rpm generates a line voltage of 460 V at 60 Hz on open circuit. The stator has 2 slots/pole/phase and 4 conductors/slot. Calculate : (*a*) the number of poles and (*b*) useful flux/pole. **(8, 32.3 mWb)**
5. Find the number of stator conductors per slot for a 3-ϕ, star-connected 50 Hz alternator for generating a line voltage of 13 kV on open circuit. The flux/pole is 0.15 Wb and speed is 300 rpm. Assume full-pitch coils and the stator to have 3 slots pole/phase. **(8)**
6. A 4-pole, 50 Hz, 3-ϕ delta-connected alternator has armature winding distributed in 36 slots each containing 16 conductors. The flux/pole is 0.04 Wb. Calculate the terminal emf on open circuit. **(818 V)**
7. A 3-phase, 16-pole alternator has a star-connected winding with 144 slots and 10 conductors per slot. The flux per pole is 30 mWb sinusoidally distributed and the speed is 375 rpm. Find the frequency, the phase and the line emf. **(50 Hz, 1600 V, 2770 V)**

MULTIPLE CHOICE QUESTIONS

1. The main parts of a synchronous generator are
(*a*) stator (*b*) rotor (*c*) exciter (*d*) all of the above
2. The frequency of a 2-pole alternator running at 3600 rpm is .... Hz.
(*a*) 50 (*b*) 60 (*c*) 1800 (*d*) 120
3. A 4-pole, 1200 rpm alternator generates emf at a frequency of
(*a*) 25 Hz (*b*) 40 Hz (*c*) 50 Hz (*d*) 60 Hz
(*Basic Elect. Engg. VTU, Karnataka 2009*)
4. The rated voltage of alternators used is power stations is usually
(*a*) 440 kV (*b*) 220 kV (*c*) 110 kV (*d*) 11 kV
(*Basic Elect. Engg. VTU, Karnataka 2009*)
5. Low-speed synchronous generators run by hydraulic turbines use a
(*a*) salient-pole rotor (*b*) smooth cylindrical rotor
(*c*) either (*a*) or (*b*) (*d*) none of the above
6. The frequency of the emf generated by an alternator depends on
(*a*) the number of poles (*b*) generator speed
(*c*) flux/pole (*d*) both (*a*) and (*b*)
7. Usually, modern alternators have short-pitched windings so as to
(*a*) increase the machine rating (*b*) improve the voltage waveform
(*c*) increase generated voltage (*d*) none of these
(*Basic Elect. Engg. VTU, Karnataka 2009*)
8. The easiest way to change the voltage generated by an alternator is to change its
(*a*) No. of poles (*b*) speed (*c*) field excitation (*d*) slots/pole
9. One basic similarity between an alternator and a dc generator is that both
(*a*) run at constant speed (*b*) have stationary armature and revolving field
(*c*) generate alternating emf (*d*) use an exciter
10. The field winding of an alternator is excited by ....
(*a*) dc (*b*) ac (*c*) both dc and ac (*d*) none of these
(*Basic Elect. Engg. VTU, Karnataka, 2009*)
11. For full-pitch coil, the pitch factor k_p is ....
(*a*) 1 (*b*) greater than 1 (*c*) less than 1 (*d*) none of these
(*Basic Elect. Engg. VTU, Karnataka 2009*)
12. The voltage regulation of an alternator can be
(*a*) positive (*b*) negative (*c*) zero (*d*) all of the above

ANSWERS

1. *d* 2. *b* 3. *b* 4. *d* 5. *a* 6. *d* 7. *b*
8. *c* 9. *c* 10. *a* 11. *a* 12. *d*.

CHAPTER 17 SYNCRONOUS MOTOR

Q. 1. Briefly discuss the construction of a simple synchronous motor. Why is this motor called a synchronous motor?

Ans. Construction of a synchronous motor is similar to that of a synchronous generator. As shown in Fig. 1, essential parts of this motor are the same as those of the aternator *i.e.* a stator and a rotor.

The stator houses 3-phase armature winding in the slots of its laminated core. This winding is similar to that used in synchronous generators and induction motors.

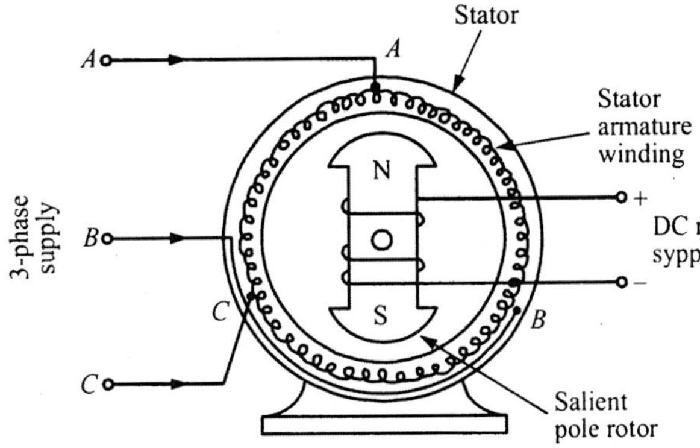

Fig. 1

The rotor has a set of salient poles that are exited (or electromagnetised) by a *dc current source*. Due to this exciting current, the rotor poles become alternately N and S poles. The number of rotor poles is made *equal* to the number of stator poles.

Modern synhronous motors often use brushless exicitation similar to that used in synchronous generators.

Most of the synchronous motors use salient-pole rotor as shown in Fig.1. A simple synchronous motor is basically not self-starting. However, it can be made self-starting with the help of *damper windings* carried by the rotor poles [Fig. 3(*b*)]

This motor is called synchronous motor beacuse its rotor runs in synchronism or in step with the synchronously rotating stator field.

Q. 2. Why is a synchronous motor not self-starting? Give reasons.

Ans. When the 3-phase winding of a synchronous motor is fed with a balanced 3-phase supply, a revolving stator magnetic field is produced. This revolving field has constant magnitude and revolves around the stator with synchronous speed $N_s = 120f/P$. It is called synchronous speed because it is *synchronised to the power line frequency*. For all practical purposes, it can be assumed as if there are actual magnetic poles which rotate around the stator. Fig. 2 shows two *stator* poles marked N_s and S_s *revolving* synchronously in clockwise direction around the rotor poles. As these stator poles rush past the rotor field poles, they pull the rotor first in *one direction* and then in *the other* at a very fast rate.

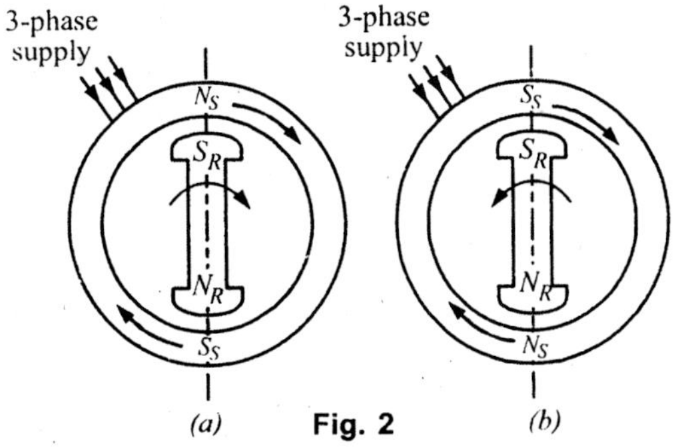

Fig. 2

Suppose at any instant, N_s happens to face S_R as shown in Fig 2(*a*). Due to mutual attraction, N_s tries to pull S_R in *clockwise* direction. However, rotor has high inertia and takes a little time to move. Meanwhile, the revolving pole N_s is replaced by S_s which repels S_R in *anti-*

clockwise direction as shown in Fig. 2(*b*). As seen, the high-inertia rotor is made to move first in one direction and then in the other direction in quick succession. The net result is that it just stays where it is. That is why a simple synchronous motor is not self- starting.

Q. 3. Explain how a simple synchronous motor can be made self-starting?

Ans. A synchronous motor cannot start by itself because it develops no starting torque at rest. However, it can be made self-starting with the help of squirrel-cage winding also known as *damper winding* or *pole face winding.* It consists of parallel copper bars embedded in the pole faces and short-circuited at both ends [Fig. 3(*b*)].

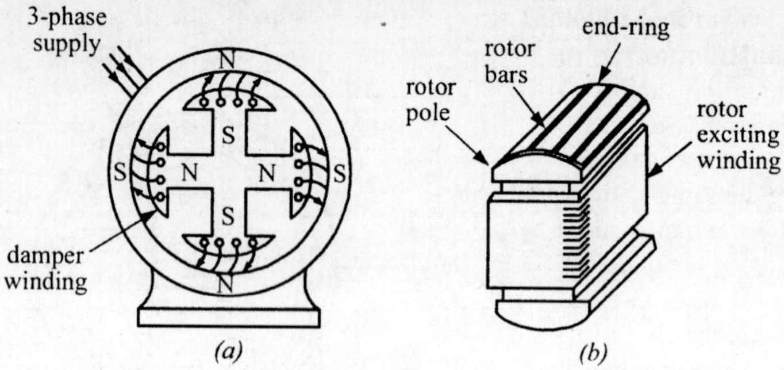

Fig. 3

The synchronous motor starts readily acting temporarily as a 3-*phase induction motor* during the starting period. How the motor runs is given below.

The supply voltage is switched on to the stator winding while the rotor pole field circuit is *left open*. The rotating stator field induces an emf in the damper winding as it passes over it. Hence, the synchronous motor starts up as an *induction motor*. When it reaches near its synchronous speed, its rotor field circuit is closed. Hence, the rotor poles are magnetised and get immediately locked with the stator poles and start rotating at synchronous speed in the *same* direction. The slip between the rotating stator poles and rotor poles becomes *zero i.e.* the stator and rotor poles run at the same speed. In fact, torque in a synchronous motor never develops at any other speed except *synchronous speed.*

Q. 4. Briefly describe the principle of operation of a synchronous motor.

Ans. When 3-phase supply is fed to stator armature winding of a synchronous motor, a revolving magnetic field is produced. This revolving field induces current in the damper winding. This current produces its own magnetic field. The stator and the rotor magnetic fields react with each other and hence produce a starting torque by mutual attraction between rotor and stator poles. As a result, rotor starts rotating in the direction of the revolving stator field just as in an induction motor. Now, the rotor excitation is *switched on.* The rotor poles become N and S poles alternately. These poles lock up with stator poles of opposite polaritly and are dragged along the stator poles. The torque developed at this stage is called *pull-in* torque. The synchronous speed of the motor is given by $N_s = 120f/P$.

Q. 5. Explain with suitable diagrams how a synchronous motor can be made to run at power factors varying from lagging to leading ones?

Ans. One of the most important property of a synchronous motor is that it can operate at lagging, unity and leading power factors. It can be easily understood by remembering that, as in a dc motor, there are *two opposing voltages* in the *stator* winding of a shynchronous motor. The two voltages are :

(*a*) the supply voltage V

(*b*) the back emf E_b induced in the *stator* winding by the rotating *rotor field.*

The net voltage E_R across the stator winding is the *vector difference* of V and E_b. Since $E_R = V - E_b$ and Z_s is the impedane of the stator winding, the stator winding current $I_a = E_R/Z_s$

and lags behind E_R by a fixed angle $\theta = \tan^{-1} (X_s/R_a)$ where R_a is resistance/phase and X_s is reactance/phase of the stator winding. The angle θ is *almost* 90° because R_a is very small as compared to X_s. If ϕ is the phase angle between V and I_a, motor power factor is $\cos\phi$ as shown in Fig 4. The power factor can be changed by changing *rotor field exitation* which changes E_b.

(*i*) Consider the case when field excitation is weak *i.e.* when motor is underexcited. In that case, E_b is lesser in magnitude than V. The armature current I_a lags behind V by an angle ϕ. Hence, motor has a lagging power factor $\cos \phi$ as shown in Fig. 4(*a*).

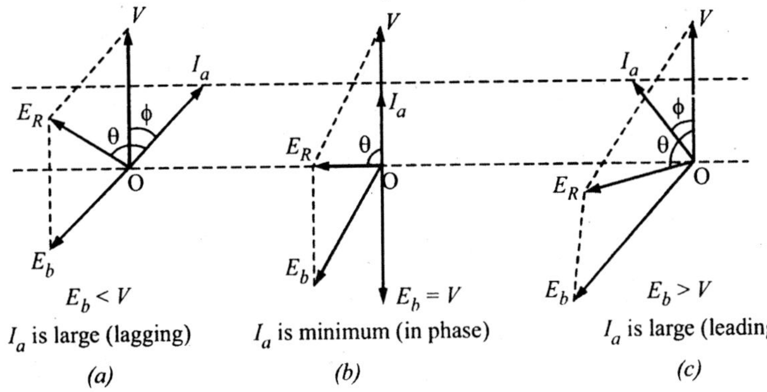

Fig. 4

(*ii*) Now, let rotor field exitation be increased so that E_b *equals* V in magnitude as shown in Fig. 4(*b*). In this case, the motor is said to have *normal* excitation. The stator armature current I_a comes in phase with V and has *minimum* value for the *given load*. The motor has a power factor of unity because $\phi = 0$.

(*iii*) When rotor field excitation is further increased *i.e.* when motor is *overexcited*, then E_b becomes greater than V in magnitude and I_a *leads* V by an angle ϕ so that motor power factor becomes *leading* as shown in Fig. 4(*c*).

The fact that a synchronous motor can run with a leading power factor makes it ideal for correcting the lagging power factor of industrial loads driven by induction motors.

Q. 6. Using appropriate phasor diagrams, derive an expression for the mechanical power developed in the armature of a synchronous motor.

Ans. When a synchronous motor runs, two opposing voltages are produced in its armature *i.e.* stator winding. These are

(*i*) the supply voltage V and

(*ii*) back emf E_b induced in the stator armature winding by the synchronously revolving pole flux of the rotor.

Whereas V is constant, E_b depends on the rotor speed and its flux. Since rotor runs at a *constant speed, i.e.* sychronous speed, the value of E_b depends *only on the flux* of the rotor poles. Rotor flux, in turn, depends on the rotor *field excitation*. The net voltage/phase in stator armature winding per phase is

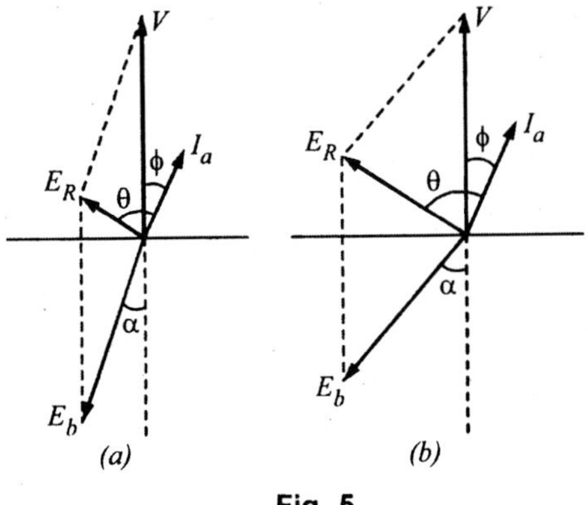

Fig. 5

$$E_R = V - E_b \qquad \text{— vector difference}$$

The stator armature winding impedance per phase is

$$Z_s = \sqrt{R_a^2 + X_a^2}$$

Therefore, stator armature current/phase, $I_a = E_R/Z_s$. Fig. 5(*a*) represents the no-load condition of an *underexcited* synchronous motor *i.e.* one in which $E_b < V$. If ϕ is the angle between V and I_a, motor power factor is cos ϕ and motor power input/phase = $VI_a \cos\phi$.

The armature current I_a lags behind E_R by a *fixed* angle $\theta = \tan^{-1}(X_s/R_a)$. Since R_a is almost negligible as compared to X_s, hence, $\theta = \tan^{-1}(X_s/0) = \tan^{-1}(\infty) = 90°$. In practice, θ is about 85° or so.

When load is applied to the motor, its rotor further shifts backward from the stator poles as shown in Fig. 5(*b*) but its speed remains the same. As a result, vector E_b further falls behind vector *V in phase* (see Q. 7). That is why load angle α is *greater* in Fig. 5(*b*) than its value in Fig. 5(*a*).

It means that when load on the motor is increased, magnitude of E_R and, hence, of I_a is increased. As a result, the motor draws more power from the supply in order to meet the extra load put on it.

The mechanical power developed in motor armature is

$$P_{mech} = E_b \times I_a \times \text{cosine of the angle between the two}$$
$$= E_b I_a \cos(\alpha - \phi) \text{ per phase.}$$

Q. 7. In what manner does a synchronous motor adjust itself to an increasing shaft load?

Ans. The stator and the rotor poles of a running synchronous motor are *magnetically locked* and, hence, both run with the same synchronous speed. But the two poles are *not in phase* with each other even when motor runs on *no load*. In other words, the centre lines of the two poles do not coincide even on *no load* as shown in Fig. 6(*a*). There is a small *backward shift* α of the rotor poles relative to the stator poles as shown. But the stator and rotor poles still keep running synchronously. This backward shift is essential for developing motor torque in order to meet no-load losses of the motor. However at not load, the torque angle α is small.

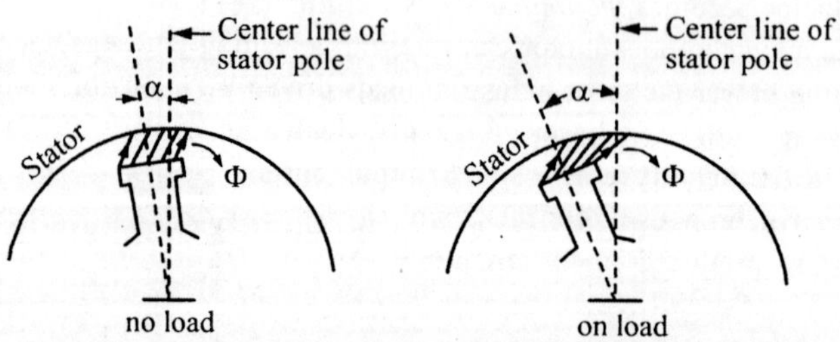

Fig. 6

As the load on the motor is increased, the backward shift of the rotor poles *increases* by a larger angle α as shown in Fig. 6(*b*). But rotor poles still continue to run synchronously.

The value of this backward shift α depends on the *load* carried by the motor. Higher the amount of load, greater the value of this angle α. It is also known as *load angle* or *coupling angle*. This angle α controls armature current I_a. Greater the value of α, higher the value of stator armature current I_a. It should be so because motor needs more input power to carry the incresed load.

If too much mechanical load is put on a sychronous motor, the rotor is pulled out of sychronism after which it comes to a stop. The maximum torque that a motor can develop without losing its sychronism is called its *pull-out* torque.

Q. 8. What do V-curves represent in terms of synchronous motor characteristics?

Ans. For a *given load,* power factor of a synchronous motor can be varied from a lagging value to a leading value. In fact, it is one of the most important characteristic of a synchronous motor particularly for power factor improvement purposes. The *V*-curves of Fig.7 show how the armature current and power factor of a synchronous motor change with rotor field excitation I_f for a given constant load.

Let us start with the case when rotor field excitation is low *i.e.* the motor is underexcited. Low field excitation reduces back emf E_b and causes the stator armature current I_a to lag. As rotor field current I_f is increased, E_b is increased, so I_a is decreased and reaches a minimum value meaning that the motor is running at unity power factor. If I_f is further increased, I_a starts increasing and the motor begins to operate at leading power factor. If we plot the graph between I_a and I_f, we get the lowest curve of Fig.7 which is for no load. *V*-curves for half load and full load are also shown in the figure.

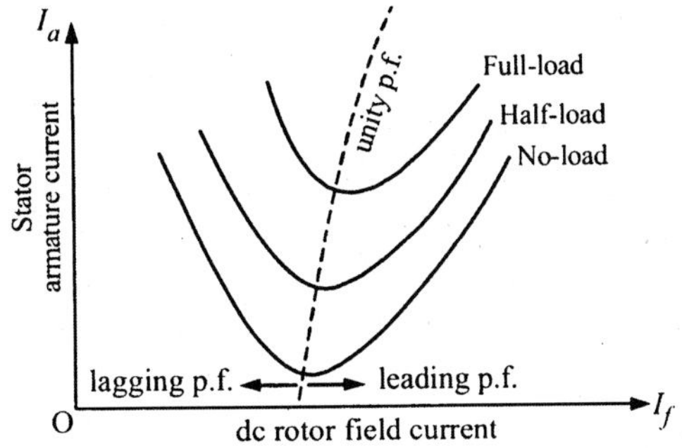

Fig. 7

As seen, *V*-curves of a synchronous motor show the relationship between stator armature current I_a and *dc* rotor field current I_f for a given load.

Q. 9. Why are synchronous motors mainly used in heavy industry?

Ans. Some of the unique electrical properties of synchronous motors which make them suitable for heavy industry are as under.

1. Most synchronous motors are rated between 150 kW and 15 MW and run at speeds from 150 to 1800 rpm.
2. They can operate at leading power factor and thereby improve the power factor of industrial plants which usually have lagging power factor.
3. They are less costly in certain kW and speed ranges *i.e.* for 35 to 350 kW rating at speeds less than 500 rpm.
4. They can be constructed with wider air gaps than induction motors which makes them better mechanically.
5. They usually run at higher efficiencies specially in the low-speed unity power factor ranges.

Q. 10. Mention some of the applications of a synchronous motor.

Ans. (1) Synchronous motors run at constant speed and have high efficiency. Hence, they are employed for loads where constant speed is desirable as in fans, blowers, pumps and compressors, motor-generator sets, continuous rolling mills, paper and cement industries.

(2) Since synchronous motors can be built for speeds as low as 120 rpm. They are well-suited for direct connection to reciprocating compressors.

(3) Pulp beaters and grinders, line shafts, steel and metal-rolling mills and rock and ore crushers are generally coupled or geared to the motor.

(4) Over-excited synchronous motors are most commonly used for improving power factor of lagging industrial loads. When used in such a role, they are called synchronous *capacitors.*

(5) They are often used for improving voltage regulation of long transmission lines.

Q. 11. Differentiate between a 3-phase induction motor and a synchronous motor.

Ans. The following table gives point-by-point differences and similarities between the two motors.

S. No.	Induction Motor	Synchronous Motor
1.	It is inherently self-starting.	It is not inherently self starting.
2.	It never runs with synchronous speed. Its speed decreases with increase in load.	It *always* runs at synchronous speed for all loads.
3.	It is called singly-excited motor because it needs just one ac supply to its armature.	It is called doubly excited motor because it needs ac supply for its armature and dc supply for rotor excitation.
4.	These motors operate only at lagging power factor	These motors operate at lagging, unity and leading power factors.
5.	It is possible to control their speed.	It is not possible to control their speed, These motors run only at synchronous speed.
6.	They are used to carry only mechanical load.	Apart from carrying mechanical load, these motors can improve the lagging power factor of the industrial loads also.
7.	Their maximum torque varies directly as the square of supply voltage *i.e.* $T \propto V^2$.	Their maximum torque varies directly as the supply voltage *i.e.* $T \propto V$.
8.	These are cheaper for speeds above 600 rpm and powers less than 150 kW.	These are cheaper and more efficient for very low-speed drives.

Q. 12. What are the disadvantages of a low lagging power factor? What is meant by power correction? Why is it desirable?

Ans. Low power factors are undesirable for several reasons.

(*a*) Suppose a load has a power factor of 0.8 or 80%. It means that only 80% of the power supplied to the load is useful. The remaining 20% does not do any useful work. Hence, for supplying 10 kW power to the load, the generator, has to produce 10/0.8 = 12.5 kW of power. It means the generator will have to produce more power and the supply lines will have to carry more current if load power factor is *lower than unity*.

(*b*) The voltage regulation of generators, transformers and supply lines is also poorer at low power factors than at unity power factor.

(*c*) Now, line and equipment losses are proportional to I^2R. These losses are also high at low power factor than at unity power factor. It is so because for delivering a given kW power, supply lines have to carry much larger current at low power factor.

Power correction means raising the power factor of the load from its low value to a higher value. Some industrial loads run at very low power factors of 0.4 or less. It is advantageous to raise this power factor to unity or nearabout. It will lead to lot of saving in power.

Any increase in power factor increase supply capacity and efficiency and, in general, improves the operating characteristics of the system. That is why power supply companies always penalise consumers for operating at low power factors.

Q. 13. What is meant by a synchronous capacitor and what is it used for?

Ans. A synchronous capacitor is nothing but an *overexcited* synchronous motor running at *no load*. Synchronous capacitors are always totally enclosed. The shaft does not extend beyond the case of the motor. They are used for correcting the power factor of a lagging load such as induction motors and transformers in a factory. They cancel the *lagging* kVARs of the induction motors with their leading kVARs. The synchronous motor is connected in *parallel* with the induction motors. They are usually placed in parallel at the *incoming power* lines to the plants as shown in Fig. 8.

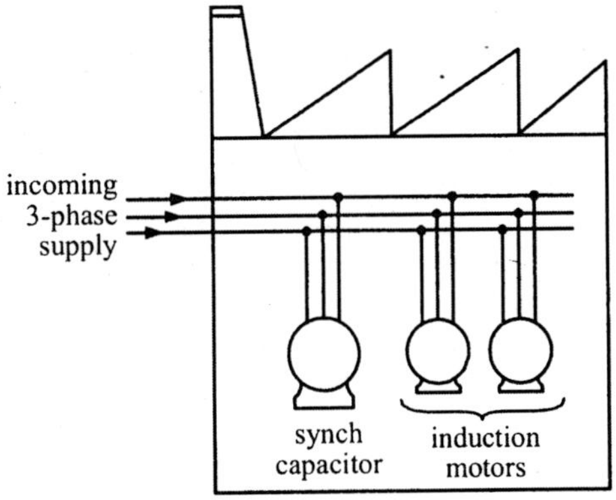

Fig. 8

The industrial loads have relatively large lagging *reactive* components of their currents. The synchronous capacitor with its large *leading* kVARs is used to cancel these laggings kVARs thus saving power.

Q. 14. An industrial plant has a load of 1500 kVA at a lagging power factor of 0.6. Calculate

(*a*) the kVA rating of the synchronous capacitor to be connected in parallel with this load to raise the line power factor to unity.

(*b*) the total kW load.

Ans. The wiring diagram and the kVA phasor diagram are show in Fig. 9.

$$\cos \phi = 0.6 \qquad \therefore \qquad \phi = \cos^{-1}(0.6) = 53°$$

$$\sin \phi = \sin 53° = 0.8$$

The 1500 kVA load has two rectangular components.

(*a*) $OA = \text{kVA} \cos \phi = 1500 \times 0.6 = 900$ kW.

It is in phase with V

(*b*) $OD = \text{kVA} \sin \phi = 1500 \times 0.8 = 1200$ kVAR (lagging)

As seen, it lags behind V by 90°.

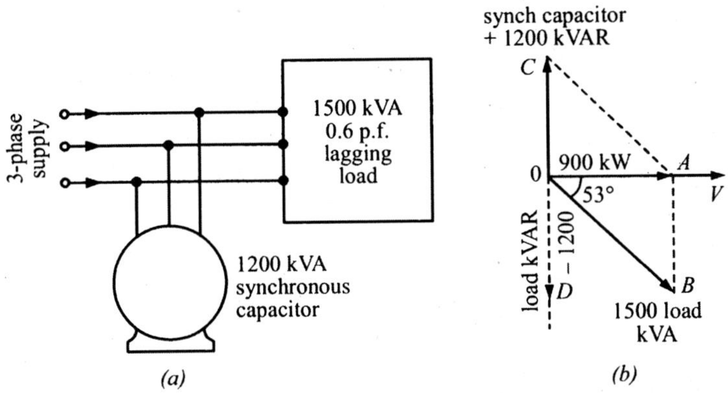

Fig. 9

The power factor will become unity if lagging kVAR of the load are neutralized by the leading kVAR supplied by the synchronous capacitor. As seen from Fig. 9(*b*), *OC* = *OD*. In other words, synchronous capacitor supplies leading **1200 kVAR** to cancel the lagging 1200 kVAR of the load.

It is seen from above that the supply rating is reduced from 1500 kVA to **900 kVA** (p.f. = 1.0) at the cost of a 1200 kVA synchronous generatror.

MULTIPLE CHOICE QUESTIONS

1. A synchronous motor is called so because
   (*a*) it runs at constant speed
   (*b*) its stator and rotor fields revolve synchronously
   (*c*) its speed depends on line frequency
   (*d*) it has unidirectional torque
2. A simple synchronous motor is not self-starting because its
   (*a*) armature current is small (*b*) rotor has high inertia
   (*c*) stator field revolves too fast (*d*) rotor poles are too weak
3. When load an a synchronous motor is incresed
   (*a*) its speed is decreased
   (*b*) rotor backward shift angle is decreased
   (*c*) armature back emf E_b is increased
   (*d*) motor load angle is increased
4. The speed regulation of a synchronous motor is
   (*a*) zero (*b*) very small (*c*) positive (*d*) negative
5. For a given load, power factor of a synchronous motor can be changed by merely adjusting its
   (*a*) armature current (*b*) armature back emf
   (*c*) rotor excitation (*d*) supply voltage
6. Change in the excitation of a synchronous motor changes its
   (*a*) speed (*b*) power factor
   (*c*) armature back emf (*d*) both (*b*) and (*c*)
7. One important feature that makes a synchronous motor different from other motors is that it
   (*a*) has a revolving magnetic field (*b*) needs two current sources
   (*c*) has a lagging power factor (*d*) has high efficiency
8. In a synchronous motor, torque is developed
   (*a*) at synchronous speed (*b*) by the stator
   (*c*) by the rotor (*d*) gradually
9. Synchronous motors are used in large power applications because of their
   (*a*) high operating efficiency (*b*) reliability
   (*c*) controlable power factor (*d*) all of the above
10. The squirrel cage winding used in synchronous motors for making them self-starting is also called
    (*a*) pole phase winding (*b*) damper winding
    (*c*) starting winding (*d*) all of the above
11. A synchronous motor is essentially a motor with
    (*a*) constant speed (*b*) leading power factor
    (*c*) unity power factor (*d*) lagging power factor

12. The armature current in a synchronous motor lags behind the resultant armature voltage by about .... degree
    (*a*) 0 (*b*) 45 (*c*) 60 (*d*) 85
13. A synchronous motor acts as a synchronous capacitor when it
    (*a*) is overexcited (*b*) has no load
    (*c*) is underexcited (*d*) both (*a*) and (*b*)
14. The synchronous speed of a synchronous motor depends on
    (*a*) supply line frequency (*b*) number of motor poles
    (*c*) motor shaft load (*d*) both (*a*) and (*b*)
15. A synchronous motor adjusts to increased load by
    (*a*) decreasing its speed
    (*b*) increasing its back emf E_b
    (*c*) decreasing its load angle
    (*d*) increasing phase between stator and rotor poles.
16. If the power factor of a load is raised from 0.8 lagging to unity, the percentage increase in the useful power supplied to the load is
    (*a*) 25 (*b*) 20 (*c*) 30 (*d*) 50
17. When a synchronous motor runs with normal excitation, its
    (*a*) $E_b = V$ (*b*) I_a is in phase with V
    (*c*) power facıor is unity (*d*) all the above
18. The V-curves of a synchronous motor show the releationship between
    (*a*) applied voltage and back emf
    (*b*) back emf and armature current
    (*c*) armature current and rotor field current
    (*d*) back emf and power factor
19. A synchronous motor is called doubly-excited motor beacuse it needs.
    (*a*) two exciting currents (*b*) two ac current sources
    (*c*) two dc current sources (*d*) one ac and one dc current source

ANSWERS

1. *b*	2. *b*	3. *d*	4. *a*	5. *c*	6. *d*	7. *b*
8. *a*	9. *d*	10. *d*	11. *a*	12. *d*	13. *d*	14. *d*
15. *d*	16. *a*	17. *d*	18. *c*	19. *d*.		

18 CHAPTER ELECTRIC POWER GENERATION

Q. 1. What are the three main types of electric generating stations? Briefly describe the type of fuel used by them.

Ans. The three main types of generation stations are:

1. thermal generating stations
2. hydropower generating stations
3. nuclear generating stations

The coal-based thermal generating stations burn fossil fuels like coal, oil, and natural gas. Coal and coke are called solid fossil fuels and are extensively used for producing heat in the electrical power stations. This heat is used to run steam turbines. Natural gas is now also being used for the same purpose.

Hydropower generating stations convert the *potential* energy of the water into electrical energy.

The nuclear power stations produce electricity from the heat released by a nuclear reaction. The fuels used are uranium-233, uranium-235 and plutonium-239. They are also called *thermal* power stations because they also use *heat* for power generation with the help of steam turbines.

Q. 2. What are the fossil fuels? Distinguish between solid and liquid fossil fuels. What are they used for?

Ans. Millions of years ago, large deposits of organic material were formed under the surface of the Earth. These deposits which provide coal, oil and natural gas are known as *fossil fuels.* Coal and coke are called *solid* fuels. They are extensively used for producing heat for electric power generation. Oil, natural gas and diesel are liquid fuels derived by petroleum processing. Natural gas is the primary fuel for electric power generation.

The fossil fuels are burnt to produce heat energy which converts water in a boiler into steam. The high-pressure steam is then used to drive a steam turbine. The turbine shaft is connected directly to a synchronous generator which converts mechanical energy into electrical energy. Almost 80 percent of electrical power prodouced today is from coal-fired systems.

Q. 3. What type of handling does coal require before it is used as fuel in a thermal power station?

Ans. First of all, ash and dust are removed from the coal. Then the size of the large *coal lumps* is reduced by grating units. These units agitate the coal to reduce it into smaller lumps. Next, coal is passed through a *pulverizer* where it is curshed into powder-fine particles. These fine particles are routed into the furnance by air pressure. The coal is held in suspension until it is ignited. It, then, releases large amount of heat energy.

Q. 4. Describe the structure and main components of a modern coal-fired thermal power station. Give its merits and demerits.

Ans. The basic structure and main components of a coal-fired generating station are shown in Fig. 1.

1. The boiler acts as a furnace which converts water into high-pressure steam by using heat produced by burning pulverised coal.
2. The steam is then used to drive a steam turbine. The mechanical energy of the turbine is converted into electrical energy because the turbine shaft is directly coupled to a synchronous generator.

3. After this, the exhausted steam is condensed to water in the *condenser.* Water from an outside surce such as a river or a lake keeps circulating through the condenser and takes away the heat. This condensed water is pumped into the boiler again
4. The voltage produced by the generator is stepped up by a transformer which is connected to the high-voltage transmission lines as shown.

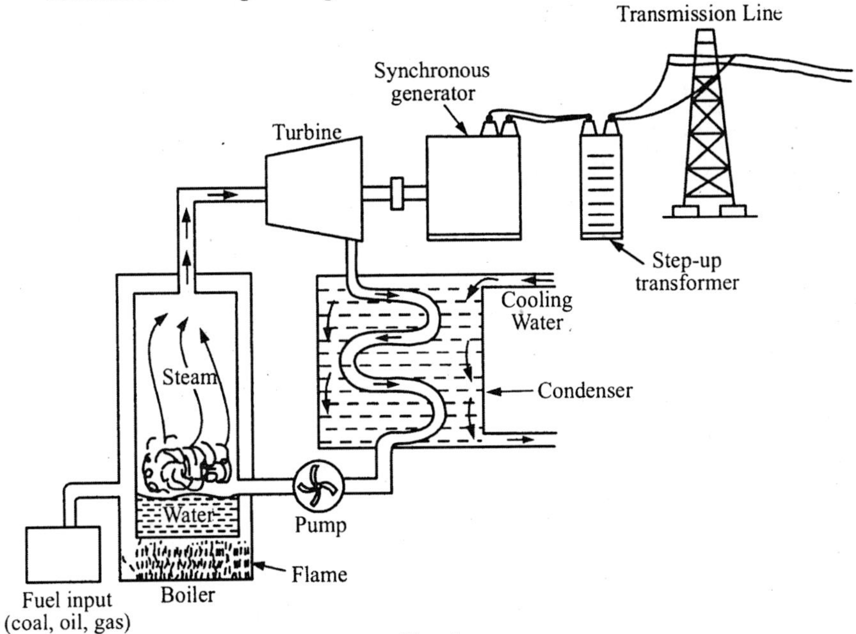

Fig. 1

Merits :

1. the fuel *i.e.* coal used is cheap
2. the initial cost of such stations is low
3. the station requires less space
4. the cost of generating electrical energy is less.

Demerits :

1. Coal is the *dirtiest* fuel because it produces atmospheric pollution. The major problems are sulphur-dioxide, fly ash and dust.
2. Efficiency of steam turbines is low *i.e.* about 40%.

Q. 5. With the help of a neat sketch, explain the operation of a hydroelectric power station. Give its advantages and disadvantages.

Ans. The hydropower generating stations convert the energy of moving water into electrical energy with the help of a hydraulic turbine coupled to a synchronous generator as shown in Fig. 2. Water reservoirs at *higher altitudes* are necessary for the production of electric power.

A concrete dam across a river bed is used to create such a water storage reservoir. Water from the reservoir is passed through a valve house which controls the flow of water. Next, water passes through penstock (huge steel pipes) which channels it to the hydraulic turbine. Here potential energy of the water is converted into kinetic energy which drives the water turbines. The mechanical energy of the turbine is finally converted into electrical energy because it is coupled to a synchronous generator. The generated voltage so produced is stepped up by a transformer which is connected to transmission lines. After water has done its work, it is allowed to leave the turbine via the tailrace into the downstream river bed.

Hydropower stations are divided into three types depending on the water head.

1. **High-head stations (head > 300m)**
   Since head is very large, the amount of impounded water is usually small.
2. **Medium-head stations (30m – 300m)**
   In this case, a large amount of water is impounded behind the dam.
3. **Low-head stations (head < 30m)**
   In this case, no reservoir is provided.

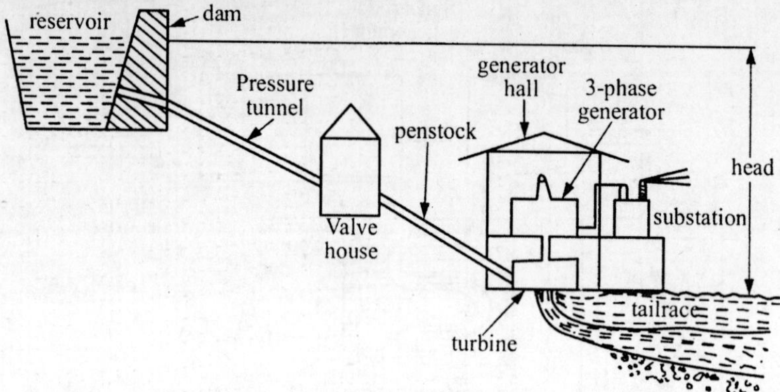

Fig. 2

Advantages :

1. no fuel is burnt
2. produces no pollution
3. running cost is less
4. construction is simple and requires less maintainance
5. easy to start up and shut down
6. the plant has longer life
7. requires less skilled persons to operate it

Disadvantages :

1. requires huge initial investment
2. power output depends on rainfall which is uncertain.
3. stations are located in the hills. Hence, transmission cost is higher.

Q. 6. (*a*) State the formula for the amount of electric power available from a hydropower station.

(*b*) A large hydropower station has a head of 300 m and average flow of 1200 m^3/s. The water reservoir covers an area of 6000 km^2. Calculate

(*a*) the available hydropower

(*b*) the number of days this power could be supplied continuously if the level of the water reservoir is allowed to drop by 1.5m.

Ans. The amount of power that can be obtained from a waterfall depends upon its

(*i*) height, h and

(*ii*) rate of flow, q.

The formula for the power available is

$$P = gqh \quad = 9.8qh$$

where P = available waterpower (kW)

q = rate of flow of water (m^3/s)

h = water head (m)

(*a*) $$P = 9.8\,qh$$

$$= 9.8 \times 1200 \times 300 = 35{,}28{,}000 \text{ kW} = \mathbf{3{,}528 \text{ MW}}$$

(*b*) Now, $1 km^2 = (1000)^2 = 10^6 m^2$. The resevoir area = **6,000** $km^2 = 6{,}000 \times 10^6 = 6 \times 10^9 m^2$. Hence, a drop of 1.5m in the water level corresponds to $1.5 \times 6 \times 10^9 m^3 = 9 \times 10^9 m^3$ of water out-flow. Because the water flows at the rate of $1200 m^3/s$, the time taken for this water to flow out is

$$t = 9 \times 10^9/1200 = 7.5 \times 10^6 \text{ second}$$
$$= 7.5 \times 10^6/3600 = 2083h$$
$$= 2083/24 = \mathbf{86.8\ days}$$

Q. 7. Briefly mention the different ways in which electrical power can be produced.

Ans. Electrical power may be produced in many ways such as from chemical reactions, heat, light or mechanical energy. However, majority of electric power produced is provided by power plants which convert heat energy produced by burning fossil fuels, the flowing of water or from nuclear reactions. The alternators at these power plants are driven by steam or gas turbines or by water turbines in the case of hydroelectric stations.

However, some other methods may be used in the future. These include solar cells, geothermal systems, wind-powered systems, nuclear fusion systems and fuel cells etc.

Q. 8. What is meant by nuclear fission? How is it achieved? What is a chain reaction?

Ans. Nuclear fission means *splitting* of uranium-235 or other nuclear fuels into two parts. This is bought about by *neutrons* moving at *very high speed.* When a neutron hits a uranium nucleus, it splits it into two parts which fly away with great speed. The amount of energy released during this reaction is extremely large. When a nucleus splits, it releases a number of neutrons. These neutrons cause more *fission reactions.* It means the nuclear fission reaction occurs as a *chain reaction* which causes huge amounts of heat energy to be given off. In nuclear power stations, the heat released by such nuclear reactions is used to drive steam turbines which, in turn, drive synchronous generators for producing electric power. In nuclear stations, the fission takes place inside a reactor which contains the nuclear fuel.

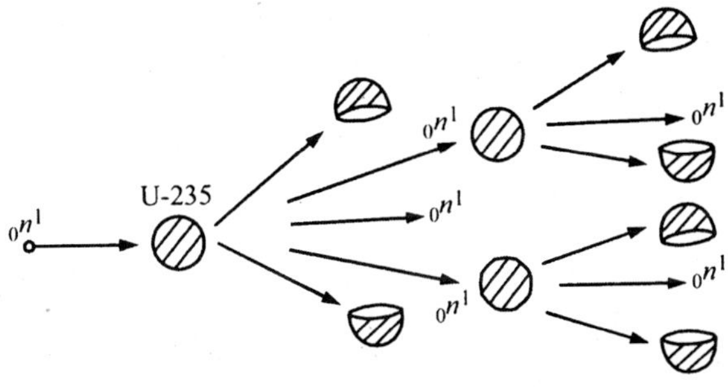

Fig. 3

Q. 9. With the help of a suitable diagram, explain the working of a nuclear power station. Give the advantages and disadvantages of nuclear power production.

Ans. As seen from Fig. 4. the nuclear power station consists of the following.

1. Nuclear reactor : It is similar to the furnance of a coal-fired power plant because uranium is 'burnt' inside it. It contains uranium-235 rods, moderator, control rods and the coolant liquid. The nuclear fission of uranium-235 atoms produces huge amount of heat which is used for raising steam.

2. Moderator : The function of the moderator rods is to slow down the fast fission neutrons in order to increase the nuclear fission of U-235 atoms. These are made of graphite.

3. Control rods : The fission reactions are controlled by cadmium rods. These cadmium rods can absorb neutrons thereby slowing down the chain reaction. If not controlled, the chain reaction can become violent and can damage the reactor.

4. Coolant : The function of the coolant is to carry away the heat from the reactor to the heat exchanger for raising steam. The coolant may be ordinary water, heavy water, liquid sodium or a gas like helium. The *hot coolant* circulates in a closed circuit between the reactor and heat exchanger.

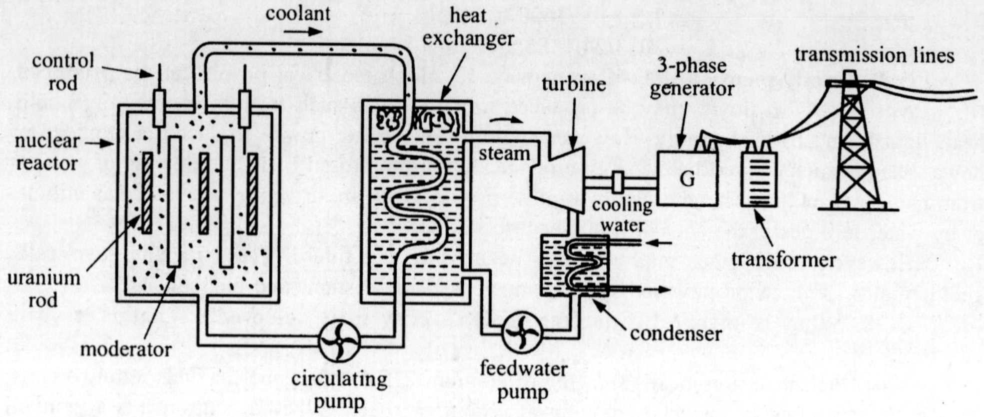

Fig. 4

As shown, the steam raised in the heat exchanger is made to drive a steam turbine which is coupled to an alternator that generates electrical power. After driving the turbine, the used steam is condensed to water by the condenser and sent back to the heat exchanger with the help of feedwater pump as shown. From there, it is pumped back to the reactor with the help of the circulating pump.

Advantages :

1. the amount of uranium fuel required *is very small.* It is found that 30 gram of uranium has roughly the same energy output as the burning of 100 tonnes of coal.
2. needs less space as compared to other power stations.
3. energy cost per unit is much lower.
4. it is a reliable source of energy because fuel resources are abundant.

Disadvantages :

1. uranium fuel is very costly
2. initial cost of the plant is high
3. it requires protective system to contain pollution
4. disposal of radioactive waste is a big problem .

Q. 10. What is a fast breeder reactor? Why is it called a breeder reactor?

Ans. It is a nuclear reactor which produces more fissile fuel than it consumes. It does not use any moderator for slowing down the neutrons.

This reactor consists of a small core of natural uranium (99.28% U-238 and 0.72% U-235) in which fast neutrons are allowed to produce fast chain reaction. Neutrons are allowed to escape from the core into a surrounding blanket of natural uranium. Here some of these neutrons are captured by U-238 nuclei for producing plutonium-239. In this way, more plutonium – 239 is produced in the natural uranium blanket than uranium-235 that is used up in the reaction. It is for this reason that this type of reactor is called a breeder reactor. After about 8 to 10 years, there is enough plutonium to start a new breeder reactor.

Q. 11. What is the meaning of solar power? Briefly describe the two most common solar energy conversion systems being used at present.

Ans. Solar power is the conversion of sunlight into electricity. The amount of *solar* energy reaching the Earth's surface is huge — almost 6000 times more than the power consumed by humans throughout the world.

There are two systems of converting sunlight into electricity.

(*i*) Photovoltaic (PV) system and

(*ii*) Concentrating solar power (CSP) system

A photovoltaic cell is also called a solar cell. It is a semiconductor device which converts sunlight into dc power by using the photoelectric effect. The first solar cell was constructed by Charls Fritts in 1883. Solar cells are often electrically connected and encapsulated as a module. The first practical application of solar modules was to power orbiting satellites and other spacecrafts but today the majority of solar modules are used for power generation.

Concentrating solar power (CSP) system uses lenses or mirrors to focus sunlight into a sharp beam. This beam is used to heat a fluid to a high temperature. The hot fluid is then used to generate steam that drives a steam turbine coupled to an electrical power generator.

Q. 12. What is a concentrating solar collector and what is its role in generating solar power? Illustrate with examples.

Ans. A concentrating solar collector performs two functions.

(*i*) it captures a large amount of sunlight from the Sun.

(*ii*) it then concentrates these rays into a sharp beam.

This powerful beam is next focussed on a small receiver for producing high temperature. The receiver uses the heat so absorbed to produce steam for operating steam turbines. These turbines are coupled to ac generators for producing electrical power.

Following four types of solar concentrators are in common use :

(*i*) Parabolic Trough Concentrator : As shown in Fig. 5(*a*), it consists of a parabolic reflector which concentrates sunlight onto a receiver placed at its focus. The receiver is a metallic tube which contains the working fluid (usually synthetic oil). The heat absorbed by the fluid is transferred to water for producing steam.

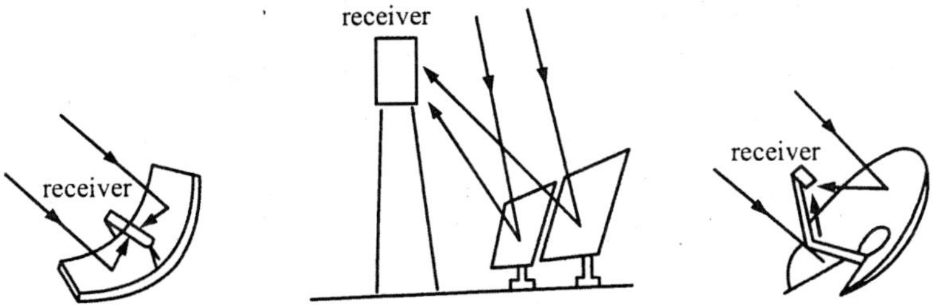

(*a*) (*b*) (*c*)

Fig. 5

(*ii*) Central Receiving Concentrator : As shown in Fig. 5(*b*), it consists of a large number of independently-moving *flat mirrors* (heliostats) which reflect the Sun's rays onto a receiver located at the top of a tower. The heliostats surround the tower from all sides. The tower supports a bundle of vertical tubes containing the working fluid. In this way, the fluid is heated by the reflected sunlight.

(*iii*) Parabolic Dish concentrator : As shown in Fig. 5(*c*), the parabolic dish concentrates the incoming solar rays at its focus. A tube containing the working fluid is placed at the focal point for absorbing the heat for final conversion into electric power.

(*iv*) Linear Fresnel Concentrator : In this case, a very large number of thin mirror strips (instead of parabolic mirrors) are used to concentrate sunlight onto the tubes containing the working fluid.

Q. 13. Give brief description of a solar power generating plant using CSP technology.

Ans. A solar power generating system using concentrated solar power (CSP) technology is shown in Fig. 6. It uses highly reflective parabolic mirrors which are shaped like a long *straight half-pipe.* A metallic tube runs along the *focal line* of the pipe. The tube is filled with synthetic oil which is heated to 400°C by the focussed sunlight. The solar rays focused on the tube are 70 – 80 times more intense than the ordinary sunlight. The heated oil transfers the heat to water which produces steam on boiling.

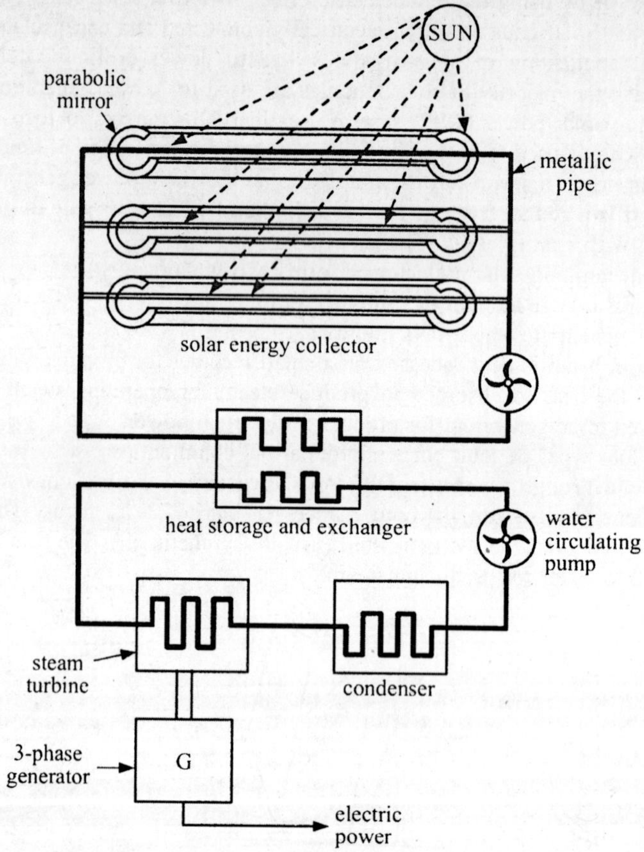

Fig. 6

The steam so produced drives steam turbines coupled to ac generators which produce electrical power.

Q. 14. What is a photovoltaic cell? Briefly explain its working. How can it be used for the generation of solar electric power?

Ans. A photovoltaic cell is also called a *solar cell**. Practically, all solar cells are photodiodes made of semiconductor materials like silicon. A solar cell works in three steps.

1. Photons in the sunlight hit the solar cell and are absorb by the semiconductor material.
2. Negatively-charged electrons are knocked off from their atoms and start flowing in the *same* direction to produce electric current.
3. An array of solar cells converts solar energy into a usable amount of *dc current.*

*Strictly speaking, solar cell is one which converts *sunlight* into electricity whereas photovoltaic cell converts any light into electricity.

Since the output of a single solar cell is very small, a large number of them are interconnected to produce solar modules or solar panels. The solar cells are connected in series for increasing the voltage. Connecting the cells in parallel gives a higher current. The modules are then connected in series or parallel or both to create a solar array with the desired dc voltage or current.

Solar cell technology is the fastest growing power generating technology in the world. It is due to the fact that solar cells with conversion efficiency of more than 40% are becoming available. It is expected that by the year 2025, solar cell system could be generating nearly 2000 GW of electric power around the world.

India's as well as Asia's biggest solar power park at Patan, Gujarat with 500 MW generation capacity was inaugurated on 30th December 2010 by the most progressive and development-oriented Chief Minister of Gujarat, Shri Narendra Modi.

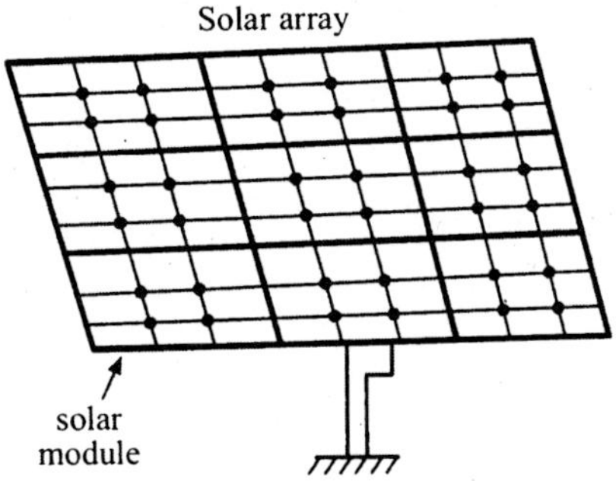

Fig. 7

Q. 15. Briefly describe the advantages and disadvantages of solar power generation.

Ans. (1) Advantage :

(*i*) Solar power is silent, limitless and free.

(*ii*) Solar power is pollution-free It releases no CO_2, SO_2 and NO_2 gases which are produced in coal-fired generating stations.

(*iii*) It does not contribute to global warming.

(*iv*) Operating costs of solar power stations are very low.

(2) Disadvantages :

(*i*) Solar electricity is not produced at night. Hence, a complementry power system is required.

(*ii*) Solar power is much reduced in cloudy conditions

(*iii*) Solar power generation depends on the amount of Sun's rays at any particular location.

(*iv*) Solar power generating stations are spread over thousands of km^2 of ground area.

(*v*) At present, solar power is very costly. It costs 4 times more than the power generated by coal-based plants.

MULTIPLE CHOICE QUESTIONS

1. In a coal-fired thermal power station, the function of a pulverizer is to
   (*a*) reduce the size of the coal lumps (*b*) crush coal into powder-fine particles
   (*c*) ingite coal (*d*) remove ash and dust from coal.
2. Steam turbines used in electric power production
   (*a*) have effieiency of about 80 percent
   (*b*) run at low speeds
   (*c*) run at a constant speed
   (*d*) run at 3600 rpm for producing 50 Hz supply
3. In thermal power plants, the function of a condenser is to
   (*a*) condense water (*b*) purify water
   (*c*) purify steam (*d*) condense used steam into water
4. The efficiency of water turbines is
   (*a*) lower than that of steam turbines (*b*) about 85 percent
   (*c*) less than 50 percent (*d*) almost 100 percent

5. A hydropower station consists of
(*a*) dams (*b*) waterways (*c*) conduits (*d*) all of the above
6. A nuclear power station consists of
(*a*) nuclear reactor (*b*) steam turbine
(*c*) heat exchanger (*d*) all of the above
7. The unique advantages of a hydropower plant are that it
(*a*) burns no fuel (*b*) produces no pollution
(*c*) is located in hills (*d*) both (*a*) and (*b*)
8. The reactor of a nuclear power plant contains
(*a*) fuel rods (*b*) moderator (*c*) coolant (*d*) all of the above
9. The function of a moderator in a nuclear reactor plant is to
(*a*) slow down fission-producing neutrons
(*b*) absorb the neutrons
(*c*) start nuclear fission
(*d*) none of the above
10. About 30 gram of uranium has roughly the same energy output as the burning of ____ tonnes of coal
(*a*) 10 (*b*) 50 (*c*) 100 (*d*) 1000
11. The coolant used for carrying away heat from a nuclear reactor to heat exchanger is
(*a*) ordinary water (*b*) heavy water
(*c*) liquid sodium (*d*) any of the above
12. The largest energy source available today is .... which supplies limitless energy
(*a*) Sun (*b*) uranaium (*c*) coal (*d*) natural gas.
13. Solar power is the conversion of .... into electric power
(*a*) light (*b*) sunlight (*c*) moonlight (*d*) none of the above
14. Which of the following cell is used for the production of solar power?
(*a*) photoemissive cell (*b*) photoconductor cell
(*c*) photovoltaic cell (*d*) photomultiplier tube
15. In the CSP system of generating solar power, which of the following concentrator is used for obtaining high temperature of the working fluid?
(*a*) parabolic trough concentrator (*b*) parabolic dish concentrator
(*c*) central reciver concentrator (*d*) all of the above
16. Solar arrays are built up of a number of solar modules connected in
(*a*) series (*b*) parallel (*c*) series-parallel (*d*) all of the above
17. The advantage of solar power is that it is
(*a*) unlimited (*b*) pollution-free (*c*) cheap (*d*) all of the above
18. The disadvanages of solar power are
(*a*) it is not produced at night
(*b*) it is much reduced in cloudy conditions
(*c*) it requires inverters to change it from dc to ac
(*d*) all of the above
19. A very remarkable thing about fast breeder reactor is that it
(*a*) uses natural uranium as fuel (*b*) produces more fuel than it consumes
(*c*) is not like pressure water reactor (*d*) uses no moderator

ANSWERS

1. *b*	2. *c*	3. *d*	4. *b*	5. *d*	6. *d*	7. *d*
8. *d*	9. *a*	10. *c*	11. *d*	12. *a*	13. *b*	14. *c*
15. *d*	16. *d*	17. *d*	18. *d*	19. *b*.		

19 CHAPTER TRANSMISSION AND DISTRIBUTION OF ELECTRIC POWER

Q. 1. Which factors determine the design of power lines? How are they classified as per their voltage?

Ans. The transmission of electrical power requires power lines to carry the electric current where it is produced to where it is used. The design of a power line depends on the following factors :

1. the amount of active power to be transmitted.
2. the distance over which the power is to be carried.
3. the cost of the power line
4. other considerations such as urban congestion, ease of installation and expected growth in load.

Power lines are of four types as per the transmission voltage.

1. Low-voltage (LV) lines : These provide power to commercial buildings, factories and houses to drive motors, electric stoves, lamps, heaters, and air conditioners. The lines are made up of insulated conductors usually made of aluminum. The consumers are connected from a local pole-mounted distribution transformer. The line may be overhead or underground operating at voltages less than 600V.

2. Medium-voltage (MV) lines : These connect load centres to one of the many substations of the supply company. The voltages are usually between 2.4 kV to 69kV.

3. High-voltage (HV) lines : These connect the main substation to generating station. The lines are either overhead wires or underground cables operating at voltages below 230 kV.

4. Extra-High Voltage (EHV) lines : These are used when generating stations are *very for away* from the load centres. Such lines operate at voltages upto 800 kV and may be as long as 100 km.

Q. 2. What are the different components of a H.V. transmission line? Describe each of them in brief.

Ans. A transmission line is composed of conductors, insulators and supporting structures as detailed below.

1. Conductors : For high-voltage lines, stranded copper conductors or steel-reinforced aluminium cables are used. The conductors are kept bare *i.e.* uncovered. Aluminium cable is preferred because it is lighter and cheaper.

2. Insulators : The job of the insulators is to support the conductors and to insulate them from ground. They are usually made of porcelain although glass and other synthetic insulating materials are also used.

Insulators offer very high resistance to surface leakage currents. They are also very thick in order to prevent breakdown under the high-voltage stresses they have to withstand. To increase the leakage path and, hence, the leakage resistance, the insulators are moulded with wave-like folds as shown in Fig. 1

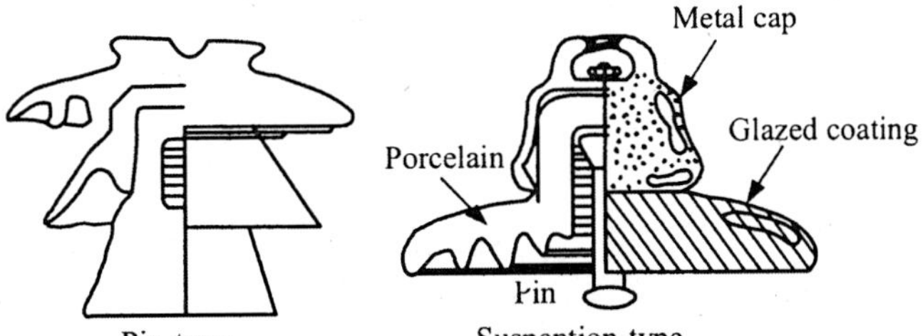

Pin-type
(a)

Suspention-type
(b)

Fig. 1

The two main types of insulators shown in Fig. 1 are (*i*) pin-type insulators and (*ii*) suspension-type insulators.

The pin-type insulator has several porcelain folds for increasing the leakage path to the ground. The conductor is fixed at the top. The insulator is bolted to the support with the help of a pin. Suspension-type insulators are used for voltages above 70 kV. The number of insulators depends upon the transmission voltage. Higher the voltage, larger the number of insulators.

3. Supporting Structure :

Supporting structure serves two purposes:

(*i*) it keeps the conductors at a safe height from the ground and

(*ii*) it keeps the conductors at a safe distance from each other.

For voltages below 70 kV, single wooden poles with cross-arms are used. For higher voltages, two poles are used to create an H-frame. The wood is treated with creosote to prevent it from rotting. For very high-voltage lines, use is made of steel towers made of galvanized angle iron pieces bolted together.

The spacing between conductors is such as to prevent arc-over under strong wind conditions. As the distance between towers and as the line voltages become higher, the conductor spacing is increased.

Q. 3. Compare the advantages and disadvantages of high voltage DC (HVDC) transmission and high-voltage AC (HVAC) transmission.

Ans. These days almost all power generation is by three-phase generators. This power is transmitted to the users by high-voltage ac transmission lines. Since high-power, high-voltage electronic converters have become available, ac power can be converted to dc power (Fig. 2) and then transmitted. DC transmission has some advantages which ac transmission does not have.

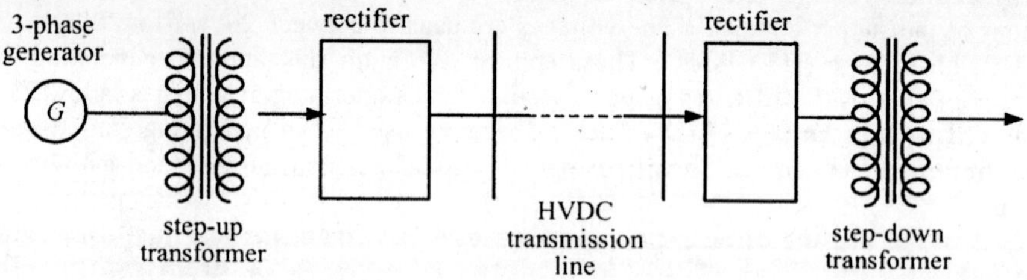

Fig. 2

The advantages are as under :

1. It requires only two conductors
2. There are no capacitive or inductive reactances. Hence, no phase differences occur at the *two ends* of the transmission line. Therefore, power can be transmitted over longer distances.
3. DC power can be controlled much more *quickly*. For example, power in the megawatt (MW) range can be reversed in a dc line in *less than one second*. It is not possible to do so in ac lines.
4. Since there is no skin effect, conductors have low resistance and hence low I^2R losses.
5. DC power can be transmitted in cables over long distances. In ac transmission, capacitance of the cable limits the range.
6. There is no dielectric power loss particularly in underground cables.
7. Over long distances, dc transmission line is economically cheaper. If a good ground return exists between the power plant and the load point, only one line conductor is needed instead of three for ac transmission. This reduces the cost of wire to one-third. Also, there will be saving on losses.

The main disadvantages of dc transmission are as under :

1. DC lines are usually point-to-point systems connecting one large generating station to one large power-consuming centre. Hence, power cannot be tapped at *different* points along a dc line.
2. DC switches and circuit breakers have their own problems.
3. The voltage step-up and step-down is not direct and simple.

The advantages of high-voltage ac transmission are as under :

1. Since ac power is transmitted at very high-voltages, its conductor size is less because it has to carry a very small current.
2. It is easier and cheaper to maintain ac substations.
3. It is very easy to step up and step down ac voltages. Hence, power can be transmitted at very high voltages with high efficiency.

The ac transmission has the following disadvantages.

1. For the same voltage, ac transmission requires more copper than dc transmission.
2. The design of ac transmission lines is more complex than that of dc transmission lines.
3. Due to line capacitance and inductance, there is a *phase shift* between current and voltage.

Q. 4. Draw a typical single-line diagram for the transmission and distribution of three-phase power.

Ans. A large network of conductors is required to connect the generating power station and the consumers. This process may be divided into two parts *i.e.* transmission system and distribution system. Each part can be again sub-divided into two.

(*i*) Primary transmission and secondary transmission

(*ii*) Primary distribution and secondary distribution.

Such a typical layout of the system is shown by a single-line diagram in Fig. 3.

1. Generating Station : Electric power is produced by three-phase alternators usually at 11 kV. Since power transmission at this voltage is not economical, it is transmitted at 230 kV or at still higher voltages with the help of step-up transformers.

2. Primary transmission : The generated power is transmitted with three-phase, three-wire overhead transmission lines. This type of transmission is called primary transmission or Extra high voltage AC (EHVAC) transmission.

3. Secondary transmission : The secondary transmission ends at a substation known as

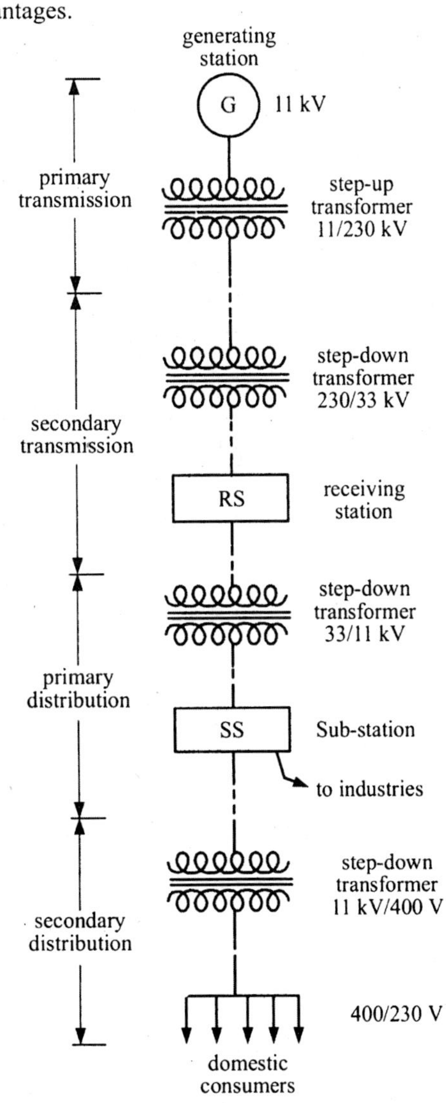

Fig. 3

Receiving Station (RS) which is outside the city. Here, voltage is stepped down to 33 kV. From here, power is transmitted to various substations located in the city. It is known as *secondary transmission.* From now onwards starts the primary and secondary distribution.

4. Primary Distribution : Here, power is stepped down from 33 kV to 11 kV or less and distributed to industrial consumers. It is called primary distribution.

5. Secondary Distribution : The primary and secondary transmission and primary distribution are carried by 3-wire, 3-phase system. However, secondary distribution is done with 4-wire, 3-phase system. This enables the consumers to use 400 V, 3-phase, 4-wire system with the help of a step-down transformer. Single-phase consumers are provided with any one phase and the neutral line. Hence they can use 400 V as well as $400/\sqrt{3}$ = 230 V. As shown in Fig. 4, secondary distribution system consists of feeders, distributors, sub-distributors and service mains. The consumers get their supply connections from the service mains. Summarizing the above, we have

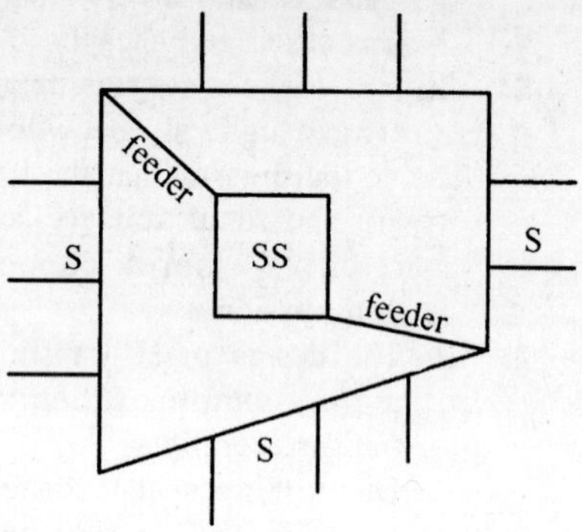

Fig. 4

1. generating voltage, 11 kV
2. primary transmission, 230 kV
3. secondary transmission, 33 kV
4. primary distribution, 11 kV
5. secondary distribution
   AC 400/230, 3-phase, 4-wire
   DC 400/230, 3-wire system.

Q. 5. How do you classify systems of power distribution? What are their usual voltages?

Ans. The power distribution systems can be classified in two main categories.

These are dc and ac systems of power distribution. They can be further subdivided as follows:

(*a*) DC systems

1. Two-wire system, 220 V
2. Three-wire system, 460/230 V,

(*b*) AC systems

1. Single-phase, 2-wire system 230 V
2. Single-phase, 3-wire system, 500/230 V
3. Three-phase, 3-wire system
4. Three-phase, 4-wire system. 400/230 V

Q. 6. Compare the overhead (OH) and underground (UG) systems of power distribution in a city.

Ans. In urban areas, *secondary* power distribution is done by UG cables. Sometimes even the primary distribution is also done by the UG cable system. The comparison between the two systems of power distribution is given below.

1. UG cable system is preferred from the point of view of public safety.
2. OH system is preferred because initial investment is many times less than for the UG cable system.
3. OH system is more flexible than the UG cable system.
4. Possibility of a fault happening in UG cable system is much less as compared to OH system.
5. The useful life of an OH system is about half that of UG cable system.
6. The maintenance cost of a UG cable system is much lower because the chances of fault in this system are much less.

7. In UG cable system, the chance of interference with communication circuits is much less as compared to OH system.

MULTIPLE CHOICE QUESTIONS

1. Which of the following factors determine th design of transmission power lines?
   (*a*) amount of power to be transmitted
   (*b*) distance over which power is to be transmitted
   (*c*) cost of the power line
   (*d*) all of the above
2. As compared to ac transmission, one main advantage of dc transmission is that it
   (*a*) has less inductive reactance (*b*) requires only two conductors
   (*c*) is a point-to-point system (*d*) none of the above
3. EHV transmission lines are used when
   (*a*) transmission voltages are higher than 1000 kV
   (*b*) generating station is very for away from load centres
   (*c*) transmission distance is short
   (*d*) none of the above
4. Which of the follwing statement is NOT true regarding transmission of ac electric power?
   (*a*) it requires more Cu than dc transmission
   (*b*) there is no phase shift between its voltages and currents
   (*c*) power can be transmitted with high efficiency
   (*d*) power can be transmitted at very high voltages.
5. In urban areas, secondary power distribution is done by underground cables because
   (*a*) chances of fault happeing are much less
   (*b*) useful life is high
   (*c*) maintenance cost is much lower
   (*d*) all of the above
6. The components of an H.V. transmission line are
   (*a*) conductors (*b*) insulators
   (*c*) suporting structure (*d*) all of the above
7. The supporting structure for very high-voltage lines consists of
   (*a*) single wooden poles having cross-arms
   (*b*) two poles forming H-frame
   (*c*) steel towers
   (*d*) none of the above
8. The main function of the insulators used in transmission lines is to
   (*a*) support the conductors
   (*b*) insulate the conductors from ground
   (*c*) to prevent arc-over between conductors
   (*d*) both (*a*) and (*b*)
9. Unlike ac lines, power in the megawatt range in dc lines can be reversed in less than
   (*a*) 1 second (*b*) 30 second (*c*) 45 second (*d*) 1 minute

ANSWERS

1. *d* 2. *b* 3. *b* 4. *b* 5. *d* 6. *d* 7. *c*
8. *d* 9. *a*.

20 CHAPTER ECONOMICS OF POWER GENERATION AND TARIFFS

Q. 1. What are the two components of the cost of producing electric energy? Which factors influence them?

Ans. Electric power systems are set up for producing electric power to be sold at a profit. Hence, every effort is made to produce power as cheaply as possible. The cost of producing electric power has two parts.

(a) Fixed Standing Cost : It does not vary with the operation of the plant. In other words, it does not depend on the number of units of electric energy produced.

(b) Running or Operating Cost : It varies with the operation of the plant. It means it depends on the number of units of electrical energy produced.

Influencing Factors :

The factors which influence the fixed cost are as under

(*i*) interest on money spent for setting up the power plant.

(*ii*) allowance for depreciation of the buildings, transmission and distribution system etc.

(*iii*) taxes and insurance.

The factors which affect the running cost are:

(*i*) most of the fuel cost

(*ii*) small part of salaries and wages

(*iii*) repair and maintenance.

Q. 2. Define the following terms in connection with an electric power system.

(*i*) average demand

(*ii*) maximum demand and

(*iii*) load factor

Ans. (*i*) Average Demand : It means the average power consumed by a consumer during some *specified* period of time of *long* duration.

$$\therefore \qquad \text{average demand} = \frac{\text{kWh consumed in the period}}{\text{hours in the period}}$$

(*ii*) Maximum Demand : It is defined as the *greatest* of all the demands which have occurred during a given period.

(*iii*) Load Factor : It is defined as the *ratio of the average power to the maximum demand.*

When average power generated or consumed is considered for a period of one month, it is called monthly load factor. If considered for a year, it is called annual load factor. Moreover, the average power may refer to either a generator or a consumer.

It is economical to obtain high load factors. If the load factor is low *i.e.* kWhs of electric energy produced are small, then charge per kWh would be high. But if load factor is high *i.e.* the number of kWhs generated is large, then cost of production and hence charge per kWh is reduced.

Q. 3. A consumer has the following conected load : 10 lanms of 60 W each and two heaters of 1000 W each. His maximum demand is 1500 W. On the average, he uses 8 lamps for 5 hours a day and each heater for 3 hours a day. Find his total connected load, monthly energy consumption and load factor.

Ans. Total connected load = 10 × 60 + 2 × 1000 = **2600 W**

Daily energy consumption is = (8 × 60 × 5) + (2 × 1000 × 3)

= 8400 Wh = 8.4 kWh

Monthly energy consumption = 8.4 × 30 = **252 kWh**

$$\text{Monthly load factor} = \frac{\text{average consumption}}{\text{maximum demand}}$$

$$= \frac{252 \text{ kWh}}{(1500 \times 10^{-3}) \times 24 \times 30 \text{ kWh}} = 0.233 \quad \text{or} \quad \mathbf{23.3\%}$$

Q. 4. A 100 MW power station delivers 100 MW for 2 hours, 50 MW for 6 hours and is shut down for the rest of each day. It is also shut down for maintenance for 45 days each year. Calcualte its annual load factor.

Ans. The station operates for (365 – 45) = 320 days in a year. Hence, number of MWh supplied in one year = (100 × 2 × 320) + (50 × 6 × 320) = 160,000 MWh

Maximum number of MWh which can be supplied per year with a maximum demand of 100 MW is

∴ = 100 × (320 × 24) = 768, 000 MWh

$$\therefore \text{ annual load factor} = \frac{160{,}000}{768{,}000} \times 100 = \mathbf{20.8\%}$$

Q. 5. A consumer has the following load-schedule for a day :

From midnight (12 PM) to 6 AM. = 200 W

From 6A.M to 12 noon = 3000 W

From 12noon to 1P.M = 100 W

From 1 P.M. to 4 P.M. = 4000 W

From 4 P.M. to 9 P.M = 2000 W

From 9P.M. to mid-night (12 P.M.) = 1000 W

Find the daily load factor.

If the tariff is 50 paisa per kW of maximum demand plus 35 paisa per kWh, find the daily bill the consumer has to pay.

Ans. Energy consumed per day. *i.e.* in 24 hours.

= (200 × 6) + (3000 × 6) + (100 × 1) + (4000 × 3) + (2000 × 5) + (1000 × 3)

= 44,300Wh

Daily average power = 44,300/24 = 1846W = 1.846 kW

$$\therefore \quad \text{daily load factor} = \frac{\text{average power}}{\text{maximum power demand}} = \frac{1846}{4000} = 0.461 = \mathbf{46.1\%}$$

Since maximum demand = 4 kW, M.D charge = 4 × 50/100 = Rs. 2.00

Energy consumed = 44.3 kWh

Energy charge = Rs. 44.3 × 35/100 = Rs. 15.50

∴ daily bill of the consumer = Rs. 2 + Rs. 15.5 = Rs. 17.5

Q. 6. What is meant by tariff? Briefly describe the different tariffs charged by electirc supply companies.

Ans. By 'tariff' is meant the *rate* at which a consumer pays for the electric energy consumed by him. The rate of charging takes into account the *fixed* as well as *running* cost of electric generation. Hence, the charging rate consists of two parts. That is why it is called *two-part* tariff and is of three types as described below:

1. Flat rate : It is not used these days because it does not take into account the fixed and running costs of producing electrical energy.

2. Block rate tariff : In this type of tariff, the fixed costs are collected by charging the first block of units at a higher rate and then reducing the rates successively.

3. Two-part tariff : As the name shows, it consists of two parts

(*i*) a fixed charge proportional to the maximum demand and

(*ii*) a low running charge proportional to the actual number of units of electric energy used.

The maximum demand is measured by a watthour meter.

This type of tariff is expressed by the equation

$$\text{charge} = \text{Rs. } A \times \text{kW} + B \times \text{kWh}$$

where A = annual charge per kW of *maximum* demand

B = price per kWh

Q. 7. Compute the cost of electrical energy and average cost for consuming 300kWh under block rate tariff as under:

First 50 kWh at Rs. 2.0 per kWh ; next 50 kWh at Rs. 1.5 per kWh; Next 50 kWh at Rs. 1.00 per kWh; next kWh at 80 paisa per kWh.

Excess over 200 kWh at 35 paisa per kWh.

Ans. Energy charge for the first 50 kWh = Rs. 50 × 2 = Rs. 100

Eneregy charge for the next 50kWh = Rs 50 × 1.5 = Rs. 75

Energy charge for the next 50kWh = Rs 50 × 1.0 = Rs. 50

Energy charge for the next 50kWh = Rs 50 × 0.8 = Rs. 40

Energy charge for the rest (300 – 200) = 100 kWh = Rs. 100 × 0.35 = Rs. 35

Ttoal charge for 300 kWh = Rs. (100 + 75 + 50 + 40 + 35) = Rs. 300

Average cost of electrical energy per kWh

= Rs. 300/300 = Rs. 1.0/kWh

Q. 8. The output of a genergating station is 4 × 10^8 units per annum and installed capacity is 10^5 kW. If the annual fixed charges are Rs. 80 per kW of installed plant and runnuing charges are 80 paisa per kWh, what is the cost per unit at the generating station?

Ans. Annual fixed charges = Rs. 80 × 10^5 = Rs. 8 × 10^6

Fixed charges = Rs. 8 × 10^6/ 4 × 10^8 = 0.02

= 2 paisa/kWh

Hence, cost at the generating station is = 80 + 2

= **82 paise/kWh**

Q. 9. A supply is offered on the basis of fixed charges of Rs 200 per annum plus 40 paise per unit or alternatively, at the rate of 60 paisa per unit for the first 400 units per annum and 50 paisa per unit for all the additional units. Find the number of units taken per annum for which the cost under these two tariffs becomes the same.

Ans. Let x kWh be the annual consumption of the consumer for which the two tariffs become equal.

Cost according to the first tarff = Rs. 200 + Rs. $40x/100$

= Rs. 200 + $2x/5$

Charges according to the alternative tariff

$$= \text{Rs. } \frac{60}{100} \times 400 + \frac{50}{100} \times (x - 400)$$

$$= \text{Rs. } 240 + \frac{1}{2}(x - 400)$$

Since charges in both cases are to be the same

$$\therefore \quad 200 + \frac{2x}{5} = 240 + \frac{1}{2}(x - 400)$$

$$\therefore \quad x = \mathbf{1400\ kWh}$$

Q. 10. What are the disadvantages of low load power factor? Fully illustrae with a suitable example.

Ans. When the load power factor of a consumer is low and lagging, it hurts the supply company in many ways.

1. It increases copper losses in the cables and machinery. It raises the *running* cost of the system.
2. It increases the excitation in the generators which reduces the efficiency of power generation.
3. The ratings of the generators, transformers, switch gears and cables is increased which means high capital charges with higher depreciation and interest.
4. Voltage drop in generators, transformers and cables is increased due to which *voltage regulation* becomes poor. Extra regulation equipment would be required to maintain voltage at the consumer's terminal within prescribed limits. It will mean extra expenditure for the power supply company.

Consider the example of a single-phase alternator which gives an output of 1500 kW at 10,000 V.

(*a*) Unity power factor

Power, $P = VI \cos\phi = VI \times 1 = VI$

Now, $VI = 1500$ kW or $10000 \times I = 15{,}00{,}000$

$\therefore$ $I = 15{,}00{,}000/10{,}000 = 150$ A

Also, in this case, kVA = kW/cos ϕ = 1500/1 = 1500

(*b*) 0.6 lagging power factor

$$P = VI\cos\phi \qquad \therefore \quad I = \frac{P}{V\cos\phi} = \frac{15{,}00{,}000}{10{,}000 \times 0.6} = 250 \text{ A}$$

kVA = 1500/0.6 = 2500

From the above, it is seen that the generator current has increasd from 150 A to 250 A. It means an increase of (250 – 150)/ 150 = 0.67 or 67%.

The kVA requirement has also increased from 1500 to 2500. As a result, a bigger size alternator would be required which means increased cost to the supply company. The increase in current will mean more I^2R losses in the generator and other equipment needed for supplying power to the customers. Hence, supply companies penalise consumers for their low power factor by charging higher tariff for maximum kVA demand in addition to normal kWh charges.

MULTIPLE CHOICE QUESTIONS

1. Cost of generating energy consists of
   (*a*) fixed cost (*b*) operating cost (*c*) variable cost (*d*) both (*a*) and (*b*)
2. One of the factors affecting the running cost of an electric supply system is
   (*a*) interest on capital investment (*b*) depereciation
   (*c*) number of kWh produced (*d*) taxes and insurance
3. Load factor of a power station is defined as
   (*a*) maximum demand /average load (*b*) average load × maximum demand
   (*c*) average load/ maximum demand (*d*) (average load × maximum demand)$^{1/2}$

4. Tow-part tariff for electric energy consists of
   (*a*) fixed charge on maximum demand
   (*b*) low charge on units consumed
   (*c*) excess charge on higher consumption
   (*d*) only (*a*) and (*b*)
5. A low load power factor for a generating station increases
   (*a*) line losses
   (*b*) size of generators and transformers
   (*c*) voltage drop in generators, transformers and cables etc.
   (*d*) all of the above

ANSWERS

1. *d* 2. *c* 3. *c* 4. *d* 5. *d*.

INDEX

D

E

F

G

H

I

INDEX

A

B

C

D

E

F

G

H

I

NOTES

NOTES

NOTES

NOTES